DATE DUE

FE 26 '00			
NO 4 '99			
OC 3 '00			
JE 9 '04			
SE 7 '04			
MR 1 9 '05			

DEMCO 38-296

Jupiter is an extraordinarily colourful and dynamic planet. Over minutes, one can watch tiny shadows cast by its moons slide over its surface; over days and weeks parades of diverse, giant swirling storms can be seen to move and evolve. It is because of this richness of visual and physical properties that Jupiter has intrigued amateur and professional astronomers and has been the goal of several space missions.

This highly illustrated volume provides a comprehensive and accessible account of Jupiter and its satellites, synthesising data from amateur and professional astronomers and space missions. It reviews systematic telescopic observations that have stretched over more than 100 years, in addition to modern observations and theories, and the wealth of data from the Pioneer, Voyager and Ulysses space missions. Many of the hand-drawings and the images from Voyager are presented and analysed here for the first time. As well as a thorough survey of the planet's atmosphere, this volume presents an up-to-date account of our present knowledge of Jupiter's satellites and magnetosphere, at a level accessible to the non-specialist.

As the first full account of Jupiter for 35 years, this volume provides the definitive account of Jupiter for advanced amateur astronomers, professional astronomers and planetary scientists.

The Giant Planet Jupiter

The Practical Astronomy Handbooks are a new concept in publishing for amateur and leisure astronomy. These books are for active amateurs who want to get the very best out of their telescopes and who want to make productive observations and new discoveries. The emphasis is strongly practical: what equipment is needed, how to use it, what to observe, and how to record observations in a way that will be useful to others. Each title in the series will be devoted either to the techniques used for a particular class of object, for example observing the Moon or variable stars, or to the application of a technique, for example the use of a new detector, to amateur astronomy in general. The series will build into an indispensable library of practical information for all active observers.

Titles available in this series

1. A Portfolio of Lunar Drawings
    by Harold Hill
2. Messier's Nebulae and Star Clusters
    by Kenneth Glyn Jones
3. Observing the Sun
    by Peter O. Taylor
4. The Observer's Guide to Astronomy (Volumes 1&2)
    edited by Patrick Martinez
5. Observing Comets, Meteors, and the Zodiacal Light
    by Stephen J. Edberg and David H. Levy
6. The Giant Planet Jupiter
    by John H. Rogers
7. High Resolution Astrophotography
    by Jean Dragesco

The Giant Planet Jupiter

JOHN H. ROGERS
British Astronomical Association

 CAMBRIDGE
UNIVERSITY PRESS

Published by the Press Syndicate of the University of Cambridge
The Pitt Building, Trumpington Street, Cambridge CB2 1RP
40 West 20th Street, New York, NY 10011-4211, USA
10 Stamford Road, Oakleigh, Melbourne 3166, Australia

© Cambridge University Press 1995

First published 1995

Printed in Great Britain at the University Press, Cambridge

ibrary

QB661.R64   1994
523.4′5—dc20   94-15303 CIP

ISBN 0 521 41008 8 hardback

Contents

Preface

As this book was being written, in 1990–1993, the British Astronomical Association (BAA) was celebrating 100 years of systematic visual observations of the planets, and the Galileo spacecraft was wending its way out to Jupiter. So this was an opportune time to summarise our knowledge of the planet to date, integrating the wealth of Earth-based records with the fascinating data from the Pioneer and Voyager spacecraft. If the Galileo mission is successful, we will soon know much more about many aspects of the jovian system, but we can expect that the largescale properties and long-term patterns described in the present account will remain valid.

A reader unfamiliar with the planet will find the overall properties of the atmosphere described in Parts II and IV. More detailed support for the generalisations therein is given in Part III, which is a complete review of the observations up to 1990 or 1991.

However, I hope that even the expert reader will find something new in this book. The author's duty of tabulating all the historical records has sometimes been unexpectedly rewarding, and I was particularly pleased to discover or rediscover the following:

long-term trends in the size and speed of the Great Red Spot, paralleling those of the South Tropical Disturbance and South Temperate white ovals, which strongly suggest that the Great Red Spot originated as recently as 1700;

the detailed Voyager record of the origin of a South Tropical Disturbance;

changes in the speed and latitude of the major retrograde jetstream, possibly related to South Tropical Disturbances;

confirmation that reddish colour is a usual sequel of intense disturbances in many latitudes;

reproducible cycles of activity in the North Equatorial Belt, involving broadening, reddening, and the formation of ovals;

a change in speed and style of outbreaks in the fastest jetstream in the 1970s, soon after the onset of a new type of coloration event in the adjacent North Temperate Belt;

confirmation of the 10-year periodicity in North Temperate Belt latitude, and explanation of it in terms of selective fading.

Conventions and abbreviations

The names and abbreviations of the belts, zones, jetstreams, and currents are given in Fig. 1.3 and Figs. 3.1&2. Other abbreviations include GRS (Great Red Spot), STropD (South Tropical Disturbance) and FFR (folded filamentary region).

A standard system of terms and units has long been used in visual reports on Jupiter. Recently, completely different terms and units have been used by space scientists. The two systems are described in Chapter 1.3 (Table 1.3). In this book, the visual conventions are used, both for historical continuity and for their intrinsic convenience. Thus all images of the planet are shown with south up. Planetary east is termed 'preceding' (p.), and planetary west is 'following' (f.). Longitudes are measured in System I (equatorial region and NTBs jetstream) and System II (all other latitudes). Speeds are given in degrees longitude per 30 days ($\Delta\lambda_1$ or $\Delta\lambda_2$). Latitudes are zenographic.

In the chapters on the satellites, as most of the data are from spacecraft, images are presented with north up unless otherwise stated.

References

Anyone of my generation learning about Jupiter must be indebted to the book by B.M. Peek (*1958*), *The Planet Jupiter*. Peek was one of the most famous Directors of the BAA Jupiter Section, and his book summarised the observations and understanding of the planet at that time. References to 'Peek' in the present text are to that book unless otherwise stated.

Sources for observations of the planet are the apparition reports by various organisations, listed in Appendix 3. For specific details, if no other organisation is specified, the information is from the BAA reports. Other organisations are abbreviated as follows: ALPO (Association of Lunar and Planetary Observers), LPL (Lunar and Planetary Laboratory of the University of Arizona), NMSUO (New Mexico State University Observatory), RAS (Royal Astronomical Society), SAF (Société Astronomique de France), SAI (Società Astronomica Italiana), SAS (Société Astronomique de Suisse).

References which are not in these series are cited in the text or in footnotes by name and date of publication (the date being in italics where necessary to avoid confusion with dates of observations). The full references are given in Appendix 4 (for Parts I–IV, the planet) and Appendix 5 (for Parts V and VI, the magnetosphere and satellites).

Acknowledgements

First, this work has depended on the many observers, of the BAA and other organisations, who have studied the planet over the years. It is their assiduous work which has made possible our present knowledge of the planet. I thank the BAA Council for permission to use BAA materials freely in this book. I am also grateful to the Royal Astronomical Society for access to their archives and for permission to reproduce illustrations from them. For my own observations, I would like to acknowledge the use of telescopes of the University of Cambridge and of the University of California at Los Angeles.

Equally, all of us interested in planets have a historic debt to the National Aeronautics and Space Administration (NASA), to the researchers responsible for the experiments on NASA spacecraft, and to the United States taxpayers who have supported the whole enterprise of space exploration. In addition to sending out the space missions themselves, NASA have (until very recently) provided spacecraft data and images either free of charge or at minimal cost, to interested researchers such as myself. Without this enlightened policy, much of this book could not have been produced. I am also grateful to the Department of Physics, Imperial College, London, for access to the NASA archive of Voyager images and maps. I have measured some of these, including stripmaps produced by the Voyager Imaging Team, to establish the drift rates of features shown in Fig. 6.4, Fig. 8.10, Fig. 10.28, and Table 12.2B.

The Voyager images were produced by the Voyager Imaging Team (leader Dr. Bradford A. Smith), and were obtained for this book as follows.

(1) Public release images from NASA.
(2) Other images of the planet were kindly provided as prints by the National Space Science Data Center, Greenbelt, Maryland, through the World Data Center A for Rockets and Satellites. (These are cited by numbers in the format '12345.12'; many are previously unpublished.)
(3) Other images of the satellites were obtained from CD-ROM discs purchased from the same source. Processing of these images, using the programs Procyon Common Lisp (Scientia Ltd.) and PhotoFinish (ZSoft Corp.), was expertly and generously done by Simon Mentha.

Original illustrations have also been kindly provided by the following: Tomio Akutsu, David Allen, Richard Baum, Barbara Carlson, S. Cortesi, Imke de Pater, Audouin Dollfus, Pierre Drossart, Marco Falorni, Robert Gaskell, Dieter Gerdes, Joseph Harrington, David Jewitt, M.J. Klein, Sanjay Limaye, A.S. McEwen, Richard McKim, Brian McLeod, Michael Mendillo, Hans-Jörg Mettig, Steven Miller, Régis Néel, P.D. Nicholson, Glenn Orton, Peter Read, Takeshi Sato, Nick Schneider, James Secosky, John Spencer, Harold Swinney, Gareth Williams.

Illustrations from journals, referenced in Appendix 4 or 5 according to citations in the figure legends, are reproduced by kind permission of the publishers, and of the authors where they could be contacted. The publishers and copyright holders are as follows: Academic Press (*Icarus*); Macmillan Magazines Ltd. (*Nature*); American Association for the Advancement of Science (*Science*); American Geophysical Union (*Journal of Geophysical Research* and *Geophysical Research Letters*); American Meteorological Society (*Journal of the Atmospheric Sciences*); Pergamon Press Ltd., Headington Hall, Oxford (*Planetary and Space Science*); American Astronomical Society (*Astrophysical Journal* and *Astronomical Journal*).

I am also grateful to the University of Cambridge for library and photocopying facilities.

Finally, I express my thanks to the following people who have kindly read drafts of chapters for this book: Peter Cattermole, Michael Foulkes, David Graham, Alan Heath, Nigel Henbest, Stephen Lewis, Richard McKim, Simon Mitton, Glenn Orton, David Southwood, Robert West, and Gareth Williams. I am particularly grateful to Dr Lewis and Professor Williams for the time they have spent trying to improve my understanding of atmospheric physics. The comments of all these people have been invaluable but the final responsibility for any inadequacies in the text is of course my own.

John H. Rogers
1993 August 5

I. Observing Jupiter

Jupiter is a world that can fascinate in many different ways.

To amateur astronomers, Jupiter can be an object of interest on every timescale. Over a matter of minutes, one can watch a moon and its shadow gliding across the edge of the planet, while one scrutinises the patterns in the clouds below. Over an hour or more, the planet's rotation brings a parade of diverse storms round to view. Over several days or weeks, these features move or evolve. Over months or years, the overall pattern of features and colours may change, often comprising great cycles of disturbance. And reviewing the observations over a span of decades, one finds that the nature of these cycles themselves has subtly changed, so that patterns may reappear that had last been seen a century before.

To physicists, Jupiter is the greatest laboratory for atmospheric dynamics. It has weather and climate and circulations that dwarf those of the Earth. Many of the largescale phenomena are still unexplained. Moreover, it has a magnetic field and radiation belts greater than those of any other planet.

To cosmologists, Jupiter may be a tiny speck, but it is the only representative that we can study of a huge class of objects: bodies intermediate in scale between the other planets and the stars. Indeed, bodies such as Jupiter may dominate the universe. As most of the mass of the galaxies cannot be seen, rival theories for the 'missing mass' range from exotic subatomic entities such as WIMPs ('weakly interacting massive particles') to more conventional bodies such as MACHOs ('massive compact halo objects'). MACHOs could be planets like Jupiter.

To lovers of landscape, Jupiter's moons offer outlandish vistas, of icy ridges or frozen oceans or rugged mountains or sulphur-spurting volcanoes.

And to art-lovers, Jupiter displays the structured but spontaneous patterns, the unpredictable symmetries, and the swirling colours, that produce an absorbing work of art.

Fig. 1.1. Jupiter is the brightest 'star' in the midnight sky. Here it stands in front of the star cluster called Praesepe in the constellation Cancer. Photograph by R. Néel on 1991 Feb. 2.

Fig. 1.2. Jupiter and its moons as seen through a telescope at low power. One can see the main belts, the Great Red Spot, the shadow of one of the moons, and the four moons themselves lying approximately in the planet's equatorial plane.

1: Observations from Earth

1.1 A VIEW OF JUPITER

To gaze at the giant planet Jupiter through a fair-sized telescope is to be transported to a wonderfully alien world. This largest planet in the solar system is partitioned by bands of variously tinted clouds, some straight and others ruffled with the turbulence of fierce winds. White and dark ovals mark atmospheric storms, some as large as the Earth. The planet rotates in front of your eyes; after no more than 10 minutes you can see that the rapid rotation is carrying the features across the disc. If you know the planet well, these features may mark out familiar patterns, but sometimes they reveal new and fascinating forms of upheaval in the clouds. Meanwhile, orbiting around the planet, you see the four 'stars' that are its four great moons. Often one of them will be seen slipping behind the planet, or gliding in front of it and casting its shadow on the clouds.

All that we see on Jupiter is clouds. The dark *belts* and the bright *zones*, and the various types of spots, are clouds of different thicknesses and colours. The belts and zones are marked out by powerful jetstreams, which are permanent winds blowing eastward or westward. Spots lying between them are storms which drift eastward or westward at slower rates.

The reason why the belts and the winds all run east–west is the planet's rotation. With a period of 9 hours 55.5 minutes, this is so fast that the poles are noticeably flattened; the polar diameter is only 14/15 of the equatorial diameter. So all the motions are channelled along lines of latitude.

The main belts and zones have been given names, and although they may change or disappear now and then, they always reappear in the same positions (Fig. 1.3). We know that the belts and dark spots are warmer than the zones and bright spots, due to heat emerging from deep in the interior. The highest and coldest feature is the Great Red Spot, which is a reddish oval circulation with twice the diameter of the Earth. The Voyager spacecraft have shown that all the 'spots' are vortices, which appear to roll between the jetstreams.

All these motions take place in an impenetrably deep atmosphere. A visible sign of the atmosphere is 'limb-darkening'[1], that is, a diffuse shading close to the edge of the visible disc, which is caused by absorption and scattering of light in the atmosphere.

The atmosphere, and indeed the whole planet, are composed mostly of hydrogen and helium, mixed with smaller amounts of ammonia, methane, and simple hydrocarbons. The clouds are believed to consist mainly of ammonia ice crystals, ammonium hydrosulphide, and water ice. There is probably no solid surface. In fact, the composition of Jupiter is believed to be much the same as that of the Sun or of the galaxy as a whole, because Jupiter, unlike Earth, has powerful enough gravity to hold onto even the lightest gases.

Jupiter orbits the Sun 5.2 times further out than the Earth. At this distance the sun's light and heat are only 1/27 as strong as at Earth, so without any other source of heat the planet and its satellites would be very cold. However, the atmosphere gets almost as much heat from the interior of the planet as it does from the Sun, and this seems to be what drives the storms. The interior heat is a result of the planet's great size; it heated up so much when it was forming 4600 million years ago, and it has such a great mass, that it has not yet cooled down.

And Jupiter is truly immense (Table 1.1). It measures 11 times the diameter of the Earth, 1323 times its volume, and 318 times its mass. In fact it has more mass than all the other planets put

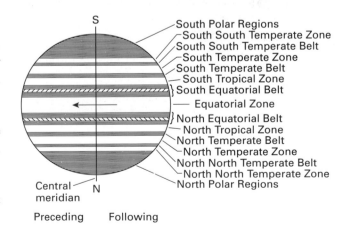

Fig. 1.3. The normal belts and zones, with the standard nomenclature. (BAA diagram.)

---

[1] The following definitions may be useful. The *limb* is the edge of the visible disc. The *terminator* is the boundary between the illuminated and darkened sides of the disc. The *phase angle* is the angle between Sun, planet, and observer; for observers of Jupiter it never exceeds 11.7°.

Table 1.1 *Physical parameters of Jupiter and Earth*

	Jupiter	*Earth*
Diameter, equatorial*	143 082 km	12 756 km
Diameter, polar*	133 792 km	12 714 km
Rotation period:†		
System I	9h 50m 30.003s	
	(877.90°/d)	
System II	9h 55m 40.632s	
	(870.27°/d)	
System III	9h 55m 29.711s	23h 56m 4s
Tilt (relative to orbit)	3.12°	23.44°
Mass	1.899 x 10^{27} kg	5.974 x 10^{24} kg
Density	1.32 g/cm³	5.52 g/cm³
Surface gravity**	2.69 *g*	1.00 *g*
Mean geometric albedo††	0.52	0.37

Data are from the *BAA Handbook (1991)* unless otherwise stated.
* The diameters of Jupiter (±8 km) are from Lindal *et al.*. (*1981*), at the
100-mbar level. Subtract approx. 50 km for cloud-top level and 100 km
for 1-bar level.
† Rotation periods are sidereal, i.e. measured with respect to the fixed
stars. The exact definition is the value in degrees per day (24 hours).
** The effective surface gravity at the equator of Jupiter is 9% less due to
the centrifugal force of rotation.
†† Albedo is a measure of the fraction of light reflected from an object.
The geometric albedo is the ratio of the object's brightness to that of one
which diffusely reflects all incident light, under vertical illumination.
Another measure of albedo, the absolute reflectivity, varies from 0.62 in
NEB to 0.76 in STropZ (Orton, *1975*).

Table 1.2 *Orbits of Jupiter and Earth*

	Jupiter	*Earth*
Mean distance from Sun	5.20280 AU	1.00000 AU
Perihelion distance	4.951 AU	0.983 AU
Aphelion distance	5.455 AU	1.017 AU
Eccentricity	0.04849	0.01671
Period (sidereal)	4332.59 d	365.26 d
Inclination	1.304°	0.000°
Mean synodic period (time between oppositions)	398.88 d	—

1 AU (astronomical unit) = 149 597 870 km.
Data are from the *BAA Handbook (1991)*.

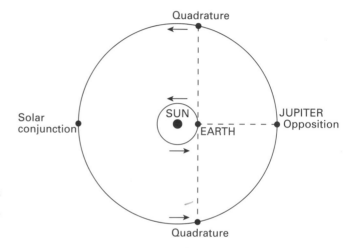

Fig. 1.4. Relative positions of Jupiter and Earth in orbit around the sun.
(Because Earth's orbit is tilted by 1.3° relative to that of Jupiter, the align-
ments at opposition, conjunction, and quadrature are not always exact.
They are defined more precisely as the times when the longitudes of the
Sun and Jupiter differ by 180°, 0°, and 90°, measured along the ecliptic
plane.)

together. In volume, it is close to the largest possible size for a gas
giant planet – as more massive planets, having stronger gravity,
would have smaller radius.

Jupiter's orbit

Jupiter takes 11.86 Earth years to complete one revolution around
the Sun (Table 1.2). As a result, it moves around the zodiac by
about one constellation per year. *Opposition* is the time when, from
our point of view, the planet is opposite the Sun in the sky, reaching
its highest altitude at midnight (Fig.1.4); the average interval
between oppositions is about 13 months. (The interval varies
slightly because both Jupiter and Earth have elliptical orbits; the
actual dates of opposition are given in Appendix 3.) *Solar conjunction* is the time when Jupiter is on the other side of the Sun from us,
and therefore cannot be seen for a few months. *Quadrature*, when
the planet is 90° away from the Sun in the sky, is when the planet
shows the maximum phase; in Jupiter's case the true phase is too
small to be detected, although the limb away from the Sun does
appear noticeably darkened.

The interval between solar conjunctions, during which we can
observe Jupiter, is called the *apparition* and lasts between 8 and 10
months. Each apparition starts with the planet rising just before the
sun as a 'morning star', proceeds through opposition, and ends with
descent into the evening twilight. The best oppositions, regrettably
but inevitably, come at the coldest time of the year, when the planet
is highest and the sun lowest in the sky; then the planet rides high
during the long winter nights. In summer apparitions, it is not so
easy to observe from latitudes as high as that of Great Britain,

because the planet is always low in the sky. Accounts of events on
Jupiter are usually dated to the apparition, which will be written in
the form 1990/91 when it bridges two calendar years.

The ellipticity of the planet's orbit also makes some difference to
observations. At oppositions when it is furthest from the Sun (at
aphelion), the magnitude is –2.3 and the apparent diameter is 44
seconds of arc. At oppositions when it is closest to the Sun (at *perihelion*), the magnitude is –2.9 and the apparent diameter is 50 seconds of arc. These perihelic oppositions fall in October, and as they
also combine high altitude in the sky with a reasonable chance of
good weather for northern hemisphere observers, these apparitions
tend to be the best observed of all.

If you have a telescope of only 5 cm aperture, you will be able to
see the main belts. A 7.5-cm telescope will reveal the largest spots
and irregularities, and the shadows of the satellites. A 15-cm telescope will show enough detail for useful observations to be made.
A 25- or 30-cm telescope will show most of the important features
of the planet.

1.2 HISTORY OF VISUAL OBSERVATION

The first person to look at Jupiter with a telescope was Galileo Galilei, in 1610 January (Fig. 1.5). His telescope was too small to show anything on the planet itself, but he could see the four moons. It was this discovery that convinced him that the Earth was not the centre of all motions in the universe. He described his discovery in *Siderius Nuncius* ('Message of the Stars'), published in 1610 March, declaring on the title page that he was:

> Revealing great, unusual, and remarkable spectacles . . . above all in FOUR PLANETS swiftly revolving about Jupiter at differing distance and periods, and known to no one before the Author recently perceived them and decided that they should be named THE MEDICEAN STARS.

(Nowadays, anyone can see them with a pair of binoculars, and they are named not after the princes Medici who were Galileo's patrons, but after the discoverer himself: the Galilean moons.) He described them thus:

> On the seventh day of January in this present year 1610, at the first hour of night, when I was viewing the heavenly bodies with a telescope, Jupiter presented itself to me; and because I had prepared a very excellent instrument for myself, I perceived (as I had not before, on account of the weakness of my previous instrument) that beside the planet there were three starlets, small indeed, but bright. Though I believed them to be among the host of fixed stars, they aroused my curiosity somewhat by appearing to lie in an exact straight line parallel to the ecliptic, and by their being more splendid than others of their size. Their arrangement with respect to Jupiter and each other was the following:

East   *   *   O   *   *West*

> [Observing them over the next three nights, he found that they moved to and fro around Jupiter, and on Jan. 13 he first saw four moons at once . . .] I had now decided beyond all question that there existed in the heavens three stars wandering about Jupiter as do Venus and Mercury about the sun, and this became plainer than daylight from observations on similar occasions which followed. Nor were there just three such stars; four wanderers complete their revolutions about Jupiter . . . the revolutions of these planets are so speedily completed that it is usually possible to take even their hourly variations. (Galileo, *1610*; translation by Drake, *1957*)

Discovery of features on the planet itself had to wait for several decades while telescope optics were gradually improved.[2]

The banded appearance of the disc was first described by Niccolo Zucchi in 1630, and soon confirmed by other Italians (Fig. 1.6). The transits of satellite shadows were first seen by G. Riccioli about 1643.

The earliest reported sightings of true spots were in 1664 by Robert Hooke in England and in 1665 by Guiseppe Campani of Rome. (Campani was famous as the best telescope maker of his day.) These spots revealed for the first time the rotation of the planet. The reports consisted simply of two paragraphs in the very first issue of the *Philosophical Transactions* of the Royal Society of London (*1665*), viz.:

> Campani affirms he hath observed by the goodness of his glasses certain protuberancies and inequalities, much greater than those that have been seen therein hitherto. He addeth, that he is now observing whether those sallies in the said planet do not change their scituation; which if they should be found to do, he judgeth that Jupiter might then be said to turn

[2] The 17th century observations were reviewed by Denning (*1899*), Antoniadi (*1926*), Chapman (*1968*), and Falorni (*1987*). The history of published observations of Jupiter up to 1878 was reviewed in detail by Hockey (*1989*). See also original descriptions by Cassini (*1666, 1672*).

Fig. 1.5. Galileo Galilei. (From the frontispiece to his collected works; by courtesy of the Institute of Astronomy, Cambridge.)

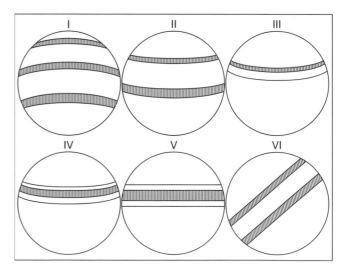

Fig. 1.6. First observations of Jupiter's belts. Drawings by Italian Jesuit priests, from *Almagestum Novum* by Father Riccioli (*1651*). I: 1639 (Fontana, at Naples). II: 1644 (Zupo, at Naples). III,IV,V: 1643 (Grimaldi, at Bologna). VI: 1648 (Grimaldi, at Bologna).

upon his Axe, which, in his opinion, would serve much to confirm the opinion of Copernicus [which however was considered to be that the moons dragged the planet round!]. Besides this, he affirms, he hath remarked in the belts of Jupiter the shaddows of the satellites, and followed them, and at length seen them emerge out of his disk.

The ingenious Mr. Hook did, some moneths since, intimate to a friend of his that he had, with an excellent twelve-foot [focal length] telescope, observed, some days before he then spoke of it (viz. on the 9th of May 1664, about 9 o'clock at night) a small spot in the biggest of the three obscurer belts of Jupiter, and that, observing it from time to time, he found that, within two hours after, the said spot had moved from east to west, about half the length of the diameter of Jupiter.

It is commonly assumed that this spot of Hooke's was the same as the long-lived dark spot followed by G.D. Cassini from 1665 to 1694. This claim was made in *Phil.Trans.* at the time and supported by the Royal Society thereafter; but this support may have owed something to Hooke's senior position, not to mention his notoriously argumentative and jealous character. The claim is inconsistent with Hooke's actual description quoted above, which seems to place the spot in the North Equatorial Belt.

The 'permanent spot', which may be identical with the present Great Red Spot, was first recognised by Giovanni Dominico Cassini in Italy in 1665. In fact the first sighting, on 1665 July 9 alongside the shadow of Ganymede, was by Cassini's friend and instrument maker (and Campani's rival), Eustachio Divini of Rome. But it was Cassini who identified it as a fixed feature and followed it on and off for many years thereafter. It was a large dark oval spot in the STropZ, but it seems to have been smaller than the present GRS (though its form cannot have been clearly observed with the primitive optics of the time). Its rotation period of just under 9h 56m was much slower than the present GRS has ever shown.

Cassini began his Jupiter researches in Italy, determining rotation periods, but carried out most of his work at Paris, as Director of the newly-built Royal Observatory of Louis XIV (Figs. 1.7–9). Among his many important researches on the solar system, he began the serious visual study of Jupiter. The telescopes of that time were bizarre contraptions with lenses of extremely long focal length, typically suspended on a mast, with the eyepiece in a separate framework that had to be guided to follow the planet. This arrangement was needed to overcome the problems of chromatic and spherical aberration that plagued lenses of shorter focal length. (At that time, neither achromatic lenses made of two pieces of glass, nor reflecting telescopes, had not yet come into use.)

Cassini discovered not only Jupiter's 'permanent spot', but also its equatorial current, its polar flattening (codiscovered by Picard), and its limb darkening (with his nephew Maraldi). He noted changes in the widths of the belts, and the birth and evolution of bright spots, and he realised that these must be clouds in a substantial atmosphere.

When the Great Red Spot was recognised in 1879, it was soon suggested that it was a rediscovery of Cassini's spot, and this has often been quoted as fact. But the identity of the two spots is by no

Fig. 1.7. Jean-Dominique Cassini. (From the Observatoire de Paris.)

means certain. There were no observations of any such feature between 1713 and 1831 – an apparently unbridgeable gap.

In the eighteenth century, few astronomers paid any attention to the surface markings of the planets. Newton's theory of universal gravitation set the agenda of astronomy for over a century; the main aim was to measure positions and movements of planets, stars, and comets with ever-inceasing accuracy, and so to work out the dynamical structure of the universe. Even when planetary surface observations were made it was mainly to determine accurate rotation periods.

William Herschel (*1781*) published sketches and timings of spots on Jupiter in 1777–1778, but abandoned the planet on finding that the rotation periods were diverse and variable; for his purpose of providing a check on the constancy of the rotation of the Earth, he turned to observations of Mars instead.

The most assiduous work was done in Germany. Johann H. Schröter, a magistrate in Lilienthal, observed from 1785 to 1797, and published his reports in 1788 and 1798 with woodcuts showing belts and dark spots (Figs. 1.10–12). He re-observed the Equatorial Current and discovered the North Tropical and South Temperate Currents, and noticed yellow colouring in the Equatorial Zone.

Samuel Heinrich Schwabe in Dessau trained as an apothecary but then became a full-time astronomer, and his decades of meticulous sunspot observations revealed the solar cycle for the first time.

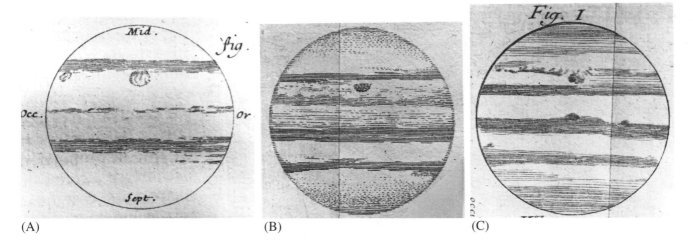

(A)　　　　　　　　　　　　(B)　　　　　　　　　　　　(C)

Fig. 1.8. Drawings of Jupiter by Cassini, including the 'Permanent Spot'. A: 1672 January; B: 1677 July; C: 1691 January. South is up. (From *Memoires de l'Académie Royale de Paris*, tome 10; reprinted from Falorni (*1987*); by courtesy of M. Falorni.)

Fig. 1.9. Paris Observatory in the early 18th century. (From *Tables astronomiques du Soleil, de la Lune, des planètes...* by J.J. Cassini (*1740*).)

He also drew Jupiter from 1827 to 1865, several times a year on average, carefully recording the surface features; it was in his drawings of 1831 that the Great Red Spot, or rather the 'Hollow' that it impresses in the South Equatorial Belt, was first recorded (Plate P1). (Most of his material is unpublished, in the archives of the Royal Astronomical Society.) In 1843, Franz von Paula Gruithuisen, a professor in Munich, published drawings showing the changes in the major belts from 1836 to 1843. In the 1860s,

Julius Schmidt, director of the Athens Observatory, confirmed the same currents that Schröter had recorded.

In England, the Rev. William Rutter Dawes and William Lassell began their noted planetary observations in the 1840s. From 1857 onwards, more observers with good-quality reflecting or refracting telescopes began to take interest in the markings on Jupiter, and many detailed and beautiful drawings began to appear in the *Monthly Notices of the Royal Astronomical Society, Astronomical*

Fig. 1.10. Johann H. Schröter. (Painted from a copperplate of 1791; by courtesy of Dieter Gerdes, Schroeter Museum, Lilienthal.)

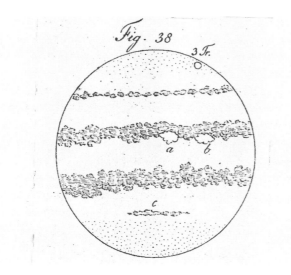

Fig. 1.11. Drawing by Schröter in 1796/97. South is up. (From Schröter, *1798*.)

Register, and later in *The Observatory* – at least in the years when the planet was at high altitudes for British observers, around 1857–1860, 1869–1872, and 1879–1883. Interest was maintained by the fortunate occurrence of striking phenomena in these favourable apparitions, so these years saw the discovery of equatorial coloration episodes, South Temperate ovals, and the NTBs jetstream. Other features recorded in these years can now be recognised as revivals of the South and North Equatorial Belts, and South Tropical Disturbances. (See Plates P1–P3 and P12.)

Most striking of all, in 1878–1881, was the emergence of the Great Red Spot. This enormous and vividly red marking, which was unlike anything seen before, attracted so much attention from astronomers worldwide that the planet was never again to be left unobserved.

There was also a technical reason for the great upsurge of observations around this time: the invention, in the mid-nineteenth century, of silver-coated glass as a material for telescope mirrors. Compared with the previous mirrors ground from the 'speculum' metal alloy, silver-on-glass mirrors (or aluminium-on-glass mirrors, as are now used) were cheaper, lighter, more reflective, and easier to maintain. Thus good telescopes came within the reach of amateurs with only moderate financial means.

The physical nature of the planet was an open question in the nineteenth century. Early astronomers tended to assume, for lack of contrary information, that the planets were like the Earth with a solid surface and atmosphere. Indeed, there was a widespread feeling that the planets, as creations of divine purpose, must be inhabited by intelligent beings. So the presence of clouds on Jupiter did not come as a surprise; the question was whether any markings represented a solid surface. At one extreme, the planet was considered cold, the white markings being attributed to snow. Conversely many writers in the late nineteenth century, considering its low density, great brightness, and general turbulence, concluded that it was very hot, and perhaps even glowing like a feeble sun. Extensive parallels with the Sun were noted: the limb darkening, fast equatorial current, changeable spots in definite bands, and cycles of activity.[3] Fortunately, these theories do not seem to have had much influence upon the serious observers. The final outcome has been a compromise. The cloud-tops are indeed frigid, as was established in the 1920s; but the turbulence of the atmosphere is indeed powered largely by internal heat, as discovered in the 1960s. E.M. Antoniadi, writing in 1926, was not far wrong:

> Thus the idea that Jupiter is a chilled sun, like the other planets, conforming to the cosmogonical hypothesis of Laplace, seems the only one admissible. Its temperature is closer to that of the Earth than to that of the Sun; and perhaps it is not totally lacking its own light. However all that does not get us much further; for, even if the material is the same on that world as this, the conditions of temperature are so different that the nature of its phenomena, like those of the Sun, is inaccessible to us. (Author's translation.)

By the 1880s, it was apparent that different latitudes showed different rotation periods, implying a regular pattern of winds, and so a major observational aim was to determine this pattern in detail. This task was made easier when longitude systems were defined and tables of central meridian longitudes published, by A. Marth (routinely referred to as Marth's Invaluable Ephemerides). The definitions of Systems I and II were stabilised by Marth in 1892.

A total of nine currents or jetstreams were known when A.S. Williams made the first systematic listing in 1896. He had himself measured many of these currents in 1887 and 1888 (published in *1889, 1909*). But measurement of currents was first done exten-

---

[3] Readers interested in the knowledge and opinions about the planet at various historical times may enjoy reading the reviews by: Green (*1887*), Hough (*1905*), Kritzinger (*1914*), Phillips (*1915*), Denning (*1923*), Antoniadi (*1926*), and Wildt (*1969*). For reviews just before spacecraft arrived, see Newburn and Gulkis (*1973*) and the special issue of *Journal of Atmospheric Science*, vol.26 (no.5.I) (*1969 September*).

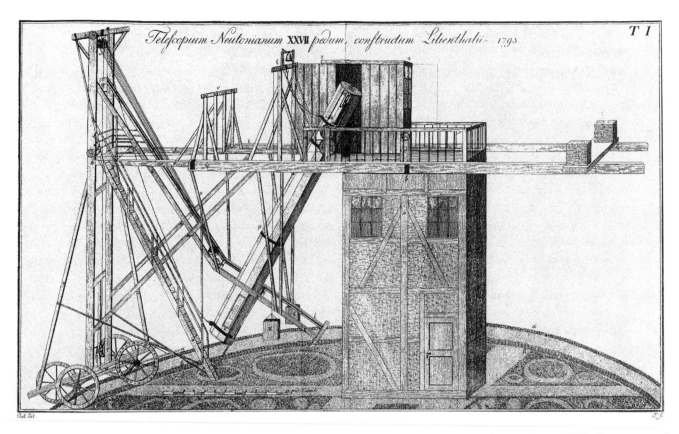

Fig. 1.12. Schröter's great reflector of focal length 27 feet (8.2 metres), used for his later Jupiter observations. (His earlier observations were made with a Herschel reflector of focal length 7 feet, 2 metres. From *Aphroditographische Fragmente* by Schröter (*1796*); by courtesy of Richard Baum.)

sively in 1898, by W.F. Denning, P.B. Molesworth, J. Gledhill, and T.E.R. Phillips. They measured longitudes simply by estimating the time at which spots crossed the central meridian – the imaginary line down the centre of the disc.

Several of these English gentlemen were memorable individuals. A. Stanley Williams was professionally a solicitor, and in his spare time a keen yachtsman. Although his observations were made with a mere 6½-inch (16.5-cm) reflector, he will figure repeatedly in this book for his systematic records both of drifts and of colours on Jupiter. William F. Denning, an accountant, persevered with astronomy despite his lack of private funds and, in later years, poor health; he laid the foundations of meteor astronomy as well as recording Jupiter's currents. Captain P.B. Molesworth (later Major; Fig. 1.13) was stationed with the Royal Engineers in Ceylon (now Sri Lanka). Viewing the planet at high elevation in the clear tropical skies with a 12½-inch (32-cm) reflector, he was the most indefatigable measurer of longitudes; in 1900 he recorded 6758 central meridian transits! And the Rev. T.E.R. Phillips (Fig. 1.14), one of a distinguished line of Anglican clergymen who became famed for their researches in natural science, would later become the greatest director of the BAA Jupiter Section.[4]

[4] One should also note the drawings from the 1880s by Nathaniel E. Green, a much-travelled professional artist and art teacher (whose pupils once included the royal family). The RAS published his Jupiter Memoir in 1887. Noting that it had once been said of him that he preferred an artistic drawing to a correct one, he replied 'I know of no difference between the two'. The 18-inch (46-cm) reflector used by Green was the same one later used by Phillips.

Fig. 1.13. Major P.B. Molesworth. (From *Memoirs of the BAA, vol. 16* (*1910*).)

Thanks to their work, by 1901 all the now-recognised slow currents had been recorded, as well as the equatorial and North Temperate jetstreams.

The distinction between amateurs and professionals was increasing in the late nineteenth century. This led to the founding, in 1890, of the British Astronomical Association (BAA), as an offshoot of the Royal Astronomical Society catering especially for amateurs. This was the start of a long, systematic, and fruitful study of Jupiter which continues to the present day. The BAA Jupiter Section's first director was the Rev. W.R. Waugh who, at the age of 72, published its first Memoir for the apparition of 1891. (The present names of the belts were defined in the second Memoir, for 1892.) The BAA Memoirs continued in an unbroken series until 1943, and reports have also been produced for almost every apparition since, either in the BAA Journal or as occasional Memoirs (Appendix 3).

In reviewing the work of observers in those years, one has to bear in mind that people differed in their abilities then as now. Some observers, though good at certain observations, may be unskilled as artists, or inaccurate in timing transits, or uncritical in analysing them. It is the function of an amateur society to encourage interest among many members, but it is the function of a director of an observing section to assess the observations and to base the published report on those that are reliable. Further comments about the skills of some observers will follow as they pertain to particular topics. In spite of these comments, one must re-emphasise that the early visual observers were great and pioneering researchers, who applied their observing skills and their scientific instincts to establish a great deal of knowledge about the planet. Their work has stood the test of time.

The first BAA Memoirs were simply compilations of separate members' descriptions and drawings. Within a few years, measurements of rotation periods grew more numerous, particularly by Denning and Molesworth from 1898 to 1903.

The Rev. T.E.R. Phillips, as Director of the BAA Jupiter Section, in 1914 began to plot more reliable drift charts by combining the transits from all the best observers. Thus he commenced the detailed and rigorous analysis that distinguished the BAA Memoirs under his directorship (1901–1934) and that of his successor, Bertrand M. Peek (director 1934–1949; Fig. 1.15). Both men were also outstanding observers, and with F. James Hargreaves they formed a group who often observed together at Phillips' observatory at Headley, Surrey, where Phillips was rector of the church. Phillips was known as a most kind and unassuming man who was the natural leader in his field. Peek was a school headmaster at Solihull, and Hargreaves was a patent agent in Coulsdon, Surrey, who eventually became a professional telescope-maker. These three colleagues observed diligently with sizable reflectors, and persevered through the vagaries of the English weather so as to produce a thorough record of the activity on the planet. Peek rounded off his observing career by writing his book, *The Planet Jupiter*, published in 1958, which ever since has been the definitive work on the visible features on the planet.

After 1942, the BAA team was broken up by distractions and disabilities and deaths. Although the Section still included keen observers (notably William E. Fox, a hydraulic engineer of Newark, Nottinghamshire, whose directorship from 1957 to 1988 occupied him during his years of retirement) they did not have the large telescopes and the intensiveness of observation to produce such detailed reports as in previous years. Some useful reports were published by European groups, and Japanese amateurs made some fine observations, but coverage was uneven. From 1949 to 1964, the most consistently good reports were from the American amateur group, the Association of Lunar and Planetary Observers, most of whose reports were compiled by Elmer J. Reese. Even these depended mainly on small apertures; so some smallscale phenomena, such as jetstream outbreaks, might have been missed during the 1950s. Reese initially worked in his family's grocery store, and used only a 6-inch (15-cm) reflector, but had a critical eye that enabled him to see much detail that was independently confirmed. In 1963/64 he joined the New Mexico State University Observatory as a professional astronomer, and inaugurated an era of professional photographic coverage of Jupiter.

1.3 METHODS OF VISUAL OBSERVATION

What can amateur astronomers do in the way of serious study of Jupiter? Much the same as they have done for a century – leaving aside for now the modern possibilities of photography and digital imaging. Visual observers can see what types of spots are present, how fast they move in the zonal winds, and what changes occur in

Fig. 1.14. The Rev. T.E.R. Phillips. (BAA.)

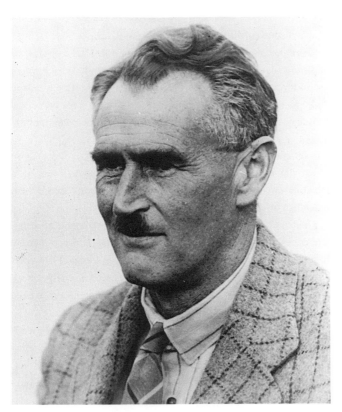

Fig. 1.15. Bertrand M. Peek. (BAA.)

the colour and darkness of the belts and zones. There is every reason to continue this coverage into the future, both to maintain a continuous and consistent record, and to be alert for new events.

This is possible thanks to the fact that the typical scale of the major phenomena is within the grasp of a moderate-sized amateur telescope. A 30-cm reflector or 20-cm refractor will show virtually all the important features, and useful observations can be made with apertures as small as 15 cm.

A large aperture has three advantages. First, resolution, which is the limiting factor in observing planets: optical diffraction restricts the resolution to 11.6 seconds of arc divided by aperture in centimetres (a formula known as the Dawes limit). Second, light grasp; this is unimportant for most planets, but is relevant for Jupiter in that it allows one to magnify the image while retaining enough light to see subtle contrasts. Third, focal length: larger aperture is often associated with longer focal length, which determines the available magnification. ('Light buckets' with large aperture but short focal length, used for looking at faint comets and nebulae, are less suitable for planets even if the primary mirror is well figured.)

The eye has a resolution of 1–2 minutes of arc, that is, one can separate two points 1–2' apart, given sufficient contrast. Therefore the telescope must magnify the image enough that details appear substantially larger than 1–2'. The magnification is determined by the focal length of the primary mirror or lens, and by that of the eyepiece. A magnification of at least 120x is needed to resolve useful detail, and one should not magnify above 20x per centimetre of aperture (e.g. 600x for a 30-cm objective) because of the Dawes limit. Within these limits, magnification is a matter of personal preference. With Jupiter, it can be as important to increase contrast between features as to increase their apparent size, and a lower power gives better contrast.

Often, though, the resolution is limited not by the telescope but by turbulence in the Earth's atmosphere – the 'seeing'. The image seems to ripple or, at worst, to boil. Some of this is due to local convection, which can be minimised by suitable siting of the telescope, and by opening up the telescope some time before observing so that it can cool down. But much of the seeing arises higher up and one just has to put up with it as best one can. The seeing is estimated on a scale named after E.M. Antoniadi, from I (perfect) to V (gross boiling) (or, in North America, on an inverse scale from 10 to 0).

The observer's notes on observations of Jupiter, as for any other astronomical object, should include the telescope used, the seeing, and the time (in Universal Time, that is, Greenwich Mean Time).

The serious amateur will want to combine his or her observations with those of others, and this is done by national amateur societies such as the BAA which organise observing sections for Jupiter. A specimen of the report form used by the BAA is given in Appendix 1.

As most telescopes invert the image, and most observers are in the northern hemisphere, the planet is generally viewed with south at the top.

Observations of longitude

Longitude measurements have always been the main work of observers. They are necessary to identify individual spots, so as to work out their drift rates and assign them to the various currents.

As there are no fixed features on Jupiter, arbitrary systems of longitude are used with constant rotation periods (Table 1.1). And as different latitudes rotate at very different rates, two such systems are used. System I, with a period of 9h 50m 30.0s that is close to the average for the equatorial region, is used for latitudes between 10°N and S; System II, with a period of 9h 55m 40.6s that is close to that of the Great Red Spot, is used for all other latitudes. Professional astronomers also use another system, System III, with a period of 9h 55m 29.7s, that is the rotation period of the interior of the planet as deduced from radio observations.

The only equipment needed to measure longitudes visually is a wrist-watch or clock accurate to better than half a minute. A spot's longitude is simply estimated from the time at which the planet's rotation carries it across the central meridian (CM) - that is, the line of longitude running down the exact centre of the disc. The observer notes the time, to the nearest minute, when the spot appears to be on the CM. This is the transit time, and can be judged to within a minute or two. The longitude is then worked out from tables (Appendix 1). (One minute of time corresponds to 0.6° of longitude.)

A long and acrimonious argument about how to determine longitudes was recorded in the *Monthly Notices* for 1904–5, between A.S. Williams, the noted English amateur, and G.W. Hough, professor at North-Western University in Evanston, Illinois. Hough maintained that accurate longitudes could only be determined by micrometer measurements, and was scathing in his attacks on 'the method of eye-estimates' used by Williams and other amateurs. While visual transit timings may indeed be subject to various forms of personal equation, Hough's micrometer results were by no means free from error; and history has judged which method has produced the bulk of our knowledge about the planet.

In modern times, some people would champion photography over visual estimates in the same way. Photography does indeed have advantages of permanence and objectivity; but if photographs are not available, visual estimates still have the necessary accuracy for most purposes.

More details of longitude measurement are given in Appendix 1.

Observations of latitude

Latitudes cannot be properly measured by visual estimation. Although a visual observer can judge the central meridian quite well, estimates of off-centre positions do not approach the required accuracy of a degree or so.

In the past, the latitudes of the belts were measured by micrometer. It was a difficult technique, as discussed by Peek in his book, and reliable only in the hands of an experienced observer. It is rarely used today. Nowadays, latitudes are measured from high-resolution photographs, as described in Appendix 2.

Observations of albedo

Important changes in the darkness of belts or brightness of zones occur on Jupiter, and one would like to have numerical estimates of the brightness (or 'albedo') of the different regions. This has been attempted in various ways. For many years observers merely listed the belts and zones in order of apparent brightness. More recently, observers have assigned intensities to them on a scale from 0 to 10, where 0 represents the brightest zone and 10 is black sky. (American observers use a reverse scale, from 10 to 0.) This is similar to the method used for intensities on Mars and Saturn. Although this is better than nothing – and the author makes such estimates himself – there are several reasons why such numbers cannot be regarded as objective.

First, the estimates must be influenced by contrast effects and by the widths of the belts. Our visual system emphasises contrasts and observers are trained to maximise this perception. So it is impossible accurately to compare (say) a polar region, lying between black sky and a dark belt, with the Equatorial Band, lying in a bright zone. Also, some narrow belts are close to the limit of resolution, so the prominence of such a belt will be a combination of its darkness and its width; the assigned intensity will depend on the telescope and the seeing.

More seriously, the whole idea of a linear visual intensity scale is flawed, because the human visual system responds in a logarithmic way. What are perceived as equal steps will, in practice, tend to represent equal ratios. (The magnitude scale of star brightness is just such a logarithmic scale.) So even on the American system, which is rather more logical in that intensity increases with brightness, there is still a problem of having a zero on the scale. So the actual scale used must be some combination of linear and logarith-

mic scales, and it is not surprising that different observers place the visible features in different ranges in the 0–10 scale.

Therefore, these estimates are not to be regarded as objective data for analysis, and will not be quoted in this book. What they are useful for, in a series of apparition reports, is to indicate trends in darkness of individual belts or zones, and to give some numerical substance to statements such as 'the belt has become rather faint'.

Observations of colour

Much the same difficulties apply to colours as to intensities. Obvious variations occur, but visual observers cannot record them other than by visual description.

There are several possible sources of spurious colour. Refracting telescopes are unsuitable for colour observations as they generally have some chromatic aberration in the lens, and even if this does not produce notable colour fringing, it may throw a violet haze around the planet; this can change the tones of browns, make belts appear purplish, and obliterate yellow colouring in the zones. The author has noticed all these effects even with a comparatively well-corrected 20-cm refractor. An eyepiece may also have some chromatic aberration. The Earth's atmosphere also has an effect, particularly when the planet is at low altitude so that the atmosphere refracts its light like a prism. At its most extreme this produces 'traffic-lights effect' – fringing with red at the top and blue-green at the bottom, in an inverting telescope – and this effect applies to individual zones as well as to the whole disc, so one must always be cautious in recording colour along the edges of zones. A neat trick is to play off the traffic-lights effect against aberration in the eyepiece, by offsetting the image from the centre of the field of view until the north and south polar regions appear equally grey – a technique first discussed by Phillips and M.A. Ainslie in 1936.[5]

Colour filters can be very useful in showing colour contrasts. With a blue filter, such as a Wratten 44A or 47, the brownish belts and any reddish or yellowish features appear darker. With a red filter, such as the Wratten 25, most of the belts appear paler but bluish spots (as on the NEBs edge) are darker. A yellow filter usually shows much the same patterns as white light, and may enhance contrast in twilight or in poor seeing.

Further discussion of colour observation and its difficulties is deferred to Chapter 4.

Drawings

Most novice observers feel an urge to draw what they see, and need some persuading that drawings with modest telescopes are rarely of value without other data, in particular transit timings. All the same, one cannot resist making drawings. And they can be useful: to provide a check on verbal descriptions of features transited, to establish their approximate latitudes when the normal belt/zone pattern is disrupted, and to confirm the existence of spots whose transits were missed. Also, 'a picture is worth a thousand words' in recording fine or subtle details (Figs. 1.16, 17).

---

[5] Phillips *et al.*(*1936*); Peek (*1937*).

Indeed, until the photographic improvements of the last few years, drawings have been the only way of capturing all the visible detail. Whereas a photograph must integrate the shimmering image for some time (typically seconds), a trained eye can catch fine details in calm moments, and thus build up a much more detailed image.

In making a drawing of the whole disc, it is best to follow an orderly scheme. First, inspect the planet carefully, not only to catch any transits before spending time drawing, but also to memorise the positions of the major features. Consciously examine the latitudes, widths, and intensities of the belts and zones, and sketch them in first. Then sketch in the other major features at a single time; the time must be recorded. Then fill in the details. This scheme will help one to avoid common errors such as drawing the belts too narrow, or drawing them too far apart so as to compress the periphery of the disc (where features are obscured by limb-darkening). And of course, it is essential to draw only what you see, not what you expect to see; a realistic blur is more useful than an imaginary fairy-castle.

Even with such advice, some observers will be more artistic and/or accurate than others, and personal differences in style occur even among good observers. The contrast of the markings has to be somewhat exaggerated to record them clearly, so individuals may represent faint or diffuse features in idiosyncratic ways. Such effects can be found out by comparing drawings among themselves or with photographs, and are one reason why a series from a single observer may be more informative than a collection from several. Some observers such as Phillips and Peek produced drawings almost as reliable as photographs, and one can make useful measurements from them. Other observers' work is no more than diagrammatic. Thus in the BAA Memoir for 1938, in Williams' last year of observation, Peek commented: 'His sketches of Jovian markings have always been diagrammatic rather than pictorial, and one can only infer that his delineation of the Red Spot region was not intended to represent the actual appearance in 1938'.

Instead of disc drawings, one can make 'strip-maps' of particular regions or of the whole planet (e.g. Figs. 1.18&19). These maps can be compiled from drawings and transits after the observing session, or with practice they can be made at the telescope. The observer extends the strip-map continuously as the planet rotates, producing a cylindrical-projection map of the planet.

Quality of observations

It is of course essential that both the observers and the person analysing their work should maintain a critical attitude to the quality of the observations. This is part of the training of any amateur astronomer but, being amateurs, we cannot enforce it. Certain observers acquire a reputation among their contemporaries for overconfidence in their discernment of detail or in their interpretation of the observations, and a conscientious recorder will allow for this and exclude doubtful observations from his analysis. But if such observations are published, a later generation of readers may find it hard to distinguish a genuinely reliable report from one that is partly the product of excessive zeal. One reason why the BAA reports of Phillips and Peek are so valuable is that both men had a

Fig. 1.16. Jupiter viewed with a typical amateur telescope, a 22-cm reflector. Drawing by Richard McKim on 1985 July 24. This shows the normal aspect of the major belts and typical disturbances, with the Great Red Spot on the p. limb.

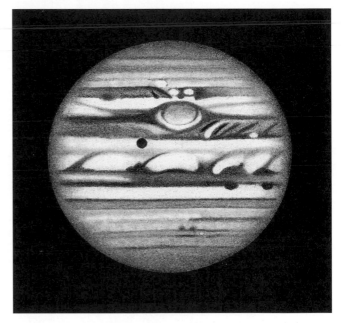

Fig. 1.17. Jupiter viewed with the 1-metre reflector at the Pic du Midi. Drawing by Richard McKim on 1986 July 16. The GRS is central, with the shadow of Io Np. it. The GRS, NEB, and NTB(S) appeared pale orange, while the dark projections on NEBs appeared deep blue. (See also Plates P19 and P24.)

thoroughly scientific approach to their own and to others' observations. More recently the work of E.J. Reese and I. Miyazaki (among others) has also been outstandingly reliable. We try to maintain this tradition in current BAA reports.

Units and conventions

The units and conventions adopted in recent years by space scientists differ from those that visual astronomers have always used.

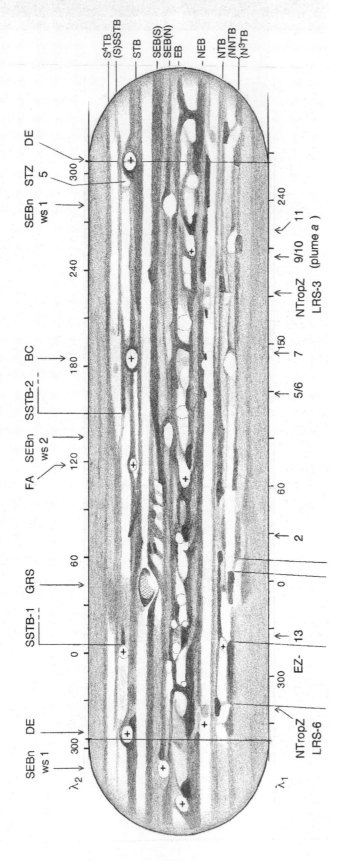

Fig. 1.18. Strip-map of the planet covering a whole rotation on a single night, 1976 Nov. 7/8, by John Rogers (300-mm refractor in Cambridge). The hemispheres at the ends were drawn at start and finish of the session. Various features are indicated including the Great Red Spot, South Temperate white ovals (FA, BC, and DE), Little Red Spots in the North Tropical Zone, and discontinuities in the high temperate belts. A new equatorial light spot seems to have appeared between the beginning and end of this session.

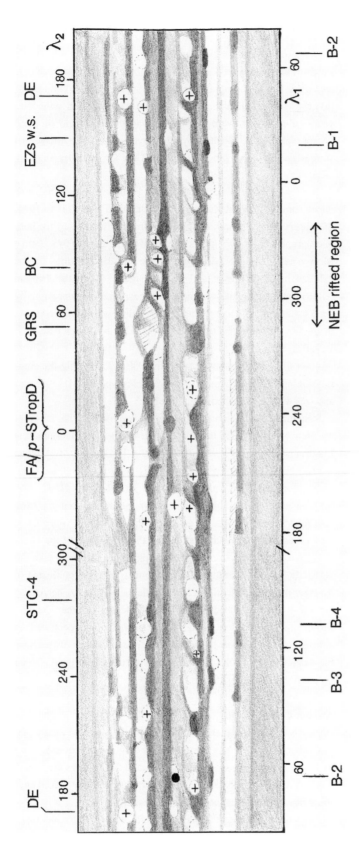

Fig. 1.19. Strip-map from two nights' observation, 1979 Jan. 7–8, by John Rogers (320-mm reflector in Los Angeles). It includes a South Tropical Dislocation p. the GRS, and four dark 'barges' in NEBn (B-1 to B-4). This was during Voyager 1 imagery; compare with Fig. 2.2.

Table 1.3 *Nomenclature*

	Visual astronomers	*Space scientists*
Image orientation	South up	North up
Directions	Preceding (p.)	East (E.)
	Following (f.)	West (W.)
Wind speeds	$\Delta\lambda_2$ (degrees per month in System II)	u (metres per second in System III)
	$\Delta\lambda_2 = -\dfrac{[2.080\,u + 8.0]}{\cos \beta'}$	$u = -\dfrac{[\Delta\lambda_2 + 8.0]\cos\beta'}{2.080}$
	Prograding(negative $\Delta\lambda_2$)	Eastward (positive u)
	Retrograding (positive $\Delta\lambda_2$)	Westward (negative u)
Circulation	Clockwise (S.hemisphere) Anticlockwise (N.hemis.)	} Cyclonic
	Anticlockwise (S.hemis.) Clockwise (N.hemis.)	} Anticyclonic

(The latitude ß' differs from the zenographic latitude ß"; see Appendix 2 for definition.)

Table 1.3 gives a translation. In this book, we mainly use the visual conventions, which were adopted for good reasons that are still valid. Thus all drawings, maps, and graphs have south up, because that is the way the planet is seen by most visual observers (and also by the Voyager spacecraft, whose pictures were turned upside-down for publication; see Fig. 2.17).

East and west are referred to as preceding (p.) and following (f.). This convention was introduced to avoid confusing east and west on the planets with east and west on the sky, which are the opposite way round; celestial east and west were actually used in some historical reports. Also, it is all too easy to confuse east with west, whereas preceding and following directions can always be identified at the telescope; they refer both to the direction in which the planet rotates, and to the direction in which it moves if the telescope drive is turned off.

Wind speeds are quoted in degrees of longitude per month in System II (or System I for the equatorial region), because that is how they are measured. It is not appropriate to report observations in metres per second, as this requires not only an accurate knowledge of the latitude (which is not always available) but also a calculation (which like any conversion of data bears some risk of error).

Circulations, however, are best described as cyclonic and anticyclonic, because this emphasises the common features of (for example) anticyclonic white ovals in both hemispheres, especially the fact that they must have high pressure in the centre while cyclonic features have low pressure.

1.4 PROFESSIONAL AND PHOTOGRAPHIC OBSERVATIONS

Professional photography

The first published photograph of Jupiter was taken in 1879 by A.A. Common. As the photographic plates of the time detected only blue light, it showed the GRS and the reddish EZ as intensely dark features with the rest of the disc light.

Photography improved rapidly, and in 1904 Earl C. Slipher began a program of systematic photography of Jupiter at the Lowell Observatory at Flagstaff, Arizona, using the 24-inch (61-cm) refractor. This program continued into the 1940s. The resolution was comparable to the best visual observations, and in the later years colour filters were used. An invaluable portfolio of these photographs was published by Slipher (*1964*). Some images of similar quality were produced at the Lick Observatory (e.g. by W.H. Wright, *1928, 1929*).

The Pic du Midi Observatory, in the French Pyrenees, has produced some of the best planetary observations. From 1941 to 1945, the observatory's 38-cm and 60-cm refractors were used for both photographs and drawings, as described by B. Lyot (*1943, 1953*). Most notable were the drawings of the surface markings on the galilean moons by Lyot and his colleagues. Later, the 106-cm reflector at the Pic gave even better images, and it has occasionally been made available to experienced amateurs, who produced some of the best photographs of all (see Fig. 1.20 and Plates P14, P15, P19, P21, P23, P24).

A set of photographs was taken in 1950–1952 during the commissioning of the 200-inch (5-metre) telescope on Mt. Palomar, and was widely published (e.g. Plate P9).[6]

Systematic professional photography resumed in the 1960s at the New Mexico State University Observatory (NMSUO) and the Lunar and Planetary Laboratory of the University of Arizona (LPL).[6] NMSUO coverage was with a 61-cm reflector, taking blue-light photographs, mainly from 1964 to 1971 by H. Gordon Solberg and Elmer J. Reese. Some of the annual reports from NMSUO were very detailed and precise, as listed in Appendix 3; they revealed both the circulation and the oscillation of the Great Red Spot. Since 1971, some photographs have still been taken at NMSUO but no reports published; summaries were published in Voyager project publications. Meanwhile the LPL team, headed by Gerard P. Kuiper, used the 154-cm reflector on the Catalina Mountains near Tucson from 1965 to 1974; they concentrated on getting the highest-resolution pictures ever obtained in narrow-band colour filters (from the infrared to the ultraviolet) and also on colour film (e.g. Plates P13 and P14).

Amateur photography

Jupiter is not an easy planet to photograph. Compared to the moon and other planets, it has a low surface brightness requiring long exposures – typically several seconds – and the features are of low contrast. Until recently, amateur photographs could not record anything like the detail that could be seen visually.

The situation changed in the mid-1980s, with the introduction of Kodak's Technical Pan film, which combined reasonably fast

---

[6] Some publications of professional photographs were as follows. Mt. Palomar: Humason (*1961*), Reese (*1962a*). NMSUO, for GRS: Reese and Smith (*1968*); Solberg (*1969*); also apparition reports listed in Appendix 3. NMSUO, pre-Voyager summaries: Beebe and Youngblood (*1979*); Terrile and Beebe (*1979*); Beebe, Orton and West (*1989*). LPL: Kuiper (*1972a, 1972b*), Fountain and Larson (*1972*), Larson *et al.*(*1973*), and apparition reports by R.B. Minton listed in Appendix 3.

speed with small grain for the first time. In the hands of the best amateur photographers around the world, with excellent telescopes and excellent seeing, this film began to record as much detail as could be seen visually. Thus some amateurs equipped with telescopes of 40–50 cm aperture, such as Donald Parker in the USA, Isao Miyazaki in Japan, and Georges Viscardy in France, began routinely taking photographs as good as those from the earlier professional programs. These photographs now provide a large fraction of the data for reports by the BAA and other amateur organisations.

One problem that is intrinsic to Jupiter is the strong limb- darkening. On a photograph printed at high contrast, the limb regions are completely lost. This can be overcome by 'dodging' the print to lighten the edges, and use of an 'unsharp mask' is a precise way of doing this. Limb-darkening is not so evident if photos are taken through a blue filter. If photographs are to be used for measurements, some such technique must be used to preserve the limb.

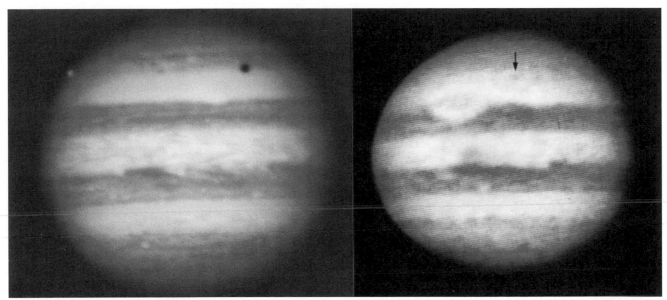

Fig. 1.20. One of the finest photographs from the Pic du Midi, by Jean Bourgeois, 1988 Dec. 29 (1-metre reflector, yellow filter). Europa and its shadow are on the SSTBn. (STB and NTB are absent.)

Fig. 1.21. CCD image by Terry Platt in Berkshire, 1989 Jan. 22 (32-cm off-axis reflector). The arrow indicates a tiny SSTBn jetstream spot near the GRS.

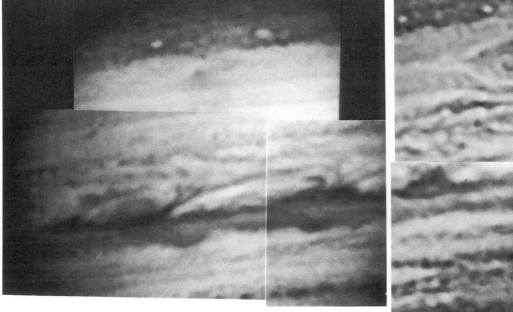

Fig. 1.22. CCD images from the Pic du Midi (1-metre reflector, red filter, processed to enhance smallscale detail); by Dr P. Laquès, F. Colas, J. Lecacheux, J. Bourgeois, C. Buil, and E. Thouvenot. Images extend from SSTB (at top, with white ovals) to NNTB (at bottom). STB, SEB, and NTB were very faint at this time. (*Left*) 1990 Jan. 22. (*Right*) 1990 Jan. 6, including GRS and N. Temperate Disturbance. (From BAA reports for 1989/90.)

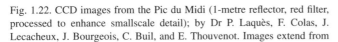

Electronic imaging

Since the late 1980s, charge-coupled devices (CCDs) have become widely available to amateurs as well as professionals. These offer two enormous advantages: sensitivity much higher than that of photographic emulsion, which allows shorter exposures that can beat the seeing; and digital recording, which allows one to enhance smallscale contrast and to suppress largescale variations such as limb-darkening. Thus CCD images are superior to photographs as well as to drawings. Observers such as Terry Platt now produce excellent images even in the frustrating skies of England (Fig. 1.21).

Meanwhile, CCDs applied to large professional telescopes are producing the best images since Voyager (Fig. 1.22, Plate P24.9&10).

With these technological advances, can visual observers still contribute anything? To the author, as one of the many observers who lack either the inclination or the resources for serious photography, the answer is certainly yes. It was visual observers who discovered most of the phenomena to be described in this book, and visual observations are as good as ever. They still provide essential data during the inevitable gaps in photographic coverage. In any case, the first principle of amateur astronomy is that it is done for enjoyment. To many of us, seeing a planet with our own eyes gives a thrill that looking at pictures cannot. As this pastime continues to reveal new phenomena and new information, so much the better.

2: Observations from spacecraft

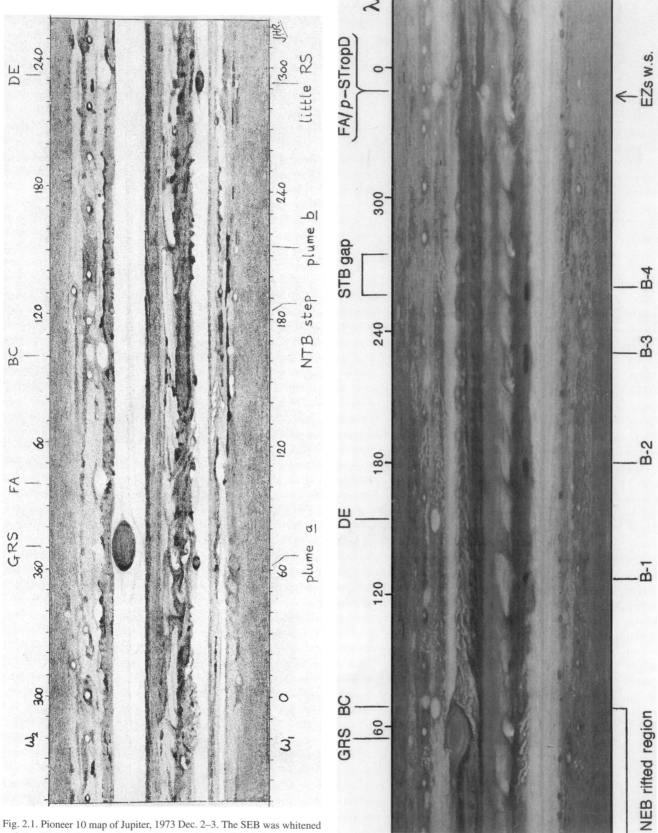

Fig. 2.1. Pioneer 10 map of Jupiter, 1973 Dec. 2–3. The SEB was whitened and the GRS dark red at the time. Features indicated include STB white ovals FA, BC, and DE; long-lived NEBs plumes *a* and *b*; and a little red spot in the NTropZ. (Drawn by the author from Pioneer images. From Rogers and Young, *1977*.)

Fig. 2.2. Voyager 1 map of Jupiter, 1979 Feb. 1. The SEB has revived and there are many spots on the SEBs and NNTBs jetstreams. Compare with Fig. 1.19. (NASA photomosaic, P-21829.)

2: Observations from spacecraft

2.1 THE PIONEER PROJECT[1]

The first two spacecraft to visit Jupiter belonged to the Pioneer series, whose main interest was not planets as such, but the particles and fields of interplanetary space. Previous Pioneers had orbited around the Sun to study the solar wind without encountering any planets; Pioneers 10 and 11 were intended mainly as the first probes of the magnetic field and radiation belts around Jupiter. But they were also provided with crude imaging systems to view the planet, and the pictures from these inevitably captured the attention of public and scientists alike.

Pioneers 10 and 11, like their predecessors, were designed by NASA's Ames Research Center, and were to be continuously spinning – both to provide all-round sampling of charged particles, and to provide a simple form of stabilisation. In this they differed from the more sophisticated and expensive planetary probes, the Mariners designed by the Jet Propulsion Laboratory. Pioneer 10 and 11 together cost about $100 million. The spacecraft were approved in early 1969 and were built in less than 3 years.

The Pioneer spacecraft

Pioneer 10 weighed 258 kg (Fig. 2.3). The core was a hexagonal box 71 cm on each side, surmounted by the high-gain antenna 2.7 metres across, which had to be pointed at Earth. A boom 6.5 metres long carried the magnetometer, and two shorter booms carried the radioactive power sources. Pioneer 11 was identical except for the addition of a second magnetometer, and weighed 260 kg. The instruments are listed in Table 2.1.

The 'camera' on the Pioneers was more properly called an imaging photopolarimeter. It consisted of a photometer with an aperture of 2.5 cm, which was scanned across the planet by the spin of the spacecraft, and shifted between each scan. Each scan line was broadcast back to Earth as soon as it was completed, and it took 25–60 minutes to build up a picture. So the raw images were very

distorted, and they were realigned by computer to create the published pictures that show the planet as if at a single instant, but still with some distortion of the surface features (Fig. 2.4). The light was split through red and blue channels. Colour images were created by combining the red and blue images plus a synthetic 'green' image that was interpolated between the two. (On Pioneer 10, the red channel was sometimes subject to some electronic interference that put a faint ripple on parts of the pictures.) The resolution was 1.7 arc-minutes (the same as the human eye), or 400 km on Jupiter in the closest images. The instrument also recorded the polarisation of the light (at a resolution of 27 arc-minutes). As this was the first opportunity to view the planet other than face-on, the variation of polarisation with phase angle would give important information on the atmospheric structure, and on the surface textures of the moons.

The other instrument for observing the jovian atmosphere was the infrared radiometer, with a $0.3° \times 1.0°$ field of view, which could also create scan images of the planet. It measured the infrared radiation in two broad bands, centred on 20 μm and 45 μm wavelength, which together account for almost the whole infrared emission. This was the first opportunity for accurate measurements of the

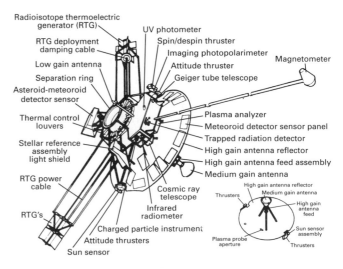

Fig. 2.3. The Pioneer spacecraft. (NASA diagram.)

[1] The Pioneer spacecraft were described in *The Pioneer Mission to Jupiter*, (NASA SP-268, *1971*), and in Martin (*1972*). General accounts of the results were given in Gehrels (*1974*) and Ashbrook (*1975*). Detailed results were published in special issues of *Science* and *Journal of Geophysical Research* listed in Appendix 4, and in the books *Jupiter* (edited by Gehrels, *1976*) and *Pioneer Odyssey* (NASA SP-349/396, by Fimmel, Swindell and Burgess, *1977*).

Table 2.1. *The Pioneer experiments*

Investigation	Principal Investigator	Objects of study	Spectral range
Magnetic fields	E. J. Smith, JPL	Magnetic field	–
Magnetic fields (Pioneer 11 only)	N. F. Ness, NASA Goddard	Magnetic field	–
Plasma analyzer	J. H. Wolfe, NASA Ames	Low-energy electrons and ions	1 eV–18 keV
Charged particle composition	J. A. Simpson, U. Chicago	Energetic charged particles	400 keV–500 MeV
Cosmic ray energy spectra	F. B. McDonald, NASA Goddard	Very energetic charged particles	50 keV–800 MeV
Jovian charged particles	J. A. Van Allen, U. Iowa	Energetic charged particles	40 keV–50 MeV
Jovian trapped radiation	R. Walker Fillius, UC San Diego	Energetic charged particles	50 keV–350 MeV
Asteroid-meteoroid astronomy	R. K. Soberman, General Electric	Photometric sighting of meteoroids near spacecraft	(Optical)
Meteoroid detection	W. H. Kinard, NASA Langley	Micrometeoriods striking spacecraft	–
Celestial mechanics	J. D. Anderson, JPL	Masses of Jupiter and moons, from spacecraft tracking	–
Ultraviolet photometry	D. L. Judge, U. Southern California	UV emissions from Jupiter and around moons	58, 122 nm
Imaging photopolarimetry	T. Gehrels, U. Arizona	Imaging Jupiter; light-scattering properties of its clouds	Red and blue light
Infrared radiometer	G. Münch, Caltech	Jupiter temperature and heat budget; H:He ratio	14–56 μm
S-Band occultation	A. J. Kliore, JPL	Structure of ionosphere and atmosphere of Jupiter and Io	–

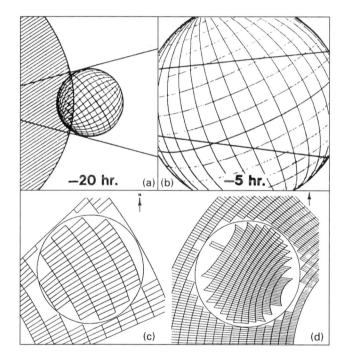

Fig. 2.4. Field of view of the imaging photopolarimeter on Pioneer, (a) 20 hours and (b) 5 hours before encounter. The field of view is 14° across. *Below:* How the scans were distorted by the planet's rotation. These diagrams indicate the disc of Jupiter with every 25th line and pixel shown. Distortion was minimal 30 hours before encounter (c), but very noticeable 4 hours before encounter (d). (NASA; adapted from Swindell and Doose, *1974.*)

temperature of the atmosphere at various levels, and of the amount of heat emitted from the day side, night side, and polar regions of the planet.

Seven instruments (eight on Pioneer 11) were to measure various aspects of the atoms, subatomic particles, and magnetic fields both around Jupiter and in interplanetary space. A set of 4 pho-

tometers, each with 20 cm aperture and 8° field of view, was to look for passing asteroids and smaller objects *en route* to Jupiter, while meteoroid detector panels (lining the underside of the great radio antenna) would count picogram-range dust impacts on the spacecraft. Finally, the tracking of the spacecraft itself constituted both the occultation experiment – to see how Jupiter's ionosphere and atmosphere affected the signals as the spacecraft passed behind the planet – and the 'celestial mechanics experiment' – to measure the masses of Jupiter and its moons, and to map Jupiter's gravitational field.

The spacecraft needed up to 106 watts of power, and the Pioneers could not use solar panels like previous spacecraft because they were going so far from the sun. Instead, they were the first spacecraft to depend entirely on nuclear energy, in the form of 'radioisotope thermal generators' (RTGs). These were not reactors; they simply consisted of four pellets of plutonium-238 whose radioactive decay provided heat, a total of 155 watts at launch decaying to 140 watts at the time of the Jupiter encounter. The heat was converted into electrical power by thermocouples. The RTGs were carried on long booms to isolate them from the spacecraft's experiments.

The high-gain antenna, broadcasting at 1024 bits per second, had a power of 8 watts, giving a signal of only 10^{-17} watts to receive at Earth. This minute signal was picked up by NASA's great 64-metre dishes in California, Spain, and Australia. (For sending signals to the spacecraft, these same antennae could emit up to 400 kilowatts.) In order to check continually whether the spacecraft's axis pointed exactly at Earth, the axis of the high-gain antenna was offset by 1° from the spin axis, so any slight mispointing would show up as a fluctuation in the signals.

For making course corrections and for re-orienting the spacecraft, there was a set of six thrusters powered by pressurised liquid hydrazine.

Finally, as the Pioneers would be the first man-made objects to

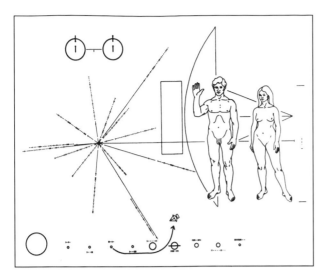

Fig. 2.5. The final 'experiment' on Pioneers 10 and 11: the plaque intended as a message to aliens. Results from this experiment are still awaited. (NASA.)

Fig. 2.6. Launch of Pioneer 10. (NASA.)

leave the solar system, they each carried a gold-plated plaque intended as 'mankind's message to the stars' (Fig. 2.5). Designed by Carl and Linda Sagan and Frank Drake, it was for the benefit of any space-faring aliens who might ever intercept the probe (Sagan, *1973*). It shows a man and a woman next to the Pioneer spacecraft for scale, and a pair of maps intended to show how to find us. The symbol at the top indicates the radio emission of the hydrogen atom as a unit of frequency; the main map is a map of the Milky Way, showing 14 pulsars relative to the Sun and the centre of the galaxy, with their pulse frequencies in binary code; and the diagram at the bottom shows the solar system, with Earth as the origin of the Pioneer spacecraft. Whether aliens ever discover the plaque or not, the decision as to how to represent us on it – and whether to send such information at all – was a bold statement of humanity's view of itself.

The flight of the Pioneers

For any mission to a planet, the minimum-energy trajectory is half of an ellipse tangential to the orbits of the Earth and the planet. For Jupiter, it takes 32.7 months to travel this path, and the opportunity to do so occurs every 13 months. Planetary probes are always sent on paths close to the minimum-energy one, so that the maximum weight can be launched, but a little extra rocket power enables the trip to be shortened. For the Pioneers, the trip took 21 months.

Pioneer 10 was launched from Cape Kennedy (Cape Canaveral), on the evening of 1972 March 2. As lightning flashed on distant thunderclouds, the ignition of the rocket engine swamped the night with fire, and the first mission to the giant planets was underway (Figs. 2.6&7). The rocket was an Atlas-Centaur with a small third stage, and in 17 minutes it boosted Pioneer directly into its Jupiter trajectory with an initial speed of 51682 km/hr – the fastest speed then attained by any man-made object. The spacecraft was then spinning 30 times a minute, but 20 minutes after launch the three booms were extended and the spin slowed to 4.8 rotations per minute – the rate that would be maintained throughout the flight.

The high-gain antenna was pointed correctly, and some of the 'particles and fields' experiments were turned on to check their performance while flying through the Earth's radiation belts. Ninety minutes after launch, Pioneer 10 left the Earth's magnetic field and was in interplanetary space.

Three course corrections were done over the next seven months, and the final arrival time was within two minutes of the planned time. The experiments were checked out during the first few months, the photometers by viewing distant Jupiter and the zodiacal light. The 'particles and fields' experiments made continuous studies of the solar wind, magnetic fields, and cosmic rays. The only anomaly was the failure of a star sensor, used for orienting the spacecraft; the orientation was maintained using the sun sensor.

The first obstacle facing the Pioneers was the asteroid belt. No spacecraft had ever entered it before. In addition to the thousands of identified asteroids, there were expected to be millions of rock-sized fragments and an unknown density of dust particles. In contrast to the dire imaginings of science fiction, the fragments were expected to be so spread out that there would be little risk of a collision – and so it proved. The Pioneer 10 meteoroid detector panels did not record any increase in dust density while crossing the asteroid belt; the impact rate fell off steadily as the craft receded from the sun, implying that the dust came from comets, not asteroids. The asteroidal photometers also found no increase in the density of millimetre-sized particles (sighted at a rate of about one a day), although the seven sightings of objects larger than 1.5 cm all occurred within the asteroid belt. In fact, the highest frequency of dust impacts was within 700 000 km of Jupiter itself, at a density about a hundred times that in deep space.

Meanwhile, Pioneer 11 was launched on 1973 April 5. One of its RTG booms failed to deploy on schedule, but it deployed itself eight hours after launch while the spacecraft was swivelling, and thereafter the flight was uneventful.

The activities of a spacecraft in a planetary encounter must be entirely pre-programmed, because of the time for radio waves to travel from the planet to Earth (46 minutes for Pioneer 10 at Jupiter). However, to save weight and cost, the Pioneers were not provided with much on-board storage capacity. Instead, each encounter was controlled by over 10 000 commands sent from Earth in real time, with no possibility of knowing if they were being obeyed. Most of the results, also, had to be broadcast back to Earth in real time.

Pioneer 10 made a final course correction on 1973 Nov. 26. The same day, its 'particles and fields' instruments recorded that it crossed Jupiter's bow shock, 7.7 million kilometres (M km) from the planet. (The time of crossing was somewhat earlier than in the speculative predictions made from Earth.) About 24 hrs later it entered Jupiter's magnetic field, 6.85 M km out. It found the outer magnetosphere to be very flattened and flexible, and on Dec.1 Pioneer found itself outside the magnetosphere again for 11 hours; evidently the boundary had contracted or buckled, probably in response to a solar flare. (Another sign of instability had come as a surprise a month or more earlier: the detection of bursts of high-energy electrons emitted from Jupiter's magnetosphere.) By Dec.3, 1.07 M km out, Pioneer 10 was in the main dipolar magnetic field and entering the densest part of the radiation belts, where electrons were much denser than predicted.

Table 2.2 *The Pioneer encounters*

Pioneer 10		
Launch	1972 Mar 2/3, 01.49 UT	
Jupiter	1973 Dec 3/4, 02.26 UT	130 000 km
Pioneer 11		
Launch	1973 Apr 5	
Jupiter	1974 Dec 2/3, 05.22 UT	42 000 km
Saturn	1979 Sep 1, 16.34 UT	21 400 km

Times are Universal Time at the spacecraft. Distances are minimum distances from the surface. There were no satellite encounters closer than 300 000 km.

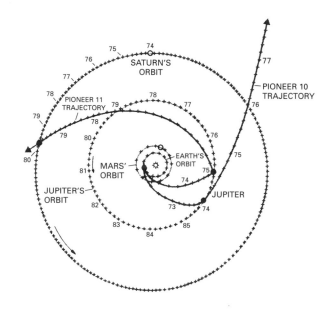

Fig. 2.7. The trajectories of the Pioneers through the solar system. (NASA.)

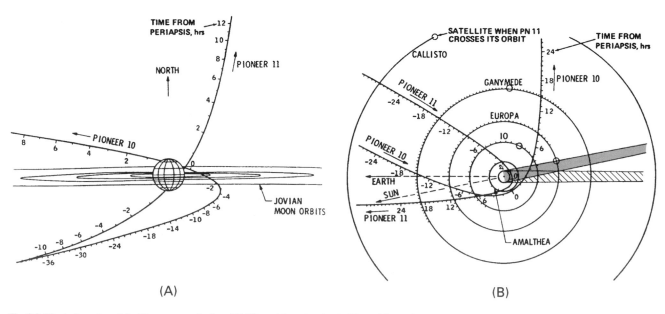

(A)   (B)

Fig. 2.8. The trajectories of the Pioneers past Jupiter. (A) Viewed from Earth. (B) Viewed from above the pole of Jupiter. (NASA.)

The flux of charged particles around Jupiter is a million times stronger than that around Earth, and the overall 'radiation dose' to Pioneer 10 amounted to a thousand times the lethal dose for a human. On the day of encounter it raised the temperature of the RTG by 5°C, changed onboard electric currents by several per cent, and created several false electronic commands, so that a few high-resolution images of the Great Red Spot and of Io were degraded or lost. The spacecraft recovered gradually, and almost completely, after the encounter.

Useful images were taken for several days before the encounter (12–26 images per day; Fig. 2.9), but only in the last 24 hrs were the images more detailed than those from Earth. The last good photo was taken 8 hours before encounter (Fig. 9.9). On that day, Pioneer 10 also took two images of Ganymede and Callisto and one of Europa, though they were not detailed enough to identify the nature of the features (Fig. 20.1); it did not come closer than 300 000 km to any of them. The satellites were also viewed by the infrared radiometer.

Closest encounter was 130 000 km above cloud-tops, with a speed of 132 000 km/hr, at 02.26 UT on 1973 Dec. 4. Pioneer 10 flew by in the same direction as the planet's rotation, at an angle of 14° to the equator. The trajectory had been carefully planned in order to have the spacecraft occulted by Io, to find out whether there is an atmosphere on that mysterious moon; this occurred at 02.41 UT for 1½ minutes. An hour later, the craft passed behind Jupiter itself for 65 minutes. Finally, Pioneer 10 looked back and revealed the view of Jupiter as a crescent for the first time (Fig. 2.10).

As Pioneer 10 moved away from Jupiter, the magnetopause and bow shock again swung back and forth over it several times within a few days, until it re-entered the solar wind. Then it cruised outwards through the solar system. The solar wind was more turbulent than expected at these great distances, and 2½ years later, when it was just beyond the orbit of Saturn and 6° above the plane of Jupiter's orbit, Pioneer 10 apparently passed through Jupiter's fluttering magnetic tail again! Then it headed for interstellar space, crossing the orbit of Neptune in 1983 June at 49000 km/hr relative to the Sun. Its magnetometer failed in 1975 December but otherwise it is in good condition. Its final velocity is 41400 km/hr, in the general direction of the star Aldebaran, though it will take 1.7 million years to get that far.

Once Pioneer 10 had achieved its goals, Pioneer 11 was free to take a different course which would achieve three exciting aims: it would overfly the polar region of Jupiter, it would pass even closer in than Pioneer 10, and it would emerge heading for Saturn. In 1974 March, as it ended its passage through the asteroid belt, its thrusters were fired for 42 minutes to put it on this course.

Pioneer 11's entry into Jupiter's domain was very similar to that of Pioneer 10. Relativistic electrons from the planet were detected 5 months before encounter. On 1974 Nov.25 the craft crossed the bow shock, again 7.7 M km from Jupiter, and the magnetopause 23 hr later. But the magnetopause and bow shock flipped back to leave Pioneer in the solar wind again for 29 hours; by then, scientists realised that the magnetosphere was pulsating like a giant jellyfish. As Pioneer plunged towards close encounter on 1974 Dec. 2/3, its close pass (42000 km) took it through an inner belt of radiation

(A)   (B)

Fig. 2.9. Processing the Pioneer images. These are Pioneer 10 images showing the Great Red Spot. (A) The raw images were distorted and 'noisy', and were received as red-and-blue pairs. (B) Processed images have missing lines interpolated and the correct outline of the planet restored, although the surface is still slightly distorted because of the prolonged scan time. (Image A50, blue.) (NASA.)

Fig. 2.10. Jupiter viewed as a crescent by Pioneer 10, with the GRS on the limb. (Image B17, blue.) (NASA.)

infrared, narrowly missed colliding with a previously unknown moon, and viewed the famous rings from the dark side. It then flew on towards interstellar space, crossing the orbit of Neptune in 1990 February.

The Pioneers are expected to be tracked until the late 1990s, when their signals will become too weak to detect. During this time they are measuring the solar wind in the outer solar system, as well as the neutral atoms and cosmic rays that flow in from interstellar space. The main prize would be detection of the heliopause – the boundary between the sun's domain (solar wind and magnetic field) and the true interstellar environment. This will be analogous to the bow shock and magnetopause around Jupiter. It could be around 50 AU (7500 M km) from the sun, or further. Pioneer 11 is heading towards Ophiuchus, in the direction of the sun's motion through the galaxy, so should encounter the heliopause first; Pioneer 10, although further out, is heading in the opposite direction. When a spacecraft passes this boundary, wherever it may be, it will be the first man-made object to enter interstellar space.

even more intense than that recorded by Pioneer 10. In the words of Robert Kraemer, director of NASA's planetary programmes, Pioneer 10 tickled the dragon's tail, but Pioneer 11 flew into its jaws. But it was travelling so fast – 172 800 km/hr, faster than any previous man-made object – that it experienced less total irradiation than its predecessor. Some instruments were deliberately shut off during encounter following spurious signals induced by the radiation, and some infrared data on the north polar region were lost, but this time no images were lost. (See Figs. 6.1 and 12.1.)

Pioneer 11 approached from some way south of Jupiter and swung by at an angle of 55° to its equator – heading in the opposite direction to the planet's rotation. It passed behind the planet for 42 minutes centred on closest approach, and then headed away to the north, taking dramatic views of the north polar regions as it departed (Fig. 2.11). Again it crossed the magnetopause several times, finally quitting it at a distance of 5.7 M km. It was in good condition except for the asteroid photometer, which malfunctioned and had to be switched off.

Heading for Saturn, its orbit around the sun was tilted at 16° to the orbits of the planets, with a perihelion of 3.5 astronomical units, so that it arched above the asteroid belt. It was the first spacecraft to go so far out of the ecliptic plane, and thus the first to enter the dipole region of the sun's magnetic field.

Pioneer 11 became the first spacecraft to reach Saturn on 1979 Sep. 1. Among its achievements, it recorded the saturnian magnetosphere for the first time, probed the atmospheric structure in the

(A)

(B)

Fig. 2.11. Pioneer's planets. (A) Jupiter from Pioneer 11; view over the north polar region (image D8). (B) Saturn from Pioneer 11; view of the dark side of the rings, with sunlight filtering through the thinner parts. (NASA images.)

The main product of the Pioneer missions to Jupiter was the first direct record of the planet's magnetic field and radiation belts (Chapter 18). Also of great importance were the infrared measurements of the planet's temperature and heat emission (Chapter 4), the photopolarimeter measurements of the heights of the clouds (Chapter 4), and the ultraviolet detection of helium around Jupiter and an ionosphere around Io. In summary, the missions were important not for revolutionary discoveries, but for precise measurements of quantities that could previously only be estimated or guessed at. And the passages through the asteroid belt and the radiation belts were important for a practical reason too: they showed that the way was clear for the next generation of Jupiter probes, the Voyagers.

2.2 THE VOYAGER PROJECT[2]

The Voyager mission also comprised a pair of fly-pasts, but the two Voyager spacecraft were in a different league from the Pioneers. Indeed, as they sent back streams of spectacular pictures of Jupiter, Saturn, Uranus, Neptune, and many of their amazing moons – live on public television – the Voyagers earned a place in history as the craft that explored the outer solar system.

Voyagers 1 and 2 had fully stabilised orientation, powerful cameras, and sophisticated on-board computers. Produced by NASA's Jet Propulsion Laboratory (JPL), they had evolved from the Mariner series that had successfully explored Mars, Venus, and Mercury. The project was originally called Mariner Jupiter/Saturn. The idea was that each craft would fly past Jupiter, pick up speed from Jupiter's gravity, and thus be swung like a slingshot into a trajectory taking it out to Saturn – as Pioneer 11 had done.

When this project was approved in 1972, it seemed a poor shadow of what NASA had been planning: a Grand Tour of all four giant planets. Once every 175 years, the four giant planets are aligned so that a single spacecraft can take a minimum-energy trip past all four of them, gaining the 'slingshot' acceleration at each one in turn. The total trip would take only 8–13 years, in contrast to 30 years for a trip direct to Neptune. The opportunity to launch a craft on this Grand Tour existed from 1976 to 1980. In 1969, JPL was designing a Grand Tour spacecraft with a more 'intelligent' computer, self-repair capabilities, and thermo-electric propulsion. But the NASA science community was split between support for the Grand Tour and support for a Jupiter Orbiter/Probe, and as the Nixon administration hacked at NASA's budget, the result was that neither mission was approved. All that was left was Mariner Jupiter/Saturn (later called Voyager). But JPL kept an option open. If Voyager 1 was successful at Jupiter and Saturn, Voyager 2 could be targeted to carry out the full Grand Tour. When Voyager 1's performance dazzled scientists, public, and congressmen alike, funding for the extended mission was assured. What was still not assured was the survival of the spacecraft; it was designed for a Jupiter/Saturn mission lasting 4 years, but it would take 12 years to reach Neptune.

The twin mission was to cost $335 million (1977 prices) as far as Saturn, and it amounted to $865 million by the time Voyager 2 had passed Neptune.

The Voyager spacecraft

Voyagers 1 and 2 were identical, each weighing 815 kg (Fig. 2.12). The overall structure had many features in common with the Pioneers. The central ten-sided 'bus', 179 cm across, was surmounted by the great dish of the high-gain antenna, 3.7 metres across. One boom carried the plutonium RTGs, which again provided the power, starting at 423 watts soon after launch. Another boom, 13 metres long, carried the magnetometer; and the two other protrusions were the two 10-metre antennae shared by the radio astronomy and plasma wave experiments. The other instruments were mounted on another boom, including a scan platform that carried the 'cameras' and the infrared and ultraviolet instruments. This scan platform would be swivelled to point the instruments during encounters without changing the orientation of the spacecraft. In the centre of the 'bus' was a tank of liquid hydrazine to power the thrusters for attitude control and course correction. Within the 'bus' were the computers.

The Voyagers had more powerful and versatile computers than any of their predecessors. They stored long series of commands transmitted at intervals from Earth, which could guide spacecraft operations for several months during cruise, or for a single day during close encounter. They also had various fail-safe programs which could rescue the mission in emergencies. Voyager also had a digital tape recorder which could store 5×10^8 bits of data, or nearly 100 images, for delayed transmission.

The signals, transmitted at 115 200 bits per second at 23 watts by the high-gain antenna, were received by NASA's three 64-metre radio telescopes. Tracking the spacecraft through occultations and through gravitational fields constituted one of the mission's experiments.

There were ten on-board experiments (Table 2.3).

The Voyager 'television cameras' were actually respectable telescopes. The wide-angle one was a telephoto lens of f/3.5, 200 mm focal length, and field of view 3.1°. The narrow-angle one was a reflecting telescope of f/8.5, 1500 mm focal length, and field of view 0.424° (Fig. 2.13). A wheel of alternative filters allowed selection of different colours ranging from 345 nm (ultraviolet) to 619 nm (a red methane absorption band), with violet, blue, green, and orange filters in between. (The methane filter might have shown up differences in cloud height, but the jovian clouds were too uniform in altitude to show any contrast in this band – unlike those of Neptune.) Usually, three or four filters were used for successive images so that a colour picture could be reconstructed. The camera used a vidicon detector which scanned the image and converted it into dots ('pixels') whose intensities were digitised into 64 grey levels for transmission; each picture had 800x800 pixels. It

[2] The Voyager spacecraft were described in a special issue of *Space Science Reviews*, vol. **21** (nos.2&3) (*1977*), and in Beatty (*1977*). Political background to the program was described by Waff (*1989*). The encounters were described in many popular articles, and the scientific results at Jupiter were described in special issues of journals listed in Appendix 4. The whole mission as far as Jupiter was described in the semi-popular book *Voyage to Jupiter* (NASA SP-439, by Morrison and Samz, *1980*).

Table 2.3. *The Voyager experiments*

Investigation	Principal Investigator or Team Leader	Primary Objectives at Jupiter	Spectral range
Imaging science	B. A. Smith, U. Arizona	High resolution reconnaissance over large phase angles; measurement of atmospheric dynamics; determination of geologic structure of satellites; search for rings and new satellites.	0.33–0.62 μm
Infrared interferometer spectrometer (IRIS)	R. A. Hanel, NASA Goddard	Determination of atmospheric composition thermal structure, and dynamics; satellite surface composition and thermal properties.	2.5–50 μm
Ultraviolet spectrometer	A. L. Broadfoot, Kitt Peak Observatory	Measurement of upper atmospheric composition and structure; auroral processes; distribution of ions and neutral atoms in the Jovian system.	40–160 nm
Photopolarimetry	C. F. Lillie/C. W. Hord, U. Colorado	Measurement of atmospheric aerosols; satellite surface texture and sodium cloud.	235–750 nm
Planetary radio astronomy	J. W. Warwick, U. Colorado	Determination of polarization and spectra of radio frequency emissions; Io radio modulation process; plasma densities	20 kHz–40 MHz
Magnetic fields	N. F. Ness, NASA Goddard	Measurement of plasma electron densities; wave-particle interactions; low-frequency wave emissions.	–
Plasma science	H. S. Bridge, MIT	Measurement of magnetospheric ion and electron distribution; solar wind interaction with Jupiter; ions from satellites.	4 eV–6 keV
Plasma waves	F. L. Scarf, TRW	Measurement of plasma electron densities; wave-particles; low-frequency wave emissions.	10 Hz–56 kHz
Low energy charged particles	S. M. Krimigis, Johns Hopkins U.	Measurement of the distribution, composition, and flow of energetic ions and electrons; satellite-energetic particle interactions.	10 keV–30 MeV
Cosmic ray particles	R. E. Vogt, Caltech	Measurement of the distribution, composition, and flow of high energy trapped nuclei; energetic electron spectra.	0.15–500 MeV
Radio science (occultation & celestial mechanics)	V. R. Eshleman, Stanford U.	Measurement of atmospheric and ionospheric structure, constituents, and dynamics; satellite masses.	–

took 48 seconds to play back each image. Each Voyager would take nearly 20 000 images during the encounter.

The infrared interferometer-spectrometer (IRIS) not only had much greater spatial resolution than Pioneer's instrument, but also made detailed spectra, sufficient to detect many absorption lines of gases in the jovian atmosphere. It used a reflector of 50 cm aperture, coupled to an interferometer which produced infrared spectra from 2.5 to 50 μm. Its field of view was 0.25° across.

Two instruments operated in the ultraviolet – the spectrometer (field of view 0.1° x 0.9°) and photopolarimeter. These wavelengths only penetrate the outer fringes of Jupiter's atmosphere, but the instruments were to observe emission from aurorae and from atoms in the radiation belts, as well as the satellites. (The photopolarimeter should have studied polarisation at both ultraviolet and visible wavelengths, but on both craft its analyser wheel repeatedly stuck and polarisation measurements were abandoned.)

The planetary radioastronomy instrument, with two 10-metre antennae pointed 90° apart, took advantage of its proximity to Jupiter to locate the sources of different types of radio emission.

The remaining five instruments measured 'particles and fields'

to extend Pioneer's survey of the magnetosphere, solar wind, and cosmic rays. One of these, the plasma wave analyser, detected plasma oscillations with frequencies in the range of human hearing; its data could be played as an audiotape so that audiences could hear the rumble of the bow shock and the whistles of lightning on Jupiter.

Even the 'message to the aliens' was much more detailed than on Pioneer. In addition to the maps and human figures, it contained 115 photographs of the Earth and the human race going about their business, and a phonograph record of earthly sounds (including whales, music from many countries, and human greetings) – and, indispensably, the names of the members of the congressional space science committees.

The flight of the Voyagers

Both Voyagers were launched from Cape Canaveral on Titan IIIE-Centaur rockets – then the largest rocket in the American fleet. The rocket put the spacecraft into Earth orbit and then, reigniting the Centaur half an hour later, into an escape trajectory. The final push towards Jupiter was done by a small final stage with a solid rocket motor, oriented by Voyager's own hydrazine thrusters. An hour

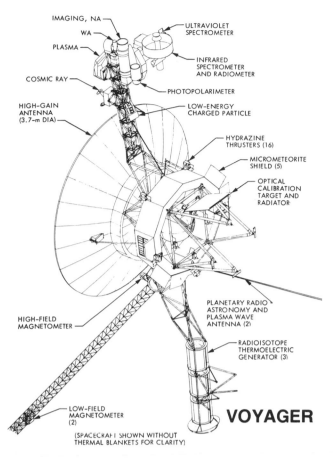

Fig. 2.12. The Voyager spacecraft. (NASA.)

after liftoff, this separated from Voyager itself, which proceeded to unfold in space.

Voyager 2 was launched first, as it was to take a slower trajectory to Jupiter, and thus arrive four months after Voyager 1. In fact, the identities of the spacecraft were undefined only eight days before the launch, when they were switched round because the one due to be launched first required a last-minute fix to its computers. Regardless, it was Voyager 2 that took off on 1977 August 20.

The launch was a harrowing one for the flight team. Anomalies within the attitude control systems had to be sorted out, and confusing data on the spacecraft systems led to fears that the scan platform had not deployed. (The boom had probably been slow in deploying, but did reach its correct position.) After one day, most of the problems had been overcome (though activation of some

IO
CLOSEST APPROACH J + 3 HRS

ISSWA

Fig. 2.13. Fields of view of Voyager's instruments, as projected on Io at closest approach to it. The square marked 'ISSWA' is for the wide-angle camera. Smaller fields within it are: (square) narrow-angle camera; (rectangle) UV spectrometer; (circle) IRIS. (NASA.)

Fig. 2.14. Launch of Voyager 1. (NASA.)

instruments was delayed), and further minor problems kept the flight team busy for several months.

Voyager 1 was launched almost flawlessly on 1977 Sep.5 – having had extra springs attached to ensure the scan platform deployed properly. It took the launch speed record with an initial speed of 54,220 km/hr. Its first course corrections were on Sep. 11 and 13. On Sep. 18, as its instruments were being checked out, it looked back from 12 M km and took the first picture of Earth and Moon together in space. It overtook Voyager 2 in mid-December. In 1978 February, Voyager 1 faced potential crisis when its scan platform stuck, perhaps because of something clogging the gears. But after three months of gingerly testing it, the problem had gone away.

Voyager 2 encountered a more serious problem in 1978 April when the primary radio receiver shut down. The problem began when mission controllers, preoccupied with Voyager 1's problem, omitted to send signals to Voyager 2 for more than a week – after which time, the spacecraft's programming made it switch to its backup receiver in case the first one had failed. When controllers did switch it back to the primary receiver, that one failed completely – and controllers had to wait another week for the craft to switch itself to the backup receiver again. But it turned out that the backup had a failed capacitor, which reduced the bandwidth for receiving command signals by a factor of a thousand. The permitted frequency would change not only with Doppler shift but also with onboard temperature, so from then on the flight team had to continuously vary the transmitted frequency to match the spacecraft's limited 'hearing range'. In case the remaining receiver should also fail, flight controllers loaded the on-board computer with a minimal program that could be performed automatically at Jupiter and Saturn, and also showed that it would be possible to transmit a few commands to the spacecraft via its planetary radio-astronomy antenna, as a last resort. Fortunately, neither contingency plan was needed.

During the flight, the 'particles and fields' experiments monitored the solar wind and cosmic rays, while the ultraviolet spectrometer made valuable observations of stars and nebulae. And the

Table 2.4. *The Voyager encounters*

Times are all when signals were received at Earth, in local time. Distances to planets are to cloud-tops. Distances to satellites are to centres; satellite encounters are only listed if closer than 220 000 km.

Voyager 1		
Launch	1977 Sep 5, 08.56 EDT	
Jupiter:		
Jupiter	1979 Mar 5, 04.42 PST	277 400 km
Io	Mar 5, 07.50 PST	20 570 km
Ganymede	Mar 5, 18.52 PST	114 710 km
Callisto	Mar 6, 09.46 PST	126 400 km
Saturn:		
Titan	1980 Nov 11, 23.05 PST	6490 km
Saturn	Nov 12, 17.11 PST	124 000 km
Mimas	Nov 12, 19.05 PST	88 440 km
Dione	Nov 12, 21.04 PST	161 520 km
Rhea	Nov 12/13, 23.46 PST	73 980 km
Voyager 2		
Launch	1977 Aug 20, 10.29 EDT	
Jupiter:		
Callisto	1979 Jul 8, 06.13 PDT	214 930 km
Ganymede	Jul 8/9, 01.06 PDT	62 130 km
Europa	Jul 9, 11.45 PDT	205 720 km
Jupiter*	Jul 9, 15.29 PDT	650 180 km
Saturn:		
1980 S26	1981 Aug 25, 21.45 PDT	107 000 km
Saturn	21.51 PDT	101 000 km
Enceladus	22.11 PDT	87 000 km
1980 S3	22.32 PDT	147 000 km
1980 S13	Aug 26, 00.28 PDT	154 000 km
Tethys	Aug 26, 00.38 PDT	93 000 km
Uranus:		
Ariel	1986 Jan 24, 08.20 PST	130 000 km
Miranda	09.03 PST	32 000 km
Uranus	10.00 PST	82 000 km
Neptune:		
1989 N1*	1989 Aug 25, (00.18 PDT	146 000 km)
Neptune	01.03 PDT	4900 km
Triton	06.17 PDT	38 360 km

*Data for 1989 N1 refer to best image, not closest approach.

imaging and infrared instruments were periodically checked by turning to view a calibration plate on the spacecraft.

As early as 1978 May, when only half way to Jupiter, Voyager 1 took pictures of the planet that were as detailed as a typical amateur view. In 1978 mid-December, a complete dress rehearsal for near-encounter was performed; photos sent then already surpassed the best views from Earth. Then the Voyager 1 'observatory phase' began on 1979 Jan. 4, 60 M km from the planet. From then on, the spacecraft made a continuous record of Jupiter, consisting of a set of images in 4 different colours every 2 hours (i.e., every 72° of jovian longitude). These images were composited into colour pictures and into the movies that showed the winds blowing across the planet. The other instruments also viewed Jupiter regularly during

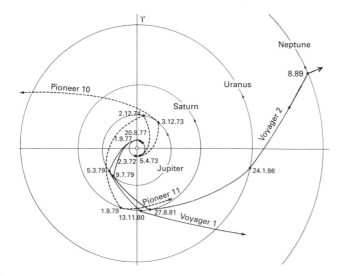

Fig. 2.15. Trajectories of the Voyagers and Pioneers through the solar system. (Adapted from NASA diagrams.)

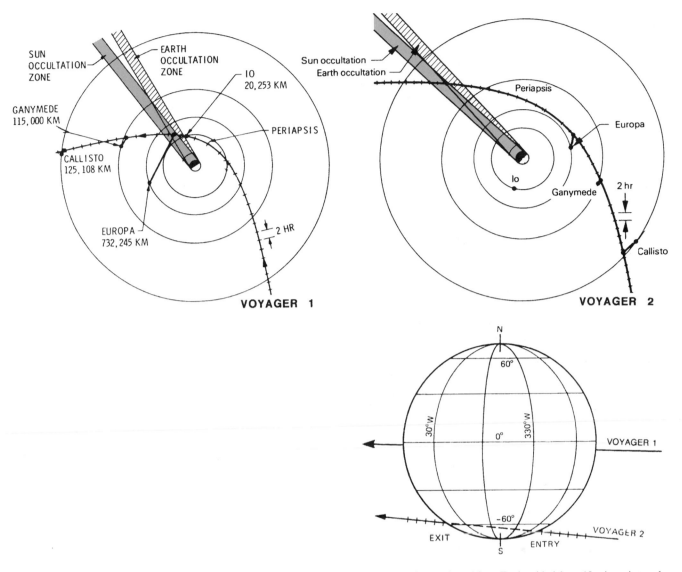

Fig. 2.16. Trajectories of the Voyagers past Jupiter. The third panel shows the paths behind Jupiter as viewed from Earth, with ticks at 10-minute intervals. (Adapted from NASA diagrams.)

the observatory phase. From Jan. 30 to Feb. 3, Voyager 1 imaged the planet every 96 seconds, recording a continuous movie of Jupiter's rotation from a distance of 35–31 M km. By mid-February, Jupiter's disk was larger than the field of the narrow-angle camera, so progressively more images were needed to maintain complete mapping. The other instruments were also scanning the planet. The final course correction was on Feb. 20.

Voyager 1 first came within Jupiter's bow shock on Feb. 28, 6.1 M km out. But the solar wind was gusty and Voyager re-crossed the bow shock (or vice versa) four times, the final entry being only 3.6 M km from the planet on March 2.

By then, daily press conferences at JPL were becoming block-buster shows at which even veteran reporters were awestruck – not to mention the scientists. There were the movies of the swirling clouds, the reports of the radiation belts and of newly-appeared sulphur emission in Io's orbit, and the psychedelic closeup pictures that were now coming in almost continuously, just 38 minutes after they were transmitted from Jupiter. The closest pictures of Jupiter,

taken a few hours before closest approach, would have a resolution of about 4 km per pixel, and only at this stage did the edges of the convoluted cloud structures finally appear diffuse. The picture that had the biggest single impact was an 11-minute time exposure taken as Voyager crossed Jupiter's equatorial plane on March 4. Fulfilling the dreams of scientists who had urged that it be taken, and confounding others who said it would show nothing, this dim, jittery image showed that Jupiter had a ring around it.

Closest approach to Jupiter was on 1979 March 5 at 04.42 PST, within the orbit of Io, at a speed of 130 300 km/hr. Three-and-a-half hours later, Voyager disappeared behind Jupiter for two hours, turning as it went in order to probe the atmosphere for as long as possible – both by beaming its radio signals to Earth, and by analysing ultraviolet rays from the sun as it moved into Jupiter's shadow. Within the shadow, it took long exposures of the dark side, which showed aurorae, lightning, and even a meteor (Chapter 17).

By this time, Voyager 1 had embarked on one of the most memo-

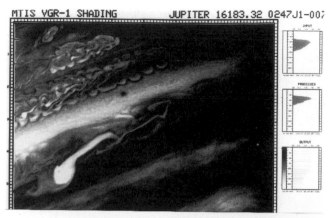

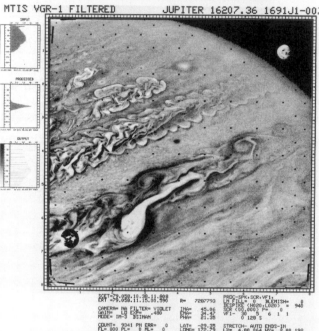

Fig. 2.17. Processing the Voyager images. Numbers at top right identify these as images of Jupiter from Voyager 1, 7 days before encounter. *(Upper:)* 'Shading' version, which preserves relative intensities. *(Lower:)* 'Filtered' version, in which local average intensity is normalised in order to maximise local contrast. (This was taken two rotations after the upper image.) 'Filtered' images reveal a grid of dots present on the camera screen.

The histograms of pixel intensities at the edges document the processing: top, original distribution; middle, after 'filtering' if any; bottom, after stretching to maximise contrast. The annotation below indicates the time the photo was taken (SCET) and received (ERT), the camera and filter used, and other data. These images were from the narrow-angle camera with ultraviolet filter (upper) and violet filter (lower).

These images show the southern hemisphere, including a mid-SEB outbreak of expanding white clouds. The lower image also shows Io (at bottom left) and Europa. South is up in the pictures as received. (NASA.)

rable days in the history of exploration. Outbound from Jupiter on March 5–6, it flew past the galilean moons, and in just 26 hours it revealed four new worlds.

Voyager had already taken many images of the galilean moons, at higher resolution than ever before, but the patterns on their surfaces remained uninterpretable until it flew right past them. As the pictures streamed in, with resolution down to 1–2 km, geologists

struggled to interpret the strange landscapes. From a distance the first satellite, Io, showed a splotchy, multicoloured disk that looked like a pizza, and even in close-up it was still puzzling. There were no impact craters, but what were these great pits, terraces, colour splashes, and isolated rugged peaks? There was no close flyby of the second satellite, Europa, but distant views showed that it too was unique: a smooth white sphere criss-crossed by mysterious lines. Then came Ganymede, different again – with heavily cratered areas, but also sets of 'tramlines' and bands of extraordinary grooves, apparently produced by global deformation of the ice crust. Finally there was Callisto – a dull icy surface covered in flattened craters, but also bearing a set of giant ripples, the frozen trace of a great impact.

As it flew past Io's south pole, Voyager was intended to pass through the 'Io flux tube' – the stream of electrical current that is believed to link Io and Jupiter. In fact it did not do so, although magnetic fields attributable to the flux tube were detected. Apparently the flux tube had snaked aside as Voyager passed. The intense radiation near Io's orbit had affected Voyager's clock and the synchrony of its computers, so photos of the satellites were being taken 40 seconds early, blurring some of the closest images.

Then on March 9, the mystery of Io was solved, when navigation engineer Linda Morabito was examining a long exposure of the crescent Io against the stars, and saw a huge faint arc protruding from the limb. It was a volcanic plume over 200 km high – and turning up the contrast on the earlier pictures of Io showed seven more such plumes reaching 100 km up. The IRIS data showed that some of the source vents were hot. Voyager had discovered the first active volcanoes beyond the Earth, and Io was popping with them.

Voyager 1 sampled the beginning of Jupiter's comet-like magnetotail before returning to the solar wind; it crossed the magnetopause on March 15 and the bow shock (several times) on March 20–22, the last crossing being 18 M km from the planet. Meanwhile, it continued to take views of the diminishing crescent for over a month after encounter (Fig. 2.18).

Voyager 2 began its observatory phase on 1979 April 24, which consisted of observing sequences like those of its predecessor. The movies showed a dramatic difference in the winds near the Great Red Spot, where the jetstream spots were now being spun round in a new circulating current.

Voyager 2's final course correction was on June 27. It crossed the bow shock on July 2, 7.0 M km out, but then crossed it another ten times as the shock oscillated to and fro. On July 7, the spacecraft took a break from observing to receive an 8–hour transmission containing its final program for the encounter – including new sequences to follow up Voyager 1's unexpected discoveries. It was now 51 light-minutes from Earth.

This time, the satellite flybys came during the inbound leg on July 8–9 – interspersed with close-up imaging of Jupiter itself. Close flybys of Callisto and then Ganymede revealed different areas than those photographed by Voyager 1. Then the pass by Europa provided the first close views of this 'crazed billiard ball' – utterly smooth, but

Fig. 2.18. Jupiter viewed as a crescent by Voyager 1. (Left) 15 days after encounter (image 16847.59, violet filter, 'filtered'). From top to bottom, it includes white oval BC, the GRS (on terminator) with turbulent STropZ, the long-lived EZs white spot, the rifted region of the NEB, and NNTBs jet-stream spots (very dark). (Right) One of the last images, 39 days after encounter (image 17566.13). It includes the shadow of one satellite on the disc, and another satellite seen as a crescent in transit over the dark side (at top right). (NASA.)

marked by hundreds of cracks in the ice. Closest approach to Jupiter was just outside Europa's orbit. Just 2 hours later, Voyager 2's thrusters burned for 76 minutes to adjust the course for Saturn.

There was no close pass by Io, but distant photos showed that many volcanoes were still active – though the largest one had shut down, and the sulphur torus in Io's orbit was weaker than before. Immediately after the closest approach to Jupiter, Voyager 2 looked back at Io's dark side for a 10-hour 'volcano watch'. This had been added to the schedule when Voyager 1 discovered the volcanoes. The resulting movie showed several of the plumes spurting continuously, lighting up in sunlight on the limb. Another new target for Voyager 2 was the jovian ring, and detailed images were taken of it while Voyager passed through Jupiter's shadow,

25–27 hrs after closest approach. Shortly before this, 22–24 hrs after closest approach, Voyager had been occulted by Jupiter, passing behind its south polar region.

The radiation belts were more intense than expected this time, and caused anomalies in several instruments. The unpredictable variations in Voyager 2's faulty receiver were a problem, and during the volcano watch the spacecraft had lost the frequency from Earth, but of course it performed its programmed tasks regardless. There was no serious loss nor permanent damage. On its way out, Voyager 2 ran further down Jupiter's magnetotail than any previous spacecraft. Nearly two years later, as Voyager 2 approached Saturn in 1981 February, it re-entered a part of Jupiter's magnetotail for a few days, as Pioneer 10 had done.

(A)

(B)

(C)

(D)

Fig. 2.19. Voyager's planets. (A) Jupiter and the four galilean moons. (B) Saturn, with Tethys (foreground) and other moons. (C) Uranus, with its moon Miranda (foreground) and its ring. (D) Neptune and its moon Triton, which has a nitrogen atmosphere and large polar ice cap. (Montages of images by the Voyager imaging team; NASA pictures.)

Beyond Jupiter, the Voyagers became the craft that unveiled the outer solar system. With Voyager 1's dazzling successes at Saturn, the way was clear for Voyager 2 to visit all the giant planets in turn. The Voyagers could not reach Pluto, but now we know that Pluto is much less massive than Earth's moon and very similar to Neptune's moon Triton, so it hardly qualifies as a major planet. And there is almost certainly no other major planet beyond it; the continuing steady course of the Pioneers, together with the limits set by the Infrared Astronomical Satellite's sky survey, leave little room for a 'Planet X'. So as Voyager 2 moved away from Neptune and Triton in 1989 August, one was left with the shocking realisation that there were no unexplored major planets left in the solar system.

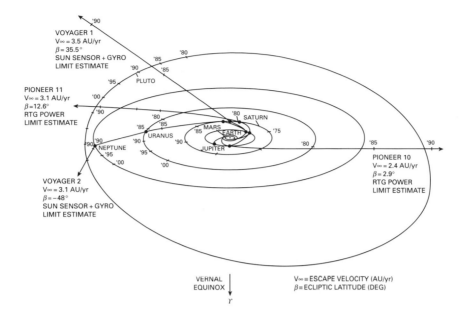

Fig. 2.20. The final trajectories of the Pioneers and Voyagers outbound from the solar system. As the Voyagers have already overtaken Pioneer 11, they will probably be the first spacecraft to reach the heliopause. (Adapted from NASA diagrams.)

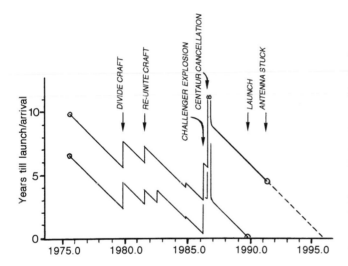

Fig. 2.21. The trials of Galileo. The graph shows how the expected times to launch (lower line) and to arrival at Jupiter (upper line) have oscillated over the 20 years of the project.

2.3 THE GALILEO PROJECT[3]

The Voyagers, for all their brilliance, constituted only the first, reconnaissance phase of planetary exploration. The follow-up phase for every planet has to be an orbiter and a lander, and for Jupiter this is the task of the Galileo project. But Galileo has had the most lengthy and frustrating history of any planetary mission. The planned time of arrival has slipped by 10 years; the cost has escalated nearly five-fold; and after all that, at the time of writing, the craft itself appears to be crippled.

[3] Political episodes in the development of Galileo are described in books by Trento (*1987*) and Murray (*1989*). The Galileo spacecraft is described in Johnson and Yeates (*1983*), and in *Space Science Reviews*, vol.60 (nos.1-4), reprinted as *The Galileo Mission* (edited by Russell, *1992*).

The trials of Galileo

The 'Jupiter Orbiter/Probe' project was formally announced by NASA in 1976, and approved by Congress in spring, 1977. That summer it survived a fierce attempt by anti-space congressmen to cut off its funds. The Orbiter and Probe were to be sent off as a single spacecraft, to be launched in 1982 January, to arrive at Jupiter in mid-1985, and to cost $295 million.

The Galileo spacecraft, as it was soon named, was to be the first planetary probe launched from the Space Shuttle, and this proved to be its undoing. As the Shuttle development encountered more and more problems, it became clear that it would not be ready in time and that the early version would not have the power to carry the whole spacecraft with its Centaur booster. So in autumn, 1979, NASA was forced to postpone Galileo for two years, and to separate the Orbiter and Probe so that they could be launched on two separate Shuttle missions with a new smaller booster, the Inertial Upper Stage (IUS). The cost had now doubled and in 1981 the Reagan administration proposed to cancel the project, only to be overruled by Congress. (In those days, even the tracking of Pioneer 10 and 11 was under threat from the administration's parsimony.) But the IUS also encountered problems during development, and in 1981–82 it was decided – after another long fight in Congress with fierce lobbying from the manufacturers – to recombine Galileo into one mission to be launched from the Shuttle by an uprated Centaur rocket. The new Centaur was itself untested, and the prospect of flying this rocket with its highly explosive liquid hydrogen fuel in the manned Space Shuttle was not universally welcomed. And the mission was delayed another two years.

By 1986 January, the mission was just four months away from launch when the ultimate Shuttle disaster struck: the explosion of the Shuttle Challenger with the deaths of 7 astronauts. Galileo and all other missions were grounded. And for Galileo, the sequel was just as damaging. In the 'play-it-safe' mood after the loss of Challenger, it was decided that Centaur could not be used on the Shuttle after all. Now, with Galileo already built as a single spacecraft, the United States did not have a rocket that could launch it to Jupiter.

The only planet they *could* launch it to, using the Shuttle and the IUS, was Venus; and that proved to be the only route of escape. A bizarre trajectory was devised that would have Galileo circle $2\frac{1}{2}$ times around the inner solar system, picking up speed by 'sling-shot' encounters with Venus, Earth, and then Earth again, before it finally aimed for Jupiter. While the Shuttle solid rocket boosters were being redesigned, Galileo was sent back to JPL to have gold foil and sunshields added so that it could survive the heat of the sun during its enforced visit to Venus. At last it was ready for launch in 1989 October; but now it would take 6 years to reach Jupiter. The cost had risen to $1350 million.

There was one final political obstacle to be overcome: the demonstrations and lawsuits by anti-nuclear campaigners, who claimed that in the event of the Shuttle blowing up on liftoff, they would be contaminated by the plutonium in Galileo's RTGs (although it had been divided into small pellets encased in iridium and graphite to ensure that there was no such risk). Fittingly for a spacecraft named Galileo, it had to run the gauntlet of anti-scientific obscurantism before it could embark on its journey of discovery.

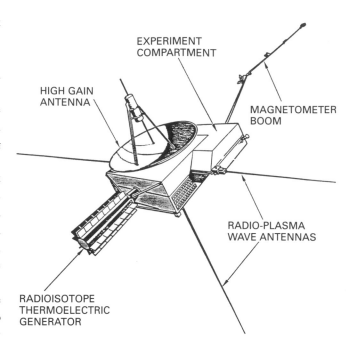

Fig. 2.22. The Ulysses spacecraft. (European Space Agency diagram.)

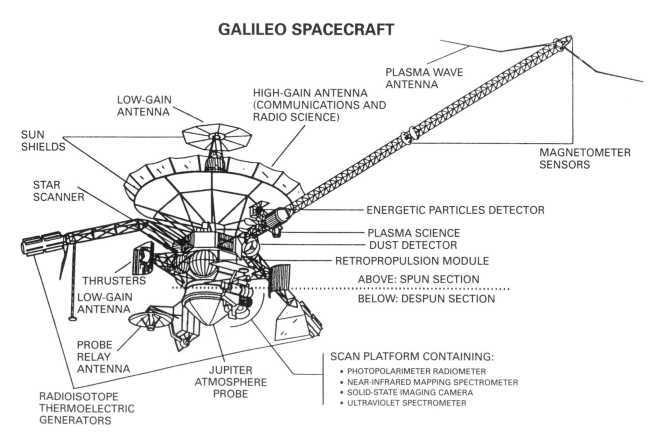

Fig. 2.23. The Galileo spacecraft. (NASA diagram.)

Table 2.5. *The Galileo experiments*

Experiment/Instrument	Principal Investigator	Objectives	Spectral range
PROBE			
Atmospheric structure	Alvin Seiff, NASA Ames Research Center	Temperature, pressure, density, molecular weight profiles	–
Neutral mass spectrometer	Hasso Niemann, NASA Goddard SFC	Chemical composition	2–150 AMU
Helium abundance interferometer	Ulf von Zahn, Bonn University, FRG	Helium/hydrogen ratio (from refractive index)	900 nm
Nephelometer	Boris Ragent, NASA Ames	Solid/liquid cloud particles (from laser scattering)	904 nm
Net flux radiometer	L. A. Sromovsky, Univ. of Wisconsin	Thermal profile, heat budget	0.3–500 µm
Lightning/energetic particles	Louis Lanzerotti, Bell Laboratories	Lightning flashes	White light
		Lightning radio bursts	0.1–100 kHz
		Energetic charged particles outside atmosphere	3–900 MeV
ORBITER (Despun)			
Solid-State imaging camera	Michael Belton, NOAO (Team Leader)	High-resolution imaging of Jupiter, moons, other planets	0.37–1.1 µm
Near-infrared mapping spectrometer	Robert Carlson, JPL	Atmospheric & surface compositions; thermal mapping	0.7–5.2 µm
Ultraviolet spectrometer (includes extreme UV sensor on spun section)	Charles Hord, Univ. of Colorado	Atmospheric gases, aerosols, etc.	54–430 nm
Photopolarimeter/radiometer	James Hansen, Goddard Institute for Space Studies	Polarimetry: reflection from clouds and surfaces	410–945 nm
		Radiometry: temperatures in atmosphere	15–100 µm
ORBITER (Spinning)			
Magnetometer	Margaret Kivelson, UCLA	Strength and fluctuations of magnetic fields	–
Energetic particles	Donald Williams, Johns Hopkins APL	Electrons, protons, heavy ions in atmosphere	10 keV–55 MeV
Plasma	Louis Frank, Univ. of Iowa	Composition, energy, distribution of ions	0.9 eV–52 keV
Plasma wave	Donald Gurnett, Univ. of Iowa	Electromagnetic waves and wave-particle interactions	5 Hz–160 kHz
Dust	Eberhard Grun, Max Planck Inst. für Kernphysik	Mass, velocity, charge of submicron particles hitting spacecraft	–
Radio science: celestial mechanics	John Anderson, JPL (Team Leader)	Masses and motions of bodies from spacecraft tracking	–
Radio science: propagation	H. Taylor Howard, Stanford Univ.	Satellite radii, atmospheric structure, from radio propagation	–
Engineering experiment: Heavy ion counter	Edward Stone, Caltech	Spacecraft charged-particle environment	6–200 MeV

But in another respect, the name was less appropriate. Of the two Jupiter-bound spacecraft then queuing up for launch by the Shuttle, the smaller one that could go direct was called Ulysses. It was Galileo that would have to make a wandering odyssey before reaching its destination.

The Ulysses mission

The joint European/American spacecraft Ulysses (Fig. 2.22) had only a passing interest in Jupiter; its aim was to explore the polar regions of the sun for the first time, and it only visited Jupiter in order to use the slingshot effect to swing it into a solar-polar orbit.

Being a 'particles and fields' mission, it carried no instruments to study Jupiter as a planet, but it probed the jovian magnetosphere in detail as it passed through it.

Ulysses was the remnant of the International Solar Polar Mission, a project that suffered agonies similar to those of Galileo. It was to have consisted of two spacecraft, one from NASA and the other from the European Space Agency. But the American spacecraft fell victim to Congress' budget cuts in 1981, leaving only the European one; and this was grounded along with Galileo when the Shuttle exploded. It was launched at last on 1990 Oct. 6, seven years late, by the Space Shuttle with the IUS final stage.

Ulysses weighs 370 kg, with nine scientific instruments, contributed more-or-less equally from Europe and the USA.

It flew past Jupiter on 1992 Feb. 8, on a north-to-south trajectory at 98000 km/hr. There was no radiation damage. It passed between the orbits of Io and Europa, and found some surprises; the magnetosphere was twice as large as expected, but the plasma in Io's orbit was patchy and was weaker than during the Voyager encounters, probably due to less activity from Io's volcanoes.

The Galileo spacecraft

Galileo consists of two spacecraft, the Orbiter and the Probe, which will separate before reaching Jupiter (Fig. 2.23). The mission is operated by the Jet Propulsion Laboratory, which also designed the Orbiter; the Probe was designed by Ames Research Center. The propulsion system and some experiments were provided by Germany. The total mass is 2.6 tonnes (2222 kg Orbiter and 340 kg Probe) – including 925 kg of fuel on the Orbiter. Both spacecraft are spin-stabilised.

The Probe will be inactive until it enters the atmosphere. It carries a detachable heat-shield and a 2.5-metre parachute. It is battery-powered, and its two transmitters will work in duplicate, each at 128 bits/second. The main task of its six scientific instruments will be to measure the temperature, pressure, heat flow, chemical composition, and cloud structure of the atmosphere through which it passes.

The Orbiter spins at 3 revolutions per minute, but includes a non-spinning ('despun') section which carries the cameras, other remote sensors, and an antenna for receiving the signals from the Probe. There is a large rocket engine for inserting the craft into jovian orbit, as well as 12 smaller thrusters. Otherwise, the components of the spacecraft are similar to those of Voyager, although improved: low- and high-gain antennae, RTGs, computers, and tape recorders. The high-gain antenna, when unfolded, should be 4.8 metres across and operate at 134000 bits/second (one picture contains 8 million bits). The plutonium RTGs produce 570 W at launch and 485 W in Jupiter orbit.

The scientific instruments also represent improved versions of those on previous probes (Table 2.5). The spun section carries five 'particles-and-fields' experiments (including the magnetometer on an 11-metre boom) and a dust detector. The despun section carries the camera, near-infrared mapping spectrometer, ultraviolet spectrometer, and photopolarimeter-radiometer, all mounted on a scanning platform. The camera consists of a telescope with 150 cm focal length that is actually a Voyager spare, so its resolution is the same; but its detector is a CCD (charge-coupled device) which is much more sensitive than previous television cameras. It should achieve a resolution of 20 metres on the galilean satellites.

The flight of Galileo

The Space Shuttle Atlantis carrying Galileo and five astronauts took off into Earth orbit on 1989 October 18. During the next 6½ hours, the astronauts opened the payload bay, raised the Galileo-IUS combination on its platform, checked their orbit and the tem-

Fig. 2.24. Launch of Galileo from the Space Shuttle. (NASA/Smithsonian.)

perature of the spacecraft, and then gently released it (Fig. 2.24). An hour later, the IUS fired to push Galileo away from Earth.

The flyby of Venus, four months later, went according to plan. It included high-resolution infrared probing of the planet's clouds, and the first unambiguous detection of lightning in them. At this stage the high-gain antenna was still furled, to protect it from the intense solar radiation, and data were sent back by the low-gain antenna. A little information, including one picture, was returned soon after the Venus encounter, but most of it was kept until the spacecraft approached Earth again so that a high transmission rate was possible. Ten months later came the first Earth flyby – which JPL treated as a full-dress planetary encounter, including probing of Earth's magnetotail, colour photography of Earth and Moon, remote sensing of the Antarctic ozone hole, and as NASA soberly announced, 'we found evidence for life on Earth.'

Then came Galileo's latest and potentially gravest crisis. On 1991 April 11, the command was sent to unfurl the high-gain antenna, which would be necessary for transmissions at high data rate from the vicinity of Jupiter. But the antenna failed to open completely. Its motors strained, but three of its 18 ribs remained stuck to the central pole – as indicated by engineering telemetry and by a slight wobble that the craft developed because of the lopsided antenna. It seems likely that the pins which held these ribs in place had become stuck during the 3-year storage and repeated transport of the spacecraft following the Shuttle disaster. Ironically (to give the Shuttle its due) a similar problem developed just 4 days earlier during the launch of the Gamma Ray Observatory: its high-gain antenna also failed to deploy, but the Shuttle astronauts were

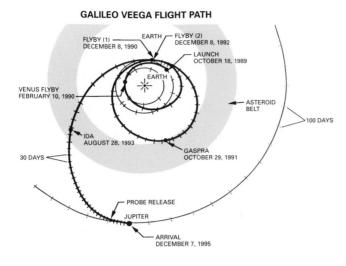

Fig. 2.25. Galileo's trajectory through the solar system – the 'Venus-Earth-Earth gravity assist'. (NASA.)

Table 2.6: *The Galileo and Ulysses encounters.*

Galileo

Launch	1989	Oct 18, 16.54 UT	
Venus	1990	Feb 10, 06.00 UT	16 120 km
Earth	1990	Dec 8, 20.34 UT	960 km
951 Gaspra	1991	Oct 29	1600 km
Earth:			
Moon	1992	Dec 8, 03.58 UT	110 000 km
Earth		, 15.35 UT	300 km
Planned encounters:			
243 Ida	1993	Aug 28	[n.d.]
Jupiter:			
Europa	1995	Dec 7	33 000 km
Io		Dec 7	1000 km
Jupiter		Dec 7	214 000 km
Europa			
Ganymede	1996–1997		*Encounters in*
Callisto			*jovian orbit*

Ulysses

Launch	1990	Oct 6	
Jupiter		1992 Feb 8, 12.01 UT	380 000 km

All times are in Universal Time, and distances are to the surface of the object.

on hand to do a 'space-walk' and free it. There was no such possibility for Galileo.

The engineers hoped that repeated heating and chilling in space might eventually work the pins free, and over the next $1\frac{1}{2}$ years of Galileo's flight, out to the asteroids and back to Earth, it was re-oriented from time to time in an attempt to achieve this – without success. This left only the low-gain antenna, which can operate at 1200 bits/second near Earth, but only at 10 bits/second at Jupiter if it is to maintain a detectable signal strength – compared to the intended transmission rate of 134 000 bits/second. This rate would be too low to carry out a useful Orbiter mission.

Meanwhile, Galileo has achieved one more 'first': the first flyby of an asteroid. This was Gaspra, a tiny fragment of rock only 12 km across. Gaspra was merely a consolation prize, and little could be learned from it; more serious asteroidal science was another casualty of the mission's delays. If Galileo had been launched in 1986, it would have flown past Amphitrite, with a diameter of 200 km, one of the largest asteroids. Nevertheless, the camera duly snapped the view as Galileo flew 1600 km from Gaspra at 29000 km/hr. Sampling of the recorded images located the asteroid right in the middle, and the relevant part of the image was returned to Earth using the low-gain antenna.

Galileo flew past Earth again in 1992 December, returning the remaining Gaspra pictures, imaging the Moon's north polar region, and skimming only 300 km above the South Atlantic to pick up speed and head for Jupiter at last. There will be one more encounter *en route*, with the asteroid Ida in 1993 August.

The last hope for freeing the high-gain antenna came with the second Earth flyby, when the craft was closer to the sun, and thus warmer, than at any other time since the problem appeared. In the two months following the flyby, mission controllers tried 'hammering' the antenna motors in thousands of brief pulses. But in 1993 February, NASA had to accept failure, and proceeded with planning for a 'low-gain antenna mission'.

Fortunately, NASA have now worked out how to achieve a hundredfold increase in the effective data transmission rate, partly by compressing the data into fewer bits, and partly by increasing the sensitivity of receivers so that a higher transmission rate can be used. For instance, a full picture would take two months to transmit at 10 bits/second, but 'only' a few hours with the new capabilities. Therefore, the experiments will still be carried out, and much of the data should be as good as intended. The tape recorder on the Orbiter has enough capacity to record not only the whole of the Probe data, but also hundreds of images or plenty of other data. The first orbit, which will be highly elliptical and last 8 months, will largely be devoted to playing back the day of arrival. The main loss will be images of Jupiter's clouds, which will have to be rather few – not a continuous survey.

Galileo at Jupiter

In mid-1995, five months before arrival, the Orbiter and Probe will separate. First the whole spacecraft will adopt the course and the spin (10 revs/minute) required for the Probe; the Probe will be released; then the Orbiter will correct its own course, and begin several months of viewing Jupiter in an 'observatory phase'.

On the day of arrival, 1995 Dec. 7, dramatic events will happen in rapid succession. The inbound Orbiter will observe Europa from 33000 km then Io from only 1000 km. (This will be its only close encounter with the volcanic moon; the radiation near Io's orbit is too dangerous for Galileo to revisit it.) A few hours later, just as the Orbiter makes its closest swing above Jupiter, the Probe will enter the upper atmosphere at 6°N, at 170 000 km/hr. It will be decelerated, first by heat shields and then by a parachute, then make a 200-km descent through the cloud layers. It should take just over an

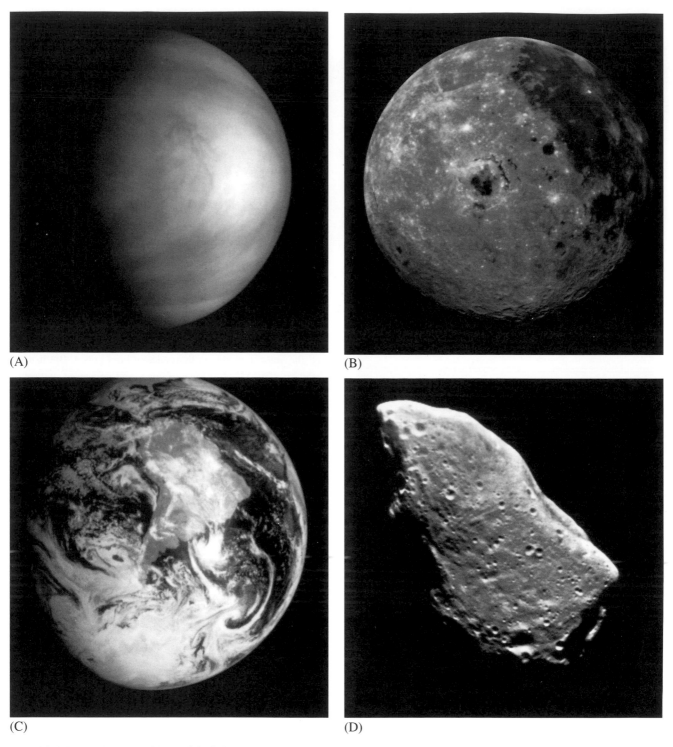

(A)

(B)

(C)

(D)

Fig. 2.26. Galileo's planets. (A) Venus (violet light). (B) Moon. (C) Earth (near-infrared, 1 μm). (D) Asteroid Gaspra. (NASA images obtained by the Galileo Imaging Team, leader M.J. Belton.)

hour to float down from the 0.1-bar level[4] to around the 25-bar level, where it is expected to expire. Its signals will be recorded by the Orbiter, 214 000 km above. About an hour later, the Orbiter will fire its main engine to go into orbit around Jupiter.

The Orbiter will broadcast the Probe's data back to Earth, and then proceed with its own mission. This is intended to last at least 2 years, during which it will make 10 orbits of Jupiter – all of them different. Almost every orbit will include a close encounter with one of the galilean moons, both to study the moons themselves, and to change the orbit through the magnetosphere. The results should be fascinating.

---

[4] One bar is approximately the atmospheric pressure at the surface of the Earth.

II. The Visible Structure of the Atmosphere

Chapters 3 and 4 summarise the observed facts about the jovian atmosphere. The observational details will follow in Part III. A reader seeking only an overall view may go straight from Part II to Part IV, in which we will revisit the topics of Part II from a more theoretical point of view.

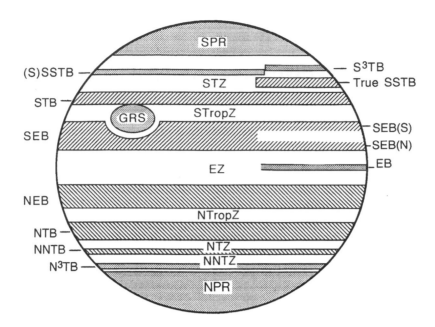

Fig. 3.1. The normal belts and zones of Jupiter's atmosphere. (See Fig. 1.3 for the full names of the belts.) At right are names of some common variations on the normal belt pattern. When a belt is double, the north and south components are abbreviated as (e.g.) SEB(N) and SEB(S). The north and south edges of a belt are abbreviated as (e.g.) SEBn and SEBs. (These abbreviations have been used by some authors to denote belt components, but we reserve them to denote belt edges.)

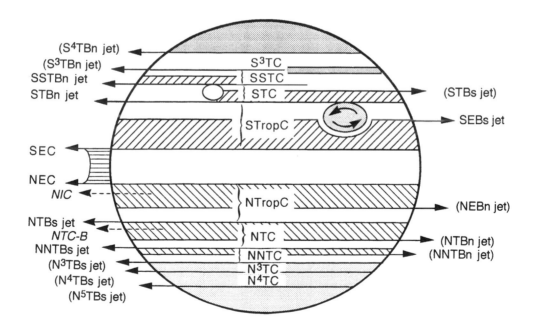

Fig. 3.2. The domains, currents and jetstreams of Jupiter's atmosphere. Domains and their slow currents are marked on the disc. Prograding jetstreams are named at left and retrograding jetstreams at right; those in brackets have only been observed by Voyager.

3: Horizontal structure: belts, currents, spots and storms

3.1 BELTS, ZONES, AND DYNAMICAL DOMAINS

The dark belts and bright zones are the most obvious features of the jovian disk. The major belts occur in much the same latitudes from year to year. Sometimes one of them may become fainter or narrower, or even disappear completely, but it always reappears. Thus there is a standard pattern of belts to which the atmosphere always reverts (Fig. 3.1).

This pattern of belts and zones corresponds to a dynamical pattern in the planet's atmosphere. Visual observers discovered a set of currents that are even more invariant in latitude than the belts, with modest wind speeds defined by the motions of large ovals and other features. Each of these 'slow currents'[1] coincides with one belt/zone pair (Fig. 3.2).

But the slow currents coexist with a pattern of much faster east-west currents, called jets or jetstreams, which seem to form the fundamental framework of the belt/zone pattern. The jetstreams sometimes show outbreaks of small spots, which revealed some of them to visual observers, but the full pattern of them was only revealed by Voyager's imaging of the smallscale cloud texture. This showed a continuous zigzag pattern of winds (Fig. 3.3), alternating between eastward and westward jetstreams that coincide with the long-term edges of the belts and zones. While the smallscale features move in this flow pattern, larger spots interrupt it. They are circulations that 'roll' between the jetstreams as they move with the slow currents.

Thus the planet outside the equatorial region is divided into a series of dynamical units, which we will call domains. These domains are clearly homologous, in the sense that they show so many different points of similarity that they must all be presumed to be set up by the same mechanism.

Each domain consists of a belt and a zone. The belt is normally dark and the zone bright, though these aspects may change from time to time. The belt is bounded by a rapid prograding (eastward)

jetstream on the equatorward side and a retrograding (westward) jetstream on the poleward side. The prograding jetstream marks the boundary of the domain. All large features in the domain normally move with a single slow current. They include white ovals, some of them very large, that lie across the latitude of the retrograding jetstream and deflect it around them. The prograding jetstream is subject to occasional outbreaks of much smaller dark spots, which move at the peak jetstream speed. The jetstream pattern defines the belts as cyclonic and the zones as anticyclonic; the stable white ovals and the jetstream spots are also anticyclonic. There may be rare exceptions to any of these generalisations, but the overall framework of the domains is fixed.

The five main domains on the planet (N. and S. Tropical, N. and S. Temperate, and N.N. Temperate) show all these properties. At higher latitudes some of the features are lost, particularly the retrograding jetstreams and the correspondence with the pattern of belts, but the domains can still be recognised up to at least 57°N (N⁵TB) and 53°S (S⁴TB).

The historical names of the belts can now be refined so as to agree with the pattern of domains. The nomenclature of the North and South Equatorial and Temperate Belts has always been straightforward, and exceptions are easily accommodated. For example, when the NEB is split, its north and south components are distinguished as NEB(N) and NEB(S), separated by a NEB Zone or NEBZ; and when there is an extra narrow belt in the NTZ, it is named the NTZ Band or NTZB. Nomenclature in higher latitudes has long been problematic, because some segments or even entire belts are often missing, and the belts that are visible are often uneven in latitude, so it is hard to define a normal pattern of belts. Those that are visible have usually been called the NNTB, NNNTB (N³TB), NNNNTB (N⁴TB), etc. in order, but they are so inconstant that this is unsatisfactory. However, the jetstream pattern as revealed by Voyager provides a natural framework by which to define these belts. The jetstream latitudes (Table 3.2) may be termed the canonical boundaries of the belts, so that a belt segment can now be named strictly according to its latitude.

Even this system is not entirely straightforward, because segments of the visible dark belts at high latitudes do not always respect the jetstream boundaries. The most important exception is

---

[1] There is a potential ambiguity between 'slow' meaning close to System II in rotation period, as in 'slow currents', and 'slow' meaning of long rotation period. As most observable winds are prograding, the two meanings usually coincide. When discussing speeds that are retrograding relative to System II, I will always specify 'rapidly retrograding' if high positive $\Delta\lambda_2$ is meant.

Table 3.1. *Average latitudes of belts and their edges.*

From Table 12.1:	(1907–1937)	(1950–1979)	(Overall)
South Polar Region:			
SPRn	–	–53.2 (±1.2; 15)	
S.S. Temperate Belt:			
(S)SSTBs	–	–46.6 (±1.3; 14)	
(S)SSTBn	–	–42.1 (±1.1; 15)	
(S)SSTB, SSTB	–41.4 (±2.3; 19)		
SSTBs	–	–41.9 (±1.8; 12)[a]	
SSTBn	–	–36.3 (±0.6; 9)[a]	
South Temperate Belt:			
STBs	–31.3 (±1.3; 7)	–33.3 (±0.9; 16)	
STB	–28.9 (±0.9; 27)	–30.2 (±0.7; 21)	
STBn	–26.0 (±1.2; 7)	–26.8 (±0.7; 17)	
From Tables 10.3 and 10.10:	(1895–1939)	(1940–1990)	(Overall)
South Tropical Band:			
STropB	–26.3 (estimate)[b]	–24.8 (±0.3; 9)	
South Equatorial Belt:			
SEBs	–18.7 (±1.1; 20)[c]	–20.7 (±0.8; 29)	
SEBn	–6.9 (±1.1; 29)	–7.0 (±0.9; 30)	–7.0 (±1.0; 59)
From Table 8.1 and 8.3:	(1913–1947)	(1950–1991)	(Overall)
North Equatorial Belt:			
NEBs	+7.1 (±1.1; 32)	+7.5 (±0.6; 27)	+7.3 (±0.9; 59)
NEBn	+17.5 (±2.7; 32)	+18.1 (±2.2; 27)	+17.8 (±2.5; 59)
North Tropical Band:			
NTropB	+22.4 (±1.0; 4)[d]	+23.3 (±0.8; 5)[e]	+22.9 (±1.0; 9)
From Table 6.2:	(1906–1942)	(1950–1991)	(Overall)
North Temperate Belt:			
NTBs	+24.5 (±1.6; 8)	+23.9 (±0.8; 21)	+24.0 (±1.2; 29)
NTB(S)	+25.1 (±0.6; 13)	+25.1 (±0.6; 13)	
NTBn	+30.6 (±1.9; 6)	+31.0 (±0.8; 15)	+30.9 (±1.2; 21)
N.N. Temperate Belt:			
NNTBs	–	+35.8 (±1.0; 18)	+35.8 (±1.0; 19)
NNTB	+38.1 (±1.0; 15)	+37.4 (±1.1; 23)	+37.7 (±1.1; 38)
NNTBn	–	+39.1 (±0.7; 12)	+39.1 (±0.7; 12)
N.N.N. Temperate Belt:			
N³TBs	–	+42.7 (±1.0; 13)	+42.7 (±1.0; 13)
N³TB	–	+44.9 (±1.0; 20)	+44.9 (±1.0; 20)
N³TBn	–	+47.0 (±1.0; 6)	+47.0 (±1.0; 6)
N.N.N.N. Temperate Belt:			
N⁴TB	+48.9 (±0.8; 5)[f]	–	+48.9 (±0.8; 5)

The table lists zenographic latitudes, given as the mean (±s.d.; N) of apparition means. (S.d. = standard deviation, including both measurement uncertainties and true variations; N = number of apparitions). Overall averages are not given for southern belts as their latitudes changed significantly during the twentieth century; also, STB gradually disappeared after 1979, being replaced by true SSTB. Suffix n or s indicates north or south edge. NTB(S) is south component; the north component is often ill-defined. Dates for some belts were more limited than the headings indicate, especially:
[a] true SSTB (1966–1989);
[b] STropB (1931–1934);
[c] SEBs (1895–1925; not listed when it was mostly within STropD in 1920s and 1930s);
[d] NTropB (1906–1943);
[e] NTropB (1962–1990);
[f] N⁴TB (1907–1916).

the most southerly of the commonly observed belts, which is usually called the SSTB by observers, but often straddles the canonical latitudes of SSTZ and/or S³TB. We will call this belt the 'southerly SSTB' or (S)SSTB, to distinguish it from the true SSTB which only appears infrequently.

The average observed latitudes of the belts and their edges are given in Table 3.1. They agree very well with the latitudes of the jetstreams in Table 3.2.

Do the belts change at all in latitude? The early observers had no reason to expect the belt latitudes to be fixed, and it was an impor-

Table 3.2 *The jetstreams: latitudes and speeds*

Jetstream	Voyager data (1979):					Earth-based data (1891–1991)					
	Lat. ß"	u (m/s)	Δλ₂ (°/mth)			Dates	Lat. ß"	u (m/s)	Δλ₂ (°/mth)	P (9h+)	(N)
S⁴TBn	53.4°S	+36.3	−129			–					
S³TBn	43.6°S	+42.1	−125			–					
SSTBn	36.5°S	+31.6	−88			(1988–1990)	36.0°S	+28 (±0)	−79 (±0)	53m 53s	(2)
STBs	32.6°S	−20.8	+42			–					
STBn	27–29°S	+44.3	−110			(1919–1991)	25.5°S	+44 (±7)	−109 (±17)	53m 12s	(14)
SEBs	20.0°S	−56.6	+117			(1928–1991)	20.7°S	−58 (±9)	+119 (±20)	58m 24s	(25)
SEC	7°S	+128	−276*	⎰ A	(1914–1974)		7°S	+105 (±3)	−228 (±5)	50m 31.5s	(23)
				⎱ B	(1914–1974)		7°S	+92 (±5)	−201 (±10)	51m 8s	(10)
NEC	5–8°N	+103	−224		(1891–1991)		7°N	+107 (±2)	−232 (±5)	50m 26s	(84)
				NIC:	*(1927–1991)*		*(11–14°N)*	*+51 (±8)*	*−117 (±16)*	*53m 2s*	*(14)*
NEBn	17.6°N	−24.3	+45			–					
NTBs	23.8°N	+163	−375*	⎧ D	(1970–1990)		23.6°N	+171 (±5)	−392 (±11)	46m 52s	(4)
				⎨ C	(1891–1990)		23.6°N	+122 (±6)	−283 (±13)	49m 17s	(10)
				⎩ B	(1928–1975)		*(26°N)*	*+45 (±10)*	*−112 (±23)*	*53m 8s*	*(7)*
NTBn	31.6°N	−32.0	+69			–					
NNTBs	35.6°N	+34.5	−94			(1929–1991)	35.4°N	+28 (±2)	−77 (±5)	53m 55s	(18)
NNTBn	39.5°N	−14.8	+31			–					
N³TBs	43.0°N	+21.8	−68			–					
N⁴TBs	48.2°N	+28.5	−94			–					
N⁵TBs	56.6°N	+14.1	-59			–					

This table lists the jetstreams that run along the belt edges, and the South and North Equatorial Currents (SEC, NEC), and two 'currents' that refer to features within the shear of the NEB and NTB, viz. North Intermediate Current (NIC) and North Temperate current B.

Voyager data are from S. Limaye (*1986*). Retrograding jetstreams are not listed for some belts as they become modest or undetectable at high latitudes. Earth-based data are from the tables at the ends of Chapters 6–11. These values are the mean (± standard deviation) of apparition means; as the speeds are usually well-determined visually, the standard deviation represents the true year-to-year variability of the mean speed. The NTBs jetstream and SEC both show multiple speed ranges denoted A, B, etc.; SEC values do not include SEB Revivals nor long-lived great white spots.

*The two fastest jetstreams were even faster according to T. Maxworthy (*1984, 1985*): SEBn, 160 m/s, -342°/mth; NTBs, 182 m/s. -417°/mth.

ß", zenographic latitude; *u*, speed in m/s in System III; Δλ₂, speed in degrees per 30 days in System II; *P*, rotation period (+ 9 hours); *N*, number of apparitions. For conversion between *u*, Δλ₂, and *P*, see Table 1.3 and Appendix 1. (System III is faster than System II by 8.0°/mth or 3.85 cos ß' m/s; System I is faster than System II by 229°/mth.)

tant discovery from systematic micrometer measurements that most belts did not migrate beyond a limited latitudinal range (T.E.R. Phillips, *1930*). However, the modern paradigm of domains implies that the edges of belts and zones should remain constant in latitude, and we should enquire whether apparent changes in latitude might be due to fading or darkening of distinct components of belts. In fact, to anticipate the following chapters, the equatorward edges of most belts are indeed almost invariant, at the latitudes of the prograding jetstreams. For example, apparent movement of the NTBs edge is largely due to fading of NTB(S) component. The only belt whose equatorward edge has shown definite latitude shifts is the (S)SSTB, though even here, spacecraft images show that variations with longitude are due to discontinuous belt components. But the poleward edges of the belts do shift over a continuous range, except for the SEBs edge which is confined by the most powerful of the retrograding jetstreams.

3.2 CURRENTS AND JETSTREAMS

The currents recorded by visual observers fall into three categories (Fig. 3.4). First, there is the great equatorial current; the entire equatorial region between about 10°N and 10°S progrades at 7–8 degrees per *day* (≈100 m/s) relative to System II, and intermediate speeds are sometimes recorded along its edges. Second, there are the nine slow currents, which govern most of the visible features outside the equatorial region, and have speeds of no more than one degree per day (≤10 m/s; Table 3.3). Third, there are the jetstreams on the edges of certain belts, only observed during infrequent outbreaks of small dark spots, which have speeds of several degrees per day (30–170 m/s). These are now known to be permanent jetstreams; some of them were initially designated as 'Currents' like the slower currents, but now that Voyager has revealed jetstreams bordering all the belts, it is more logical to give them systematic designations, as in Table 3.2.

The jetstreams are detectable from Earth only when they are disturbed by distinct spot outbreaks. All the larger features move with the slow currents. Even the rather high prograding speeds of white ovals in the SSTC, and the retrograding speeds of dark streaks in the NTC, have been shown by Voyager to be much less than the speeds of the jetstreams in those regions; these are indeed slow currents that lie between the jetstreams.

(A)

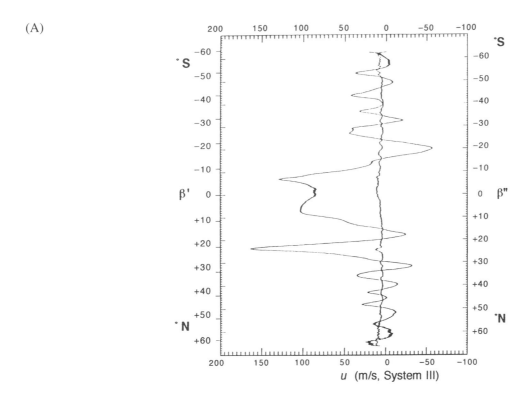

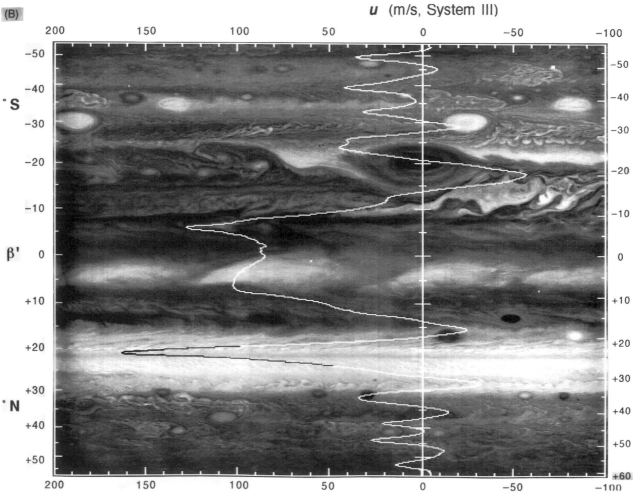

Fig. 3.3. Mean east-west wind speeds determined by Voyager. The east-west wind speed (*u*, in m/s in System III), averaged around the planet, is plotted against zenographic and zenocentric latitude. (A) From Voyager 2, violet-filter maps. (B) The mean profile is superimposed on a photo-map of the planet. (Note however that this mean curve will not apply to the GRS and STropD which are shown in this sector of the map.) (From Limaye, *1986*; these versions by courtesy of S.S. Limaye.)

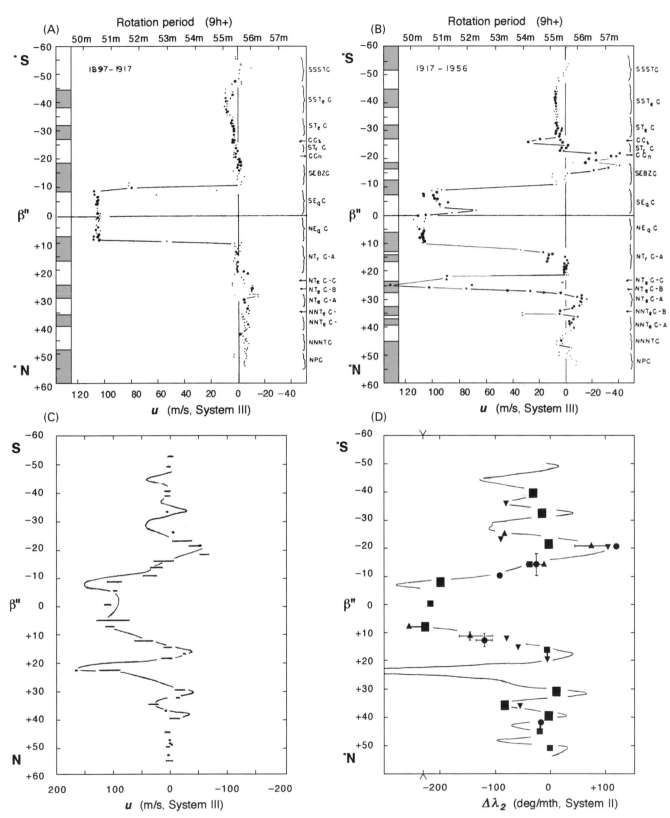

Fig. 3.4. Mean east-west wind speeds determined from Earth:
(A) 1897–1917; (B) 1917–1956. Both are from BAA and other historical data, showing the mean speed at each zenographic latitude. Averaging over multiple apparitions has reduced the apparent peak speeds of some jet-streams. (Adapted from Chapman, 1969.)

(C) 1974–1979, from NMSUO data, matched to a preliminary version of the Voyager profile. (From Beebe & Youngblood, 1979.)
(D) 1986–1989, from BAA data, matched to the Voyager profile of Fig. 3.3. Symbols denote different apparitions. (From Rogers, 1990.)

Table 3.3 *The slow currents: latitudes and speeds*

Current	(Old name)	Lat.range	(Interval)	P	(9h 55m+)	Δλ₂		u		(N; n)
S³TC	(SPC)	45–52°S	(1900–1991)	29.4s	(±7.8s)	−8.3	(±5.7)	+0.1	(±1.9)	(14; 2.6)
SSTC		38–42°S	(1888–1991)	6.4s	(±2.6s)	−25.1	(±1.9)	+6.6	(±0.7)	(72; 4.9)
STC		29–35°S	(1880–1940)	20.2s	(±2.4s)	−15.0	(±1.7)	+2.9	(±0.7)	(43; 5.5)
			(1940–1991)	5s → 24s		−26 → −12		+7.5 → +1.5		(decelerating)
STropC	(–)	13–25°S	(1831–1991)	40.5s	(±3.2s)	−0.1	(±2.3)	−3.6	(±1.0)	(see below)

Components of STropC:

—GRS			(1831–1880)	35.9s		−3.5		−2.0		(18; 1.0)
			(1884–1898)	41.3s		+0.5		−3.8		(13; 1.0)
			(1898–1939)	38.9s		−1.3		−3.0		(37; 1.0)
			(1940–1991)	42.0s		+1.0		−4.0		(46; 1.0)
—SEB(S)			(1911–1939)	39.7s	(±5.7s)	−0.6	(±4.1)	−3.4	(±1.9)	(12; 1.9)
			(1940–1991)	45.3s	(±7.9s)	+3.3	(±5.7)	−5.2	(±2.6)	(8; 2.3)

Current	(Old name)	Lat.range	(Interval)	P	(9h 55m+)	Δλ₂		u		(N; n)
NTropC		14–22°N	(1887–1991)	28.3s	(±7.4s)	−9.1	(±5.4)	+0.5	(±2.5)	(80; 10.9)
NTC	(NTC-A)	25–34°N	(1899–1991)	64.1s	(±6.0s)	+17.0	(±4.2)	−10.6	(±1.8)	(48; 3.6)
NNTC	(NNTC-A)	37–42°N	(1900–1991)	40.1s	(±4.4s)	−0.3	(±3.2)	−2.9	(±1.2)	(52; 4.4)
N³TC	(NNNTC)	44–47°N	(1900–1991)	19.8s	(±5.6s)	−15.2	(±4.1)	+2.5	(±1.4)	(25; 4.1)
N⁴TC	(NPC)	49–55°N	(1900–1991)	42.5s	(±3.9s)	+1.4	(±2.8)	−2.9	(±0.9)	(16; 3.5)

For each of the nine slow currents, the table lists the zenographic latitude range, then the mean speed in the three commonly used forms: P, rotation period (+9h 55m); $\Delta\lambda_2$, speed in degrees per 30 days in System II; u, speed in m/s in System III. Data are from the tables in Chapters 6–11. These values are the mean (± standard deviation) of apparition means; as the speeds are usually well-determined visually, the standard deviation represents the true year-to-year variability of the mean speed. All values (including those for the GRS) are as derived within apparitions, and are thus slightly too slow due to phase effects. To convert to true speeds, the correction is: −0.9s, −0.6°/mth, +0.3 m/s. In the last column, N = number of apparitions used, n = mean number of features in these apparitions (since 1914, as the individual spots were not always listed before then).

The slow currents

The nine slow currents (Table 3.3) had all been recorded by 1901. Since then, they have shown only temporary variations in speed.

The different speeds of the nine currents are unexplained. They are not obviously related to the type of feature they carry, nor to the patterns of jetstreams around them, nor to the mean windspeed for the same latitudes in Voyager imagery.

In a given domain, the same slow-current speed is shown by both dark and bright spots, cyclonic and anticyclonic spots, and medium and large spots. For example, in the NTropC, bright (anticyclonic) and dark (cyclonic) ovals of equal size in slightly different latitudes all move with the same slow current, whether they are isolated or clustered. They may show individual speeds within the current, and sometimes even collide, but there is no systematic difference between bright and dark spots. (It is slightly further south in the NEB that one encounters the steep gradient of speed that merges into the equatorial current.) In some domains, either bright spots or dark spots are more common, and one might suppose that the motion of the dominant type could entrain the other features in that latitude. This may have been the case in the STC when all features in the current changed their speeds with the formation of three great white ovals in the 1940s; but even as the white ovals have shrunk so that other STC features are remote from them, all these features still share the same STC speeds.

The uniformity of the slow currents is important because bright and dark spots are now known to have opposite circulations, and lie on opposite sides of the retrograding jetstreams, which are commonly deflected round them (section 3.3). The larger spots therefore are embedded in waves on the retrograding jetstreams.

This explains an apparent paradox: that the slow currents are represented mainly by features at latitudes where Voyager detected retrograding jetstreams, and these jetstreams are not detected from Earth, except in the case of the SEBs. The slow-current features observed from Earth are mostly large bright ovals or dark oblongs, of sizes comparable to the distances between jetstreams. The prime example is the Great Red Spot, whose circulation makes it seem to roll between the SEBs and STBn jetstreams; its centre (22°S) lies at almost the same latitude as the SEBs jetstream (20°S), which is deflected around the Spot. Similarly, Voyager revealed that the STB white ovals deflect the STBs jetstream. Flow patterns of this sort explain how the slow STC can govern motions not only of dark patches within the STB, but also of the STB white ovals on the south side of the STBs jetstream, and (in 1986) of a little red spot north of the STBn.

In other words, the alternating flow pattern revealed by Voyager governs very small spots. Medium-sized stable spots tend to develop at fixed latitudes on either side of the retrograding jetstream where the local flow equals the slow-current speed. Larger ovals or dark oblongs move at the same slow-current speed, but come to be centred closer to the latitude of the retrograding jetstream so that it is deflected around them.

The speeds of the slow currents do fluctuate – most notably the NTropC in the 1960s and 1970s, and the STC when the great ovals formed. Smaller changes in the overall speeds of the currents can occur within a few months. Individual large spots can change their drift rates within a week or two.

Occasionally, one or two features in a domain move with speeds more typical of the next domain towards the pole. In the STropZ, the 1986 Little Red Spot was an example of this, as are the large dark structures called 'South Tropical Disturbances'. There are several records of such 'invasions' of the STC into the S. Tropical domain, and of the SSTC into the S. Temperate domain, and of the N^3TC into the N.N. Temperate domain. These may be special instances of a tendency for unusual large features confined to zones to move slightly faster than other spots in the domain. This tendency has also been seen for the rare Little Red Spots in the NTropC, and possibly for spots in the NNTC and the S^3TC, but for the small spots tracked in these high latitudes, the currents may be less distinct.

In each hemisphere, most of the slow currents form a progression of increasing speed with increasing latitude. This progression becomes even more striking if one includes the infrequent faster drifting features of some zones. One then sees a regular gradient, which in the southern hemisphere consists of:

STropC – (S.Tropical Disturbances) –

STC – (STBs ovals) – SSTC

and in the northern hemisphere:

NTC – NNTC – (NNTZ spots) – N^3TC.

(However, the highest-latitude current in each hemisphere, S^3TC and N^4TC, is slower.) Do these symmetrical gradients hint at an underlying control of the slow currents which is unaffected by the intervening jetstreams? We will return to this question in Chapter 15.1.

The jetstreams

The fastest jetstream on the planet, on the NTBs, was discovered during an outbreak of spots in 1880, but as there were few currents then known, the exceptional nature of this jetstream was not evident until later. Three other rapid currents, on the SEBs, STBn, and NNTBs, were discovered during outbreaks in the 1920s, and the pattern began to emerge that these were special jetstreams on the edges of belts – prograding on the equatorward edges of three belts (like the great equatorial current) and retrograding on the poleward edge of the SEB. (The fifth jetstream to be detected from Earth, on the SSTBn, was not observed until 1988.)

This pattern suggested that the jetstreams might be permanent features of the atmosphere, and that all the belts might be bounded by prograding and retrograding jetstreams in the same way. This hypothesis also fitted well with evidence on the altitudes of the clouds (Chapter 4), leading to the theory that the belts are low-pressure cyclonic regions with thin clouds, while the zones are high-pressure anticyclonic regions with thick clouds (Chapter 15).

Our view from Earth is selective because the jetstreams are only detectable during outbreaks which may indicate an unusual state of the atmosphere. However, spacecraft can observe the jetstreams even when they are undisturbed. The Pioneer 'snapshots' showed the white NTropZ and STropZ to be marked by faint streaks which suggested that the major jetstreams were still present.

It was the Voyager spacecraft that revealed the full pattern of the jetstreams. The winds were identified by tracking small cloud features, and can be seen directly in the Voyager movies. The most precise analysis was published by Sanjay Limaye in 1986. He used a computer to align pairs of successive Voyager strip-maps and, taking each latitude in turn, to find out what displacement gave the best correlation between the maps. This was done for many pairs of maps, and the resulting plot of displacements against latitude (Fig.3.3) revealed a smoother pattern than had been detected before. Jetstreams were detected on all the major belt edges, and the pattern continued far into the polar regions. There was virtually no change between the two Voyager encounters.

For the five jetstreams that have been observed from Earth, the latitudes and speeds determined by Voyager were essentially the same as those determined from Earth. Two of them (SEBs and NNTBs) were undergoing outbreaks of spots during the Voyager encounters, but the others were quiet, so the Voyager data implied that their positions and speeds do not change during outbreaks.

In the Pioneer and Voyager photographs of the quiet white zones and some temporarily whitened belts, a dense texture of streaks forms a herring-bone pattern slanting towards the jetstreams, rather than running along lines of latitude in the direction of motion. So these are streak-lines, not stream-lines. This implies that clouds are continually being formed and pulled apart by the latitudinal shear within these 'quiet' zones.

Retrograding jetstreams become modest at high latitudes, and exceptions to the general pattern are seen in the S.S. Temperate and N.N.N. Temperate domains (at 40°S and 45°N), where the prograding 'slow currents' observed from Earth correspond to the slowest currents observed by Voyager. In other words, no retrograding jetstreams were reported for SSTBs nor N^3TBn. However, slower and probably retrograding velocities were present in the Voyager data: the SSTC features were shown to be an alternating series of cyclonic ovals (38°S) and anticyclonic ovals (40.5°S), whose circulations must entail locally retrograding speeds at 40°S. Perhaps the absence of a continuously retrograding jetstream in these domains is related to the high density of ovals in them, and/or to the narrowness of the domains themselves (Table 3.2).

Several other intriguing asymmetries stand out in the jetstream pattern (Fig.3.3). These and other asymmetries will be discussed in Chapter 13.2.

The constancy of the jetstreams

The prograding jetstreams seem to be extremely stable, showing no variations in latitude and only slight variations in speed. The retrograding jetstreams may vary more, in the course of rearrangements into 'circulating currents' as will be described in section 3.3. However, these rearrangements have always been transient.

The latitudes of the jetstreams, except for one, have not changed significantly over the period of Earth-based observations (Fig.3.4), nor over the four months between the Voyager encounters (Limaye, *1986*). Taking the jetstreams individually (see relevant chapters for observational details):

NNTBs, NTBs, and STBn (prograding): These are the jetstreams repeatedly observed from Earth, and their spots have always appeared in the same latitudes. Between outbreaks, the average latitudes of the belt edges have adhered closely to the latitudes of the jetstreams.

SEBs (retrograding): This jetstream has apparently shifted 2° south since the great S. Tropical Disturbance of 1901–1939, according to measurements of the belt latitude (Chapters 10.2 and 14.4).

STBs (retrograding): Although it has not been directly observed from Earth, its position in 1940 can be inferred from the way in which the STBs white ovals began to form at that time (Chapter 11). The latitude of the STBs edge in 1940–1945 (31.3°S), which became the retrograding north edge of the ovals, was almost the same as that of the jetstream in 1979 (32.6°S). The difference may not be significant but it may be analogous to the shift of the SEBs jetstream.

SSTBn (prograding): A similar argument for stability could be made in relation to the south edge of the STBs ovals, but the observations are less clear. However, the existence in many apparitions since 1914 of a narrow 'STZB' or SSTBn edge at 36°S to 37°S is evidence for the permanence of this jetstream, and it was directly observed in 1988–1990.

S⁴TBn (prograding): Although the S³TB is not consistently related to the Voyager jetstreams, there has been a sharp boundary at 53.2°S (sometimes coinciding with a narrow S⁴TB) in many apparitions since 1950; this latitude is indistinguishable from that of the Voyager jetstream.

NTBn, NEBn (retrograding), and SSTBs (velocity minimum): Although these jetstreams are never observed from Earth, Voyager pictures showed that they are intimately related to characteristic slow-moving dark or bright ovals, whose mean latitudes and speeds have not changed during recorded history.

The structure of the domains makes it unlikely that the jetstreams could shift without altering the behaviour of the slow currents, so the constancy of the nine slow currents is further evidence for the fixity of the jetstreams. (Indeed, the great S. Tropical Disturbance was a longitudinal sector in which the SEBs edge was temporarily shifted, and both the slow current and the SEBs jetstream had altered speeds within it (Chapter 10.7).)

Regarding the jetstream speeds, we must try to distinguish variation in the true speed of the jetstream from variation in its influence on a given spot. Certainly spots can be more or less entrained by the jetstream. The Voyager movies showed one or two anticyclonic rings nearly stationary in the STropZ, apparently identical to those which were retrograding at the full speed of the adjacent SEBs jetstream, and they also showed that spots on the SEBs and NNTBs jetstreams frequently overtook one another and merged. But in years when all the jetstream spots have travelled faster than the peak speed recorded by Voyager, the jetstream speed itself must have changed.

The prograding jetstreams on NNTBs and NTBs have not varied much. There can be variations in speed within a single outbreak of spots, but the speed of the spots *en masse* remains remarkably constant (Tables 6.4 and 7.1). A dramatic increase in the observed speed of NTBs outbreaks from 1970 onwards was probably a change in behaviour rather than speed; the slower type of outbreak persisted concurrently with the new faster type, the latter represent-

ing the full speed of the jetstream (Chapter 7.4). The prograding STBn jetstream has changed its speed, being 20% slower in the Voyager data than in 1932; this may have been related to the great S. Tropical Disturbance, which recirculated the SEBs and STBn jetstreams.

There have been more frequent and substantial changes in the speed of the SEBs jetstream, which has sometimes retrograded 30% faster than when Voyager measured it. These changes do not seem to be related to SEB outbreaks but may well be a rapid (≈6–month) response to S. Tropical Disturbances (Chapters 10.7 and 14.4).

3.3 SPOTS AND STREAKS

Visual observers see a variety of 'spots' on Jupiter's belts and zones, some of them impressive and long-lived, others barely resolvable. The spots in the equatorial region, which are often the most conspicuous, are a species apart and will be dealt with in Chapter 9. The spots in all other latitudes are the subject of this section. One can group them into several categories visually, and spacecraft images have simplified the picture still further.

White oval spots are the most familiar; many are stable features while others are rapidly expanding clouds. Dark streaks or oblongs are also common. The particularly well-defined ones that sometimes occur on the NEBn edge are often described as 'barges'. When this name was first suggested, the BAA President, Major P.H. Hepburn, commented: 'I feel rather shy of the term 'barges'; it is introducing canal language into Jupiter, and I do not know that I can congratulate Commander Ainslie or Dr. Steavenson on the invention of the name.'[2] But the name has endured as it is such an apt description. A more questionable example of canal language, more used by less experienced observers, is the word 'bridges' to describe dark wisps that appear to cross a zone. There really are such features in the Equatorial Zone, but in other zones they are rare; probably the only true 'bridges' are the borders of circulations such as the great ovals. An optical illusion of a 'bridge' can be created by an unresolved spot in the zone, or by the close alignment of two spots on opposite sides of the zone, or by contrast effects around a bright area. Of course, it would be quite wrong for observers to let such suspicions deter them from drawing what they see, and a feature recorded as a 'bridge' may well be consistently tracked and only later recognised as a tiny spot.

The Pioneer photos, although they could not track circulations, showed that small spots had the same oval shapes as the familiar large ovals, and even small 'bridges' or 'bulges' recorded from Earth mostly turned out to be tiny dark rings near the edges of the belts. The Voyager images proved that all these spots were well-defined atmospheric circulations, all rolling like ball-bearings between the jetstreams.

---

[2] BAA meeting report, *1922*. M.A. Ainslie had introduced the term 'barges' in 1917/18 (BAA report). B.M. Peek, who could have been at the 1922 meeting, gave a different account in his book (p.59), attributing to Ainslie the remark that 'on no account must we allow canal language to become associated with Jupiter!' Canal language was, of course, notorious in respect of Mars.

Table 3.4. *The characteristic stable spots: mean latitudes*

Description	Current	Sense	Latitude	
SSTB w. ovals	SSTC	A	40.6°S	(±0.4)
STBs w. ovals	STC	A	32.9°S	(±0.6)
Great Red Spot	STropC	A	22.3°S	(±0.3)
SEB(S) d. ss.	STropC	C	16.7°S	(±0.5)
NEBn d. barges	NTropC	C	15.9°N	(±0.6)
NTropZ w. ss.	NTropC	A	19.2°N	(±0.5)
NTropZ red ss.	NTropC	A	19.1°N	(±0.3)
NTBn d. streaks	NTC	C	30.9°N	(±0.9)
NNTBn d.streaks	NNTC	C	38.5°N	(±0.2)
NNTBn w.ovals	NNTC	A	41.2°N	(±0.6)
N³TBn w.ovals	N³TC	A	46.4°N	(±1.0)

For each class of spot, the table lists: description (w., white; d., dark; ss., spots); current/domain; sense (anticyclonic or cyclonic); and mean zenographic latitude (± standard deviation). Data from Appendix 2 (q.v. for details).

Details of how the circulations may work can be left to Chapter 14; to appreciate the observations, all we need to know is that spots are either cyclonic or anticyclonic. As in the atmosphere of Earth or any other planet, a cyclonic circulation is one that blows clockwise in the southern hemisphere or anticlockwise in the northern, and the area within it must be at low pressure. An anticyclonic circulation has the opposite sense and surrounds a high-pressure area.

So the spots can now be placed in just four categories. The most conspicuous, which are commonly seen on the poleward edges of belts, are bright anticyclonic spots and dark cyclonic spots. Bright cyclonic spots are seen within belts. And dark anticyclonic spots, always tiny, are characteristic of outbreaks on the jetstreams. The large spots, like the jetstreams, lie in fixed latitudes regardless of the ebb and flow of belts around them (Table 3.4).

In the Voyager analyses, anticyclonic spots were far more common than cyclonic ones. However, this may not be a general rule. The Voyager total included a large number of jetstream spots, which are not always present, and excluded the more complicated cyclonic regions, which would be regarded as spots by Earth-based observers.

White and red anticyclonic spots

These occur on or near the poleward edges of the belts (Table 3.4; Fig. 3.5, A1 to A6; Fig. 3.6). First among them is the planet's best-known feature, the Great Red Spot. It has existed for centuries. It is a vast oval nestling in the SEBs edge, that rolls between the SEBs and STBn jetstreams. In its circulation and its high altitude, it appears to be a 'super-zone' – an enhanced portion of the South Tropical Zone within which it sits.

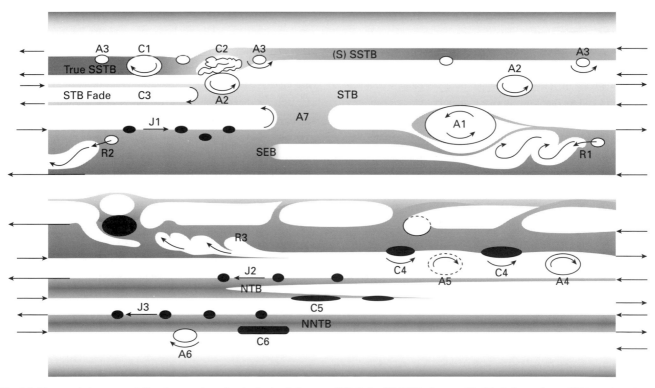

Fig. 3.5. Characteristic spots and disturbances. A, anticyclonic circulations: A1, Great Red Spot; A2, STBs white oval; A3, SSTB white oval; A4, Little Red Spot; A5, NTropZ white oval; A6, NNTBn white oval; A7, S. Tropical Disturbance. C, cyclonic circulations: C1, SSTB oval; C2, SSTB FFR; C3, STB Fade; C4, NEBn barges; C5, C6, dark streaks on NTBn and NNTBn. J, jetstream spot outbreaks: J1, on SEBs; J2, on NTBs; J3, on NNTBs. R, 'rift' outbreaks in cyclonic belts: R1, following GRS; R2, mid-SEB outbreak; R3, NEB rifted region.

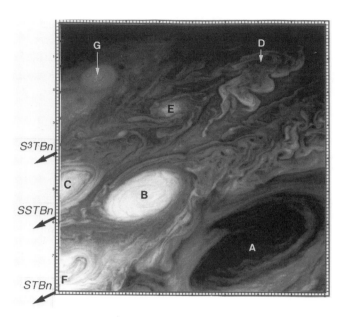

S³TBn

SSTBn

STBn

Fig. 3.6. Voyager image showing anticyclonic ovals (anticlockwise circulation) in four consecutive domains, from SEB to S³TB. A, GRS; B, STBs white oval BC; E, G, smaller ovals in latitudes of SSTB and S³TB. These alternate with cyclonic (clockwise) features: C, oval in SSTB; D, FFR in SSTB; F, f. end of STB Fade. At left, approximate latitudes of prograding jetstreams are indicated. South is to upper left. (Voyager 1 image 16303.32, violet, unfiltered. NASA.)

Three great white ovals on the STBs edge, which have existed for over 50 years, are very similar to the Great Red Spot in shape and in position relative to a belt, and Voyager showed that they had identical circulation patterns as well. They are the homologues of the Great Red Spot in the next domain to the south (33°S). In the next domain again, another set of homologous ovals was revealed by Pioneer and Voyager, forming a strikingly regular chain of tiny white ovals in the SSTBs latitude (40.5°S). Earth-based photographs show that these tiny white ovals often appear at this latitude and can last for several years, although they rarely form the long chains that the spacecraft saw.

In the northern hemisphere, anticyclonic white ovals commonly occur on the NEBn (19°N), NNTBn (41°N), and N³TBn (46°N) (though not on the NTBn edge for some reason). They are homologous to those in the south but are smaller and shorter-lived; even those on the NEBn rarely last for more than a year or two.

All these ovals commonly have dark borders, which are more prominent for the smaller ones. The spacecraft pictures reveal still smaller anticyclonic rings in which the dark collar is the dominant element, so they are essentially grey or brown rings. Pioneer found them on the NEBn edge, and Voyager found them in high latitudes; the jetstream spots (see below) appear identical except for their motion.

Of course the Great Red Spot is, sometimes, red rather than white. Similar red ovals occasionally appear in the homologous position in the NEBn/NTropZ (19°N), although these Little Red Spots are much smaller and have seldom lasted for more than a year. It is not clear why they are different from white ovals in the same latitude. There are also similar little ovals which are brown. Modern photography has also revealed two Little Red Spots in the STropZ (23°S; 1986/87 and 1990/91), and one in the NTZ (33°N; 1992). Each of these was created out of interactions involving pre-existing spots, and lasted less than a year.

As well as the distinct white ovals described, visual observers commonly see more diffuse white 'ovals' that span most or all of a zone (STZ, STropZ, NTropZ, NTZ). The Voyager images showed some of these as anticyclonic ovals with diffuse borders, but some of the visually conspicuous bright areas were not clearly shown; they may be regions of thicker cloud without distinct edges or circulations.

Dark cyclonic spots

These too are seen near the poleward edges of belts, and appear as dark streaks or oblongs (Table 3.4; Fig. 3.5, C4–C6). The most striking examples are the very dark brown 'barges' that are sometimes seen on the NEBn (16°N). Similar features are commonly seen on the NTBn (31°N); they are smaller than the NEBn oblongs and sometimes have more tapered ends, but they too can be extremely dark. Similar though still smaller features are occasionally recorded in the latitudes of SEBs (17°S), SSTBs (38°S), and NNTBn (38.5°N). Sometimes they persist in a zone when the adjacent belt has faded.

Four of the NEBn oblongs were viewed by Voyager and were found to have closed cyclonic circulations, that is, they circulated in the same sense as the belt within which they were embedded. Indeed, one cannot make a sharp distinction between the dark streaks on NEBn, NTBn and NNTBn and the longer sections of dark belt component that sometimes occur in the same latitudes.

These features appear to be sectors of 'super-belt', having a circulation and colour that are exaggerations of those of the surrounding belt, just as the white ovals appear to be sectors of 'super-zone'. On belt edges which are marked both by dark cyclonic oblongs and by white anticyclonic ovals (NEBn, NNTBn, SSTBs), the former do tend to lie further into the belt than the latter, as the former belong dynamically to the belt and the latter to the zone.

Light cyclonic spots

These features, which occur within the belts, are not fully documented by Earth-based observations because some of them are of low contrast, but spacecraft showed that they include a wide range of forms (e.g. C1, C2, R1, R2, R3 in Fig. 3.5). The following descriptions are largely from Voyager data.

At one extreme are well-formed stable white ovals, notably in the SSTB (Fig. 3.6). Other vaguely oval regions in the SSTB, which are more obscure and indistinct as photographed from Earth, are 'folded filament regions' (FFRs), which were shown by Voyager to be a more common form of cyclonic circulation (Fig. 3.7). They appeared highly turbulent in the Voyager images, being filled with bright and dark filaments which only last for a day or so,

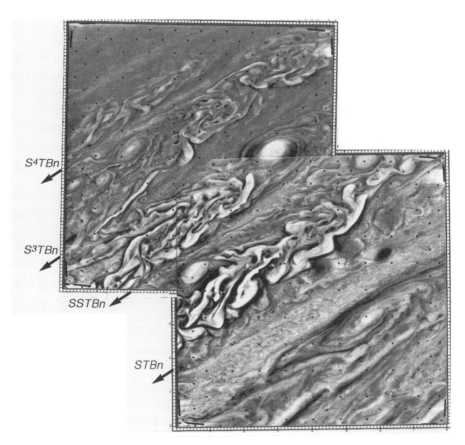

S4TBn

S3TBn

SSTBn

STBn

Fig. 3.7. Voyager images showing cyclonic FFRs (folded filamentary regions; clockwise circulation) in four consecutive domains, from STB to S⁴TB. The FFRs fill almost the whole width of the image. The northernmost, in STB, was an isolated FFR midway between ovals DE and FA. At left, approximate latitudes of prograding jetstreams are indicated. South is to upper left. The oval at bottom right is a SEBs jetstream spot, with anticyclonic circulation. (Voyager 1 images 16310.44,46, violet, filtered. NASA.)

and tiny bright spots erupt within them. They are increasingly common towards higher latitudes and are very extensive in the S³TB, N³TB and N⁴TB.

Cyclonic white ovals, dark oblongs, and FFRs may be interconvertible. They all have a narrow bright outer border, within which is a wavy or braided inner border. In the NNTB there seemed to be intermediate stages between FFRs and dark oblongs, and in the SSTB, a cyclonic white oval was observed to turn into a FFR during the Voyager encounters. Cyclonic white ovals and FFRs together formed a regular chain alternating with the SSTBs anticyclonic ovals, and Pioneer 11 saw dark oblongs in the same latitude.

Less confined regions of cyclonic disturbance occur in the STB following the great white ovals, and in the SEB f. the Great Red Spot. The turbulence f. the Great Red Spot is on a large enough scale to be conspicuous to visual observers, who record brilliant white spots appearing and prograding towards the Red Spot, often changing from day to day. There are even longer turbulent regions in some belts, which are not associated with anticyclonic ovals. Examples have lasted for a year or so in the STB (e.g. 1974, 1979) and up to 4 years in the NEB (1977–1981, gradually expanding). Even in the SSTB, some disturbance from the FFRs was spilling along adjacent jetstreams in the Voyager movies, and this effect was more pronounced in the STB and SEB. In fact, almost all the tiny spots that were seen to be newly created during the Voyager 2 movie were thrown out from cyclonic FFRs (M. MacLow and A. Ingersoll, 1986); they included both cyclonic and anticyclonic spots at high latitudes, and also the main jetstream spots on SEBs and NNTBs.

The extreme of unconfined cyclonic activity is reached in the conspicuous outbursts of white spots or 'rifts' that occur in the NEB and SEB. The vigorous 'boiling' and cyclonic eddying of these turbulent regions was dramatically shown by the Voyager movies, but they are also commonly observed from Earth, as clusters of white spots which burst out and are then pulled into oblique streaks by the shear across the belt. The most dramatic of all cyclonic disturbances is the classical revival of the SEB, which begins with an outbreak of very bright and dark spots in the belt latitudes, from which numerous spots are emitted on the jetstreams in both directions (Chapter 10.3).

Dark anticyclonic spots (jetstream spots)

Outbreaks on the jetstreams, as observed from Earth, consist mainly of tiny dark spots (Fig. 3.5, J1–J3). Voyager viewed outbreaks on the SEBs and NNTBs jetstreams and found that the dark spots on both were similar and characteristic, all being anticyclonic rings or eddies, moving at the peak velocity of the jetstream (Figs. 3.7, 3.8). The jetstream spots are normally dark grey, although some on NTBs have been distinctly blue. In Voyager pictures,

some appeared reddish-brown and some had white cores. The
'bays' between them, which sometimes appear bright to visual
observers, did not show distinct structure but some may have had a
whiter cloud cover.

Visual observations show that the dark jetstream spots are sometimes created from distinct disturbances (notably in the SEB), but
sometimes they form over a wide range of longitudes. The Voyager
pictures showed some of the spots forming, both on SEBs and
NNTBs; indeed, some were ejected from the areas of cyclonic turbulence within the belts, while others materialised along long
stretches of belt edge.

The dark jetstream spots looked much the same as other tiny anticyclonic rings in the spacecraft images, and like them they are warm
in the infrared (Chapter 4). However, several lines of evidence indicate that they extend to a high altitude, unlike other dark features
(Chapter 14). So these appear to be eddies that become very strong
and stable, perhaps associated with an updraft of warm air.

3.4 THREE TYPES OF DISTURBANCE

Much of the activity in Jupiter's atmosphere consists not of isolated spots, but of largescale patterns of disturbance, which affect
whole domains in a stereotyped manner. They fall into three distinct categories: circulating currents, jetstream spot outbreaks, and
coloration episodes.

Circulating currents

'The Circulating Current' was the name given to a phenomenon
that astonished observers in the 1920s and 1930s, in connection
with the dark obstruction of the STropZ that was the great South
Tropical Disturbance (STropD; Fig. 3.5, A7). Spots retrograding
on the SEBs jetstream were deflected southwards when they
arrived at the concave p. end of the Disturbance, and ended up
going back the way they had come on the STBn jetstream! Now we
know that the SEBs and STBn jetstreams are permanent, and that
the STropD was a temporary reconnection of them. Thus the region
of STropZ excluded by the STropD, which shrank as the STropD
grew longer during its 38–year life, constituted an anticyclonic circulation in the same sense as the Great Red Spot. Other STropDs
have been observed since, the greatest of which developed during
the Voyager encounters and lasted for two years. The development
of its circulating current was photographed in great detail. The circulation may typically be incomplete: the opposite end of the circulation, from STBn to SEBs, was never definitely observed with
the great STropD and was not occurring with the Voyager STropD;
and the Voyager movie showed retrograding SEBs flow continuing
past the STropD, even though the SEBs spots were being skimmed
off at the STropD.

Similar phenomena probably occur in other zones, although
largescale recirculation has not been directly observed. A 'South
Temperate Disturbance' from 1918 to 1920 had the same appearance. When three similar features appeared across the South
Temperate Zone around 1940, they progressively expanded until
the intervening sectors of STZ became the three STBs white ovals.
Clearly these 'South Temperate Disturbances' had divided the

zone into three anticyclonic circulating currents. Both South
Tropical and South Temperate Disturbances always drift faster
than the normal slow current for that zone.

Similar dark disturbances are sometimes seen spanning the
North Temperate Zone; they too have been suspected of blocking
the passage of jetstream spots (on the NNTBs), and of growing
longer with time, but they have not been shown to involve a circulating current. Voyager also recorded examples of spots recirculating in the NTZ and NNTZ, although these may have been one-off
events.

Cyclonic circulating currents, in belts, are equally distinct phenomena. The fading away of sectors of the South Temperate Belt
involves whitening of a belt rather than darkening of a zone, and
Voyager revealed that one such STB Fade consisted of a cyclonic
circulating current (Fig. 3.5, C3). It reconnected the STBn and
STBs jetstreams alongside the STBs white ovals at each end of the
faded sector. Just like the smaller cyclonic ovals in the SSTB,
described above, the STB Fade was enclosed by a beautifully
braided ribbon surrounded by a bright border. Such Fades are regular occurrences on the STB, and Voyager revealed a similar circulation in one sector of SSTB.

Long, closed cyclonic circulations in higher-latitude belts
appeared similar except that the interiors were filled with folded
filaments or dark material as described above. It appears that recirculation between jetstreams is common in high latitudes, but constitutes a special event in the lower-latitude domains that can be
observed visually.

It seems to be characteristic of anticyclonic circulating currents
to contract with time (in the STropZ, and the STBs white ovals).
Conversely, cyclonic circulating currents expand with time (STB
Fades). (Cyclonic FFRs in the NEB also expand with time, viz.
'rifted regions'.)

Interactions between circulating currents in adjacent latitudes,
which give rise to 'dislocation' phenomena, will be discussed in
Chapter 13.

Jetstream spot outbreaks

In this category fall the outbreaks of small dark spots travelling on
jetstreams (which are the only way the jetstreams can be detected
from Earth), and also the more extensive eruptions within the SEB.
The latter span the entire width of the belt as well as producing jetstream spots.

The outbreaks on a given jetstream occur at infrequent intervals,
sometimes of decades, and typically last for between three months
and three years. There have been a few longer episodes of activity
on SEBs and (on a small scale) on NNTBs, but these may have
depended on renewal at 3– to 5–year intervals.

Although the small size of the spots makes them difficult to
observe, the outbreaks that are detectable visually do seem to be
different from the much smaller-scale activity shown in Voyager
images and, in recent years, in good Earth-based photographs.
Voyager's discovery of the distinctive nature of jetstream spots
(see above) confirmed that these outbreaks are distinct events. No
other distinct largescale features move at the peak jetstream
speeds, except for extremely bright, high-altitude clouds which

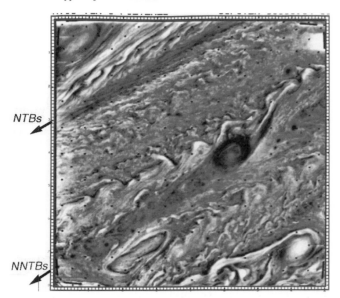

NTBs

NNTBs

Fig. 3.8. Voyager image showing the NTBs and NNTBs prograding jet-streams. Streaks in the upper (southern) half indicate the NTBs jetstream although this part was mostly white at the time. The dark oval is a typical NNTBs jetstream spot. Also shows a well-formed cyclonic circulation in NNTB domain at lower left. (Voyager 1 image 16318.34, violet, filtered. NASA.)

appear during the most energetic outbreaks on NTBs.

Outbreaks on the prograding jetstreams (STBn, NTBs, NNTBs) are usually restricted to the jetstream itself, except that they are sometimes followed by a diffuse yellow or reddish coloration of the adjacent zone (see below). Outbreaks on these jetstreams do not seem to be regularly associated with revival of the relevant belts. Even though the STB, NTB, and NNTB do sometimes fade away, they usually revive peacefully. Occasions when NTBs jet-stream outbreaks have led to a revival of the belt are not typical.

The most spectacular of the outbreaks are those which occur during revivals of the SEB, when a localised source pours out dark spots onto the SEBs jetstream as well as bright, dark and coloured spots (some of which may be quite large) onto the equatorial jetstream and into the SEB latitudes. But similar activity can also occur without prior fading of the belt. These are divergent cyclonic eruptions. Outbreaks of white spots or 'rifts' also occur in the NEB and appear to be similar phenomena, although they are less distinct in time and place, and there is never any disturbance visible on the NEBn jet-stream. These outbreaks in both SEB and NEB often lead to the revival of the belt after a period when it has been faint or absent, and a striking reddish colour often develops across the newly restored belt – another parallel with other jetstream outbreaks.

There is some evidence that jetstream outbreaks in different lati-tudes, as well as coloration episodes, tend to be correlated within a larger pattern of disturbance. This theory of 'global upheavals' will be discussed in Chapter 13.

Coloration episodes
Although a general discussion of colours on the planet belongs to the next chapter, there is a type of coloration episode that seems to form a distinct category of disturbance, with some relation to jet-stream outbreaks. These episodes comprise strong yellowish or reddish coloration of a zone. They develop over a few weeks or months, and last for months or years.

These episodes are most characteristic of the Equatorial Zone. The colour is most commonly described as ochre, but a real range has been observed from a rather pure yellow through a coppery shade to a deep ruddy colour. Similar episodes occur more rarely in other zones.

The coloration is always seen all round the planet; although the intensity may vary broadly with longitude, no local source for the colour has ever been identified, and its longitudinal spread has never been tracked. Infrared observations indicate that the colour is at high altitudes (Chapter 4), where the winds could be very differ-ent from those governing the visible spots. Indeed in the Voyager images at least one white haze front, pushing against the orange material, could be tracked as it blew rapidly across the underlying dark patches (Chapter 9.2).

EZ coloration is often associated with episodes of SEB fading prior to revival (Chapter 13), although it also occurs separately. In other zones, coloration episodes are almost always associated with adjacent jetstream outbreaks or belt revivals. In the STropZ, col-oration has always been accompanied by SEB reddening (six appari-tions, usually following a SEB Revival) and/or STBn jetstream activity (three apparitions) (Chapter 10.6). In the NTropZ (Chapter 7.4 and 8.2), a light yellow or orange tint usually develops within a few months before or after a NTBs jetstream outbreak. Other NTropZ colourings, more like a dull brownish-yellow, have appar-ently spilled over from a reddish, newly-revived NEB. In the temper-ate zones, strong coloration has only been observed twice in the NTZ (continuous with the NTropZ coloration in 1912 and 1980/81) and once in the STZ (continuous with the STropZ coloration in 1928/29).

In 1928/29, during a global upheaval which included SEB and STBn jetstream activity, strong reddish or tawny colour spread over at least some longitudes in three separate regions of the planet: the NTropZ and northern NEB; the southern EZ, later extending to the reviving SEB; and the southern STropZ, later extending over the whole of the STropZ and the STZ. The STB was described as a grey belt with an 'overlying red veil'. The BAA inferred that the colour changes 'were caused by the extension and drift of large masses of reddish vapour of some kind. This vapour, however, appears to have been largely transparent, permitting the surface details to be seen through it though modifying their apparent colour.'

It is a little-known fact that these coloration episodes show little respect for the boundaries of the belts and zones (Fig. 3.9). Although they are usually described as affecting a zone, they sometimes affect only half the width of a zone, and they have occa-sionally spread over two adjacent zones. And they sometimes affect an adjacent belt as well; when the NEB or SEB becomes red-dish after a revival, the colour sometimes spills across the adjacent NTropZ or STropZ as mentioned above.

Many EZ episodes and almost all other zonal colorations have shown this disregard for zone boundaries. One should remember that there is a strong tendency for visual observers to refer the colour simply to a zone, first because a light orange colour is more difficult to detect if it overlies a dark belt, and second because the human visual system has rather poor resolution for colour alone and unconsciously tends to identify colour boundaries with adjacent brightness boundaries. So the variation in latitude may be underestimated in Fig. 3.9. The NTropZ entries for 1970 and 1971 depend on professional photographs, but the example for 1985 is from visual observations.

A coloration episode wanes gradually, but as it does so the colour often changes. In the EZ, and in the NTropZ in 1985, it can change from a reddish-brown to a more yellowish colour, then gradually brighten to white. On the other hand, EZ coloration can change from rather pure orange or yellow into a browner colour, developing into a massive brown Equatorial Band; then this EB may gradually become grey. This development poses problems with regard to the altitude of the clouds, as described in the next chapter, since orange colour is higher than the white zones, whereas brown and grey belts are if anything lower. It suggests that the coloration episode is only a part of some more vertically extensive disturbance of the EZ.

Finally, we should mention that just as zones sometimes darken, so belts sometimes whiten or 'fade'. We can tentatively distinguish two types of belt fadings. One, seen in sectors of the STB and probably in belts further south, has a sharply limited extent and is part of a circulating current phenomenon as described above; it is often accompanied by compensatory darkening of an adjacent belt or zone (Chapter 13.1). The other, which is most familiar in the SEB and NTB, involves whitening of a broad region with variable boundaries. This may have some analogy with the coloration episodes, except that the cover is white. In the case of the SEB, the fading may affect both components or only the SEB(S), and any South Tropical Disturbance alongside fades as well. (The SEB fading is usually accompanied by coloration of the EZ and reddening of the GRS.) SEB fadings and the subsequent revivals have sometimes recurred at 3–year intervals, sometimes less frequently. The NEB has also sometimes undergone fading and revival like the SEB, also with a temporary 3–year periodicity. In the case of the NTB, again the two very different components often fade at about the same time; occasionally sectors of the NNTB have also faded and revived along the NTB, although this could be coincidence. NTB fadings occur rather regularly, every 10 years. The causes of these events are still entirely unknown.

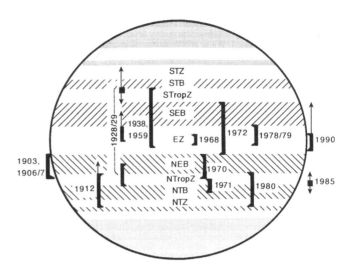

Fig. 3.9. Extent of some major colorations. Arrows show where some colorations spread to later.

4: Vertical structure: colours and clouds

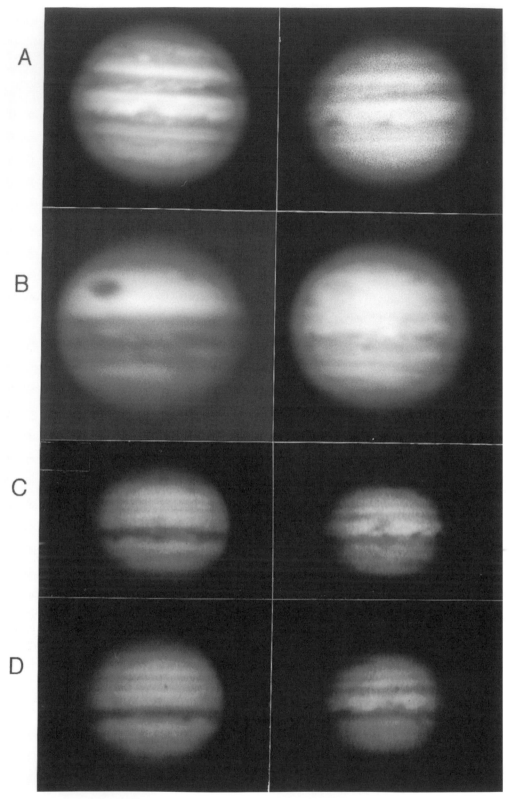

Fig. 4.1. Images of Jupiter in various colours. (South is up.)

(A) White and red; 1985 Aug. 2 (Donald C. Parker, Florida, 32-cm reflector); (left) ω_2 108, white light; (right) 50 mins later, W29 red filter (600–900 nm). The main bluish features (dark in red) are the NEBs dark projections and N. Tropical Band. NTropZ was ochre. Also shown are oval DE and a mid-SEB 'rift' outbreak.

(B) Blue/violet and red; 1990 May 21, ω_2 58 (Isao Miyazaki, Okinawa, 40–cm reflector); (left) B-390 blue/violet filter; (right) R-60 red filter. There is strong 'reddish' or yellow colour in the GRS and EZ (with SEB faint) and in NTropZ (after a NTBs jetstream outbreak). But some 'cold grey'

spots are dark in red, in STB, SEB (Np. GRS), and NEBs. (For other blue/red pairs from the 1989/90 global upheaval, see Figs. 7.8, 9.18, 10.24, and colour Plate P24.)

(C,D) Blue and near-infrared; 1989 Aug. 14, ω_2 299–303 (C), and 1989 Aug. 16, ω_2 236 (D) (Mark Kidger, La Palma, 1-metre Jacobus Kapteyn Telescope, CCD detector); (left) Johnson-B blue filter; (right) Gunn-Z 1-μm filter. At 1 μm, the EZ is always bright (this was taken before the 1989/90 coloration developed). The NEBs projections and a 'NTZ Belt' (at lower right in C) are especially dark at 1 μm.

4: Vertical structure: colours and clouds

4.1 THE COLOURS OF THE CLOUDS

Jupiter is famous for its colours and their changes. The different colours – white, brown, grey, bluish, yellow, or reddish – represent clouds with different compositions and different vertical structures. In this chapter, we look at the vertical profiles of the atmosphere in terms of pressure, temperature, and cloud structure. The conclusions will be summarised in section 4.7. The chemistry of the clouds will be dealt with in Chapter 16, although we should note one definite fact now: that the white clouds of the zones are made of ammonia ice. Much of this chapter will be a tour of the infrared spectrum of Jupiter, which reveals different atmospheric levels at different wavelengths. First, though, we consider the colours in the clouds that the human eye can see and the camera can photograph.

Visual observations of colour

First, we need to take a cautious look at the problems inherent in the visual records of colour. Using the human eye, observers can only judge colours in a relative and subjective manner, even if they are wise to possible instrumental and atmospheric illusions (Chapter 1.3).

Some visual observers record the basic colours of the disc as grey and white (with some belts brownish); others normally see them as brown and yellowish or cream-coloured (with some belts reddish). The planet certainly does have an overall yellowish-brown cast, and spectra show that this applies to the zones as well as the belts, although it is stronger in the belts. However, observers who can subconsciously discount this tint, and record variations from grey, are better placed to record real colour contrasts than those who cannot. These personal differences could be due to the physiological phenomenon of 'colour-constancy', whereby everyday objects are perceived in their true colours regardless of the tone of the illumination, as long as the full spectrum is present. The physiological basis of this is uncertain but it apparently involves unconscious averaging of colours from different parts of the visual field to establish the current standard of whiteness. One might therefore expect to see maximum colour contrasts on Jupiter if the observatory were dark or subject only to red (or sodium) light, whereas one should see Jupiter in its true but murky shades of brown in the presence of white light or twilight. The author does not know of any experiments on this matter, and it would be interesting to try some.

The colours on Jupiter are often subtle and rarely pure, so 'bluish' or 'yellowish' are much more common than blue or yellow. In assessing visual colour reports, it is essential to compare each description with the full range of colours seen by the same observer. Apart from the variations in the basic perceived tint, there are marked variations in the perceived range of colours, some observers reporting only slight shades of grey while others report the most exotic shades of blue and red. Observers who recorded every single belt as reddish-brown will be ignored in summarising colours for this book! And reported 'creamy' tints in zones will be equated with white, reserving 'yellowish' or 'reddish' to describe the coloration episodes that were agreed to be unmistakable.

Colours generally appear clearer in twilight, and this cannot be explained by colour-constancy.

Large telescopes give a strong advantage in seeing colours, because they give both high light levels and high magnification. It is remarkable how a narrow 'grey' belt can reveal strong colour when viewed in a large telescope. Magnification is important because our visual system has lower resolution for colour than for brightness. Although both are detected by the cones in the retina, it takes three cones to detect colour, and the blue-sensitive cones are comparatively sparse. In fact, our visual system tends to perceive colour boundaries as coinciding with nearby brightness boundaries, even if they do not really match.

Numerical colour estimates have not really been successful. A.S. Williams made a 58-year series of estimates of the 'redness' of the equatorial belts and zone, on a 0–10 scale, from 1878 to 1936; but his figures did not always agree with the collective descriptions in the BAA reports, nor with the numerical summaries of those descriptions that Phillips published in 1915. This may be partly due to different sections of a belt having different colours, which is common, but it illustrates the caution that must be taken with visual reports. We had a salutary lesson in 1990: when the SEB turned orange, visual observers (including the author) thought the

NEB changed from brown to grey, but photographs showed the same brown throughout. The apparent change in the NEB was entirely a contrast effect. B.M. Peek, who himself made many notes of colour, ended up very sceptical of the changes documented by such subjective means, and it would be very difficult to reduce different observers' estimates to a really uniform scale.

In principle, it should be possible to record colours be making numerical intensity estimates (as described in Chapter 1.3) through different colour filters. Alan Heath of the BAA has done this for many years, and the author has tried it occasionally, using a Kodak Wratten 25 red and Wratten 44A or 47 blue filter. The blue filter always gives higher contrast, because of the generally brownish colour of the belts, and local colour contrasts can be clearly seen; a yellow zone or the Great Red Spot looks darker in blue, and the bluish-grey dark patches on NEBs look darker in red (Figs. 4.1 and 4.2). But attempts to record these differences numerically, by estimating intensities on the 0–10 scale through each filter and subtracting one from the other, give results that seem to depend on darkness as well as on colour. This may be because of the non-linearity of the intensity scale, and also because the uncertainties in intensity estimates are doubled in comparing two sets; moreover, the dimmer image (especially in the blue filter) makes minor belts more difficult to see regardless of their colour. Such estimates undoubtedly encode real colour information, but more study will be needed before they can be interpreted clearly.

All this means that subtle colour changes from year to year are hard to document because they depend on observers' qualitative memory. So it would be unwise to put much weight on reports of long-term variations in the range between grey and brown, although such variations surely occur. But changes to distinctly reddish or yellowish colours are conspicuous and often confirmed by many observers. It is these changes which are considered the most important, and the historical record of them can be taken as generally reliable.

Photographic observations of colour

Photographs should give more objective records of colour, and splendid colour photographs have been produced from large telescopes (Plates P13 and P24). However, even these must be interpreted with caution, as the colour balance of film may vary.

For more precise work, monochrome photographs are taken through different colour filters. This takes advantage of the higher resolution of monochrome film and eliminates any atmospheric dispersion, and in principle the relative intensities can be measured on the negatives to produce truly objective colour information. Filter photography is difficult for amateurs because of the dimness of the image, particularly in blue. Not until 1989 did amateurs regularly produce both red and blue images with good resolution.

Colour photos can be reconstructed by superimposing a red-light image through a red filter, a green-light (or yellow- or white-light) image through a green filter, and a blue-light image through a blue filter. The pictures in Plate P21.5&6 were made in this way. The latter shows quite realistic colours, but the former, made from higher-contrast monochrome prints, shows how the colours can be

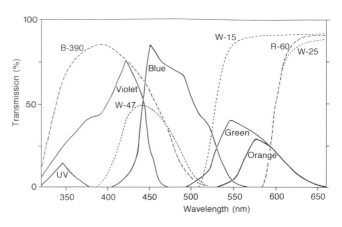

Fig. 4.2. Transmission spectra of some filters used for Jupiter. Solid curves are for filters on the Voyager narrow-angle camera. Dashed curves are for Kodak Wratten filters (W-47, W-15, W-25) and Hoya B-390 and R-60.

greatly exaggerated using this method.

Similar variability can occur with spacecraft images. The Pioneer 'photographs' were made using only blue and red filters, the middle wavelengths being interpolated; nevertheless, the published images had quite realistic colours (e.g. Plate P16.1). The Voyager pictures, constructed from multiple colour-filter images, should in principle contain the most accurate colour information of all, but this is not the case in the publicly-released versions of them. They were printed with somewhat exaggerated colour relative to true white, which resulted in most areas appearing strongly reddish, and true red/brown contrasts became hard to see. A few were processed to exaggerate colour contrasts (e.g. Plate P17.4), which are highly informative though not realistic.[1]

To measure relative colours precisely, photometry of photographs has been the best method until recently, although not often used. Relative spectra obtained by this method are shown in Fig. 4.3, and scans from Voyager imagery in Fig. 4.4. Nowadays, CCD images record intensities digitally, so these should be an excellent source of accurate colour data in the future.

Despite all the caveats, there is no doubt that real colour contrasts exist. So let us now see what colours Jupiter does have to offer.

An overview of colours

The '*white*' clouds dominate the zones and white ovals. In the Voyager pictures, they presented an almost unbroken cloud deck, though with puffy or streaky textures indicating convective activity.

In the belts and some spots, at least five other colours can be distinguished. When more varied descriptions are quoted in the following chapters, it will be as well to consider them simply as impressions of tones within these five basic categories. As we shall see in section 4.2, the different colours are associated with different

---

[1] The Voyager colour images were described by Owen and Terrile (*1981*). The true colours were illustrated by Young (*1985*). Direct photometry of belts and zones at the telescope had earlier been done by Pilcher *et al.*(*1973*) and Orton (*1975*).

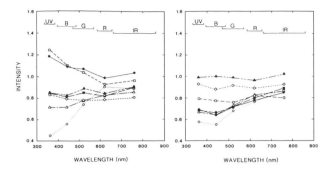

Fig. 4.3. Relative colours of belts, zones, and GRS. These visible-light spectra are from micro-densitometry of photographic plates taken with different filters, in 1973 September (left) and 1979 January (right). Intensities are given relative to NTropZ = 1.0; the slopes of the other regions in 1973 suggest that NTropZ was actually somewhat 'reddish' then. Symbols: ▲ NTB, ■ NEB(S), △ SEB(N), ● SEB(S), □ STropZ, ◇ STB, ○ Little Red Spot in NTropZ (1973) and Great Red Spot (1979). (From Beebe and Hockey, *1986*.)

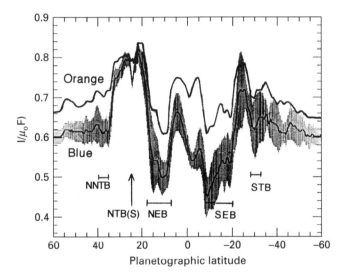

Fig. 4.4. Global average north-south profiles of reflectivity from Voyager 1 orange and blue images. (The range of the blue values is also shown.) Note that belts are 'redder' than zones except for EZ(S) which was 'reddest' of all. (Adapted from Gierasch, Conrath & Magalhaes, *1986*.)

infrared temperatures which may indicate different vertical arrangements for the clouds. Reddish and yellowish colours are above the other clouds. Brownish-grey, grey, and bluish-grey clouds are at almost the same altitude as the white clouds, but are thinner and associated with thinner underlying cloud layers.

Bluish-grey is normally seen only in the dark patches on the NEBs. It can also occur in the SEB(N) or other faint features around the GRS when the SEB has faded, and along the SEBs jetstream, and in occasional North Tropical Bands. All these features can also be plain grey.

Grey is the usual colour of the STB and NTB.

Grey-brown or brown is typical of the equatorial belts.

The commonest colours for dark features are grey or brown or anything in between; these shades apply to the polar regions and to the SEB and NEB. 'Purple' tint is sometimes recorded, but at least some such impressions are attributable to chromatic aberration. The SEB and NEB usually fluctuate between grey and brown, and

sometimes become strongly 'reddish' – that is, orange-brown. Dark spots on the NEBn edge are also often strongly reddish-brown. The various shades of grey and brown may be due to varying densities of a yellow or reddish haze that overlies a basically greyish belt.

Yellow or ochre appears in distinct episodes in the EZ and occasionally in other zones. Sometimes the yellow or orange colour appears almost pure, but more often it is better described as yellowish-brown, brownish-yellow (ochre), or orange-brown (tawny or coppery). The exact tint often changes during a coloration episode, and there may be discrepancies, in its location and its timing, between different visual observers and photographs. We will discuss this in Chapter 9.4, with the suggestion that the colour has a variable spectrum. Yellowish-brown tint is also sometimes seen along the NEBn edge and in the NNTZ. Reddish and yellowish colours seem to form a continuum, and all such colours are sometimes referred to as 'reddish'.

Reddish colour in the colloquial sense is famous in the Great Red Spot. At its strongest, the colour is brick-red; it is never a pure red. Similar colour is shown by occasional Little Red Spots, and by the NTB(S) (along the fastest jetstream). In recent years, photographs have sometimes revealed similar colour flanking the STB white ovals.

It is intriguing that reddish colours are associated with many regions notable for their strong winds or turbulence, from the largest to the smallest scales. The GRS is the largest vortex; the EZ and NTB(S) are the fastest jetstreams; colour episodes in zones are associated with jetstream outbreaks, and colour episodes in the SEB and NEB follow turbulent revivals. On a much smaller scale, Voyager pictures revealed diffuse spots of red haze over the cusps of tiny eddies in turbulent cyclonic regions in the temperate belts (Plates P17 and P20). Similarly, in the rifted SEBZ following the GRS, spots on the turbulent dark filaments were reddish-brown – and in the Space Telescope photo of 1991 March (Plate P24.10), the SEB redness after a Revival appeared to be most concentrated in small dark red spots on the cusps of wisps.

Green is almost never seen, but in 1929/30, F.J. Hargreaves, M.A. Ainslie and T.E.R. Phillips agreed that some dark markings on SEB(N) were green. This might have been due to a bluish-grey feature seen though the semi-transparent orange haze which suffused the whole region. Hargreaves also noted a NEBs projection in 1941/42 as 'unmistakably blue-green'; the NEB was then reddish-brown so this could have been a similar case. Voyager photographs showed the NEB as a complex mixture of streaks of blue-grey, brown, yellowish, and white, some of which seemed to cross each other at different altitudes (Plate P16).

Photographs through red and blue filters not only record the relative colours, but also emphasise detail at different heights in the atmosphere. This is not due to the intrinsic colours of the clouds themselves, but to a 'reddish' (violet-absorbing) haze that seems to hang over the belts (see below). Red or orange filter photographs see through this haze to record fine details in the clouds beneath. Blue or violet photographs show the belts as largely featureless and very dark, while white clouds (with less haze above them) stand out clearly.

Near-ultraviolet photographs (330–360 nm) give an exaggerated version of the blue images, with 'reddish' features appearing even darker, as the haze absorbs even more strongly in the ultraviolet (Figs. 4.5, 9.12, 9.16, 9.18). In Voyager scans at 240 nm, only the highest levels in the atmosphere are seen, and the belt structure is smoothed out and altered; the STropZ and NTropZ are still bright but the S. half of the SEB is even brighter, while the polar regions are very dark (C. Hord and colleagues, *1979*). Wavelengths shorter than 230 nm are strongly absorbed, by ammonia, methane, and acetylene.

Near-infrared photographs give an exaggerated version of the red images; the GRS is the brightest feature (Fig. 4.10). This holds for wavelengths up to 1 µm, except for the methane absorption bands (see below). At wavelengths around 1 µm, where methane absorption sets in, the EZ is especially bright due to high-altitude haze (Fig. 4.1C,D).

The violet-absorbing haze

This haze is actually present all over the planet, and accounts for the steep absorption that increases from green through violet into the ultraviolet (Fig. 4.13A).[2] It is due to an aerosol at high altitude. There appears to be more of this haze over the belts, either because it is denser or because the belt cloud-tops lie deeper. It may well be the same haze, in more concentrated form, that gives the colour to the Great Red Spot and to reddish/ochre episodes in zones and belts. There is reason to suspect that the absorption has a variable spectrum, extending to longer wavelengths when the colour is more intense (Chapter 9.4).

4.2 THE HEIGHTS OF THE CLOUDS: VIEWING AT AN ANGLE

What are the relative heights of the clouds? This is a contentious question, to which different techniques give a variety of answers. First, we consider the information imprinted in visible and near-visible light waves by their passage through the layers of the atmosphere: scattering, polarisation, and absorption. Each type of data alone can be interpreted in various ways, and only by putting them together can one home in on the single true description of the atmospheric structure – something that has not yet been fully achieved. Descending through the atmosphere, we will encounter at least three layers. First, there is the haze of micron-sized particles around the 100-millibar level[3], probably the same as the violet-absorbing haze; second, the white ammonia clouds at 300–700 millibars; and third, deeper cloud layers which can only be detected at infrared wavelengths.

Limb-darkening and limb-brightening

A conspicuous feature of Jupiter's visible disc is the darkening towards the limb. This is a sign that light is being lost above the main cloud layers, the more so when it travels through the atmos-

Fig. 4.5. Near-ultraviolet image from Voyager 2. The limb is bright due to high-altitude haze, the polar region is dark, and a NEBs plume core is very bright as it projects above much of the UV-absorbing haze. Southern EZ appears dark because it was covered by ochre UV-absorbing haze. (Image 20130.29, 18 days before encounter, unfiltered. NASA.)

phere at a shallow angle. In principle it could be due to scattering or absorption by gas or by particles. The colour and polarisation of the limb-darkened light, and measurements of methane absorption bands, can help to distinguish these possibilities. The limb-darkening on Jupiter is strongest in red and infra-red light, and weaker in blue.

Absorption by methane gas was considered as a possible cause, as there are weak methane bands in red light, but in fact these have no perceptible effect on Jupiter (unlike Uranus and Neptune, where these bands are responsible for the blue colour of the planets). Rather, the limb-darkening is now attributed to high-altitude haze, which produces some absorption at all wavelengths, and also produces some forward-scattering of light.[4]

The properties of this haze were somewhat clarified by Pioneers 10 and 11, whose photopolarimeter measured the limb-darkening at high phase angles for the first time.[4] When the planet was viewed as a crescent (150° phase angle), the atmosphere was bright, which implied at least a thin layer of forward-scattering, micron-sized haze particles around the 100–mbar level (though the vertical extent was not well-defined). This haze must cover both zones and belts, and lie above most of the gas above the cloud-tops. So it is presumably the same as the violet-absorbing haze layer.

However, in ultraviolet light, the effect is offset by broader scattering of rays at very high altitude which would otherwise be absorbed by the haze below, so one sees limb-brightening (Fig. 4.5).[5] Blue light shows an effect intermediate between red and ultraviolet, with only slight limb-darkening. Scattering can again

[2] The high-altitude violet-absorbing haze was first proposed by Axel (*1972*). Detailed analysis by Tomasko *et al.*(*1986*).

[3] Units of pressure: 1 bar = 1000 millibars (mbar) = 10^5 Pascals = 10^6 dyne/cm^2 ≈ mean atmospheric pressure on Earth at sea level.

[4] Smith (*1986*); West (*1979b*). For the Pioneer analysis, see Tomasko *et al.*(*1978*).
    Phase angle is defined as the angle between the incident sunlight and the observer.

[5] Fountain (*1972*). Voyager data were from a photopolarimeter which operated at 240 nm: Hord *et al.*(*1979*), West *et al.*(*1981*).

be by gas molecules or by haze, and the ultraviolet limb-brightening on Jupiter is attributed to molecular scattering by hydrogen (Rayleigh scattering). The amount of scattering by this process varies as the inverse fourth power of the wavelength, so ultraviolet and blue light is affected much more than red; this is what causes the sky on Earth to be blue. It also causes the scattered light to be polarised (see below). On Jupiter, Rayleigh scattering sends back ultraviolet light which would otherwise be absorbed by the violet-absorbing haze. Its strength indicates that this haze lies around 150–200 mbar, with little absorption above 100 mbar over most of the planet, but almost complete absorption above 400 mbar.[5]

But in the polar regions, which are extremely dark in ultraviolet, the absorber must extend much higher, above the 40-mbar level (see section 4.7).

The Pioneers looked for one other phase-dependent effect, the rainbow (or cloudbow) that would arise if there were spherical droplets in the clouds. Such a cloudbow can be detected on Earth and Venus, but there was none on Jupiter. This confirms that the clouds are made of solid crystals, not liquid droplets.

Polarisation of light

Polarisation of light contains a lot of information about the atmosphere through which it has passed. Polarisation is induced both by pure gas (by Rayleigh scattering) and by some kinds of haze or cloud particles. So it depends on the depth of the gas above the cloud-tops that reflected the light, and on the size, shape, and refractive index of any haze particles in that region. To sort out the contributions of these factors, one needs to know how the polarisation varies with wavelength and with phase, and to compare the measurements with other data.

[6] Lyot (1929). These and later observations were reviewed by: Dollfus (1957, 1961); Gehrels et al. (1969); Kuiper (1972a,b); Carlson & Lutz (1989).

The first observations of polarisation were by B. Lyot in the 1920s.[6] A weak polarisation is seen at the limb. It differs at the east and west limbs, implying that sunlight affects the structure between dawn and dusk. However, the most intense polarisation is over the polar regions, at all visible wavelengths.

This pattern of polarisation is nicely shown by modern imaging (Fig. 4.6 and Plate P21.2&3). The pattern of belts and spots is barely visible on these images, though sometimes there is slightly higher polarisation in the belts. The strong polar polarisation has a sharp boundary at about 58°N or S, and reaches a maximum of about 60% polarisation. But these polar caps are variable in size and intensity.

If the polarisation were due to Rayleigh scattering, it would increase steeply with phase angle. As it does not, it must be produced not merely by gas but by a thin haze of particles, ≈1 μm in size. This important point could not be finally proved by Earth-based data because the phase angle from Earth never exceeds 11.7°. So a major goal of the Pioneer spacecraft was to measure polarisation at high phase angles (Fig.4.7), up to 150°. The detailed analysis of the Pioneer data was presented by P.H. Smith and M.G. Tomasko (*1984*). In the Pioneer maps, polarisation in red light rises abruptly towards the limb and the terminator. Smith and Tomasko interpreted this as due to a partially transparent haze of polarising particles at about 120 mbar, which may well be the same as the violet-absorbing haze at the same level; it may extend down to the main cloud-tops at about 320 mbar.

If there is any polarisation which can be attributed to Rayleigh scattering, it should show up more strongly in blue light, and it should show whether there is any height difference between belts and zones. Earth-based observations had shown no difference between belts and zones; the Pioneer data showed only a small difference in polarisation between belts and zones in blue light, and none in red light. D.L. Coffeen (*1974*) (Fig. 4.8) analysed a Pioneer 10 scan across the crescent Jupiter, on the assumption that the

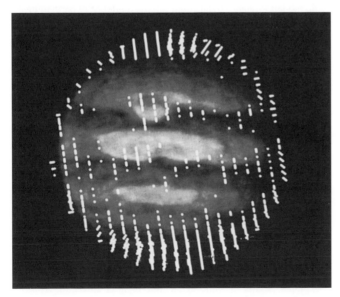

Fig. 4.6. Map of polarisation of light from Jupiter, from Observatoire de Meudon, 1989 Jan. 27, phase angle 10°. Linear polarisation is indicated by lines on an orange-light image. Note the intense polarisation in the polar regions, and more modest polarisation around the limb (radial to disc) and in the belts and GRS. (From Dollfus, *1990*; by courtesy of A. Dollfus.)

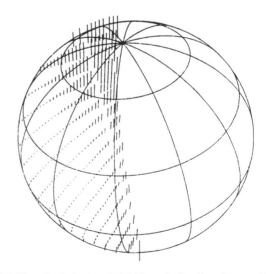

Fig. 4.7. Map of polarisation of light from Jupiter, from Pioneer 11, 1974 Dec. 2. This 'image' was made over 67 mins at a mean phase angle of 98°, in red light. Degree of polarisation is indicated by the height of the lines; orientation is not indicated. (From Smith and Tomasko, *1984*.)

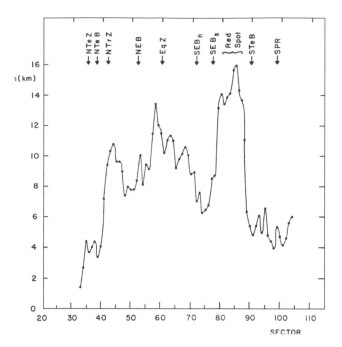

Fig. 4.8. Model of the cloud-top altitudes, based on Pioneer 10 polarisation measurements. This was derived from a single north-south scan of the crescent planet on 1973 Dec 4, including the GRS (similar to Fig. 2.10). Polarisation was interpreted as due to Rayleigh scattering. This may be incorrect and much of the variation may not be due to altitude differences, although the height of the GRS is confirmed. (From Coffeen, *1974.*)

polarisation was due to Rayleigh scattering. He inferred that the GRS was the highest feature, several kilometres above the whitened SEB and other zones; and the polar regions, being most highly polarised, were the deepest. But in fact, the difference between belts and zones may not be due to altitude at all. Smith and Tomasko (*1984*) showed that it can all be attributed to the dark brownish tint of the belts: light scattered back from the cloud-tops is depolarised, and as there is less of it in the blue range, it does not so greatly swamp the polarised light that was scattered at higher altitudes. Simple models indicate that both belts and zones have cloud-tops at about 320 mbar which are virtually opaque in visible light. If the visible cloud-tops were deeper, as suggested by some infrared studies (see below), the light should be more polarised – almost 30% in blue if we could see down to 2 bars. Instead, the visible difference between belts and zones must be due to the colour of the clouds at the 320–700 mbar level, i.e. the ammonia clouds.

Nevertheless, the polarisation *is* different above the GRS and EZ, implying that their cloud-tops are higher than other zones (P.H. Smith, *1986*). And the intense polarisation in the polar regions is *more* than can be accounted for by Rayleigh scattering; here too, haze particles must contribute. Smith & Tomasko (*1984*) proposed that this polarisation is caused by the same haze that absorbs ultraviolet light.

Near-infrared methane absorption (0.89 μm)

Methane is one of the most abundant gases in the atmosphere after hydrogen and helium (Chapter 16), and its absorption bands give the most direct evidence on the heights of the visible clouds. It absorbs at several wavelengths in the infrared, the first major band being at 0.89 μm. Thus Jupiter is much darker at 0.89 μm than at

neighbouring wavelengths, but some of the sun's radiation does penetrate to the cloud-tops. So the 0.89-μm picture represents not the albedo of the clouds, but their depths: the higher they are, the brighter (Figs. 4.9&10).[7]

Typically the GRS is the brightest feature, the zones are bright, and the belts are dark. There are variable bright polar hoods (Chapter 5). The EZ, NTropZ, and NTB(S) were especially bright in years when they were orange. However, there is not a complete correlation with visual brightness or colour. Some white zones (and ovals) were not noticeably bright in the methane band, including the NTropZ and STZ, and the whitened SEB before a revival. The mismatches in high temperate latitudes may be because these are colder than the tropical regions so the white ammonia cloud-tops lie lower in the atmosphere (R.B. Minton, *1972a*). But mismatches in the equatorial region suggest that some dark clouds lie at the same level as lighter ones. For example, there is no extra methane absorption over the dark patches in the NEB nor the festoons.

There are also weaker methane absorption lines in red light. But there is virtually no difference between pictures in these bands and broad-band red images, either from Earth or from Voyager (which had a filter for one of these lines at 619 nm). This implies that red light does not penetrate below the ammonia cloud level, neither in the belts generally nor even in small gaps (R.A. West and colleagues, *1985*).

A detailed analysis of Earth-based methane-band images, taken in 1977 February in the 890, 725, and 619 nm bands, was done by West (*1979a,b*) and West & Tomasko (*1980*). At that time the SEB and EZ formed a very broad, dusky brown belt after a Revival, but were not strongly coloured visually – though reddish SEB colour developed later that year. The whole EZ and SEB(N) were bright at 0.89 μm, even though dusky in visible; was this due to high-altitude reddish haze?

West did photometry of the limb-darkening in each absorption band, added Pioneer 10 data, and fitted a model of two cloud layers with overlying haze (Fig. 4.11). The model indicated that upper (ammonia) cloud-tops were about 1 km lower in the SEB(S) than the STropZ, and the EZ and the GRS (diffusely) extended ≈10 km higher. In the NTZ and STZ, the upper cloud-tops may be 3 km lower than in tropical zones. These results agree well with Coffeen's (*1974*) polarisation analysis.

When the SEB became white visually in 1989/90, as in 1970, it remained dark at 0.89 μm (D. Kuehn and R. Beebe, *1993*). Thus these events involve whitening of the belt clouds but not a higher-altitude covering as in the true zones.

At wavelengths from 1 to 3 μm, methane and ammonia absorption becomes very strong, and only the high-altitude haze over the EZ and polar regions remains visible (Figs. 4.12&13, and Plate P21.4). Indeed, at 3–4 μm the planet is virtually black, except for the glow of aurorae (Fig. 4.12C and Chapter 17). But beyond 4 μm, the planet begins to shine again, due to heat emerging from the interior.

---

[7] Kuiper (*1972b*), with appendix by S.M. Larson. The same volume contains other images at 0.89 μm that were taken at LPL from 1968 onwards: Fountain & Larson (*1972*), Fountain (*1972*), and Minton (*1972a*).

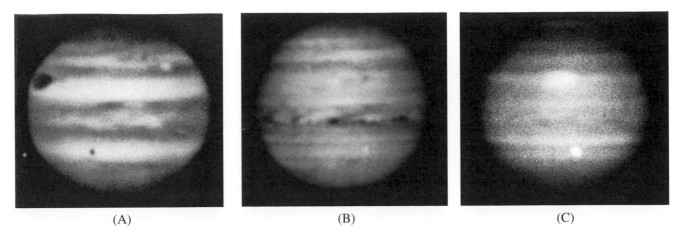

(A)   (B)   (C)

Fig. 4.9. Images of Jupiter in violet, near-infrared, and 0.89 μm methane absorption band; 1970 May 16 (LPL, 154–cm reflector). (A) 07.25 UT, violet (0.33–0.40 μm). (B) 05.41 UT, near-infrared (0.70–0.88 μm). (C) 05.51 UT, methane (0.89 μm). In methane, the brightest features are the GRS (central) and Europa (on NTZ). In violet, the GRS is very dark (near p. limb) and Europa is off the p. limb with its shadow f. it. SEB is dark in methane even though it is whitened at visible wavelengths. Note 'bluish' NTB(S) (dark in infrared and methane), and orange patches in EZ(S) (dark in violet). (Published by Fountain and Larson (1972). By courtesy of the Lunar and Planetary Lab., University of Arizona.)

(A)   (B)   (C)

Fig. 4.10. Images of Jupiter in blue, near-infrared, and 0.89 μm methane absorption band; 1971 July 10 (LPL, 154-cm reflector). (A) 04.16 UT, blue. (B) 04.25 UT, near-infrared. (C) 04.38 UT, methane. In methane, the brightest features are the GRS (near p. limb) and Io (central on NTropZ; dusky in blue). SEB is still dark in methane; the SEB Revival has begun with a vigorous source near central meridian, which shows various colours. NTB has also revived and is strongly orange (dark in blue). Also note bright polar hoods in methane image. (Published by Minton (1972b). By courtesy of the Lunar and Planetary Lab., University of Arizona.)

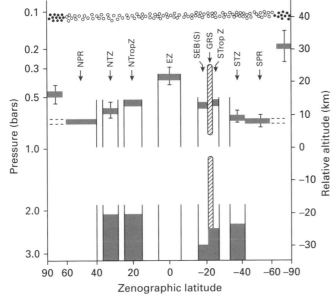

Fig. 4.11. Model of cloud structures, derived from methane absorption photometry of images in 1977 February. Belts were not modelled, except for SEB(S). The model includes an upper haze layer (dark and back-scattering over the poles, shiny and forward-scattering elsewhere), an upper cloud layer (presumably ammonia cloud), and a lower cloud layer. These layers may be vertically wider than shown. The optical density of the upper cloud layer is between 1 and 3, i.e., between 35% and 5% of light penetrates this cloud layer, revealing the deeper cloud beneath. (Adapted from West & Tomasko (1980).)

(A)

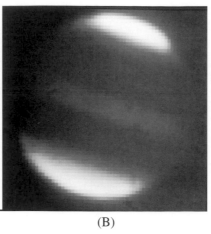
(B)

(C)

Fig. 4.12. Images of Jupiter in near-infrared, 2–4 μm, where solar radiation is strongly absorbed by methane and ammonia in the atmosphere. (A) 2.16 μm; (B) 2.34 μm. The reflective haze above the EZ, also seen at 1.0 μm (Fig. 4.1), is bright at 2.16 μm but barely visible at 2.34 μm. But both images show reflection from the high-altitude haze in the polar regions. (By courtesy of David Allen; copyright Anglo-Australian Telescope Board, 1991.)

(C) 3.55 μm. This wavelength is almost totally absorbed; the polar emission is from H_3^+ in aurorae. The bright spot near the centre is Ganymede – bright because it is free from methane absorption. Taken with the ProtoCAM instrument at the NASA Infrared Telescope Facility, Mauna Kea, Hawaii (Baron *et al., 1991*). (This image by courtesy of Steven Miller, Gilda E. Ballester, Jonathan Tennyson, and Robert D. Joseph.)

4.3 THERMAL RADIATION

Thermal emission from Jupiter

Jupiter emits almost twice as much heat as it receives from the sun, implying that it is warm inside. This heat is emitted as infrared radiation.

Exactly how much heat is emitted was difficult to determine from Earth, partly because of the difficulty of measuring far-infrared fluxes, and partly because we can only see the sunlit side of the planet. The dark side could be presumed to have the same temperature as the sunlit side, because the atmosphere is much too thick (at the relevant levels) to heat up during the jovian day. But the flux from the polar regions could only be estimated. So the first accurate measurments came when Pioneer 11 flew over the north polar region, and found that it had the same effective temperature as the equatorial region, to within 3°. (In fact these measurements refer to different depths in the atmosphere, but in each case it is the effective temperature that measures the net amount of heat being lost.) The Pioneers also confirmed that the day and night sides had the same effective temperature: 125 (±3)°K.

The current best value, from the Voyager spacecraft, is 124.4 (±0.3)°K, which implies that the total heat emitted is 1.67 times the heat absorbed from the sun (Table 4.1).[8]

Quoting a single temperature is an oversimplification, as the observed emission temperature depends on the wavelength and differs between belts and zones. The belts appear 3–4° warmer than the zones, but this is attributable to their lower albedo. It is remarkable that the poles are not colder; possible reasons will be discussed in Chapter 15.

Table 4.1 *The heat budget of Jupiter*

Data from Hanel *et al.* (1981).

	Over whole planet	Per unit area (projected on sunlit disk:)
Incident solar energy	7.64×10^{17} W	50.8 W/m²
Reflected solar energy	2.62×10^{17} W	17.4 W/m²
Ratio = Bond albedo = 0.343 (±0.032)		
		(average over planet:)
Absorbed solar energy	5.01×10^{17} W	8.2 W/m²
Thermal emission	8.37×10^{17} W	13.6 W/m²
Ratio = 1.67 (±0.09)		

The infrared spectrum of Jupiter is a complicated curve which contains a wealth of information (Fig.4.13). The Earth-based spectrum contains gaps due to absorption in the Earth's atmosphere, and large uncertainties at the longer wavelengths. A more complete and accurate spectrum was provided by Voyager (Fig.4.14). Whereas the radiation out to ≈3 μm, like visible light, consists of reflected and scattered sunlight, the radiation longer than 3 μm is *emitted* from Jupiter because of its warmth.

The emission around 8 μm is special, coming from methane in rarefied warm layers in the stratosphere (section 4.4). Otherwise, the emission comes from the troposphere, from different levels for different wavelengths. A 'window' at 5 μm allows radiation through from particularly warm and deep levels reaching to 260°K, well below the visible clouds. At 9–20 μm, the emission represents a temperature of about 125°K; this comes from quite high levels because of strong ammonia and hydrogen-molecule absorption deeper down. At longer wavelengths still, the emission temperature is 140–155°K; this is the temperature in or just below the white clouds of the zones. At these far-infrared wavelengths, the

[8] Pre-spacecraft results were reviewed by Kuiper (1972b). Pioneer results were summarised by Ingersoll et al.(1976) and Ingersoll (1976b). Voyager results were given by Hanel et al. (1981).

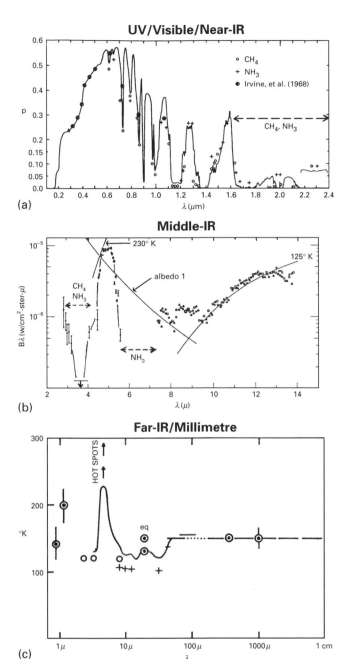

UV/Visible/Near-IR

○ CH₄
+ NH₃
● Irvine, et al. (1968)

Middle-IR

Far-IR/Millimetre

(a)

(b)

(c)

Fig. 4.13. The spectrum of Jupiter: Earth-based spectra in three ranges. Adapted from Kuiper (*1972b*). Note that the ordinate is defined differently in each waveband, for convenience in interpretation. (a) Visible and near-infrared. This consists of reflected sunlight. Methane and ammonia absorption bands are indicated. The 'shoulder' in the ultraviolet is where absorption is offset by Rayleigh scattering. Scale is geometric albedo. (b) Middle infrared. There are no data around 6–7 μm and 14–17 μm because these wavelengths are absorbed by water vapour and CO₂ respectively in Earth's atmosphere. Dashed curves indicate flux that would be observed from black-bodies emitting at 230°K (5-μm level) and 125°K (ammonia cloud-tops), and from a planet that reflects all incident sunlight (albedo = 1). Scale is surface brightness. (From Gillett *et al.,1969*). (c) Far infrared to microwave. Scale is radiation temperature.

main opacity is due to hydrogen itself. Spectra of the infrared emission allow the complete temperature profile of the atmosphere to be reconstructed from about 1 to 400 mbar (section 4.4).

Mid-infrared thermal emission (5 μm); the heights of the clouds (continued)[9]

At wavelengths of 4.5–5.5 μm, the jovian atmosphere is exceptionally free of molecular absorption, and the disc has an average radiation temperature of 206–217°K, with hot spots up to 255–260°K. The temperatures beautifully match the visible cloud patterns (Figs. 4.15&16). A picture of Jupiter at 5 μm looks like a negative of the picture in visible light. Belts and dark spots are bright at 5 μm, while zones and white ovals are dark. But the best correlation is with the visible colour.

Both white and reddish/ochre regions are cold, below 200°K; the actual temperature in the zones may be 180–195°K, representing a level somewhat below the visible white clouds. The belts and 'barges' emit at 220–230°K, and these emission temperatures are thought to be very close to the true temperatures. The blue-grey dark patches on the NEBs are hot-spots with radiation temperatures of 240–260°K, but as they get dimmer towards the limb (unlike the main belts), the radiation must be partially absorbed above the emitting level; the actual cloud temperatures are probably about 290°K.

The correlation with colour holds good down to small scales and through temporal changes. Cold areas have included not only the zones and white ovals but also the GRS, as well as belts that were temporarily orange in 1972 – the NTB and parts of the fading SEB. While the visible colour and brightness of the EZ and SEB underwent great changes throughout the 1970s, the correlation with 5-μm brightness always held true. And the warm areas include all dark grey-brown features, even on small scales: NEBn 'barges', jetstream spots, the rim of the GRS and a narrow S. Tropical Band, and even the dark rims of ovals in the SSTB.

Owen and Terrile (*1981*) did note a few exceptions to the colour-temperature relationship in the Voyager images. The turbulent SEBZ following the GRS, full of white and brown swirls, showed no contrast at 5 μm; possibly the dark material was drafted upwards in this region, or the mixture of reddish, brown, and white material was not fully resolved. Also, the fine bluish streaks in the NTropZ were not warm, unlike the bluish NEBs patches.

Given the striking correlation of the 5-μm temperatures with the visible cloud layers, the conclusion seemed irresistible: that the same clouds were being viewed at 5-μm and visible wavelengths. The resulting model[10] has become widely popularised. However, this seductively simple model, with its separate, 'all-or-none' cloud layers at different altitudes, is in conflict with other evidence (described in the previous section) on the altitudes of the clouds. It is still generally accepted that orange cloud (violet-absorbing haze) is the highest visible cloud material, and white clouds next. But the

---

[9] The 5-μm emission was discovered by Gillett, Low & Stein (1969). The first detailed 5-μm maps were by Keay et al. (1972, 1973). Jupiter was subsequently monitored from 1974 onwards by R. Terrile and colleagues with the Mt. Palomar 5-metre telescope. For later 5-μm maps, see: Westphal et al.(1974); Armstrong et al.(1976); Terrile & Westphal (1977a); Terrile & Beebe (1979); Terrile et al. (1979a,b); Owen & Terrile (1981); Beebe, Orton & West (1989).

[10] Terrile and Westphal (1977a); Owen & Terrile (1981).

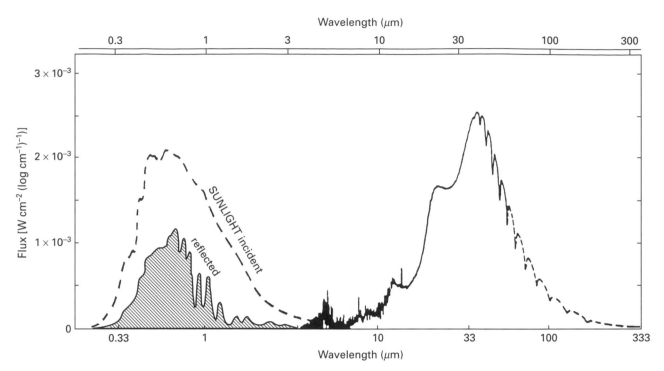

Fig. 4.14. The spectrum of Jupiter, from the ultraviolet to the millimetre range. Earth-based results are combined with those from the Voyager IRIS. The hump at left is reflected solar radiation centred on the visible range; the dashed curve is the solar radiation received by the planet. The hump at right is thermal emission as measured by Voyager; the dashed part is hypothetical. (From Hanel *et al., 1981*.)

idea that brown and grey clouds lie at deeper, warmer levels, and the blue-grey hot-spots deepest of all, can no longer be maintained. In fact, both in the zones and in the belts, the 5-μm radiation probably comes from below the visible clouds (Fig.4.17).

The recent analysis of the 5-μm radiation, using Voyager data for the equatorial region, was by B. Bézard and colleagues (*1983*). They concluded that the cold 5-μm emission from the EZ could come from a thick cloud layer at 180–195°K and 2 bars pressure. This postulated cloud layer must always go with overlying white ammonia clouds, to maintain the visible/5-μm correlation; this is consistent with the pattern of 45-μm emission (see below), which comes from almost the same level. The white ammonia clouds and the 190°K/2-bar clouds may always form together in updrafts in anticyclonic regions.

In the belts, the disparity between the visible and 5-μm altitudes is even greater. The old model of the 5-μm data agreed with other indications that belts lie deeper than zones: the dynamical evidence that the belts are cyclonic areas in which gas was thought to be sinking; the greater detail seen in belts in red light than in blue; the Pioneer polarimetry results; and the 0.89-μm methane absorption. But these data only imply height differences between belts and zones of 1–3 km, whereas the 5-μm temperatures differ by ≈60°K or ≈60 km. Moreover, the 0.89-μm methane absorption is not well correlated with visible colour; the polarisation differences can be explained by albedo differences alone; and light from the belts is not polarised enough to be coming from great depths. So there may be slight differences in altitude or transparency, but the visible clouds cannot be at the 5-μm-emitting levels.

The bluish-grey NEBs hotspots pose the greatest conundrum. Their 5-μm emission is from clouds at 260–290°K/4–6 bars (possi-

bly water clouds; Chapter 16). Owen and Terrile proposed that their blue colour was due to Rayleigh scattering of light at this depth, like the blue sky on Earth. But this is ruled out by the absence of strong polarisation and the absence of strong methane absorption in them.

What remains true is that the deep cloud layers are tightly correlated with visible colour at the cloud-tops. We will look at these models again in section 4.7, after considering the rest of the infrared spectrum.

4.4 THE UPPER ATMOSPHERE: STUDIES FROM INFRARED EMISSION

Infrared spectra and the temperature-pressure profile

Above the visible clouds, except for the diffuse violet-absorbing haze, the atmosphere is clear and chemically uniform. It is characterised by the profile of temperature and pressure shown schematically in Fig. 4.18, which also shows the major divisions of the atmosphere. Pressure must fall off exponentially with altitude, as measured by the 'scale height' – the height over which the pressure changes by a factor of 2.72 (called *e*). The scale height on Jupiter is 20–30 km (section 4.5).

The temperature-pressure profile is well-established up to the top of the stratosphere at about 1 mbar. It has been established by three methods.[11]

---

[11] Hunten (*1976*); Wallace (*1976*); Orton and Ingersoll (*1976*). Radiative-convective modelling was first done by Trafton (*1967*) and refined since by Wallace *et al.* (*1974*) and others.

(A)

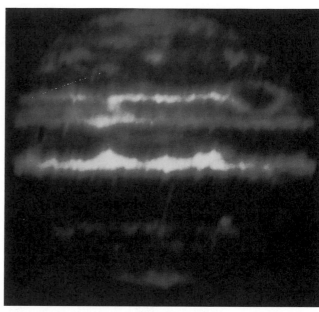

(B)

Fig. 4.15. Image of Jupiter at 5 μm (B), compared with a visible-light image from Voyager 1 (A) taken one hour earlier; 1979 Jan 10. Most visibly dark features are bright at 5 μm, including the rim of the GRS (at far right), S. Tropical Band, and NNTBs jetstream spots. However, orange areas are dark at 5 μm (the GRS itself, and the EZ(S)). (NASA image P-20957, published in Terrile and Beebe *(1979)*.)

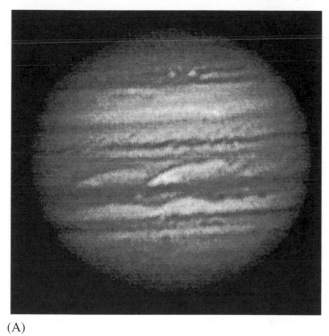

(A)

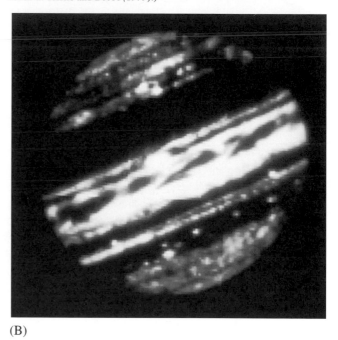

(B)

Fig. 4.16. Image of Jupiter at 5 μm in 1992, compared with a visible-light image. At this time the STB was absent and SEB(S) was faint and yellowish visually. Prominent spots in the 5-μm image include NTBs and NNTBs jetstream spots, and patches flanking the tiny white ovals in the SSTB. (A) Visible-light CCD image by D.C. Parker, 1992 Feb 15, ω_2 226. (B) Image at 4.9 μm, 1992 Feb 9, ω_2 239. This is a mosaic of 9 images, obtained with the ProtoCAM instrument at the NASA Infrared Telescope Facility, Mauna Kea, Hawaii, by J. Harrington, R.L. Baron, T. Owen, T.E. Dowling, and C. Kaminski. (By courtesy of J. Harrington.)

(1) Radiative-convective modelling. Knowing the atmospheric composition (which determines the scale height) and the overall heat budget, a theoretical model of the atmosphere is constructed taking account of the absorption and emission of heat at each level, to match the effective temperature of the planet as measured by infrared radiation. These models can predict the infrared spectrum of the planet and can be adjusted according to the observed spectrum.

(2) Inversion of the infrared spectrum. This method starts with the observed infrared spectrum, and works back from it to deduce the temperature-pressure profile. Typically it involves comparing observations in the 7.8–μm region (a methane emission band) with

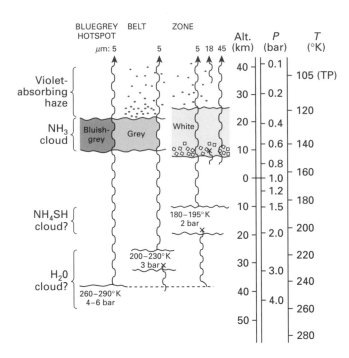

Fig. 4.17. Model of cloud structure, relating the visible clouds to 5-µm emission. This replaces the earlier model which identified the visible clouds with the 5-µm-emitting levels. Vertical wavy arrows indicate infrared emission; waves ending in X indicate radiation that is absorbed within the clouds. The origin of 18-µm and 45-µm radiation is also shown; the 45-µm radiation is partially absorbed in the zones but not the belts, due (hypothetically) to a layer of larger cloud particles at the base of the ammonia clouds. For more details, see Fig. 4.28. (TP = tropopause.)

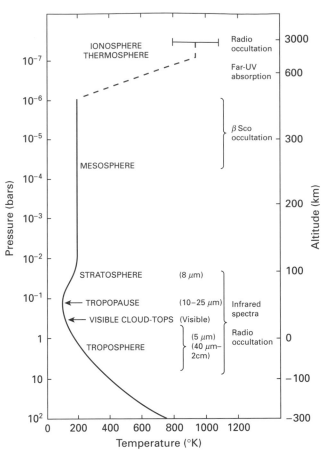

Fig. 4.18. Overview of the layers of Jupiter's atmosphere, showing how temperature and pressure vary with height. (Adapted from Hunten, *1976*.)

longer wavelengths that are dominated by hydrogen opacity. Only crude attempts could be made from Earth-based observations. The Pioneer infrared radiometer yielded a more complete though low-resolution profile. The Voyager IRIS yielded yet more reliable profiles, for the whole planet and for local regions (Fig. 4.19).

(3) Occultations of spacecraft radio signals. This completely independent technique, which has much better vertical resolution but is at risk from potential errors, will be described in section 4.5.

The major divisions of the atmosphere are the same as on the Earth: the troposphere, which is defined as the region that is stirred by convection, and the stratosphere, in which there is little convection. Above them, more poorly known, are the mesosphere and the thermosphere, the latter being the region of lowest pressures and highest temperatures.

In the lower, cloudy parts of the *troposphere*, one can assume that convection is the main way by which heat is transported upwards. In that case the temperature should fall off with altitude with a gradient of 2°K/km, which is the 'adiabatic lapse rate' for an atmosphere of Jupiter's composition, and indeed this is found at ≈1 bar and deeper. With this gradient, gas will cool on rising or warm on sinking without changing its total energy content – a characteristic of a convectively stirred atmosphere. The temperature at 1 bar is about 165°K. The white ammonia clouds occupy the region ≈300–700 mbar. Higher in the troposphere, heat will be lost upwards by radiation as well as convection.

At a level called the *tropopause*, convection almost ceases, and above it is the *stratosphere*, where thermal radiation is dominant.

Heat is absorbed both from above (the sun) and from below, and is re-radiated upwards. The balance between absorption and emission depends on the infrared wavelength and on the density and temperature of the atmosphere at each level, so different wavelengths come from different levels in the atmosphere. The tropopause is defined by a temperature inversion; above it, the temperature rises with altitude because solar heat (mostly in the 1–4 µm range) is absorbed in the stratosphere. The gases there are so thin that they cannot radiate away absorbed solar heat as readily as at lower altitudes. The temperature minimum is 105°K at a level of 100–160 mbar. The upper stratosphere reaches about 170°K at 1 mbar.

Temperatures at higher altitudes have been estimated from occultation measurements (see next section). The temperature at 10^{-5} bar is also ≈170°K, but above that there is a much hotter region called the *thermosphere*. This overlaps with the *ionosphere* (Chapter 17); temperatures here vary from ≈850°K to ≈1300°K. It may be heated by material waves propagating upwards, and by magnetospheric particles plunging downwards.

Detailed temperature profiles from Voyager
The Voyager data showed how the temperature profiles differed over various parts of the planet (Figs. 4.19 & 20). The tropopause is warmest and lowest at latitude ≈15°S, and cooler and higher at high latitudes; thus the stratosphere is coolest over the south polar region. There is only a slight difference (≈2°K) in upper atmos-

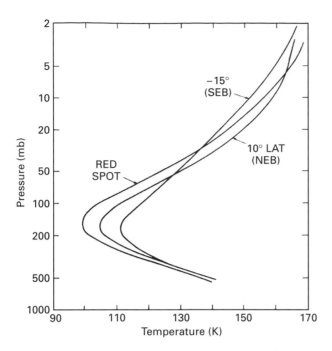

Fig. 4.19 Temperature pressure profiles from the Voyager IRIS, obtained by inversion of infrared spectra, for the NEB (10°N), SEB (15°S), and GRS. (From Hanel *et al., 1979a.*)

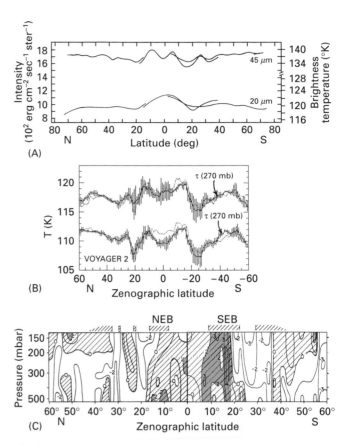

Fig. 4.20. Temperature profiles from pole to pole, from Pioneer and Voyager data. (A) From the Pioneer infrared radiometers. This shows the brightness temperature and flux in two broad-band channels centred at 20 μm and 45 μm; each one records about half the total infrared flux. Profiles are from Pioneer 10 at low latitudes and Pioneer 11 at high latitudes. Viewing angle 41°; for near-vertical viewing, less complete data gave similar profiles but 2–3° warmer. (Adapted from Ingersoll *et al.,1976.*) (B) From the Voyager IRIS. Infrared spectra were converted into temperature at two pressure levels, 270 mbar (just above cloud-tops) and 150 mbar (just below tropopause). (From Gierasch *et al., 1986.*) (C) Cross-section through the atmosphere, compiled from Voyager IRIS curves like those of Fig. 4.19, showing deviations from the average vertical profile in °K (shaded areas are warmer). Warmer areas include the SEB, NEB, and NTB (in spite of NTB being whitened); zones are colder. Data may be inaccurate below the 300-mbar level, i.e., within the visible clouds. (From Pirraglia *et al., 1981.*)

pheric temperature between the belts and zones. The atmosphere is also cooler over the GRS (by about 5°K) and over the STB white ovals, and warmer over a NEBs dark hotspot and a NEBn barge – in each case, up to 10–20 mbar, where the temperature difference disappears. These profiles are consistent with the presumed updraft in the anticyclonic ovals and downdraft in the cyclonic NEB spots.

Far-infrared thermal emission (10–50 μm): maps of the troposphere

Maps of Jupiter at mid- and far-infrared wavelengths are maps of the temperature at the levels where the atmosphere becomes opaque to those wavelengths. Like the radiation in the 5-μm 'window', radiation around 45 μm and longer (all the way to the microwave range at ≈2 cm) comes from below the visible clouds, from a level of ≈800 mbar (Fig. 4.21).

Thus pictures at 45 μm resemble the 5-μm image in showing the zones dark and the belts bright; the Pioneer 11 image (Fig.4.21A) is a negative of the visible image. In the Voyager 45-μm maps of the whole planet[12], the temperate regions were lost in strong limb-darkening, but the SEB and NEB were warm belts with many hotspots. These maps correlated quite well with the orange-light albedo. They showed features such as an EZn plume, the EZs white spot, and the STB Fade, as opaque (cool) locations whose drift rates agreed with those observed visually. But the correlation did not hold everywhere, as other EZn plumes were not obvious on these maps, and other comparatively opaque (cool) regions coincided with NEBn barges.

[12] The Voyager 45-μm maps (Hanel *et al., 1979b*) were transformed into maps of optical depth at 45 μm (Gierasch *et al., 1986*; Magalhaes *et al., 1990*).

In contrast, wavelengths in the range 10–25 μm come from levels around 150 mbar, well above the visible clouds. Images of the planet in this range have always shown the same picture, from early scans to modern spacecraft or Earth-based pictures (R. Beebe, G. Orton and R. West, *1989*). There is a general banded structure which, like that at 5 μm, is a negative of the visible pattern; but it is more diffuse (Fig. 4.22). These images confirm the general difference between belts and zones that was shown by the spectra (Fig. 4.20).

Do any changes occur at this level? In the crude scan images from Pioneer, with a resolution of ≈3% of the planet's diameter, the only change between Pioneer 10 and Pioneer 11 was that the whitened SEB became much warmer at 20 μm in 1974. There was no change in its visual appearance, but perhaps this was a forerunner of the SEB Revival that occurred the following year? Scans by

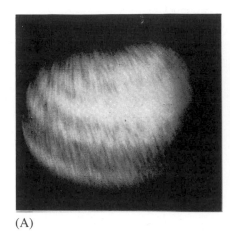
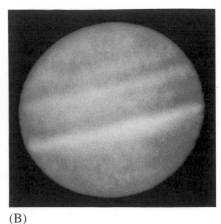

(A)   (B)

Fig. 4.21. Images of Jupiter at 45 μm and 2.0 cm, showing emission from below the visible clouds; belts are warmer (brighter). South is up. (A) At 45 μm, from Pioneer 11, in 1974. This is an inbound image showing the southern hemisphere; there is no difference between day and night sides. (Compare with the optical image in Plate P16.1.) The scan took 2½ hours so any local features were smeared out. The major dark band corresponds to the visibly-bright STropZ-SEB. (From Orton *et al., 1981.*) (B) At 2.0 cm, from the Very Large Array in New Mexico, in 1983. This shows the thermal emission at 0.5 to 1 bar. (From de Pater (*1986*); by courtesy of Imke de Pater.)

Fig. 4.22. Images of Jupiter at 18 μm, showing emission from the upper troposphere. (North is up.) (From Beebe, Orton & West, *1989.*)

Glenn Orton's group at 17.8 μm (Beebe *et al., 1989*) showed essentially the same belt structure from 1979 to 1984, except for a slight seasonal variation in hemispheric asymmetry, and a slight cooling of the EZ (which concurrently became brighter visually).

The Voyager IRIS viewed the planet not only at close encounter, but also 2–3 days before and after the encounters, when it produced maps of the whole planet at low resolution (≈5–10% of the disc diameter). Later, the spectra from each point were analysed to produce 'four-dimensional' maps of the atmosphere, showing the temperature at each pressure level on maps of the surface.[13] These maps did not even resolve belts and zones fully, but did show plenty of longitudinal structure in the warm NEB and SEB as well

as the cold spot over the GRS. In the maps showing the temperature at the 270-mbar level – just above the visible clouds – the NEB at ≈15°N showed a series of hot-spots 30–50° apart in longitude, in the same longitude range as the visible dark 'barges' and moving at similar speed, but not coincident with them. The hot-spots in the belts were not obviously related to visible features.

Comparison of several Voyager maps gave a surprising impression of largescale thermal features that were moving only sluggishly, with a curious asymmetry between the two hemispheres. J.A. Magalhaes and colleagues (*1989*) compared the maps of temperature at 270 mbar, made by Voyager 1 two days before and after encounter, and by Voyager 2. Because they could not trace individual features with confidence on just these three maps, they resorted to computer correlations between the three maps at each latitude, to produce a profile of the probable wind speeds. The technique does work; on visible-light and 45-μm maps, it showed up the known pattern of cloud-level currents. But the 270-mbar map showed vir-

---

[13] Voyager maps included raw maps of brightness temperature at 16.6 μm/≈150 mbar (Hanel *et al., 1979b*), and later processed maps of the real temperature at the 270 mbar level (Gierasch *et al., 1986*; Magalhaes *et al., 1990*).

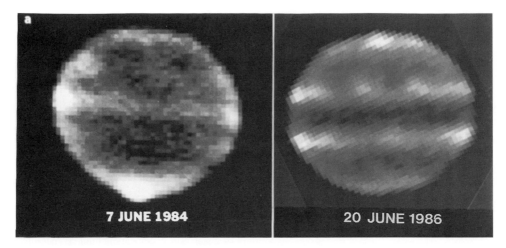

Fig. 4.23. Images of Jupiter at 7.8 μm, showing emission from the stratos-phere. (North is up.) Note change from bright EZ (1984) to dark EZ (1986), and variable polar hotspots. (From Beebe, Orton & West, *1989*.)

tually no motion; the displacements were only tens of degrees lon-gitude even over 4 months between the two encounters. The whole southern hemisphere was stationary in λ_2 (u = -3 to -4 m/s in λ_3), while the northern hemisphere from 5°N to 44°N was moving at $\Delta\lambda_2$ = -13 (u = +2 to +4 m/s in λ_3), and the equatorial region moved at an intermediate rate. There was no fast equatorial current.

What could produce these slow motions? Magalhaes and col-leagues suggested that they were controlled by deep levels in the atmosphere – though whether deep fluctuations could be transmit-ted to the upper troposphere, without producing visible effects in between, seems uncertain. It seems more likely that they are con-nected with visible spots, whose influence extends surprisingly far in latitude. The motion reported for the southern hemisphere was identical to that of the Great Red Spot, and the correlation – both there and over the equator – mainly consisted of the entire hemi-sphere centred on the GRS being almost 1° cooler than the other hemisphere. The motion for the northern hemisphere was very sim-ilar to that of the conspicuous NEBn 'barges', as well as that of the N.N.N. Temperate Current. What these results may have done is to define a new level of high-altitude motions in the atmosphere.

Ground-based observations have also shown structure in these wavebands. D. Deming and colleagues (*1989*) mapped the 8–13 μm emission using the 3-metre dish of the NASA Infrared Telescope Facility on Mauna Kea, Hawaii. (Large telescopes are used for such long-wavelength imaging because of the limitations of diffraction; at 13 μm, even for a 3-metre telescope, the Dawes limit is 1.0 arc sec-ond.) They revealed wave-like patterns with a 'wavelength' of ≈35° in longitude, both at the equator (presumably related to the EZn plumes) and at 20°N (the NEBn). The NEBn pattern seems to be peculiar to the upper atmosphere; it was not 'spillover' from the EZn plumes as it was not rapidly moving, but there was no such pattern in visible light at that time (late 1987). The motions of these features will be discussed in section 4.6. Imaging at these wavelengths has not yet reached sufficient resolution to say much about the features being recorded; future results should be very interesting.

Mid-infrared methane emission (7.8 μm): the stratosphere[14]

Between 7.4 and 8.0 μm, methane shows very strong absorption and emission, so the atmosphere is not transparent to these wave-lengths until one gets up above the tropopause, where the methane appears in emission. Images at these wavelengths show the stratos-pheric temperatures at the 10–20 mbar level. Such images show some remarkable structures (Fig. 4.23):

Variable polar hotspots, possibly associated with auroral activity (Chapter 5).

Two major warm 'belts'; but they are centred near 23°N and S, extending further poleward than the visible NEB and SEB. The N. and S. 'belts' had an asymmetry which reversed in 1982 and again in 1986, which might be a sea-sonal effect (Fig. 4.24).

Limb brightening, especially of the warm 'belts', which often show bright spots at the limbs.

Occasional patchy structure in the belts. In 1985 it was promi-nent and perhaps wave-like, with warm patches separated by ≈30° longitude in the South Tropical region and 45° in the North. More subdued patches in subsequent years sometimes changed brightness on timescales of ≈20 days.

Marked variations in the EZ temperature, possibly with a 4-year cycle (Fig.4.24). The EZ was the warmest band in 1980, 1984, and 1989–90, but a cool zone in 1982 and 1986–87.

According to the relation between temperature gradients and winds (section 4.6), this equatorial cycle must entail a cyclic change in stratospheric wind speeds; when the equatorial stratos-

[14] Limb-brightening at 7.8 μm was discovered in the first scans by Gillett & Westphal (*1973*), and polar hotspots by Caldwell *et al.* (*1980, 1983*). Glenn Orton's group at the Jet Propulsion Lab. has made central meridian scans over 1979-1983 and images over 1984-1990 with a resolution 4% of the disc diameter, using the 3.0-metre NASA Infrared Telescope Facility on Hawaii. Their results are in: Beebe, Orton & West (*1989*); Orton *et al.* (*1991*); Leovy *et al.* (*1991*).

phere is cool, the speed must decrease with altitude as jetstreams usually do, but when it is warm, the speed must increase with altitude. Leovy and colleagues (*1991*) interpret this cycle by analogy with Earth's quasibiennial oscillation, which is also an equatorial cycle in east-west wind patterns, at an altitude of about 25 km. On Earth this oscillation is driven by convective equatorial waves travelling eastward or westward in alternate years, which propagate vertically; perhaps something similar is happening on Jupiter.

An intriguing coincidence, not previously pointed out, is that two of the years when the equatorial stratosphere was warmest (1980 and 1989–90) coincided with EZ coloration episodes, which consist of enhanced UV-absorbing haze in the upper troposphere. One can suggest that this haze traps heat by absorbing short-wavelength light, either in the upper troposphere, or perhaps in the stratosphere itself if it extends that high. But in 1984, the other warm year, there was no visible EZ colour.

4.5 THE UPPER ATMOSPHERE: STUDIES FROM OCCULTATIONS

A direct way of examining the upper atmosphere is to follow the light of a star as it disappears behind (is 'occulted' by) the planet. The radio emissions of a spacecraft can be followed in the same way. These occultations are important as they give the only direct measurements of the density scale height.[15]

The fading of the star or spacecraft at occultation is mainly caused by differential refraction in the density gradient of the atmosphere. The intensity is reduced by half when the light or radio waves are bent through an angle equal to that subtended by the scale height at the distance of the observer or of the spacecraft. Therefore, occultations of stars viewed from Earth probe very small angles of refraction, in the rarefied upper atmosphere; occultations of spacecraft radio waves probe much deeper.

The data yield a vertical profile of refractive index, and thus of gas density; this gives the density scale height. As the pressure scale height can be assumed from the known atmospheric composition, the two heights can be compared to give a profile of temperature with altitude. Alternatively, by assuming the temperature/pressure profile derived from infrared spectra, one can use occultations to determine the mean molecular weight and thus the gas composition (Chapter 16).

Occultations of stars

Occultations of stars brighter than 6th magnitude have been recorded on only two dates: σ Arietis on 1952 Nov. 20, and the ß Scorpii multiple-star system on 1971 May 13.[16] These occultations probe the outer fringes – the 'mesosphere' – where the pressure is ≈2 μbar and the density ≈7×10^{13} molecules/cm^3. The most impressive feature of an optical occultation is the flashing of the star as it fades, like greatly exaggerated twinkling. This is presumably due to refocussing of the light by small fluctuations in density, either by static layers of gas or by turbulence. But it cannot be usefully

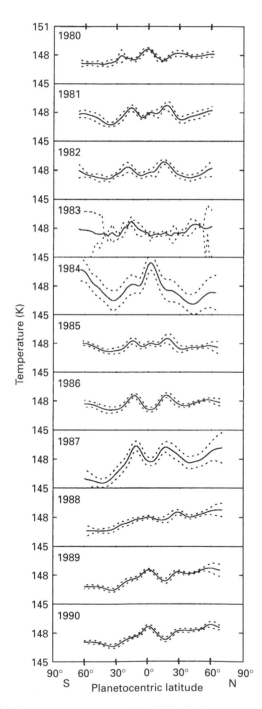

Fig. 4.24. Scans of Jupiter at 7.8 μm, 1980–1990, showing the cyclic warming and cooling of the stratosphere over the EZ. The curves show temperatures at 20 mbar (or 10 mbar in 1984 and 1987), globally averaged, with dotted curves representing standard deviations. (From Orton *et al.*, *1991*; by courtesy of Glenn Orton.)

[15] Reviewed by Hunten and Veverka (*1976*). The scale height is the height over which pressure or density falls by a factor of 2.72. For pressure, it is

$$H = RT/\mu g$$

where R is the gas constant, T the temperature, μ the mean molecular weight, and g the force of gravity.

[16] Reviewed by Hunten and Veverka (*1976*) and Wallace (*1976*). For _ Ari: photometric analysis by Baum and Code (*1953*); film showing flashes by Pettit & Richardson (*1953*); visual reports in Haas (*1953*). For ß Sco: photometric analysis by Hubbard *et al.* (*1972*) and Veverka *et al.* (*1974*); photos in Larson (*1972a*), Dragesco (*1972*), and Combes *et al.* (*1972*).

analysed. More important is the overall rate of fading, from which the scale height and temperature can be deduced.

The occultation of σ Arietis (magnitude 5.5) gave an apparent scale height of about 8 km, inconsistent with the mainly hydrogen atmosphere, but this may have been due to imprecision in the data. ß Scorpii, a much brighter star, gave much better data (Fig. 4.25). Both ß Scorpii C (magnitude 4.9) and ß Scorpii A (magnitude 2.6) were occulted, and ß Scorpii A is itself double, with a separation of 1.5×10^{-3} arc-seconds, or 3 km at the distance of Jupiter; this duplicity was actually visible as a doubling of the flashes at occultation (Fig. 4.25B). The flashes were not symmetrical between ingress and egress. The ingress, in high southern latitudes, lasted nearly 8 minutes, with flashes during most of this time. The best values for the scale height were 25–30 km, and the deduced temperature was $170\pm30°$K at 14 μbar, with no difference between the limbs at sunset (ingress) and sunrise (egress). Although vertical temperature gradients were also deduced, they were far from certain.

The Voyager spacecraft also observed jovian occultations of stars, notably Regulus (α Leonis), and of the sun. But these observations were made with the ultraviolet spectrometer and recorded hydrogen absorption, not refraction; these short wavelengths are fully absorbed above the level at which refraction would have started. They gave temperatures of 200°K at 10^{-6} to 10^{-7} bar, and $\approx$1100 ($\pm$200)°K at $\approx$1500 km above the cloudtops (S.K. Atreya and colleagues, *1979, 1981*).

Occultations of spacecraft

The radio signal from a spacecraft is distorted at two levels: first in the ionosphere, by electrons (Chapter 17), and then in the atmosphere, by gaseous refraction. In addition to the fading, the Doppler shift gives extra information. As the spacecraft is close to the planet, the angle of refraction is large and probes from ≈1 mbar down to levels near the visible cloud-tops.

All four Jupiter space-probes passed behind the planet. The first results from the Pioneers gave high temperatures which could not be reconciled with the infrared profiles of the atmosphere. It turned out that there are many problems in analysing occultation data which were not fully realised until after the Pioneer encounters. The main error was to ignore the oblateness of the planet. When this was included in the analysis, the Pioneer curves agreed well with other information below the 10-mbar level, although they were essentially undetermined above.[17]

For the Voyager occultations, the false assumptions had been sorted out, and there were several other features that made the data more reliable. The signals were stronger; two different frequencies were used (wavelengths of 3.6 and 13 cm); and the trajectory and the pointing of the spacecraft were planned to give the clearest results. The Voyager 1 occultation was almost across the greatest diameter of the disc, from latitude 11–12°S to 0–1°N (Fig. 2.16). (In contrast the Voyager 2 occultation was behind the S. polar region, latitudes 67°S to 50°S, and the radio signals never disappeared; at

[17] Pioneer occultations: Hubbard *et al.* (*1975*); von Eshleman (*1975*); Kliore & Woiceshyn (*1976*). Voyager occultations: Lindal *et al.* (*1981*).

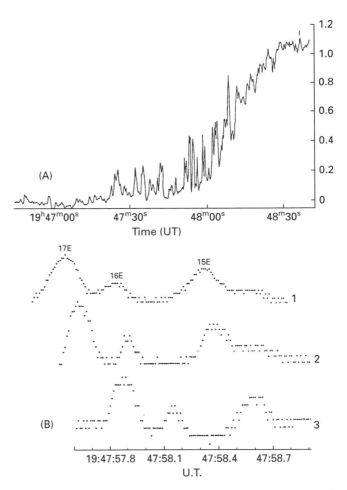

Fig. 4.25. (A) Light-curve of the occultation of ß Sco A by Jupiter, 1971 May 13. This shows the gradual emergence over one minute, with superimposed flashes. The lower panel (B) enlarges a one-second portion with two double flashes, in three colour channels (1, broad UV; 2, near-UV; 3, red). The time delay is due to atmospheric refraction. (From Veverka *et al.*, *1974*.)

mid-occultation they were refracted by 0.3°, and they showed 'flashing' due to atmospheric and ionospheric irregularities.) The resulting profiles agreed well with the infrared profiles (Fig. 4.26), including the temperature minimum (tropopause) of 110°K at 140 mbar, and an increase to 160 ±20°K in the 1–10 mbar region of the atmosphere. (These highest regions are the least certain because one has to assume a temperature at the top and work downwards.)

The Voyager 1 profiles showed a remarkable vertical waviness in the stratosphere. As the initial Voyager reports stated, this could have been due to non-uniform layers of gas absorbing heat at those levels, or it could have been due to material waves ('inertia gravity waves') propagating upwards. The main fluctuations could be interpreted as waves with an amplitude of 10°K and a wavelength of about 3 scale heights. M. Allison (*1990*) noted that these were not present in the high-latitude Voyager 2 data, and suggested that they were propagated up from the same planetary-scale wave system that controls the EZn plumes (Chapter 14). These ideas should be testable by the Galileo entry probe.

Eclipses of satellites

An 'occultation' of the Sun by Jupiter as observed from a jovian moon is similar in geometry to the occultations of spacecraft. As

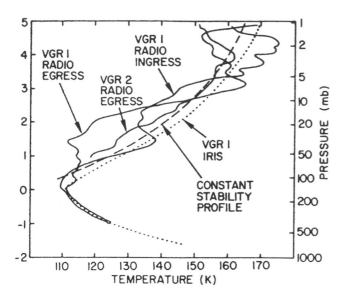

Fig. 4.26. Temperature/pressure profiles from the Voyager radio occultations. Pressure in millibars at right; altitude in scale heights at left. The latitudes were: 11–12˚S (Voyager 1 ingress); 0–1˚N (Voyager 1 egress); 57–73˚S (Voyager 2 egress); see Fig. 2.16. The three occultation profiles are compared with an IRIS profile recorded close to the Voyager 1 ingress point (vertical resolution ≈ 2 scale heights), and with a theoretical ('constant stability') profile derived from radiative-convective modelling. The large fluctuations in the two Voyager 1 profiles may be vertically-propagating waves. (From Allison (*1990*), after Lindal *et al.* (*1981*).)

the sun and the moon are large objects, the only useful observation is that the moon sometimes remains faintly visible several minutes after the expected total eclipse, due to sunlight refracted or scattered in the jovian atmosphere. This was observed by G. Kuiper (*1947, 1961*) and analysed photometrically by several groups.[18] They concluded that the light is scattered by the high-altitude haze, and confirmed that the haze particles are violet-absorbing and ≤0.4 μm in diameter.

Likewise, when Voyager 2 was immersed in Jupiter's shadow, it could still see a glow along the planet's limb (Fig. 19.2). This glow was not resolved, but must have been due to scattering of light way up in the stratosphere. According to R.A. West's analysis (*1988*), the fine haze of sub-micron particles responsible must extend up to levels around ≈1 mbar, and varies with latitude.

4.6 HOW HIGH DO THE WINDS EXTEND?

The upper atmosphere
The general profile of winds at different altitudes can be calculated from the maps of temperature profiles described in the previous sections. This is a straightforward consequence of the physical laws governing atmospheres, as expressed in the 'thermal wind equation'. A horizontal temperature gradient connotes a vertical wind gradient, such that where temperature increases with latitude, eastward winds decrease with altitude.[19]

[18] Smith *et al.* (*1977*); Greene *et al.* (*1980*); Smith (*1980*). Early results were reviewed by Harris (*1961*).

From the Voyager IRIS maps (Figs. 4.20), the belts are a few degrees warmer than the zones, from ≈20 mbar to ≈1 bar (even when they are whitened, like the NTB during the Voyager encounters). The result is that all the jetstreams (eastward and westward) should become weaker at greater heights, disappearing over 2–3 scale heights (Fig. 4.27). The same is true for major circulations, both anticyclonic (the GRS and white ovals) and cyclonic (the NEBn barges). All the circulations must disappear in the mid-stratosphere, around the level of 10–20 mbar where the temperature contrasts fall to nothing.

The visible clouds
The only direct observations of differential motions in the visible clouds have been in the EZ, where several rare and diverse observations support the prediction that higher clouds should move more slowly. Speeds intermediate between Systems I and II have been detected for at least one white/orange 'front' in the Voyager images, for brilliant 'irradiating spots' in 1918/19, and for bright and dark spots that spill across into the EZ from SEB Revivals (Chapter 9.3). All these features are likely to be above the main ammonia cloud level.

The thermal patterns at 270 mbar above the EZ seemed to be moving even slower, at System II speeds (section 4.4). These may have represented either very deep perturbation or high-altitude diffusion from the GRS and NEBn barges, but as the level concerned is only just above the ammonia clouds, such a striking contrast in speed will have to be confirmed at higher resolution before any explanation can be established.

As red light seems to penetrate deeper into the belts than blue light, one might hope to find a systematic difference in speed between features observed in red and blue. Magalhaes and colleagues (*1990*) aligned sets of Voyager maps in orange and violet, and did indeed find that at four latitudes the violet-light maps showed less extreme speeds (averaged around the planet). They showed lower prograding speeds on SEBn and NEBs, and lower retrograding speeds on SEBs and NEBn, by ≈10 to 30 m/s (20 to 50°/month) in each case. This could correspond to an altitude difference of 0.3–0.6 scale heights (6–12 km). However, the difference may have been due to the maps being dominated by different types of features, rather than due to an altitude difference. On the SEBn, the difference arose because a slow-moving great white spot dominated the violet maps, whereas faster-moving smallscale features dominated the orange maps. The great white spot was so well-organised and long-lived that it must have affected all visible layers; its unique slow drift cannot have been due simply to alti-

[19] The thermal wind equation is:

$$\frac{\partial u}{\partial z} = \frac{-g}{2T\Omega r.\sin \varnothing} \frac{\partial T}{\partial \varnothing}$$

where $\partial u/\partial z$ is the gradient of eastward wind speed with altitude, T is the temperature, Ω is the planetary rotation rate, r is the planetary radius, $\varnothing$ is the zenographic latitude, and $\partial T/\partial \varnothing$ is the gradient of temperature with latitude at a given pressure. It was applied to Jupiter by: Ingersoll and Cuzzi (*1969*); Pirraglia *et al.* (*1981*); Conrath *et al.* (*1981a*); and Gierasch *et al.* (*1986*).

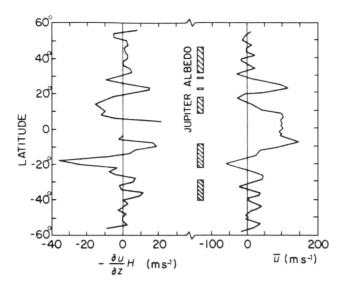

Fig. 4.27. Jetstreams decay with altitude above the cloud-tops. These profiles from north (top) to south (bottom) compare the cloud-top wind speeds (*u*, from Voyager images, preliminary version) with the vertical wind shear above the clouds at the 140-mbar level (*-(du/dz)H*, calculated from Voyager IRIS temperatures according to the thermal wind equation). (The shear values are lower limits because of the limited resolution of the IRIS; shear appears strongest at the SEBs because this is one of the broadest jetstreams.) The almost perfect correlation between speed and shear indicates that all the jetstream motions lessen with increasing height. (From Pirraglia *et al.*, *1981*.)

tude. No-one has yet examined the other latitudes in detail to see whether the differential speeds there have similar explanations. In both Voyager and visual observations, white clouds erupting in the SEB and NEB indeed appear to billow above the belt, but they are rapidly drawn out into streaks which follow the prevailing winds at each latitude.

4.7 SUMMARY : THE VERTICAL STRUCTURE OF THE ATMOSPHERE

The equatorial, tropical and temperate regions[20]
The vertical temperature/pressure profile seems to be well-established from 1 mbar to 1 bar (Figs. 4.19&26). What remains to be discussed is how the various cloud layers fit onto this profile.

The topmost layer is the violet-absorbing, polarising, forward-scattering haze, which is encountered at 120–200 mbar, close to the tropopause; it may extend down to the visible cloud-tops. This thin haze is present over both belts and zones. As the reddish and yellowish colours over the Great Red Spot and over coloured belts and zones are at similar high altitudes (according to polarisation, 5 µm emission, and methane absorption), they are probably due to thickenings of this haze. The composition of this haze is unknown; one possibility is ammonia crystals, merging into the thicker ammonia clouds below, but coated with a coloured substance at the higher altitudes.

[20] This model is largely from Smith & Tomasko (*1984*); West, Strobel & Tomasko (*1986*); Smith (*1986*).

The next layer comprises the visibly opaque white clouds of the zones. The top is diffuse, at 200–320 mbar. This is indicated by both polarisation and methane-absorption measurements; it also agrees with the expectation of an ammonia cloud layer at this level. So this cloud layer consists largely of ammonia ice crystals (Chapter 16), and this implies that it extends down to 630–700 mbar, the level at which ammonia reaches saturating concentrations in the atmosphere.

Below this, the situation is uncertain. The 5-µm radiation comes from deeper levels, and until recently it was thought that the matching visible clouds were at the same levels (section 4.3). But the polarisation and methane data are not consistent with this; they indicate that the visible clouds are at almost the same level, the ammonia ice level, in belts and zones. Even in the zones, the 5-µm radiation probably comes from below the ammonia clouds, at about 1.5–2.0 bar/180–195°K – although it may be partly absorbed by large particles near the base of the ammonia clouds in the zones. (These particles at the 700-mbar level, ≈3–100 µm in size, are also required to explain the 45-µm opacity of the zones.) In the belts, the 5-µm radiation also comes from below the visible clouds, from 2–4 bar/220–230°K in the brown belts, and 4–8 bar/240–290°K in the blue-grey hotspots on the NEBs edge (Figs. 4.17&28). Alternatively, the belts' 5-µm emission may be a mixture coming from both the ≈190°K and the ≈260°K layers.

So this new model states that the ammonia cloud covers belts as well as zones. In the zones it has larger particles at 700 mbar, and there is also a thick layer of cloud around 2 bar. In the belts the ammonia cloud is mixed with darker colours, and may be thinner, and there is also less cloud at 2 bar (Fig. 4.28). (A possibility proposed by West, Strobel and Tomasko (*1986*) is that the dark cloud particles form nuclei on which white ammonia ice can condense if the temperature falls or the ammonia concentration rises, for example during whitening of a belt.) The pattern of the deeper layers must be almost invariably coupled to the visible colour in the ammonia layer. A physically reasonable model has the ammonia cloud largely opaque to visible light yet largely transparent at 5 µm, with cloud particles either micron-sized or larger.

The polarisation and methane absorption data indicate that the ammonia cloud-tops are slightly deeper in the belts than the zones, but only by 1–3 km – or the clouds may just be thinner in the belts.

The model does not explain why great detail is seen in the belts in red light, as compared to blue light; this seems to show that red light penetrates deeper in the belts. However, if the optical density is as low as 1.5 in the belts, which is allowed by the present models, 20% of the light does get through in spite of being scattered and so losing its polarisation; so this may be sufficient to reveal the detail in or even below the ammonia cloud layer.

The polar regions
Between about 40° and 60° latitude, there is a transition to a very different atmospheric structure over the polar regions. The polar regions show strong polarisation in visible light, setting in above 40–48°N or S, and most of all above 58°N or S. These regions are also very dark in ultraviolet and in 0.89-µm methane absorption

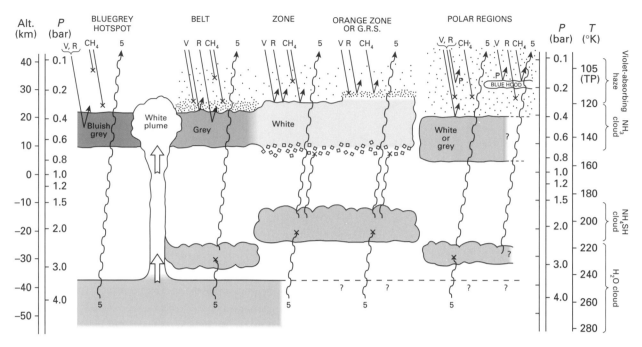

Fig. 4.28. Model of the cloud structures, synthesising data from this chapter. The diagram shows typical ray-paths for violet (V), red (R), 0.89-µm (CH₄), and 5-µm (5) radiation. (For 18-µm and 45-µm radiation, see Fig. 4.17.) Waves ending in X indicate absorption of radiation; scattering with P indicates that polarisation is induced.

Table 4.2 *Summary of data on altitudes of clouds in different regions.*

Feature	Visual colour	5 µm emission	0.89 µm methane	Optical polar'n
GRS ⎫ Coloured EZ ⎬ NTB(S) ⎭	orange or ochre	dark (cool)	very bright (high)	weak (high)
Main zones	white	dark (cool)	mostly bright (high)	N
Main belts	dark brown or grey	bright (warm)	dark (low)	N
NEBs patches & festoons	dark bluish-grey	v. bright (v. warm)	N	N
Polar regions	dark grey	bright (warm)	dark (low)	strong (*)
S. polar hood	blue	bright (warm)	bright (high)	strong (*)

'Low' and 'high' denote the inferred altitude of clouds. For polarisation, N indicates 'no contrast', and * is a reminder that polarisation does not simply indicate altitude.

(apart from the 'polar hoods' described below). Here, the polarising and violet-absorbing haze must extend up to about 40 mbar. However the top of the main cloud layer is lower than elsewhere, around 400 mbar (West *et al.*, *1986*). Thus the strong polarisation comes from a combination of Rayleigh scattering and particulate scattering. This structure also accounts for the general dimness and low contrast at visible wavelengths. The overall colour however is grey (relatively bluer than lower latitudes). Consistent with the depth and darkness of the clouds, the polar regions are warm at 5 µm.

At the highest latitudes (≥68°S), a remarkable blue polar hood is sometimes photographed (Chapter 5). It is highly reflective in blue and near-ultraviolet, and also in the methane 0.89-µm band, indicating that it is very high altitude (Figs. 4.9 & 10). According to the 1977 methane images studied by West and Tomasko (*1980*), the south polar hood lies at 140 mbar, and the north polar hood at 400 mbar. This suggests that it consists of some special condensate above, or a coating on, the violet-absorbing haze particles. This coincides with the most highly polarised part of the disk so the polarisation is presumably due to scattering by these particles. But the pole is also warm at 5 µm so the polar hood must be transparent to this wavelength, and not associated with thicker ammonia clouds like the zones. At 2 µm – a wavelength which is almost completely absorbed by gas over most of the planet – the poles are also notably reflective, and S.J. Kim and colleagues (*1991*) concluded that the scattering haze extended thinly all the way up to 5–15 mbar.

The nature of this condensate is unknown. The first suggestion, by G. Münch and R.L. Younkin (*1964*), was frozen methane. This would require the polar atmosphere to be much colder than elsewhere (64°K at 50 mbar) which is unlikely. The Voyager 2 occultation crossed the edge of this region, and no such low temperatures were apparent (Fig. 4.26). Another substance is probably needed to explain the polar hoods – perhaps something produced by auroral chemistry (Chapter 16.1).

III. The Observational Record of the Atmosphere

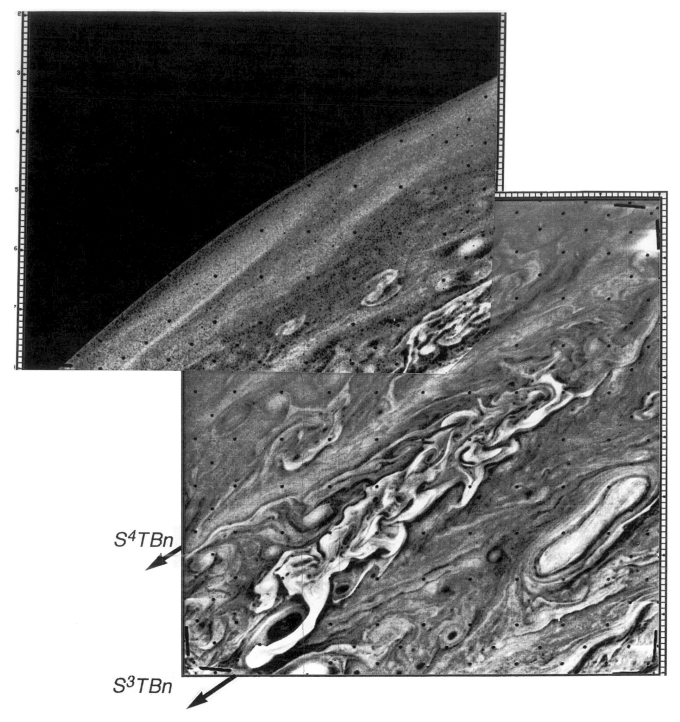

S^4TBn

S^3TBn

Fig. 5.1. Voyager closeup of the south polar region. The south polar hood, at top (bright in violet), is resolved into oblique diffuse sheets. The bright 'dumbell' at lower right is in SSTB latitudes. Cyclonic filamentary structures can be seen in four consecutive domains from S.S. Temperate to S.S.S.S.S. Temperate, even though the S^4TBn jetstream (indicated) marks the canonical boundary of the SPR. (Images 16314.34&36, violet, filtered. NASA.)

5: The Polar Regions (>57°N and >53°S)

5.1 VISUAL OBSERVATIONS

For visual observers, the north and south polar regions are considered to be the shaded areas, usually featureless, that extend down from the poles to the first zone or belt that can be discerned. They often extend down to latitudes of ≈40° to 48°, at the edge of the NNTB and SSTB, or of the NNTZ and SSTZ. However, the highest-resolution observations occasionally show latitudinal structure and regular currents extending further, up to about 57°N and 53°S, which are the latitudes of the uttermost prograding jetstreams that were clearly defined by Voyager. Therefore, this chapter only considers the latitudes beyond those limits. The boundary at ≈57°N is not usually visible, but a belt there would be the N⁵TB. The boundary at 53°S is often sharp (Chapter 12), and when there is a belt there it is defined as the S⁴TB.

Definite colours are rarely if ever seen in the polar regions. There have often been reports of colour differences between the north and south polar regions, but these are attributable to the 'traffic lights effect' of atmospheric dispersion. Refraction in Earth's atmosphere causes the inverted image to be fringed with red at the top and blue-green at the bottom (Chapter 4.1). As the polar regions of Jupiter generally have a gentle gradient of increasing darkness towards the poles, the traffic lights effect applies to every infinitesimal latitude band, causing the whole south polar region to be slightly reddish and the north polar region slightly bluish (for northern-hemisphere observers). This is indeed the colour difference that has often been reported. It can be neutralised by making use of chromatic aberration in the eyepiece (Chapter 1)! Therefore, colour estimates of the polar regions should be treated with great caution.

However, it was fairly common in early visual reports for *both* polar regions to be described as bluish-grey, which agrees with the impression on the Voyager pictures.

There are probably real variations in the darkness of the polar regions. For example, the north polar region was considered particularly dark in 1905/06 and 1930–1932, and light in 1932/33 and 1936.

Generally, if any detail is seen in the polar regions, it consists of ill-defined dusky features. In apparitions when observers have

reported an exceptional amount of belt or spot structure in the 'polar regions', it seems to have been mostly in what we are now defining as the high temperate latitudes (Chapter 6), but it sometimes extends into the polar regions proper. From 1938 to 1942 some belts were recorded unusually far north, one of them at an estimated latitude of 60°N, and a belt was measured at 59.5°N in 1967/68 (NMSUO); this would have been a N⁵TB. There seem to be no records of belts further north. In 1905/06 and again in 1987/88, there was a general increase in detail in the high northern latitudes including light ovals in the polar region; the disturbance was estimated as extending up to ≈65°N in both apparitions, and several light ovals were photographed centred at 60–62°N in 1987/88. In the south polar region, short belt segments are occasionally seen beyond the S⁴TB; there was a visual estimate of 60°S for one in 1911, and a photographic measurement of 58°S in 1965/66 (NMSUO). A few small bright ovals have been photographed at about 60°S, by NMSUO (see below) and by J. Dragesco in 1983 (SAF report).

No drift rates have ever been determined visually in the polar regions, apart from one, perhaps doubtful instance in 1911 (see below). The so-called 'North Polar' and 'South Polar' currents appear to be the characteristic currents of distinct latitude bands, and are here re-named 'N.N.N.N. Temperate Current' (N⁴TC; Chapter 6) and 'S.S.S. Temperate Current' (S³TC; Chapter 12).

5.2 PROFESSIONAL AND SPACECRAFT OBSERVATIONS

Earth-based photographs rarely show any detail in these regions, although they have occasionally shown isolated small bright spots. A 'South Polar Belt' at ≈65°S was reported by NMSUO and LPL from 1968 to 1972, but this may have been a broad region limited by the blue polar hood (see below).

The Pioneer and Voyager photographs also showed rather little detail in these regions; the impressively convoluted patterns of whorls and bands were mainly in the high temperate latitudes. However, the Voyager pictures did show some similar types of features – light ovals likely to be anticyclonic, and nests of folded fila-

Table 5.1 *Observations of the polar hoods*

A. Visibility

1968–1971	Blue	SPH	LPL
1968–1971	0.9 μm (CH₄)	SPH	LPL
1971–1972	0.89 μm (CH₄)	SPH, NPH	LPL
1974, 1979	5 μm IR	SPH, NPH	Terrile & Beebe(*1979*), Beebe *et al.*(*1989*)
1976	Blue	SPH	Beebe *et al.*(*1989*)
1979–1981	Blue	Weak SPH?	Beebe *et al.*(*1989*)
1979	Blue, violet	SPH	Voyager
1986	0.72 μm (CH₄)	SPH	Carlson & Lutz(*1989*)
1986, 1989	Blue	Weak SPH	Dragesco, Akutsu (BAA)

B. Zenographic latitude of SPH north edge

1968/69	0.89 μm (CH₄)	–67.9 (±1.3)	Minton (*1972a*)
1969/70	0.89 μm (CH₄)	–69.4 (±1.2)	Minton (*1972a*)
1971	0.89 μm (CH₄)	–70.2 (±0.7)	Minton (*1972a*)
1971	Blue	–68.4 (±0.9)	NMSUO
1972	0.89 μm (CH₄)	–68.4 (±1.3)	Minton (*1972a*)
1979	Violet	–66.4 (±1.0)	Voyager (author's measurements)
Average:		–68.4	

SPH, South Polar Hood; NPH, North Polar Hood.

ments likely to be cyclonic – up to ≈65°N and S (Figs. 5.1&2, Fig.6.5, Fig.12.3&4). There was one white oval as far north as 73.5°N. These features were sparser and had lower contrast than those in the temperate latitudes, as if the polar regions were dimmed by haze.

The colour of the polar regions was grey in professional photos (LPL, 1969–72) and in the Voyager images, contrasting with the browner aspect of most other latitudes. The distinctive polarisation, infrared, and ultraviolet properties of the polar regions were described in Chapter 4.

The blue polar hoods

Although the polar regions are generally dark, photographs in some wavebands sometimes show bright polar caps or 'polar hoods', particularly over the south pole down to latitudes 66–69°S. The south polar hood is quite often seen in the highest-resolution blue photos (Table 5.1A), and was very distinct on the Voyager blue and violet photographs (Fig. 12.3 and Plate P19.1), It is never visible in green or red light. It is most distinct in the methane infrared absorption bands at 0.89 μm and 2–3 μm (Figs. 4.9–12). In these bands the south polar hood is remarkably bright, and a weaker north polar hood is also recorded.

However, the north polar hood has never been convincingly shown in blue light, and the south polar hood is certainly variable. For instance, it was not visible on the blue images from Pioneer 10 nor from large Earth-based telescopes in 1973.

The optical properties of the polar hood appear contradictory in terms of the paradigm that relates colours and altitudes over the rest of the planet. The hood is blue and warm, and yet highly polarising and at high altitude. As discussed in Chapter 4.7, it must be a high-altitude haze or coating which reflects and polarises in the blue and near-infrared, but is transparent in the middle infrared.

The latitude of the edge of the south polar hood is listed in Table 5.1B. R.B. Minton (*1972a*) noted a variation of a few degrees over a few years, in line with the tilt of the planet, such that the apparent width of the bright hood remained constant. Seasonal variations could perhaps produce such an effect. Voyager photos revealed irregularities in the edge of the hood and oblique bands within it (Fig. 5.1).

The magnetic polar hotspots

Both polar regions exhibit aurorae, detectable in the ultraviolet and at 3.5 μm in the infrared (Fig. 4.12C and Chapter 17). They may be responsible for several remarkable features of the polar stratosphere.

Infrared images at 7.8 μm, which show stratospheric thermal emission, reveal hotspots close to the north and south magnetic poles (Chapter 4.4). The north polar hotspot remained fixed at $\lambda_3 = 180 (\pm10°)$, 60°N ($\pm10°$), over several apparitions, within the auroral zone. The south polar hotspot was sometimes seen near $\lambda_3 \approx 0$, but was variable in intensity and rotation period (R. Beebe, G. Orton and R. West, *1989*). These hotspots are thought to be due to heating of the stratosphere by the large flux of charged particles descending in the aurorae.

The north polar hotspot shows increased infrared emission from methane and acetylene, which could be attributed to a warmer stratosphere. But there must also be different chemistry there, as traces of other hydrocarbons are detected there and nowhere else, while ethane emission is not increased (T. Kostiuk and colleagues, *1989*; Chapter 16.1).

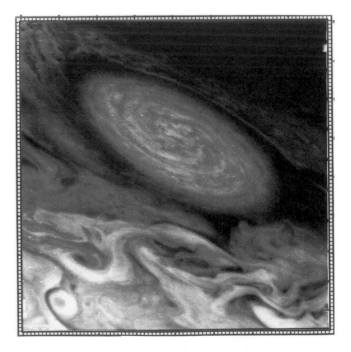

Fig. 5.2. Grand spiral in south polar region at about 60°S, 5000 km across. This is anticyclonic and lies in the putative S.S.S.S. Temperate domain. Photographed one day before Voyager 1 close approach. South is to top right. (Image 16373.12, green, unfiltered. NASA.)

East-west currents

The only Earth-based velocity measurements relate to the south polar region at 60°S, and are listed in Table 12.2. One was visual (two dark spots in 1911; $\Delta\lambda_2 = +7.3$). The other two were by NMSUO, for single small bright yellow spots in 1968 ($\Delta\lambda_2 = -22$) and 1970 ($\Delta\lambda_2 = -19.4$). The 1968 spot was remarkable in showing steady acceleration from $\Delta\lambda_2 = -9$ to $\Delta\lambda_2 = -34$, while it also moved in latitude from 58.6°S to 61.1°S ($\pm0.3°$) over 8 weeks. The NMSUO observers noted that the change in speed was not enough to conserve angular momentum. Instead, it could have been due to the spot drifting into a previously undetected prograding jetstream at ≥61°S.

Indeed, the Voyager images and correlation analysis (S.S. Limaye, *1986*; Fig. 3.3) did suggest the existence of one more domain between 53°S (the prograding S^3TBn jetstream) and 61°S (the next prograding jetstream). The images showed an anticyclonic light oval at 59°S with grand spiral structure (Fig. 5.2), and large cyclonic folded filaments spanning 53–57°S (e.g. Fig. 5.1). The speeds of these cyclonic features (measured by the author; Table 12.2B) ranged from $\Delta\lambda_2$ -3 to -13, and along with the NMSUO speeds these may represent a 'S.S.S.S. Temperate Current' for this domain. These typical domain structures suggested a retrograding jetstream at 57°S, and the Limaye analysis did indeed show one, albeit weak ($\Delta\lambda_2 = +14$).

In the north polar region, the Voyager data showed a broad, modestly retrograding jetstream at 61°N and a modestly prograding jetstream at 65°N (Limaye, *1986*; Fig. 3.3). Four features measured by the author in the north polar region were all tiny bright spots at about 61°N or 62°N (Figs. 6.4&5), and were all moving with the retrograding current, at $\Delta\lambda_2 = +10$ to +27 (N^5TBn).

Thus no spots have been tracked beyond 62°N or S, and all the speeds recorded in the polar regions may refer to distinct currents in domains similar to those of the high temperate latitudes.

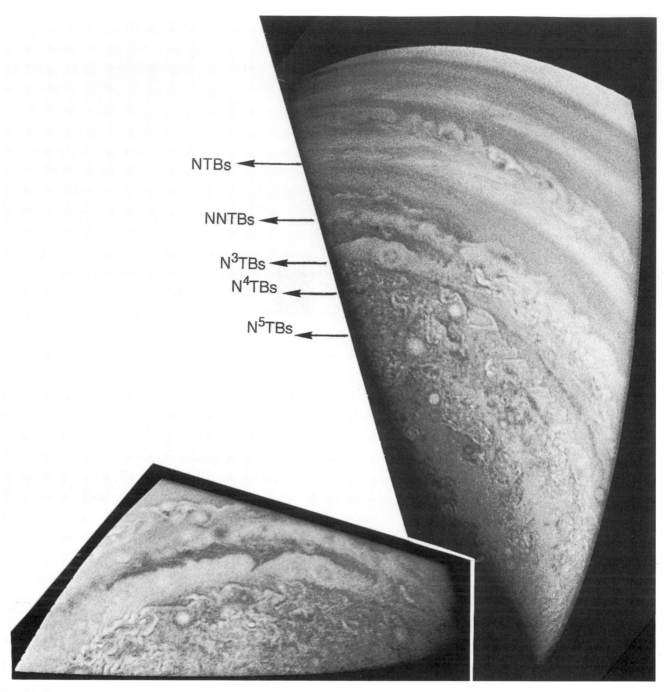

NTBs ←

NNTBs ←

N³TBs ←
N⁴TBs ←

N⁵TBs ←

Fig. 6.1. Pioneer 11 closeups of the far northern regions. Latitudes of pro-grading jetstreams are indicated approximately. The two images overlap, showing a dark segment of NNTB, and turbulent 'folded filament' regions in NTB, N³TB, and further north. (Images D2 and D4, combined red and blue. NASA.)

6: North North Temperate Regions (57°N to 35°N)

6.1 OVERVIEW

The chapters to come will each deal with a single dynamical domain; but in this chapter, we deal with three such domains which have many characteristics in common, not least the difficulty of observing them. They are the North North Temperate domain, containing the N.N. Temperate Belt (NNTB) and governed by the N.N. Temperate Current (NNTC); the N.N.N. Temperate domain (N³TB, N³TC); and the N.N.N.N. Temperate domain (N⁴TB, N⁴TC).

These regions are very variable in appearance, and the N³TB and N⁴TB are seldom present as continuous belts. Visual observers usually see nothing but grey shading extending from the pole to the NNTZ or NNTB, or even to the NTB. If higher-latitude belts are present, they generally appear only as short segments embedded in the general polar shading. So the boundaries of the high-latitude domains are defined by the jetstreams revealed by the Voyager spacecraft, which are not detectable from Earth, and belts are named accordingly (Chapter 3). In fact, there is evidence that the three domains under discussion are not as well-demarcated dynamically as those at lower latitudes, and the densely convoluted texture in the Voyager images suggests that features may interact between the domains. However, Earth-based observers can sometimes see small subtle markings, and occasionally the region seems to clear for a few years so that features can be seen which define the three characteristic currents.

There are five common problems which make it difficult to establish drifts in this region. First is the general obscurity of the region, which is affected by limb-darkening and foreshortening. Secondly, most of the features that exist seem to be small and of low contrast; in this region, unlike most of the planet, the dominant features are at or below the limit of resolution of most observers, even with large telescopes. Thirdly, very similar spots or streaks can belong to entirely different currents, in adjacent latitudes. Fourthly, many of the features change rapidly, over a matter of weeks. Fifthly, it seems that similar but transient belt segments tend to move through a region in succession ('serial behaviour'), so that it is difficult to keep track of individual features. The fourth and fifth points have often been suspected by the author, and were clearly demonstrated by several examples in the well-observed

apparitions of 1985–1989 (see below). It has not been possible to analyse 'serial behaviour' in any detail and one can only suspect that it may arise from belt segments changing as they pass some fixed region of disturbance, just as segments of STB have darkened or faded when passing the Great Red Spot (Chapter 11). Overall, the N.N. Temperate regions are not only less structured than their southern counterparts but also less stable.

Sometimes there appear to be fairly large dark markings, but these do not usually last long even if they appear to be well-formed, and they may be merely temporary alignments of belt segments passing in different currents. In 1986–1989 this was certainly so, and it seemed that light ovals rather than dark patches were the more stable features of the region.

The parameters of the three domains under discussion are laid out in Table 6.1.

This Table also demonstrates the curious displaced symmetry between the northern hemisphere domains and the southern hemisphere domains that are offset by one unit. Although there are some obvious differences, notably the greater speed of the South Temperate Current and the greater size of the three ovals in it, the overall correspondence is remarkable and unexplained. Perhaps all the correspondences might stem from the fact that the N.N.N. Temperate and S.S. Temperate domains happen to be rather narrow and lack perceptible retrograding jetstreams. Or perhaps the pattern is determined by the stronger jetstreams nearer the equator – the south equatorial current in the south but the unique NTBs jetstream in the north. Or perhaps there is a deeper connection – literally so, if the model of the atmosphere as a set of concentric rotating cylinders is correct (Chapter 15).

The boundaries of the belts

The measured latitudes of the belts in this region, along with the NTB, are shown in Fig. 6.2 and Table 6.2. The values for the NNTB and N³TB have been consistent over the years – surprisingly so, given the frequent discontinuities seen in these belts. The reader should be aware, though, that many of these measurements probably refer to belt segments that were not visible at all longitudes, and may represent a selected minority of well-defined aspects, particularly for the N³TB and N⁴TB; the number of indi-

Table 6.1 *Parameters of the N.N. and S.S. Temperate regions*

	North				South			
	Name	β"	Δλ₂ (Voyager)	Δλ₂ (Earth)	Name	β"	Δλ₂ (Voyager)	Δλ₂ (Earth)
Prograding jetstream (Uttermost jetstream readily detectable from Earth in spot outbreaks)*	NNTBs	35.6°N	−94	−77	STBn	27–29°S	−110	−109
Belt (Uttermost regular belt; subject to occasional fading of sectors)	NNTB				STB			
Retrograding jetstream	NNTBn	39.5°N	+31		STBs	32.6°S	+42	
Slow current (Subject to occasional 'invasions' by N³TC/SSTC)	NNTC			−0.3	STC			−15.0
Anticyclonic white ovals (number in 1979:)	(2)	41.2°N			(3)	33.2°S		
Prograding jetstream	N³TBs	43.0°N	−68		SSTBn	36.5°S	−88	
Belt (Not always present; may extend further poleward)	N³TB				SSTB			
Velocity minimum (No retrograding jetstream)	N³TBn	45.3°N	−4		SSTBs	39.8°S	−17	
Slow current (Rather fast)	N³TC			−15.2	SSTC			−25.1
Anticyclonic white ovals (number in 1979:)	(6–8)	46.4°N			(12)	40.5° S		
(number in 1987:)	(≥6)				(7)			
Prograding jetstream	N⁴TBs	48.2°N	−94		S³TBn	43.6°S	−125	
Belt (N⁴TB usually absent, S³TB usually in anomalous latitude)	N⁴TB				S³TB			
Retrograding jetstream (Modest speed)	N⁴TBn	51.6°N	+33		S³TBs	49.3°S	+17	
Slow current (formerly N. and S. Polar currents; slow, variable, rarely observed; uttermost current detectable from Earth)	N⁴TC			+1.4	S³TC			−8.3
Prograding jetstream (Uttermost jetstream clearly shown by Voyager)	N⁵TBs	56.6°N	−59		S⁴TBn	53.4°S	−129	

Jetstream data are from Limaye (*1986*) (see Table 3.2). Data are zenographic latitude (ß") and speed (Δλ₂, degrees per month).
* Other similarities between the NNTBs and STBn jetstreams (section 6.4) are the simultaneous motion of belt segments on both of them in 1987, and the fact that spots in the jetstream have been observed to vanish on reaching a long-lived dark feature in the adjacent zone ('NTZ Disturbance' or South Tropical Disturbance). However, there is no historical tendency for activity on the two jetstreams to coincide.

vidual measurements for these is much less than for other belts. The main change with time is that the early years included few examples of the true N³TB. Instead, from 1907 to 1916, there were consistent measurements of a belt further north that must now be called the N⁴TB. It seems possible that the N³TB and N⁴TB tend not to be present simultaneously, just as the SSTB and S³TB tend not to occur together (Chapter 12); but the N.N. temperate region is too obscure for any such correlation to be established at present.

Since the Voyager spacecraft established the pattern of jetstreams on the planet, the latitudes of belt edges now seem more significant than the latitudes of belt centres. Accurate measurements of the edges of these low-contrast features can only be obtained from high-resolution photographs, and the limited number of published values has been supplemented by the author's

measurements (Appendix 2). The histograms on Fig. 6.2 show the results. The values for the south edges of belts are clearly clustered about 35.8°N (NNTBs) and 42.7°N (N³TBs); these means are very close to the corresponding jetstreams as observed by Voyager. The values for the north edges of belts also look well clustered, but this is probably because they could only be measured when the belt appeared reasonably regular; there were years when the north edge was too indistinct to be measured, or obviously extended into another domain. Of the north edge values that could be used, the averages were 39.1°N (NNTBn) and 47.0°N (N³TBn), which do agree well with the boundaries expected from Voyager jetstream data. Further north, the measurements are few and widely scattered, with no sign of permanent belt features.

The averages exclude some anomalous belt segments that

Table 6.2 *Latitudes of the belts*

	(1906–1942)	(1950–1991)	(Overall)
North Temperate Belt:			
NTBs	+24.5 (±1.6; 8)	+23.9 (±0.8; 21)	+24.0 (±1.2; 29)
NTB(S)	–	+25.1 (±0.6; 13)	+25.1 (±0.6; 13)
NTBn	+30.6 (±1.9; 6)	+31.0 (±0.8; 15)	+30.9 (±1.2; 21)
N.N. Temperate Belt:			
NNTBs	–	+35.8 (±1.0; 18)	+35.8 (±1.0; 19)
NNTB	+38.1 (±1.0; 15)	+37.4 (±1.1; 23)	+37.7 (±1.1; 38)
NNTBn	–	+39.1 (±0.7; 12)	+39.1 (±0.7; 12)
N.N.N. Temperate Belt:			
N^3TBs	–	+42.7 (±1.0; 13)	+42.7 (±1.0; 13)
N^3TB	–	+44.9 (±1.0; 20)	+44.9 (±1.0; 20)
N^3TBn	–	+47.0 (±1.0; 6)	+47.0 (±1.0; 6)
N.N.N.N. Temperate Belt:			
N^4TB	+48.9 (±0.8; 5)*	-	+48.9 (±0.8; 5)

The table lists zenographic latitudes, given as the mean (±s.d.; N) of apparition means. (S.d. = standard deviation, N = number of apparitions). Data are those in Fig. 6.2. NTB(S) is south component; the north component is often ill-defined. Dates for some belts were more limited than the headings indicate, especially: *N^4TB (1907–1916)

clearly occupied the normal latitudes of a zone, which will be described in section 6.2.

White ovals

Distinct though tiny white ovals are seen occasionally, and shown more commonly on high-resolution photographs. The Voyager pictures showed two at 41°N (plus one or two less distinct light ovals), and 6 to 10 at 46°N (some being smaller than others)(Fig. 6.5). All the latitudes measured for these ovals are listed in Appendix 2. The values from the 1950s seem to be scattered, but the later values fall into four distinct clusters. In relation to the Voyager jetstreams, these clusters lie in homologous positions in the canonical latitudes of NNTZ (41.2°N), N^3TZ (46.4°N), N^4TZ (53.0°N), and N^5TZ (61.4°N). Although these latitudes do not usually show visible bright zones, they are anticyclonic relative to the jetstreams, and Voyager showed that the ovals too were anticyclonic, according to motions tracked around them or spiral structure within them (Fig. 6.8).

These bright ovals were about 4500 km across at 41°N and about 3000 km across further north, according to the Voyager images. They were actually circular, appearing oval only due to foreshortening. Most of them had dark collars. They were reported to be very bright in 0.89 μm (methane) images, indicating the high altitude of their white cores (R. Beebe, G. Orton and R. West, *1989*). Thus in every respect they resembled the more noticeable tiny white ovals in the SSTB at 40.5°S.

Although they quite often lie on the northern edges of the belts, these bright spots seem to be fixed in latitude regardless of the pattern of belts and zones around them. In the Voyager images, there was little belt/zone structure north of the NNTB; and in 1987/88, the 46°N bright spots appeared as notches in the south edge of a belt at some longitudes, but as sizable ovals breaking up the N^3TB at other longitudes.

These white ovals generally move with the characteristic current for their latitude, although the motion can be very variable. The 41°N ovals in 1986 moved with the N^3TC rather than the NNTC, and in the Voyager maps one such oval suddenly accelerated from NNTC ($\Delta\lambda_2 - -4$) to a N^3TC speed ($\Delta\lambda_2 = -16$) (Fig. 6.4). In 1988/89, several of these tiny, bright, dark-collared ovals appeared for a few months and their NNTC motion varied considerably; one swung from $\Delta\lambda_2 = -8$ to $\Delta\lambda_2 = +4$, while two others converged and may have merged. According to A. Sanchez-Lavega and J. Quesada (*1988*), the ovals at 47°N in 1975, about 3000 km across, also had very diverse speeds which may have been influenced by a prograding jetstream. The 62°N ovals in the Voyager maps moved with the retrograding stream in that domain.

Before 1973, there were very few records of the white ovals at 41°N, but several professional photographs instead showed very dark streaks or oblongs on the NNTBn edge (see below), which have not been so conspicuous recently.

The three slow currents

In view of the observational difficulties in this region, it was pleasing to discover that the three currents detected by visual observers correspond almost exactly to the three domains defined by the Voyager jetstreams. The NNTC is commonly observed, for features in the NNTB or NNTZ, but the N^3TC and N^4TC are only infrequently recorded.

The NNTC (formerly NNTC-A) is plotted in Fig. 6.3. Its mean speed over the last century has been $\Delta\lambda_2 = -0.3°/mth$ (±3.2).

The NNTC has a reputation for variability, and indeed single spots

[*continued on page 91*

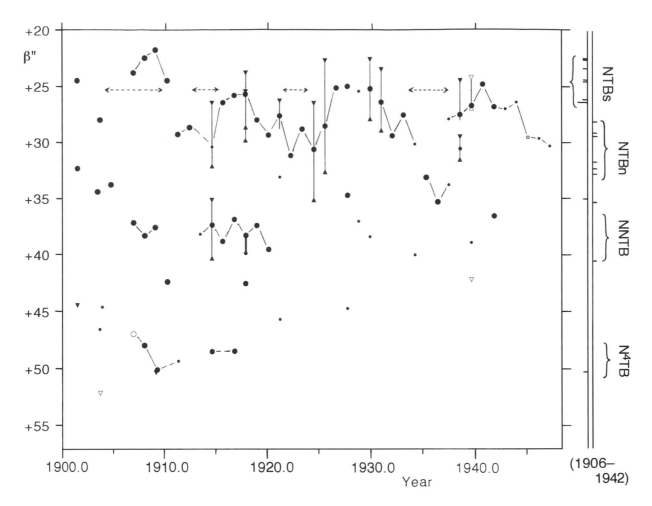

Fig. 6.2 (above and on facing page). Latitudes of the belts (NTB to N⁴TB). The chart shows apparition means, from sources listed in Appendices 2 and 3. When there are data from different sources for an apparition, they are either averaged or plotted separately. (Some segments were only present at certain longitudes.) Circles, belt centres; triangles, north and south edges of belts. Vertical lines are drawn to indicate the extent of the belts, for clarity; these are not error bars. Small symbols are single micrometer values, and open symbols are data of lower precision, which are not used for statistics. Dashed horizontal lines indicate times when NTB was faint or absent. At the sides are histograms of the latitudes of the belt edges. Open arrows at far right indicate latitudes of jetstreams from Voyager (Table 3.2); they agree closely with the visible belt edges. Statistics of these data are in Table 6.2.

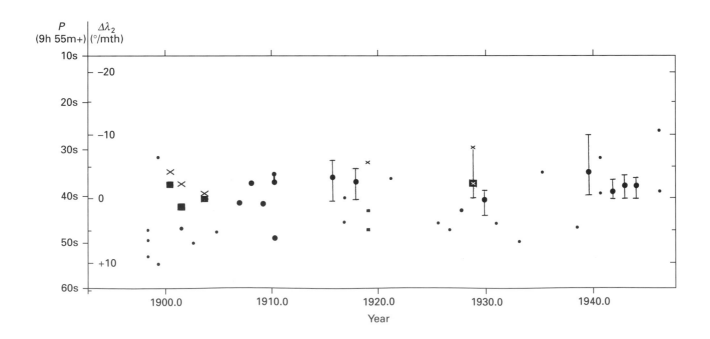

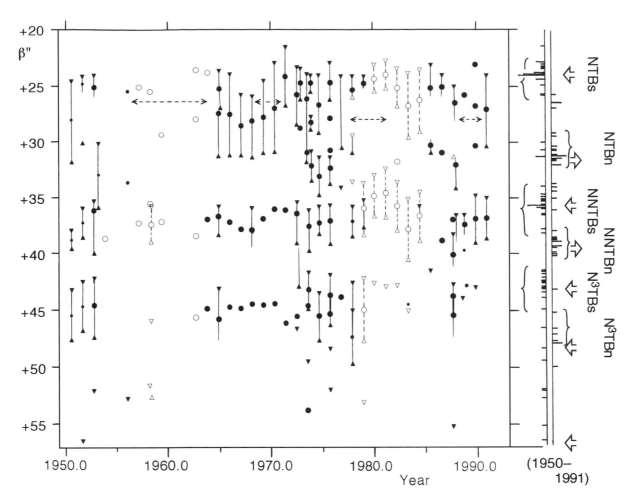

Fig. 6.3 (below and on facing page). Speed of the N.N. Temperate Current, 1898–1991. From data listed in Appendices 1 and 3. Speed is shown as P (rotation period), as $\Delta\lambda_2$ (degrees per month), and as u (m/s). Large symbols are apparition means of ≥ 3 features, and vertical bars indicate the range of speeds (not published before 1914). Small symbols represent single features. The mean NNTC speed is $\Delta\lambda_2 = -0.3°$/month (± 3.2, standard deviation of apparition means; omitting the infrequent speeds faster than -12 or slower than $+14$, which are attributed to N³TC and NTC respectively).

For 11 apparitions, light spots in NNTZ (X) are plotted separately from

dark features in NNTB (squares); note that the former usually drift faster. The histogram at right shows the speeds from just these 11 apparitions; ⊠, NNTZ light spots; ■, NNTB dark features. Small squares represent individual features (BAA in 7 apparitions); large squares, apparition means (Molesworth in 1900, 1901, 1903; SAF in 1984; the reliability of these values is uncertain but they agree well with the BAA individual features). The mean $\Delta\lambda_2$ is $+0.2$ for NNTB dark features and -3.0 for NNTZ light spots, a difference of $3.2°$/month. Speeds faster than $\Delta\lambda_2 = -12$, typical of N³TC, are omitted from statistics; the mean difference would be $4.9°$/month if they were included.

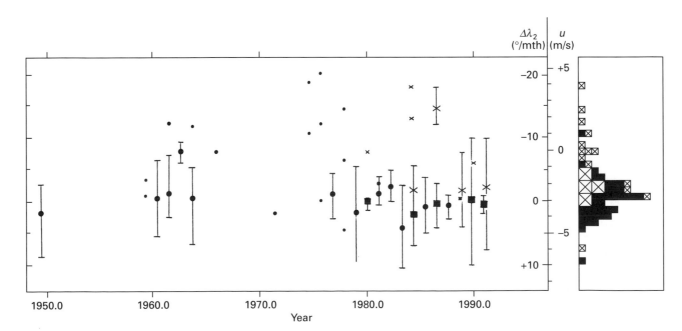

Table 6.3 *Records of N⁴TC and N³TC*

Appar'n	N⁴TC				N³TC				N³TC in NNTZ/NNTB				Source
	P	$\Delta\lambda_2$	N	lat.ß"	P	$\Delta\lambda_2$	N	lat.ß"	P	$\Delta\lambda_2$	N	lat.ß"	
1881	42.4s												Peek
1892	38.9s												Peek
1900	–				19.1s		4	43					Mole.
1902	44.3s		1	NPRs									Phil.
1903	*21.5*		*11*	*52**									*Mole.*
	42.0s		4	50–55									Denn.
1906/7	40.6s		2	NPR									Phil.
1908/9	44.6s		3	50									Phil.
1909/10	45.1s		4	NPRs									Hawks
1917/18	52.0s	+8.0	3	52	29.3s	–8.3	1	42.5					BAA
1918/19	42.0s	+1.0	1	–									BAA
1919/20	42.7s	+1.5	1	–									BAA
1927/28	43.9s	+2.5	4	NPR									BAA
1928/29	*23.9s*	*–12.2*	*2*	*NPR**	20.1s	–15.0	2	N³TB					BAA
1929/30	39.0s	–1.1	6	NPR	26.5s	–10.4	1	N³TB					BAA
1930/31	33.6s	–5.4	4	NPR									BAA
1931/32					10s	–22.3	1	NPRs					BAA
1939/40	40s	–0.2	1	NPRs									BAA
1941/42					19s	–15.6	5	N³TB					BAA
1942/43					24.0s	–12.3	2	N³TB?					BAA
1945					10s	–22.3	1	N³TB	*4s*	*–26.5*	*1*	*NNTB**	BAA
1946									*26s*	*–10.5*	*1*	*NNTB**	BAA
1951					17s	–17	?	–					Reese
1952/53	39.5s	–0.9	1	NPRs	23s	–12.7	5	N³TB					ALPO
					23s	–13.0	3	43					BAA
1961/62					19s	–16.1	8	N³TB	*24s*	*–12.0*	*2*	*NNTB**	ALPO
1962/63					22s	–13.8	13	46					ALPO
1963/64					15s	–18.7	7	44.5	*25s*	*–11.6*	*1*	*NNTB**	ALPO
1962–					24.1s	–12.1							
1970†	*11.5s*	*21.3*	*?*	*55.0**	23.1s	–12.9	?	45.9–					NMSUO
					11.6s	–21.2		42.8					
1974**					9.3s	–20.3	1	46.5	*21s*	*–14.7*	*2*	*37**	BAA, LPL
1975	*11s*	*–22*	*2*	*55.2**					*20s*	*–16*	*2*	*42**	BAA, SL&Q
1976/77					24.5s	–11.8	1	44					BAA
1977/78					27s	–9.8	1	47	*21s*	*–14.2*	*1*	*38**	BAA
1978/79					13s	–20	1	42*	*13s*	*–20*	*1*	*42**	BAA
					24s	–12	1	45.2	*19s*	*–16*	*1*	*41.0**	Voyager
	(V.diverse)			54–56	11.2s	–21.5	5	46.5					Voyager
1985									*13s*	*–20*	*1*	*41**	BAA
									≈6s	*≈–26*	*3*	*NNTZ**	SAF
1986					23.6s	–12.4	3	N³TB	*21s*	*–14.7*	*3*	*41.9**	BAA
1987/88	43.6s	+2.1	9	52	18s	–16.3	9	43–46	*20s*	*–15.0*	*1*	*40**	BAA
1988/89	45.0s	+3.1	8	51	19s	–16.2	6	45,48					BAA
1989/90					24s	–11.9	2	45.8					BAA
1990/91	42.7s	+1.5	1	NPR									BAA
Average	42.4s	+1.3	(N=12)		20.2s	–15.0	(N=9)						
1900–46	(±4.4s)	(±3.2)			(±6.6s)	(±4.8)							
Average	42.7s	+1.5	(N=4)		19.6s	–15.2	(N=16)						
1951–91	(±2.3s)	(±1.7)			(±5.2s)	(±3.5)							
Average	42.5s	+1.4	(N=16)		19.8s	–15.2	(N=25)		19.0s	–15.9	(N=11)		
1900–91	(±3.9s)	(±2.8)			(±5.6s)	(±4.1)			(±7.0s)	(±5.1)			

[Unweighted averages of apparition means (± standard deviation).]

Notes to Table 6.3:

P = Rotation period (9h 55m+). $\Delta\lambda_2$ = drift (deg/mth). N = number of spots. ß" = zenographic latitude. (Most latitude values are estimates, except when a decimal place is given, and after 1980, which are measured.) Sources are listed in Appendix 3. Early observers include Molesworth, Phillips, and Denning. Voyager data in 1978/79 are from Fig. 6.4.

Among early determinations, A.S. Williams' value for 1888 has been omitted; he recorded dark spots aligned on three northern belts, all moving together, which was probably an optical illusion. Scriven Bolton's values for the early 1900s have been omitted for the same reason, and some by Molesworth and Phillips have been omitted because the intervals between observations were excessively long. (See Appendix 1 for further discussion.)

* N³TC speeds in anomalous latitude, shown in italics. Of the values in N⁴TB latitudes, the early ones are included in the N³TC statistics because the latitudes may have been imprecise, but later ones were definitely anomalous: $\Delta\lambda_2$ = –21 to –25 at 55°N for spots in the 1960s (NMSUO), in 1975 (Sanchez-Lavega and Quesada, *1988*), in 1979 (N⁴TC spot 6 in Voyager data, Fig. 6.4), and possibly in 1988/89 (suspected by BAA, not listed). The 'N³TC' values in NNTB/NNTZ latitudes are tabulated separately in the right-hand section. Again, some (but not all) may actually have been in the N³TB domain due to imprecision in latitudes. This was the case for the white spot tracked at 42°N (NNTZ) by the BAA in 1978/79; Voyager images showed that it was actually a cyclonic oval in the N³TB domain.

† NMSUO values deduced from summary by Reese (*1971a*); dates of original observations were not given.

**In 1974, Minton (LPL) recorded a white spot in N³TB; his period and drift values are slightly discordant. The BAA recorded N³TC for two dark features of NNTB.

can show marked variations as mentioned above. The exact limits of the current cannot be absolutely defined because occasional features in this domain move with slower speeds typical of the NTC or faster speeds typical of the N³TC. However, given that we have excluded these clearly unusual speeds, the NNTC is actually no more variable than other northern-hemisphere currents (Table 3.3).

Exceptionally slow speeds on the NNTB occurred in 1934, when one dark spot on the NNTB(N) (40°N) and one on the main NNTB showed $\Delta\lambda_2$ = +17.5 and +16 – speeds typical of the NTC. At the other extreme, the p. end of a darker segment of NNTB in 1945 moved at $\Delta\lambda_2$ = -26.5 (section 6.4).

B.M. Peek noted that light spots in the NNTZ tend to drift slightly faster than dark features in the NNTB, and this has been confirmed in several apparitions recently. For the 11 apparitions where both light and dark features were reported, they are plotted separately in Fig. 6.3 and in a histogram on the right-hand side. The NNTZ light spots have more diverse speeds, but on average they do drift 3.2°/mth faster than the NNTB dark features – even if the fastest ones, with full N³TC speeds, are excluded. This is the only domain where the normal anticyclonic ovals of the zone commonly show a different drift from features in the belt. This differential drift has sometimes been observed for adjacent spots, for example in 1979/80 and 1989/90 (BAA reports). There was also an example in early 1979, during the Voyager encounter, when dark patches of NNTB at 39°N had $\Delta\lambda_2$ = +1 to +11 (BAA) while an adjacent white oval at 41°N had $\Delta\lambda_2$ = +1 (Voyager). Indeed, this velocity gradient appeared to extend into the N³TC domain, as the same Voyager photos (Figs. 6.4&5) showed a cyclonic spiral at 45°N with $\Delta\lambda_2$ = -12, while the white ovals at 46–47°N were moving at -16 to -28. The appearance of a velocity gradient spanning two domains may be merely a coincidence, but there is other evidence that the NNTC-N³TC boundary is often blurred.

Some NNTZ light spots (and occasional NNTB features) do move with full N³TC speeds, $\Delta\lambda_2 \approx$ -16 (Table 6.3, right-hand section). Some of these cases could be due to imprecision of latitude measurement, but one cannot explain them all away in this manner, particularly a Voyager measurement. It is not clear whether there is really intermingling between the NNTC and N³TC, or whether these events represent specific 'invasions' as will be described in the South Temperate region.

All the observations of the N³TC and N⁴TC are listed in Table 6.3. These are all the drifts ever recorded north of 42°N, with the exception of a few whose reliability is doubtful. In fact, the slow current north of 48°N was named the 'North Polar Current' by T.E.R. Phillips and by Peek; however, all the measured latitudes for features in it are south of 57°N, between Voyager's N⁴TBs and N⁵TBs jetstreams, so 'N⁴TC' is the logical name for the current. When the latitude of features moving in this current was not measured, they were described as being in the NPR, which certainly covers these latitudes as far as visual observers are concerned. Almost all the spots tracked were dark features, although in recent years the tiny light ovals have been found to be more reliable tracers.

The N⁴TC was apparently observed in 1881 and 1892 (even before the NNTC), but the discovery of the N³TC, and its separation from the N⁴TC and NNTC, were due to P.B. Molesworth's detailed observations from Ceylon in the apparitions 1900–1903. In those years, an exceptional number of far northern spots was visible, and Molesworth established numerous drift rates which had a mean $\Delta\lambda_2$ of -16 (N³TC) in 1900 and -1 (N⁴TC) in 1901. In retrospect his achievement may seem surprising. Peek criticised Molesworth's drift charts for being overcrowded and thus at risk of confusion between spots, and indeed the large number of spots tracked (21 in 1901), sometimes with long intervals between observations, does raise some concern on this score. Moreover, in 1903, Molesworth obtained $\Delta\lambda_2$ = -14 (N³TC) for his 11 spots while W.F. Denning obtained $\Delta\lambda_2$ = +1 (N⁴TC) for his four, all being at 50°-55°N; no-one reported both currents in the same apparition. Even when these two currents were first defined and tabulated in the BAA Memoir for 1929/30, the evidence for their latitudinal separation was incomplete – assignment often being made by drift alone in the absence of latitude measurements. But in spite of these uncertainties, Molesworth's results have stood the test of time, and since the 1960s the N³TC and N⁴TC have repeatedly been observed in their characteristic domains.

The only discrepant feature of Table 6.4 is that fast drifts typical of the N³TC have several times been recorded in the N.N. Temperate domain (see above) and in the N.N.N.N. Temperate domain. The latter discrepancy is hard to prove, at such a high latitude, but the table shows two definite records of N³TC speeds at 55°N from professional photography (in the 1960s and 1975), and several suspected instances.

In an attempt to clarify these currents, the author has measured the drifts of spots on the Voyager 1 strip-maps (Figs. 6.4&5). The N³TC clearly controlled the chain of tiny white ovals at 46°-47°N and a cyclonic spiral at 45°N (moving slightly slower). However, three features in the expected N⁴TC latitudes had extremely diverse speeds, with one moving even faster than the N³TC and one suddenly acquiring a retrograding drift. These spots were very small and may perhaps have been caught up by the surrounding jetstreams, or diverted by local circulations. The somewhat steadier N⁴TC recorded from Earth in 1987/88 was marked by large light ovals which may themselves have imposed an unusual degree of order on the region (Fig. 6.6).

6.2 VISUAL AND PHOTOGRAPHIC OBSERVATIONS

The N. N. Temperate Belt

The NNTB is usually a distinct belt, although uneven in latitude and intensity, or it may merge with the north polar shading. Sometimes there is only a narrow south component which may even lie in the latitudes of the NTZ (see next chapter). For example, from 1984 to 1986 there was a constant arrangement in one sector, consisting of a narrow NNTB(S) carrying tiny dark spots or streaks at NNTBs jetstream speeds, alongside a broad clear NNTZ, whose f. end ran into a broader NNTB or a complex shaded region.

It is difficult to observe the colour of such a minor belt, but it seems that the NNTB really does adopt a variety of colours. It was rich brown or reddish-brown in 1907/8, 1946, and 1989/90; but at other times it has been plain grey. During the Voyager encounters, when it was the main northern belt all round the planet, the author saw it as mostly grey but with some brown segments.

Markings on the belt are commonly just dark patches or discontinuities, but at high resolution, as described above, small white ovals or dark bars may occasionally be seen on the NNTB north edge. From 1941 to 1943, when most of the belt was very faint, the best observations resolved two segments of a very narrow, very dark north component, which persisted through two apparitions with $\Delta\lambda_2$ between 0 and -4 (NNTC).

Distinct, very dark bars were photographed alongside a clear NNTZ in 1939, 1962, 1964, 1967, and 1968 (the main NNTB being extremely faint in the last two apparitions). They were at 38.6°N (see Appendix 2) and so it is likely that they were cyclonic 'barges', like those sometimes seen on the NEBn or NTBn. Similar dark streaks or belt segments at the same latitude, though less con-

spicuous because they were embedded in the NNTB, were photographed in the 1970s – and also viewed by Pioneer (Fig. 6.1) and Voyager (Fig. 6.7).

Sometimes the NNTB is very faint, as in 1923–1926; in 1924 it disappeared completely. It was also very faint at most longitudes in 1930–1934, 1941–1943, and 1967, and in some other apparitions. The revival of the belt is usually a quiet affair, as a few well-documented examples show.

1909/10: At first, NNTB was scarcely visible; then a series of dark condensations appeared at three separate longitudes; then it became a broad diffuse belt, light brown; the latitude of 42°N suggests that it covered NNTZ. At the same time the NTB also revived quietly; was there any connection?

1975: NNTB initially consisted only of a narrow NNTB(S) at 37°N, but a section of narrow but dark NNTB(N) developed gradually in the bright NNTZ; at first its ends were indistinct but the p. end became continuous with NNTB(S) at an inflexion moving with NNTC, while the f. end became a sharp step-down to the N³TB adjacent to a tiny bright oval at 41°N. In 1976/77, the most conspicuous belt segments in the region were mergers of NNTB and N³TB, moving with NNTC.

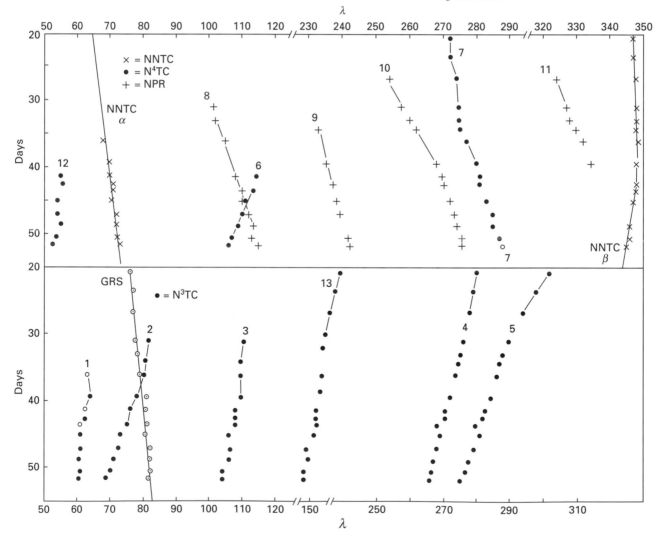

Fig. 6.4. Drifts of spots measured from Voyager strip-maps. Voyager 1 maps produced by the Voyager Imaging Team were measured by the author. The longitude scale is arbitrary and differs from λ_2 by 6°/month (established by measurements on other features that were tracked visually). The time scale is in days from 1979 Jan. 1 (close encounter was on day 64). Upper panel: x, white ovals in NNTC, 41°N; ●, white spots in N⁴TC, 54–56°N; +, tiny white spots further north, 62°N. Lower panel: ●, white spots in N³TC; no.1 was a cyclonic spiral at 45°N, the others were anticyclonic ovals at 46–47°N. Some of the features tracked are labelled on Fig. 6.5. Deduced speeds were as follows ($\Delta\lambda_2$ in °/month). NNTC: α, +1; ß, $-4 \rightarrow -16$. N³TC: no.1, -12; 2, -28; 3, -16; 4, -20.5; 5, -27; 13, -16. N⁴TC: 6, -31; 7, +1.5 $\rightarrow$ +14; 12, -13. NPR: 8, +14; 9, +10; 10, $+27 \rightarrow$ +12; 11, +19.

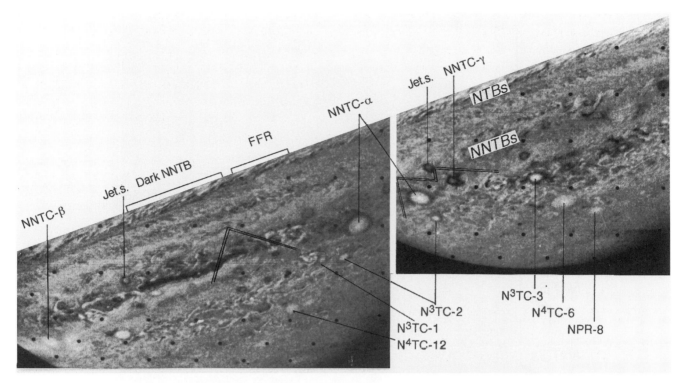

Fig. 6.5. Views of the northern hemisphere from Voyager 1, 14 days before close encounter. The NTBs and NNTBs jetstreams are indicated. Labels above indicate features of the NNTB, including jetstream spots (jet.s.), a cyclonic folded filamentary region (FFR), a cyclonic dark segment, two anticyclonic white ovals (NNTC-α, ß), and an anticyclonic dark oval that was tracked from Earth (NNTC-γ). Labels below indicate some of the features tracked in Fig. 6.4, mostly anticyclonic white ovals although N³TC-1 was a cyclonic spiral. Corner outlines indicate the approximate positions of Figs. 6.7 (left) and 6.8 (right), 11 days later (by which time the prograding drifts of N³TC-3, N⁴TC-6, and a NNTBs jetstream spot, had brought them into the frame of Fig. 6.8). The same region can be seen on many published Voyager images as oval α was at the same longitude as the Great Red Spot. (Images 16005.01 and 15982.42, orange, filtered. NASA.)

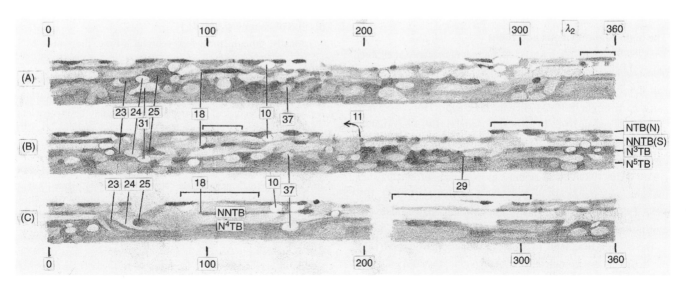

Fig. 6.6. Strip-maps of N.N. Temperate region in 1987, from amateur photographs. (A) Aug. 14–23; (B) Aug. 30 – Sep. 3; (C) Sep. 6–8. Sectors marked with a line above were filled in from lower-resolution drawings. Numbered features were tracked, as follows: no.10, NTC; 11, NNTBs jetstream; 18, NNTC; 23,24,25,29, N³TC; 31,37, N⁴TC. (From BAA report.)

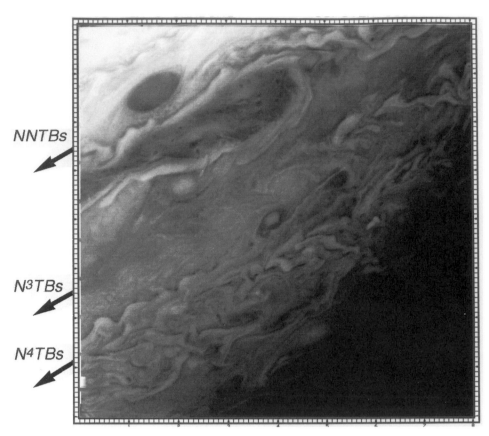

NNTBs

N3TBs

N4TBs

Fig. 6.7. Voyager 1 close-up of NNTB. This shows a dark segment of NNTB with a distinct cyclonic circulation around its f. end. A NNTBs jetstream spot is passing along its S. edge. Further north is a vast area of cyclonic folded fila-ments. (See Fig. 6.5 for location. Image 16314.55, violet, unfiltered. NASA.)

Figs. 6.7, 8, & 9 were all taken 3 days before close encounter and are at roughly the same scale. Jetstream latitudes are indicated approximately.

1990: in 1988/89, half the length of the NNTB had faded away at the same time as the NTB; reviving in 1989/90 along with the NTB(N) (see next chapter).

The higher-latitude belts and zones

The NNTZ is quite often observed, but not necessarily at all longi-tudes; for example, in 1979 and 1980/81, it was present around just half the planet. Brighter patches are sometimes seen in it. It was particularly bright in 1914 (when the NTZ and NTropZ were heav-ily shaded) and in 1975. In 1974 R.B. Minton (LPL) reported it to be 'light reddish' in filter photos, as was the NTZ. The author has often seen the NNTZ as slightly yellowish in recent years (1978–1980; 1989/90), in contrast to its greyer surroundings, and the same appearance has been recorded previously (1908/9, 1913, 1952, 1957/58, 1962/63). This yellowish colour may be more com-mon as the region is so difficult to observe clearly.

The N³TB is detectable in perhaps half the apparitions on record, but is usually not complete. (For example, it was seen only occa-sionally in 1973 and 1974 although it was quite well-defined in 1973 professional photographs. But in 1975 it was a distinct, con-tinuous belt while the NNTB was absent, and the N³TB remained a continuous belt until 1978, accompanied by fragments of N⁴TB.) Often, though, the belts in these high latitudes do not fit into a regu-lar pattern.

Further north, a dusky N³TZ and perhaps a streak of a higher-lat-itude belt may be visible, but there is usually little to see.

Episodes of enhanced detail

Occasionally, though, for a few years at a time, the whole N.N. Temperate region displays more detail. This could be because a veil has cleared, and/or because there really are more large fea-tures. The rest of this section will describe those episodes which have been well recorded; however, the record may be incomplete, as the following accounts (from BAA reports) all depended on high-resolution observations.

1900–1903: The currents were discovered in these years, mainly thanks to dark spots in the southern part of the 'North Polar Region' (Plate P4.9). In 1903 there were dusky spots or streaks on N³TB and N⁴TB, with wisps around them; some remained in 1904. But these were not prominent; the region was still generally 'faint and nebulous' to Molesworth.

1905/06: According to Denning (in *MNRAS*): 'There were signs of consid-erable disturbance in the region far north, about 40–65°N, and extensive irregularities in the N³TB and North Polar Cap [NPR] were visible. These consisted of large white spots with very dark areas adjoining.' Other observers also saw them, along with large white spots in the NNTZ, while the NPR was decidedly dark.

1941/42: Distinct N³TB, north of narrow NNTB with jetstream spots; N³TC was shown by 4 dark spots and a gap in N³TB. In 1942/43, a

'N³TB(S) dark streak' was a string of 'tiny knobs' and elsewhere there was 'a mass of fine detail' in the N³TB (F.J. Hargreaves).

1985–1989: Dusky shading, with irregular detail down to the limit of resolution, developed across a sector of NNTZ and NTZ in 1985. In 1986, features in a generally dusky sector included: conspicuous though variable segments of N³TB; a conspicuous but transient merged NNTB-N³TB; and later a persistent NNTB(N) dark bar. There were also at least 6 small white ovals at 42°N moving with the N³TC. Some of them were at the ends of segments of NNTB(N) or of N³TB, but they did not maintain a fixed relationship with such belt segments.

In 1987/88, there was even more structure (Fig. 6.6). It was surprising and gratifying to find that all the enigmatic shapes represented discrete features which were moving on the historically-defined currents, regardless of the surrounding belt patterns. There were assorted features in the NNTC (40°N), and bright ovals or bays in the N³TC (45°N) and N⁴TC (52°N). The N⁴TC ovals were spaced ≈20–30° apart in a semiregular series around the whole planet (Fig.6.6), and were so large (typically ≈ 9000 km across, 11° x 7°) that they touched the N³TC ovals. At some longitudes, there were segments of N³TB or N⁴TB or N⁵TB at various latitudes, interrupted by light bays or ovals; in one sector (possibly the same dusky sector as the previous year), everything north of the NTB [*sic*] constituted a bizarre field of irregular patches and shadings. Some light ovals were centred as far north as 60–62°N. The more conspicuous dark patches changed rapidly (within weeks) and were generally formed as irregular dark spaces between differentially drifting light ovals in N³TC and N⁴TC.

By 1988/89, these regions were not as obviously disturbed. Light ovals were still present but they seemed more subdued; there was also a set of small dark blocks at 51°N in the N⁴TC. Then the region gradually sank back into its usual murkiness.

6.3 SPACECRAFT OBSERVATIONS

Many of the spots and currents recorded by Voyager have already been described. What remains is to describe the fine texture of the region, which is dominated by turbulent cyclonic regions consisting of 'folded filaments' in the canonical latitudes of the NNTB, N³TB, and N⁴TB.

The baroque loops, swirls, and scalloppings of the region were first revealed by Pioneer 11 (Fig.6.1), and shown in more detail by Voyager (Figs. 6.7–9). The region contained not only anticyclonic white ovals (section 6.1) but also, according to the Voyager team:

> Surrounding these features are large areas of puffy unorganised clouds and other regions of folded ribbon-like clouds. A zonally elongated cyclonic flow stretching tens of degrees in longitude (40°N)[*NNTB*] is observed.... Long currents are observed around these regions..... These recirculating cyclonic currents are similar to some of the ribbon-like flow patterns in the southern hemisphere *[e.g. the STB Fade]*, although it is interesting that they are not associated with alternate anticyclonic spots as in the south. At least one small (2000 km diameter) feature (45°N, 55°W) showed very rapid cyclonic rotation *[= N³TC spiral no.1 in Figs. 6.4&5]*. [Mitchell *et al., 1979; notes in italics by the author.*]

These cyclonic circulations in the NNTB, centred at 38.5°N, adopted several forms. Some were complex folded-filament regions (FFR); others were plainer closed regions, whose interiors ranged from light (one simple oval) to dark (some long belt seg-

ments, e.g. Fig. 6.7). During the Voyager 1 approach in 1979 February, one of the cyclonic FFRs in the NNTB seemed to be darkening to form a similar short segment; this was one of several dark patches which the author had observed to form within the NNTB in 1979 January (Plate P17). At their p. ends, some cyclonic regions were drawn into long turbulent oblique strips; NNTBs jet-stream spots were appearing from one of these (see below).

The Voyager colour closeup (Plate P20.2) shows a large cyclonic FFR in the NNTB. At the cusps of the tiny whorls in it, there are diffuse red patches that appear to be spread across the underlying white and grey filaments.

In among the turbulence, several tiny dark spots could be seen running along the various jetstreams. Many of the appearances and interactions of spots in the Voyager 2 movies[1] were in this region. All the new spots, whether bright or dark, appeared from cyclonic FFRs. Other events included:

> *Appearance event 6:* An anticyclonic dark-haloed white spot condensed from a cyclonic FFR, and moved at $\Delta\lambda_2 = +30$, i.e. on the NNTBn jetstream.
>
> *Interaction event 1:* A small fragment from a FFR was whirled half way round a nearby anticyclonic spot.
>
> *Interaction event 11:* An anticyclonic bright oval prograding on N³TBs came close to a cyclonic oval in the NNTB, which deflected it Sf. onto the NNTBn; there it moved away in the retrograding direction.
>
> *Interaction event 7:* An anticyclonic spot on the NNTBn jet-stream ($\Delta\lambda_2 = +30$) overtook an anticyclonic dark-haloed white oval that had been formed by merger of two such spots.

There were also several instances of small spots on the N³TBs jet-stream passing other spots without merging.

Lightning

Finally, these were the latitudes covered by the Voyager 1 long exposures of the dark side, which revealed jovian lightning flashes. All of the flashes were in the turbulent cyclonic latitudes, with 82% being at 49–53°N ('N⁴TB'), while a few were at 45°N ('N³TB') and 57–60°N ('N⁵TB') (Chapter 17.1).

6.4 THE NNTBs PROGRADING JETSTREAM

On the south edge of the NNTB, outbreaks of tiny dark spots have occasionally been recorded, whose latitude (35–36°N) and rapidly prograding motion are the same as those of the jetstream observed by Voyager. The data are summarised in Table 6.4.

The NNTBs jetstream was discovered in 1929/30. The rapid motion of the spots was first noticed by Hargeaves and was designated NNTC-B by Peek. Two similar-looking spots had been recorded on drawings in 1926, without drift determinations. Antoniadi's fine drawings in 1928 (Plate P6) also appear to show some, possibly representing minor activity such as has been seen in recent years. It was notable that the outbreaks of 1926, 1929, and

---

[1] These events were described by MacLow & Ingersoll (*1986*); their event numbering is used.

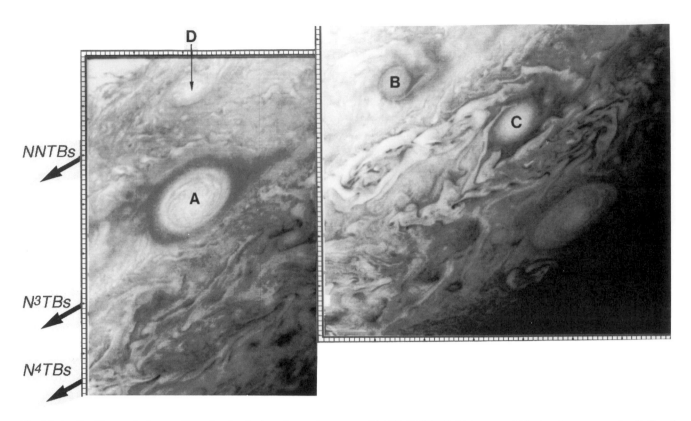

Fig. 6.8. Another Voyager 1 close-up. Several anticyclonic ovals are shown (labelled in Fig. 6.5): A, NNTC-α (41°N); B, NNTC-γ; C, N³TC-3 (46.5°N); north of it, N⁴TC-6 (55.5°N – a grand spiral with remarkably rapid drift); D, NNTBs jetstream spot. The two images are not well aligned because they were taken 25 min apart. (Images 16316.12&43, orange. NASA.)

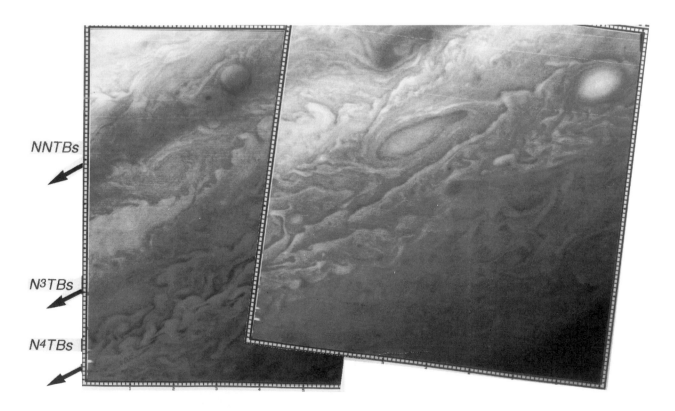

Fig. 6.9. Voyager 2 close-up. NNTB is absent except for a diffuse dusky NNTB(S), which carries a jetstream spot. Note weird sheets of cloud in NNTZ, a cyclonic 'fish' swimming in N³TB latitude, and an anticyclonic N³TBn white oval following it (at top right). The two images are not well aligned because they were taken 25 min apart. (Images 20572.32 and 20573.03, violet, unfiltered. NASA.)

1940 onwards, all coincided with similar outbreaks on the NTBs jetstream.

1929/30: The first spot appeared in 1929 October, and over a few months spots proliferated more than half way round the planet. They appeared as small dark projections from the NNTBs edge, but in the best seeing they were suspected of being detached spots within the NTZ. They had all disappeared by the next apparition.

1940–1945: The next outbreak began in 1940 September-October, when three spots were tracked. From 1941 to 1943 the spots were abundant, $\geq 20°$ apart, mostly moving with similar speeds (e.g. Plate P8.3). The NNTB was often very faint, except for narrow dark segments in its north half; the jetstream spots showed no tendency to arise near these segments nor to interact with them as they passed.

During 1942 and 1943, there was another, slightly larger dark spot on NNTBs with an unprecedented steady motion that persisted for two apparitions: $\Delta\lambda_2 = -16$ to -18. Two NNTBs jetstream spots apparently disappeared after conjunction with it; a third jetstream spot may possibly have survived its encounter.

In 1944/45, there was another NNTBs dark spot with $\Delta\lambda_2 = -29$ (Plate P9.3), very like the one of 1942–1943, and the final spot in the jetstream outbreak did not survive an encounter with it. There was also a segment of darker NNTB whose p. end moved with $\Delta\lambda_2 = -26.5$. Was this unique drift an after-effect of the NNTBs activity, like similar intermediate drifts in the NTB (Chapter 7.3)?

Substantial variations in speeds, which were documented in later outbreaks, were not reported for the outbreaks before 1945, but in some of those years the intervals between observations for some spots were rather long; although there were enough well-observed spots to establish the general flow of the jetstream, questions might be raised about a few spots.

No further outbreaks were reported for 20 years. This may have been partly because of poorer coverage. A photograph from Mt. Palomar in 1951 showed tiny spots on the NNTBs but they would have been too small to detect visually.

But from 1965 onwards, when professional and then amateur photography was undertaken in earnest, activity has been reported more frequently. Some of these episodes involved spots too small for most visual observers, so the definition of an 'outbreak' became somewhat blurred. The episodes of 1968–1970, 1972, 1978–1979, and 1988–1991 were substantial outbreaks, even though the first two were only recorded photographically. The detailed observations of these four outbreaks, including views from Voyager, revealed important new aspects. First, the spots were appearing around particular longitudes. Second, their speeds were definitely variable. Third, they tended to appear between the five-yearly NTBs jetstream outbreaks, not simultaneously with them as before.

1965/66 (NMSUO): Two spots were recorded between 1965 December and 1966 February.

1968–1970 (NMSUO): The next NNTBs jetstream spot recorded was an isolated one which seemed to oscillate. From $\Delta\lambda_2 = -75$ in 1968 January, it accelerated in February to -94, then decelerated again in March, as if oscillating with period 66 days and amplitude 3°. The latitude did not vary by more than 0.6°.

Another spot appeared in 1968 December, followed by large numbers in 1969 and 1970. Virtually nothing was left in 1971.

Table 6.4. *Records of the NNTBs jetstream.*

Apparition	$\Delta\lambda_2$	(range)	(N)	Lat. ß"	Source
1929/30	−78	(−66 to −92)	(7)		BAA
1940/41	−72	(−67 to −80)	(3)		BAA
1941/42	−77	(−68 to −82)	(6)+		BAA
1942/43	−80	(−74 to −79)	(5)		
		(and −100)	(1)+		BAA
1943/44	−79		(1)*		BAA
1944/45	(−80)		(1)+		BAA
1965/66	−80		(1)*	+35.1	NMSUO
1967/68	−85	(oscil.)	(1)	+35.5	NMSUO,LPL
1968/69	−79		(12)		NMSUO,LPL
1970	−79		(9)		NMSUO,LPL
1972	−74**		(5)*		LPL
1978/79	−74	(−59 to −91)	(6)	+35.2††	BAA
Voyager	−70	(−57 to −80)	(5)	+34	MacLow & Ingersoll (*1986*)
1984	−84	(−80,−88)	(2)*	+35.7††	BAA
1985	(−75)		(1)	+35.8††	BAA
1986	−83		(4)	+35.3	BAA
1987/88	−76	(−68 to −86)	(6)	+36	BAA
1988/89	−69	(−42 to −84)	(16)	+36	BAA
1989/90	−78	(−67 to −87)	(7)+	+34.7	BAA
1990/91	−73	(−61 to −82)	(9)+	+34.9	BAA
Average:	−77		(N=18)	+35.4	
	(±4.6)			(±0.5)	
Voyager:	−94		(global)	+35.6	Limaye(*1986*)

* More spots were present but were not tracked.
+ Plus single slower–moving spots: $\Delta\lambda_2 = -15.7$ (1941/42), -15.6 (1942/43), -29 (1944/45), -44 and -27 (1989/90), -45 and -38 accelerating to -60 (1990/91).
**Mean speed given by Minton (LPL) was -49, range -21 to -75, but Reese suggested that identifications should be revised to give a mean of -78, range -70 to -87. The best–observed pair had $\Delta\lambda_2 = -74$. However, speeds certainly varied in 1988–1990, so perhaps the question of the 1972 speeds must be left open.
††Author's measurements on photos.

These spots all first appeared in a restricted 65° span of longitudes, centred on $\lambda_2 \approx 60$ in 1969 and $\lambda_2 \approx 0$ in 1970. E.J. Reese (NMSUO) deduced a speed of $\Delta\lambda_2 = -5$ for the source region. R.B. Minton (*1973*)(LPL) connected these observations with those of 1972; the same source region, still with $\Delta\lambda_2 = -5$, could account for them all. He suggested that it was a subsurface source, but the speed would not be unusual for a NNTB feature moving with the NNTC, like the source regions in 1979 and 1988/89. Minton also pointed out that the NNTB was discontinuous during the 1968 and 1972 outbreaks, as in 1941–1943 (and also as in 1988/89); but none of the visible features coincided with the source region for the spots.

1972 (Minton, 1973; LPL): The typical dark spots were seen again throughout the 1972 apparition (April to October). Again they were born in a restricted longitude range, λ_2 180–280. A little way downstream was a dark step from NTB(N) onto NNTBs, retrograding with the North Temperate Current (see next chapter), and the jetstream spots all disappeared at or before their encounter with the step.

1974–1977? (minor activity): In 1974, a pair of jetstream spots was tracked photographically by M. Watanabe (in Sato, *1980*). Some photos showed tiny spots in this latitude in 1976 and 1977 (Sanchez-Lavega and Quesada, *1988*), and several observations in 1976 suggested that a little spot in the NTZ had $\Delta\lambda_2 = -66$ (BAA).

1978/79 (BAA): A major outbreak was first detected in 1978 November (see Fig. 1.19). Voyager images showed it clearly, through to the Voyager 2 encounter in 1979 July, when activity was declining (see below). Activity had ceased by the next apparition.

1984–1987 (minor activity: BAA): Tiny spots were photographed which did not constitute a major outbreak. Several were recorded by photographs in 1984 May-June, strung along a narrow NNTB(S) in the absence of a substantial NNTB. In 1985, on the same narrow band, only one tiny streak could be tracked. In 1986 and 1987, the jetstream was marked by several tiny dark spots in the 'NTZ'. Their latitude was in fact normal, but according to Pic du Midi photgraphs in 1986 (Plates P19.6&7, P21.5), the zone had a reddish extension to the north, reminiscent of the NTB(S). In 1987, these spots in the NTZ were apparently cast loose from a segment of NNTB(S) further f. which also carried tiny jetstream spots.

The jetstream gave rise to two remarkable observations in 1987. First, one of the jetstream spots was seen silhouetted against a pre-existing NTZ white oval that was retrograding in the North Temperate Current (Fig. 6.10). Both the dark spot and the oval continued on their respective currents thereafter. Second, the activity was terminated in 1987 October when a long section of darker NNTB(S) extended itself along the same jetstream. A belt segment moving with a jetstream is very unusual in itself (although a similar segment was probably seen in 1985); but what was extraordinary was that a segment of belt in the South Tropical Zone ran along the STBn jetstream at exactly the same time and the same longitudes! (Plates P22.5, P23.2). It is hard to imagine that this was anything more than a chance coincidence, although it fits into the pattern of offset symmetry in Table 6.1.

1988/89–- (BAA): There was a profusion of dark spots on the NNTB, representing the largest outbreak since 1979. Some of the spots showed uneven motion and unusual size (curiously, as did concurrent spots on the SEBs jetstream). Pic du Midi photos (Plate P21.6) showed that the jetstream spots were dark grey in contrast to the brown NNTB.

Numerous dark spots were all appearing near $\lambda_2 = 60$ and prograding at $\Delta\lambda_2 = -78$. They were crowded ($\leq 10°$ apart) and some were rapidly varying. At $\lambda_2 \leq 340$, they slowed down irregularly and became sparser and larger (mean $\Delta\lambda_2 = -53$). A well-observed example halted for a week before regaining jetstream speed (reminiscent of the oscillating spot in 1968). The NNTBs spots were generally not seen p. a faint 'NTZ Disturbance' (NTD; Chapter 7.2) that spanned the NTZ at $\lambda_2 \approx 205$, very like the 'step' of 1972, and they may have disappeared near there. But in late January, one big spot did come through that region, and smaller jetstream spots then came through behind it (although the NTD was still present). Also, the spots were no longer decelerated at higher longitudes (mean $\Delta\lambda_2 = -76$ throughout).

It was not clear how these spots arose. At λ_2 60, there was a nearly stationary f. end to a NNTB dark segment, with little visible f. it. One possibility is that most of the dark spots were created at λ_2 60, perhaps because a cyclonic structure at the f. end of the NNTB distorted the jetstream and led to instability in its flow. Another possibility is that the jetstream flow was continuous, but that white cloud cover overlay both the NNTB and most of the dark spots at $\lambda_2 > 60$ – possibly the same white cloud cover that had hidden the NTB at that time.

In late 1989, the jetstream had only a few tiny spots (at least one of which penetrated the NTD), plus one large dark spot with $\Delta\lambda_2 = -27$ (as in 1945). But in 1990 Jan.-Mar., many full-speed jetstream spots appeared again – apparently formed by the breakup of a long stretch of belt. (This coincided with a NTBs jetstream outbreak.) In 1990/91, activity continued, and this time spots passed through the NTD with their speeds undiminished – becoming large and dark as they did so. There were again some slower-moving spots.

Spacecraft observations[2]

A major NNTBs jetstream outbreak was viewed by the Voyager spacecraft, which revealed the true nature of the spots for the first time. They were all small, sharply defined, anticylonic rings, about 3000 km across. Some were distinctly orange, and some had white

---

[2] Described and illustrated by Mitchell *et al.* (*1979*) and Smith *et al.*(*1979a*). The interactions shown by Voyager 2 were studied in detail by MacLow & Ingersoll (*1986*); those described here are their events I2 to I6, I10, and A4 with A5.

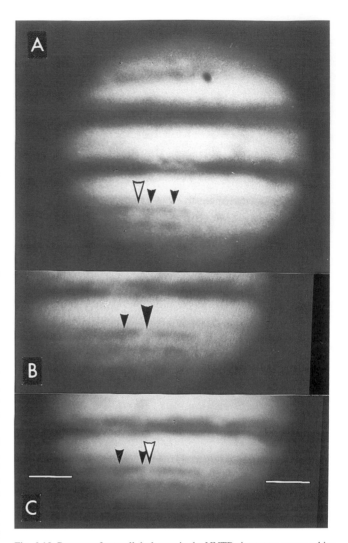

Fig. 6.10. Passage of a small dark spot in the NNTBs jetstream over a white oval retrograding in the NTZ, in 1987. (A) Sep 7d 01h 36m, ω_2 170 (G. Viscardy). (B) Sep 12d 00h 08m, ω_2 149 (G. Viscardy). (C) Sep 13d 06h 13m, ω_2 160 (D. Parker). White lines in (C) indicate the latitude of these spots in NTZ. Small black arrowheads indicate a pair of the small dark jetstream spots and a large open arrowhead indicates the white oval. (From BAA report.)

cores. Two are shown in close-up in Plate P20.2. They were spaced 10–15° apart around most of the planet, but were absent in the remaining 90°, as if obscured by a higher level of diffuse clouds. The spots did not all move at the same speed and sometimes they overtook each other and merged. As in other interactions between spots, the pair of interacting spots would spiral round each other before merging.

The genesis of some of these spots was observed. Three were recorded thrown out from cyclonic FFRs in the NNTB, one in the Voyager 2 movies, and two others in the Voyager 1 strip-maps. The latter two spots, with white cores, were created from a long oblique cyclonic disturbance where streamers from its p. end were fluttering past a 41°N white oval. (Perhaps a similar feature might account for the localised genesis of jetstream spots in 1968–1970, 1972, and 1988/89.) But another group of six dark jetstream dark spots materialised simultaneously over a 70°-long stretch of fairly quiet NNTBs. And at one longitude during the Voyager 2 encounter, two such spots appeared in a transient circulating current within the NTZ: they were created at $\lambda_2 \approx 240$, moving southwards in a clockwise (anticyclonic) curve to retrograde in the whitened latitudes of the NTBn jetstream, but after 13 days they interacted in an anticyclonic manner at $\lambda_2 \approx 270$, and ended up as a single fuzzy dark anticyclonic spot prograding on the NNTBs jetstream.

MacLow and Ingersoll (*1986*) recorded many more disappeances (mostly by mergers) than appearances; the outbreak appeared visually to be declining at that time.

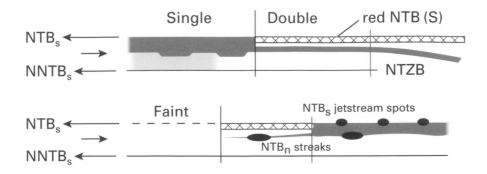

Fig. 7.1. Typical aspects of the N. Temperate region. The latitudes of the jetstreams from Voyager are indicated at left.

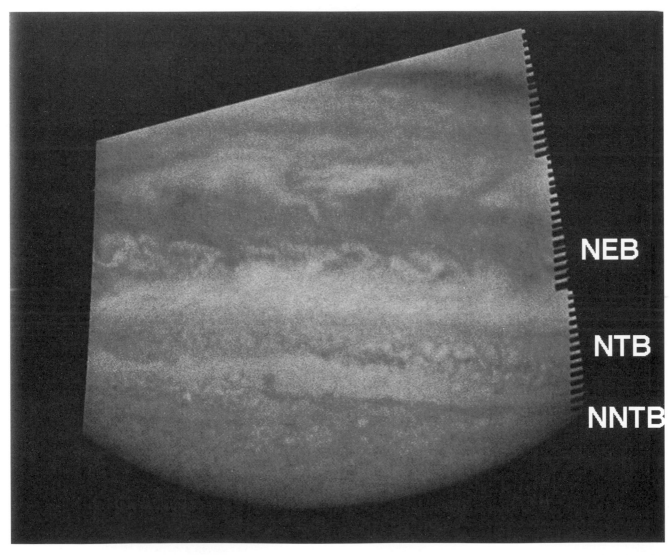

Fig. 7.2. View from Pioneer 1O. The Pioneer spacecraft revealed oblique sectors of smallscale turbulence within the NTB that, by analogy with other belts, must have been cyclonic disturbances. (Image A8, blue plus red. NASA.)

7: North Temperate Region (35°N to 23°N)

7.1 OVERVIEW

We now enter the regions of the planet where the dominant features are within the grasp of ordinary amateur telescopes, so that the historical record of activity can be taken as reasonably complete. Indeed, for the North Temperate region, the Earth-based record is almost all there is; during the Voyager encounters, the entire region was covered with white cloud.

The main belt in the region is the North Temperate Belt (NTB). It is usually described as grey or bluish-grey. When the belt is present, the south edge is usually straight while the north edge is often irregular. Often the belt is double, and then the NTB(S) component tends to be narrow and straight, while the NTB(N) component is quite variable in latitude and often carries well-defined dark streaks. Sometimes the NTB(N) swerves into the NTZ as a narrow band (NTZB) that may itself be continuous with the NNTB(S).

Sometimes the NTB fades away for several years. Over the last hundred years it has been very faint, or even absent, for about 30% of the time.

The latitudes of the belt and its components were tabulated in Table 6.2 and Fig. 6.2; the mean latitudes of the edges (24.0°N, 30.9°N) coincide with the NTBs and NTBn jetstreams revealed by Voyager. But there are marked fluctuations in the measured latitudes. The fluctuations of NTBs can probably be explained by occasional fadings and episodes of coloration (section 7.3). The fluctuations of NTBn are probably real and may even be underestimated because of observational selection effects.

The most remarkable feature of the NTB is the jetstream on its south edge, which is much faster than any other jetstream on the planet (the equatorial currents included). The occasional outbreaks on this jetstream are striking events that will be described in section 7.4.

Although the NTB usually appears unremarkable, detailed study has revealed some interesting temporal patterns in its behaviour. First, there appear to be cyclic changes in its latitude with a period of about ten years, coupled to the recurrent fadings of the belt. The NTBs jetstream outbreaks also tend to occur at intervals of five or ten years, although they do not seem to be related to the other cyclic changes. (After fadings, the belt usually recovers quietly; a

jetstream outbreak is not required, unlike in the SEB.) Furthermore, the latitude of the NTB seems to be related to that of the NEB north edge. These periodicities will be considered in section 7.3.

Also, a longer-term change seems to have occurred in the 1960s in the behaviour both of the NTB and of the NTBs jetstream, as follows.

From about 1890 to 1960, the NTB was usually a single belt, and bluish-grey in colour. Sometimes the colour was merely grey. There were definite records of brown or reddish colour in only three apparitions, and even then many other reports were of grey colour. (The apparitions were 1918/19, 1927/28, and 1940/41; these did not coincide with colour in NTropZ, but did twice follow NTBs jetstream outbreaks. Reports by a few observers who tended to see *all* belts as reddish-brown have been ignored.) The NTBs jetstream outbreaks in this era, typically coming at intervals of ten or more years, consisted of small dark spots like those in NNTBs jetstream outbreaks, and all had rotation periods of about 9h 49m ($\Delta\lambda_2 \approx$ -289; Peek's 'North Temperate Current C'). One or two years after these outbreaks, an intermediate period of about 9h 53m to 9h 54m ($\Delta\lambda_2 \approx$ -106; 'Current B') used to be recorded for segments of the NTB.

However, in 1964, after a prolonged fade, the NTB reappeared with a vivid orange colour. This colour was noticed by E.J. Reese in 1963/64 while the belt was still faint; then in summer 1964, the belt revived, dark and narrow, remarkably reddish visually and on photographs (Plate P13.1). It remained so until 1966. Major revivals of a visibly orange NTB(S) recurred in 1971 and 1985, along with more extensive NTropZ coloration. This belt was observed in the infrared in 1971 and 1972 and was at very high altitude, according to its coldness at 5 μm wavelength, and its brightness in the 0.89 μm methane absorption band (Chapter 4; Fig. 4.10). These three orange revivals may also have been associated with NTBs jetstream outbreaks (see section 7.4), and so may have been part of more extensive disturbances which lofted coloured material above the jetstream.

The orange colour of the narrow, featureless NTB(S) component has persisted in most of the other years since 1964 according to high-resolution photographs, even though it is sometimes very

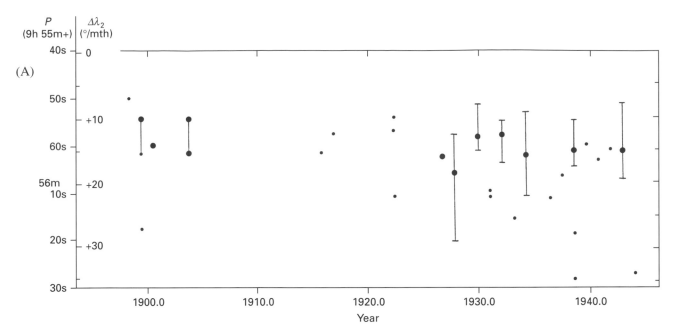

Fig. 7.3. History of the N. Temperate Current. From data listed in Appendices 1 and 3. Speed is shown as *P* (rotation period), as $\Delta\lambda_2$ (degrees per month), and as *u* (m/s). Large symbols are apparition means of ≥3 features, and vertical bars indicate the range of speeds (not published before 1914). Small symbols represent single features; open circles represent imprecise values. The mean NTC speed is $\Delta\lambda_2$ = +17.0°/month (±4.2, standard deviation of apparition means).

faint. (In contrast, the NTB(N) and its streaks tend to be grey.) Although the orange colour is not obvious visually outside the major revivals, the author has occasionally been able to see it, so the fact that it was never mentioned between 1890 and 1964 may point to a significant change.

The other major change since 1964 has been in the behaviour of NTBs jetstream outbreaks. From 1970 to 1990, they have usually occurred every five years; they have included large, brilliant white spots as well as dark spots; and they have shown rotation periods of 9h 47m ($\Delta\lambda_2 \approx$ -392), fully two minutes faster than the previous type of outbreaks, in what might be called 'Current D'. And they have not been followed by 'Current B' activity. Thus in the 1960s, the belt that used to be cold grey became liable to vivid orange revivals, and from 1970 onwards it displayed a faster speed than had ever been seen before.

But before we conclude that these changes were certainly related, we should note that the orange belt in 1964/65 carried a reddish spot that marked the last outbreak of the Current C type; the first Current D outbreak did not occur until five years later.

We have not so far mentioned the behaviour before 1890. There was at least one cycle of activity more like those of recent years, with a Current D outbreak in 1880, followed by the formation of a revived reddish NTB in 1881 (e.g. Plate P12.8); NTB was also drawn as a massive orange belt in 1883 (N.E. Green, *1887*). This colour would not be seen again for 80 years.

The North Temperate Current

The characteristic slow current of the region is the North Temperate Current (NTC). Peek defined it as North Temperate Current A, in contrast to Currents B and C, which are the much

faster speeds associated with jetstream outbreaks; but we will use 'North Temperate Current' to mean the slow current, for consistency with other latitudes.

The NTC is most commonly displayed by the dark spots and streaks on the NTB(N) component or NTBn edge, but also controls segments of the whole NTB, and sometimes also spots in the NTZ. Occasionally a NTC feature extends all the way to the NNTB as described in section 7.2. Distinct white ovals are very rare, though diffuse white spots are sometimes tracked in the NTZ.

The NTC is the only one of the slow currents to have a substantial retrograding drift, with a mean $\Delta\lambda_2$ of +17.0 and a range of about +10 to +25 (Fig. 7.3). A few dark spots have retrograded as fast as $\Delta\lambda_2$ = +30 to +34. However, this current is clearly different from the NTBn retrograding jetstream, which was only detected by Voyager, with a speed of $\Delta\lambda_2$ = +69 at latitude 31.6°N.

7.2 THE NTB(N) AND NORTH TEMPERATE ZONE

North Temperate Zone

The NTZ is very variable in apparent width and in brightness.

Sometimes it appears to share the tone of the NNTZ or NTropZ. This does not reflect any consistent relationship between these zones, but these appearances can be quite striking. For example, when the NTB is very faint, the NTropZ and NTZ often appear as a single broad bright zone with the tenuous NTB(S) in the middle of it (e.g. 1979, 1989). At other times, the NTZ and NNTZ appear to merge likewise, when the NNTB is reduced to a narrow NNTB(S) (e.g. 1975, 1985). In 1985 the bright white NTZ contrasted strikingly with the dull brownish NTropZ, as was also the case in 1928/29. Sometimes, both the NTZ and the NNTZ are subsumed in

(B)

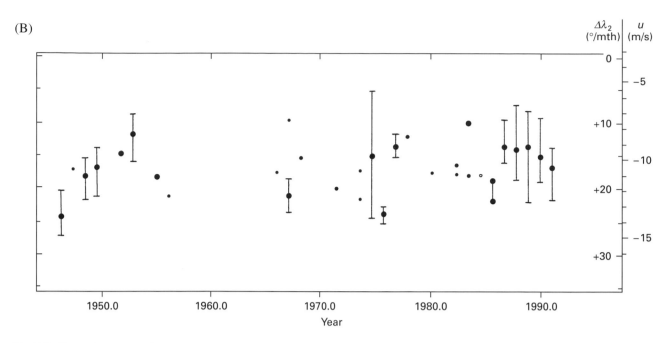

Fig. 7.3B. For caption see opposite

the dusky polar shading (e.g. 1925); at other times, the NTZ and NTropZ are shaded together while the NNTZ is bright, as in 1914 (grey) and in 1912 and 1980/81 (dusky yellow).

Complex disturbed sectors of NTZ were observed in 1927/28, 1982, and 1985. In 1927/28, the spots and shadings all moved with the normal NTC. In 1985 there was a tangle of irregular structures: these appeared to include large features moving with the NNTC, but they were changing from week to week and were not fully resolved. There is no evidence that these were coherent disturbances.

NTZ Bands and Disturbances

It is not uncommon for a narrow belt to run through the NTZ, sometimes obliquely. Depending on the existence of shading north or south of it, such a band may appear to be NTB(N), NTZB, or NNTB(S). It is often not possible to identify a narrow belt segment until one has measured its drift and latitude, which determine its correct name. Thus a narrow band at 36°N with prograding jet-stream spots is NNTB(S), regardless of the surrounding albedo patterns.

Sometimes these bands really do tilt across the canonical boundaries. For example, in 1935–1937, the main belt in the region was a 'NTZB' at latitude 32°N to 35°N, dark and bluish-grey and sometimes quite narrow; its p. end moved with the NTC in 1936, but further f. it trended N. to become the NNTB. Similarly in 1975–1978, a 'NTZB' had a p. end arising from NTB(N), and trended northwards with kinks and inflexions (one dark segment in 1975 moving with NTC) to form a tenuous belt at ≈34°N which merged with the NNTB further f. (Fig. 1.18).

In several apparitions there have been more substantial dark belt segments spanning the NTZ; we can call these North Temperate Disturbances, by analogy with South Tropical Disturbances. They move with the NTC even though they extend to NNTBs latitudes.

The records of them are as follows.

1924 (BAA): NTB extremely broad, to 35°N.

1949 (ALPO): For a length of about 60°, NTB and NNTB were merged to form a wide dark belt.

1952/53 (BAA and ALPO; Plates P8.7&8, P9.5&6): There was conspicuous dark streak about 40° long, again spanning the latitudes from NTBn to NNTBs. Its p. end lay in the NTZ but its f. end lay along NNTBs, drifting with the NTC ($\Delta\lambda_2$ = +17) and slowly lengthening. In good seeing it appeared as a chain of dark spots. It may have been derived from some short dark streaks on NTBn in 1951/52, which would have coalesced and drifted northwards to create the Disturbance. It became rather difficult and possibly fragmentary in early 1953.

1972–1975 (BAA and LPL; Fig. 7.4, Plate P14.8): This feature generally appeared as a dark step across the NTZ, and seems to have been the f. end of a dark segment of NTB(N) which, however, sometimes lay so far north as to contact the NNTBs edge. However, it had a variable appearance and relationship to the surrounding belts. Spots and streaks were commonly seen in the NTZ flanking it.

It was first observed by Minton (LPL) in 1972 as a dislocation between NTB(N) and NNTB(S), and NNTBs jetstream spots disappeared on or before reaching it. In 1973, photographs showed it as the f. end of a dark, 'cold grey' NTB(N) which contrasted with the browner NNTB Nf. it. In the Pioneer 10 and 11 images, also, the 'step' was the f. end of a NTB(N) component, divided from the NTB(S) by a turbulent white stream. These photographs apparently missed the periods when the 'step' most clearly joined the NNTB. In 1974, NTB(N) was occasionally seen as many small spots and streaks (LPL). In 1975 Aug.-Sep., the 'step' may have been represented by a broad forking of the NTB, but it was not seen thereafter.

Its drift ($\Delta\lambda_2$) varied between +19 and +22 throughout its life, typical of the NTC. The latitude of the NTB(N) segment p. the step was 28.8°N in 1972, 32.2°N in 1973, and 33.1°N in 1974 (LPL; see Fig. 6.2); the NNTB f. the step was at normal latitudes of ≥36°N.

1988–1992 (BAA; Fig. 7.5; Plates P23.3&6 and P24.11): This Disturbance started as an almost exact replica of that of 1952. When the NTB faded in 1988, the best photographs revealed that faint features remained visible on

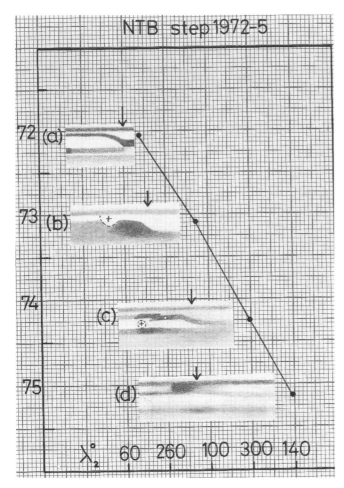

Fig. 7.4. The 'NTB Step', 1972–1975. The drawings show the region from NTB(S) to NNTB. Four representative observations are shown: **(a)** 1972 June; **(b)** 1973 Aug.; **(c)** 1974 Oct.; **(d)** 1975 Aug. **(a)** by Minton (LPL); **(b-d)** by Rogers (BAA). (This chart drawn by R. McKim for BAA report.)

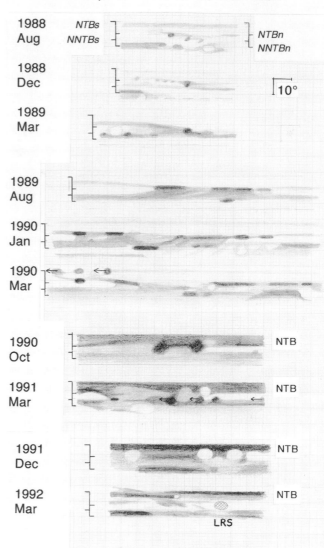

Fig. 7.5. The N. Temperate Disturbance, 1988–1992. These strip-maps were made from high-resolution photographs (e.g. Plates P23.3&6 and P24.11). Bars at left indicate the latitudes of the NTBs, NTBn, NNTBs, and NNTBn jetstreams from Voyager. Arrows indicate spots moving on the jetstreams. The main NTB was very faint from 1988 to 1990, when a NTBs jetstream outbreak led to its revival. The NNTB was variable but was usually disrupted alongside the Disturbance, though dark bars there retained NNTC drifts while the Disturbance moved with the NTC. The last map shows a Little Red Spot in the NTZ after the Disturbance broke up.

the extrapolated NTC tracks of a pair of NTBn dark spots, constituting a diffuse streak or chain of spots across the NTZ. This Disturbance, like that of 1972, seemed at first to block the motion of NNTBs jetstream spots, although later several did run past it (Chapter 6.4).

In mid-1989, as NTB(N) began to revive, the Disturbance had grown into a massive 'cold grey' belt segment spanning 30–38°N, with a length of 60° (later growing to 70–90° long) (Fig. 4.1C). Its p. end appeared distinct and concave at high resolution (Fig. 1.22) but the apparent p. and f. ends of the Disturbance sometimes shifted from one dark streak to another, though it still moved approximately with NTC. A few NNTBs jetstream spots were detectable moving into it although they were slower than usual. This huge 'NTZ Disturbance Belt' gradually broke up in early 1990, particularly when a NTBs jetstream outbreak ran past it. The debris included typical NTB(N) streaks at 29.5°N, some of which might have been revivals of those of 1988.

However, the Disturbance was still present in 1990/91, now shortened to a dusky wedge extending from NTBn, 29° long. A notable feature was a bright white oval on NTBn at its f. end, at 29.5°N and therefore presumably cyclonic. NNTBs spots were now passing it and becoming large and dark as they did so. In 1991, other regions of NTZ alongside these spots became dusky, and the Disturbance was subsumed in a general shading of the NTZ interrupted by light ovals. In late 1991 this structure was still present, the Disturbance being a very dark block flanked by 2–3 white ovals at 29.8°N. But in 1992 Feb., it suddenly faded away to leave only a faint streak.

But it had an unexpected sting in its tail. A Little Red Spot appeared in

the NTZ at 33°N just f. the erstwhile Disturbance in late February – the first such spot ever recorded in the NTZ. It was a dark oval in blue-light photos but a tiny bright spot in red light. It may have incorporated a NNTBs jetstream spot which suddenly halted; however, one NNTBs jetstream spot later ran past it. It lasted for one month.

The N. Temperate Disturbance had a typical NTC drift throughout, averaging $\Delta\lambda_2 = +15$ from 1988 to 1990/91, then +12 to 1992.

Whether these 'North Temperate Disturbances' involve circulating currents like those of South Tropical Disturbances has not been established. The existence of such anticyclonic circulations is suggested by the disappearance of NNTBs jetstream spots in the vicinity in 1972 and 1988, and by the distinct concave p. and f. ends shown in some of the highest-resolution views (Plate P9.5&6; Fig.

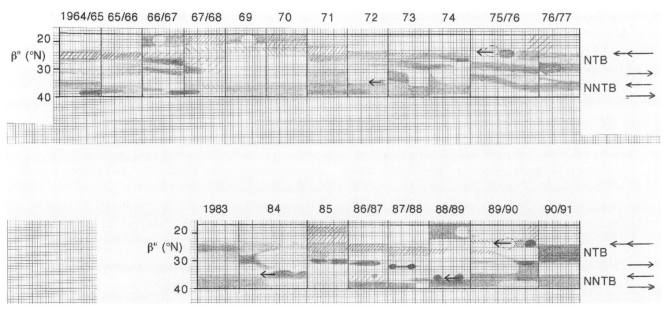

Fig. 7.6. A quick history of the N. Temperate region, 1964–1991, showing variations in latitude and colour. Data are mainly from photographs as reported by LPL (1964–1973) and BAA (thereafter). Colours are shown by shading: //// yellow or orange; \\\\ bluish. ß" is zenographic latitude. The NTBs and NNTBs jetstreams are indicated by arrows. The chart includes: NTB fading of 1968–70 and 1984; orange NTropZ-NTB of 1971 and 1985; N. Temperate Disturbances of 1973–75 and 1989/90; NTBs jetstream outbreaks of 1975 and 1990, and also expanded NEBn of 1966–70 and 1988/89.

1.22), and by the appearance of the Little Red Spot. Contrary evidence, however, is that the ends are sometimes variable or indistinct, and that some NNTBs jetstream spots clearly managed to move past the last Disturbance, though they interacted with it.

Dark spots and streaks on NTBn

The NTBn dark spots and streaks were very notable in 1986 and 1987, when they lay along a tenuous, incomplete NTB(N) at 31–32°N. Of five streaks present in 1986, some may have persisted into 1987, but there were then many more of them; some arose by subdivision of longer, patchy dark segments of NTB(N). They presented a remarkable spectacle in good seeing (Plate P22.6), as some were small but extraordinarily dark, looking like miniature satellite shadows. Some spots were seen to disappear suddenly, within a week or so, having persisted stably for months. Also, two merged to form one large dark spot, which was one of the last to disappear in 1988. Usually the dark streaks appeared grey visually, and Pic du Midi photographs showed most of them as greyish but some as distinctly brown.

The only previous occasions on which such features had attracted any special comment were in 1900 and 1951, when dark streaks were conspicuous, and in 1931/32, when two NTBn spots were described as minute and black.

7.3 PERIODIC BEHAVIOUR OF THE NTB

Periods of either 10 years or 11.9 years (one jovian year) have been reported for NTB fadings, NTB latitude shifts, and NTBs jetstream outbreaks, and in recent times the jetstream outbreaks have shown a period of five years. The latitude shifts have also been noted as being concordant with those of the NEBn, which itself has an even shorter timescale for fluctuation. Exactly what these changes consist of, and how they may be related, will be the subject of this section.

Surveying the historical record, the most consistent of all the periodicities is that of the fadings of the NTB. The intervals have ranged from 8 to 13 years throughout recorded history, except for one interval of 22 years; and if that is taken as a double interval, the mean period is 10.1 years. The times of these fadings, and of the belt's recovery from them, are listed below. (Occasionally there are briefer, less complete fadings, which are not listed.) The major fadings last between 1½ and 8 years, during which time the belt is faint and narrow, or even completely absent. Both the fading and revival are usually gradual affairs taking up to a year.

The fluctuation in latitude has the same 10–year period (Fig. 6.2; von E. Schoenberg & W.D. Heintz, *1955*), and one must therefore ask whether it is related to the fadings. The same question arises on theoretical grounds. In view of the relationship between belts and jetstreams, and particularly the two-component structure of the NTB in recent years, are the latitude shifts really due to migration of the belt, or to selective fading of one of the components?

It turns out that northward displacements of the belt closely precede fadings of it, and at least since the 1960s, they have been due to the orange NTB(S) fading earlier than the grey, streaky NTB(N). This was the case with the northward displacements of 1987–1989 and 1966–1968 (Fig. 7.6). In 1966–1968, LPL photographs showed an extremely faint orange strip of NTB(S) remaining on the S. edge of the main grey belt, which was still visible in a 1969 photo even after the main belt had disappeared. (In 1978, another major fading was not recorded as a latitude shift because the NTB(N) faded most completely, leaving the tenuous orange NTB(S) that was photographed by Voyager.) In 1988, there was successive fading of NTB(S), NTB(N), and finally the last NTBn dark spots, until the belt disappeared.

For the era before 1964 – before anyone perceived the orange NTB(S), and before high-resolution colour photography – it is difficult to determine if the same process was occurring. Latitudes of belt edges were not measured, and the latitude of the belt as a whole could only be measured by micrometer when it was reasonably dark. One can however suggest the hypothesis that northerly displacement[1] was due to splitting of the belt leaving a S. component which faded before the N. component. There is some evidence to support this, as seen in the following account (and see Fig. 6.2).

This chronicle lists all the major northward displacements, fadings, and revivals that have been observed.

> *Faint, 1896–1899:* In 1899, dusky NTB(N) gradually revived, followed by a separate NTB(S).
>
> *Faint, 1904–1909:* (Measured latitudes were for a N. Tropical Band; see next chapter.) In 1909/10, NTB revived as a broad, vague, grey band, with two narrow but 'hazy' components at some longitudes.
>
> *Northward displacement, 1911–1912:* There was probably a very faint NTB(S) in 1911 while the measured latitude was for NTB(N), which also faded away in 1913. *Faint, 1913–1914:* Both the NTZ and NTropZ were heavily shaded, having been dull yellowish in 1912; one or both NTB components disappeared within this shading. (NTB revived during solar conjunction.)
>
> *Northward displacement, 1918–1920:* No evidence for doubling. *Northward displacement, 1922:* Faint NTB(S) occasionally drawn, while the measured latitude was for streaky and incomplete NTB(N). *Faint, 1922–1923:* NTB(N) streaks still present at some longitudes. (NTB revived during solar conjunction.)
>
> *Northward displacement, 1931–1934:* Usually two uneven components. Both faded in 1934. *Faint, 1934–1937;* in 1935 and 1936 the major belt was in normal NTZ latitudes. Revival in 1937 was again quiet.
>
> *Northward displacement, 1945–1947:* No visual evidence for doubling, but one Pic du Midi photo in 1945 showed two narrow components, NTB(S) being faint, so the measured latitude may have been for NTB(N). The belt did not fade as a whole, however.
>
> *Faint, 1956–1963:* (No precise latitude data.) The gradual revival was the first appearance of the massive orange NTB, as described in section 7.1.
>
> *Northward displacement, 1966–1968: Faint, 1969–1970.**
> *Faint, 1978–1981.**
>
> *Northward displacement, 1987–1989: Faint, 1988–1990.**
>
> *The onset of these recent major fades has already been described above and sketched in Fig. 7.6. The recovery from each of these fades overlapped with NTBs jetstream outbreaks, and so will be described in section 7.4.

As early as 1898, Denning noted that NTBs jetstream outbreaks had occurred at 10-year intervals, and a subsequent 10-year recurrence was noted by Peek. More recently the intervals have been exactly five years (section 7.4). However, these intervals do not fit consistently with the 10-year intervals between fadings of the NTB. If anything, jetstream outbreaks have tended to occur when the belt looked normal, although the D current outbreaks of 1880, 1970, 1980, and 1990 were major exceptions that occurred while the belt was faint.

It was also suggested that the jetstream outbreaks tended to occur near perihelion, implying a mean period of 11.9 years[1]; but since then, the five-yearly jetstream outbreaks have not maintained any relationship with the jovian year.

The relationship in latitude between the NTB and NEBn was first noticed by T.E.R. Phillips, who wrote:

> In each hemisphere the direction of [latitude shift] shown by contiguous features is commonly sympathetic, e.g., the displacements of the NTB clearly have some relation to the extensions and recessions of the NEBn edge. (Phillips, *1930*.)

The same tendency was noted by E.J. Reese (*1971a*) from data in the 1960s; NTBs moved north (due to fading) just as NEBn also moved north.

In comparing the latitude charts of the NTB (Fig.6.2) and NEB (Fig.8.14), one can indeed see a tendency for such coordinated shifts throughout history, although given that the NEBn latitude fluctuates much more often than the NTB latitude, it is not surprising that there have been plenty of years when the NTropZ really was very wide or very narrow. No physical connection can be discerned that could explain the correlation, but we will encounter similar correlations in other latitudes (Chapters 11–13).

The extreme southward displacements of the NTB seem to have a variety of causes. In recent decades, a strong southerly NTB(S) has been formed either by the dark bluish belt that appears in D current outbreaks, or by the strongly orange component that may also be associated with jetstream outbreaks and may cover part of NTropZ. In the era before 1964, whenever the NTB was substantial and southerly it was always bluish-grey or plain grey. From 1906 to 1909, an extreme southerly latitude was attributed to the presence of a strong bluish N. Tropical Band (see next chapter).

7.4 THE NTBs PROGRADING JETSTREAM

This is the fastest jetstream on the planet. To Voyager, it was detectable only by the motions of streaks in the white clouds, which peaked at 23.8°N, precisely at the south edge of the narrow faint orange NTB(S) (Fig. 3.8; Plate P20.2).[2] (The blandness of this component in the Voyager images may suggest that it was a high-altitude covering, as was the more conspicuous orange NTB(S) in 1971; Fig. 4.10) The jetstream defines the NTBs edge, and has not changed its latitude noticeably over more than a century. Details are summarised in Table 7.1.

---

[1] Reese and Smith (*1966*); Favero *et al.*(*1979*).
[2] A detailed description was given by Maxworthy (*1984*). The mean speed recorded by Voyager was $\Delta\lambda_1 = -188$ (period 9h 46m 19s) (Maxworthy, *1984*) or $\Delta\lambda_1 = -163$ (period 9h 47m 15s) (Limaye, *1986*).

Table 7.1. *Records of the NTBs jetstream and intermediate currents*

Current:	D		C		B			
	Period	$\Delta\lambda_1$ (N)	Period	$\Delta\lambda_1$ (N)	Period	$\Delta\lambda_1$ (N)	Lat.β"	Ref.
1880	[48m 0s]	−112]*(2)						See text
1891			49m 19s	−53 (4)	[54m 31s	+178]* (1)		See text
1892			(49m+)				(+22.4)	Hough
1926			49m 3s	−65 (1)				BAA
1928/29					53m 3s	+113 (3)		BAA
1929/30			49m 17s	−54.5 (21)				BAA
1930/31			49m 10s	−60 (5)				BAA
1931/32					52m 46s	+101 (7)		BAA
1932/33					53m 46s	+145 (1)		BAA
1939/40			48m 57s	−69 (11)				BAA
1940/41			49m 11s	−59 (13)				BAA
1941/42					52m 58s	+110 (1)		BAA
1942/43			49m 6s	−63 (2)	53m 29s	+132.5 (2)		BAA
1943/44					53m 38s	+139 (1)		BAA
Average (1891–1944)			49m 9.0s (±7.7s)	−60.4 (N=7)	53m 16.7s (+24.3s)	+123.4 (N=6)		
1964/65			49m 18.5s	−53 (1)			+24.2	NMSUO
1970	47m 3s	−155 (1)					+23.8	NMSUO
1975:								
Main:	46m 57s	−159 (3)						BAA
Main:	47m 18s	−144						SAF
Main:	46m 50s	−164 (1)	49m 35s	−41 (4)	52m 19s	+81 (2)	+23	SL&Q†
Second:	[48m 47s	−77]*						SAF
Second:	[48m 19s	−98]*(5)					+23	SL&Q†
1980	46m 33s	−178 (1)						NMSUO
1990	46m 50s	−165 (2)	49m 57s	−25 (9)			+23.8	BAA
Average (1964–1990)	46m 51.9s (±13.9s)	−163.4 (N=4)	49m 36.8s (±19.3s)	−39.6 (N=3)			+23.6 (±0.5°)	
Average (overall)	46m 51.9s (±13.9s)	−163.4 (N=4) (±10.7)	49m 17.4s (±17.4s)	−54.2 (N=10) (±12.6)	53m 8.4s (+31.1s)	+117.2 (N=7) (+22.5)	+23.6 (±0.5°)	
Voyager	47m 15s	−146 (global)					+23.8	Limaye (*1986*)
Voyager	46m 19s	−188 (global)					+23.7	Maxworthy (*1984*)

For each apparition, the Table gives the observed speed in two forms: P (rotation period, 9 hrs+), $\Delta\lambda_1$ (speed in System I in degrees per 30 days). (N is the number of spots with drifts recorded; see text for total number of spots in outbreak). 1892 values are approximate and not included in averages.
* Intermediate drift rates not included in averages.
†Sanchez-Lavega & Quesada (*1988*). Their speeds were approximate, ≈±30°/mth. Their latitude was for D and C current spots; B current spots were at 26.5°N. See text for descriptions of features.

We have already discussed the general features of the jetstream, and noted the striking difference in speed between recent outbreaks that match the Voyager speed (Current D), and earlier outbreaks that were slower (Current C). In this section, we describe the recorded outbreaks in detail. Speeds are given in longitude System I. As the outbreaks are often accompanied or followed by several months of yellow or orange colour in the NTropZ, we include notes on such colorations.

NTBs outbreaks in the nineteenth century
The first outbreak recorded was in 1880. W.F. Denning (*1898b*) suggested that there had been comparable outbreaks every ten years since 1850. Indeed, several drawings showed irregularities or distinct dark spots on the NTB between 1870 August and 1871 January, which could well have been jetstream spots. However, the drawings he referred to in 1850 and 1860 seem to show features in the NEB rather than the NTB (Chapter 8.5).

1880 (Fig. 7.7A; Plates P3.3&4 and P12.7): The NTB was faint and narrow and 'blue' at the time. Two black spots 20° apart were detected on it on 1880 Oct. 17. Over the ensuing months, many tiny dark spots formed and the length of the outbreak grew from about 30° on Oct. 29 to 120° on Nov. 23. The tiny dark spots were so changeable that most of them could not be tracked, but Denning found that the two chief ones, near the centre of the expanding outbreak, had a period within a few seconds of 9h 48m 0s ($\Delta\lambda_1$ = -112). Such a rapid speed, which may not have been the fastest in the out-

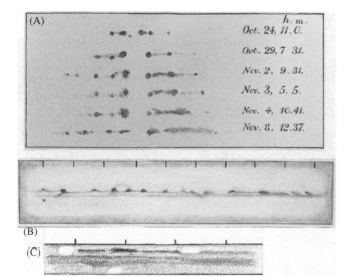

Fig. 7.7. Strip-drawings of NTBs jetstream outbreaks. Marks are at 20° intervals. (A) 1880 Oct.-Nov., by W.F. Denning (*1880*). (B) 1929 Nov.14, by T.E.R. Phillips (from Peek, *1931, 1958*). (C) 1975 Oct.11, by D. Gray (from BAA archives).

break, would not be seen again for almost a century. By the middle of 1881 January, the spots had completely encircled the planet to create a new dark NTB(S), within which the individual spots lost their distinctiveness. The new NTB was described as red by E.E. Barnard (and in Plate P12.8).

1891–1892 (Plate P3.9): The first dark spot (called *c* below) was seen on the NTBs edge by A.S. Williams on 1891 May 14, and the outbreak was extensive by August. Denning noted that one spot was as dark and prominent as a satellite shadow. There were 'a great many small blackish spots', forming a broken chain all round the planet, and unlike the spots of 1880, they appeared at many longitudes without having any particular centre.

Barnard reported a mean period of 9h 49m 3s ($\Delta\lambda_1$ = -65) from his observations, and Williams reported periods of 9h 49m 27.2s, 1.6s, and 5.5s ($\Delta\lambda_1$ = -47, -66, -63) for three well-observed spots up to September (*a,b,c*). (Spot *a* was 93° p. spot *c*. The mean of these values is 9h 49m 9.3s, $\Delta\lambda_1$ = –60.) However, in mid-September the three spots all suddenly decelerated to 9h 49m 44s, 33s, 19s ($\Delta\lambda_1$ = -34, -42, -53) respectively. Later, the f. spot (*c*) accelerated again, and as it caught up with the two leading ones they accelerated likewise, until the mean period had returned to about its original value.

At the same time there was a feature on the NTBn edge, described as 'a large protuberant mass', which had a steady period of 9h 54m 31.3s ($\Delta\lambda_2$ = -51) in Sep.-Oct. but then suddenly halted in System II. It is unclear whether this was the first manifestation of the B current, or of the NNTBs jetstream, or whether it was unique.

By 1892 January, most of the spots had dwindled or disappeared, and the NTB had become very broad and dark. However, some spots were still drawn on the belt in the 1892 apparition. The NTropZ remained bright throughout.

NTBs outbreaks 1929–1965

Two major and two minor outbreaks all consisted of tiny dark spots with periods of about 9h 49m (the C current), just as in 1891. In each outbreak, individual spots were tracked for more than three months.

It may seem surprising that there were intervals of 37 years and 21 years with no definite NTBs activity. It would be possible for observers to miss a short-lived outbreak during solar conjunction,

or one with only a few small spots, and larger outbreaks could conceivably have been missed in the 1950s. However, an outbreak as extensive as those of 1891 or 1929 or 1975 would have been detected in any year.

1915–1917?: In these three years, there were a few records of condensations on the NTBs edge, but they were not considered as a probable outbreak when Phillips later reviewed the observations.

1926: Only one spot was tracked, for three months from June 15, although one or two others may have existed. The NTropZ was very bright but had an orangish tint. The NTB(S) was dark and bluish-grey or cold grey; but in late 1927 some longitudes appeared clearly ruddy. Then in 1928/29 the NTB was dark and double with occasional dark spots between the components, which moved with the B current.

1929–1931 (Fig. 7.7B; Plate P6.3; BAA and Peek (*1931*)): Many spots were seen from 1929 Sep 27 onwards. They were first identified by A.S. Williams, who had then been observing for over 50 years, and with only a 16.5–cm reflector was able to recognise a phenomenon that he had not seen for 37 years. The spots broke out around two-thirds of the circumference within no more than a few weeks. Those at the f. end moved somewhat faster so the outbreak did not spread into the remaining longitude sector. The peak of activity probably came in 1929 December, with spots spaced ≈14° apart,; thereafter the number of spots diminished, but in the next apparition over 200° of longitude was still disturbed on a smaller scale.

In 1928/29, the NTropZ had been narrow and had a striking dull yellow or reddish colour; but this had disappeared by the time of the NTBs outbreak. In 1929/30 and 1930/31, the zone was bright, and the yellow tint recorded by some observers may have been nothing out of the ordinary.

The B current was observed from 1931 December, for dark streaks and light spots in NTB(S), and continued until 1933.

1939–1943: The first spot was seen on 1939 Sep 23, and by the end of October, there were many of them around half the circumference of the NTB (Plate P9.1). Spots then appeared in the remaining half as well, until the outbreak covered the entire NTB by the end of the year. It was still intensely active up until 1940 December, when the dark spots became fewer and weaker. The NTB sometimes appeared brown in 1940/41.

A very important observation was achieved by Ainslie, Hargreaves, and Phillips on 1939 Dec. 27. One of the tiny dark jetstream spots was due to encounter a large white NTropZ spot which extended over the same latitude; and on that night, all three observers agreed that the jetstream spot was projected upon the northern part of the slow-moving white spot.

In 1941/42, there were hardly any NTBs spots, but the B current was observed for a vague p. end to a section of NTB. In 1942/43, there was a slight but definite revival of NTBs jetstream spots, while the B current still operated on a segment of NTB.

In most of these years, the NTropZ was bright and 'white or yellow', so again, these reports may not indicate any abnormal colour. In 1941/42, the NTropZ was dull yellowish and very narrow following an expansion of the NEB. In 1944 March, the NTropZ had a dull 'rosy' tint (Hargreaves).

1952/53?: Several BAA drawings show isolated NTBs dark spots at various longitudes in 1952 December and 1953 February. The 'fluffy' appearance of the NTB(S) from Mt. Palomar in 1952 October (Plate P9.5) may also be a sign of activity. No drifts were obtained. The NTB(S) was grey.

1964/65 (NMSUO): While the strong orange NTB(S) was intensifying, a single dark spot on its south edge was photographed (Reese and Smith, *1966*). It was tracked from 1964 July 7 to 1965 April 1. It was ≈6000 km across, too faint for visual observers, and was unusual in being frankly red like the NTB(S) itself. It faded gradually and could only be recorded in ultraviolet photographs after January. Its speed fluctuated irregularly, and was 2°/mth (3 s) slower than the mean until November and a similar amount faster after December.

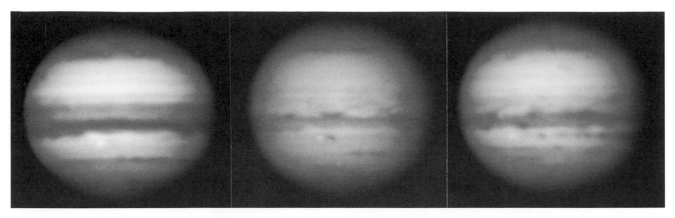

Fig. 7.8. The 1990 NTBs jetstream outbreak, photographed by I. Miyazaki. **(Left)** 1990 Feb. 15, ω_1 78 (blue/violet filter); **(Middle)** 2 min later (red filter). These show the initial white spot of the outbreak, followed by a deep blue spot (dark in red). **(Right)** 1990 March 18, ω_1 255 (white light). The leading white spot is followed by a chain of dark spots.

The spot was twice photographed during conjunctions with large slow-moving white ovals that spanned the NTropZ, and in both cases it appeared to encroach on the white ovals, suggesting that the red jetstream spot was higher. Another indication that it was at high altitude was that it darkened relative to the NTB as it rotated towards the limb.

NTBs outbreaks 1970–1990

The next outbreak, in 1970, was again detected only by professional photography, and it was of a novel kind: the single spot was bright white, and had a rotation period fully two minutes faster than the usual jetstream C current. This was a forerunner of a series of major outbreaks with the same characteristics. The main features are white spots; they have periods of about 9h 47m (D current), which is more-or-less the same as the peak jetstream velocity detected by Voyager; they occur at intervals of five years; and they are not followed by visible B current activity.

1970 (NMSUO): The white spot was tracked from 1970 Aug 12 to Sep 21, on an extremely faint NTB. It was especially bright in blue and ultraviolet, and when near the limb, implying that it was at high altitude.

By 1971, the NTB(S) had revived with a vivid orange colour that extended as far south as 21.4°N (Figs. 4.10 and 7.6). But the best LPL photos suggest that there was a slightly darker, browner NTB(S) in its normal latitudes plus orange haze that overlapped the northern NTropZ. The southern NTropZ (which had been orange in 1970) was now white again. If any further NTBs spots had occurred during conjunction, there was now no sign of them.

1975 (BAA, SAF, and Sanchez-Lavega & Quesada (1988); Fig. 7.7C): This spectacular outbreak began with one bright white spot on 1975 Sep. 14. By Sep. 23, two more white spots and several dark spots had appeared within a span of 60° longitude, and a secondary outbreak (one white and one dark spot) appeared 70° following. Disturbance also extended p. both outbreaks and covered 170° by late October. The white spots, about 8° long, became less conspicuous after a few weeks. At high resolution, the dark spots between them were seen to be very dark chunks of NTB(S).

Visual observations showed drifts of $\Delta\lambda_1 = -159$ for the white spots in the main outbreak, and somewhat slower for the secondary outbreak. Professional photographs (Sanchez-Lavega & Quesada, 1988) confirmed this, but also showed that the *dark* spots in the main outbreak were moving much slower – with approximate $\Delta\lambda_1$ ranging from -24 to -53, at the familiar speed of the C current! In the secondary outbreak, there was a wide range of speeds for dark and bright spots ranging from $\Delta\lambda_1 \approx -61$ to -125.

And tiny white spots at 26.5°N (between the NTB components) were moving with a typical B current speed. Thus the white spots of the D current seemed to be superimposed on an outbreak that otherwise had many of the usual features!

Equally remarkably, a large white spot in the slow North Tropical Current was apparently created in the NTBs outbreak in mid-October (BAA).

The main outbreak revived briefly in mid-October, and again in early December with some new white spots, but by late December all but a few vague dark spots had faded out.

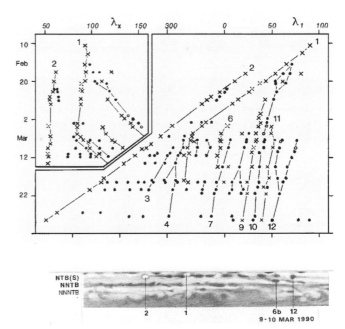

Fig. 7.9. Chart of the motions in the 1990 NTBs outbreak, plotted in System I longitude. x, white oval; ●, dark spot. **(Inset:)** The early stages plotted in a longitude system moving at -5.0°/day relative to System I. **(Below:)** Strip-map by Miyazaki from his photos on 1990 March 9–10, to the same scale. Feature numbers correspond to those on the chart, except no. 6b which is a NTBn dark streak near the p. end of the N. Temperate Disturbance. (From the BAA report.)

A broad dark grey NTB had already existed, but after the outbreak the S. component, at 24°N, ended up much stronger than before. The NTropZ, which had been bright white, was left unevenly but strongly shaded and yellowish, and slow-moving white spots that had been in it had mostly disappeared. It remained somewhat yellowish until mid-1976.

1980 (NMSUO, BAA): The NTB was virtually absent when a jetstream outbreak was first photographed by NMSUO on 1980 May 11. One very bright white spot was tracked briefly, but this was too late in the apparition for extensive observations. A series of dark spots followed the white spot, and the NTB thus revived in some very dark segments. However, it faded again within a few months (during solar conjunction). The author had noticed a slight yellow colour developing in the NTropZ several months *before* the outbreak; by late 1980, the NTropZ and NTZ were both dusky yellow. During 1980/81, first NTB(S) then NTB(N) gradually darkened again, while NTropZ became white.

1985? (BAA): The NTB had been rather faint in early 1984, and did not revive fully until 1985. No NTBs disturbance was then present, but a striking orange-grey or dull pink colour developed across the NTropZ, spreading from the northern half from 1985 June onwards, and also affecting sectors of the NTB(S). This may indicate jetstream-related activity.

1990 (BAA, and Sanchez-Lavega *et al.(1991)*; Figs.7.8&9, Plate P24.4): The whole NTB had disappeared in 1988, though much of NTB(N) was reappearing in 1989/90. The jetstream outbreak was discovered photographically by I. Miyazaki on 1990 Feb. 10, as a small white spot, which became large and brilliant by Feb. 15 (Fig. 7.8). On that date it was followed by a dark blue spot (invisible in blue light!), and over the next few days two more white ovals appeared p. and f. it, along with numerous, changeable small dark spots. One of the white ovals was also very brilliant at 0.89 μm, indicating high altitude. The two leading white spots (nos.1 and 2; Fig. 7.9) moved with $\Delta\lambda_1$ = -169 and -160; and no.2 showed a possible oscillation with a period of about 35 days. But after a few weeks it became

evident that the trailing white spot, and all the dark spots, moved with $\Delta\lambda_1 \approx$ -25 instead, so that the outbreak lengthened at 130°/month! Thus the D and C currents were again active simultaneously. In fact, white spot no.1 suddenly switched from one to the other, a deceleration of 5°/day in just two days, and survived for a week thereafter! By late March the dark spots were fairly stable and uniformly spaced at intervals of 13–16°. Apart from a few vivid blue segments at first, these spots were grey. By early April, the outbreak had spread all round the planet, the dark spots were merging to form a continuous dark grey NTB, and the NTropZ developed 'warm'-coloured shading for a few months.

But the revived belt did not turn orange; it remained very dark grey for several years. Less than two years later came a new surprise: a typical C-current outbreak, consisting of four small dark NTBs spots with $\Delta\lambda_1$ = -57 (1991 Nov. onwards, with one such spot persisting into early 1993).

The behaviour of these outbreaks raises many challenging questions. Why is this jetstream so fast? How can the B and C currents exist when they are so much slower than the D current of the jetstream itself? How can they all apply simultaneously to neighbouring spots? Are they at different altitudes? Did the speed of the jetstream change between 1965 and 1970, or did the earlier outbreaks not represent the full jetstream speed? Have the outbreaks now reverted to the pre-1970 style? How can spots be created and maintained at such enormous speeds? What is the instability that can produce either a single spot, or many spots at different longitudes, and/or a diffuse yellow/orange coloration? What controls the five- and ten-year periodicities, and how do they relate to the jovian year and to global upheavals?

8: North Tropical Region (23°N to 9°N)

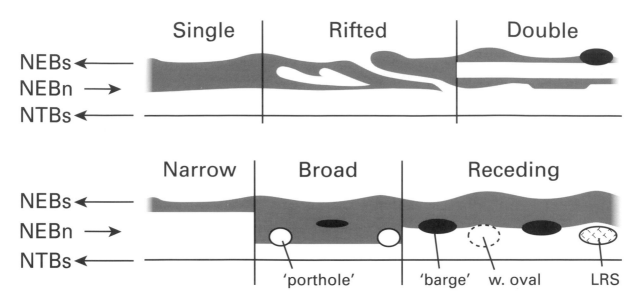

Fig. 8.1. Sketches of typical aspects of the NTropZ and NEB. The NEB commonly evolves as sketched from left to right, (top row) during an episode of rifting, or (bottom row) during a typical cycle of activity – broadening, reddening, and creation of stable ovals. W., White; LRS, Little Red Spot. (Detailed forms of NEBs disturbances are not shown; see Fig. 9.1.)

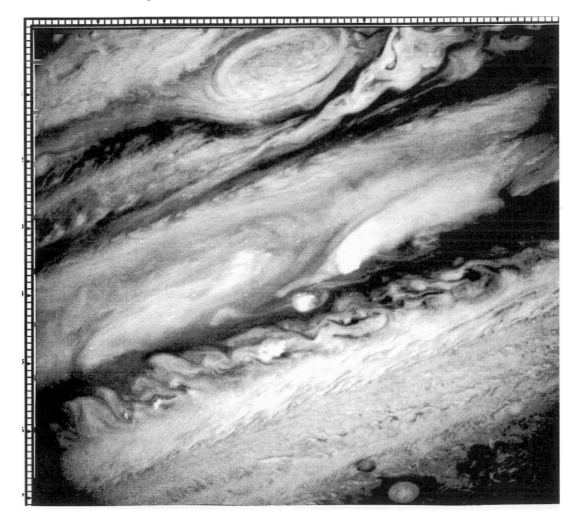

Fig. 8.2. Voyager 1 view of the NEB and EZ, showing the turbulent 'rifted region'. At centre, a NEBs plume with a brilliant white core is overtaking another bright white spot that has just erupted in the southern NEB. The GRS is at top. This was taken on 1979 Feb.24, 10 days before close encounter; see also Fig. 9.2. (Image 16105.46, J1–10d, orange. NASA.)

8: North Tropical Region (23°N to 9°N)

8.1 OVERVIEW

The North Tropical region is perhaps the most regular domain on the planet. Indeed, the North Tropical Zone (NTropZ) and North Equatorial Belt (NEB) show many of the planet's features in their most typical forms.

The NTropZ is usually bright and white, while the NEB is almost always a substantial, very dark belt. The boundary between them, the NEBn edge, fluctuates from year to year, most commonly lying between 14°N and 20°N although it has an extreme range from 11°N to 22°N (Table 8.1). The NEBs edge is usually irregular, with conspicuous projections and plumes extending into the Equatorial Zone, but its mean latitude never deviates much from 7–8°N.

The NEB sometimes appears to be double, but this can be due to several different kinds of activity. Sometimes it is due to accumulation of 'rifts' (bright streaks) in the middle of the belt, but it is possible that doubling may sometimes occur quietly, and sometimes it is mimicked by independent accumulation of great dark oblong spots on both edges of the belt. Sometimes the NEB even appears triple, when it has extended unusually far north and a 'North Tropical Band' has been incorporated as an extra NEB(N) component.

The NEB frequently plays out a distinct pattern of activity, to be defined for the first time in this chapter. A typical cycle consists of a northward broadening of the belt, often with rift activity, followed by increased redness and by the appearance of an array of bright and dark ovals on NEBn. Between 1890 and 1915, these cycles occurred every 3 years and involved wholesale fading of the NEB followed by turbulent revival. It will be shown here that the same pattern has continued in subsequent years, the only difference being that the belt does not disappear before the onset of activity. These cycles are the northern equivalent of the famous SEB Revivals. Details of these cycles, and a chronicle of the width, colour, and activity of the NEB, will be deferred to section 8.5.

Various types of conspicuous spots are common in the North Tropical region, as well as many less distinct irregularities which go down to the limit of visual resolution. In a given apparition, spots of the same type(s) tend to be present at many longitudes. Listed from north to south, they can include the following: white spots or red ovals in the NTropZ; minor irregularities on NEBn; dark 'barges'; active regions of northern NEB, liable to outbreaks of white 'rifts'; white rifts in mid-NEB; and dark bluish patches and white plumes on NEBs. (Details are in sections 8.4 and 8.5.) The 'NEBn' white ovals and dark barges may lie in the NTropZ or within the NEB, depending on the width of the belt, but regardless of their surroundings, all spots move at speeds characteristic of their latitude, from System II in the NTropZ to System I on the NEBs edge.

Currents in the region

Most of the large spots in the domain move with the North Tropical Current (NTropC), including all the NTropZ white and red ovals at 19°N and the NEBn dark barges at 16°N (section 8.4), as well as many smaller features. Because this current has shown interesting changes during its history, it will be described fully in section 8.3.

There is a retrograding jetstream at 17.6°N, discovered by Voyager, but it has a modest speed and has never been observed from Earth. High-resolution Earth-based photos have revealed no more than a tendency for tiny features at 17–18°N to move slightly slower than the rest of the NTropC (sections 8.3 and 8.4).

Active regions of northern NEB at ≈15°N tend to drift somewhat faster than the normal NTropC. From this latitude, a steep gradient of velocity leads up to the Equatorial Current (System I). Distinct white rifts in mid-NEB (latitudes 10–14°N), arising either within the active regions or in isolation, can move at any speed between System I and System II, but they mostly drift in the range $\Delta\lambda_2 = -80$ to -150 ($\Delta\lambda_1 = +150$ to $+80$); this range has been named the North Intermediate Current (NIC; section 8.5). The stretching of these rifts in the veloc-

Table 8.1. *Latitudes of the NEB edges.*

	(1913–1947)	(1950–1991)	(1913–1991)
NEBs	+7.1 (±1.1; 32)	+7.5 (±0.6; 27)	+7.3 (±0.9; 59)
NEBn	+17.5 (±2.7; 32)	+18.1 (±2.2; 27)	+17.8 (±2.5; 59)

These are averages (±s.d.; *N*) of apparition means (where s.d. = standard deviation; *N* = number of values). The data are those plotted in Fig. 8.14 (excluding imprecise values).

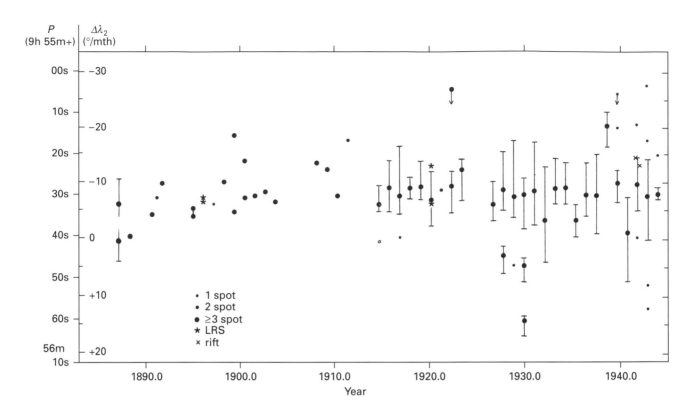

Fig. 8.3. History of the N. Tropical Current (NTropC) from 1887 to 1991. From data listed in Appendices 1 and 3. Speed is shown as *P* (rotation period), as $\Delta\lambda_2$ (degrees per month), and as *u* (m/s). Large symbols are apparition means of ≥3 features, and vertical bars indicate the range of speeds (not published before 1914). In some apparitions, two averages are plotted, representing either different sectors of longitude (see text), or different sources of data. Arrows indicate spots with marked deceleration.

The mean NTropC speed from these data is as follows. (1887–1948:) $\Delta\lambda_2$ = -8.4°/month, period 9h 55m 29.2s (±5.4s, standard deviation of apparition means); (1949–1991:) $\Delta\lambda_2$ = -10.0°/month, period 9h 55m 27.1s (±9.3s); (1887–1991:) $\Delta\lambda_2$ = -9.0°/month, period 9h 55m 28.3s (±7.4s).

Some slow-moving NEB rifts are included (crosses); faster-moving rifts are shown on Fig. 8.13.

ity gradient can sometimes be watched over a few days. The NEBs edge, at 7–8°N, moves with the N. Equatorial Current (Chapter 9).

These major features are summarised in Table 8.2.

8.2 NORTH TROPICAL ZONE

Coloration episodes

The NTropZ is always present and usually has the typical white or cream colour of the jovian zones. Its brightness varies, but there have been few years in which it was heavily shaded.

Most coloration episodes in NTropZ have been in association with NTBs jetstream outbreaks and were described in the previous chapter. Typically, the NTropZ remains quite bright but looks yellowish or orangish after (or occasionally just before) the outbreak. More rarely, there have been episodes of *dark* brownish-yellow coloration in the NTropZ, usually associated with marked revival or expansion of the north half of the NEB which was strongly reddish at the same time (section 8.5). So these are NEB expansion events in which the usual reddish colour extends at least part way across the NTropZ as well.

The following is a list of all NTropZ coloration episodes.

1903 (P.B. Molesworth): NEB very red and NTropZ brownish-yellow except for a bright white strip on N. edge.

1906/7: Red colour spanning the same latitudes as in 1903, again just after a NEB Revival, though this time there was also a NTropB (Plate P5.2).

1912–1914: In 1912, *before* the NEB Revival, NTropZ and NTZ were merged and all dull yellowish. Same colour ('brownish-yellow') over much-narrowed NTropZ in 1913, continuous with reddish half of NEB after Revival, just as in 1903 and 1906/7. In 1914, narrow NTropZ (and NTZ) were again heavily shaded at times (Plate P5.5) though no colour was noted.

1926: Bright but orangish, coinciding with minor NTBs outbreak. Heavily disturbed sector in 1927/28 (see below).

1928/29: Dull yellow or reddish (as was much of the recently-expanded NEB north half).

1930/31: Yellowish, coinciding with NTBs outbreak.

1932/33: Slightly shaded, occasionally drawn tawny; associated with red NEB?

1941/42: Very dull yellow or brownish (again associated with great broadening of NEB – and with a NTBs outbreak though it had started two years earlier).

1944: Dull rose-coloured according to F.J. Hargreaves (after NTBs outbreak).

1970–1971 (LPL): In 1970, photos showed southern NTropZ yellow-orange while NEB had become ochre (*before* the NTBs jetstream spot appeared) (Plate P13.4; Chapter 7.4). Colour shifted to NTB(S) in 1971.

1975 (BAA): Yellowish after NTBs outbreak.

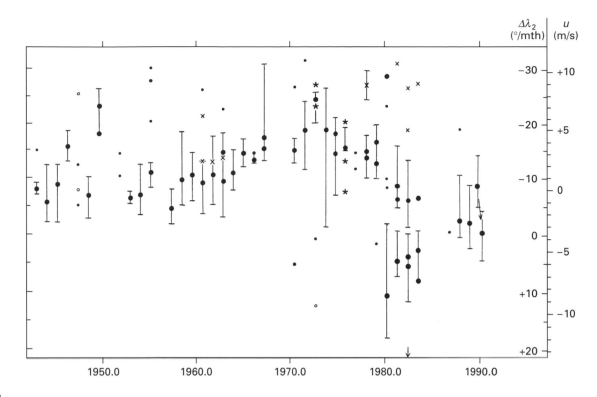

Fig. 8.3B.

Table 8.2. *Typical N. Tropical spots and currents*

Name	Latitude	Current
White spots and Little Red Spots	19°N	NTropC
Tiny features	17–18°N	Slow NTropC to NEBn retrograding jetstream
Dark barges	16°N	NTropC
Active regions	15°N	Fast NTropC
White rifts in mid-NEB	11–14°N	NIC
Dark patches, projections, white spots	7–8°N	NEC

Table 8.3 *The North Tropical Band*

Apparition	Description	Lat. β''
1906/7	NTropB, becoming bluish-grey	23.8
1907/8	NEB(N), strongly blue	22.4
1908/9	NEB(N), bluish-grey (*NTropC*)	22.8 or 20.4
1942/43	NTropB	22
1962/63	NTB(S) or NTropB	23.6
1963/64	NTB(S) or NTropB (*NTropC*)	23.8
1967/68	NTropB, faint blue (also 1969)	23.9
1985	NTropB, bluish (*NTropC*)	[19.0]
1988/89	NEBn, brown or bluish-grey	22.0
1989/90	{ NTropB (reddish or grey)	[20.8]
	{ NTB(S) (blue)	23.1
Average:		22.9
		(±0.9) (N=9)

1980/81 (BAA): NTropZ yellow colour commenced in 1979/80 *before* NTBs jetstream outbreak; after it, NTropZ and NTZ were both yellow.

1985 (BAA): Dark pinkish or dirty orange colour, subsiding through yellowish-brown to white. This was the only episode not associated with NEB revival nor NTBs outbreak, but a possible relationship to NTBs outbreaks was discussed in the previous chapter.

1990 (BAA): 'Warm' colour after NTBs outbreak.

Spots in the NTropZ will be discussed in sections 8.3 and 8.4. A sector of extensive disturbance with many spots and shadings was present in 1927/28, but it does not seem to have been a coherent phenomenon.

This table includes all definite observations of a narrow bluish line, whether it appeared to be NEBn, NTropB, or NTB(S). The band may also have been present as the north edge of an expanded NEB in several other apparitions. Latitudes were obtained as follows: (1906–1909) micrometer measurements by Scriven Bolton in 1906–1909 and Phillips in 1908/9; (1942/43) estimated relative to the measured latitudes of the NEB and NTB; (1962–1964) medium-precision photographic measurements by ALPO; (1967–1989) measured directly from high-resolution photographs.

The two values in square brackets, which seem anomalous, are omitted from statistics.

'NTropC' indicates when spots on the Band were tracked. In each case, they moved with the NTropC, viz: 1908/9 (BAA: two spots with $\Delta\lambda_2$ = –10.5); 1963/64 (ALPO: two dark condensations with $\Delta\lambda_2$ = –12.5 and –15); 1985 (BAA: two tiny white spots with $\Delta\lambda_2 \approx$ –8 and 0).

North Tropical Band

Sometimes there is a very narrow bluish band in the NTropZ, at about 23°N (Table 8.3). In some years it lies in the middle of clear white NTropZ, and a similar bluish band at 19°N was in the middle of dull orange NTropZ in 1985 (Fig. 4.1). It formed an extra NEB(N) component during the great expansions of the NEB in

1906/7 and 1988 (Plate P23). Conversely, sometimes it has been difficult to distinguish from a very faint NTB(S), and the very dark bluish NTB(S) that follows some NTBs jetstream outbreaks is in the same latitude. However, when features on the NTropB can be tracked, they move with the NTropC (Table 8.3).

In Voyager colour photographs, a very faint bluish fringe was visible alongside the faint orange NTB(S), and the closeups (Plate P20.2) revealed that it was formed by blue streaks between the highly sheared white streaks of the zone; these may be thinnings in the white cloud cover. Perhaps the blue NTropB appears when the white clouds are disrupted still further.

8.3 VARIATIONS IN THE NORTH TROPICAL CURRENT

The North Tropical Current (NTropC) controls all the visible features between 22°N and 15°N, that is, in the NTropZ and northern NEB. Although there are variations in the speed of the current, it applies to all major spots whether bright or dark, and it is unaffected by the ebb and flow of the NEBn edge across these latitudes.

The current was first recorded in 1787 by J.H. Schröter ($\Delta\lambda_2 = -5$, period 9h 55m 34s). It was recorded again in 1835 ($\Delta\lambda_2 = -12$, period 9h 55m 24s), in 1866 ($\Delta\lambda_2 = -17$, period 9h 55m 18s), and 1881 ($\Delta\lambda_2 = -4$, period 9h 55m 35s) (Peek, *1958*). Since then, it has been recorded in most apparitions up to the present (Fig. 8.3). Its average speed is $\Delta\lambda_2 = -9$ (period 9h 55m 29s).

The NTropC is the only slow current, apart from the STC, which has shown a long–term change in speed during recorded history (Fig. 8.3). In the 1880s and 1890s, it was unusually slow, and in the late 1960s and 1970s, it was unusually fast. In the intervening decades the mean speed was fairly constant although there were fluctuations for a year or two at a time.

The changes in speed were not associated with any other visible changes in the region. There is no way of telling whether the NEBn retrograding jetstream has changed its speed, although it has probably not changed its position, as there has been no change in the typical forms and latitudes of the spots in this region.

Distinct sectors with different drift rates have been notable in the following apparitions, at times when a prominent array of alternating white and dark ovals allowed speeds to be determined all round the planet.

1887 (A.S. Williams): Different speeds in two hemispheres.

1899–1900: Different speeds in two sectors, separated by a sharp boundary. In 1899, spots did not cross the boundary, but in 1900, some spots did; they suddenly accelerated to cross the fast sector, 90° long, in which the NEBn edge was generally indistinct and spots appeared veiled. For example, one white spot in 1900 (tracked from 1899) appeared 'nebulous' while rushing through the fast sector, then distinct again on decelerating into the slow sector (T.E.R. Phillips).

1915–1917: Speeds varied almost sinusoidally with longitude.

1927–1930: In 1927/28, one hemisphere moved exceptionally slowly. Only one slow spot was recorded in 1928/29, but in 1929/30 there were two slow sectors alternating with two fast sectors. The mean speeds were $\Delta\lambda_2 = -7.9$ (3 spots), +4.5 (6 spots), –8.2 (10 spots), and +14.1 (4 spots). Each longitude sector contained both white and dark spots, but the four very slow spots in the last sector were all white, and crossed the tracks of faster dark spots.

1939/40: One white spot moved very fast ($\Delta\lambda_2 = -20$) as it crossed an inactive stretch.

1942/43: Again a barren stretch through which one dark spot moved very fast ($\Delta\lambda_2 = -27.5$), having apparently moved northwards to reach the NEBn edge.

1959–1960 (ALPO): Fast and slow spots were interspersed in 1959, and in 1960 the speeds varied sinusoidally around the planet.

1974; 1980–1984: The wide ranges reported in these years were due to tracking of very small features apparently influenced by the retrograding jetstream (see below).

1990/91: Soon after a stationary array of barges had faded away, the only features visible were 4 white spots moving fast ($\Delta\lambda_2 \approx -9$) along what had been a 60°-long barge-free sector. Then, after a large NEB rift had passed, the white spots slowed down, and a new array of stationary dark barges developed between them.

The fastest speeds in the NTropC are shown by the rare Little Red Spots in the NTropZ (see section 8.4), and by spots crossing 'blank sections' in a few apparitions (see above), and by active or rifted regions in the northern NEB (see section 8.5). Because these regions merge into the North Intermediate Current, one cannot put an upper limit on the speeds of the NTropC.

The slowest speeds in the NTropC, apart from a handful of spots described below, have been $\Delta\lambda_2 = +7.5$ in several apparitions. R.B. Minton (LPL) noted that the slowest spots in 1973 and 1974 were those between 17° and 18°N; these latitudes coincide with the retrograding jetstream subsequently discovered by Voyager. A very small dusky spot in 1971 (NMSUO, 18.4°N, $\Delta\lambda_2 = +5.1$) seems to belong with this group. The high precision of the LPL data may thus have revealed a minority of spots that are partly influenced by the jetstream, whereas most sizeable spots are centred north or south of it.

Fig. 8.3 shows that a few spots have been reported with retrograding speeds up to $\Delta\lambda_2 = +17$, particularly in slow-moving sectors in 1929/30 and 1981. Reports of even slower speeds, $\approx+25$ in 1939/40 (BAA) and +28 in 1983 (SAF), were not thoroughly documented.

The way in which the NTropC reduced its speed in the early 1980s was quite remarkable, as revealed by the analyses of amateur photographs by the Société Astronomique de France (SAF). The change apparently began with a phenomenon seen in the Voyager images in 1979, that is, the long train of wave-like shadings at 10° intervals on NEBn following a long-lived NEB rifted region (section 8.4 and Figs. 8.7–11). These were not clearly observed from Earth, but in 1980/81, when the rifted region had expanded around much of the NEB, a similar 'wave-train' was observed: the SAF reported the NEBn to be very disturbed with a series of 7 or 8 small projections about 18° apart, moving with $\Delta\lambda_2 = +10.5$ (range 0 to +17) while the other NEBn features had negative drifts. The picture was the same in 1982, with 6 projections $\approx12°$ apart with $\Delta\lambda_2 = +4.4$ (range –1 to +7.5).[1] In 1983, the SAF tracked as many as 38 features in the current; with so many, there must be some risk of confusion, but the published SAF chart shows

---

[1] Sanchez-Lavega and Quesada (*1988*) confirmed this picture in 1982, 1983 and 1984, reporting up to 7 projections $\approx12°$ apart with $\Delta\lambda_2 = +1$ to +18.

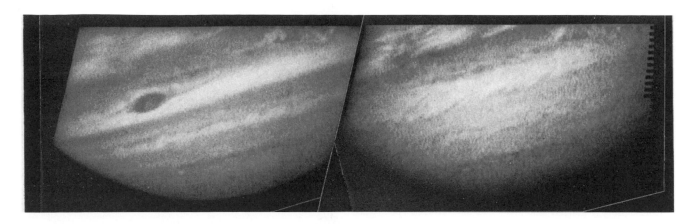

Fig. 8.4. Pioneer 10 view of the Little Red Spot in the NTropZ. (Left) Blue image; (right) red image. (Image A16. For colour version see Plate P16.3. NASA.)

that the enormous range of reported speeds really existed. The NEBn was divided into four sectors with alternately positive and negative drifts, each sector containing dark *and* bright spots (mean $\Delta\lambda_2$ = –3.1, +4.0, –7.5, +5.1), and one of the slow (positive $\Delta\lambda_2$) sectors was again neatly following a persistent rift. In 1984, the SAF reported even more spots, and agreed with the BAA's conclusion from just four features: that the slow drift rates now applied to the majority of features in the NTropC, for the first time in recorded history. The NEBn was then blank for several years, but when major spots reappeared from 1987 onwards, their drifts ($\Delta\lambda_2$ ≈ 0) were still unusually slow. Thus a change that began with tiny wave-like features retrograding from a disturbed sector had developed into a general deceleration of the whole NTropC.

One wonders whether the unusually slow drift in 1940/41, when Peek described the NEBn edge as having a series of northward steps followed by a 'sawtooth' pattern of small dark spots, was perhaps a similar phenomenon.

8.4 SPOTS ON THE NEB NORTH EDGE

Conspicuous spots are common on and around the NEB north edge. The more distinct ones fall into several classes which tend to occur in characteristic latitudes (Table 8.2 and Appendix 2). The best-known are white ovals (at 19°N) and dark barges (at 16°N); there are also rare Little Red Spots (at 19°N). These spots are only the most distinct types from a wide range of features that are commonly seen on the NEBn. However, spacecraft photos suggest that many of the less distinct features are similar though fainter or smaller.

Little Red Spots

Thanks to a well-observed example during the Pioneer 10 mission, it is now recognised that the NTropZ sometimes harbours small red ovals that are miniature sisters of the Great Red Spot. They are, inevitably, named Little Red Spots (LRSs). Although such spots had been recorded before Pioneer, they had not been clearly distinguished from the other dark spots in the region, particularly the NEBn barges

which can also be distinctly reddish. But Pioneer 10's LRS (Fig. 8.4; Plate P16.3) was different from typical barges in its colour (red rather than reddish-brown), shape (a lozenge, not a 'barge'), and latitude (19°N rather than 16°N) (R. Beebe and T. Hockey, *1986*). These characteristics have been used in a review of the visual records to determine how often LRSs have occurred in the past.

It turns out that they have only been clearly recorded in four apparitions: 1895/96, 1919/20, 1973, and 1976/77. In each of these apparitions, there were at least two such spots, 70–120° apart in longitude, at 19°N latitude (listed in Appendix 2). They must have a mean lifetime of about a year. All the LRSs were detached dark spots in the NTropZ, sometimes bordered by a recess in the NEBn edge. (It is a moot point, however, whether a LRS would be discernible if NEB material did surround it.)

Details of the spots in these four apparitions are as follows.

1895/96: After the NEB narrowed, several small very dark spots remained in the NTropZ, and two of them were seen as red by several observers ('intensely red' to Antoniadi; Plate P4.3). From the shape and colour which they both shared, they were named the 'Violin' and 'Garnet' spots; perfect seeing revealed a shape like o°. Their speed ($\Delta\lambda_2$ = –7) was faster than for other NTropC spots. They lasted for most of the apparition then faded away.

1919/20: As the NEBn edge receded irregularly, a striking set of spots appeared that included dark barges and white ovals and two LRSs; $\Delta\lambda_2$ = –13 (the fastest spot in NTropC) and –6.5 (Plate P5.8&9). Both were very dark oval spots, and one had a minute white core in the best seeing. This LRS was probably re-identified in 1920 Nov.–Dec. although it had decelerated.

1973 (LPL, BAA): The main LRS (LRS1) was well shown in Earth-based and Pioneer photos (Fig. 8.4, Plates P15.4 and P16.3); one observer also saw it as red. It was 12° long with tenuous bands sometimes connected to the ends. LRS2 was smaller, also reddish according to photos, but darker than LRS1 in yellow light. LRS2 formed from a NEBn projection in May and disappeared in October. They had $\Delta\lambda_2$ = –26 and –23, being the fastest spots in the NTropC. As LRS2 disappeared, 'LRS3' formed ($\Delta\lambda_2$ = +1), alongside the NEB rifted region (see below, and Fig. 2.9B). But it was grey-brown, not red.

1976/77 (BAA, UAI, NMSUO): Three LRSs were seen. One was described by Beebe and Hockey (*1986*); it was 10.8° long, 5.3° wide, and had the same red colour as the GRS, being as bright as the STropZ at the infrared wavelength of 0.8 μm. To visual observers, this LRS was dusky and only occasionally seen as pink; the second was darker and more widely seen as red. They had $\Delta\lambda_2 = -20$ and -13, the former being the fastest spot in the NTropC. (See Fig. 1.18.) As in 1973, there were adjacent segments of NEB(N) which were also strongly reddish. A third similar spot was visible for only one month and was only once noted as red by a BAA observer but was confirmed as red like the other two spots by the UAI ($\Delta\lambda_2 = -8$).

The colour is difficult to observe visually except with large apertures, and so is subject to some uncertainty. Thus in most cases, we cannot be sure whether the colour was different from that of NEBn barges. There is little information on the visual colours of spot 1 of 1919/20 (only a general statement that several such spots were very dark red), on LRS2 of 1973 ('red' in this LPL report included brown), and on spot 1 of 1976/77 (only once seen as red by the BAA). So the record of LRSs may well be incomplete.

Note that LRSs tend to move faster than other features in the NTropC; they did so in 1895/96 and 1919/20 when the NTropC was slow, and still maintained an edge in 1973 and 1976 when the whole current was faster.

In addition to *bona fide* Little Red Spots, other features occur with similar shape and latitude which are not necessarily red. (They are also listed in Appendix 2.) Outside the 4 'LRS' apparitions, such features were recorded clearly only when powerful telescopes were used; they were too small and/or faint to be notable to visual observers. Such minor features are probably common, and do not necessarily move fast.

No LRS has been observed while forming, but there are records of the formation of similar brownish spots which were later photographed by Pioneer 10 and Voyager 2. According to Minton (LPL), the grey-brown 'LRS3' of 1973 formed alongside an active rifted region of northern NEB; he suggested that 'a large eddy in the NEBZ may have contributed to the rapid formation of LRS3 in the adjacent NTropZ.' A spot in 1974 was similarly situated. He also reported that a non-red NTropZ oval in 1974 began as a NEBn spot that shifted from 16.3°N to 18.7°N, but in retrospect, this would have crossed the NEBn jetstream (then undiscovered). It is possible that all these spots arose in a similar manner to one observed by Voyager that is described below.

White ovals and dark barges

These are the well-known spots of the NEBn edge, shown by spacecraft to be anticyclonic and cyclonic respectively. The white ovals are always slightly further north than the dark barges. The typical patterns of these spots have been observed throughout history, irrespective of the long-term changes in NTropC speed and NEB width. (Many examples are shown in the Plates.)

The white ovals occur at 19°N and are sometimes very bright. They are actually circular white spots spanning 5–7° (6000–8000 km). Depending on the width of the NEB, they may be entirely within the NTropZ (when, lacking dark rims, their borders are hard to define), or they may form bays in the NEBn edge, or they may be completely enclosed by NEB material. The last aspect can be striking, as in the very broad NEB of 1914 or 1989, producing an appearance of 'portholes' in the belt (Plate P5.5).

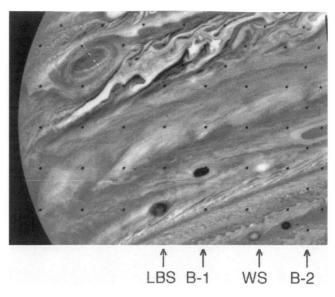

Fig. 8.5. Voyager 2 view of the NEB and EZ, showing various ovals. In this violet image, red and brown features appear dark. Features are: LBS, little brown spot that appeared in NTropZ since the Voyager 1 encounter (anticyclonic); B-1, barge no.1 (cyclonic); WS, white oval (anticyclonic); B-2, barge no.2, obviously paler than no.1 (cyclonic). The line of the NEBn retrograding jetstream can be followed, snaking between these alternating circulations. LBS and B-1 are also shown in Plates P18.4 and P19.2. (Image 20120.41, J2–18d, violet, filtered; 1979 June 21. NASA.)

The dark streaks or oblongs or 'barges' mostly lie around 16°N, and again may lie off the NEBn edge or completely within the NEB. They are typically 2.5° to 3° wide in latitude. They are often strung along a narrow NEB(N) component and less conspicuous versions are just indistinct condensations of it. When the NEB is very broad, some of the 'barges' can extend further north than usual and a few seem to be condensations of belt segments up to 19°N; these cannot be clearly distinguished from the lozenge-shaped, LRS-like spots. The name 'barges' was first used in 1917/18 by M.A. Ainslie (see Chapter 3.3), as these dark oblongs were left isolated by the retreating NEBn. He described them then as 'chocolate red'. The barges often have a striking reddish-brown colour, which has also been described as ruby-coloured (1928–1930, Phillips), black-red like clotted blood (1928/29, Peek), roseate brown (1928/29, Antoniadi), copper-red (1946, Reese), ox-blood red (1950, BAA), or rich chocolate brown (1978/79, Rogers). But there is a real range of colour; one large and outstandingly dark barge in 1957/58 was definitely not reddish according to most observers, and those of 1990 were merely the same dark brown as the NEB itself.

These spots usually occur in sets; there may be half-a-dozen white ovals, or dark streaks, or both, scattered around the planet. For example, in 1974 there were mostly dark spots, and in 1975 mostly white spots; while in 1988/89 there was approximately one dark spot every 45° and one white oval every 90°. The spacing is not usually periodic, although it may tend to become so when there are many of these spots. As A.S. Williams first noticed, there tends to be 'persistence of type' over several years: the pattern of about 8 dark barges and 9 white ovals that he recorded in 1887 and 1888 had developed over several years, following several years in which the NEBn had been featureless. Similar conspicuous arrays of bright and dark spots were recorded in 1914–1920 (Plate P5);

1928–1930 (Plate P7); 1931–1933; 1942/43 (Plate P8); 1953–1954; 1961; and 1964–1966. Almost all these arrays developed within a year following a major northwards expansion of the NEB, and no complete arrays developed while the NEBn latitude was stable in the 1970s – although there were still some dark barges *or* white spots that looked typical. It may be significant that the more restricted array of four dark barges in 1978/79 was in the hemisphere following an active rifted region (see below). A classic array of bright *and* dark spots was not observed again until 1988/89 (Plates P23 and P24.3).

Surprisingly little is known about the genesis of these spots. The following are the recorded instances and they do not reveal any common pattern. A notable advent of dark spots was observed in 1890, when six small black spots appeared on a narrow NEB(N), sometimes joined to NEB(S) by fine curving wisps (Plate P3.7). As described by E.E. Barnard, they were at first black and round like satellite shadows, but later became dark reddish and elongated; by 1891, they had a deep red colour. White spots were observed to appear in 1914, beginning as 'small but intensely brilliant nuclei' within pre-existing dark spots on NEB(N). In 1975, a white spot formed from the NTBs jetstream outbreak; it lay slightly further north in the NTropZ than others. Both dark and white spots arose in a different way in the very broad NEB in 1988/89, when the dark barges condensed gradually along the central NEB component, and the round white spots materialised equally unremarkably in the north half of the belt. In 1990/91, again, barges began as dark streaks which progressively thickened, soon after a large NEB rift had passed.

An unusual interaction occurred in 1942/43. A dark streak in mid-NEB moved northwards to the NEBn; another dark spot appeared p. it and grew into a streak, and a white spot appeared between them, forming a triplet.

The lifetimes of these spots cover a wide range. Many remain visible for no more than one or two months, but many last for a full apparition, and some have been tracked for two years. More could undoubtedly be followed from year to year were it not that their large numbers, and the possibility of sudden changes in their visibility or their drift rates, make identification uncertain after a lapse of several months. A.S. Williams and W.F. Denning claimed to have tracked two brilliant white spots from 1885 and 1887 up to 1891, and although some of their identifications may be uncertain, the persistence of such spots over this period would not be implausible. There are few other documented examples, viz.: two dark spots tracked through 1890 and 1891 (Barnard); one white spot tracked through 1899 and 1900 (Phillips); another 1941–1943 (BAA); another 1961–1962 (ALPO). Of the four barges viewed by Voyager in 1979, only one persisted to the first half of next apparition, giving them an average life of about a year. But of the many dark and white ovals that appeared in 1988/89, almost all could be tracked through the following apparition as the NEBn receded around them, then disappeared in summer, 1990.

Given the range of speeds that sometimes coexist in the NTropZ, one must ask whether spots sometimes collide. Merging of dark spots or streaks in the north half of NEB is probably not uncommon. Such mergers were recorded in 1927/28 (BAA), 1955/56 (BAA), and 1974

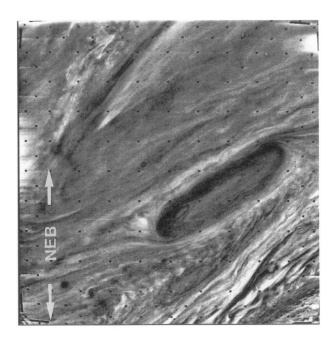

Fig. 8.6. Voyager 1 closeup of NEBn barge no.2, 10° (12,000 km) long. The cyclonic circulation is obvious. Also note a diffuse pattern of N.-S. bands representing transverse waves on the NEBn retrograding jetstream around the N. edge of the oval, with a wavelength of 0.8° (940 km). (Image 16320.09, J1–3d, violet, filtered. NASA.)

(LPL). Two white ovals in the same latitude merged in 1966/67 (NMSUO), but mergers between white ovals seem to be rare.

Occasionally, the regular arrays of spots include striking pairs or triplets of adjacent white and dark spots, centred about 10–12° apart. The formation of a triplet in 1942/43 was described above. In three other years, groups have formed when a fast-moving spot comes up against a slower-moving one from the following side and they then remain together:

1928/29: There were two striking and colourful spot triplets, each consisting of two very dark ruby-coloured streaks separated by a brilliant white spot. The first of these already existed (Plates P6.8, P7.2); the second formed as rapidly-moving white and dark spots caught up with another dark ruby-coloured spot.

1967/68 (NMSUO): There was one large, very dark brown bar, which was approached by successive faster-moving white spots. The first, with a small dark spot, formed a transient triplet with the barge (Plate P13.3). Then another white spot moving even faster ($\Delta\lambda_2 = -31$), and a mid-NEB 'rift' moving faster still, approached until all these bright features merged into a single white spot pressing against the f. end of the big dark barge. The barge disappeared suddenly a few weeks later.

1988/89 (BAA): An initially stationary white spot made jerky prograding movements (period ≈3 months) until it came to rest adjacent to a dark spot.

Given the opposite circulations of the dark and bright spots revealed by spacecraft, these triplets must fit together like cogwheels, and such an arrangement may be stable irrespective of how it forms.

Spacecraft observations of NEBn spots

The Pioneer photos showed that several visually-observed little blips on NEBn were in fact small circular brown rings, much like the anticyclonic rings seen in other domains (J. Rogers and P.

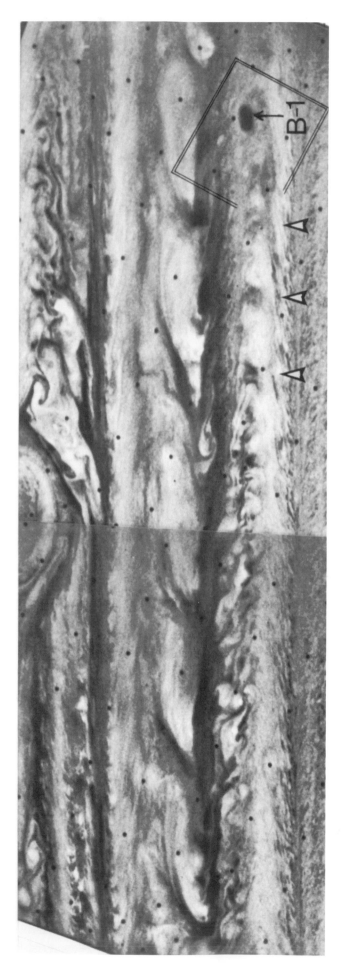

Left:

Fig. 8.7. Voyager 1 photos forming a strip-map of the equatorial and N. Tropical region. Includes the NEB rifted region (left half of map) and the NEBn wavetrain f. it (arrowheads) leading up to the small barge no.1 (at right). The dusky streak Np. barge no. 1 is the 'circulating current' that became the Little Brown Spot. Box outlines area of Fig. 8.8, 10 days later; also see Plates P18.4 and P19.1&2. The GRS is at top centre. (Image 16005.01 plus 16007.30, J1–13d; orange, filtered. NASA.)

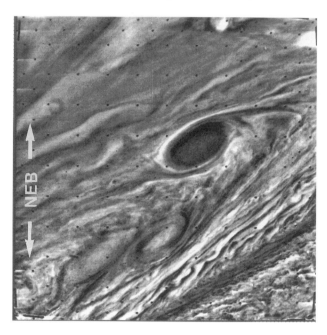

Fig. 8.8. Voyager 1 closeup of the small barge no.1 (very dark oval) and the incipient anticyclonic circulation Np. it (below it; this became the Little Brown Spot in Fig. 8.5). (Image 16318.36, J1–3d, violet, filtered. NASA.)

Young, *1977*). The Little Red Spot and the similar brown spots must also have been anticyclonic features, by comparison with the velocity profile later revealed by Voyager.

Voyager discovered that the white and brown ovals in the NTropZ have anticyclonic circulations and lie on the north side of the retrograding jetstream, while the dark barges in the NEBn have cyclonic circulations and lie on the south side of the jetstream. The jetstream snakes between these two sets of spots (Fig. 8.5).

Four great dark barges were present during the Voyager encounters (e.g. Plate P16 and Fig. 8.6). The closed cyclonic circulation around and within the largest of them (no.3) was mapped in detail by A. Hatzes and colleagues (*1981*). The clockwise flow at 20 (±6) m/s gave the spot a circulation period of 10–20 days. The average radial velocity gradient was 3.6 m/s per 100 km ('vorticity' 3.6×10^{-5} s^{-1}), three times that of the surrounding belt. The barge was oscillating with a period of 15 days, not in longitude but in length and width, maintaining a constant area; its dimensions oscillated between 16400×2600 km ($13.6° \times 2.2°$) and 13600×3200 km ($11.2° \times 2.8°$). No external cause of this oscillation could be seen, and although there were some bulges circulating around the perimeter, they could not have caused the oscillation or the period

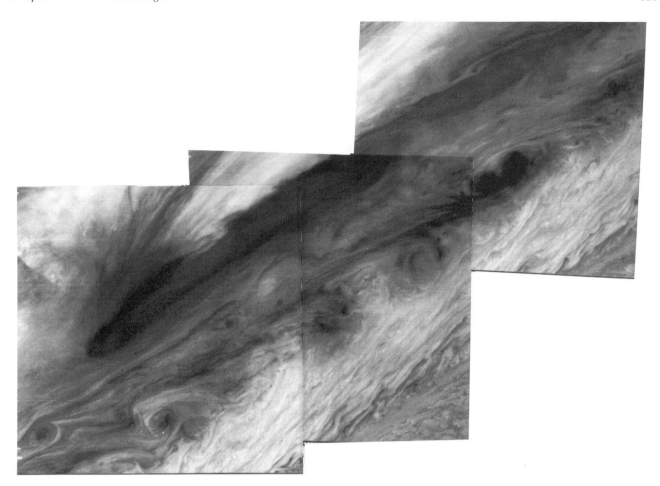

Fig. 8.9. Voyager 2 closeup of the NEBn wavetrain f. the NEB rifted region. The 'waves' (arrowed) are diffuse dusky areas without any regular relation to the underlying cloud streaks. A new barge is forming towards the f. (right-hand) side. (20587.06, 18, 26, J2–3d, orange, unfiltered. NASA. A colour version of the right-hand frame is Plate P16.5. A colour-enhanced version of the left-hand frame was on the cover of *Science* (Smith *et al.*, *1979b*); this overlaps Fig. 9.13 which shows more of the NEBs projection.)

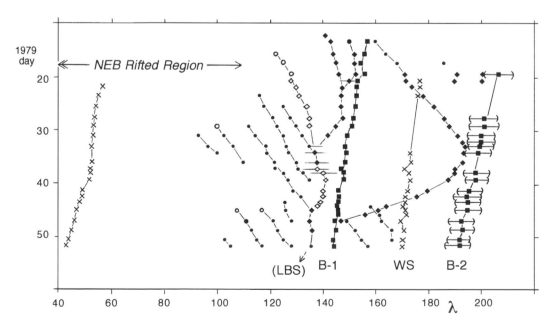

Fig. 8.10. Drifts of spots in NEBn and NTropZ in the wake of the rifted region, from Voyager data. Voyager 1 maps produced by the Voyager Imaging Team were measured by the author. The longitude scale is arbitrary and differs from λ_2 by 6°/month (established by measurements on features that were tracked visually). The time scale is in days from 1979 Jan. 1 (close encounter was on day 64). Symbols: X white oval NTropZ; ■ dark barge NEBn; ● dark 'waves' NEBn; ◆ dark lozenge NTropZ; ○ ◇ similar features with white cores (several of the dark 'waves' were first seen around the tips of white streamers from the rifted region).

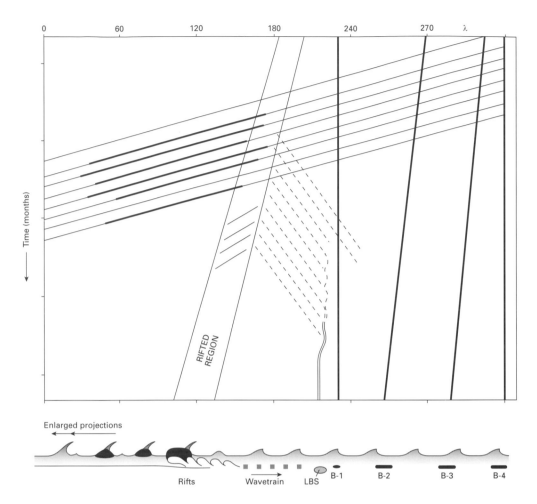

Fig. 8.11. Arrangement of the NEB during the Voyager encounters in 1979. This is a schematic drift-chart, using a longitude system moving with the NTropC at $\Delta\lambda_2 = -10$. Note similarity to SEB Revivals (Fig. 10.12 *et seq.*), including: 'retrograding branch' (Voyager's NEBn wavetrain; dashed lines), 'central branch' (the NEB rifts), 'prograding branch' (NEBs dark projections, intensified on passing the rifted region; Chapter 9.2), and quasi-stationary features (the barges and Little Brown Spot). Also note resemblance to NEB Revival (Fig. 8.15).

would have been half the circulation period. At the time of the Voyager 2 encounter in 1979 July, bright yellow-white cloud streaks from the NTropZ were pushing over this same barge. But this barge remained visible from Earth until 1979 December, when it was overtaken by the active region of NEBn and was lost.

The barges were also warm in the infrared (5 μm and longer wavelengths).[2]

When these major kinds of spots have been accounted for, the NEBn still has a diffuseness and raggedness in the spacecraft photographs, particularly alongside and following the 'rifted' region of the NEB. This active region of cyclonic turbulence will be described in the next section; here we focus on the disturbance that followed it on the NEBn edge (Figs. 8.7–11).

Some of the Voyager photographs showed a remarkable series of dark patches on the NEBn edge at 17°N which looked like a train of waves, spaced 10° apart (Fig. 8.7&9). Their colour was grey, in contrast to the reddish-brown NEB(N) to the south. As shown in Fig. 8.10, from Voyager 1 maps, they were retrograding at speeds of about $\Delta\lambda_2 = +34$. Thus they were moving with the NEBn jetstream (peak at 17.6°N, $\Delta\lambda_2 = +45$). Their spacing indicated that they

[2] Owen & Terrile (*1981*); Conrath et al. (*1981a*)

were produced from the active region once every 5 days. Several of these dark patches were first seen as dark rims around the Nf. ends of white streamers or rifts in the active region, and then detached from them. Thus they were retrograding NEBn spots emitted from the NEB active region, just as retrograding SEBs spots are emitted from SEB outbreaks. But unlike the SEBs spots, the NEBn features were diffuse squarish patches without strong circulation. Closeups did show turbulent streaks with hints of anticyclonic eddying motions underlying some of these diffuse patches (Fig. 8.9). The Voyager 2 movie likewise showed bright 'tails' at the Nf. ends of some of the rifts, flapping to N. and S. and setting off waves in the retrograding jetstream.

Some of these wave-like features had substance, as shown by the behaviour of a group during the Voyager 1 encounter (Fig. 8.10). These continued as fairly compact dark spots at 18–19°N for some distance, until they drew level with the p. end of a barge. Then one, encountering barge no.2, was deflected to the north (to 21°N) and underwent an amazing acceleration to $\Delta\lambda_2 = -185$, clearly under the influence of the NTBs jetstream. (It stretched into a streak as it squeezed past the north edges of other NTropZ spots.) A similar NEBn spot, encountering barge no.1, also reversed its more modest drift, and merged with the next spot in the wavetrain; then it did the same again. The oscillatory motion of this spot seemed to be due to successive mergers of 'wave' features that were bounced back from the longitude of barge no.1, creating an anticyclonic circulating current (Fig. 8.8).

It must have been this circulation which evolved into a distinct brown lozenge at 20°N, with the shape and position of a Little Red Spot, by the time of the Voyager 2 encounter (Fig. 8.5; Plate P19.1,2; also drawn in Plate P18.4). Although this 'Little Brown Spot' was anticyclonic, it was warm at 5 μm wavelength, in keeping with its colour (Owen & Terrile, *1981*).

Not all the NEBn wave features were absorbed by this circulation; in both the later Voyager 1 and the Voyager 2 coverage, some could be seen continuing to retrograde past barge no.1.

The barges may also have been formed as a result of the turbulent 'rifting'. A new barge seemed to be forming in the wake of the rifted region in the Voyager 2 closeups (Fig. 8.9; Plate P16.5). This may have been one of several new barges observed visually in 1979/80. Indeed, the chart of these apparitions (Fig. 8.12) suggests that all the barges arose f. the rifted region, and in some cases persisted until they were overrun by it again a year or so later.

It seems possible that arrays of dark barges and bright ovals arise from spontaneous meanderings of the NEBn jetstream, developing as eddies on the inner sides of the curves. Perhaps these instabilities were triggered by the large rifted region in 1979. The more extensive arrays of barges and ovals that arise after great NEB expansions may likewise be triggered when the NEBn jetstream is made unstable, either by a change of speed or by the surrounding turbulence.

8.5 ACTIVITY IN THE NORTH EQUATORIAL BELT[3]

Although the NEB is not subject to distinct jetstream outbreaks like other belts, it has marked variations in internal activity. We have already discussed the arrays of NEBn barges and ovals. Three other aspects of NEB activity remain to be described: the NEBn latitude variations, the colour of the belt, and the white spots and 'rifts' that sometimes erupt within it. These four aspects of activity tend to occur together as part of a cycle that involves broadening, reddening, creation of stable NEBn ovals, and sometimes rifting.

NEB Rifts and the North Intermediate Current

Below latitude 15°N or so, largescale features no longer adhere to the uniform domain of the NTropC, but are caught up in the steep gradient of speed that leads to the great equatorial current on the NEBs edge.

'Rifts' in the NEB are white spots or streaks near the middle of the belt. They are often small and have varied, rapidly-changing shapes, with detail at the limit of resolution. They are usually oriented from Sp. to Nf., sheared by the velocity gradient. They clearly represent vigorous turbulence in the cyclonic region of the

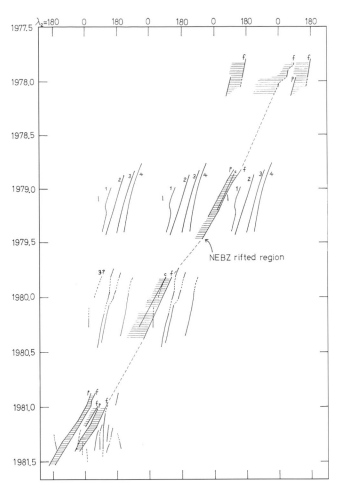

Fig. 8.12. Chart of the NEB rifted region and dark spots on NEBn, from BAA data, 1977–1981. Longitudes are System II. The rifted region is shown shaded with lines for p. and f. ends and centre, if tracked. (The tracks for 1977/78 were uncertain but probable.) Tracks of dark spots in NEBn/NTropZ are shown as lines (dashed if imprecise). In 1978/79, the four barges are numbered. In 1979/80, one NEBn barge may have been a continuation of no.3; the feature f. it was a dusky spot in NTropZ (like the Little Brown Spot?), not a barge. (Drawn by R. McKim; from BAA report.) Note the steady drift and progressive expansion of the rifted region. Also note that NEBn barges tended to appear f. the rifted region and were destroyed when it caught up with them again.

NEB. The more distinct spots are often teardrop-shaped, and sometimes appear in chains, which sometimes consist of most intricate curls – a beautiful sight visually[4], and amazing in spacecraft photographs. Individual white spots typically last only for a few days or, at most, weeks, before being dragged apart by the velocity gradient. But longer rifts have in recent years been tracked for several months, gradually lengthening and continually producing new small white spots within their turbulent borders. Sometimes, rifts repeatedly arise from a long-lived, slower-moving active region (see below).

The term 'rifts' reflects the predisposition of many visual observers to see dark markings as positive features on a bright background. In fact, the bright features are often the more substan-

---

[3] Aspects of NEB activity have been discussed by Mackal (*1966*) and Wacker (*1973*). Much of section 8.5 is adapted from Rogers (*1989b*), which includes a complete survey of colour and rift activity. The account in this chapter is expanded to include a complete listing of major episodes of broadening of the belt and of major arrays of NEBn ovals. For examples of typical NEB cycles, see Plates P4, P6, and P23.

[4] Several examples are shown in the Plates, especially Plate P18. Individual rifts in the 1980s were described by Rogers (*1988*), Sanchez-Lavega and Quesada (*1988*) and Sanchez-Lavega et al.(*1990*), as well as in BAA reports on those years.

tial ones, especially in the case of 'rifts' which are probably chains of white clouds over convective plumes.

Rifts do not usually impinge on NEBn barges, being slightly further south. But one important and unique event was observed in 1946, when a small bright spot in mid-NEB overtook a typical barge that overlapped the same latitudes. According to Peek: 'The two were in conjunction on May 28, when the author was fortunate enough to obtain a few moments of seeing that was good enough to reveal two little dark spots, one on either side of the northern half of the white one; these were the two ends of the dark streak, the middle of which was obscured.' The white spot then disappeared. The dark streak was seen again, perhaps partly obscured, on May 31 and June 2, but it was not recorded thereafter as the apparition was ending.

Although Voyager showed a continuous gradation of speed across the NEB, most visual records of spots or rifts in these latitudes fall into two distinct currents (Fig.8.13 and Table 8.4). Some spots have speeds at the fast end of the range of the North Tropical current ('fast NTropC'), with $\Delta\lambda_2$ around –30 or rarely up to –60. Others move with speeds intermediate between Systems I and II, in what we shall call the North Intermediate Current (NIC), with $\Delta\lambda_2$ between –80 and –150.

NEB rifts and NIC: spacecraft observations

A novel discovery in the 1970s was that short-lived white spots in mid-NEB (moving with the NIC) were arising within long-lived 'active regions' on NEBn, which moved more slowly with 'fast NTropC' speeds. These are analogous to 'folded filamentary regions' seen in other cyclonic belts. One was viewed by Voyager and another by Pioneer.

The rifted region observed by Voyager in 1979 had arisen in 1977 and it persisted and spread over several years. It moved with 'fast NTropC' speed at $\Delta\lambda_2 \approx$ –25 to –33 (mean –28) from 1978 to 1981.[5] During the Voyager 1 encounter in early 1979 it was about 60° long. Within this region, individual white spots moved much faster with the NIC, for example $\Delta\lambda_2$ = –103 in 1979 February and –98 in 1980 March (Rogers). The Voyager films showed this activity dramatically (Figs. 8.2, 9.2). New white spots were arising as small brilliant spots near the f. end of the disturbance with $\Delta\lambda_2$ = –60 to –100, then accelerating to NIC speeds of –98 to –148, meanwhile expanding and eddying cyclonically in the strong wind shear of the NIC.[5] The average speed of spots in the Voyager 1 coverage was about –123.

During the Pioneer encounters in 1973 and 1974, a similar rifted region was present. In those years Minton (LPL) tracked a single active region of NEB up to 90° long moving with fast NTropC (Table 8.4) ($\Delta\lambda_2$ = –35 initially, decelerating to –18 in 1974). He reported that small white and red spots within it had lifetimes from two to eight weeks, apparently fixed relative to the active region itself. But Pioneer 10 photographed the active region as a series of beautiful swirls, which over several days moved with the NIC (Rogers and Young, *1977*; Plate P16.2; Fig. 2.1).

[5] Terrile and Beebe (*1979*) and BAA reports; Figs. 8.12&13; Table 8.4. For Earth-based views see Fig. 1.19 and Plates P18 and P17.3; for Voyager views see Figs. 2.18, 8.2, 9.2, 9.12, and Plate P17.1&2. Detailed Voyager maps were presented by Smith et al.(*1981a*); speeds quoted here were measured from these maps by the author.

Table 8.4. *Records of the 'fast NTropC' and the North Intermediate Current: Drifts of features in mid-NEB*

Apparition	Av.$\Delta\lambda_2$ (NIC)	Av.$\Delta\lambda_2$ (fast NTropC)	Duration	Source
1927/28	–117 (8)	–55 (4)	7–34 d	BAA
1931/32		–50 (4)	1–2 mth	BAA
1932/33		–41 (2)	12 d	BAA
1941/42		–30 (1 + AR)	(var.)	BAA
1944/45	(≈–81)(–)			BAA
1946	(≈–103)(–)			BAA
1948	–88 (1)			BAA
1949	–123 (2)	(–52)(3)	1–2 mth	ALPO
1951	–104 (–)			ALPO
1952/53	–133 (2)		17,26 d	ALPO
1960	–122 (2)		33 d	ALPO
1965/66		(≈–36)(*)		BAA
1967/68	–113 (2)		11 d	NMSUO
1972		–26 (*)	1–5 mth	BAA
		(up to –39)(*)	(var.)	LPL
1973		–44 (2)	3–7 wk	BAA
		–35 (AR)	(yrs)	LPL
1974		–18 (AR)	(yrs)	LPL
1978/79	–115 (5)	–26 (AR)	5 d/(yrs)	BAA
		–28 (1)		SAF
1979/80	–106 (2)	(–26)(AR)	6–13d/(yrs)	BAA
1980/81		–28 (AR)	(yrs)	BAA
		–33 (1)		SAF
1982		–31 (2)		SAF
1983		–23 (2)		SAF
1984		–27 (–)		SAF
	–115 (1)		8 wk	BAA
1985	–146 (4)		10–42 d	BAA
1986	–111 (*)		1–6 mth	BAA
1987/88	–141 (8)		10–58 d	BAA
1988/89		–60 (5)	6–38 d	BAA
1989/90	–125 (5)		2–5 wk	BAA
1990/91	–97 (3)		1–5 mth	BAA
Average	–117.1 (±15.6) (N=15)	–35.0 (±13.1) (N=13)		
Voyager:	–123			Rogers

This table lists apparition means of the data plotted in Fig. 8.13 (q.v. for individual points). Speed is given as $\Delta\lambda_2$ (°/month). (N) is number of white spots or points on rifts included in the average; other features were as follows: (AR) active region; *1965/66, 2 red sections; *1972, 2 small dark spots accelerating (LPL) and 2 dark streaks (BAA); *1986, p. and f. ends of 2 rifts, one with variable speed over 6 mth, individually ill-defined but giving a reliable average. For data on 1987/88, see also Sanchez-Lavega et al.(*1990*). The column 'Duration' gives the length of time some of the features were tracked.

NIC and 'fast NTropC' are arbitrarily separated at $\Delta\lambda_2$ = –80 (see the histogram in Fig. 8.13 which shows this to be the approximate division). 'Fast NTropC' is separated from ordinary NTropC around $\Delta\lambda_2 \approx$ –20, but the separation is not clear: we have included one long-lived rifted region which travelled slightly slower, but we have excluded mid-NEB white spots from the 1890s and early 1960s which travelled slower, as well as NEBn dark spots which travelled slightly faster. (These are plotted with the ordinary NTropC in Fig. 8.3.) Thus the mean value for 'fast NTropC' is of little significance.

Some imprecise values are given in brackets and not included in means. There were other apparitions not listed here when observations were sufficient to detect NIC motions over a few days (just quoted as 'intermediate between Systems I and II'), but not to quote a worthwhile drift rate.

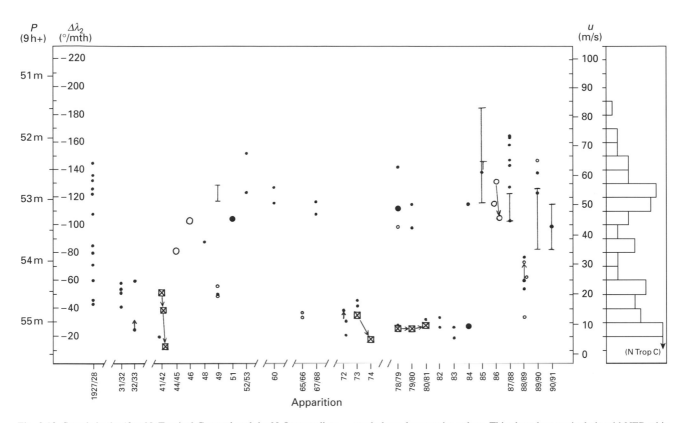

Fig. 8.13. Speeds in the 'fast N. Tropical Current' and the N. Intermediate Current (NIC). The chart shows the individual records of speeds in these currents, from data listed in Appendices 1 and 3; for averages and descriptions, see Table 8.4. Symbols: • single spot or rift (all white except in 1965/66 and 1972); ⊤ ⊥ p. and f. ends of long rifts (often ill-defined); ⊠ active region; ● apparition average; ↓ ↑ decelerating or accelerating. Open symbols are less precise values. This chart does not include mid-NEB white spots which moved with the ordinary NTropC in the 1890s and 1960s. Below $\Delta\lambda_2 = -30$, these fast currents merge into the ordinary NTropC (Fig. 8.3).

At right is a histogram of these data, giving equal weight to all precise points, and omitting imprecise points.

NEB rifts and NIC: historical record

Peek listed NIC speeds ($\Delta\lambda_2$ faster than –80) for only 1927/28, 1944–46, and 1948. More recently, the NIC has been recorded much more often, as listed in Table 8.4 and Fig. 8.13. High-resolution observations can reveal NIC drifts in most apparitions, and – at least within the well-observed recent era – NIC drifts have been quite distinct from fast NTropC drifts.

A greater frequency of rifts presumably implies a high level of activity in the NEB, perhaps analogous to jetstream outbreaks in other belts (especially the SEB). One might expect NEB disturbance to be reflected in enhanced activity on the NIC, by analogy with prograding white rifts in SEB disturbances.

However, the years when the NIC has been recorded are not well correlated with the other indicators of activity in the NEB as chronicled below. Detection of the NIC depends more on the frequency and resolution of the observations. There are several reasons why NIC drifts may be missed, particularly when activity is at a high level, even if there are enough good observations. Spots in this current may simply be too small and short-lived. They may be arising repeatedly within an active segment that shows fast NTropC motion. They may merge into a continuous NEBZ in which no motions can be detected. Or they may be sheared so much that they really are pulled apart before they have attained NIC velocities.

Could a better impression of the level of rift activity be gained by inspection of drawings and photographs? Unfortunately, it is difficult to identify the NIC white spots by their visual appearance alone. All the other types of NEB spots may vary independently to produce a false impression of rifting. Another criterion of serious rift activity at the limit of resolution may be visible doubling of the NEB, which has often been the end result of such activity, as in 1981 and 1986; but it is possible that doubling may occur due to unrelated causes. Conversely, some active periods may have been missed due to poor observing conditions. Thus the limits of visual resolution, and the problems of confusion with other types of activity, mean that the historical record of NEB rift activity is inevitably rather uncertain.

NEB colour

The NEB is variably and sometimes strikingly coloured. Typically it varies from brownish-grey to reddish-brown. When there are colour differences within it, the north component is usually the redder one, sometimes appearing frankly red. Sometimes the north edge appears to have a yellowish fringe, and although this colour

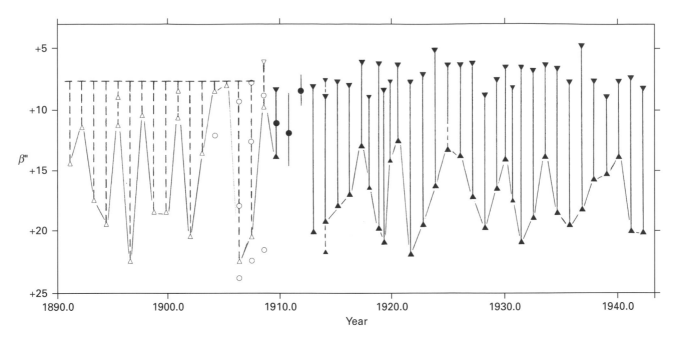

Fig. 8.14. Latitude of the NEB edges, 1891–1991. The chart shows apparition means, from sources listed in Appendices 2 and 3. When there are data
from different sources for an apparition, they are either averaged or plotted
separately. Vertical lines are drawn to indicate the extent of the belt. Open
symbols are data of lower precision, which are not used for statistics. Up to
1908, the data consist mainly of estimates of the width of the belt, from
Phillips (1915); these are plotted as latitudes on the assumption that the
NEBs edge was fixed at 7.5°N. Note the huge fluctuations in NEBn latitude
with three-year periodicity up to 1915, more irregular fluctuations thereafter, and stasis in recent decades. Statistics of these data are in Table 8.1.

could sometimes be an illusion caused by 'traffic-lights effect' of
atmospheric dispersion, it was certainly real during the Voyager
encounter. The dark patches on NEBs are usually grey or bluish-
grey. However, the NEB(S) can be a rich brown colour. On rare
occasions when the NEB(N) has appeared bluish, it was apparently
because the belt had extended so far north that the bluish NTropB
had been incorporated as an extra northern component (section
8.2).

A vivid range of colour descriptions by careful observers may be
found in the BAA Memoirs. For example, in the apparition
1928/29 in which the NEB was very active and variegated in
colour, Phillips wrote: 'Towards daylight and sunrise the colours
were splendidly seen in the 18–inch [reflector]…NEB a beautiful
pinkish-red or rose colour, except at S. edge where it is grey.'
Hargreaves (with a 14.5–inch reflector) reported a similar vivid
contrast in that year and also in 1941/42, when his observations
were summarised thus: 'On November 17 he called the belt as a
whole dull brick-red or brown, but noted at the same time that the
dark projections at its southern edge were grey…In many, though
quite definitely not in all, there was a striking blueness of tint. For
instance…on March 23 he wrote that the contrast between the
bright blue of one of these objects and the deep red of the NEB(N)
was striking.' (Quotations from BAA reports.) But these are only
vivid examples of a colour contrast that is frequently observed.
Sometimes a rift will partly isolate a bluish-black mass on NEBs
from the brown main belt. Occasionally the belt is triple and any of
the components may have a distinctive colour. It is also not uncommon for different longitude sectors of the belt or its components to
have different colours; for example, a long oblique rift may mark

the f. end of a reddish N. component and the p. end of a grey S.
component.

The Pioneer and Voyager spacecraft revealed even finer texture
of colours within the belt; in the Voyager colour close-ups (Plate
P16), the belt seems to contain a superposition of streaks representing the different colours of the white EZ(N), blue-grey NEBs
patches, brown NEB proper, and yellow/orange NEBn fringe –
presumably at different altitudes. Such mixtures of streaky cloud
layers could be responsible for the wide continuum of colours that
are observed visually.

The tendency of the NEB to be especially reddish a year or so
after 'revivals' or 'expansions' will become evident from the following chronicle.

NEB latitude variations

The latitudes of the NEB edges over recorded time are plotted in
Fig. 8.14. It is obvious that the north edge has fluctuated greatly,
especially around the turn of the century. The rather stable width in
the late twentieth century has been historically unusual. Though
the average NEBn latitude is 18°N (Table 8.1), it often retreats to
14°N or even lower latitudes, before rapidly expanding to
20–22°N. These expansions are the most distinct element in the
typical pattern of NEB activity.

In the rest of this chapter, we chronicle the NEB activity in detail,
including records of expansions, colorations, and rift activity.

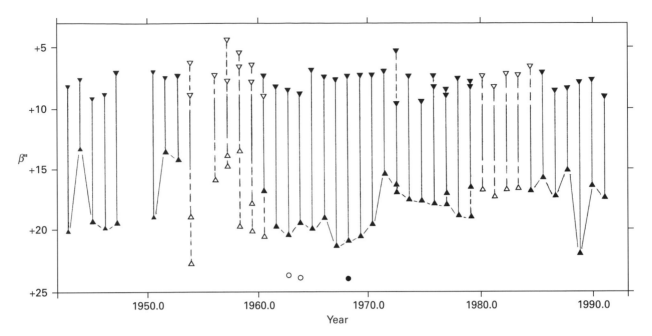

Fig. 8.14B

NEB Revivals in the mid-19th century

First, we should recall one 'prehistoric' oddity, from 1860, which turned out to be the closest NEB analogue of a SEB Revival on record.

1860: NEB had been reduced to a very narrow S. component throughout 1859, and on 1860 Feb.29 and Mar.2 a distinct oblique streak was seen connecting it to the NTB (Plate P2.1). This curious streak spanned only 7° longitude, but it rapidly stretched out (and also darkened) until it spanned over 180° by April 9, and lapped itself by May 6. On April 20 it was described as a bluish-black band with condensations, and thereafter looked like a regular belt.

This was the account published by J. Long (*1860*). The Sp. end had $\Delta\lambda_2 = -153$ (period 9h 52m 13s, later accelerating). Denning (*1898b*) later interpreted it as a NTBs jetstream outbreak, but the drawings show that it was clearly too far south for this, and must actually have belonged to the NIC. However, the Nf. end did appear to be on the NTB, and had $\Delta\lambda_2 = +15$ (period 9h 56m 1s), which is typical of the North Temperate Current. All this suggested that the streak spanned the NTBs jetstream but was unaffected by it!

But other drawings in the manuscript collection of the Royal Astronomical Society show clearly that this was a uniquely well-organised NEB Revival. The NTB feature was irrelevant, and the elongating streak was a reviving NEB component, just like the 'central branch' of SEB Revivals (Chapter 10.3). The outbreak was first recorded by S. Schwabe on Feb. 26 as a NEBn projection. Two drawings on March 2, by W. Huggins (Plate P2.2) and Capt. Noble, show the initial streak as two spots, in NTB and NEB latitudes. The former was probably there by chance; the latter was presumably the source of the streak elongating to recreate the NEB. Schwabe showed the NEB streak lapping itself on April 14–17, and drawings by Huggins showed it very turbulent throughout April (e.g. Plate P2.3). By May, according to these observers, the NEB had reappeared as a broad disturbed belt. By 1861, there was a broad though incomplete NEB(N) marked by classic 'porthole' spots (Plate P2.4).

NEB fadings and revivals seem to have been frequent in the 19th century. Major revivals were recorded in 1837 (F.v.P. Gruithuisen, *1843*), in 1856 (S. Schwabe, unpublished), in 1860, and in 1871/72 (J. Gledhill in *The Astronomical Register*, and others; Plate P2.9).

NEB Revivals: outbreaks every three years from 1893 to 1915

At the turn of the century there occurred a remarkable though almost-forgotten era in which NEB activity reached extreme levels, even though NIC was never detected. The NEB, just like the SEB in the 1950s, displayed violent Revivals approximately every three years.

Three of the Revivals attracted attention at the time, in 1893, 1906, and 1912.[6] A typical outbreak proceeds like that of 1893 (Fig. 8.15). First the NEB(N) fades to invisibility, leaving only a narrow residual NEB(S). Then the Revival begins with very dark spots appearing in the NEB(N) latitude, and violent activity spreads both Sp., creating spots in the equatorial jetstream, and p. and f., re-creating the NEB(N). The whole NEB is highly disturbed, and the whole affair resembles a SEB Revival, except that there is no distinct source and no retrograding jetstream to carry the spots round the planet; they just proliferate messily in the NTropC. The revived NEB, or at least its N. component, is often red. Plate P4 shows two complete cycles of this type.

Although a typical onset of Revival activity was observed only in the three years mentioned, it is evident from the original BAA Memoirs that Revivals actually occurred in each of the following years, with the possible exception of 1909. The date of onset is given. (It is also worth noting that the appearance of a striking array of NEBn dark spots in 1890 came three years before the first of these Revivals.)

1893: Classic Revival (Fig. 8.15, Plate P4.1). It started in 1893 September with several small dark spots in 'NTropZ', with dark streaks sloping Sp. to NEB(S), on which many spots were appearing in System I – reminiscent of a NTBs jetstream outbreak. Small dark spots also proliferated on streaks of

---

[6] Peek (*1958*) and Wacker (*1973*) reviewed them. The typical Revival of 1893 was described by Waugh (*1893*) and in the BAA report. The triennial pattern was described by Rogers (*1978, 1989b*).

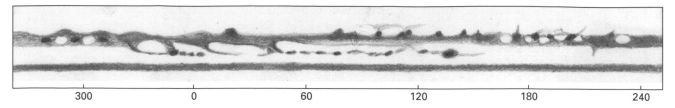

| | 300 | 0 | 60 | 120 | 180 | 240 |

Fig. 8.15. Strip-map of NEB during its Revival in 1893, from E.M. Antoniadi's drawings on Oct. 19–20. The map suggests that the source of disturbance was at $\lambda_2 \approx 45$. The spot at $\lambda_2 = 140$ was very dark red; it is unclear whether it was cyclonic (a barge) or anticyclonic (a Little Red Spot). NTB at bottom. (Drawn by the author from drawings in Waugh (*1893*) and the BAA report.)

new NEB(N). One larger and darker NEB(N) spot was deep red, and moved irregularly in the NTropC. By 1894/95 the belt was still broad and disturbed, with a coppery colour. In 1895/96, NEB(S) was still reddish, but NEB(N) faded away leaving very dark Little Red Spots in the NTropZ (Plate P4.2).

1896: Another great Revival; it started during solar conjunction but the belt then became very broad and active (Plate P4), and although it was neutral grey in 1897 Feb.–Mar., it became markedly reddish in April–May. In 1898, it broke up into a grey-brown NEB(S) and fragmentary NEB(N) again.

1899 (early): A mild Revival: NEB broader with disturbance, but colour neutral or purplish. There was a striking array of bright and dark ovals on NEBn in two velocity sectors (see above). NEB was similar in 1900 but almost disappeared in 1901.

1902: A slow but active Revival, in which NEB(N) progressively strengthened and acquired an array of dark and bright spots. NEB was brownish but by 1903 became very red, when NTropZ was also coloured brownish-yellow. These colours disappeared and NEB again faded to a diffuse grey line in 1904/5.

1906: Classic Revival, starting on April 10 with an extremely dark spot near NTBs from which dark streaks were pouring Sp. into the NEB (Plate P5.1). This was late in the apparition, but by 1906/7 the belt was extremely broad, and ruddy, covering the canonical NTropZ. In 1907/8, the orange belt material between the components began to fade, and the northern component became a blue NTropB, while the southern (main belt) was still brick-red, becoming brown. The belt looked similar in 1908/9.

1909?: Possibly a mild Revival during solar conjunction, as the belt was slightly broader and then narrowed again, but no disturbance was noted and colour was merely brown; possibly NEB(N) was still absent. In 1911, NEB(S) was reddish and NEB(N) fragmentary. By 1912, NEB was just a tenuous southerly line, with a vast yellowish-grey 'zone' extending to NNTB.

1912: Classic Revival starting with a dark marking on August 23, followed by rapid spread of dark projections along NEBs. This was at the end of the apparition. By 1913, NEB(S) was grey with very disturbed S. edge, and NEB(N) red though fading again, merging with brownish-yellow NTropZ. In 1914, the striking array of 'porthole' spots appeared along NEB(N) (Plate P5.5).

1915 (mid): A less well observed and atypical Revival, as the belt had not faded away. A disturbed dark (S) component and stable NEBn spots were already present, but NEBZ darkened and showed some rifts, and the belt became reddish. Much of NEB(N) was brick-red in 1916/17.

1918?: A possible final event in the series: broadening and increased disturbance, and possibly slightly redder. But the belt was continuously dark, reddish-brown, and disturbed (with persisting arrays of dark and bright NEBn ovals), from 1915 to 1919 (A.S. Williams, *1930*).

It seems surprising that the NIC was never recorded during these great NEB Revivals. One can only assume that these events unfolded in a way similar to the smaller disturbances of recent years, with the sources of spots lying in the NTropC, giving rise to mid-NEB disturbances whose NIC velocities were missed. There may be a clue in the latterday analogue of these Revivals that occurred in 1988/89 (see below). In 1988/89, just after the NEB broadened dramatically to the north, there were only a few small and short-lived rifts visible; but when their positions were plotted in System II, it became clear that there were persistent sources, giving rise to two or three tiny white rifts in succession, and moving at $\Delta\lambda_2 \approx -60$. This speed is very rarely observed, but both the speed and the latitude (15°N) corresponded to those recorded by Voyager for the very beginnings of rifts, before they expanded and accelerated into the NIC. So it seems that in 1988–89, incipient rifts were appearing at a normal latitude and speed, but were failing to flourish. It is likewise possible that no coherent features moved with the NIC during the NEB Revivals at the turn of the century.

Subsequent history of NEB activity

Since 1918, the variations in width, doubling, darkness, and colour of the NEB have been much less than before. Indeed, apart from a spasm in 1926–28, the belt has never been very narrow nor very split nor very red, and there have been no classic Revivals. Activity has consisted of more modest broadenings, and gradual proliferation of white or dark spots in various regions of the belt.

Nor has there been any strict periodicity since the 1920s. Episodes of broadening to the north occurred at intervals of 3–6 years until the 1960s.

The following listing demonstrates that all the various parameters of activity tend to occur together in a cycle that lasts about two years; the classic Revivals were an extreme form of this. The most distinct event in this cycle is the broadening to the north, which is followed by the appearance of an array of dark and white ovals in the NEB north half and, over a year or so, by reddening of the belt; occasionally, yellow/orange shading may extend over the NTropZ as well. NEB rift activity may accompany the broadening or may occur after it; this seems to be only loosely associated with the rest of the cycle.

Colour outside the years listed below was generally grey or brown; for a full listing see Rogers (*1989b*). 'White and dark ovals' denotes round white spots at 19°N and dark spots or barges at 16°N.

1922: Expansion very far north during solar conjunction. Some white ovals appeared in N. Colour merely changed from grey in 1922 to brown in 1923, when the belt became double.

1926–28 (Plate P6): NEB had been narrow in 1925. In 1926 June–July, NEB(N) reappeared in a classic Revival, very disturbed, becoming redder. NEB was still broadening and very red in 1927/28, when there was also intense disturbance in mid-NEB (the numerous and beautiful rifts revealed the NIC for the first time), NEBn (white and dark spots in contrasting velocity sectors), and NTropZ. In 1928/29 there was a grey-blue NEB(S) but NEB(N) was still reddish with some prominent spots, and yellow/orange shading extended over NTropZ. All this coincided with a 'global upheaval' (Chapter 13).

1931–33: In 1930/31 there were some rifts. In 1931/32 there was a general increase in width, redness, and disturbance all across the NEB, and an array of white and dark NEBn spots. All this persisted in 1932/33, when internal activity was decreasing, but a white spot with $\Delta\lambda_2 = -59$ was recorded.

1935: NEB broadened in 1935, with varied colours; reddish in 1936 and 1937.

1939–42: In 1938, NEB(N) was very faint except for two dark streaks (Plate P7.7&8). In 1939, the N. half was very disturbed with dusky loops into NTropZ and some rifts; probably similar in 1940/41 when it had changed from brownish to reddish; activity had developed slowly. Not until 1941 did NEB expand northwards. In 1941/42, parts were triple, generally purplish, with a very red central component; there were rifts in S. half (a single active region showed $\Delta\lambda_2 = -51$, decelerating), and many white and dark spots in N. half (Plates P8.1–3, P9.2). In 1942/43, NEBn spots persisted but the belt was brownish and quieter.

1944–46: After another major broadening in 1944 (Plate P9.3), for 2 years there were outbreaks of numerous white spots or rifts within the belt, moving with NIC, and a modest array of spots on NEBn.

Data are incomplete for the 1950s, and what data there are do not uphold the previous correlation between broadening, rift activity, ovals, and colour. The most 'rifted' appearances, in 1955/56, did not coincide with the major broadenings. The colour was usually recorded as reddish-brown, although real changes might have been missed.

1951–54: The ALPO reported substantial NEB rift outbreaks in late 1951, succeeded by dark patches which may have been barges. A year later (around the end of 1952), a major broadening set in, with appearance of many white spots and rifts and dark streaks. In 1953/54, NEB was broad with an array of white and dark ovals.

1955/56: Double; disturbed in middle and on S edge.

1959: Broad again, and double, but not many spots.

1961–64 (Plates P10.5&6, P14.1–3, P13.1): NEB was broad throughout these years, although its N. half was sometimes eroded. It was often disturbed, with many rifts, but these apparently persisted for months and moved with the NTropC, not NIC ($\Delta\lambda_2 = -13$ in each year: ALPO). In late 1963 this activity became intense. There had been an array of NEBn white ovals and dark streaks in 1961, and a similar array was prominent from 1964 to 1966. Red colour and activity diminished in 1964/65.

In the next 23 years, there were no major broadenings, and only modest examples of colour changes and of NEBn spot patterns, but there was still waxing and waning of rift activity. These half-hearted displays of different forms of activity were not perceptibly correlated.

The improved coverage in these decades revealed many details of the rifts, notably the fact that short-lived NIC rifts were appearing within long-lived, slower-moving active regions (described above).

1966–69: NEB was even broader than before, and was entirely double from the end of 1965 onwards, with some disturbance. Photographs by LPL revealed much detail in the white NEBZ which suggested that it was formed by vigorous rift activity (Plate P14.4–7). But NIC was only measured for two white spots in 1967/68 (NMSUO). Otherwise, only fast NTropC was recorded for mid-NEB. The rifting could have been moving with NIC but was too detailed to track visually.

1970: Reddish colour covered NEB and southern NTropZ (Plate P13.4).

1971–72: In 1971 NEB was quite narrow and contained some white rifts. In 1972 it was even more disturbed, with large white and dark spots in Systems I and II spanning both S. and N. halves, but only a moderate broadening occurred. Within the belt, nothing was tracked faster than $\Delta\lambda_2 -39$. The belt had little colour until 1975.

1975–78: There had been an active region in 1973 and 1974 (observed by Pioneer; see above), but it never amounted to a widespread disturbance. But in 1975, during the great global upheaval (Plate P15), most of NEB was double, with some distinct clusters of white spots; it was also very reddish, especially NEB(N). In 1976, activity had partly subsided, but NEB(N) was still reddish and there were still some NIC rifts [though not accurately tracked]. In 1977/78, most of NEB was again double with at least two large rifted sectors.

1980–82 (Fig. 8.12 and Plate P18). One of the active regions of 1977 persisted, gradually expanding (observed by Voyager; see above). By 1980 it was quite extensive so several sections of NEB were again double or rifted, but the rest was normal. By 1981 it was ≈120° long, and by 1982, rifts could be seen at any longitude. Over the next three years there were only a few rifts, although a few were tracked in the NIC. There was still little colour.

1986–87 (Plates P22.5–7, P23.1&2): Many sections were double or rifted in 1986 and 1987/88. One major rift region 30–50° long persisted throughout the 1986 apparition and was the most long-lived feature that has ever been tracked in the NIC. In late 1987, several rifts were recorded in their early stages and most were in mid-NEB at about 12°N.

In early 1988, rift activity increased and led to the first major broadening since the 1960s – a classic example of the typical cycle.

1988–90 (Plate P23): In 1988, the NEB broadened dramatically to the north; this began with increased 'rift' activity and dark streaks extending into the NTropZ, but the main broadening occurred during solar conjunction. In mid-1988, the best photographs showed a subtle reticulate pattern (Plate P23.3), which was soon largely replaced by a triple belt structure: only a few small short-lived rifts were visible. These were arising from persistent though inconspicuous active regions, which gave rise to two or three tiny white rifts in succession, all moving at $\Delta\lambda_2 = -60$. Then a striking array of dark and bright ovals appeared in the usual latitudes, and persisted in 1989/90 as the NEBn edge gradually faded and receded around them; then occasional large rifts appeared again in the NIC. The belt was not reddish as in previous cycles but probably a richer brown than in the early 1980s.

1990/91: The largest single rift on record was tracked in the NIC from 1990 Aug. to 1991 Jan., expanding halfway around the planet. In spring, 1991, a wave-like array of NEBn barges and white bays gradually developed with a precise 'wavelength' of 24° longitude. But there was no NEB expansion.

This last episode may have been 'the exception that proves the rule' concerning the origin of arrays of NEBn barges and ovals. No sooner had the author realised the linkage between NEBn expansions and arrays of spots, and described it in the first draft of this chapter, than this striking array of spots appeared without any NEB expansion! I propose that all such arrays arise from wave-like instabilities in the flow of the NEBn retrograding jetstream (Chapter 14.5). In most cases, the jetstream is destabilised as part of the whole cycle of activity that includes broadening of the belt. But it can also be destabilised in the wake of a large rift or active region, and this may be supposed to have happened in 1979 and in 1990/91.

Similarities between activity in NEB and SEB

An important question is whether the patterns of activity in the NEB resemble those in the SEB. The pattern of Revivals in the SEB is well-known and is well-organised on a grand scale (Chapter 10), whereas the pattern of NEB activity had been little-recognised until recently.

NEB rifts seem to be only loosely associated with the overall cycle of activity; nevertheless, the NEB rifts are clearly similar to, although smaller than, the white spot outbreaks that occur in the SEB. Thus:

(1) They emerge from sources in the poleward half of the belt, prograding only slowly if at all relative to System II.

(2) The white spots spread across the belt, stretching towards System I.

(3) Voyager films show that this shearing motion introduces dramatic cyclonic eddies.

(4) The rifts sometimes intensify long-lived System I features as they pass, viz. the SEBn white spot and the NEBs dark blue patches (see next chapter).

(5) The rifts in NEB and SEB expand at similar rates and can be explained by the same meteorology (see Chapter 14).

The similarity in overall pattern is evident if one compares Fig. 8.11, showing the pattern of disturbances in the NEB during the Voyager encounters, with the charts of SEB Revivals in Chapter 10.3.

The larger patterns of activity in the two belts are also similar. The triennial NEB Revivals were clearly analogous to classical SEB Revivals, and the subsequent cycles of NEB activity have been similar to NEB Revivals in almost all respects. Detailed parallels include the following.

(6) The whitening of the poleward half of the belt, followed by Revival.

(7) Turbulent rift activity.

(8) Within a year, strong reddish colour over the belt and sometimes over the adjacent zone.

(9) Three-year periodicity in some historical eras.

(10) Although SEB Revivals do not usually create large stable ovals analogous to NEBn barges and ovals, some SEB Revivals have indeed formed small slow-moving SEBs spots.

The major differences between the SEB and NEB patterns are as follows.

(1) SEB outbreaks are more organised and more conspicuous than NEB cycles. In contrast, small NEB rifts are often present and it is sometimes difficult to define distinct outbreaks.

(2) The strong SEBs jetstream adds a third, retrograding branch to SEB outbreaks.

(3) There are different patterns of stable spots. The Great Red Spot controls a region of localised continuous SEBZ activity. In the north, there can be many stable dark spots on NEBs and NEBn, which can appear as part of the cycle of activity.

SEB outbreaks are a major element in 'global upheavals'. Major episodes of NEB activity do not tend to coincide with SEB activity, but the possibility that they are an alternative element in some global upheavals will be discussed in Chapter 13.

9: Equatorial Region
(9°N to 9°S)

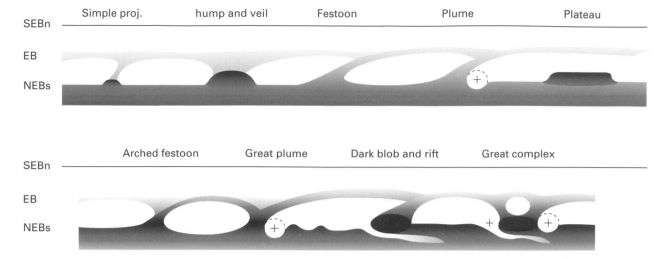

SEBn	Simple proj.	hump and veil	Festoon	Plume	Plateau

EB

NEBs

SEBn	Arched festoon	Great plume	Dark blob and rift	Great complex

EB

NEBs

Fig. 9.1. Diagrams of typical forms of projections on NEBs: simple forms at the top, more complicated forms below. All sorts of intermediate and combined forms also occur. The names allude to the usual aspect with south up. Bright white spots are marked +.

Fig. 9.2. The NEB and EZ from Voyager 1, showing the NEB rifted region and a NEBs plume with bright white core (top right). This complex core was formed by partial, cyclonic merger of the plume core with another bright spot that it was overtaking. See Fig. 8.2 (59 hrs earlier). Other features of this image and Fig. 8.2 are: the pattern of streaks in the EZ; wavelike wisps on SEBn (at top); and a small round light spot on NEBs (at centre of frame), which was fixed relative to the plumes. (A third view, 20 hrs later, is in McKim and Rogers (1992). This is image 16178.44, J1–8d, orange, filtered. NASA.)

9: Equatorial Region (9°N to 9°S)

9.1 OVERVIEW

The equatorial region consists of the Equatorial Zone (EZ), with its boundaries the NEBs and the SEBn. These edges rarely stray more than 1° from their average latitudes (taken globally), which are 7.3°N for NEBs and 7.0°S for SEBn (Figs. 8.14, 10.7; Tables 8.1, 10.3). The region differs strikingly from the rest of the planet.

Firstly, the whole region moves with the great equatorial current or jetstream, prograding at some 8° per day relative to the stable features in other latitudes. This is the greatest current in area, and is almost always marked by major spots. This current can readily be detected by only a few days' observation in an amateur telescope, by the motion of features on NEBs relative to spots on NEBn or to the GRS.

Secondly, the typical shapes of features are different here. The most obvious ones are large dark patches or projections from the NEBs, often bluish in colour and outstandingly warm at 5 μm wavelength. They often stream out in the Sf. direction into curving grey bands called 'festoons'. They are often accompanied by white spots, which can also adopt plume-like shapes. So the NEBs features completely mould the EZ(N). These shapes reflect the fact that cloud systems close to the equator are not much affected by Coriolis forces.

Thirdly, the most obvious type of largescale disturbance is a strong ochre coloration of the Equatorial Zone, colour which is less common in other zones on the planet.

The EZ is often split by a dusky Equatorial Band (EB). The EB is typically on or close to the equator, but is very variable in latitude and appearance. Sometimes it is absent, sometimes it is broad and dark, but most often it is an irregular grey band formed largely from the festoons of NEBs projections.

As the equatorial current is so fast, all features in the equatorial region are referred to System I of longitude, which has a rotation period five minutes shorter than System II: 9h 50m 30s.

The current was discovered by J-D. Cassini, but curiously, all measurements of it seem to be imprecise until the late nineteenth century. Cassini gave a rotation period of about 9h 51m, for 1689–1691. William Herschel tracked a complex bright patch in the NEB in 1779, and was surprised to find discordant periods of 9h 51m 45s and 9h 50m 48s over successive five-day intervals; this may have been a complex interaction of a NEBs spot with a NEB rift. J.H. Schroeter, in 1785–1787, obtained a period of 9h 50m 30s. W.F. Denning, in 1882, quoted 9h 50m 17s for a dark patch on the equator and a surprising 9h 51m 47s for a white spot in EZ(N); again these values were probably imprecise. Various observers from 1879 to 1885 gave accurate values but these all referred to a single 'great white spot' on the SEBn, which was moving very fast but decelerating (section 9.5). It was not until 1887 that A.S. Williams reliably measured the rotation periods for the whole array of spots in the equatorial region. These speeds have been measured in almost every apparition since.

Features in the northern and southern halves of the region usually drift independently, and are therefore referred to separate northern and southern 'branches' of the current. Occasionally, features on the EB can also be tracked and define a central branch. These three branches are named the North, Central, and South Equatorial Currents (NEC, CEC, SEC), and will be described in detail in sections 9.2, 9.3, and 9.5.

The aspect of the equatorial region was reversed before 1910

In the late twentieth century, we are familiar with an array of conspicuous projections and white spots along the NEBs, with few if any features on the SEBn. But the reverse pattern obtained in the nineteenth century: the SEBn highly disturbed, and the NEBs quiet (Plates P2–P4). This pattern was widely drawn in 1870–1872, and again in 1880–1882. (In the intervening years, more complex disturbance was drawn throughout the EZ.) Again, from 1887 to 1908, there was a regular array of spots or projections on the SEBn, and little if anything on the NEBs (e.g. Williams, *1889, 1909*). These SEBn features were very like the present-day ones on the NEBs. The dark spots were projections or distinct spots, sometimes almost like a satellite shadow, sometimes larger and more diffuse or doubled. From them, grey wisps either formed a symmetrical pattern of loops enclosing white ovals, or formed festoons streaming Nf. towards the equator. The white ovals or bays were often bright and sometimes had small bright off-centre nuclei. Although these features often formed a regularly-spaced series, they were also liable to large and rapid changes in intensity. The only notable difference from today's NEBs features was the spacing. Numerous nineteenth-century drawings and maps show a typical spacing of 17–25° for the SEBn spots, in contrast to the typical spacing of 30–35° for the modern NEBs features. When NEBs features were present then, they were much smaller and sparser.

The change came between 1908 and 1913. In 1908, the disturbance on the SEBn abruptly ceased, and the EZ became very bright white. The zone remained quiet from 1908/9 to 1912, although there were still many ill-defined white spots and faint wisps. Then the NEB Revival in late 1912 (Chapter 8.5) led to the abrupt onset of major NEBs disturbance, first tracked in 1913. At the same time, the speed of the N. Equatorial Current suddenly increased (see below). In contrast to previous NEB Revivals, the NEBs activity from this one persisted indefinitely – to the present day.

9.2 THE NEB SOUTH EDGE

The NEBs edge and its projections

The dark projections on the NEBs edge, and white spots associated with them, are often the most conspicuous things on the planet. Indeed, the author's study of Jupiter began in earnest with curiosity about these features. In the absence of any published information about their behaviour or their nature, and having access to the 30–cm refractor of the University of Cambridge during the summer of 1973, I observed for several weeks in order to find out how the NEBs features were changing. Only an eager student would have so underestimated the time needed for such a study! In fact, although these features sometimes show patterns in their behaviour, more often they metamorphose through various forms with no apparent regularity.

There are typically 11 or 12 dark projections spaced every 25–35° around the NEBs. Their average lifetime is several years. This is implied by the fact that most of them persist throughout a given apparition. However, as they are liable to change their drift and their appearance quite suddenly (within only a few weeks or even days), it is rarely possible to re-identify them individually from one apparition to the next. Only in the early 1960s, when American observers made special efforts to observe close to solar conjunction, was it possible to follow most of the projections over several years (Fig. 9.3).

Some of the forms that these features commonly take are shown in Figure 9.1, along with the names that visual observers have given them: projections, plateaux, plumes, festoons, and great complexes. The term 'plume' is used for dark features by some authors and for white features by others; the author reserves it for the combination of a dark projection and festoon with a bright white spot at the f. side. During the Voyager encounters, at any one time, three or four of the 11–13 projections had active white cores that qualified them as plumes in this sense. Minor dark projections are quite common f. large plumes.

Sometimes these forms are stable for months, but sometimes a single feature can transform between these or other aspects in a matter of weeks or even days. Sometimes a dark projection or

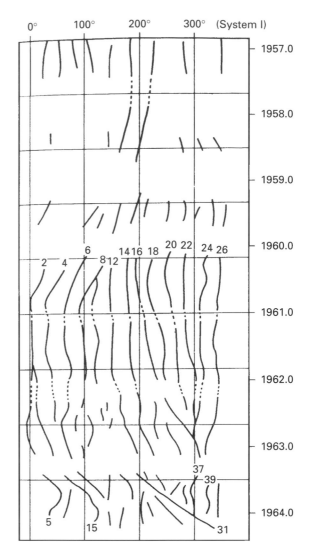

Fig. 9.3. Longitude chart of NEBs dark projections from 1959 to 1964, from ALPO data. (From the 1963/64 ALPO report in *The Strolling Astronomer*.)

white oval splits, or two of them merge. Sometimes a projection subsides and is replaced by another one 10–20° p. or f. it.

The NEBs projections have a distinctive colour, plain grey or bluish-grey, which often contrasts with the browner NEB proper. This was first noticed in 1916/17, when several observers recorded the projections being plain grey against the reddish-brown NEB. The contrast was again noticed in 1927/28/29 and in 1941/42, some projections being distinctly blue-grey. The same has been reported in many apparitions since, including 1949 and 1958 (by the ALPO), and especially from 1979 onwards (e.g. J. Olivarez, *1984*, and BAA reports). The bluish tint is mainly seen in the largest and darkest projections, and is best seen with large telescopes, probably because a large area is required for the human eye to recognise the colour. Photographs show that it is often present in most dark projections. But there is probably genuine variation between plain grey and bluish-grey; certainly the Voyager images showed different projections having subtly different colours.

The EZ bright spots are as important as the dark features,

although they are often less distinct visually.[1] The light areas between the dark projections are sometimes roughly oval, but often there is a distinct bright spot immediately p. or f. a projection, indenting the NEB or even forming a rift into it. These are often most notable on the f. edges of projections, where the white spot can vary in size and brightness over a few days, often appearing as a small, brilliant nucleus of a plume.

In some apparitions, all the NEBs features change their appearance simultaneously. Thus in 1972, the major features were huge white gashes that extended far into the NEB, whereas in 1973 and 1974, the main features were plumes, and in 1976, there was chaotic activity around half the circumference of the NEBs, with dramatic changes occurring in only a few days. (The years 1974–1976 showed great activity associated with two long-lived plumes, as described below.)

The NEBs projections and white spots have also appeared especially active in three other periods, as listed below. (By 'active' I mean unusually large, complex, intense, and/or variable. This is only a matter of degree, and the judgement of what is exceptional is admittedly subjective.) In each of these three periods, the onset coincided with strong coloration of the EZ (section 9.4). Coloration tends to outline the white spots, and to produce ochre veils interspersed with the usual festoons, so it may perhaps give a false impression of enhanced activity. However, the increase in NEBs activity in each case impressed experienced observers and persisted long after the colour faded, so it was probably real.

1938–1942 (BAA) (apart from a lull in 1941). In 1938 (as also in 1975) the intense NEBs activity developed while there was even more intense disturbance within the SEB. The BAA recorded many examples of spots merging or splitting in these years, but found no general pattern to the behaviour.

1961–1963 (ALPO): In 1961, both coloration and disturbance in the EZ were considered to be the greatest since 1938. The ALPO report described splittings and mergers of dark projections, and disappearances of some bright ovals. In 1962 the ALPO reported: 'The EZ was churned by tremendous turbulence; Clark Chapman noted on July 15, 1962 [using a 32–cm reflector and seeing many tiny white spots], that changes could be noted in intervals as short as 15 or 30 minutes.' One very bright oval with a dark projection f. it moved at $\Delta\lambda_1 = +6°$/mth and obliterated several stationary features. The EZ became quieter and lighter in 1963 January.

1978–1984 (BAA): Intense activity began in 1978 November. However, two very large complexes had remained stable for 2 or 3 years (1976/77 to 1978/79). They formed a symmetrical pair on opposite sides of the planet, and each typically consisted of a large very dark mass (sometimes blue-black) with a bright white plume at the f. edge and other white spots on the p. and S. edges, with associated wisps. They often looked much like other NEBs features and showed similar motion and variations, but were distinguished by their greater stability in size and motion. In 1979 January, one of them dwindled and was replaced by other projections shortly f. its position. (In Fig. 1.19, they are at $\lambda_1 = 80$ and 280.)

In 1978/79, any NEBs projection was liable to turn into an enormous blue-black mass or a brilliant white-cored plume, often within a matter of

days. Many such changes were due to the passage of NEB rifts (see below). Similar activity was maintained until 1984.

Another interesting pair of great complexes arose in 1982, and was tracked for three years by the BAA. These two were adjacent. One of them (here named Q) arose in 1982 April 20° f. the other (named P), with prominent disturbance between them. They progressively separated by 1–2°/month, but otherwise shared the general NEC drift, except that one or more faster-moving projections f. them drifted towards complex Q but disappeared before reaching it (becoming flatter and diffuse: BAA, SAF). For many months in 1983, P and Q spanned the entire width of the EZ! (Fig.9.4). Complex Q then included a stable light oval lying within the EB, as well as shading across the EZ(S). Complex P temporarily became more diffuse and smaller while a new projection grew 20° f. it. Q was still a classic great complex in 1984, and was then balanced by a similar complex 160° away, as in 1976–79; the regions following these two complexes were the most disturbed and variable in the EZ.

In 1985, there were still about 11 dark projections, but they were flatter than usual, and the long-lived ones could not be certainly identified. As the projections subsided, the EB also became faint and tenuous.

The subsequent years (1986–1989) have also revealed some interesting phenomena, which may not be unusual but which were covered in unprecedented detail by amateur photography. The implications – interactions with NEB rifts, exceptionally fast and

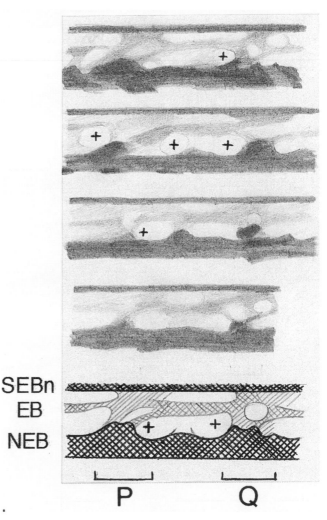

Fig. 9.4. The pair of long-lived great complexes in 1983. From top to bottom: 1983 May, early June, mid-June, September, and a diagram showing the typical features including a stable light oval on the equator in complex Q. Bright white spots are marked +. (From the 1983 BAA report).

---

[1] There used to be some debate about whether the white spots were just spaces between the dark features, or vice versa, and whether the curving dark festoons or loops were just contrast effects at the edges of bright ovals. The best visual observers were never in much doubt that all these features are real, and the Voyager images have proved as much: both dark and bright features are distinct entities and they are not always adjacent.

slow drift rates, correlation of form with motion in NEBs features, and independent motions on the equator – will all be discussed in subsequent sections. First we list some of the curious changes in the aspects of the NEBs/EZ(N) features.

1985: Passage of a NEB rift had dramatic effects on two projections (see below) (Fig. 9.5A).

1986: Grey shadings filled a large fraction of the EZ, and the characteristic appearance of the NEBs dark features evolved through four stages, each lasting a couple of months. Several features went through most or all of these stages individually (Fig. 9.5B). Stage I: Large projection or plume with diffuse shading on the p. edge. At least some of these appearances were probably induced by passing NEB rifts. Stage II: Long 'plateau' with diffuse shading at the p. end and a more distinct projection or plume at the f. end. Most of these appearances were unrelated to rifts, but represented large projections which 'collapsed' and rapidly lengthened (especially in the p. direction) before vanishing. Stage III: Huge blue-black oval patch, contrasting with the main NEB in colour and separated from it by a sharp border or bright rift around the p. side. These appearances were all induced by passage of a long-lived NEB rift, which intensified the dark patch and left a persistent rift draped around the p. end. Stage IV: The projection all but vanished, leaving only subtle wisps and shadings.

1987/88: Many regions were still almost blank until the autumn, when projections reappeared, but they remained unusually variable. At least one of them 'collapsed' like similar features in 1986, simultaneously merging with a stationary feature p. it, and also being disrupted by chaotic interactions with a NEB rift (Fig. 9.5C). The rapid drift of the p. ends of the collapsing projections in 1986 and 1987 ($\Delta\lambda_1 \approx -18$ to -29) may have general implications for the nature of the North Equatorial Current (see below).

1988/89: Now the NEBs features were unusually small, typically consisting of a white notch in a dark brown NEB(S), with a small grey projection at the

f. side instead of the p. side (Fig. 9.5D). This coincided with a great northward expansion of the NEB, and with a sudden deceleration of the NEC.

Although such variations in drift and in appearance have been laboriously observed and charted over the years, they have revealed little in the way of general patterns or principles. (For this reason, full details of many apparitions remain unpublished, although extensive strip-maps and charts were published in some reports, especially by the BAA in the 1970s and 1980s.) However, long-term patterns have emerged from at least some of the records. First, a pair of plumes showed unique behaviour and longevity from 1963 to 1976. Second, dark projections were found to be greatly influenced by the passage of turbulent 'rifts' in the NEB.

The long-lived plumes, 1963–1976[2]

During the Pioneer missions to Jupiter, the most conspicuous features in the EZ were two elegant and brilliant plumes, on opposite sides of the planet, which were advancing with almost unprecedented speed through the forest of ordinary NEBs projections. Elmer Reese and Reta Beebe (*1976*) worked back through the NMSUO records and found that one of these plumes (called spot a), and possibly the other (b), had been present for an entire jovian year (Fig. 9.6). Plume or spot a existed for 13.5 years in all, from mid-1963 until late

[2] These were described by Reese and Beebe (*1976*). Charts for 1963-1966 were also published by the BAA, and detailed descriptions and charts were also given in the LPL and BAA reports for 1973-1976, and by Fountain *et al.*(*1974*), Rogers and Young (*1977*), and Sato (*1980*). All these sources agree, even in many small details of the behaviour of the plumes.

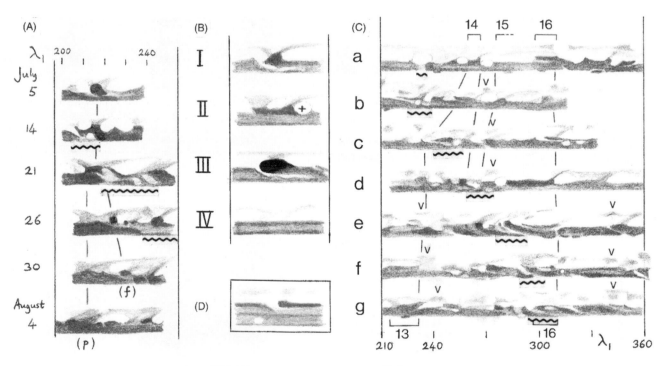

Fig. 9.5. Changing aspects of NEBs projections, 1985–1989.
(A) A dark projection inflated and disrupted by a passing NEB rift in 1985, drawn from photographs. The dark patch splits into p. and f. parts. Here and in subsequent figures, the NEB rift is zigzag-underlined. (From Rogers, *1988*, redrawn from the 1985 BAA report. Also see photographs by D. Parker in Plate P19.1, Fig. 4.1, and Fig. 20.3e.)
(B) Typical forms in 1986. From top to bottom, stages I (July), II (August),

III (Sep.), IV (Nov.).
(C) A disturbed part of the NEB/EZ(N) in 1987, showing a NEB rift passing (zigzag underline), NEBs projections collapsing and becoming chaotic, and differential motion of blobs on the EB (arrowheads). NEBs projections are numbered 13–16. Dates: Aug 21 (a), 22–23 (b), 30 (c), Sep 6–7 (d), 12–14 (e), 18 (f), 21–22 (g). (From the 1987/88 BAA report.)
(D) A typical NEBs feature in 1988/89. (Also see Plates P21.6 and P23.)

1976; plume or spot b existed from 1963 to 1966, and from 1970 to 1975, but its continuity between these periods was uncertain.

The structure and speed of a and b varied dramatically during their lifetime (Figs. 9.6&7). When first seen in midsummer, 1963, spot a was only one of several features with similar slow motion (positive $\Delta\lambda_1$), but in the autumn the others speeded up to $\Delta\lambda_1 \approx 0$, while a pursued its previous course, ploughing through a series of features f. it. So in 1963/64, both features were moving remarkably slower than the other System I features. They appeared as brilliant white spots (embedded in the coloured shading of the EZ), each forming a bay in NEBs with a very dark grey projection on its f. side, as though the dark grey material was piled up on the advancing edge of the plume (Fig. 9.8).

In 1964/65, a and b retained the same shape and drift, ploughing through the other EZ(N) features which had opposite $\Delta\lambda_1$. In 1965/66, they were still outstandingly bright, but the other features had changed their drift to agree with that of a and b. Then the whole array suddenly changed again to become nearly stationary in λ_1. (Fig. 9.6. The continuity of all features, even a, is uncertain at this point.) In 1966/67/68, the main features in the EZ(N) were white ovals, some very bright and with diverse motions; spot a was no more distinctive than others.

In the next few years, published data were scanty, but a remarkable aspect in 1972 was well documented. Then, the two long-lived spots and all other NEBs features were moving fairly fast, and all appeared as large white areas sprawling over EZ(N) and NEB, without large dark projections.

From 1973 to 1975, the two plumes were moving much faster than other features, and had a white core with a dark grey or bluish projection and festoon on the p. side – again on the leading side relative to other NEBs features (Fig. 9.9). The bright core was typically 4–8° in diameter, centred at 7°N, with the N. edge fixed, but with variable expansion to the S.

One is tempted to infer that the white core was the centre of the structure, and that the dark projection developed like a bow-wave against the surrounding flow. (However, this theory would not account for the similar plume shapes often observed among ordinary NEBs features that have no relative motion; see below.)

In 1974, as the two plumes accelerated while the EZ(N) in general decelerated, they began to collide with and consume the other projections that lay in their path (Fig. 9.10). Just before the first observed collision, in 1974 October, the large white core of plume b dissolved into a mess of white spots and dark festoons, from which a renewed 'standard plume' re-emerged in November, about 10° behind its original course. In subsequent collisions, a and b usually retained their shape, but decelerated slightly at the projected time of impact. As plume a underwent repeated collisions of this sort in 1974 and 1975, the decelerations caused its drift to oscillate with a period of 6–8 weeks. Sometimes the 'target' projection faded away before impact, sometimes it merged with the dark projection of the long-lived plume.

On the f. side of each plume, new projections began to be ejected – first behind plume b (which was moving the fastest) and in 1975, behind plume a as well. The first such ejection, in 1973, was

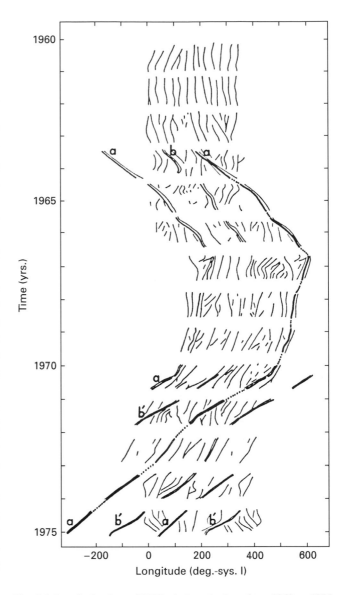

Fig. 9.6. Longitude chart of NEBs dark projections from 1960 to 1975, showing the two long-lived plumes, from NMSUO data. (From Reese and Beebe, *1976*).

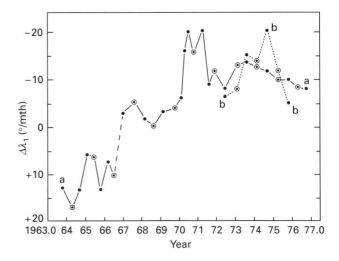

Fig. 9.7. Changes in the speed of the long-lived plume a. Data are replotted from Reese and Beebe (*1976*) up to 1973, and from BAA from 1973 onwards (plume b also shown; LPL data were very similar). No correction has been made for phase effects. ● within an apparition, ⊙ between apparitions.

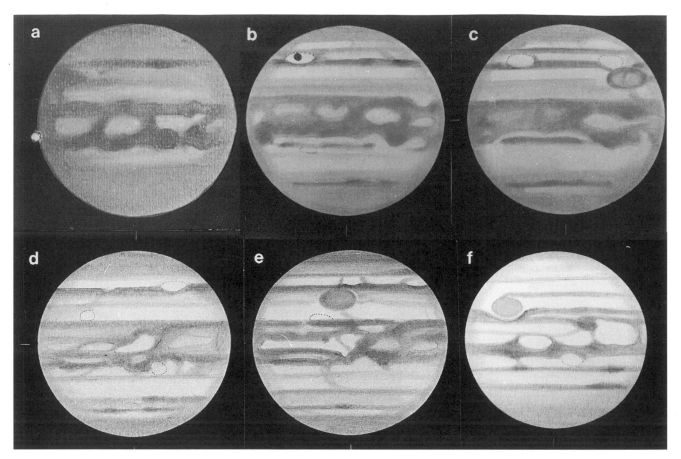

Fig. 9.8. Drawings by Jean Dragesco, 1961–1965, showing the largescale changes in the equatorial region. (a) 1961 July 26, ω_1 62, ω_2 204 (GRS near p.limb); SEB replaced by a disturbed 'EZs belt'. In (b-e), SEB is still faint, EZ is mostly very dark and NEB is highly rifted. (b) 1962 Sep. 22, ω_1 332, ω_2 149 (satellite shadow on STB white oval). (c) 1962 Sep. 23, ω_1 131, ω_2 335 (including GRS). (d) 1963 Oct. 21, ω_1 252, ω_2 301; the large white spot at centre of the disc is long-lived spot a. (e) 1963 Nov. 29, ω_1 273, ω_2 26; again spot a is at centre of disc. (f) 1965 Aug. 31, ω_1 143, ω_2 58; now the SEB is normal and EZ is clear again; spot a is the large white oval on the p. side.

Fig. 9.9. Pioneer 10 image, 1973 Dec.3, showing long-lived plume a. (NASA.)

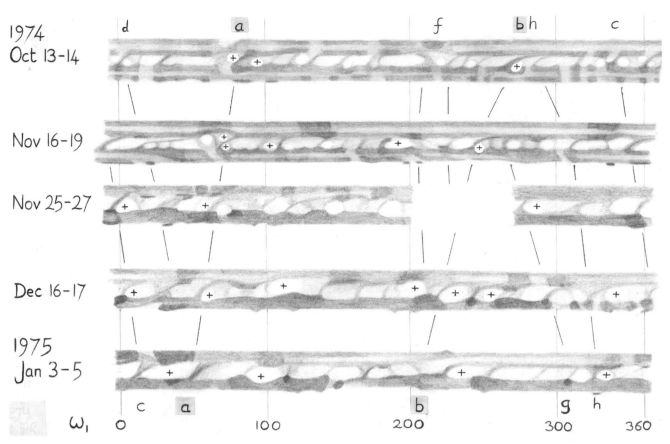

Fig. 9.10. Strip-maps of the equatorial region in 1974/75, by J.H. Rogers and P.J. Young (*1977*). Note long-lived plumes a and b colliding with other NEBs projections and b ejecting new ones. Bright white spots are marked +. For completeness, the October map was supplemented with observations from Oct.6 at λ_1 200–260, and the December one from Dec.11 at λ_1 0–40. (For detailed strip-maps in 1975/76, see the BAA report.)

described by the author (BAA report): 'Between July 23 and August 9 the diffuse f. border of the core of plume b formed into a new plume around a new, gradually brightening core. The new plume then moved off in the f. direction, while the diffuse f. border of the core of plume b re-formed.' Ejections were more rapid in late 1974, when in five months plume b ejected four projections which became established, and several more transient ones. Among the latter, the LPL report noted the rapid ejection of a train of five small 'dark blue spots' with an average speed $\Delta\lambda_1 = +73°$/mth, or 94°/mth (45 m/s) relative to the plume; they included the slowest velocities ever observed in this latitude (Table 9.2). This serial ejection occurred just after a NEB rifted region passed, so this was an example of the rift interactions described below. In the BAA data, each ejected feature typically showed an initial large positive $\Delta\lambda_1$ but those features which survived gradually accelerated towards a more normal drift rate. Plume a as well was seen to eject projections in 1975, when it produced four or five in Aug. and Sep. alone. (The last of the ejections from plume a is shown in Plate P15.7, when the small white core of the plume was brilliant, but detached from its accompanying dark festoon.)

Thus it was clear that the two great plumes had some persistent driving force that survived superficial disturbances. But eventually, in one week in 1975 October, both great plumes suddenly broke up and lost their white cores. This occurred when their repeated collisions with other projections had reduced their speeds to those of their surroundings. (In 1975, a and b had average $\Delta\lambda_1 = -10$ and -5

overall, but $\Delta\lambda_1 = -18$ before collisions, and sometimes positive $\Delta\lambda_1$ after collisions.) After the breakup, the sites of the plumes were chaotic, with many short-lived or variable plumes and white spots and dark shadings. But plume a seems to have re-formed in 1976, with a very chaotic region f. it. This plume began its final collision with another dark projection in 1976 December, and thereafter only complicated and minor features were seen. The plumes had finally expired.

Interactions with NEB rifts[3]

The turbulent mid-NEB rifts often cause sudden intensification or disruption of the NEBs dark patches as they pass. This was first noticed in the BAA analysis of the 1985 apparition, but was confirmed by going back to the author's detailed observations in earlier years, including the time of the Voyager encounter.

In 1985, a fairly small rift passed several dark features in succession. Two of these dark features underwent dramatic transformations (Fig. 9.5A). First, each expanded considerably as the rift passed, forming a very large dark bluish patch. Next, this patch was torn apart: a dark spot formed at the f. end and was dragged f. at $\Delta\lambda_1 \approx +30$, before disappearing, while at the original λ_1 of the dark patch, only a diffuse grey veil remained. Finally a dark 'projection' re-formed on its original site within a few days.

[3] This section is adapted from Rogers (*1988*).

The following year, similar but more extensive interactions led to some impressive intensifications and a few splittings of dark features (see above).

Did such interactions occur in previous years? In the early BAA Memoirs, the summaries of NEBs activity are too brief to reveal them. One suggestive example is reproduced in Plate IV of Peek's book, which shows strip-maps of a beautiful NEB rift in 1927. On 1927 Oct.5, Peek particularly noted an 'intense blue-black' spot in NEBs; this spot is shown in the Plate mentioned, and was not so dark a few days before or after, so it was a transient intensification that immediately followed the passage of the rift. Another probable example was reported by the LPL in 1974, when one of the long-lived NEBs plumes ejected five dark spots rapidly retrograding (see above), just as a rifted region of NEBZ passed this plume. Subsequent ejections f. the two great plumes may have been similar but the rifts were not sufficiently well tracked to be sure.

The most consistent series of rift-induced events was found in the author's observations for 1978/79, when there was a persistently rifted region of the NEB about 60° long (Chapter 8.5). I had made detailed observations covering the Voyager period, and noted many times when a dark projection became strikingly larger or darker than before, but did not realise that these events might be related to the passage of rifts. Seven years later, when I laid the track of the rifted region over the chart of the NEBs (1978 Dec. to 1979 Feb.), it was obvious that of 12 observed intensifications of NEBs dark patches, 9 or 10 began during or immediately after the passage of the NEB active region.

The most common result of interaction in 1978/79 was that the NEBs feature became much larger and darker, without splitting. The first five correlated events, which happened to five successive NEBs features, are shown and described in Fig. 9.11. It is remarkable that the active region had such a dramatic effect on the NEBs even though it was then quite small and northerly in the belt. The last well-observed event in this apparition affected one of the same projections again (g); since the event in Fig. 9.11b, it had completed one circuit of the planet relative to the NEB active region, which had become much more elaborate in the meantime. On 1979 Feb.8, g was a long dark bluish 'plateau' with the NEB rifts just past it. On Feb.10–13, its f. end had grown into a large blue-black 'hump'.

This event was well covered by Voyager 1 photographs (Fig. 9.12 and Plate P17). In agreement with the visual observations, a new dark patch appeared and expanded, and also retrograded; it comprised a mixture of white and blue-grey streaks roughly parallel to the NEB (complex g–g'). It is notable that this event occurred f. a brilliant white plume, which was also visibly active (see below), so that the interaction resembled the ejections from white plumes that were seen in 1974–1975. The Voyager team reported that the NEB active region was 'interacting strongly with the flow of the equatorial plumes' (though they did not describe the nature of this interaction), but that interactions had ceased at the time of the Voyager 2 encounter in mid-1979 (B.A. Smith and colleagues, *1979a,b*).

However, in 1979–80, at least one dramatic event confirmed that NEBZ-NEBs interactions were continuing (Plate P18.5).

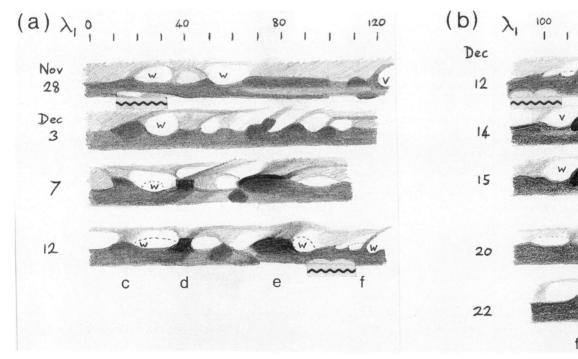

Fig. 9.11. Five successive intensifications of NEBs projections in 1978 December, caused by a NEB rifted region (underlined with zigzags) which was still small and often barely visible. w, bright white spot. (Drawn by the author; from Rogers, *1988*).
(a) Projection c intensified Dec.1 ('big very dark block'; similar Dec.3). Projection d intensified Dec.7 ('like a volcano with dark smoke dropping ash'). Projection e was seen as a long bluish NEB(S) segment on Nov.28, was more active on Dec.3 and intensified further on Dec.7 ('now massive again, very dark blue-black'). By Dec.10 and 12 it was a typical great complex with remnants of the NEB rift draped around it.
(b) Feature f was a long-lived plume. On Dec.14 it acquired a new extremely dark spot and brilliant white spot p. it. Both were fading on Dec.15. On Dec.20, a similar black and white pair (g) appeared, and they were probably fading on Dec.22.

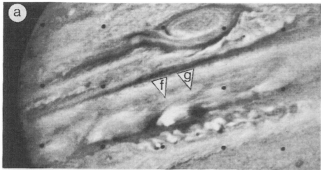

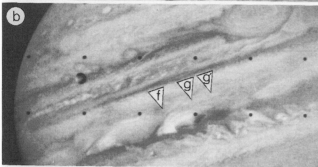

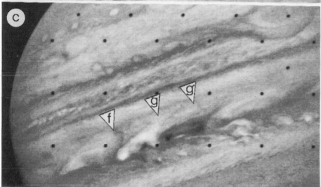

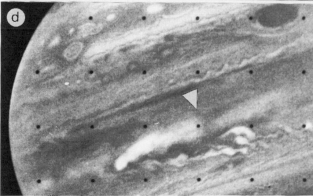

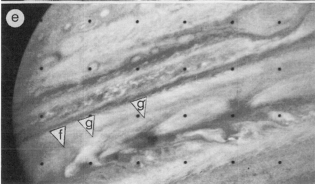

It may seem surprising that the NEBs white rifts can have such strong effects on the NEBs dark patches, when the latter are much more conspicuous and long-lived than the former. But this analysis shows that such interactions are common. Interactions sometimes happen even if the rifts are small features in the north of the NEB, apparently not in contact with the NEBs patches, as in 1974 and 1978. The rifts can have either of two effects on the dark patches. They may cause them to expand and darken, as happened many times in 1979 and 1986, or the rift may cause the f. edge of the patch to eject a new dark spot, which is dragged f. after the rift for a few days or weeks, as happened in 1974, 1980 and 1986. The two events of 1985, and the one observed by Voyager 1, combined both effects in succession.

Although major intensifications of NEBs patches sometimes occurred independently, it seems that at least half of such intensifications can be attributed to the passage of NEB rifts.

How do the NEB rifts intensify NEBs dark patches? A detailed meteorological interpretation would be beyond the scope of this chapter, but two general hypotheses can be outlined. The first invokes mainly vertical (convective) gas motions, as suggested by atmospheric models (Chapter 14), and proposes that a downdraft that forms the dark patch is enhanced by the proximity of convective updrafts in the white spots of the rift. The second hypothesis invokes mainly horizontal (wave) motions, as suggested by the Voyager images, and proposes that the passing of the NEB rift sends waves into the EZ which break up the white cloud layer, revealing more underlying dark material.

It is notable that a dark NEBs patch, after being torn apart by the passage of the rift, as in 1985, often re-forms at its original System I longitude within a few days. This shows that the dark patches must have stable roots which can survive separation both from the associated white spot and from the superficial visible structures.

Fig. 9.12. Intensification and expansion of dark patch g in 1979 Feb, viewed by Voyager 1. These photographs are in orange light so that bluish features appear darker, except for (d). They were 'filtered' to enhance small-scale contrast; this did not significantly alter the relative intensities of the NEB/EZ features. The images contain a grid of calibration dots. Times are in UT at spacecraft. (From Rogers, *1988*. See also Plate C4. NASA images.)

(a) Feb.3d 18h. Note bright plume (arrowed f) and the small dark bluish patch g.

(b) Feb.6d 05h. A second dark bluish patch g', has appeared, and the white plume core has been divided. (Io is in front of the SEB).

(c) Feb.8d 17h. Patches g and g' have expanded into a large streaky complex. Patch g' was pulled f. by 12° ($\Delta\lambda_1 \approx +80$) from Feb.6–11, just like the features in 1985. Some of the bluish streaks seem to have intensified individually, and they may also have been slowly drifting Sf. away from the disturbed edge of the NEB.

(d) Feb.8d 17h. (ultraviolet light; note that the bluish complex is invisible while the orange EZ(S) and GRS are dark). A white/orange front, which crossed the bluish complex, is arrowed. The white core of plume f had brightened around Jan.29 as seen by Voyager, and from Feb. 2 onwards a front of white material against orange material pushed away from the core with $\Delta\lambda_1 \approx +105$.

(e) Feb.11d 04h. The complex g-g' is fading as white streaks develop among the bluish streaks.

It is less clear whether the white nuclei of NEBs plumes are generally affected by NEB rifts. During the Voyager encounters, the white cores were repeatedly seen to brighten suddenly (see below). The initial accounts (G. Hunt and colleagues, *1981, 1982*) reported that these brightenings did not occur at any particular longitude, but R. Beebe, G. Orton and R. West (*1989*) said that they did tend to occur alongside the NEBZ rifted region. They showed one example in a set of strip-maps. The brightening of plume f just before Fig. 9.12 was another example, and Fig.9.2 shows a third example.

Spacecraft observations and physical interpretations

The NEBs projections are distinct 'hot-spots' at 5-µm wavelength (Chapter 4), having thinner clouds than other areas so that 5-µm radiation escapes from very deep levels. This confirms that they are distinct structures, not just extensions of the NEB. But they are not particularly warm at the cloud-tops, according to Voyager IRIS temperature maps, which show only the adjacent bright plumes (or patches within them) as distinctly cold features (Hunt, Conrath & Pirraglia, *1981*).

Voyager's cameras revealed a densely streaked texture in the clouds of the NEB and EZ. At least four colours were present – brown of the NEB, blue-grey of the NEBs dark patches, white of the plume cores, and ochre of the southern EZ – and these were intricately mixed in some places. The expanding NEBs dark patch described above was a mixture of bluish-grey and white streaks, and when it later faded, this involved an increase in the white streaks at the expense of the blue-grey ones (Fig. 9.12 and Plate P17).

Voyager confirmed that smallscale streaks in the EZ(N) run parallel to the projections and festoons seen visually, and give a powerful impression of rapid wind circulations along these lines. However, few small features could actually be tracked. On the closeups of dark projections (e.g. Figs. 9.2&13), the streaks, sometimes curve into the p. edge of the projection as in visual drawings of 'looped plumes', and a few white cloud flecks flicker to and fro from one day to the next along these streaks, but do not reveal any distinct flow pattern.

So what are the NEBs features really? While physical theories will be deferred to Chapter 14.6, let us see what can be inferred directly from the observations.

The 5-µm data show that the dark features are hot-spots with thin clouds while the bright features have cold, thick upper-level clouds. From their typical forms, it would be tempting to conclude that they are respectively cyclonic (blue-grey patches circulating anticlockwise) and anticyclonic (white spots circulating clockwise). This would be consistent with the circulations of spots elsewhere on the planet, including the mirror-image white spot on the SEBn (section 9.5). But it is doubtful if Coriolis forces could maintain typical cyclones and anticyclones so close to the equator, and no such circulations have been reported in the Voyager imagery of NEBs plumes. (Indeed, when a bright NEB rift did graze past a plume core, they interacted not anticyclonically but cyclonically, in keeping with the velocity shear within the NEB; Figs. 8.2 and 9.2.)

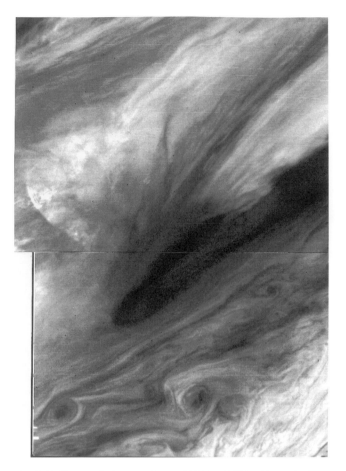

Fig. 9.13. Voyager 2 closeup of part of NEB and EZ, with a quiet NEBs projection and festoon. An irregular white cloud (at left) lies transversely across the p. edge of the festoon, with a few waves parallel to its edge. Small cyclonic vortices in the NEB (at bottom) are in the wake of the rifted region. The picture is about 16° (20 000 km) wide, extending from SEBn (top left) to NTropZ (bottom right). It overlaps Fig. 8.9. (Images 20587.02, 06; J2–3d; orange, unfiltered. An enhanced-colour version was published on the cover of the Voyager 2 issue of *Science* (Smith et al., *1979b*). NASA images.)

Also, the shapes of typical plumes suggest that the EB should be moving much slower than the NEBs, by several degrees per day (given that the timescale for expansion in the plume cores, according to Voyager, is a few days). The EB does sometimes move slower than the NEBs, but only by several degrees per month (see below) – not nearly enough to account for the shapes of the plumes.

Thus the dark blue-grey NEBs features are still unexplained. However, the shape of the white plumes is accounted for by vertical, not horizontal, motion.

It is predicted from the thermal wind equation (Chapter 4.6) that the equatorial jetstream will diminish at higher altitudes, so when white clouds rise in the plume core and expand at higher altitudes, they are drawn out in the f. direction like a comet's tail. Voyager images confirm that this is the explanation of the white plumes.

Several of the white plume cores brightened suddenly while Voyager was watching, and these events were analysed by G. Hunt, J-P. Müller and P. Gee (*1982*). The brightening was due to the rapid formation and expansion of new nuclei, 2000–4000 km across,

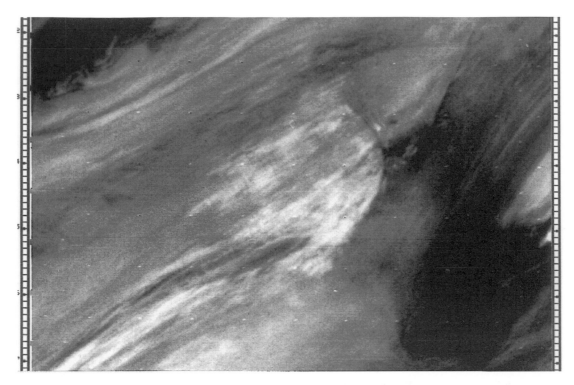

Fig. 9.14. Voyager 1 closeup of an irregular white cloud on the equator overlying the p. edge of an inactive dark projection. The cloud is marked by numerous north-south waves of wavelength 240 km; traces of similar waves can be dimly discerned outside the cloud. The picture is about 17° (21 000 km) wide, extending from SEBn (top left) to NEBs (bottom right). (Image 16324.06, J1–3d, orange, unfiltered. NASA.)

made of puffy white clouds like cumulus with individual sizes of 100–200 km or less. The nuclei were sometimes irregular or even double. A typical bright nucleus, at maximum rate of expansion, doubled in area daily and spread out over several days. These measurements were consistent with the hypothesis that the clouds formed at the top of a rising convective column, probably driven by water condensation deep down like a terrestrial thunderstorm (Chapter 14.6). The column would measure 40–60 km from its base to its top.

The plume material streaming away from the convective nucleus, according to Hunt and colleagues (*1981*), appeared to move in a slower current of ≈80–100 m/s, i.e. $\Delta\lambda_1 \approx +13$ to $+55°$/mth. One plume front was tracked by Rogers (*1988*) as it happened to coincide with the dark patch expansion described above (Fig. 9.12 and Plate P17). On or about Feb.2, white cloud from the core of plume f began spreading very rapidly f. in the EZ, forming a front between the white plume material and the orange EZ material. This 'white/orange front' appeared in the photographs of Feb.5–11 as a north-south arc of orange haze which cut directly across the white and blue-grey streaks. It apparently crossed the expanding dark patch g without interacting with it. Presumably the front was at a higher altitude. The front moved at a speed of $\Delta\lambda_1 = +105°$/mth. This speed was in a range never before detected in the EZ; it almost exactly matched the NIC motion of the adjacent rifts in the southern NEB, although there was no visible connection. The Voyager pictures include other examples of these fronts (Plate P20.1).

Further observational evidence for this high-altitude shear will be described in section 9.3.

The Voyager images also revealed several packets of medium-scale waves in the EZ, lying north-south, that is, transversely across the streaklines.[4] Some were in the tails of white plumes from NEBs, while others were in the general field of the EZ. In Fig.9.14, the waves are most conspicuous in the irregular white cloud patch (whose edge is parallel to the waves), but the original image shows faint traces of them over much larger areas, suggesting that the waves were widespread but the white cloud rendered them visible in one area. Different packets had wavelengths from 70 km to 600 km; the most common wavelength was 300 km (0.24°), and the packets consisted of 10–30 waves. They lasted for up to 6 days. The packets had the same NEC or CEC drifts as nearby cloud features, but the phase speed of the waves, unfortunately, could not be determined from the Voyager photos.

Hunt and Müller (*1979*) interpreted them as gravity waves, which are caused by an air flow rising and cooling, then sinking and warming, in a regular oscillation. Gravity waves in Earth's atmosphere often occur downstream of mountains or islands, and also downstream of convective plumes such as thunderstorms. So those in a jovian plume tail could be triggered by the upthrusting motions in the plume core. Flasar and Gierasch (*1986*) suggested that these gravity waves are trapped just below the white ammonia-cloud layer, and that they might have a period of one jovian day, being excited by diurnal convection cycles in the ammonia clouds.

[4] Hunt & Müller (*1979*); Flasar & Gierasch (*1986*).

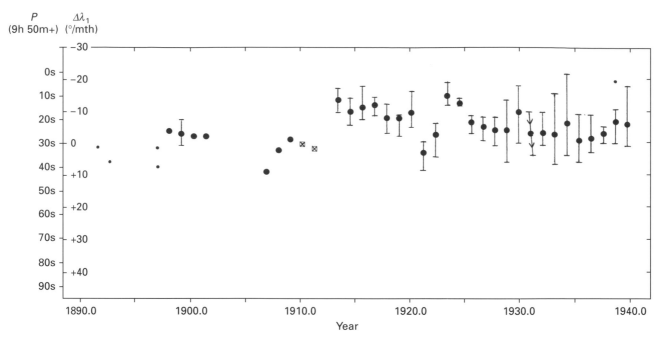

Fig. 9.15. History of the North Equatorial Current. Speed is shown as P (rotation period), as $\Delta\lambda_1$ (degrees per month), and as u (m/s in System III). Large symbols are apparition averages of >3 features, and vertical bars indicate the range of speeds (not published before 1914). Small symbols represent single features. Diamonds, long-lived Reese-Beebe plumes. Squares, apparition averages omitting the Reese-Beebe plumes. Note that the vertical scale of speed differs from that used for charts of the non-equatorial currents.

Earlier values, not included in the chart, were for 1884/85, 9h 50m 9.2s (G. Hough), and 1888, 9h 50m 24s (A.S. Williams, *1909*). Values in 1882 (W.F. Denning) and 1887 (Williams) were of doubtful reliability. See also list of sources for SEC, Fig. 9.20; however, there were few observaions of the NEC in the nineteenth century. From 1892 to 1948, the chart uses values taken from BAA Memoirs. Values in 1910 and 1911 were for all equatorial spots. From 1949 to 1964, we use values from ALPO or BAA or the average, depending on the quality of the data in each report. From 1964/65 onwards, data are from the BAA, except for 1971 (NMSUO) and 1972 (average of BAA and LPL data).

The North Equatorial Current

The mean speed of the North Equatorial Current (NEC) is very close to System I (Table 9.1). The history of the NEC is presented in Fig. 9.15.

There were few observations before 1913, as the NEBs was mostly quiet, but the average speed was consistently close to System I.

The NEC was strikingly faster from 1913 onwards, when the now-familiar array of NEBs disturbances first appeared within it. Since then, the speed has shown a very slow, fluctuating decline, punctuated by occasional marked decelerations that last only a year or so. These decelerations are undoubtedly real; they amount to 10–20 seconds in rotation period which comfortably exceeds the uncertainty in the mean period for an apparition.

It is frustrating that the variations in speed of the NEC are still unexplained, in spite of the vast number of observations and the importance of the question for atmospheric physics. Attempts have been made in the past to correlate the NEC speed with NEB activity, EZ coloration, or the jovian seasons, but all without success. The only possible correlation that seems worth mentioning is between NEC decelerations and the jovian seasons.

Of six such decelerations recorded, five have occurred at the season of maximum solar heating of the NEB. These were in 1906/7 (T.E.R. Phillips), 1953/54, 1965/66, 1975/76, and 1988/89; all occurred within one year of the greatest northerly latitude of the sun, and 1–2 years after perihelion. However, this may not be a firm correlation, as the deceleration of 1921 came at a different season, and three other perihelia have not been followed by NEC decelera-

Table 9.1. *Average speed of the North Equatorial Current*

	$\Delta\lambda_1$	P (9h 50m+)	(N)	u
1891–1939	−4.0	24.6s (±7.4s)	(37)	+107.7
1940–1991	−2.1	27.3s (±7.0s)	(47)	+106.7
1891–1991	−3.0	26.1s (±7.3s)	(84)	+107.2
Voyager (1979)	+5	37s (global)		+103

These are the means (± s.d.; N) of apparition means as plotted on Fig. 9.15, but including the Reese–Beebe plumes. (S.d. = standard deviation; N = number of apparitions). Speed is given as $\Delta\lambda_1$ (deg/mth), as P (rotation period), and as u (m/s in System III).

tions. In any case it is odd that the current should slow down when energy input is greatest, and that the decelerations are quite sudden and transient. So if solar heating is responsible, the connection must be indirect. (Reese and Beebe (*1976*) also tried to account for the speed changes of the long-lived plumes by solar heating, but the same mechanism can hardly account for both the long-lived plumes and the rest of the NEC, as they often diverged markedly!)

The 1906/7, 1953/54, and 1988/89 decelerations coincided with sudden expansions of the NEB to the north, but the others did not, and in general the NEC speed is not correlated with expansions of the NEB, nor with the degree of activity on the NEBs. The 1975 deceleration was unique in that it was largely due to small projections 'ejected' f. the long-lived plumes, which themselves were moving very fast, as described above.

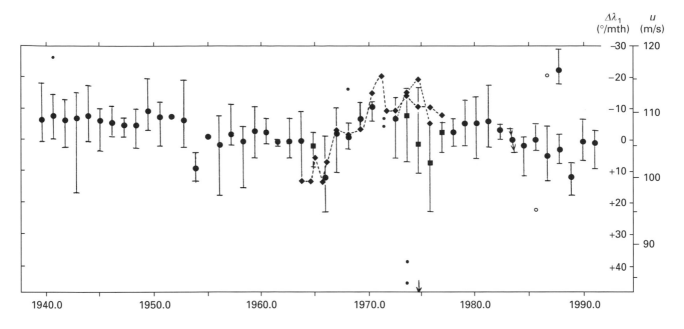

Fig. 9.15B

Global changes in the NEC have sometimes been followed in detail: for example, in the published ALPO chart for 1963/64, and the BAA chart for 1974, and the SAF chart for 1983 (confirmed by the BAA). In each case, features throughout one hemisphere changed speed within ≈2 months, catching up with the other hemisphere which had changed several months earlier.

The most extreme speeds reported for individual features are listed in Table 9.2. The slowest ones are mostly spots 'ejected' f. the Reese-Beebe plumes, or jolted by interaction with NEB rifts, as described above; most did not maintain their speeds for more than 2 weeks.

The fastest ones include the Reese-Beebe plumes themselves in the 1970s, and 'collapsing' projections as described above in the 1980s, and several short-lived white spots about which few details were published. The fastest of all ($\Delta\lambda_1 \approx -50$, in 1942/43) was a projection followed by a white spot which, in its one-month rampage along the NEBs, eliminated several near-stationary projections that lay in its path.

Finally, can anything more be inferred about the overall structure of the NEBs features in relation to the underlying NEC flow? It is intriguing that in 1973, when all features were moving fast, they tended to be classic plumes with the dark projection p. the white core, whereas in the 1988/89 deceleration, the NEBs features adopted an unusual form, the dark humps being small and f. the white spots. We noted above that the shape of the Reese-Beebe features may have depended on their drift relative to the other NEBs features; so one wonders whether the common NEBs features are influenced in a similar way, by their drift relative to an underlying flow that is roughly equal to System I. Perhaps NEBs features develop into plumes when they exceed this speed, and collapse into the reversed type of feature when they fall behind (as in 1988/89).

Unfortunately, this speculation rests on very shaky ground. One cannot tell whether the proposed relationship is general, because not enough apparitions have been analysed with close attention to

the shapes of the NEBs features; in pre-1970 reports, the published drawings are often heterogeneous and there are few detailed descriptions. Even the apparitions that are well-documented provide counter-examples. Thus classic plumes were common in 1979, and small 'reversed' structures were common in 1968, even though in both years the NEC speed was very close to System I.

Even if true, such a relationship would raise further problems, as it would suggest that the NEC bulk flow remains stable while the whole array of NEBs spots can drift collectively one way or the other relative to it.

A different (and equally speculative) hypothesis about the NEC was suggested by Voyager observations of the SEC, in which the bulk flow was found to be faster than the major white spot and the historical SEC (section 9.5). Michael Allison (*1990*) suggested that the same might be true of the NEC: that the visible NEBs features might be waves embedded in an invisible, faster jetstream. The Voyager images showed no evidence for such faster motion; for example, the small round spot in Fig. 9.2 did not move relative to the main projections. The only evidence that may support faster motion was the tendency of projections in 1986 and 1987 to 'collapse' with $\Delta\lambda_1 \approx -21$.

Thus many important questions about the NEBs activity are still unanswered. However, answers may come soon, as the Galileo probe is due to descend into the atmosphere at this latitude, measuring the gas composition, temperature, wind speeds, and vertical motions directly.

9.3 FEATURES ON THE EQUATOR

The Equatorial Band

Although the EZ usually has the same white or creamy colour as other zones, it is always marked by various shadings – sometimes very faint, sometimes massively dark. Most commonly they include an EB, and projections and festoons from the NEBs (which join the EB), and more diffuse wisps and shadings in the northern

Table 9.2. *Extreme speeds of spots in the North Equatorial Current*

Δλ₁ ≤ −20 (P < 9h 50m 4s)					Δλ₁ ≥ +22 (P ≥ 9h 51m 0s)				
Appar'n	Source	$\Delta\lambda_1$	P (9h+)	Description (duration)	Appar'n	Source	$\Delta\lambda_1$	P (9h+)	Description (duration)
1934	BAA	(−22)	(50m 0.5s)	W.s flanked by d. projs. (2 mth)					
1940	BAA	−26	49m 55s	One spot (5 wk)					
1942/43	BAA	(−50)	(49m 23s)	Proj. followed by w.s. (1 mth)					
1949	ALPO	−28	49m 53s	One w.s., accelerated from Δλ₁ = −8					
[1952/53	BAA	−19	50m 4.5s	Fastest of the usual range]	1952/53	BAA	(+33, +48)	(51m 15s, 35s)	Two features (10 d, 7 d)
					1963/64	ALPO	+33	51m 14s	One spot (1 mth)
					1965/66	BAA	+23	51m 1s	Slowest of the usual range
1970	NMSUO	−20	50m 3s	Plume *a* (late in appar'n)					
1971	NMSUO	−20	50m 3s	Plume *a* (early in appar'n)					
[1973	LPL	−15	50m 9s	Plume *b*]	1973	LPL	+41, +51	51m 22s, 31s	Two 'blue spots' ejected f. plume *b* (19 d)
1974	BAA	−21	50m 2s	Plume *b* (late in appar'n)	1974	LPL	+69 to +83	51m 59–80s	Four ditto (12 d)
						BAA	+30, +63	51m 10s, 55s	Two ditto (9 d, 2 mth)
					1975	BAA	+23	51m 1s	Two slowest of projs. ejected f. plume *a* (6–12 d)
					1979/80	JHR	(+58)	(51m 48s)	Small d.proj. jolted by NEB rift (Plate P18.5)
					1985	BAA	(+22)	(51m 0s)	One small w.bay (2 wk)
1986	BAA	(−21)	(50m 2s)	P. end of 'collapsing' proj. (4 wk)					
1987/88	BAA	−29 to −18	49m 51s to 50m 6s	P. ends of 3 'collapsing' projs.(3–4 wk)					

(Values in brackets are imprecise.)
W.s., white spot; d.proj., dark projection from NEBs.

and southern halves of the zone. All these shadings are typically grey or bluish-grey, except during distinct coloration episodes (section 9.4). Occasionally they become dense enough to render most of the EZ(N) or EZ(S) grey.

The EB varies considerably, sometimes being absent, and sometimes being a broad dark belt. Parts may be double. The most common appearance is a faint irregular band, which seems to be largely formed from the dark festoons from NEBs, and distorted by EZ(N) white spots. Pioneer images showed a massive EB made of up to five components, bluish in the north where some of them were continuous with festoons from NEBs, and increasingly redder to the south.

Features on the EB are usually no more than dusky, indistinct streaks. The detailed photography in 1986–89 gave some insights. In these years the EB was tenuous and incomplete but showed numerous dusky 'blobs', often drifting more slowly than the adjacent NEBs disturbances (see below). Some of them looked as if they might be formed from EB material compressed by the expansion of EZ(N) white clouds in the Sf. direction. However, some blobs looked more complicated. It may be significant that several of the blobs were first seen due south of NEB projections to which they seemed to be connected, and three such EB/NEBs complexes were distinctly blue, consistent with the EB blobs arising from the NEBs projections. In 1988/89, with NEBs quieter, the blobs and streaks were grey rather than blue according to photographs.

The Central Equatorial Current

Only rarely are features on the EB distinct and long-lived enough to be tracked. The available data are listed in Table 9.3. In most of these apparitions, there was no systematic difference from the NEC. Until recently, the most detailed BAA analysis was in 1939/40, when the NEBs was extremely disturbed. There were many spots on the EB: some evanescent, some white spots coupled to NEC features, and some dark streaks which could be tracked independently for 1–5 months. The difference in speed between the latter features and adjacent NEBs features ranged from −7 to +8°/mth, but the average difference was zero.

These results differed from the Voyager measurements of the smallscale cloud texture, which showed a velocity minimum on the equator, with Δλ₁ = +44°/mth.

However, this velocity minimum has been detected in the detailed observations of 1987–1990, when EB 'blobs' were tracked over 3–5 weeks and had Δλ₁ = +4 to +28°/mth. In 1987/88, these drifts were distinctly slower than those of the adjacent NEBs disturbances (Fig. 9.5C). In 1988/89 the NEBs features had decelerated to the same speed (though they were not necessarily connected to the EB), but in 1989/90, though the EB blobs were short-lived, differential motion was again observed. It may also have been detected in 1949, when the ALPO reported (without comment) that two out of three EB dark features, observed over

Table 9.3. *The Central Equatorial Current*

Appar'n	Source	$\Delta\lambda_1$	P (9h+)	(n)	Notes
(A) Typical drifts:					
1900	Molesworth	−3	50m 26s	(13)	
1909/10	Hawks (BAA)	−1.5	50m 28s	(11)	
1927/28	BAA	−1	50m 28.5s	(6)	
1928/29	BAA	−2	50m 27.7s	(7)	
1932/33	BAA	+1.5	50m 32s	(1)	
1934	BAA	0	50m 12s, 37s, 40s		
1938	BAA	−33	49m 46s	(1)	
1939/40	BAA	−4	50m 24.5s	(6)	
1942/43	BAA	−7	50m 20s	(2)	
1949	ALPO	0	50m 30s	(3)	
1961	ALPO	+1	50m 29s	(16)	
1983	SAF	+5	50m 36s	(6)	
1987/88	BAA	+13	50m 48s	(6)	
1988/89	BAA	+17	50m 52s	(7)	
1989/90	BAA	+24	51m 2s	(1)	
Average:		+0.7	50m 30.6		$u = +105$
		(±12.6)	(±17s) (N=15)		
Voyager (1979):		+44	51m 30s (global)		$u = +85$
(B) Exceptional observations of slow drifts:					
1918/19	BAA	+120	53m 12s	Irradiating w.ss.	
1928/29	BAA	+78	52m 15s	SEB Revival*	
1979 (Voyager):					
Hunt et al.(*1981*)		+13 to +55	50m 48s to 51m 44s	EZn plume clouds	
Rogers (*1988*)		+105	52m 52s	White/orange front	
Magalhaes et al.(*1989*)		+221	55m 30s	Infrared patterns	

Notes: This Table lists drifts for features on the equator that were not obviously connected to NEBs or SEBn. Many other drift values for streaks on the EB were published in BAA Memoirs from 1899 to 1904/5, by Scriven Bolton and P.B. Molesworth, but they cannot be relied on for reasons given in Appendix 1. Only Molesworth's value for 1900 may be worth recording, as the average interval between observations was only 14 days. In 1909/10 and 1911, Phillips (BAA) listed all EZ spots as 'equatorial' but his values are plotted with the NEC in Fig. 9.15.
*(There were also slow drifts in EZ(S) in other SEB Revivals; see also 1938, 1958, 1971, etc. in the SEC.)

1–2 months, drifted +6 to +20°/mth slower than adjacent NEBs features.

Slower winds at higher altitudes

It is believed that the whole equatorial current becomes slower with increasing altitude, in accordance with the thermal wind equation (Chapter 4.6), and with the shapes of the NEBs plumes (section 9.2). But is there any observational evidence for this belief? A few exceptional observations do support it (Table 9.3B).

1. 'Irradiating spots' of 1918/19. In 1919 January, very bright spots were repeatedly seen, shining at or even projecting from the limb of the planet. Each display lasted no more than half an hour, and these white spots could not be discerned when near the central meridian. As Phillips noted in the BAA report, they were probably high-altitude white clouds floating above most of the limb-darken-

ing. (Their prominence may have been enhanced by contrast with a high-altitude violet-absorbing haze which was then developing, which may have increased the overall limb-darkening.) By plotting the times of the limb displays, the BAA concluded that there were two equatorial bright spots with periods of about 9h 53m, and a third (slightly north of the equator) with the System I period of ≈9h 50.5m. However, the positions of the latter spot were very scattered, and can alternatively be interpreted as due to two or three spots in succession with a period of 9h 53m 12s ($\Delta\lambda_1$ = +120), which also fits the earlier two spots.

2. Spots from SEB Revivals. The energetic SEB Revival process (Chapter 10.3) often flings great dark and bright spots into the EZs, and they generally move slower than System I. There were prominent examples in 1938, 1958, and 1971; but the most extreme example was in 1927/28 when one pair of spots on SEBn had P = 9h 53m 20s ($\Delta\lambda_1$ = +133), and passed right over a pre-existing dark mass there, temporarily obscuring it. These were part of a small cluster of slow-moving spots, including two right on the equator which for a week displayed P = 9h 52m 15s ($\Delta\lambda_1$ = +78). These SEB-derived spots must surely be at high altitude to spread so far in latitude. However, that cannot be the whole story. A great white spot that appeared in 1976 was also a slow-moving spot derived from a SEB Revival, but it became one of the most long-lived and well-formed features on the planet (section 9.5); it must certainly have been a deep-seated disturbance, not just a drifting cloud.

3. Plume fronts in Voyager pictures. These observations, showing speeds up to $\Delta\lambda_1$ ≈ +105, were described in section 9.2 (p. 143).

4. Infrared patterns in Voyager IRIS maps. As described in Chapter 4.4, J. Magalhaes and colleagues (*1989*) found that the pattern of temperature at the 270–mbar level, just above the cloud-tops, showed no equatorial current: it appeared to be stationary in System III. This surprising result is not yet understood.

9.4 COLORATIONS OF THE EQUATORIAL ZONE

Overview

The most remarkable disturbances of the EZ are episodes of reddish or yellowish coloration. The colour has been described in many ways, most often as ochre (yellowish-brown) or tawny (orange-brown), but both the colour and the darkness certainly vary from one event to another, ranging from pure yellow to dark brown or coppery red. It is more yellow or orange than the typical brown of the belts, but never as deeply red as the maximally coloured GRS.

A complete listing of these coloration events follows; but first, as these events have not been analysed so systematically before, let us summarise the general conclusions.

Coloration has been present for 41% of the time from 1859 to 1992. However, this figure includes years of waning when it had declined into a faint yellowish tint or a dark grey EB; there has been conspicuous colour for only ≈30% of the time. The colour has occurred in 18 episodes as listed below, with an average and typical duration of 3.0 years. The shortest episodes have lasted for one year or slightly less (1925, 1943–44, 1958–59). The longest one listed was 7½ years (1878–84) but this, like other long durations, is an arbitrary figure because the colour may have been renewed more than once. Thus, a waning episode is occasionally rejuvenated (1884?, 1949, 1964).

Some of the following generalisations are derived from only a minority of episodes, as listed in brackets, but they may also have been true for other episodes. In many episodes there were not enough data, or data were from various observers who might interpret colours differently, or there was a more complicated succession of phenomena.

The colour typically develops over a few months, simultaneously at all longitudes. Sometimes the colour fills the whole EZ, and sometimes it is more restricted in latitude; for example to an EB (1897, 1968, 1989/90). This confinement to an EB may be more common than visual observers can perceive; the 1968 and 1989/90 examples of coloured EB were defined photographically, so in 1989/90, violet-filter photos showed a broad dark EB while visually the whole EZ(N) appeared yellow. This may be largely due to the low resolution of the human visual system for colour (Chapter 4.1).

Sometimes the colour begins as a pure yellow, but evolves within a few months to a darker, less pure colour such as ochre or tawny or merely brown (1869?, 1918–19, 1977–78?, 1989–90). Details of the most recent and best-observed episode will be discussed after the observations; this evolution may perhaps have been overlooked in the early stages of previous episodes, which were often not widely noticed at first. Sometimes the evolution continues as the coloured material concentrates over a year or two into a massive dark belt, EB(S), passing from yellow/orange through brown to grey (1938–39, 1961–63, 1968–71, 1972–76).

However, some episodes have begun with dull grey or brown colour and only later become more vivid (1878–84, 1927, 1946–48). And some episodes end with a gradual brightening back through yellow to white (1928–30, 1979–82).

The colour may be less obvious at some longitudes than others, but it never shows any distinct structures, apart from those impressed upon it by the usual EZ features. Activity on the NEBs is not diminished and may even be increased. The usual white areas of EZ(N) (or, before 1908, of EZ(S)) often stand out dramatically as white ovals against the surrounding colour. The grey projections and festoons from NEBs can produce an astonishing filigree of multicoloured filaments and veils. An evolving coloration sometimes becomes displaced from EZ(N) by white spot activity. (The 'white/orange front' described from Voyager data on p. 143 and Plate P17 was an illustration of this process.)

EZ coloration events are clearly associated with whitenings ('fadings') of the SEB, or with the SEB Revival that follows (Chapter 10.3). Of the 18 EZ coloration episodes listed below, ten occurred while the SEB southern half was faint and continued until the SEB Revival, and three others began during a SEB Revival (1859?, 1943, 1958). (Only two SEB Revivals occurred with no coloration of the EZ: 1952 and 1955.) Often the EZ colour spreads over the SEB soon after the Revival. There may also be an association with jetstream outbreaks in temperate latitudes (Chapter 13). However, there is no clear association with cycles of NEB activity.

A complete table of EZ colour and shading from 1877 to 1986

was given by Rogers (*1989b*). This need not be repeated here; instead, the individual coloration episodes will be described in more detail (sometimes from records that are not easily accessible). This account differs from the earlier one by inclusion of 'coloration' in several more apparitions (1878–1881, 1918/19, 1934, and 1958), which the author now regards as significant, partly due to further research in the literature, and partly due to our experience with the coloration of 1990 (see below). This showed clearly that positive reports by one or two experienced, colour-sensitive observers should not be discounted even if many visual observers do not report any colour. In the following accounts, the author readily admits to having selected some of the more impressive published descriptions. Having given due regard to the reliability and reproducibility of the visual observations, as well as to my own experience, I believe that the descriptions quoted represent good discrimination rather than good imagination!

Equatorial coloration events in the nineteenth century[5]

The first records of yellowish-brown colour in the EZ were by Schroeter in 1786, and by several observers in 1790–92. Another event, in 1839, was recorded by Professor Franz von Paula Gruithuisen (*1843*). This flamboyant gentleman had earlier claimed to see a city on the Moon, but his series of drawings of Jupiter look realistically modest, so the ruddy-brown shading drawn in 1839 was probably real. These were only occasional observations at a time when Jupiter was seldom observed. Only from 1858 onwards can the record of coloration episodes be taken as reasonably complete.

1859–60: In 1859/60, EZ was shaded and notably brown or somewhat tawny, with light ovals in it. (This was just after a SEB Revival, and the NEB was absent.) W. Huggins recalled a similar tint in 1858/59. The EZ was then white from 1860/61 to 1869.

1869–72 (Plate P12.1&2): This was the most striking episode recorded in the nineteenth century; it was reviewed in detail by T. Hockey (*1992*). (The SEB was partly faded.) The EZ coloration was first discovered by John Browning, who waged a long fight in the RAS to get more erudite but deskbound astronomers to accept the reality of the colour change. In 1869 October, he found the formerly white EZ developed a strong yellow colour. Within one or two months a yellow, tawny, or ruddy colour was confirmed by other observers. In 1870 January, Browning found the colour more dusky and temporarily displaced from the SEBn by a white strip.

The 'ochrish or tawny' colour across the whole EZ was even stronger in 1870/71, and remained very intense for three years. The colour was also confirmed by an early photographic plate; being blue-sensitive, the plate registered the EZ as totally dark. The colour faded during 1872 (after a SEB Revival), and by 1873 there was only a slight, variable rosy tint left.

However, some degree of coloured shading was seen in many apparitions up to 1888, probably uneven and fluctuating. Indeed, during the late nineteenth century when the terms 'belt' and 'zone' had not yet been given their modern definitions, the EZ was so often filled up with coloured shading that the whole span from NEBn to SEBs was often referred to as 'the great equatorial zone' or 'equatorial belt'.

1874–76: The colour reappeared in 1874, though not as intensely as before. A bronze-yellow or coppery tint of the EZ was notable in 1874 May and

---

[5] In addition to the numerous contemporary records, the best reviews of this topic are by Williams (*1920*) and Hockey (*1992*).

into 1875. In 1876, beatiful drawings by G.D. Hirst in Australia (Plates P12.5&6 and P3.1) showed the EZ as dark and ochre as the NEB, SEB, and GRS. It was white again in 1877.

1878–84 (Plates P12.7–9 and P3.3–5): This period is interesting because it included the most extreme whitening of the SEB observed until then. EZ colour was not widely reported until 1882/83, but extensive brown or brownish-grey streaks and shadings were present from 1878 to 1884/85, along with marked disturbance along both edges. These unusual shadings, often as dark as SEB(N) or NEB, probably represented a coloration episode.

In 1878 and 1879, much of the EZ was still white, but much was occupied by greyish-brown shadings. In 1880 and 1881/82, it was more shaded and disturbed, with a distinct EB, and some observers coloured the EZ dull ochre or coppery.

In 1882/83, shading and disturbance persisted, but the EZ was more clearly yellow. (The SEB had just revived and was ruddy.) Thereafter the EZ(N) was strongly shaded, grey in 1883/84 but brown in 1884/85 with a strong EB. (This clearing from the S. corresponds to the clearing from the more active N. that has been observed in more recent years.) From 1885/86 onwards, the EZ was clear and white again.

During the era of frequent NEB Revivals (1893–1915), when there were no SEB Revivals, there was only one full coloration event, and two episodes of shading which had little if any colour.

1897–99 (Plate P4.4–8): In spring, 1897, A.S. Williams observed an intensely pink EB and a strong rosy tinge over EZ(S) and SEB. Other BAA observers confirmed an orange colour, which soon diminished. But the whole zone remained shaded and strongly orange-yellow throughout 1898 and 1899. It was mottled with many wisps and light patches, especially white ovals along the S. edge – a wonderful sight. In 1900 it was only dull yellow (with white ovals along both edges), and thereafter it was brilliant white.

(1905–6: This was merely a bluish-grey shading which filled the EZ(N) during the absence of the NEB in 1905/6. In 1906/7 the shading became a light 'dingy yellow' (possibly an extension of the reddish colour of the newly revived NEB). It was never very strong, and it faded back to normal during 1907.)

(1913–14: After the NEB Revival of 1912 and the onset of long-term activity in the EZ(N), during 1913 the EZ(S) gradually acquired grey shading, which also outlined the brilliant white ovals in EZ(N) (Plate P5.4). It was similar in 1914 but then lightened. The shading was notable in contrast to the brilliant whiteness of the preceding years, but was not very intense, and its greyness contrasted with concurrent coloration in NTropZ (1913) and STropZ (1914).)

Equatorial coloration events in the twentieth century

When SEB fadings and Revivals resumed, so did EZ coloration episodes. At the time, interest in EZ coloration episodes centred mainly on a search for periodicity. A.S. Williams (*1920, 1922, 1936a,b*) proposed a period of 11.95 years, while R.A. McIntosh (*1936*) proposed one of 7.35 years. Both periods satisfied many of the observations to that time, and indeed coloration episodes have always tended to recur at intervals in this range, but subsequent history has shown that there is no sustained periodicity. In a footnote to McIntosh's paper, Peek made a connection which has proved more robust: '...a close correlation between the colour of the EZ and the state of the SEB is rather more probable than a true periodicity in the EZ itself.'

This connection will be explored further in Chapter 13.3; for

Fig. 9.16. Ultraviolet image of Jupiter from LPL, 1968 March 22 (ω_2 209) (0.33–0.38 μm), showing the intense ultraviolet absorption in the orange EB. (Compare with Plate P14.4&5. Image by the Lunar and Planetary Lab. of the University of Arizona, from Fountain and Larson, *1972*.)

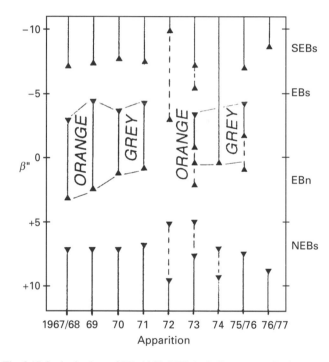

Fig. 9.17. Latitude chart of EB, 1967–1977, including onsets of coloration in 1967/68 (EB) and 1972 (whole EZ). Data from NMSUO (1967–71), LPL (1972–73), BAA (1972–77), and SAF (1973–76). Values from different sources are plotted separately when they differ noticeably – sometimes by several degrees due to the irregularity or diffuseness of the belt edges. Triangles indicate S. and N. edges of belts and lines indicate the belt extent.

now, we will merely note the state of the SEB in passing. The first three SEB Revivals, in 1919, 1928, and 1938, were each preceded and accompanied by a major EZ coloration episode. But from 1943 to 1962, when SEB Revivals occurred every three years, EZ colorations were less well synchronised with them. Then there was a prolonged series of colorations in the 1960s and 1970s. The synchrony with SEB activity was re-established with the well-marked episodes of 1978 and 1990.

Fig. 9.18. Onset of EZ coloration, 1989 Oct. 30: filter photographs by T. Akutsu (Okinawa, 32–cm reflector). (Left) Red R-60 filter, 16.25 UT, ω_2 5. (GRS on f. side.) (Middle) Blue/violet B-390 filter, 18.04 UT, ω_2 65. (GRS on p. side.) (Right) Ultraviolet U-360 filter, 11 min later. Absorption was already strong in ultraviolet although it did not become obvious visually for another 2 months. (For other colour images of this epissode, see Plate P24, Fig. 7.8, and Fig. 4.1B.)

1918–21: A light but definite yellow colour covered the EZ in 1918/19, at least as early as 1918 August (Williams, *1920, 1922*) and December (Ainslie, in BAA Journal). (The SEB was then fading, while the NEB had just broadened and reddened.) In 1919/20, the colour was remarkable: a strong yellow-orange or tawny shading, of uneven intensity. In late 1920 this colour was still present, but less intense, and it weakened to light brownish and then almost to white by 1921 June.

1925: The EZ was dull during 1925, noted as brownish-yellow by Phillips in June; there were white spots along NEBs. But in 1926, while weak shading persisted, its colour was cold grey. (In 1926, the SEB faded and the NEB began a Revival.)

1927–30: In 1927/28, EZ shading persisted but had become orangey in EZ(N). (SEB(N) and NEB were also very reddish; the rest of the SEB was faint.) In 1928/29, during the colourful SEB Revival and with intense disturbance throughout the EZ, strong ochre shading persisted in EZ(S), and later spread over the SEB. Filter photographs from Lick Observatory (Wright, *1928, 1929*) showed little coloration in 1927 Oct. but confirmed its presence in 1928 Oct.–Dec. Some strips and spots remained white. In 1929/30 the EZ was generally light yellow, though sometimes grey. In 1930/31, slight yellow colour remained, brightening to white during the apparition.

1934–35: In these years, when the planet was observed only at low altitude, colour notes were conflicting. In 1932/33, orange colour was reported only by A.S. Williams (fulfilling his own long-standing prediction; Williams, *1936a,b*) and C.F.O. Smith (whose colour reports were unconfirmed), and they disagreed as to the timing. In 1934, these two observers agreed that the EZ had a strong orangey colour throughout; there were no other colour notes to disagree with them in that year, so these positive reports should probably be given the benefit of the doubt, although drawings showed no shading. In 1935, the EZ was definitely yellow early, becoming white later.

1937–39: In 1936 (with the SEB faded) the EZ was shaded, especially in the S. Most observers saw the colour as merely off-white (even Williams), so it is difficult to accept the report by McIntosh (*1936*) that the colour was distinctly yellow. But in 1937, the dusky EZ(S) was certainly yellowish, while there were numerous white spots in EZ(N). The coloration in 1937 and the following year was confirmed by Lowell Observatory filter photographs (Slipher, *1964*). In 1938, following the SEB Revival, the whole EZ was amazingly disturbed and colourful, on a background of continuing yellow or brown or tawny colour, which spread over the SEB. In 1939, EZ(S) had evolved into a massive dark grey belt. By 1940/41, this had gone and the EZ was very bright.

1943–44: There was no EZ coloration during the next SEB fading, but during the next SEB Revival, in 1943 April, new orange or yellow shading appeared in EZ(S) (while EZ(N) remained strongly disturbed). In 1943/44, EZ was slightly shaded in places and was yellow or tawny, until 1944 March (Plate P12.10).

1946–49 (BAA, ALPO): In 1946, the zone was merely greyish-white until July, when E.J. Reese found it dull ochre. During 1947 the dull ochre colour (with some white ovals) evolved into a lighter yellowish-white. In 1948 it was yellowish or yellow-white throughout.

In 1949, colour re-intensified: the EZ was dull and yellow-ochre to ruddy, except for a grey EB and white areas along the very disturbed NEBs. In 1950, though a strong grey EB remained, the EZ was mostly white. It remained white for nine years.

1958–59 (ALPO and others): The first report of colour was again by Reese who noticed increasing ochre shading in late 1958 (after the SEB Revival). This was confirmed by others in the next apparition, from 1959 January to July, when it was very striking. Reese, and Japanese observers (in the ALPO Journal), described the EZ as dusky deep orange, and it was described as pink by observers in Brazil, Italy, and New Zealand (in locally printed bulletins). The coloration also covered the SEBZ, with a more ochre tint. Bright ovals often stood out in the EZ(N). Japanese observers found the EZ somewhat brighter from 1959 July, and in 1960 the colour had disappeared.

This was considered the most intense coloration for many years, though it was short-lived. But the next one, which started less than two years later, was very prolonged and was probably the darkest on record.

1961–65 (ALPO, SAI, BAA) (Fig. 9.8): This event began in 1961 when the SEB faded and the SEB(N) was replaced by an unprecedented belt occupying the southern one-third of the canonical EZ latitudes (Plates P10.5&6, P14.1). This 'EZs belt' had many spots and projections on its south edge, moving with the SEC, which confirmed its unique latitude. It was very dark but flagrantly orange. The remaining EZ was largely shaded, perforated by light ovals, and was deep yellow or orange.

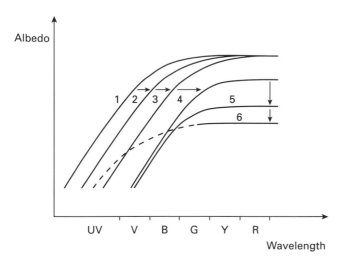

Fig. 9.19. Hypothesis for progressive change in the spectrum of the EZ during a coloration episode. This is purely diagrammatic, based on qualitative colour impressions. (1) White; (2) visually white but with stronger absorption in UV and violet; (3) yellow; (4) ochre or tawny; (5) brown; (6) grey (massive EB). There are two processes, reddening and darkening, and their relative intensity determines whether the maximum colour is more yellowish (ochre) or reddish (tawny). When the episode ends with a dark grey EB, presumably the UV/violet-absorbing haze has cleared but the underlying zone has darkened right across the spectrum.

In 1962, the 'EZs belt' migrated southwards until by August it had resumed the normal position of SEB(N), though it was still orange. But the remaining EZ was then so dark that the whole span from SEB(N) to NEB appeared as a single massive belt (Plates P11.1–3, P14.2&3). The shading was rich yellow or warm brown, and even the light ovals were yellow, both visually (they contrasted with true white spots in the northern NEB) and photographically. In blue light, the whole equatorial region was solidly dark, and in violet light it was the darkest 'belt' on the disk. (See the filter photographs from the Pic du Midi in the 1962 SAF report and in de Callatay and Dollfus (1974).) High resolution revealed tremendous disturbance and complicated texture – even before the SEB Revival, which spread to EZ(S) late in 1962.

Clearing began in 1963 January. In 1963/64, EZ was still very dark and disturbed but had lost much of its colour. In the S., the shading had condensed into a brown EB(S), while EZ(N) was grey and whitish.

But the coloration was vividly renewed in 1964 (during the next SEB Revival). While the dark SEB(N)/EB(S) persisted, the ALPO reported orange colour spreading in from the N. and S. edges from 1964 August onwards. High-resolution photographs in late 1964 showed the whole EZ intensely coppery red (Plates P12.12, P13.1). S. Cortesi reported a notable yellow colour (again mixed with many grey veils and streaks) from October which persisted until early 1965. In 1965/66 and 1966/67, the EZ was white again.

1968–71 (LPL and others): The next event, unusually, affected only the middle of the EZ at first. It was a pure orange EB, which appeared in 1968 January. This event was well covered by the Arizona LPL photos (Plate P13.3&4; Plate P14.4&5; Fig. 9.16&17). In ultraviolet photos, this EB was the darkest belt on the planet! Being quite narrow, its colour was not widely noticed visually, though the SAF reported ochre colour.

This EB persisted for 4 years and gradually lost its colour (Fig. 4.9&10; Plate P13.5). By 1969, it was ochre or orange-brown and was broadening to the south as the SEB faded. In 1970 it was dark, brownish, and more southerly. In 1971 this massive EB(S) was grey (LPL, NMSUO). Then it was disrupted by the spectacular and colourful SEB Revival (Fig. 10.20&21).

1972–76 (LPL, BAA): In 1972, coloration revived: yellow-orange colour covered the whole EZ and SEB (LPL, BAA; Plate P13.6). There were large 'white' clouds on EZn and EZs but even these were light yellow in LPL photos; those on EZs slowly blended into the ochre of the revived but fading SEB. The new coloration again condensed into a massive dark EB(S) (Fig. 9.17) which persisted until 1976/77. It was brown or grey-brown in 1973, 1974, and 1975, similar to the adjacent SEB(N), but plain grey in 1976/77 (BAA). (Those of us who began observing in the early 1970s, when most of the SEB was white, came to regard this massive belt as normal, but historically it must be regarded as a persistence of the coloration episode.)

1977–82 (BAA and others): A yellow EZ was first noticed by the author in 1977 December; the same month, a blue-light photo showed a broad dark 'EB' (Sanchez-Lavega and Quesada, 1988). In 1978/79, the colour was obvious in EZ(S); the author found it browner than in 1977/78, typically ochre, while others described it as yellow or orange, and it appeared strikingly orange in the Voyager pictures (Plates P16, P17, P20). EZ(N) was also coloured in late 1978, but a white strip along the EZn edge expanded until EZ(N) was mostly white. It was also extremely disturbed. Thus the zone presented a bizarre and spectacular mixture of strong ochre colour, grey EB and festoons, and huge blue-black NEBs projections and white spots.

During 1979 the EZ(S) reverted to yellowish, and this colour persisted for another three years, gradually fading, according to the author and some other BAA observers. (Meanwhile, EZ(N) remained disturbed but uncoloured, and the grey EB persisted.) Sectors of EZ(S) were still light yellow up to 1982 March but not thereafter.

(1986: Extensive shading of EZ(N) was grey, associated with complicated NEBs activity; there was no coloration.)

1989–92 (BAA) (Figs. 9.18, 7.8, 4.1B, 10.24; Plate P24): After the SEB had faded away, traces of colour began to appear in autumn 1989. As described below, this began with a strongly violet-absorbing EB, progressed to a golden yellow colour across the whole EZ(N) in 1990 Jan., and became more brownish-yellow over the next two months. Once again the yellow colour was mingled with a tangle of grey festoons and wisps, and with white or yellowish-white patches on the north side, providing a weird spectacle.

In 1990/91, colour spread south over the revived SEB, but cleared from EZ(N), apparently displaced by large white areas associated with NEBs activity, and the grey NEBs festoons formed a grey EB. In 1991/92, EZ(S) remained shaded but had a duller brown colour, and the EB was grey, in contrast to the SEB which remained ochre. Thus the colour had shifted southwards during this episode, as also happened in 1882–83 and 1927–29.

The colour in EZ coloration episodes

In several episodes the colour has evolved from pure yellow or orange to a browner colour. Detailed observations of the most recent episode suggest that this is only part of a progressive change in spectrum, whereby absorption slowly spreads in from the ultraviolet through the visible range.

The first blue-filter photos, in 1989 October, already showed a strong dark EB and dusky EZ(N), absent from red-filter photos. An ultraviolet photo showed the EB even more strikingly, almost as dark as the NEB (Fig. 9.18). However, direct colour photos in the autumn (Plate P24) showed the zone only slightly cream or faintly yellowish, and visually, yellow colour was only suspected in November. In 1990 January, the colour became striking; it was described as 'golden yellow' by Rogers and R. Néel, and photos routinely showed a distinct yellow or 'clotted cream' colour all

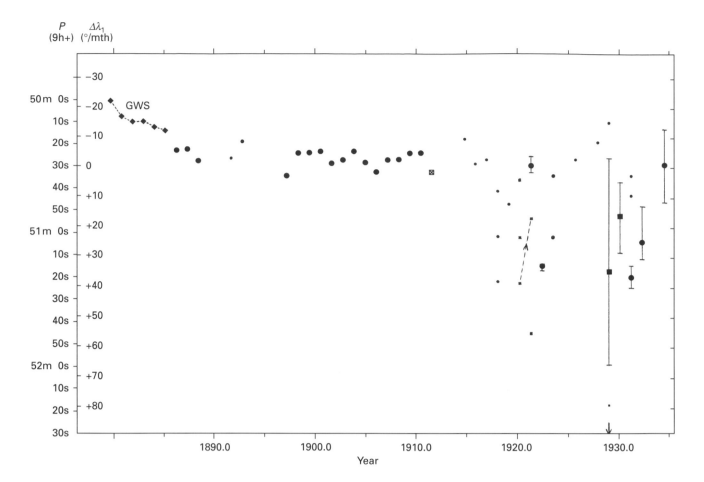

Fig. 9.20. History of the South Equatorial Current. Speed is shown as P (rotation period), as $\Delta\lambda_1$ (degrees per month), and as u (m/s in System III). Note that the vertical scale of speed differs from that used for charts of the non-equatorial currents. Symbols: diamond, Great White Spot; square, spots from SEB Revival; circle, other spots. Large symbol, average of three or more spots; small symbol, one spot or pair. Vertical bars indicate the range of speeds (not published before 1914).

See Appendix 3 for details of sources. Values up to 1907 are taken from compilations by Williams (*1896, 1897, 1909*) and from other reports in MNRAS. From 1905 to 1948, data are from BAA Memoirs. From 1949 to

1964, data are mostly from the ALPO, but the second point in 1952 and 1958 represents the BAA average for the same apparition; the large difference (and range) in each of these SEB Revivals could be due to a real range of speeds, or perhaps to misidentifications of some spots. (For 1928/29 and 1960, . indicates features before the SEB Revival began.) Data in 1968 and 1971 are all from NMSUO (all the spots in the 1971 SEB Revival had periods > 9h 51m 50s). Points for 1972 to 1974 are averages of BAA and LPL data. Points for 1975 onwards are BAA values, except for the circles for 1980–1984, which are for numerous elusive light areas reported by the SAF.

over the EZ(N). However, some visual observers still did not perceive more than a slight 'warm' tone at this time. From Feb./Mar. onwards, the colour seemed to become more brownish yellow, and also more visible to observers who had not recorded the earlier, purer yellow colour.

I therefore suggest that the spectrum evolves progressively, as sketched in Fig. 9.19. First, the universal ultraviolet absorption spreads into the violet where it is detectable in filter photographs but not visually. Then it spreads further into the blue to produce a pure yellow colour.

In spite of the striking intensity of this colour, some observers find it surprisingly hard to see. The differences between observers in 1990 were reminiscent of the debates in 1869–1871 on this question. In both episodes, the colour then changed to a duller, browner tone, and eventually everyone could see it. The browner colour must be due to the blue/violet absorption becoming augmented by absorption right across the visible spectrum.

It is not known what generates the colour absorption, but such a

change might be caused either by increasing density of the absorbing substance, or by increasing chemical complexity as the event progresses.

The subsequent evolution of EZ coloration often continues the trend, but cannot be due to a single haze layer. Often it evolves into a dull brown or even dark grey shade, concentrated in a massive southerly EB. Whereas the initial colour is at high altitude, the brown or grey colour is typical of thin clouds which allow warm 5-μm emission to penetrate from deep levels (Chapter 4). This was directly shown for the EB in the 1970s. So the event is not simply a high-altitude yellow veil that spreads above the EZ; it is part of a general change in cloud structure that gradually replaces part of the EZ with a dark belt.

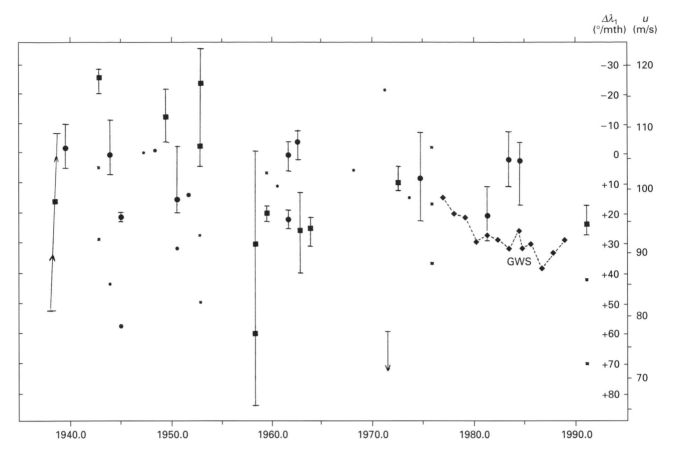

Fig. 9.20b. For legend see opposite

9.5 THE SEB NORTH EDGE

In the eighteenth century, as described in section 9.1, the EZ(S)/SEBn was as active as the EZ(N)/NEBs is nowadays. That activity ceased in 1908. Since then, notable disturbances on SEBn have only been seen in two circumstances.

Firstly, extensive disturbance often develops during SEB Revivals (Chapter 10.3). These outbreaks disgorge spots onto SEBn which may be very large, very bright, very dark, or multi-coloured. The SEBn spots often last for a year or so.

Secondly, we must record the long-lived great white spot or rift on SEBn which was a familiar feature from 1976 to 1989. When most developed, it was a bright white oval, interrupting the SEB(N), and connected by a rift to a stretch of northerly SEBZ f. it. This defined a sharp p. end for a dark section of SEB(N) f. the spot. Before describing it in detail, though, we should note that it was not the first feature of its type. Similar but short-lived features have been conspicuous in several apparitions; and in the nineteenth century, one such feature was very conspicuous and lasted for at least six years.

The South Equatorial Current

The average speeds are in Table 9.4, and the historical record is plotted in Fig. 9.20.

The initial deceleration in the 1880s, which impressed observers

Table 9.4. *Average speeds of the South Equatorial Current*

Interval	$\Delta\lambda_1$	Period	(±s.d.)	(N)	u (m/s)
1879–1885 (GWS only)	−15.0	9h 50m	9.8s (±4.7s)	(6)	+112.9
1886–1909 (all points)	− 3.0	9h 50m	26.1s (±3.9s)	(17)	+107.2
1914–1974 (branch A)*	+ 1.1	9h 50m	31.5s (±9.9s)	(23)	+105.2
1914–1974 (branch B)*	+27.8	9h 51m	7.6s (±13.8s)	(10)	+92.4
1976–1989 (GWS only)	+27.5	9h 51m	7.0s (±6.3s)	(13)	+92.6
Voyager:					
(Limaye, *1986*)	−47	9h 49m 27s			+128
(Maxworthy, *1985*)	−113	9h 47m 59s			+160
(Beebe et al., *1989*)	−93	9h 48m 26s			+150

Notes on Table 9.4: The Table lists the mean (±s.d.; N) of points plotted in Fig. 9.20, some of which are apparition means, while others are single spots or differentially–moving clusters of spots. They were assigned a weight of one for a single spot or pair, and three for an average of three or more spots. Spots attributable to SEB Revivals were excluded: i.e. these averages include only points shown as diamonds (GWS) or as circles (other averages) on Fig. 9.20. For spots plotted as squares (in SEB Revivals), which are often very slow, see Table 10.1. Speed is given as $\Delta\lambda_1$ (deg/mth), as P (rotation period), and as u (m/s in System III).
*For 1914–1974, the points are arbitrarily divided into two 'branches', A and B, following Peek and Reese. Branch B includes the points with P> 9h 50m 50s; these are from only 10 apparitions (1917/18, 1922, 1923, 1930/31/32, 1943/44/45, 1950, 1961, 1973) having excluded the drifts from SEB Revivals.

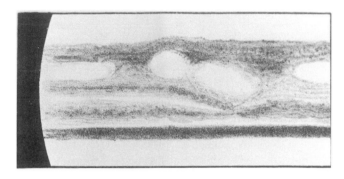

Fig. 9.21. The Great White Spot on 1880 Oct 18; it was temporarily double. Drawing by W.F. Denning (*1880*).

Fig. 9.22. A short-lived Great White Spot on 1950 Sep 8. Unpublished drawing by W.E. Fox.

at the time, concerned only the single Great White Spot (GWS), described in detail below (indicated by diamonds in Fig. 9.20). And most of the values since 1976 refer to the more recent Great White Spot or rift, which also decelerated throughout its long lifetime – even though it started off moving much slower than the one a century earlier.

Only from 1886 onwards have multiple features been tracked in the SEC. The change in character of the chart after 1909 is due to two factors. Firstly, only averages were given up to that date; but also, that was when the long-running SEBn activity ceased (section 9.1). So the stable SEC shown before 1909 was certainly real. Since 1909, there have been few distinct features to track, except during SEB Revivals (square symbols on the chart) when they cannot be expected to adhere to a stable underlying current. As the SEBn activity from a Revival often lasts a year or more, the assignment of symbol for the year following a Revival is rather arbitrary. These SEB Revival spots often drift quite slowly at first but accelerate to System I speeds. (This was very clearly observed in 1938 as indicated by arrows on the chart.) But three SEB Revivals produced very fast speeds (1943, 1949, and 1952). The fastest features on record were two spots reported by the BAA during the 1952/53 SEB Revival, with $\Delta\lambda_1 = -68$ (P = 9h 48m 59.5s, lasting 3 weeks) and -36 (9h 49m 41.5s, lasting 7 weeks). The slowest features, described in section 9.3 and too slow to include in the chart, were in the SEB Revivals of 1927/28 and 1971 (see Chapter 10.3).

The remaining features, plotted as circles, presumably represent the intrinsic speed of the SEC. Peek noted that the SEC speeds fall into two groups: half the values are close to System I ('branch A'), the other half are considerably slower ('branch B'). Peek divided them arbitrarily at a period of 9h 51m. Reese later preferred a division at 9h 50m 50s, which is also used in the present analysis (Table 9.4); but all these authors have admitted that the division is only arbitrary. A histogram of the points in Fig. 9.20 shows that branch A is a tight cluster, whereas branch B is simply a continous spread ranging down to mid-SEB speeds that we have not included. So the 'mean value for branch B' does not mean much.

To summarise, the SEC includes a component that adheres closely to System I, but many features move much more slowly, both transient ones from SEB disturbances, and well-formed great white spot structures.

Voyager gave a strikingly different result. The global average speed of the SEC was much faster than suspected from visual data: $\Delta\lambda_1 = -47$ (P = 9h 49m 27s) or even faster (Table 9.3B). This value refers to the fine texture of the clouds, and implies that the GWS, and previously-recorded SEC features, represent waves with speeds much slower than the bulk flow. (However, the bulk speed had been tentatively detected by A.S. Williams in the vicinity of the earlier GWS in 1881/82, as noted below!).

The great white spot on SEBn, 1879–1885[6]

This great white spot (GWS) in the southern EZ, nestling in the SEBn edge, was first observed by G. Hough in 1879 September, and remained very conspicuous for the next six years (Fig. 9.21). Most often it was a well-defined, small but very bright oval, often with a diffuse plume to the f. side; sometimes it was large and bright, sometimes double, sometimes diffuse, and occasionally invisible. The brightness could change within a few days. Dark material often flanked it, particularly on the f. side. Neighbouring features generally moved with the GWS, but Denning and Williams, corresponding in *The Observatory*, believed that in 1881/82 dark material was repeatedly erupting from a source ≈35° f. the GWS, and streaming towards it with a rotation period of ≈9h 49m 14s. (This old and rough estimate is surprisingly close to the rotation period measured for this latitude by Voyager: 9h 49m 27s.) These dark patches typically became more diffuse over a few days but nevertheless obscured the GWS as they crossed it. Another dark spot, just p. the GWS, appeared in 1882 October and moved along with it.

The rotation period of the GWS was fairly steady and remarkably fast, averaging 9h 50m 7.4s from 1880–1882, 9h 50m 12.2s from 1882–1884, and 9h 50m 9.8s from 1880–1885, according to A. Marth's compilations. It gradually decelerated during its lifetime, as shown in Fig. 9.20. But in the short term it could show striking changes in motion. In 1880 October, having drifted at 9h 50m 0.7s for several months, it shifted 19° f. within two weeks, and thereafter drifted at 9h 50m 7.3s. In subsequent years it showed similar sudden shifts in the f. direction, for example in 1883 January (10° f.), 1883 summer (17° f.), and 1885 January (12° f.). (Two months before this last shift, it had shown an unprecedented acceleration to 9h 49m 38s, which brought it 40° ahead of its previous track – only partly corrected by the 12° shift back.) As Denning pointed out, the shifts could have been caused by alternating activity of multiple bright nuclei within the complex.

[6] This was described in many contemporary papers, notably by Denning (*1880, 1883, 1885*) and Williams (*1910*).

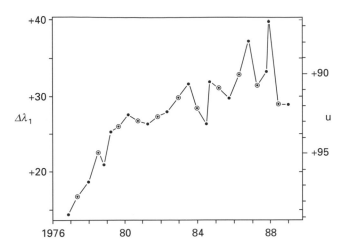

Fig. 9.23. Changing drift rate of the GWS/rift, 1976–1989. (From BAA report for 1987/88, with later BAA data added. Data for 1976–1982 are the author's, but complete BAA results in Fig. 9.20 do not differ significantly.) Drifts are plotted within each apparition (dot; after subtraction of the phase effect of 0.6°/mth), and between apparitions (dot in circle; average value from one opposition to the next). The latter values show less variability.

Passages of the GWS past the GRS occurred every 45 days or so, and were carefully measured, but there were no reports of either spot changing its appearance at these passages. The first longitude shift described above coincided with one such passage, but the last shift did not.

After the last shift, the GWS period from 1885 January to April was 9h 50m 15s, and it was not definitely identified thereafter. At that time it was as brilliant and distinctive as ever; but in the 1885/86 and 1887 apparitions, there were many other white spots along the SEBn and the rotation period had increased, so although Denning and Williams made tentative identifications of the GWS, it had to be abandoned as lost.

Other great white spots and rifts on SEBn

Minor features of this sort, resembling the GWS/rift of 1976–1989, were common in the 1890s. They consisted of a small rift curving into the SEBn from a light spot in the EZ(S), and were known as 'trumpet-shaped rifts'. Presumably their drifts were included in the SEC averages for those years.

Unusually slow drifts were recorded for groups of spots on SEBn in each apparition from 1917/18 to 1923, and several of these had the classic white-spot-and-rift structure. Thereafter, the only notable example has been in 1950.

1917 Aug–Dec: One such feature was tracked for 4 months, $\Delta\lambda_1$ = +23.5 (period 9h 51m 1.6s).

1922 April–June: There was a pair of these features only ≈20° apart, $\Delta\lambda_1$ = +33.5 (period 9h 51m 15s).

1923: Similar appearances and drifts were recorded for several features, but they only lasted a few weeks.

1950 Jun–Oct (ALPO, BAA [unpublished], and SAS): This was an impressive great white oval, coupled to a very conspicuous sharp p. end of dark SEB(N), though it did not open into SEBZ (Fig. 9.22 and Plate P9.4). When passing the Red Spot Hollow (then white), it surpassed it in brightness and matched it in size; the white spot was 26° long, though very variable, especially at the p. end. $\Delta\lambda_1$ = +19 (period 9h 50m 55s), lasting 4 months.

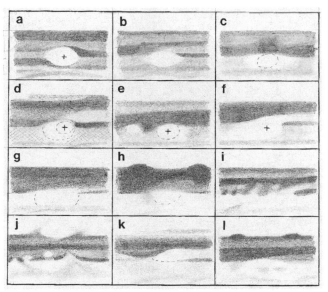

Fig. 9.24. Typical appearances of the GWS/rift, 1976–1989. Bright white spot is marked +. (a) 1976 Nov., (b) 1977 Nov.-Dec., (c) 1978 Jan.-Feb., (d) 1979 Jan.-Feb. (after passing GRS), (e) 1981 Mar., (f) 1980 Dec. (after passing GRS; before e); (g) 1982 March; (h) 1984 April-May; (i) 1986 July; (j) 1987 Aug. and Oct.; (k) 1987 Sep. (after passing GRS); (l) 1988 Sep. and Dec. Each panel shows the full width of the SEB, with EB at bottom. Drawn from visual and photographic data.

It remains a mystery why these 'great white spots' vanished within a few months, whereas the apparently identical one that appeared in 1976 was to last for 12½ years.

The great white spot or rift on SEBn, 1976–1989[7]

This GWS appeared in the aftermath of the 1975 SEB Revival. It was first seen on 1976 August 4 by W.E. Fox, when it was already a prominent bright white oval. (Sanchez-Lavega and Rodrigo (*1985*) claimed that it actually appeared at the very start of the SEB Revival on 1975 July 7, but BAA observations showed no evidence for such a feature in 1975, and the 1976 drift seems inconsistent with such an origin.) In fact there were two such features in 1976 November, but the second one soon disappeared.

Over the next 12 years the GWS evolved both in speed and in appearance.

Its speed diminished, rapidly and steadily at first, and more irregularly in its later years (Fig. 9.23).

Its appearance was initially a simple white oval, but this acquired the familiar rift into a trailing SEBZ in 1977 December, and in later years the white spot became less and less conspicuous while the rift became the most stable feature. Typical appearances are shown in Figs. 9.24 and 10.52.

The initial white oval was ≈13° long, prominent in 1976/77 when EZ(S) was shaded (Fig. 1.18), but less so in 1977/78. In 1977 December, the bright rift into a trailing SEBZ first made its appearance.

Conjunctions with the GRS occurred every 7 weeks at first. Up to the end of 1978, these encounters had no noticeable effect on the GWS. But in 1979 January, the GWS became faint or invisible for two weeks before passing the GRS, then after the conjunction it suddenly reappeared in full

[7] Previous reviews have been given by Sanchez-Lavega and Rodrigo (*1985*) and in the BAA report for 1987/88.

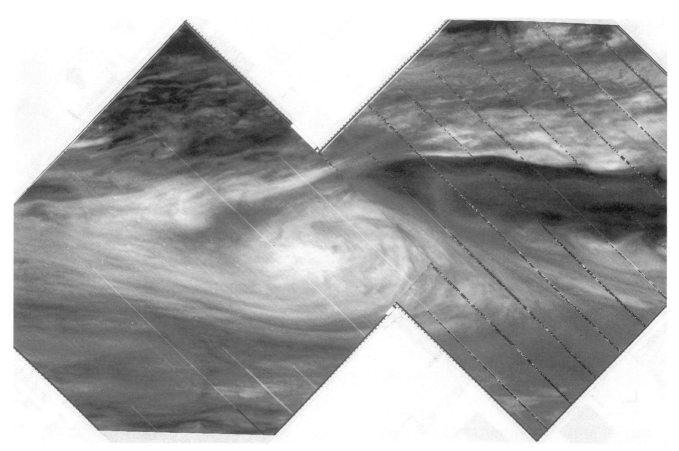

Fig. 9.25. Voyager 2 closeup of the GWS. It had recently passed the GRS and is very bright. The S. Tropical Disturbance f. end complex (Chapter 10.7) is off the top of the picture. South is up. (Image 20590.22, 26; J2–3d; green, unfiltered. NASA.)

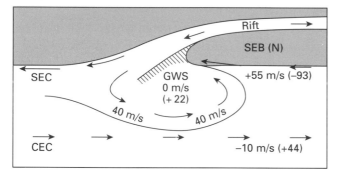

Fig.9.26. Circulation of the GWS from Voyager data, according to Beebe, Orton, and West (*1989*). Wind speeds are given in m/s relative to the GWS, and (in brackets) in deg/mth relative to System I.

splendour as a white spot and rift (compare Fig. 1.19 with Plate P18.3). Over the next few years, the spot continued to vary between these extremes: usually it was a white oval, of varying brightness, often with an attached rift into a SEBZ, but sometimes it was faint, especially when approaching the GRS. Although observations were not so frequent in the early 1980s, there was evidence for the spot brightening after it passed the GRS in 1980 Mar.-Apr. and in 1983 Aug.

The dramatic effects of the GRS conjunction in 1979 Jan.–Feb. were well shown in Voyager movies (see below; also Fig. 2.18, Fig. 10.58, and Plate P19.2). These suggested that the important interaction was not with the GRS, but with the turbulent mid-SEB rifts f. it, which were conspicuous throughout the lifetime of the GWS. However, in visual observations the GWS rift rarely if ever formed a largescale connection to these mid-SEB rifts as it passed.

The very dark SEB(N) segment f. the GWS/rift was the same colour as the rest of the SEB(N) in 1976/77. But in 1979 Feb., and sometimes in 1979/80, it was ruddy-brown in contrast to the greyer belt p. the GWS. No colour contrast was noted in later years.

From 1982 to 1985, the feature was seldom prominent, and visual sightings became rare, though it was still tracked photographically. Although the classic form was sometimes seen, more often it was merely a light area in EZ(N) attached to the p. end of a SEB(N) segment. In 1985 and 1986, the SEBn rift was maintained but there was usually no white spot, even after passing the GRS. In 1986 and 1987, Isao Miyazaki often noted what looked like ripples on the SEBn extending up to 60° p. the long-lived rift. They were most extensive in 1987 Oct., when the rift's speed was temporarily its slowest ever ($\Delta\lambda_1 = +40$ for three months). At that time the mouth of the rift in EZ(N) was subject to faint, variously tinted shadings. However, two GRS conjunctions in 1987 Aug.–Sep. and Oct.–Nov. again rejuvenated it, for the first time in many years. Each time the whole feature was invisible while passing the GRS, but a week later the rift temporarily opened up anew. However, another conjunction that Christmas did not enhance the rift. In 1988/89, although its drift had accelerated, it was feebler than ever – just a subtle 'crease' in SEBn (Fig. 9.24l) with an indistinct light strip of EZ(S) alongside it. The bright rift was no longer present, as mid-SEB and SEBs disturbance had also ceased at the same time. The last traces were photographed in 1989 January, after a lifetime of 12.4 years.

Spacecraft observations and physical interpretations

Movies from both Voyagers 1 and 2 spectacularly showed the GWS brightening as it passed the GRS region, as was also noted visually. The brightening occurred as a narrow stream of white material (alongside the Red Spot Hollow) actually connected the GWS to the SEBZ rifts.[8] This gave the impression that energy, and perhaps white clouds, were being transferred from the SEBZ to the

GWS. In contrast, Plate P16.4 shows the GWS when it was f. the GRS, dim and covered in yellow haze.

A Voyager closeup of the GWS is in Fig. 9.25. (This was shortly after passing the GRS.) The picture gives a strong impression of anticyclonic (anticlockwise) circulation and streaming, around the edges of the GWS and out of the SEB along the rift. T. Maxworthy (*1985*) tracked cloud features near the GWS but detected only the latitudinal winds, not circulating motions, as the clouds within the GWS were too ill-defined to follow. However, Beebe, Orton and West (*1989*) did describe anticyclonic circulation, at about 40 m/s. Thus the GWS rolls between the fast SEC and the slow CEC, like higher-latitude anticyclonic white ovals (Fig.9.26). According to Beebe and colleagues, orange chevrons in the rapid SEC jetstream were flowing round the south edge of the GWS. When the GWS passed the GRS, a wavy white streamer stretched Np. from the mid-SEB rifted region to the Sp. edge of the GWS, and small brown-eyed eddies moved along the same track to enter the GWS. They followed the anticyclonic flow two-thirds of the way round the GWS, expanding and fading until the underlying cloud streaks at the f. side could be seen through the now-diffuse eddies.

Physical models of the GWS as an anticyclone were presented by Maxworthy (*1985*), who interpreted it as a 'solitary wave', and by Sanchez-Lavega and Rodrigo (*1985*), who suggested that the circulation was due to the transfer of white clouds from the SEB

every time it passed the GRS region. This would imply that the long life of this GWS was entirely due to the persistence of the mid-SEB disturbance f. the GRS. (However, it is unclear whether such disturbance was totally absent earlier in the century when GWSs were short-lived.)

Infrared data on the GWS were paradoxical, implying an unusual cloud layering, according to Earth-based and Voyager observations in 1979.[9] The white colour implies a high-level ammonia cloud cover, but the GWS appeared as a 5-μm hot-spot at $\approx$253°K, implying that the deeper clouds normally associated with white features were absent. One suggestion was that that the white cloud cover was transferred from the SEB rifts to overlie the warmer EZ(S).

In shape, the SEBn GWS closely resembled the common white spots on the NEBs, particularly those that lie at the p. edges of dark projections, with a rift curving into the belt on the f. side. The slow-moving NEBs white bays seen in 1988/89 (Fig. 9.5D) were especially close copies, although smaller. This suggests that the shape in each case was due to the spots retrograding relative to the bulk flow: certainly the SEBn GWS was doing so, and we have speculated (section 9.2) that the 1988/89 NEBs features may likewise have retrograded relative to an underlying System I flow.

Likewise, the brightenings of the GWS after passing the mid-SEB rifted region may be analogous to the NEBs plume-brightenings and patch-darkenings induced by passing NEB rifts.

---

[8] Smith *et al.*(*1979b*); Beebe, Orton and West (*1989*), in their Figs.146-147; and Fig. 10.58.

[9] Bézard et al.(*1983*); Sanchez-Lavega and Rodrigo (*1985*).

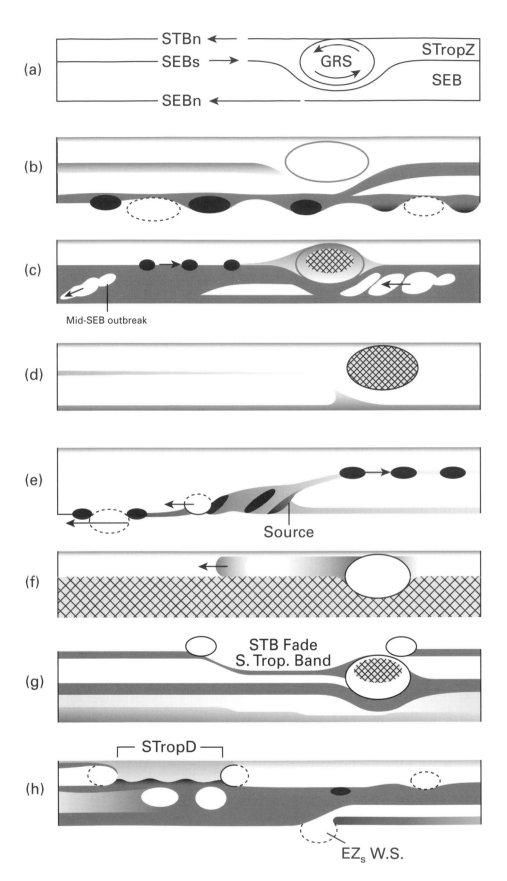

Figure 10.1. Diagram of typical aspects of S. Tropical region. (a) Jetstreams and abbreviations. (b) With SEB quiet (late 19th century). (c) With SEB active. (d) With SEB faint. (e) SEB Revival (GRS not shown). (f) Aftermath of a SEB Revival; cross-hatching indicates reddish colour in the belt. (g) South Tropical Dislocation. (h) Various long-lived spot types, including South Tropical Disturbance (STropD) (GRS not shown).

10: South Tropical Region (9°S to 27°S)

10.1 OVERVIEW

In the domain of the South Tropical region, we encounter the most famous and impressive features of the entire planet. Here is the Great Red Spot, and here occur the violent South Equatorial Belt Revivals, the grand South Tropical Disturbances, the intricate South Tropical Dislocations, and all the wonderful phenomena associated with them. Here too is the fastest retrograding jetstream on the planet, which underlies many of these special phenomena. We know now that many of the features of this region have smaller-scale parallels in other domains on the planet; but in their size and longevity, the features of the South Tropical region achieve a degree of organisation that transcends such comparisons, and justifies the unique status that they have always had for visual observers.

All these features require detailed description in later sections, but as they all interact in important ways, we begin with an overview of the whole region.

The South Equatorial Belt (SEB) is usually the broadest belt on the planet and sometimes the darkest, being typically as prominent

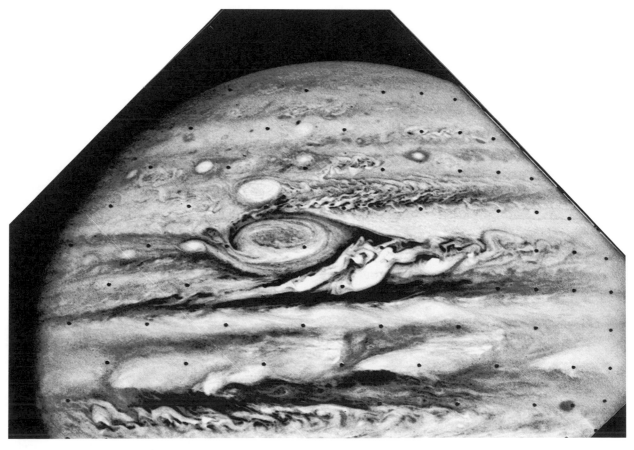

Figure 10.2. Voyager 1 view including the GRS, on 1979 March 3d 22h. On the p. side is a S. Tropical Dislocation with SEBs jetstream spots. On the f. side is the turbulent SEBZ. Also see Fig. 3.6 (March 2d 15h), Fig. 12.4 and Plate P17.4 (March 4d 7h), and Fig. 10.37 (March 5d 4h). (Image 16342.06, J1–2d, wide-angle, green, filtered. NASA.)

as the NEB. Sometimes it is split along part or all of its length by a division termed the SEB Zone (SEBZ). The South Tropical Zone (STropZ) vies with the NTropZ as the brightest and most conspicuous zone on the planet.

The Great Red Spot (GRS) is the archetype of anticyclonic ovals. It is a huge oval that spans the whole width of the STropZ, indents the SEB south edge, and deflects its jetstream; this indentation, the Red Spot Hollow, is always present and is often more prominent than the GRS itself. The Spot only rarely has an obvious red colour. Sometimes it is a diffuse grey or fawn-coloured patch, or a light area enclosed by a dark ring, but its oval form persists, even though only part of it may be visibly outlined. It influences the structure of the SEB around half the circumference of the planet, and controls the origin of South Tropical Disturbances, Dislocations, and Bands.

The SEB has cycles of activity that are more obviously organised than in any other belt. The climax is the classical SEB Revival, which is not merely a revival of the dark belt but also a pattern of disturbances on several currents that affects all the structures in the region. It involves enormous cyclonic turbulence and jetstream outbreaks. Such cycles have occurred 14 times in the 20th century up to 1991. The cycle begins with fading of the SEB, leaving barely a trace of SEB(S) and a narrow SEB(N), so that the white STropZ appears immensely broad. Other dark features, such as a S. Tropical Disturbance if present, also disappear, but the GRS becomes darker and often redder than before. After one or two years with the belt in this 'pregnant' condition, the Revival begins at a single point. From this source, intense disturbance pours out in the form of bright and dark spots, on three distinct currents. These are: a southern branch, retrograding in the SEBs jetstream and reconstituting the SEB(S); a northern branch, prograding in the South Equatorial Current and sometimes breaking into the southern EZ; and a central branch, prograding more slowly in the mid-latitudes of the SEB, filling it with turbulent spots and, ultimately, with the dark material of the restored belt. The southern branch sooner or later runs into the GRS, which thereupon fades; the GRS ends up as a pale oval in a dark ring. The activity lasts several months overall, and is followed by a flush of reddish colour across the revived SEB.

In some decades, these cycles have ceased, and the SEB has instead shown periods of continuous activity lasting for several years. This type of 'standstill' activity has been familiar in the 1980s. It is not as obviously energetic as a classical Revival, but includes some of the same elements: retrograding spot activity on the SEBs jetstream, and eruptions of prograding bright spots and streaks from a distinct source within the belt. These 'mid-SEB outbreaks' also resemble rift activity in the NEB.

The mid-SEB spots or 'rifts', whether in classical Revivals or in mid-SEB outbreaks, constitute the most dramatic examples of turbulent cyclonic disturbances on the planet. Similar cyclonic turbulence, evident visually as bright white spots, occurs almost continuously for several tens of degrees following the GRS, even when the rest of the belt is quiet.

A South Tropical Disturbance (STropD) is a coherent sector of shading and disturbance that spans the STropZ. A famous one existed from 1901 to 1939, but at least six smaller ones have occurred since. They have several characteristic features, of which the most important (though the most difficult to observe) is apparently the 'Circulating Current' – the reconnection of the SEBs and STBn jetstreams, which results in retrograding spots on the SEBs jetstream being deflected onto the STBn jetstream so that they end up going back the way they came. This behaviour was seen in 1920 and in the 1930s, and was revealed again by Voyager in 1979.

In recent years a more complicated phenomenon has repeatedly occurred, the South Tropical Dislocation. This is triggered when a faded sector of STB passes the GRS; as it does so, the STropZ p. the GRS darkens, so that the normal pattern of belts and zones is reversed. Eventually a structure resembling a STropD emerges from the p. end of the GRS. The second such event, in 1979, produced a STropD that lasted for two years. Conversely the fourth S. Tropical Dislocation, in 1985, produced a Little Red Spot.

There are other varieties of darkening of the STropZ p. the GRS, which do not appear to involve any circulating current, and have drifts intermediate between a STropD and the STBn jetstream. General darkening of this sector sometimes occurs during a SEB Revival; and sometimes (generally at the end of an episode of SEB activity) a distinct narrow belt emerges from the p. end of the GRS, termed a South Tropical Band (STropB), which lies on or close to the prograding STBn jetstream.

The two Voyager spacecraft photographed Jupiter at a most fortunate time, when they were able to record continuous SEB activity, a mid-SEB outbreak, a S. Tropical Dislocation, and the origin of a major STropD.

Currents in the South Tropical region:

(i) Currents in mid-SEB

A great range of currents operates in this domain, and they are all summarised in Table 10.1 and shown in Figs. 10.3–6 and Fig. 10.61. The fast speeds to the south, including the STBn jetstream (Fig. 10.61) and STropDs (Fig. 10.6) will be covered later in this chapter. Here we consider the slow current of the domain, the South Tropical Current (STropC; Figs. 10.5&6); the retrograding jetstream on SEBs (Fig. 10.4); and the gradient of speed across the SEB, from the SEBs retrograding jetstream to the equatorial prograding jetstream (Fig. 10.3).

To visual observers, this gradient across the SEB is sometimes revealed in the behaviour of white spots and streaks during Revivals or other disturbances; they arise from sources stationary in System II, in the centre or (at highest resolution) in the south half of the belt, and some are rapidly stretched in the Np. direction to flow with velocities approaching System I. However, most white spots in mid-SEB (including the SEBZ branch of Revivals, mid-SEB outbreaks, and the perennial activity f. the GRS) prograde in a range of intermediate speeds, $\Delta\lambda_2 = -5$ to -70. The speeds are quite diverse and the spots are obviously jostled and distorted by the intense turbulence in these active regions. As these drifts overlap the range of the S. Tropical Current (see below), they do not define

Table 10.1. *Summary of drifts in the South Tropical region*

Type of feature	Lat. β''	$\Delta\lambda_2$: (1898–1939)	(1940–1990)	(1898–1990)
SEB(N) in Revival	7–11°S			−197.2 (±32.6)
SEBZ in Revival ⎫				
Mid-SEB outbreaks ⎰				
— Leading edge	11–16°S			−53.6 (±20.3)
— Most spots	11–16°S			−33.4 (±21.3)
— Source*	14–17°S			+0.7 (±2.2)
SEB(S) spots*	17°S	−0.6 (±4.1)	+3.3 (±5.7)	+1.1 (±5.2)
SEBs jetstream:	20–21°S			
(inc. SEB(S) in Revival)				
— STropD absent		–	+104.9 (±14.3) ⎫	+119.1 (±20.2)
— outside STropD		+138.2 (±8.3)	+126.6 (±18.6) ⎰	
— inside STropD		+56.8 (±17.1)	–	–
GRS:*	22°S			
— outside STropD		−0.1 (±1.4)	+1.0 (±0.7) ⎫	0.0 (±1.9)
— inside STropD		−2.7 (±1.9)	– ⎰	
STropZ (misc.,slow):	21–25°S			
— outside STropD*		+2.2 (±6.9)	−0.4 (±6.3)	+0.6 (±6.6)
— inside STropD		−9.5 (±7.9)	–	–
STropDs	21–25°S	(−16 → −3; see Fig 10.6)	−10.2 (±5.7)	–
STropZ (misc.,fast)	21–25°S	–	−32.1 (±4.7)	−32.1 (±4.7)
STBn jetstream (outside STropD)	25–26°S	−117.8 (±12.1)	−103.1 (±17.5)	−109.3 (±17.0)

This is an abstract of data in Figs. 10.3–6 and 10.61, and Tables 10.2 and 10.10. All drifts are given as $\Delta\lambda_2$ (degrees per month): mean (±s.d.) of apparition averages, weighted according to number of features up to a maximum of 4. All drifts are as given within apparitions, not corrected for phase effect. A few anomalous drifts were omitted (e.g. SEB Revival sources in 1971 and 1975). β'' is zenographic latitude.

*These four categories appear to define a *South Tropical Current* (STropC), analogous to the slow currents of other domains, with an average speed of zero in System II. The speeds of the SEB(S) spots (1911–1939) were little affected by the great STropD; average $\Delta\lambda_2 = +1.1$ outside the STropD and −2.1 within it. The difference of 3°/mth is the same as for the GRS, but it is not statistically significant given the scattered speeds of the SEB(S) spots. The speeds of STropZ spots within the STropD presumably include features moving with the STropD as a whole.

a distinct 'South Intermediate Current'. They are most notable in the SEB(N) and SEBZ branches of SEB Revivals, which will be considered in more detail in Section 10.3.

In the quiet years before the great Revival of 1928, surprisingly slow drifts were quoted for single white spots in the SEB ($\Delta\lambda_2 = -4$ in 1911, −4.5 and +0.4 just p. the Red Spot Hollow in 1915, and −13 in 1922). Drifts quoted in earlier years were unreliable because the spots were not observed often enough (mean interval >15 d). Slow drifts quoted for mid-SEB spots in 1929/30 and 1955/56 have not been included in Fig. 10.3 as they were associated with STropDs.

Currents (ii) The SEBs retrograding jetstream
This unique jetstream is one of the most important features of the region. It is by far the most rapidly retrograding jetstream on the planet, and is the only one that ever carries distinct spots that reveal it to Earth-based observers. It outlines the north edge of the Great Red Spot; it is a major contributor to the spectacle of SEB Revivals; and its occasional recirculation into the STBn jetstream, producing the 'Circulating Current', is the main characteristic of S. Tropical Disturbances. Voyager revealed that the spots on it were anticyclonic rings just like those on prograding jetstreams.

The jetstream was not discovered until 1920, when two large round dark spots in the SEB Revival made it evident, but even they

(being quite southerly) probably did not represent the full jetstream speed; that was not observed until 1928. Because the jetstream defines the S. edge of the SEB and the N. edge of the GRS, it must have been always present. The absence of earlier records may be because there was no activity, or because observers were not aware that such extreme speeds were possible.

The speed of the SEBs jetstream shows substantial variations (Fig. 10.4). A great range can be seen even within a single SEB Revival, and this may be because some spots are not fully entrained by the jetstream; but there are also definite variations in the mean speed between years. The two Voyagers found a speed of $\Delta\lambda_2 = +118$, essentially the same as that observed visually in the same year, but in other years the average speed has been as high as +150, with individual spots attaining +168 (period 9h 59m 31.5s) – far outside the range observed by Voyager.

What causes these speed variations? The speed does not depend on the season of the jovian year. It shows no tendency to increase during episodes of SEB disturbance. (In fact, it slowed down during each of the three episodes of continuous activity in the 1980s, and in SEB Revivals the SEBs spots in the first group are usually among the fastest; but there are also examples where later spots moved faster.)

The most promising explanation is that the faster speeds reflect the presence of a STropD, as would be predicted on theoretical

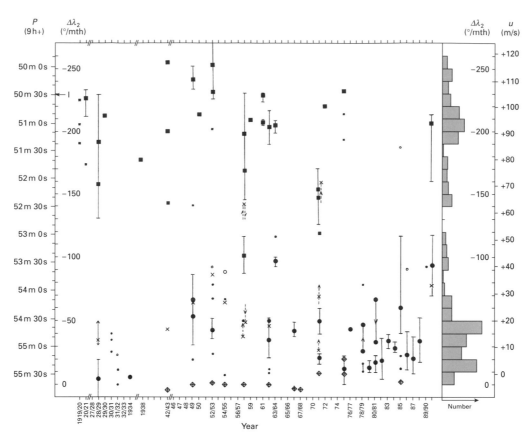

Figure 10.3. Speeds recorded in the SEB(N) and SEBZ, spanning the full range between Systems I and II. SEB(N) is only shown during SEB Revivals (see Fig. 9.20 for ordinary S. Equatorial Current). From sources listed in Appendices 1 and 3. Speed is shown as P (rotation period), as $\Delta\lambda_2$ (degrees per month), and as u (m/s in System III). Symbols: ■, SEB(N) in N. branch of SEB Revivals; ●, SEBZ (central branch of SEB Revivals, mid-SEB outbreaks, etc.); ×, leading edge of central branch of SEB Revivals; ⊕, source region for SEB Revivals or mid-SEB outbreaks (expanded in Fig. 10.5). Arrows show changing speed of individual spots. Large symbols are apparition means of ≥3 features, and vertical bars indicate the range of speeds. At the edge is a histogram of the data (omitting the source regions and imprecise points; weight 1 for single features, 4 for apparition means). Average speeds are listed in Table 10.1.

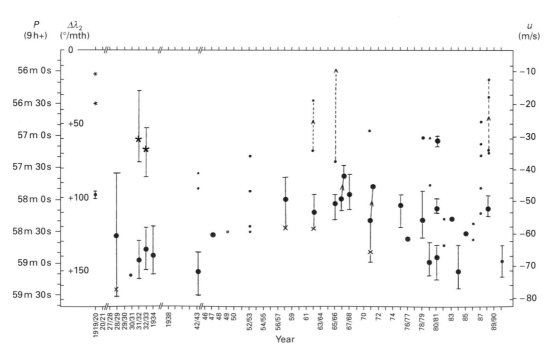

Figure 10.4. Speeds recorded in the SEBs jetstream, including S. branch of SEB Revivals. From sources listed in Appendices 1 and 3. Speed is shown as P (rotation period), as $\Delta\lambda_2$ (degrees per month), and as u (m/s in System III). Symbols for SEBs spots: ★ within STropD; × leading spot(s) in SEB Revivals; ● others. Arrows show changing speed of individual spots. Large symbols are apparition means of ≥3 features, and vertical bars indicate the range of speeds. Also see Table 10.2.

Table 10.2. *Records of the SEBs jetstream (including S. branches of SEB Revivals)*

Appar'n	$\Delta\lambda_2$ (normal)		$\Delta\lambda_2$	
	(+STropD)	(−STropD)	(in STropD)	Source
1919/20	[+101,+96]		+16.8,+35.5	BAA
1928/29	+126 (18)			
1930/31	+153 (2)			
1931/32	+143.5 (9)		+61 (16)	
1932/33	+136 (32)		+68 (15)	
1934	+140 (9)			
1942/43(A)	+84,+94			
(B)	+151 (3)			
1947	+126 (3)			
			(odd, near GRS)	Source
1949		[+124?]		ALPO
1952/53		+103 (4)		BAA
1957/58	+101 (6/9)			ALPO,BAA
1962/63		+110 (12)	+68 → +34 (from STBn, f.GRS, before Rev.)	ALPO
1965/66		+104 (12)	+75 → −11 (*sic*; p.GRS, before others)	NMSUO
1966/67		+101 (56) →+85 (13)		NMSUO
1967/68		+98 (18)		NMSUO
1971	+115.5 (24)		+54.5 (f.GRS, before Revival)	NMSUO
1975		+105 (3n)	+18 (jet.s.slowed, then Dislocation)	BAA
1976/77	+128 (4)			UAI
1977/78		[+67 or +110?]		BAA
1978/79		+115 (20)	+59 (2) (p.GRS, before others)	BAA
1979/80	+144 (16)		+59,+91 (f.STropD, p.GRS)	BAA
1982	+132,+114			BAA
1983		+114 (group)		BAA
1984	+150 (4n)			BAA
1985	+124 (4)			BAA
1986		+117,+128		BAA
1987/88		+77 (5)		BAA
1988/89		+107 (7)	+67 → +31 (p.GRS); +69 → +19 (p.GRS)	BAA
1990/91	+143 (?)			BAA
Average: (±s.d.)	+130.2(N=14) (±16.6)	+104.9(N=12) (±14.3)	–	–
Voyager	–	+118 (global)	–	Limaye (*1986*)

The Table gives the mean $\Delta\lambda_2$ for each apparition in which the jetstream was recorded. (In brackets, number of spots; for range of speeds, see Fig. 10.4.) Column 2, when a STropD was present within the previous year (these spots are outside the STropD). Column 3, no STropD recently present. Column 4 (upper), spots within the great STropD; (lower), unusual speeds close to the GRS. (This table omits some short–term interactions with the GRS, of which there are examples in Figs. 10.35&62.). Values in square brackets are uncertain or anomalous so not included in averages. The overall average speed is +119.1 (±20.2) degrees/month.

grounds. Most of the extreme speeds have occurred when a STropD is present or has only just disappeared; as shown in Fig. 10.4 and Table 10.2, the mean speed has been +130 when a STropD has been present within the previous year, and +105 when there has been no STropD. The speed variations from 1978 to 1986 were most suggestive, with the retrograding speed increasing in 1979/80 and again in 1984 within less than 8 months of the origin of a STropD, and slackening again within a similar period after

each STropD had disappeared. (The allowance of a year's persistence, arbitrarily given in compiling Table 10.2, may be too generous; the intermediate speeds seen in 1982, 1985, and 1958, each 6–12 months after a STropD had disappeared, suggest that the effect really falls off in less than a year.) However, there are also a few counter-examples to the trend, notably in the 1943A and 1990 SEB Revivals. The hypothesis that the STropD pumps up the SEBs jetstream will be considered in more detail in Chapter 14.4.

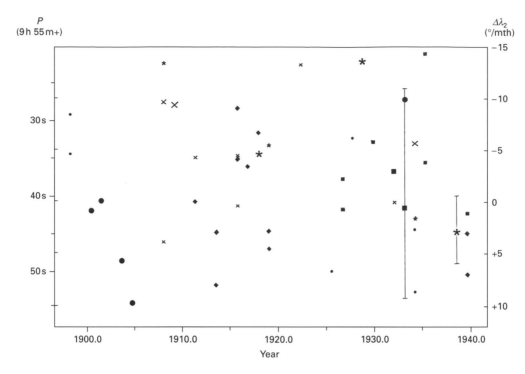

Above and opposite:

Figure 10.5. Speeds recorded in the South Tropical Current (STropC), i.e. slow spots in SEB(S), SEBs, and STropZ; see also Fig. 10.6 for the GRS and STropDs. From sources listed in Appendices 1 and 3. Speed is shown as P (rotation period), as $\Delta\lambda_2$ (degrees per month), and as u (m/s in System III). Symbols: × white spot in SEBZ; ◆ dark spot in SEB(S), not in STropD; ■ ditto, in STropD; ● spot(s) in STropZ and features on SEBs; ★ ditto, in STropD; ⊕ source of SEB Revival. Large symbols are apparition means of ≥3 features, and vertical bars indicate the range of speeds. Arrows show changing speed of individual spots (including oscillating spots in 1940–42; see section 10.9). STropZ white spots A and B in the 1980s persisted for several years. Not included are averages for 1903 ($\Delta\lambda_2$ = -16) and 1909/10 ($\Delta\lambda_2$ = -10) as it was not stated whether the features were inside or outside the STropD.

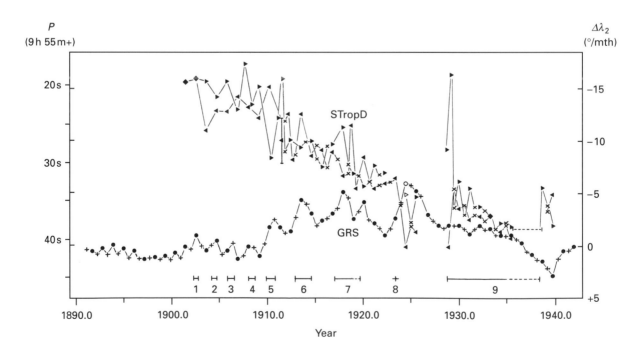

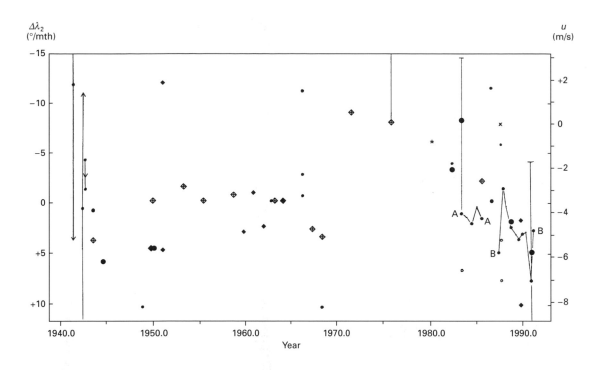

Below and opposite:

Figure 10.6. Speeds recorded for the GRS and the S. Tropical Disturbances. Symbols: ● GRS; ▶ p. end STropD; ◀ f. end STropD. These values exclude times when the STropD ends were passing the GRS. (Bars indicate the 9 con   junctions between STropD and GRS.) Crosses, mean speed from one opposition to the next (adjusted for phase effect by addition of 0.6°/mth, to match the intra-apparition speeds). (For GRS speeds before 1890, see Fig. 10.31.)

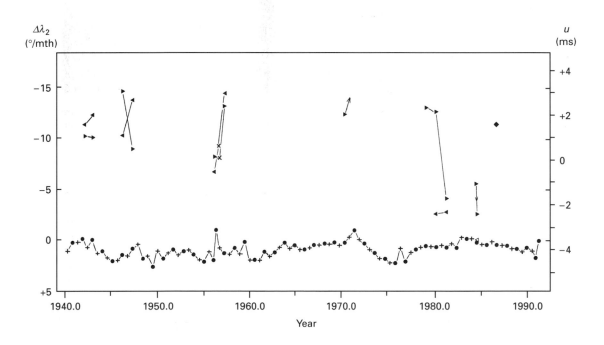

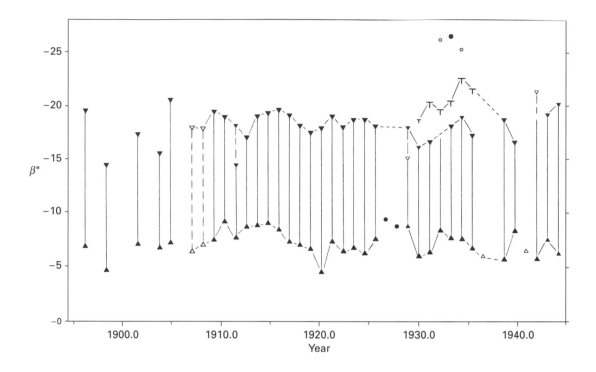

Figure 10.7. Latitudes of the SEB edges, 1895–1991. The chart shows apparition means, from sources listed in Appendices 2 and 3. When there are data from different sources for an apparition, they are either averaged or plotted separately. Vertical lines are drawn to indicate the extent of the belt. Small symbols are single micrometer measurements, and open symbols are data of lower precision, which are not used for statistics. ▼ ▲ S. and N. edges of SEB; ●, narrow band (including S. Tropical Bands); T, SEBs edge within the great STropD (▼ in 1929–1935 is SEBs outside STropD). Statistics of these data are in Table 10.3.

There also seems to have been a long-term shift in the latitude of the SEBs jetstream, as judged by the latitude of the visible SEBs edge. When the great STropD existed, the SEBs edge moved northwards within the Circulating Current, but was further south within the expanding STropD, and has remained quite southerly since the great STropD disappeared.

So far we have not referred to SEBs spots that were occasionally seen within the boundaries of the great STropD. They were retrograding at a much more modest rate, averaging $\Delta\lambda_2 = +26$ during the 1919/20 Revival and $\Delta\lambda_2 = +65$ in 1932 and 1933. Thus the SEBs jetstream seems to have been much less vigorous in the lee of the Circulating Current. Likewise, in 1979/80, more modest speeds were recorded for two small spots f. the STropD, that is, beyond the point of recirculation.

Table 10.2 also lists several other SEBs spots which had exceptionally weak retrograding speeds, all immediately p. or f. the GRS. The one f. the GRS in 1962 had come from the STBn jetstream, and some SEBs spots have slowed down drastically as they approached the GRS. However, most spots maintain their retrograding speeds or even increase them as they swing round the Red Spot Hollow (NMSUO reports; e.g. Figs. 10.35&62b).

Currents (iii) The South Tropical Current
Peek did not define a 'South Tropical Current', but the motions of occasional spots in the SEB(S), as well as the GRS, do seem to define a steady South Tropical Current almost stationary in λ_2, analogous to the slow currents of other domains. The GRS dominates this domain, obviously entraining the rifted regions p. and f. it, and often it is the only feature to be tracked with a slow drift. So it could be that, as in some other domains, the largest oval entrains the motion of other features in the domain. However, sometimes there are spots far away from the GRS which move with similar speeds. The most uniform drifts are shown by a class of dark spots or streaks that occasionally appear in SEB(S) when the belt is quiet. These seem to define a steady if infrequently observed current, with $\Delta\lambda_2$ about 2°/month slower than the GRS (Table 10.1 and Fig. 10.5). Their average drift from 1911 to 1939 was $\Delta\lambda_2 = -0.6$, while the GRS had average drift -1.9; and their average drift from 1940 to 1990 was $+3.3$, while the GRS had average drift $+1.0$. These spots or streaks are in SEB(S), not on its south edge. In many ways they resemble the NEBn barges although they are smaller. The examples in 1989 were at latitude 16.7°S – on the equatorial side of the SEBs jetstream. Voyager photographed tiny dark lozenges with virtually the same speed and latitude, and it was within one of these that the mid-SEB outbreak began (16.3°S, $\Delta\lambda_2 = +5$) (section 10.4). Thus it is relevant that the mean drift of sources in SEB Revivals and other outbreaks is almost the same ($\Delta\lambda_2 = +0.7$). These features do appear to define a South Tropical Current.

Drifts on the SEBs edge and in the STropZ are more varied. Some white and dark spots retrograde slightly, like the SEB(S)

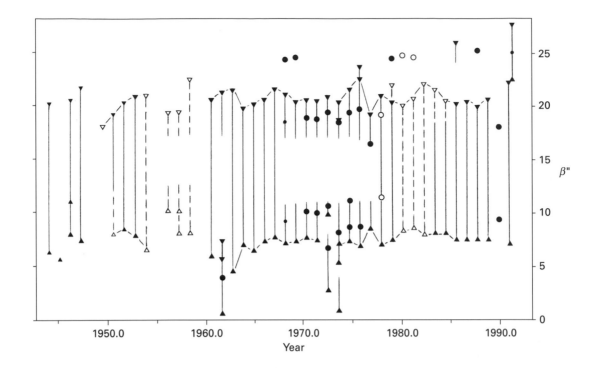

spots, and some may be physically associated with the SEB(S) spots (as in 1942/43 and 1989); others have prograding drifts more like the STropDs, or any drift in between. The average speed of all these spots is zero. STropDs have their own, faster, drift rate, similar to the South Temperate Current, and must be considered separately (section 10.7). Some amorphous dark shadings or bands move even faster (section 10.8). The smaller dark spots resembling jetstream spots may move at any speed between the SEBs jetstream and the STBn jetstream, and some change speed or even oscillate (section 10.9). Their motion probably depends mainly on their latitude; this was clearly true of those observed by Voyager, which were small rings like jetstream spots whose drifts were determined strictly by the latitudinal gradient of winds between the jetstreams.

small brilliant nuclei are not characteristic of the white spots between the components of the SEB, nor in general are rotation periods found here that are markedly shorter than that of System II' (mean 9h 55m 27s, $\Delta\lambda_2 = -10$, 1907–1934). The situation has been markedly different since 1975, in that SEBZ sectors have only been short, and often highly turbulent; they do show 'small brilliant nuclei' and they do prograde rapidly relative to System II, as described above (Fig. 10.3). Their velocities and forms display the gradient of wind speeds across the belt. This splitting is particularly common immediately p. and f. the GRS.

The SEBZ f. the GRS represents the accumulation of turbulent white spots in the cyclonic disturbance there. From 1977 to 1988 the f. end of this rifted region lay 40°–60° f. the centre of the GRS.

10.2 THE SOUTH EQUATORIAL BELT

The most memorable aspects of the SEB are linked to the patterns of activity to be described in the next two sections; but a brief account is needed of the appearance of the belt when it is doing nothing in particular.

The edges of the SEB are almost constant in latitude at 7°S (coinciding the the S. Equatorial jetstream) and 20°S (coinciding with the SEBs jetstream) (Table 10.3 and Fig. 10.7). However, there seems to have been a small long-term shift in the SEBs edge, related to the great STropD, as described above.

In recent years the SEB has usually been a solid dark belt at most longitudes; but up to the 1960s the belt was usually double, split by a SEBZ which was often broad and white. Peek (*1958*) summarised this aspect as follows: 'Except during the great revivals, the light space between the components is normally tranquil; white spots and transverse grey streaks do appear, however, and may resemble superficially some of the mid-NEB markings, though

Table 10.3. *Latitudes of the SEB*

	β''	(±s.d.)	[N]
SEBn			
1895–1939	−6.9	(±1.1)	[29]
1940–1990	−7.0	(±0.9)	[30]
1895–1990	−7.0	(±1.0)	[59]
SEBs			
1895–1925*	−18.7	(±1.1)	[20]
1930–35 (outside STropD)	−17.5	(±0.8)	[5]
(inside STropD)	−20.9	(±1.1)	[6]
1940–1990*	−20.7	(±0.8)	[29]

Figures are means of apparition means. In brackets, standard deviation, and number of apparitions. From data shown in Fig. 10.7. (SEBn does not include anomalous values in 1928, 1961, 1972; SEBs does not include 1897/98. Open symbols in Fig. 10.7, representing medium–precision values, are also not included in averages.)

*These SEBs values were all outside STropD if present.

Typically, white spots arise at the f. end and prograde towards the GRS (Fig. 10.8). Sometimes spots arise about once a week so there are several such spots jostling one another; sometimes they arise less than once a month, and there are only one or two bright spots in an apparently bland SEBZ, although greater resolution is likely to reveal smallscale turbulence at such times. Curving dark columns often separate the teardrop-shaped white spots, and sometimes the whole f. end of the disturbance will break away and prograde as a dark step on the SEB(N), while new white spots may form a new f. end behind it. New white spots can also erupt within the length of the disturbed SEBZ. Because of the speed of drift and the rapid changes in appearance of the spots, it is often difficult to track them with confidence, but some can be followed as they are squeezed north of the Red Spot Hollow, where they disappear.

This turbulence f. the GRS was impressively shown in the Voyager photographs and movies, which are described further in section 10.4 (Fig. 10.9).

The rifted region f. the GRS was first drawn by W.R. Dawes in 1857 (Plate P1.4). It was not notably active in the late 19th century, although the SEBZ often opened to the EZ at that point, but it was notable in most years from 1910 to 1917/18, and also during the episodes of continuous SEB activity in 1931–34, 1966–67, and 1976–88.

Unfortunately, drift rates in the early years do not seem to have been well established. P.B. Molesworth reported drifts for white spots in the SEBZ that extended all round the planet, but his values of around 9h 51½ m in the years 1900 to 1903 were totally at variance with those now recognised as typical. Molesworth's tables indicate that the average intervals between transits were 20–27 days, which must be too long to be sure of identifications of such unstable spots; in 1900, for instance, he noted that there were very rapid changes both in the white spots and in the SEBZ around them. His value for 1899, though equally insecure, was much more plausible at 9h 54m 45s ($\Delta\lambda_2 = -41$; 5 white spots); and his description in that year, of the white spots being traced down to the f. shoulder of the Red Spot Hollow where they disappeared, is also consonant with recent experience.

In 1903, in this turbulent region f. the GRS, Molesworth made one of the very few visual observations of a change occurring within the few hours in which a feature crosses the disc of the planet. He saw a small brilliant spot appearing in the SEBs edge after the region had crossed the central meridian, and within 5 minutes it had grown into a bright oblique rift (Fig. 10.10; Molesworth, *1905*). The only other reliable record of such a rapid change was also in the SEB, during the 1928 Revival. As such a feature must span several thousand kilometres in order to be seen at all, one cannot infer that it grew from nothing within 5 minutes; but a growing cloud must at some time cross the threshold of visibility, and this must be what Molesworth observed.

In 1899, according to Molesworth, new white spots were also being budded off from the p. end of a broad sector of SEBZ *preceding* the GRS. This sort of activity has also been seen recently, in 1966–68 (section 10.4).

Another segment of SEBZ, lying further north, was familiar

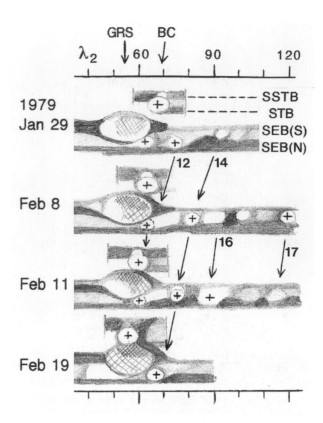

Figure 10.8. The rifted region f. the GRS in 1979 Jan.-Feb. (drawings by the author). The prograding motion of white spots labelled 12, 14, 16, 17, is evident. (Only a short segment of the STB-SSTB is shown, around oval BC. Preceding the GRS is the S. Tropical Dislocation.) Bright white spots are marked +.

from 1976 to 1988, dragged along in the train of the long-lived EZs white spot (Chapter 9.5). Typically its p. end connected to this great white spot, and the SEBZ tapered away f. it. When it approached the GRS, this SEBZ segment would disappear and did not merge visibly with the region of turbulent SEBZ; but after passing the GRS, the EZs white spot would sometimes break through to the permanent SEBZ p. the GRS and thus re-establish its own strip of SEBZ. Photographs in 1988 showed that part of this strip consisted of a remarkable chain of tiny white spots, spaced 4° apart.

The SEB never develops prominent 'barges' like those of the NEB, but several smaller, very dark spots or streaks were visible just inside the southern edge between 1911 and 1919, unrelated to the STropD. Similar streaks have occasionally been seen since, when the SEB is quiet (see above, and Fig. 10.5). Sometimes they persist as the SEB fades before a Revival (1926, 1989). One in 1913 and two in 1989 were deep reddish-brown like NEBn barges; of the two in 1989, one was on the f. edge of the Red Spot Hollow and one on the p. edge of a long-lived STropZ white spot. A dark reddish spot on SEBs in 1885, about half the size of the GRS but short-lived (N.E. Green, *1887*), may have been a similar feature.

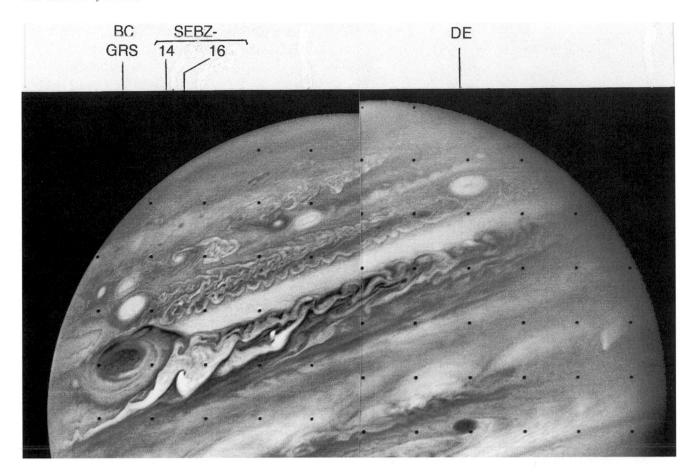

Figure 10.9. Voyager 1 view of the rifted region f. the GRS on 1979 Feb. 17. Includes white spots 14 and 16 of Fig. 10.8, rolling together north of the GRS. (Images 15920.50 and 15921.02, J1–16d, violet, filtered. NASA.)

The colour of the SEB, like that of the NEB, varies from cold grey to strongly reddish. This rich orange-brown, reddish, ochre, or tawny colour usually develops in the year following a SEB Revival. All such instances are listed in the chronicle of SEB Revivals in section 10.3. The belt was also notably reddish in 1895–99, 1905–7, 1912–14, 1925, 1933, 1948, and 1977/78. More varied colours are often seen just before and during SEB Revivals.

More usually, the SEB is brown or greyish-brown.

In the rifted region f. the GRS in 1977–1979, the author found both components to be less reddish than the remainder of the belt, and they remained grey or only slightly brown along with the rest of the belt in 1979–1981. The SEB(N) just p. the GRS was also greyer than the main belt according to the author. The segment of SEB(N) f. the long-lived EZs white spot was sometimes strongly brown but sometimes grey. During episodes of continuous activity, the narrow SEB(S) carrying jetstream spots sometimes contrasts with the main belt. It was reddish-brown in 1932/33, grey to the author in 1977–1980, mixed grey and brown on Voyager photographs, and strikingly blue in Pic du Midi photographs of 1986 (Plate P21.5).

Voyager images of quiet sectors of the SEB showed a wealth of streaks and wavelike patterns, as shown in Fig. 10.11.

10.3 REVIVALS OF THE SOUTH EQUATORIAL BELT

The pattern of a SEB Revival

There have been 14 SEB Revivals between 1919 and 1990, and by good fortune, almost all of them were observed from the beginning; only three began during solar conjunction. They are listed in Table 10.4. The typical pattern of activity is as follows.

The SEB fades over a matter of months, and remains faint for between one and three years before the Revival begins. If the SEB is not already double, the fading begins in the middle of the belt (1949, 1956, 1989), creating a diffuse SEBZ, but soon the S. component virtually disappears; this can occur within 2–3 months (1919, 1989) or over a much longer period. Some fadings in the 'triennial' era did not go to completion. The N. component may remain as a dark though quite narrow belt, or it may fade as well (1937, 1951, 1989). If a STropD is present, it too fades away (1919, 1926, 1936, 1942, 1957, 1971). The fading of the belt may not be entirely uniform; well-defined, very dark streaks of SEB(S) may linger for several months (1926, 1959–1960, 1989). A faint narrow line usually remains to mark the SEB(S) and/or the Red Spot Hollow.

The Pioneer spacecraft showed that the whitened SEB latitudes had exactly the same cloud cover as the STropZ, as if the bright

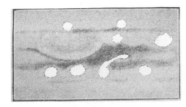
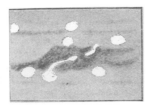

Figure 10.10. Sudden appearance of a rift and bright spot in SEB f. the GRS on 1903 Dec 17 (by P.B. Molesworth). Left, 01h 45m; right 02h 05m. (From 1903/4 BAA Memoir).

Figure 10.11. Voyager 1 closeup of the SEB, 1979 March 2/3, from STBn (upper left) to SEBn (lower right). At upper left, the large oval is a SEBs jetstream spot showing anticyclonic circulation. Note many periodic streaks in the belt. (Image 16325.10, J1–3d, orange, unfiltered. NASA.)

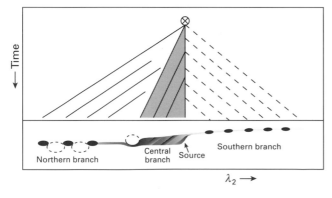

Figure 10.12. Diagram of a SEB Revival. (Upper:) Drifts of spots in System II longitude. (Lower:) Map of the expanding disturbance in the SEB (south up).

white STropZ had broadened to cover the belt. However, in 1961 the faded SEB appeared bluish, and in 1971 the LPL showed that the faded SEB was the brightest region on the planet in the ultraviolet at 0.3 µm wavelength, much brighter than the STropZ proper

(Kuiper, *1972b*). The residual SEB(N) sometimes develops a faint bluish-grey colour (1928, 1936, 1941/42, 1952, 1989/90); sometimes there is a distinct bluish or grey triangle on SEB(N)s Np. the GRS (1942/43, 1957, 1962, 1974, 1989/90, etc.). Bluish colour is thus common. As this colour is rarely seen at other times, it is evidence that the fading is not merely a covering of the belt with white clouds. (LPL photos in 1968 also revealed a bluish band in SEBZ Np. the GRS even though the SEB had not faded – Plate P13.3.)

Reddish colour is only seen in the faded belt when the coloration of the previous SEB Revival lingers during the next fading (1949, 1959, 1972), or when the SEB(N) is involved with strong coloration in the EZ (1937, 1962, 1974). A yellow or reddish coloration in the EZ almost always develops while the SEB is faint (10 out of the 14 cycles); and if it has not already appeared, it sometimes does so after the Revival when the SEB itself becomes reddish (1943, 1959, 1972). (See Chapter 9.4 for more details.)

As the SEB fades, the GRS intensifies, both in darkness and (often) in colour, to become a prominent dark reddish oval floating in the broad white zone (12/14). Sometimes, rather faint bluish shadings develop around it (e.g. Plate P16.1 and Fig. 1.22). The GRS also usually decelerates (section 10.5).

Sometimes the SEB(S) shows signs of diffuse darkening just before the real outbreak begins (1937?, 1958, 1962?, 1971), but these may merely be coincidental fluctuations.

The Revival always begins with a dark spot or streak lying across the latitudes of the SEB (Fig. 10.14). Often it is associated with a bright white spot on its p. side, and the white spot can even appear a few days before the streak, in which case it is in the S. component or on the SEB(S)n edge (latitude 13–17°S) (1943B, 1949, 1971A,B, 1975A,B; 'B' indicates a secondary outbreak). Possibly the white spot always appears first, but is less likely to be noticed against the bright zone.

From this source point, intense disturbance and/or dark belt material then spreads from the source in three currents or 'branches' (Fig. 10.12).

(1) The southern branch, in the retrograding SEBs jetstream, usually consists mainly of dark spots, which may be small dark spots like those on other jetstreams, or may be larger; when colour has been recorded it is grey (1928) or bluish (1952, 1971). They typically run in chains with a spacing of 12–20°. The range of speeds is shown in Fig. 10.4 and Table 10.2. Often, though, the S. branch becomes so turbulent that few spots can be tracked.

(2) The central branch, with forms an expanding belt segment whose p. end progrades down the SEBZ while the f. end remains stationary at the original source longitude. Sometimes this appears to be simply a dark belt segment, often starting off with a dark bluish-grey colour (1952, 1955, 1958, 1990). But in many Revivals it is very turbulent, with brilliant spots and dark 'bridges' continuing to appear for several months at the original source, or at lower longitudes. It is possible that higher resolution would reveal vigorous turbulence in it every time (as in 1990).

Central-branch speeds are shown in Fig. 10.3. The leading edge of the central branch is usually one of the fastest features in it, with initial $\Delta\lambda_2$ ranging from –27 to –86 (average -54); but it may accelerate or fluctuate in speed (1958, 1962/63, 1971). The central branch spots (excluding the source) generally move with $\Delta\lambda_2$ between -5 and -70 (Fig. 10.3). (Spots in mid-SEB outbreaks and in the rifts f. the GRS have drifts in the same range.) The speeds in the central branch show a slight tendency to fall into two ranges centred around $\Delta\lambda_2 \approx -46$ and -16; the overall average is -33.

The source of an outbreak often remains identifiable for several weeks, and is almost always stationary (average $\Delta\lambda_2 = +0.7$, excluding 1971 and 1975).

(3) The northern branch, spilling into the S. Equatorial Current, is variable; sometimes it contains the most spectacular spots of all (e.g. 1938, 1971); but sometimes it does not appear at all until 1–2 months into the Revival (1962, 1975, 1990), although the track of the leading edge can then be extrapolated back to imply that it started from the initial source. Sometimes the spots are visibly deflected northwards and/or accelerated as they pass the Red Spot Hollow (1928, 1952, 1958, 1971, 1990), and they may spread across part of the EZ.

The northern branch spots may move at any speed between $\Delta\lambda_2 = -140$ and -260, the latter extreme being faster than System I (Fig. 10.3). The average is $\Delta\lambda_2 = -197$ ($\Delta\lambda_1 = +32$). There may be a great range within a single Revival. Often the disturbance on SEBn is seen to accelerate up to the usual System I speeds within a year, but sometimes the slower northern-branch speeds persist for up to a year after a Revival. Visible disturbance tends to persist longer in the northern branch than in the other two branches, sometimes being recognisable in the following apparition.

Vigorous new eruptions sometimes occur close to the leading edge of the northern or the central branch. The spots in the Revival can include some of the brightest and some of the darkest that ever occur on the planet. There can also be a mixture of strong colours, so much so that photographs in red and blue filters can show quite different patterns across the SEB(N) and EZ (1938, 1971, 1990). The different colours may represent clouds at different altitudes.

Sometimes a second source appears just like the first (1943, 1971, 1975, 1990); in 1975 there were four such sources. In 1975 and 1943, these appeared just as the leading edge of the northern branch was passing that longitude.

The reviving SEB is often so turbulent that it is difficult or impossible to recognise spots after only 2 or 3 days, which is typically the minimum interval at which the same longitude is presented for observation. So the best-observed Revivals are not necessarily the ones for which the most speeds are reported.

After the northern and southern branches have overlapped, the SEB is essentially restored, but disturbances may continue for several months; new spots and rifts continue appearing close to the original sources, or at other longitudes.

When the southern branch hits the GRS, the GRS almost always fades, beginning within a few days or weeks (in 11 out of the 14 Revivals). Sometimes (1928, 1943?, 1952, 1971, 1990?) the dark jetstream spots from SEB(S) can be seen running round the curve of the Red Spot Hollow, and some apparently penetrate onto the SEB(S) following it, although no single spot has ever been tracked for any great distance before and after passing the GRS – perhaps

because of difficulties in identification, but probably also because the turbulent region just f. the GRS renders them unstable. Occasionally (1928, 1962, 1975), at the height of the disturbance, the SEB(S) appears to run straight across the fading GRS; then the reddish material is confined to the southern half of the Spot.

The breakup of the red material of the GRS is often patchy, and it often recovers temporarily. Some observers interpret these changes in terms of white clouds spreading across the Spot. In any case, within a few months the GRS becomes a light oval with a dark ring around it (11/14). The dark ring appears to be derived from SEB material.

The final stage of the Revival is an 'orange flush' that spreads across the revived SEB (10/14). It is often continuous with the persisting coloration of the EZ. Although this SEB coloration is such a common sequel, it had been largely forgotten in recent years, partly because it was not mentioned by Peek in his book, and partly becase it was absent in most recent Revivals (it was absent in 1919/20, 1962/63, 1964/65, and 1975/76). But it was noted by W. Wacker (*1973*) and R. Doel (*1978*). And it was well enough known in 1955 that Guido Ruggieri wrote to the BAA Jupiter Section Director, comparing that year's Revival with that of 1952: 'In both disturbances at the beginning there was little colour. Afterwards there developed that characteristic dark blue tint which remains an enigma. It is easy to foresee that in the next few months the blue tint will change to brown, then to red…. These naturally are my views on which you and other members of the Section may not be in accord.' (See Plate P12.11.)

Sometimes there is also darkening or coloration of parts of the STropZ in the aftermath of the Revival, mainly p. the GRS. This may be reddish or yellow coloration, probably extending from the SEB colour (1928/29/30, 1938, 1943, 1953, 1959); or it may consist of grey shadings or a dark grey S. Tropical Dislocation, possibly derived from grey SEBs jetstream material (1947, 1952, 1962/63, 1971?, 1975/76, 1990/91).

Some Revivals are clearly more vigorous than others, and the relative strengths of the three branches also vary. The most spectacular Revivals on record were those of 1928 and 1975, while those of 1955, 1962, and 1964 were the weakest. The northern branch was particularly active in 1938 and 1971, but almost undetectable in 1962 and 1975. The central branch can be dominated by one dark belt segment (1964, 1990, etc.), or by chains of white spots (1949), or by intense turbulence (1928, 1975). There may be a weak relationship between the strength of the outbreak and the time interval preceding it; at least, outbreaks coming less than 3 years after the previous one (1955, 1964) have been anomalously weak. There is no relationship between the strength of the outbreak and subsequent events; the SEB has faded again immediately after some of the strongest outbreaks.

Many SEB Revivals coincide not only with coloration in the EZ, but also with jetstream outbreaks in other latitudes. These episodes are called 'global upheavals' (Chapter 13.3).

The following sections give the circumstances of all recorded SEB cycles, listed according to the date the revival began (Table 10.4). Most of them are illustrated in the Plates.

SEB Revivals in the nineteenth century

Until 1919, the phenomenon of SEB Revivals was unknown. But in retrospect, it is clear that the SEB did undergo major fading and revival several times in the nineteenth century, even though violent disturbances and rapid motions were not detected.

1859: SEB(S) was faint (except for a pair of dark bars) in 1858 December and 1859 spring. There must have been a SEB revival during solar conjunction; in 1859 autumn, SEB(S) was present again (N.E. Green) and almost the whole length of STropZ was a massive dark 'belt' (S. Schwabe; e.g. Plate P1.8&9). Whether the revival was violent or peaceful is unknown, but it was accompanied as usual by fading of the GRS; and in 1860 March, several observers remarked on yellowish-red colour in the SEB.

1871: SEB(S) was rather tenuous throughout 1869 and 1870, and absent preceding the GRS, which appeared as an isolated ring with a white interior. In 1871 Jan., J. Birmingham drew SEB(S) thick and spotty at some longitudes, but he used a small telescope and this 'revival' was not confirmed. In 1871 Dec. and 1872 Jan., observers drew a dark spot and streaks on SEB(S) p. the GRS (e.g. Plate P2.8&9). These developed into a great oblique dark band sloping Nf. from the STB to the revived Red Spot Hollow. Also, on 1872 Jan. 14, J. Gledhill noted the SEB(S) as a prominent belt with its S. edge 'thrown into waves'. This was evidently a SEB Revival, and by 1873 both SEB components were broad and dark. However, the strengthening of SEB(S) was only a matter of degree, and the belt did not excite nearly as much comment as the concurrent disturbances in EZ and NEB. Oddly, the GRS became darker than before, although its oval outline was disrupted.

SEB(N) was very disturbed throughout these years.

1882: The unprecedented prominence of the GRS from 1879 onwards was accompanied by fading of the SEB(S), which was absent in drawings of 1880 and 1881. When the next apparition started in autumn 1882, the GRS had become extremely faint, and the SEB was reviving to be the most conspicuous feature on the planet. According to Green's RAS Memoir, in 1883 February and March, 'the coppery hue [of the EZ] ... migrated gradually southwards, till in 1883 it appeared in full force in the southern belt, rivalling and at last surpassing the colour of the red spot itself.' Green's drawings show that this colour was mainly in the massive SEB(S) p. and f. the GRS, and also that the SEB(N) was riven by large bright and dark spots (Plate P3.5). The whole picture resembled that of 1938 or 1990/91, and may likewise have been the late stages of a violent Revival that started during solar conjunction.

It is worth noting that the 1859, 1871, and 1882 Revivals were preceded by coppery coloration of the EZ, and the 1882 Revival was preceded by an outbreak on the NTBs jetstream; thus these were components of global upheavals. There were probably no other SEB fadings and revivals in the second half of the 19th century.

The quiet period, 1883–1918

After 1882, there was no SEB Revival for the next 37 years. Whether this was a spontaneous fluctuation in activity, or was related to the development of the great STropD from 1901 onwards, we cannot tell.

The Red Spot Hollow had a strange form from 1886 to 1895. Its p. edge had been very dark and reddish in 1884. During 1884–86, the SEB(S) gradually faded away for ≈80°–120° p. the GRS, until it was faint or absent in that sector, although it remained strong elsewhere (Plate P3.6). The GRS intensified in colour and form

during 1891, becoming a distinctly pink oval again. It then faded during solar conjunction, and was very faint (although unevenly so) by autumn 1892. Also in 1892, P.B. Molesworth, A.S. Williams and others noted that SEB(N) was unusually disturbed with white rifts; but this was only an enhancement of activity that was obvious throughout the 1890s, and the SEB(S) remained absent p. the GRS until 1895. Then the SEB became notably broader and redder (though with no evident disturbance), the Hollow was restored, and for the next two years the SEB was ruddy and the GRS virtually invisible.

From 1886 until 1908, except for the sector p. the GRS, the SEB was almost always split into two comparably dark components. The SEBn edge often had many projections into the EZ, sometimes accompanied by rifts through to the SEBZ, resembling the more recent aspect of the NEB. The SEBZ was sometimes dusky, sometimes brighter and containing bright spots, but (as discussed in section 10.2) these spots were not distinct enough to be widely observed. The SEBs edge appeared to be smooth and quiet until 1900, and there was no 'arch' over the GRS. Thus the belt really was free of either classical or mid-SEB outbreaks.

From 1900 to 1906, the SEBs edge appeared scallopped or humped, and as one of the humps developed into the great STropD, it is of interest to ask whether these small humps represented retrograding jetstream activity. (Molesworth was the only observer to detect them in most of these years, and perhaps such minor features could have been detected earlier if someone with an equally good telescope and skies had been active then?) Rotation periods were published for many of them by Molesworth, who found them to be stationary in System II; the points for 1900 to 1905 in Fig. 10.5 represent his values, for (e.g.) 12 dark spots and white bays in in 1900 and 23 spots in 1901. Perhaps there may be some question of identification when so many spots are tracked. Nevertheless, Molesworth's identifications were made from transits with a mean interval of 9–13 days, which should be quite adequate; and in 1903 Denning also observed that these features were moving at exactly the same speed as the GRS. Similar nearly-stationary projections remained visible each year until at least 1906.

From 1910 to 1912 (and intermittently to 1918), there was considerable disturbance in the SEBZ f. the GRS, and of SEB(S) within the longitudes of the STropD. This may have comprised an episode of continuous SEB activity (section 10.4), but it may have been a special slower-moving disturbance related to the STropD; unfortunately no drift rates could be obtained for the most disturbed regions.

SEB Revivals in the twentieth century

The first Revival to be observed in detail, revealing the remarkable motions, occurred in 1919/20.

1919: The Red Spot Hollow and STropD began to fade in 1919 Jan., then the SEB(S) faded away completely, except for a few very faint fragments. Only the periphery of the GRS darkened (though it had a large f. cusp) and it was always grey.

The outbreak was first seen on Dec.8 as a group of dark humps on SEB(N) near λ_2 230 (Fig. 10.13). The initial longitude was not recorded

Table 10.4. *SEB Revivals*

Date started		Discovered by	λ_2		Reese source λ_2	(O–C)	GRS λ_2	Drift of leading edge ($\Delta\lambda_2$): C.branch	Source	S.branch	Drift rates from:
1919	Dec 8	Phillips (UK)	230	(227)*	A: 223	(+4)	322	–	–		
1928	Aug 10	Taffara (Italy)	128		A: 131	(–3)	315	–35 (6)	–	+163 (3)	BAA
1938	(early)	–	–								
1943	Feb 7	Peek (UK)	20		B: 29	(–9)	172	–43	+4		BAA
–(2)	Feb 27		288		C: 278	(+10)					
1946–47		–	–								
1949	Jul 23	McIntosh (NZ), Hare (USA), Murayama (Japan)	155		B: 146	(+9)	241	–64	0		ALPO/ McIntosh
[Ws:	Jul 19	McIntosh (NZ)	163]								
1952	Oct 20	Reese (USA)	204	(208)*	B: 201	(+7)	269	–86	–1.5		ALPO/BAA
1955	Feb 4	Ruggieri (Italy)	229		C: 240	(–11)	294	–66	0		ALPO/BAA
1958	Mar 30	Miller (USA)	40	(47)*	B: 53	(–6)	315	≥–39†	–	+121 (1)	ALPO/BAA
1962	Sep 23	Bornhurst (USA)	234	(241)*	C: 246	(–5)	14	–46	0	–121 (2)	ALPO
1964	Jun 14	Reese (USA)	232	(245)*	[B: 192]	[+53]	19				
1971	Jun 21	Fox/Minton (USA)	79		A: 79	(0)	10	≥–27†	–10	+137 (6)	NMSUO/
[Ws:	Jun 18	(Hawaii,Flagstaff)]									LPL/UAI
–(2)	Jul 20 [Ws: Jul 18]		144		C: 141	(+3)					
1975	Jul 5	Reese (USA)	55		A: 60	(–5)	42	–	–20		BAA
[Ws:	Jul 2	(NMSUO)]									
–(2)	Aug 2 [Ws: Jul 29]		206		B: 228	(–22)		–	–8		
–(3)	Aug 13		158					–	–8		
–(4)	Aug 17		119		C: 120	(–1)		–	–19		
1990	(Jul)	–	–	(270)*	B:273–265	(≈0)	28	–77	–		BAA
–(2)	Sep 4		123		[C: 150]	[–27]					
1993	Apr 6–7	GEA (Spain), Balella (Italy)	21		B: 14	(+7)	42				

Columns 1, 2: Date and discoverer of first observation of dark streak. [Note in square brackets indicates when a white spot was seen earlier.]

Column 3: Initial longitude of dark streak. *(In some cases the initial value may not have been accurate, or the outbreak probably began a few days before it was observed. Longitude in brackets was given by Reese as a better value for the source position over the first few days. These values are from Chapman and Reese (*1968*) and Reese (*1972a*), except for 1990 which is from Japanese and BAA reports.)

Column 4: Predicted λ_2 positions of sources A, B, and C are from Reese (*1972a*), with later ones calculated from his formula:

$$\lambda_2 = L - 0.256265(JD - 2442301)$$

where JD = Julian date, L(A) = 137.3, L(B) = 312.3, L(C) = 206.5.

Column 5: Difference between observed and calculated longitude [in square brackets for values >22°].

Column 6: λ_2 of GRS is given for opposition in that apparition.

Columns 7–9: Drift rates ($\Delta\lambda_2$) for the leading edge or leading spot-group in the central and southern branches (in brackets, number of features included), and for the source. For the overall speeds of the northern and central branches, see Fig. 10.3; for the overall speeds of the southern branch, see Fig. 10.4 and Table 10.2 (SEBs jetstream). As these speeds are often very diverse, it would be misleading to quote a single average per Revival for comparative purposes.

† 1958: Leading edge accelerated from –39 to –142 then decelerated to –49. 1971: Leading edge accelerated from –27 progressively to –159.

Column 10: Source of drift data.

precisely but from several early observations it must have been in the range λ_2 220–230. The first classical Revival then developed. According to the BAA Memoir: 'H. Thomson wrote on 1920 Feb 27: "The SEB is a most extraordinary spectacle. It consists largely of round dark dots and white spots." Moreover, the changes of aspect were so rapid that it was almost impossible to identify the markings after so short an interval as a couple of days.' Everything was turmoil and confusion; but the drawings apparently include a typical central branch prograding towards the GRS, and in late March the Red Spot Hollow and the ends of the great STropD became visible again. Few drifts were obtained; two dark projections on SEBs just f. the source had $\Delta\lambda_2 = +17$ and +35.5, which could represent a very weak retrograding jetstream within the boundaries of the STropD. The jetstream proper was brought to observers' attention in 1920 Feb. when two isolated SEB(S) streaks shortly f. the GRS developed into a remarkable a pair of very dark round spots in the STropZ, about 60° apart, with the unprecedented drifts of $\Delta\lambda_2 = +101$ and +96. It was these same two spots that first revealed the Circulating Current (Plate P5.8&9; section 10.7).

Later, SEB(N) became broken up and indistinct, merging into the ochre equatorial shading. In 1920/21 there was still a group of spots on it which included drifts much slower than System I.

1928:[1] The SEB(S), Red Spot Hollow, and STropD all disappeared during solar conjunction in 1925–26. Two short streaks of SEB(S) nick-named 'Remnant' and 'Fragment' persisted until late 1926 (Plate P6.3). In late 1927, there was a faint southerly S. Tropical Band in some longitudes, with a loop northwards into the middle of the STropZ that was nicknamed the 'False Red Spot', and also an outbreak of tiny spots prograding on STBn. All these SEBs and STBn features were within the boundaries of the (invisible) STropD, and had disappeared by 1927 December. By 1927, SEB(N) was also quite tenuous and reddish. Meanwhile the GRS intensified,

[1] See Plates P6 and P7, and Figs. 10.14&15. It is well worth reading the original reports on this remarkable apparition, not only in the BAA Memoir, but also in: Hargreaves, Peek & Phillips (*1928*) (q.v. for detailed maps); Wright (*1929*) (multispectral photos from Lick Observatory); Hargreaves (*1929*) (the irradiating spot); Hargreaves (*1939*) (the Circulating Current). Many excellent photographs from the Lowell Observatory are in Slipher (*1964*).

1919 Dec. 8. The Director.

1920 Jan. 1. Thomson.

1920 Feb. 6. The Director.

Figure 10.13. SEB Revival of 1919/20: Strip-maps from STB to NEB; bracket indicates width of SEB. (Top:) First sighting of the source, by T.E.R. Phillips. (Middle:) The source developing, by H. Thomson. (Bottom:) Intense turbulence in the reviving SEB, by Phillips. (From the BAA Memoir.)

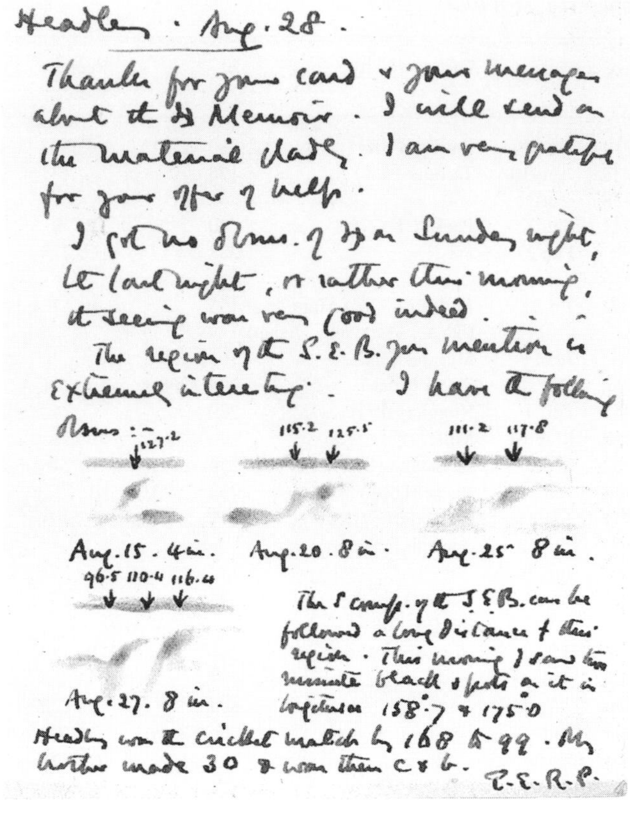

Figure 10.14. SEB Revival of 1928: Drawings by Phillips of the initial source, on a postcard that he sent to Peek. The last two lines read: 'Headley won the cricket match by 168 to 99. My brother made 30 and was then caught and bowled. -T.E.R.P.' (From the BAA archives.)

becoming a prominent reddish oval with a lighter centre and pointed, reddish ends, which tapered into narrow ruddy wisps of southerly S. Tropical Band (Plate P6.5). These colours were confirmed photographically.

The subsequent Revival was one of the best-observed of all, and apparently one of the most energetic ever, on all three branches. It did not begin until 1928 Aug. 10, when the source was discovered by L. Taffara at Catania Observatory, Italy; Peek also discovered it on Aug. 12 (Fig. 10.14). This source, at λ_2 128, appeared as a dark spot in the latitude of SEB(S) with a wisp connecting it to SEB(N). Within 3 weeks, a conspicuous chain of dark spots was retrograding on SEB(S) at speeds up to $\Delta\lambda_2 = +168$. They were dark grey in the Lick photographs and were 12–20° apart. Meanwhile a second chain of spots began rapidly prograding on SEB(N), although there was such chaos that it was difficult to track individual spots. (The published maps, and Fig. 10.15, include a striking 'sawtooth' pattern approaching the GRS in the central branch; this cluster was spreading out with $\Delta\lambda_2$ from –30 to –11.) The southern and central branches overlapped near the GRS, and thereafter the central latitudes of the SEB became intensely turbulent, even surpassing the chaos of 1919/20 (Fig. 10.15). On Oct.14, Hargeaves recorded definite changes in less than an hour, as illustrated in Peek's book.

The colour changes were among the most outstanding features of the Revival, and being part of a global upheaval, they spread far beyond the latitudes of the SEB. In 1928 Sep., in longitudes as yet undisturbed, there was a striking colour pattern: the southern EZ was tawny, SEB(N) was bluish-grey or striking blue, SEBZ was bluish-white, and STropZ was strongly reddish, with a sharp colour border at the latitude of the still-absent SEB(S). (The narrow reddish strip that had been developing alongside STBn before

the Revival had spread over STB and STZ in August, and in September it spread northwards to cover the whole STropZ, in the hemisphere p. the GRS, with an intense orange tint.) During October the blue colour disappeared, and the ochre/orange tint of the EZ spread progressively southwards; SEB(N) and SEBZ became strongly pinkish-red, a colour which remained prominent for many months.

Peek summarised the behaviour of the retrograding southern branch thus: 'As the spots proceeded on their course, the S. component of the belt seemed to be developing with them; and although the leader apparently faded before reaching the Red Spot, those following it were deflected northwards along the S. edge of the Hollow, which was beginning to reappear, becoming flattened out into short streaks during the process. Many of them survived the passage but all were lost after proceeding a few degrees beyond the f. end of the Hollow. The point of disappearance turned out to be close to the position where the p. end of the reviving STropD was ultimately identified.... Soon after the passage of the first few spots around the Hollow, the Red Spot began to fade and the whole region to lose its characteristic features.' (Peek, *1958*. See Fig. 10.15.) Thereafter some SEB(S) spots, which also became very stretched out, were carried straight on across the N. half of the GRS without deflection. These also were lost a few degrees further on; by this time the whole region of the GRS and immediately f. it was a chaos of spots and streaks, in which it was impossible to tell whether recirculation was occurring. The GRS temporarily recovered in late October but then faded again.

The northern branch was no less striking. SEB(N) was broken up into many dark and bright spots which, being deflected northwards past the GRS, spread as far as the equator. They had diverse and variable speeds ranging from $\Delta\lambda_2 = -90$ to -230 ($\Delta\lambda_1 = +139$ to 0); some groups converged, others diverged and new spots formed within them. These motions were much slower than usual for these latitudes, and applied even at the equator, where for one week in October at least two spots moved with $\Delta\lambda_2 = -151$

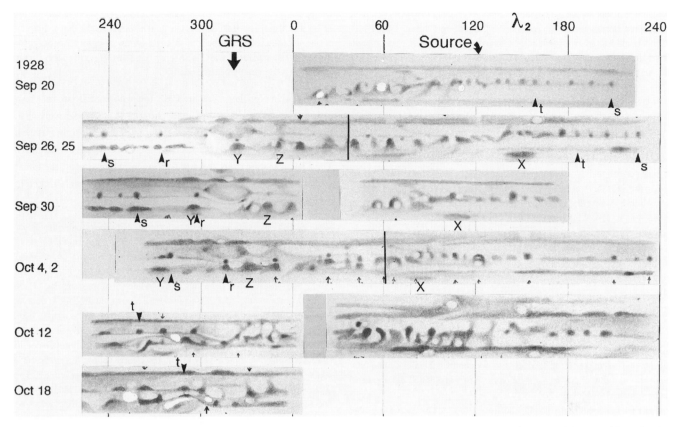

Figure 10.15. SEB Revival of 1928: Original strip-maps by Phillips, from his notebook. The GRS is at λ_2 315, p. end of STropD at 345, source of SEB outbreak at 125. Representative SEB(S) and SEB(N) spots are indicated. Spots r, s, t, are dark spots in the southern branch. X marks the great dark 'blob' on SEBn; from Oct. 12, it was obscured by the slower-moving SEB(N) activity as it approached the GRS. Y and Z are probable identifica-tions of spots around the turbulent leading edge of the central branch. The GRS becomes confused as SEB(S) spots hit it and is unrecognisable by Oct. 18. (These maps, from the BAA archives, were aligned by the author following Phillips' annotated longitudes. Strip-maps for earlier and later dates were published in Peek (*1958*). A more complete set of strip-maps was published in Hargreaves et al.(*1928*) and the BAA Memoir.)

($\Delta\lambda_1$ = +78)! At least some of these spots were at high altitudes, as shown by two remarkable phenomena in October.

Firstly, a dark and a bright spot with the very slow speed of $\Delta\lambda_2$ = –96 ($\Delta\lambda_1$ = +133) passed right over a pre-existing dark 'blob' on SEB(N), obscuring it as they passed. (This large dark spot had existed before the Revival, with $\Delta\lambda_1$ = –15, and was gradually decelerated as it caught up with the northern-branch spots, persisting to January with $\Delta\lambda_1$ = +26.)

Secondly, one SEB(N) spot was so brilliant that, in late October, it remained bright far from the central meridian and even appeared to project beyond the limb of the planet. This 'irradiating spot' appearance was pre-sumably a contrast effect produced by a cloud that was so high in the atmosphere that it was unaffected by the usual limb-darkening. In fact, the irradiation must have been produced when the spot was a little over 15° longitude from the true limb. From the times at which irradiation was observed at the p. and f. limbs, it was possible to calculate the longitude of the spot responsible, and to identify it with features mapped under more normal illumination (Hargreaves, *1929*). It corresponded to an exceedingly bright spot in SEB(N), moving with $\Delta\lambda_2$ = –188; this was especially con-spicuous in the Lick Observatory ultraviolet photographs (Wright, *1929*) which emphasise high-altitude features, and showed a vast bright cloud whose brightest part coincided with the longitude of the visible irradiating spot. This spot expanded rapidly across SEB(N) before it faded and broke up. ('Irradiating spots' were also seen in the SEB Revival of 1943, in the SEBZ f. the GRS in 1963 (ALPO), and in the EZ in 1918/19 (Chapter 9.3).)

The SEB was almost back to normal by the end of 1928, and the GRS was a whitish oval surrounded by the shading of the STropD. In 1929/30 there was still a very disturbed section of SEB(N). The rosy-red colour of the belt also persisted in 1929/30, especially in the southern half; and the

STropD was reddish while the rest of STropZ was yellowish. These colours faded away during 1930.

The belt did not fade again until 1936, but from 1930 to 1934 there was an episode of continuous activity in the SEB, to be discussed in section 10.4.

1938: During 1936 and 1937, the SEB(S) was again absent or extremely faint, and the GRS very conspicuously dark brick-red (probably the reddest it had been since the 19th century). SEB(N) in 1936 was weak and bluish, and in 1937 was very faint and merged with the yellow-brown southern EZ. Phillips recorded a distinct, very southerly SEB(S) in one sector on 1937 Aug. 31, and the GRS seemed to fade somewhat thereafter. (Reese (*1962b*) speculated that this marked the actual Revival, but this seems very unlikely; his suggested onset near λ_2 285 was merely conjecture.)

The belt and the STropD revived during solar conjunction, some time between 1937 Dec. and 1938 May. By May, the GRS had already com-pletely faded within a dark ring, and no disturbance remained in the main latitudes of the SEB.

However, there was massive disturbance in southern EZ with typical northern-branch motion – clearly the aftermath of a SEB Revival (Plate P7.7–9). It included many dark grey masses with $\Delta\lambda_2$ = –177, which smoothly accelerated to $\Delta\lambda_2$ = –229 ($\Delta\lambda_1$ = 0). To visual observers, these dark grey masses separated the yellow-brown EZ from a notably reddish SEB. Lowell Observatory photographs with colour filters (Slipher, *1964*) confirmed a strong reddish colour over most of the SEB, and a remarkable multicoloured disturbance of SEB(N)/EZs, with large, irregular, very dark blue spots mingled with reddish veils and lighter areas. There were also many tiny spots on the STBn jetstream p. the STropD, which may have originated from the SEB Revival via the Circulating Current (Plate P7.7). The STropZ in the hemisphere p. the GRS (the STropD) was dusky and reddish, but it was white f. the GRS. The SEB(S) and SEBZ remained reddish through 1939.

1943: SEB(S) faded gradually in 1939/40, but the process was arrested, and for 3 years both components remained visible. SEB(S) was very narrow and faint, and so was SEB(N) from 1941/42 onwards; in that apparition the whole belt had a cold grey or lavender tint. Although the Red Spot Hollow was almost invisible, the GRS did not darken, never appearing as more than a pale 'doughnut'. Nor did the EZ acquire any colour, until 1943 March when the SEB Revival had already started.

At the λ_2 where the short-lived STropD of 1941/42 disappeared, from 1942 Dec. to 1943 March, there was a small grey projection or 'step-up' on SEBs with $\Delta\lambda_2 = 0$ to +2. This was where the Revival began; on Feb.7, this projection had darkened and a new dark spot had appeared in the SEBZ immediately Nf. it, at λ_2 20. By Feb.11, a new white and dark spot had appeared just p. this pair, and a vigorous Revival developed, with the usual retrograding and prograding branches (Fig. 10.16).

Meanwhile, the 'step-up' on SEBs was replaced in April by a 'step-down' on the same track ($\Delta\lambda_2 = +1$), marking the f. end of a dusky stretch of STropZ alongside the reviving SEB. In retrospect these features, looking like the p. and f. ends of a STropD, might have represented the rim of a single bright oval in the STropZ, which was not contrasted enough to be fully visible. On a few nights in April the very disturbed SEB(S) did appear to outline several such ovals (Fig. 10.16).

On Feb. 27, a second source appeared at λ_2 288, just as the leading edge of the original northern branch was passing this longitude. The new outburst first consisted of a brilliant white spot on SEB(S)n edge; by March 2, there were two such spots lying on the p. edge of a very dark streak across the SEBZ, and by March 4, new spots were beginning to emerge from this source in a new southern and northern branches, the latter being distinguished from the original northern branch by having a much faster speed ($\Delta\lambda_2 = -257$ instead of the previous -136).

In contrast to 1928, according to Peek, 'there was far less confusion and rapid change among the spots and markings that appeared in the middle of the belt', and the SEBZ remained clear where the southern and northern branches overlapped. But SEB(S) was more turbulent, so that few spots on it could be tracked far.

The SEB(S) spots began to encounter the GRS at the start of April, and although observations were intermittent, several sightings of dark streaks N. of and f. the GRS suggested that the spots were retrograding around the curve of the Red Spot Hollow.

As in 1928, an 'irradiating spot' was observed shining apparently at the planet's f. limb (actually the terminator), three times in March (in the SEBZ) and three times in April (in the SEB(N)). This was probably one of a pair of very bright spots in the same latitudes ($\Delta\lambda_2 = -142$), within the northern branch, which were conspicuous on the disc in March (Fig. 10.16).

During the Revival, colours in the SEB varied between bluish-grey and light brown, and were never striking. But colour had developed by the 1943/44 apparition (Plate P12.10). The GRS was a completely white oval (perhaps brighter than ever before), outlined against a lightly shaded STropZ which sometimes appeared yellowish or pinkish. The SEB itself was notably orangish or reddish, and strong red colour persisted in the SEB(S) into 1945.

The 1943 Revival was the first of a series which occurred at three-yearly intervals until 1958.

1946: In 1946, SEB(S) was rather faint, and the GRS was a reddish 'doughnut' and sometimes quite conspicuous. However, there was still some shading across the SEBZ, and there was a new dark STropD which did not fade away. By 1947, SEB(S) was darker than SEB(N) and the GRS was entirely

white; and two small dark projections were detected retrograding in the SEBs jetstream. The data are sparse, but it seems likely that a Revival occurred during solar conjunction, some time after 1946 July. According to Reese, in 1947, SEB(S) was bluish-grey but SEB(N) was reddish-brown and the SEBZ gradually became ochre.

We should record here that in 1948 the whole SEB became reddish-brown or orange-ochre, and parts of the STropZ were also tinted brown; the STropZ contained large dark and bright patches, and there were dark masses framing the white or yellowish GRS. The SEB was then quiet, except for typical white spots f. the GRS, so this was a rare example of strong coloration not closely linked to a Revival.

1949: British observers were again unfortunate with the timing of the 1949 Revival; the planet was so far south that very little could be seen. Fortunately, R.A. McIntosh in New Zealand was able to carry out a single-handed coverage of the disturbances using a 36-cm reflector (McIntosh, *1950*), and observations from the USA and Japan were compiled into a thorough report by the ALPO. Because these reports are difficult to find, an extensive synthesis of them is given here. This outbreak seems to have been intermediate in style between classical Revivals and mid-SEB outbreaks.

Before the outbreak, there was a broad clear SEBZ which developed in 1949 April–May, becoming very bright and clearly yellow, but the SEB(S) was by no means faint, and it had well-marked ends abutting the GRS. The GRS was merely pale pink and sometimes parts of it were already faint. The SEB(S) carried several small dark spots with slow STropC drifts, of which one ($\Delta\lambda_2 = +17$; λ_2 157 on July 12) coincided with the initial outbreak in the SEB.

The outbreak began with a small brilliant spot in the SEB(S) at λ_2 163 on July 19 (McIntosh). By July 23, this was accompanied by the usual 'bridge', at λ_2 156, attached to a very dark spot on SEB(N)s at λ_2 154 (E.E. Hare, USA; S. Murayama, Japan). By July 28 and 29, the white spot had expanded to fill the SEBZ and had a dark 'bridge' on each side. By August 4, the white spot was prograding at $\Delta\lambda_2 = -48$, and a second one had appeared f. it; the Revival was under way.

Most of the activity seemed to be in the central branch, in which there were many bright spots separated by dusky columns or dark patches, moving with mean $\Delta\lambda_2 = -66$ (McIntosh). This was not merely an idiosyncrasy of McIntosh's observing style, as the ALPO showed a similar aspect on strip-maps and reported similar drifts (mean $\Delta\lambda_2 = -54$), although they did not always agree on drifts for individual spots. The p. end of the central branch had $\Delta\lambda_2 = -82$, later decelerating (McIntosh), or -64 overall (ALPO). The f. end of the central branch remained fixed, and new spots continued to arise there until the end of October. The strip-maps indicate that one white spot was created every 6 days.

A southern branch developed with numerous small dark spots by late August, but by this time it had already encountered the GRS, and only a few SEB(S) spots were ever seen beyond the GRS. (McIntosh and Reese (ALPO) suggested different drift rates for the SEBs spots, but both authors admitted that other interpretations of their charts were possible; in Table 10.2 and Fig. 10.4, the tentative ALPO value is entered.) The GRS retrograded by some 7° in September, as the Red Spot Hollow re-formed, but returned to its original position by October. The GRS lost its colour according to most observations from Oct.3 onwards, becoming a pale grey ring encircled by dark material; it remained so in 1950.

The northern branch was not tracked precisely, but dark material on SEB(N) passed the GRS without incident and returned to the original source longitude on about Oct. 23. According to McIntosh, this SEB(N) did not mingle with the original disturbance, but formed a separate, darker SEB(N) advancing alongside it. But at this time the source apparently switched off and a solid dark SEB section began to advance with $\Delta\lambda_2 = -46$ during November. By that time, most of the SEB had become dark.

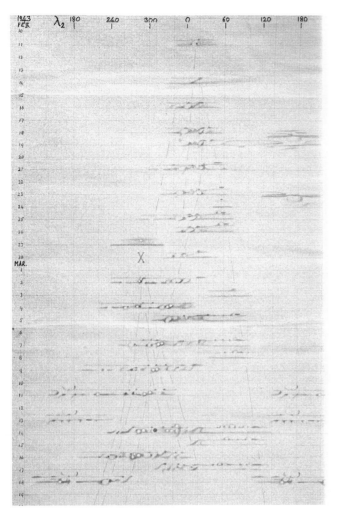

Figure 10.16. SEB Revival of 1943: BAA strip-maps. X marks the secondary source. (Drawn by F.J. Hargreaves from observations by himself, Phillips, and Peek. Previously published in the BAA Memoir and in Peek (*1958*).)

The northern branch probably affected the EZ also: McIntosh noted a large bright area in EZ opposite the SEB source on August 3, crossing the EB, and various greyish festoons in EZ(S) thereafter. The ALPO thought that many of the dark spots tracked in the SEC ($\Delta\lambda_1$ –4 to –22) were probably associated with the SEB disturbance.

The only colour notes were from Reese, who found the SEB orange-reddish the following summer.

1952: During 1951, the SEB(S) faded until it was practically invisible and the SEB(N) was also very faint, just a dusky blue-grey band; never before had the SEB been so thoroughly obliterated (Plate P9.5&6). At the same time the GRS became very prominent, often with an orange tint, though the interior remained somewhat lighter.

The Revival that followed was very energetic and was well observed. The initial outbreak was recorded by Reese on 1952 Oct. 22, as a classic 'bridge' across the SEBZ latitudes with a very dark and conspicuous round spot at its midpoint (observed independently by K. Komoda in Japan on Oct. 24). The usual branches set off at once, most conspicuous being the central branch which consisted of a dark 'cold grey' or bluish belt segment.

The chain of dark spots of the southern branch (also bluish according to G. Ruggieri) soon encountered the Red Spot; they were apparently deflected around it and the Spot faded patchily in November. Its appearance fluctuated for several months; eventually it became a simple ring. Many

SEB(S) spots were seen f. the GRS although the identification of individual ones that might have passed it, as in the BAA Report, seems very speculative.

By late November the belt was filled with intense disturbance, 'like a stormy sky of cumulonimbus cloud' (Ruggieri), who thought that activity at the original source had intensified at this time. According to the BAA Report: 'In 1953 January, the SEB became exceedingly active, and observers commented on the beautiful colours seen. Creamy white spots on a roseate background were very prominent, one "being so brilliant as to appear to shine by its own light". F.M. Bateson recorded this particular spot on 1953 Jan 13 "like a satellite as it moved off the disk", while on Jan. 29 he commented on the "bewildering mass of detail, especially on the SEB(N)". On the same date W.E. Fox noted two extremely dark spots as being the "blackest markings I have ever seen on the planet".' According to the ALPO, when the prograding disturbance reached the GRS in late January, bright clouds were deflected over the SEBn and accounted for some rather slow motions in the South Equatorial Current thereafter.

Reese and others reported increasing reddish colour in the SEB from November to January, particularly an orange or red-brown colour in the advancing central branch and N. component; but from the source region to the GRS, the SEB was only a dull brownish or tan; and f. the GRS, the still-separate components were still grey or even bluish grey.

Preceding the GRS, large clouds of dark material apparently spread from the SEB(S) into the STropZ, producing a dark zone resembling a STropD by late December, and the outline of the GRS was sometimes rendered

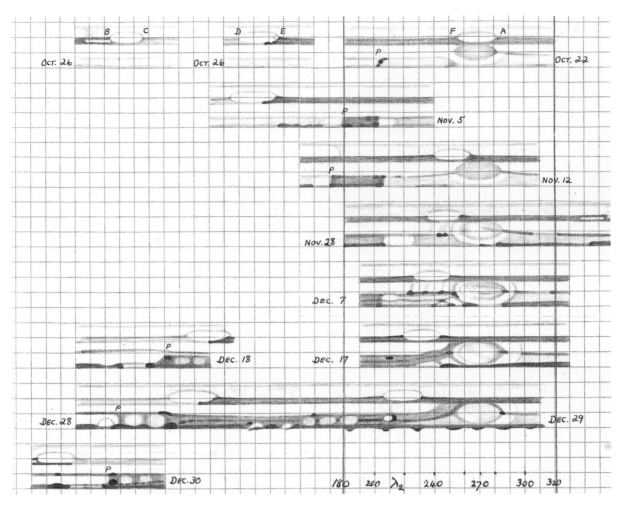

Figure 10.17. SEB Revival of 1952: Strip-maps of the early stages, by E.J. Reese. 'P' marks the initial source on Oct. 22 and the leading edge of the central branch thereafter. (See also Plate P8.8&9.) (From BAA archives.)

indistinct. Over the next few months, the whole of the southern hemisphere p. the GRS seemed to become veiled with a dusky haze. Reese commented in February that 'the whole region from the SEB(N) to the STB appears as one wide dusky belt'. It seems to have had little colour, although there were a few reports of a tan tint.

In the next apparition, SEB was a rich orange-brown, while STropZ began dull and yellowish then brightened (ALPO).

1955: During 1954, SEB(S) became faint though not absent, and the GRS darkened in 1954 November, to become a dull reddish-brown oval. The Revival started only 2.3 years after the previous one. Although the published observations were rather few and made with small telescopes, it was clear to experienced observers that this disturbance really was one of the weakest on record.

The outbreak began with the classic dark bridge seen by Ruggieri on 1955 Feb. 4 (also see Plate P10.1). It soon generated a typical massive dark central branch whose leading edge, an oblique dark streak associated with spots and a rift in SEBn, advanced with $\Delta\lambda_2 = -64$ (ALPO). The f. end remained stationary. There were plenty of dark spots preceding it on the SEB(N), but they were generally small. The reviving SEB(S) seems to have been just a narrow dark belt, with no spots reported on it. The source was again p. the GRS, and in late March the Red Spot Hollow revived and the GRS started to fade as usual; it was a white Hollow by June.

Ruggieri noted that the advancing central branch was dark bluish-grey (Plate P12.11). Reese (unpublished notes) recorded SEB(S) orange-brown before and during the Revival, and the whole belt orange-brown in 1955/56.

1958: (ALPO, BAA, and others.) In 1956 March, a narrow white SEBZ developed, which was the first stage in fading of the SEB that proceeded during solar conjunction. In 1957 the whole SEB was almost invisible, while the GRS became a striking dark orange or reddish oval. A dusky triangle did remain on the Np. side of the Red Spot Hollow. A STropD also disappeared in 1957 July – near the same longitude where the Revival later began.

The SEB began to darken in 1958 March (R. Mourão; McIntosh); but the actual outbreak was first noticed by J.S. Miller (Los Angeles, ALPO) on March 30, as the usual oblique bridge to a small dark spot on SEB(S). By extrapolating the tracks of the ends of the central branch, Reese estimated that it started about March 27, λ_2 47. Soon there were classic chains of dark spots on the northern and southern branches, and many streaks and rifts in the central branch, while the source remained fixed (Plate P10.4). (The most detailed strip-maps were published by Chapman and Reese, *1968.*) The dark SEB(N) in the central branch was a striking bluish colour according to unpublished reports by T. Sato (Japan) and McIntosh (New Zealand).

The Red Spot Hollow formed in dramatic fashion, as the leading edge of the central branch, with a bright gap in SEB(N), closed up from the f. side at $\Delta\lambda_2 = -38$ (BAA) or -42 (ALPO); this closed on May 23 to form a very tight Red Spot Hollow, which caused a sudden decrease in longitude of the f. end of the GRS by 5°, though it then returned to its original position. Then in June the SEB(S) spots reached the GRS, which at once began to be veiled by whitish clouds and within a week or two faded from brick-red to a diffuse pale grey; it disappeared in early 1959.

When the Red Spot Hollow formed, much of the dark material of the central branch was deflected onto SEB(N) so the leading edge appeared to accelerate into the northern branch, having $\Delta\lambda_2 = -142$ for the month of June, but then it suddenly decelerated again to $\Delta\lambda_2 = -49$ (ALPO).

By July, much of the SEB was a scene of chaotic activity, and by August the dark belt had largely revived. In August and September at the end of the apparition, there were a few reports of dark clouds flanking the faded GRS.

By early 1959, a very extensive ochre or orange shading had developed, covering the whole EZ, SEB, and STropZ! SEB(S) was already fading again , but the GRS remained an obscure hollow ring. SEB(N) was dark reddish-brown and still carried many spots and gaps. By 1960, the coloration had all gone except for a reddish SEB(N).

1962: (This account is mainly from the detailed ALPO reports.) In 1961, the SEB(S) was faint, both it and the SEBZ appearing as if covered by a light, bluish haze, and 'SEB(N)' was exceptionally far north and disturbed (Chapter 9.4). By summer 1962, SEB(N) was more normal in latitude but markedly orange-brown, merged with the intense reddish EZ coloration, with a few humps on it still. In 1961 the GRS was a vivid dark orange, perhaps more so than ever before this century, with a lighter centre. There were still traces of the Red Spot Hollow, especially a dusky triangle of varying darkness on the Np. side.

One spot was retrograding on SEBs *before* the outbreak, having originated on STBn (section 10.9), but it died out in August. SEB(S) may already have been darkening in late July; but essentially it was still faint at that time.

The typical dark bridge across SEBZ was first seen on Sep.24 by L. Bornhurst (USA). (See also Fig. 10.18.) A chain of dark spots on the southern branch and a series of dark 'bridges' in the central branch soon appeared, but no northern branch. As usual, there was a brilliant white oval on SEB(N)s marking the leading edge of the central branch, $\Delta\lambda_2 = -46$. On Nov. 23, this oval was replaced by a large and brilliant cloud covering SEB(N)s, which moved off with $\Delta\lambda_2 = -204$; this was the first sign of a northern branch. Curiously, this track extrapolated back to the position of the original outbreak on Sep. 24; so a submerged northern branch may indeed have formed then, but did not emerge until it lapped the central branch on Nov. 23. This brilliant cloud broke up into smaller 'bays' over the next few weeks, with very dark spots between them and connections to the EZ, and the cloud was suspected to have dissipated into the EZ.

The retrograding spots (Plate P11.1&2) could not be traced past the GRS. Some drawings showed the SEB(S) running straight across the Spot, but this was not widely confirmed and could have been a temporary 'canal effect'. The GRS did seem to be eroded at the edges after mid-November, and encircled by dark STropZ material, but it never really faded, and the Hollow never fully revived; there were just white spots there.

From late October to January, dusky prograding material accumulated p. the GRS in the STropZ, which became much darker than the adjacent SEBZ, and even darker than the reviving SEB further p.! Reese (ALPO report) speculated that the retrograding SEB(S) spots, subjected to much turbulence as they approached the GRS, 'were torn asunder and sent swirling into the STropZ where they spread out into dusky masses which

were carried along ... in the opposite direction from whence they came.' In the light of the Voyager movies, this description no longer seems fanciful. This duskiness persisted until the end of 1963, 'cold grey' in tone, especially along STBn.

This Revival seemed to be weaker than its predecessors, with a minimal northern branch, and a southern branch that produced a lot of smoke but little progress. SEBZ was still active in 1963 February. But SEB(S) never fully revived and the GRS did not disappear.

1964: By mid-1963, SEB(S) was again almost invisible and SEBZ white, and the GRS remained very dark and vivid pink or orange. In 1963/64 there were still some small spots prograding in SEBZ, and a very bright oval just f. the GRS 'irradiated' near the f. limb (ALPO).

Reese reported a classic outbreak first seen on 1964 June 14. This was very early in the apparition, but several unpublished observations (J. Dragesco, A. Heath, J. Olivarez, T. Sato; Plate P11.4&5) confirm that a weak central branch set off near this longitude and prograded towards the GRS, where it formed the Red Spot Hollow in late September. All this was described and shown on whole-planet maps by European observers (Fig. 10.19).

Otherwise there was no visible disturbance. SEB was grey and generally spot-free, in contrast to the dark orange EZ. The SEB(S) finally revived quietly, undergoing 'an insidious simultaneous darkening at all longitudes' between 1964 Sep. and Nov. But the GRS still did not fade.

This last, atypical revival was the end of the roughly triennial series. A session of continuous SEB activity occurred from 1965 to 1968. Two well-observed cycles then took place in which the classical Revivals were very violent, with multiple sources, and associated with global upheavals.

1971: This violent but unusual Revival remains the only one to have been thoroughly covered by professional photography, mainly from Arizona and New Mexico.[2] Indeed, the initial outbreak was discovered by W.E. Fox of the BAA on 1971 June 21 while he was a guest observer at the LPL 1.5 metre telescope.

The SEB(S) faded in 1969 and disappeared in 1970, while the weak grey-brown SEB(N) and a massive southerly EB formed a belt resembling the usual double SEB. The GRS was as dark and red as ever. By 1971 June, Fox noticed an increasing bluish tint in the residual SEB(S) which suggested that a Revival was imminent. The tiny white spot that began the outbreak was photographed on June 18 from Hawaii and Flagstaff, and by June 21 it was larger (λ_2 80, β'' 14°S) with a classic oblique bridge f. it; it was the brightest feature on the disc at all wavelengths. This was only 70° f. the centre of the GRS. LPL photographs showed the 'bridge' and adjacent dark material as

---

[2] See Minton (*1972b*) and Larson (*1972b*) (LPL), and Reese (*1972a*) (NMSUO). The 1971 Revival is illustrated by LPL photographs in Fig. 4.10, Fig. 10.21, and Plate P13.5; and by Japanese drawings in Fig. 10.20 and Plate P15.1&2.

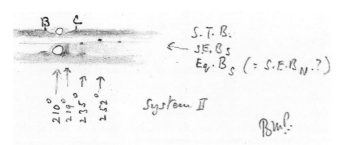

Figure 10.18. SEB Revival of 1962: An early view of the source region by B.M. Peek, on 1962 Oct. 7. He had discovered it independently on Sep. 25. This was the last Revival he observed. STB oval BC is marked. (From BAA archives.)

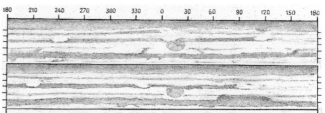

Figure 10.19. SEB Revival of 1964: Strip-maps from observations by members of the Société Astronomique de Suisse, by S. Cortesi. (Upper:) 1964 Aug. 5–18. (Lower:) Sep. 8–30. From a source near λ_2 220, a dark SEB segment prograndes towards the GRS. The pre-existing dark belt includes southern EZ. (From the SAS report in *Orion*; by courtesy of S. Cortesi.)

Figure 10.20. SEB Revival of 1971: Drawings by Japanese observers. (a) 1971 June 26, 12.43 UT, ω_2 54 (I. Hirabayashi); the initial outbreak. (b) Aug.15, 11.14 UT, ω_2 311 (I. Hirabayashi); N. branch developed p. GRS.

(c) Aug.27, 09.47 UT, ω_2 260 (R. Horiguchi); intense turbulence. (Also see Fig. 4.10, Plate P13.5, and Plate P15.1&2.)

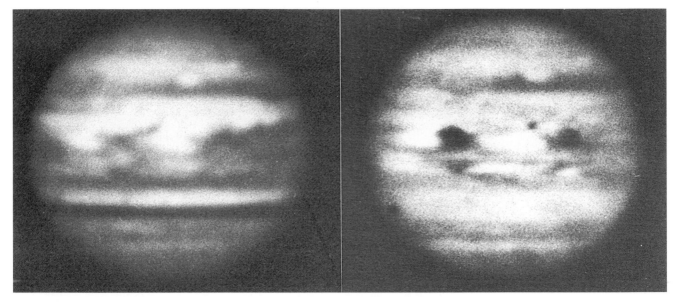

Figure 10.21. SEB Revival of 1971: Images in blue light (left) and near-infrared (right), showing two spectacular bright clouds flanked by dark spots and shadings of contrasting colours. 1971 Aug. 7, 03.03–10 UT, ω_2 255. (For colour version see Plate P13.5. Lunar and Planetary Lab. of the University of Arizona; from Larson (*1972b*).)

cold grey on June 23 but changed to yellowish-brown by July 7. According to S. Larson (LPL), subsequent dark 'bridges' showed the same colour changes.

By July 10 a vigorous Revival was underway with five small dark bluish spots already retrograding on the darkening SEB(S), ≈15° apart. These bluish spots continued to be dispensed from the original source, one every three days. The source had an unusual drift of $\Delta\lambda_2 = -10$, close to System III. The initial white spot had become a large, brilliant white oval in SEB(N)s, with turbulence around it. It had moved at $\Delta\lambda_2 = -27$ (Larson) or –50 (Minton) up till July 7, but was moving north and accelerating. Later, after it had passed the GRS, it crossed over the SEB(N) into the EZ, moving with $\Delta\lambda_2 = -159$. Other white spots then developed on SEB(N) p. and f. it, all with similar speeds. Between July 24 and 26, the appearance of a new bright oval and very large dark patch in SEB(N), immediately p. the leading white spot, was described by Reese as one of the most violent and spectacular events ever seen on Jupiter.

A secondary outbreak began on July 18 (white spot) at λ_2 144, ß" 13°S. Again it was the brightest feature at all wavelengths; rapidly expanded; and was accompanied after 2 days by a dark bridge. The original southern branch had already passed this longitude but the northern branch had not.

The new source produced a slow northern branch with appearance and speed similar to the original one.

So this Revival, uniquely, was dominated by a northern branch with a slow drift, perhaps due to interaction with the GRS as was observed in 1958. This was the most vigorous northern branch observed since 1928, and it spilled over extensively into the EZ. The SEB(N) became a complicated mass of detail surrounding the huge and brilliant prograding white spots, which sprawled as far as the equator. Most dramatic was the profusion of colours, including strong blue in the very dark SEB(N) patches and strong orange spreading around them (Fig. 10.21 and Plate P13.5).

SEB(S) spots reached the GRS in late August. Some disappeared, and some passed round the N. side of the Red Spot Hollow, but the GRS did not fade. Around that time, shadings were also reported in the STropZ, but the apparition was nearing its end. By 1972, there were eight gaps or white spots in SEB(N), probably remains of the 1971 activity, but now moving almost at System I speed ($\Delta\lambda_2 = -219$) and still accelerating; by late September they were merely gaps or faint sections of belt, and they disappeared in 1973.

But by 1972 the southern 2/3 of the SEB had already faded again, with a SEBZ tinged with orange colour similar to, but fainter than, the orange coloration that had then spread over the EZ (Plate P13.6). As in 1959, the orange coloration of one cycle apparently overlapped the fading at the start of the next.

1975: SEB(S) was quite faint in summer 1972, and for 3 years it continued to fade until barely visible. A massive dark brown southerly EB had formed, and again paired with a very ruddy SEB(N). The GRS remained very dark and strikingly red throughout, sometimes with dusky blue shading Np. or f. it. This very long 'pregnant state' ended with one of the greatest Revivals ever recorded (Plate P15 and Fig. 10.22).

The first warning of the SEB Revival came on 1975 July 2, when Fox saw a curious shading just f. the GRS (BAA), and NMSUO photographed a brilliant white spot there, at 17°S (R. Terrile and R. Beebe, *1979*). The classic dark bridge was seen alongside this white spot by Reese on July 5. This source was immediately f. the GRS as if it was part of the Hollow, and this may be why the initial development of the Revival was rather slow, consisting only of a rather faint chain of dark spots f. it on the SEB(S), which was also darkening diffusely in all longitudes.

On July 29, T. Broadbank saw a brilliant white spot 150° away in longitude, and on August 2/3 this had developed into a second source of the Revival with a classic dark bridge. Then, in an unprecedented development, two more sources appeared in rapid succession: no.3 on Aug 12/13, then no.4 on Aug 16/17 (BAA; UAI; Sato, *1980*). All four sources appeared close to the same System I longitude and, as it turned out, within 26° of the extrapolated track of the first feature to appear in the northern branch – a huge dark block spanning SEB(N) and EB(S) that was first seen on August 16 alongside source no.3. The BAA concluded that this marked a submerged northern branch stemming from the original outbreak, with $\Delta\lambda_2 = -215$, which had triggered the three secondary outbreaks as it passed critical System II longitudes (see below), and only then became visible.

Otherwise, there was hardly any northern branch, perhaps because the Red Spot Hollow lay immediately p. the initial source. Indeed, this source and subsequent bridges slowed down and faded as they moved into the Hollow. Unusually, the four sources had prograding drifts ($\Delta\lambda_2 = -8$ to -20). Nor was the usual central branch drift detectable: there was intense disturbance but its motion ($\Delta\lambda_2 = -14$) was similar to that of the four sources. These became lost around mid-September in the general profusion of chaotic spots and streaks which occupied the SEBZ up to $\lambda_2 \approx 230$.

The SEB(S) had now broken up into spots forming a typical southern branch, whose rather indistinct leading edge reached the GRS in mid-September. The northern half of the GRS soon faded, while the southern half of the SEB(S) became visible running straight across the middle. The Spot also moved f. by 5° in two weeks. During November and December, the Spot was merely a light ring, with the split SEB(S) surrounding it. The light interior was still reddish, particularly in the southern part, and was probably changing rapidly. All this occurred even though the southern branch was dramatically halted p. the GRS in Sep.–Oct. by the development of the first S. Tropical Dislocation (section 10.7).

Activity subsided during Oct.–Nov. The SEBZ became heavily shaded all round the planet, and SEB(N) re-formed, though some bridges and bays remained in 1976 January. Belt components underwent complicated latitude changes, particularly alongside the S. Tropical Dislocation. From December onwards, the SEB(S) began to fade again, starting f. the GRS, and dark red material reappeared in the GRS. But by summer 1976 it was evident that the SEB had, at last, definitively revived. A few SEBs jet-stream spots then remained (UAI) but activity then stopped. Unusually, the SEB had no strong reddish coloration (until 1977/78).

After these cataclysmic events, SEB Revivals ceased for the next 14 years. The SEB remained broad and dark, and the GRS remained generally faint, with continuous rifting activity in the SEBZ f. it, while the long-lived SEBn white spot that appeared in 1976 repeatedly circled the planet. A spell of continuous SEB disturbance began in 1978 and continued, with some fluctuations, to 1988, as described in the next section. Then in late 1988, a change came over the region. The continuous SEB disturbance stopped

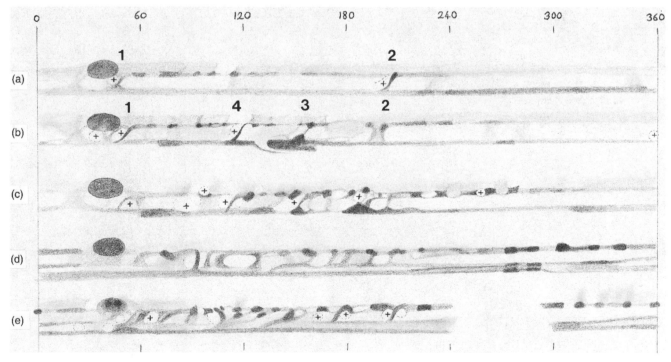

Figure 10.22. SEB Revival of 1975: Strip-maps by the author, starting with the onset of the secondary outbreak. Four sources, numbered 1–4, appear in succession. (a) 1975 Aug. 1–4; (b) Aug. 15–18; (c) Aug. 26–28; (d) Sep. 12/13; (e) Sep. 14/15. The GRS begins to fade on Sep. 14 as the SEB(S) spots arrive. Bright white spots are marked +.

rather suddenly, not only on the SEBs, but also in the SEBZ f. the GRS; the long-lived white spot or rift on SEBn gradually faded away; and the drift of the GRS gradually changed from negative to positive. Then, within no more than $2\frac{1}{2}$ months during solar conjunction, the SEB faded and a new cycle of activity began.

1990: The fading was discovered in mid-July 1989, when both components were still visible and two segments of SEB(S) were still dark and reddish-brown (f. the GRS and p. a long-lived STropZ white spot). The fading progressed until by the new year SEB(S) was virtually invisible and SEB(N) was also very faint (and bluish-grey). As the STB had disappeared earlier, this left an unprecedented white expanse from the SSTB to the equator, in which the dull brick-red GRS floated in solitary majesty; and around the new year, the northern EZ developed a vivid yellow colour, which completed a quite extraordinary picture (Plate P24.3&4). A blue streak was present Np. the GRS in alternate months; in 1990 Feb., photos revealed a 'micro-Revival' which began with a white spot due N. of the GRS and produced a blue streak prograding with $\Delta\lambda_2 = -92$ for a few days.

The true Revival began during solar conjunction, as was discovered in 1990 August, but it was possible to extrapolate back and deduce that it started the previous month near λ_2 270. The S. branch had already hit the GRS, which was already faded, and the reviving N. component reached the GRS soon after, but the central branch was well-observed as a great dark bluish-grey front with a white spot p. it. (The Revival is illustrated in Plates P22.9 and P24, and Figs. 10.23&24.) On Sep.4 a secondary disturbance began, photographed by I. Miyazaki as a dark blue streak bridging the components; it may have been close to the leading edge of the initial southern branch, although both components were darkening already. Great disturbance ensued in this sector.

The N. branch behaved oddly. Its p. end became arrested at the Red Spot Hollow, where a very dark bluish spot developed in September-October. Then the preceding sector of SEB(N) all darkened, and broke up into dark and light spots which became very prominent in October.

The yellowish-brown colour of the EZ and EB spread southwards across the whole SEB, parts of which were a striking orange-brown in November. By December, dark grey segments of SEB components appeared to be embedded in a general yellowish-brown murk, which also tinted some lighter 'rifts' in the belt, although a few short-lived spots were still white.

By then the GRS was a pale ochre oval with a dark rim, and a massive grey S. Tropical Band began to cover the whole STropZ (section 10.7). From mid-March, 1991, the SEB became quiet with a dark reddish-brown SEB(N). The ochre colour over the rest of the belt persisted even into 1992, when the belt was gradually fading again.

Origins of SEB Revivals

Why do SEB Revivals start at such localised sources? Can we identify any triggering factors in any of the recognised longitude systems?

System I: Secondary sources were apparently triggered by the primary northern branch in 1943 and 1975 (though not 1971 and 1990), but there have never been any evident triggering factors in System I for the primary outbreaks.

System II: The sources are stationary in System II once they have appeared, but there have only rarely been visible precursors at similar longitudes. The 1928 Revival began near a white spot at the f. end of

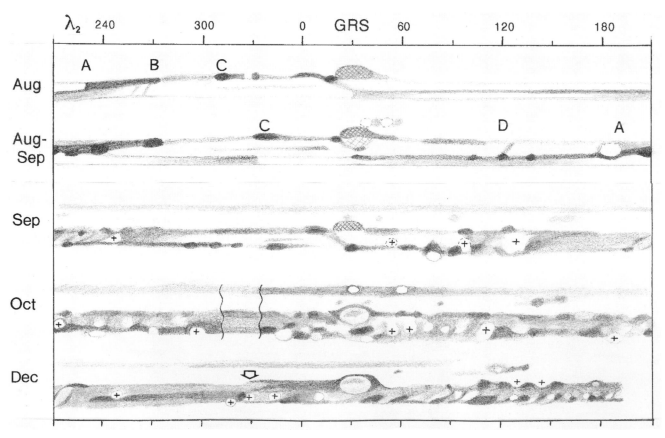

Figure 10.23. SEB Revival of 1990: BAA strip-maps. From top to bottom: Aug. 19–25; Aug.27–Sep.4; Sep. 21–24; Oct. 18–22; Dec. 4–7. The Revival had already started during solar conjunction and the GRS was rapidly eroding. Features indicated are: A, leading edge of central branch; B, source region; C, retrograding spot in SEB(S); D, secondary outbreak; thick arrowhead, p. end of S. Tropical Band appearing p. the GRS. The series is continued in Fig. 10.56 showing how this developed into a S. Tropical Dislocation. (Also see Plate P24. Drawn by the author from BAA observations, mainly photographs. From the BAA reports.)

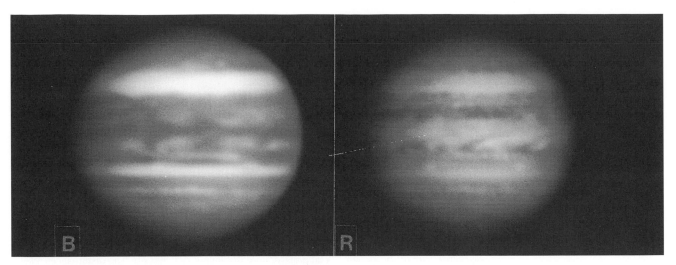

Figure 10.24. SEB Revival of 1990: Photographs in blue/violet (left) and red (right), showing multicoloured spots and belts. 1990 Nov. 19, ω_2 296. (Isao Miyazaki; from BAA report.)

the STropD, and the 1943 Revival near a similar feature. Otherwise there is no consistent relationship to STropDs, nor to the position of the GRS. (A majority have begun 205°–295° f. the GRS.) It is interesting that the 1943 and 1949 Revivals appeared adjacent to small dark slow-moving spots on SEB(S), as the Voyager images showed a mid-SEB outbreak to begin in just such a spot at 16.3°S (section 10.4).

System III: The Reese 'uniformly rotating source' hypothesis. Revival sources are so localised and energetic as to suggest that they may be not merely ephemeral. E.J. Reese therefore compared the longitudes of the initial outbreaks in order to see whether they might arise from one or more permanent sources, which would be fixed to some underlying layer within the planet. He first plotted the source longitudes in System II, and found only rather loose relationships between them.[3] But when Reese (*1972a*) plotted the source longitudes in System III, the longitude system of the planet's core, he did indeed find that they fell onto three straight parallel lines (Fig. 10.25 and Table 10.4). These were interpreted as representing three permanent loci which drift only slowly and steadily, with a rotation period of 9h 55m 30.11s ($\Delta\lambda_2 = -7.7$, $\Delta\lambda_3 = +0.3°$/mth). These are sites of potential instability, at which SEB eruptions tend to be triggered. This hypothesis was strengthened by the 1975 revival, which had sources within 2° of Reese's loci A and C, and another source 23° p. his locus B, and by studies of mid-SEB outbreaks (section 10.4), some of which appeared on the same tracks. (The agreement is best if locus B is shifted 20° p. after 1958.) Moreover, locus A emerged as the source of four of the strongest classical Revivals (1919, 1928, 1971, 1975), while locus C was responsible for one of the weakest (1955) and for secondary outbreaks. In years when there were both primary and secondary outbreaks (1943, 1971, 1975) they appeared on the lines in the order: A, B, C.

After 1975, faith in the Reese hypothesis was weakened by the mid-SEB outbreaks, most of which did not appear at his loci (section 10.4), and by the 1990 SEB Revival. The 1990 primary source lay on the original line B, but it thus conflicted with more recent

outbreaks 20° p. that line, and the secondary source did not fit any of the lines. However, as this chapter is being finalised in 1993 April, we are just witnessing the start of a new Revival, again coinciding with Reese locus B – a striking confirmation of the hypothesis! The source is all the more remarkable in being immediately p. the Red Spot Hollow, in a region where the surface currents are distorted by the Red Spot circulation. That this Revival has started here, and produced typical northern and central branches, is further evidence for the very deep-seated nature of these disturbances.

However, the drifts of Reese's hypothetical sources are not usually displayed in the drifts of the observed sources during a Revival, apart from the exceptional cases in 1971 and 1975.

From our present knowledge of the planet, the Reese sources could not be 'volcanoes' as originally envisaged. They might perhaps be long-lived circulations or waves or even floating objects at a deep level.

All these observations can be fitted together into the following tentative model for SEB outbreaks. (The physical mechanisms will be discussed in Chapter 14.6.) While the SEB is whitened, an unstable situation builds up below the cloud layer at all longitudes, and can only be released in a classical Revival. Whatever the Reese sources may be, the instability always breaks first over them, just as clouds on Earth first form over mountains. The instability can be triggered by the passage of spots on SEBs or SEBn past the Reese sources, or if necessary it will break out spontaneously. Then the eruption begins with a billowing white cloud which, if the Voyager observations of mid-SEB outbreaks are taken as representative, appears in one of the small cyclonic spots at 16–17°S. The classic dark 'bridge' across the SEBZ forms at the border of this white cloud. From this moment on, the outbreak becomes self-sustaining, fixed in the S. Tropical Current, but creating the three spreading branches as it destabilises the neighbouring jetstreams and pours masses of clouds into them.

The mid-SEB outbreaks to be discussed in the next section have many similarities to classical Revivals. There are some differences. One is that mid-SEB outbreaks have been separated by as little as 4 months, whereas classical Revivals do not recur within less than 2–3

[3] This unsuccessful theory was first proposed by Reese (*1953a*); also Reese (*1955, 1962b*), and Chapman and Reese (*1968*).

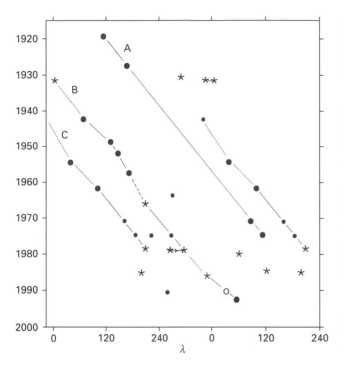

Figure 10.25. E.J. Reese's 'permanent sources' of SEB outbreaks. Longitudes of outbreaks are plotted in 'old' System III (period 9h 55m 29.37s) against time. The present System III period (9h 55m 29.711s) is actually closer to the drift of Reese's three loci (period 9h 55m 30.11s). Filled circles are initial sources of classical revivals (smaller symbols for secondary outbreaks); stars are initial sources of mid-SEB outbreaks. Lines A, B, and C mark the hypothetical permanent sources. (This is an extended version of the chart of Reese (*1972a*), adapted from Rogers, (*1989a*).)

years. A second difference is that mid-SEB outbreaks do not obviously produce retrograding SEBs spots, but occur when the SEBs jetstream already shows activity. However, it is still possible that mid-SEB outbreaks and classical Revivals are essentially the same phenomenon. Let us suppose that mid-SEB outbreaks tend to erupt at intervals of 1–3 years, but that the atmospheric changes involved in the whitening of the SEB tend to delay their appearance and to block the development of SEBs jetstream spots. Then, when enough energy has built up to produce a really vigorous mid-SEB outbreak, some trigger sets it off and it breaks through the cloud cover, thus initiating all the branches of the disturbance at once. The turbulence on all the branches disrupts the white cloud as it advances and permits the revival of the belt. If this hypothesis is true, there is no necessary connection between the fading and the subsequent Revival of the SEB. SEB activity would occur anyway, but the prior whitening process focusses it into a single great eruption.

10.4 LONG-RUNNING SEB ACTIVITY AND MID-SEB OUTBREAKS[4]

In several historical periods, the cycles of SEB Revivals have ceased, and long-running SEBs and SEBZ activity has developed

[4] This type of activity was described by Rogers (*1989a*), from which this account is taken. A report by one ALPO observer of mid-SEB outbreaks in 1978 is not included as it conflicted with independent observations that showed no such activity.

instead. The retrograding SEBs spots, spaced 15–20° apart, do not have any particular source; their origin was documented in the Voyager observations as described below. There is usually a region of continuous prograding disturbance within the SEB following the GRS (section 10.2). And once every year or two during these episodes, substantial outbreaks of white rifts appear at longitudes remote from the GRS, which we refer to as mid-SEB outbreaks. (They are called 'bright streak disturbances' by the ALPO.)

Each of these mid-SEB outbreaks begins with a small white spot in the southern half of the SEB. Typically, within a few days, this white spot expands in the Np. direction, and white spots and streaks prograde rapidly to form a highly disturbed SEB Zone. New white spots continue to appear near the original site for several months, until the source switches off. The disturbances resemble the continuous disturbance f. the GRS, and they prograde similarly until they disappear north of the Red Spot Hollow, although they do not actually mix with the more proximal disturbance. The best-observed such outbreak, which seems to have been typical, was in 1985 (Fig. 10.26). Another example is in Fig. 10.52 (1980/81). The outbreaks are listed in Table 10.5.

Visual observations

1930–34: After two quiet years, SEBs jetstream spots reappeared in 1931 March. For the next 3 years the jetstream was intensely active with many such spots. These spots were recirculated onto the STBn by the STropD (section 10.7). The activity switched off during solar conjunction in late 1934.

Mid-SEB outbreaks were not identified at the time, but the BAA Memoirs contain clear evidence of three such outbreaks. The first apparently started even before the SEBs activity, on 1930 Dec. 14, which was the first recorded sighting of a mid-SEB white spot with $\Delta\lambda_2 = -25.5$ (see Plate P7.4). The second outbreak was first clearly recorded on a drawing on 1932 Feb. 4, but a drawing of 1931 Nov. 11 seems to show the outbreak already active. In the 1932/33 apparition, a broad SEBZ split the entire belt, and no mid-SEB drifts were recorded. (A minor episode of ochre coloration affected the SEB and STropD later in 1933, but this colour is not typical of the SEB in such periods.) In 1934 the SEBZ was similar but there does seem to have been another outbreak within it, whose origin was not observed. Drawings in 1934 March and April show intense rifting of SEB already underway (e.g. Plate P7.6), just p. the STropD and ≈160° p. the GRS.

1965–68: SEBs activity began suddenly in 1965 Dec., and numerous jetstream spots were observed up till 1968 (especially by NMSUO) (Plate P11.6&8). Activity actually started in 1965 Nov. with four spots on STBn, and one rather slowly retrograding spot on SEBs which drifted from 21.4°S to 23.1°S as it came to a halt 50° p. the GRS (Fig. 10.62). In 1966/67, the SEBs spots were appearing 30–90° f. the GRS. The later SEBs spots retrograded up to the Red Spot Hollow without change of speed, and some were tracked around the GRS; these spots permitted the discovery of the circulation within the GRS (section 10.5). Within the broad clear SEBZ, no definite mid-SEB outbreaks were identified, but the ALPO reported an outbreak of white spots in the SEBZ p. the GRS (P. Budine, *1968*), the first record being on 1966 Nov. 24. Likewise S. Cortesi (Société Astronomique de Suisse) reported that small bright prograding spots appeared in the SEBZ from 1967 Jan. 1 onwards, resembling a SEB Revival. Throughout the next apparition (1967/68), NMSUO reported white spots forming and prograding from a source 25° p. the GRS with $\Delta\lambda_2 = +4$.

There was also SEBZ activity f. the GRS (described in the 1966/67 NMSUO report and 1967/68 BAA report).

1978–88: After the great Revival of 1975, activity on the retrograding jet-stream was re-activated in 1978 and remained high until late 1980; there were three mid-SEB outbreaks in 1979 and 1980. The first sign of new activity on SEBs was a brilliant white spot retrograding in 1978 Jan.–Feb. (BAA; Plate P18.2). Within weeks, much of the SEB(S) was disturbed. No definite drifts could be obtained until 1978 Nov.–Dec., when several features were moving towards the GRS with $\Delta\lambda_2 = +60$. But they were soon overtaken by faster-retrograding spots with $\Delta\lambda_2 = +110$ to $+120$, which were appearing $60°–100°$ f. the GRS and running all the way up to Red Spot Hollow where they disappeared. Occasional mergers of spots were seen. The dark spots were separated by white bays which occasionally brightened suddenly. The speed continued to increase and averaged $+144$ in 1979/80, while a STropD was present (section 10.7 and Plate P18.5). SEBs activity declined in 1980/81, showing a great range of speeds, and ceasing in 1981 April. (As it did so, there was again evidence for recirculation at the STropD; sections 10.7&9.)

On 1979 Feb. 21, a mid-SEB outbreak was seen nearly $100°$ f. the GRS; and in 1979 Nov, a new outbreak was active at longitudes $30°$ higher still. This second outbreak actually started on 1979 May 31 according to Voyager 2 photos. Activity reappeared $140–150°$ f. the GRS in 1980 Nov., and remained strikingly turbulent at least until 1981 March, extending p. towards GRS (see Fig. 10.52). (This was recorded in detail by the SAF.) In early 1981, white spots were arising $100°$ f. the GRS at the rate of one every 7 days, typically with $\Delta\lambda_2 = -66$, decelerating to about -33 after 1–3 weeks. The f. end of the disturbance prograded towards the GRS after the source switched off.

After several lean years, and the development of the third S. Tropical Dislocation which included stationary SEBs spots in 1983, intense retrograding activity resumed in 1984 (Plate P19.3). There followed three mid-SEB outbreaks in 1985 and 1986. (The first is shown in Figs. 10.26, 4.1A, and 20.3e, and in Plates P19.5 and P22.3).

The number and retrograding speed of spots on SEBs gradually dwindled until there was little activity left in 1986 and 1987. SEBs activity and retrograding speed revived for a third time in 1988/89, but declined rapidly; no new spots were formed in the SEBs jetstream, nor in the perennial disturbance f. the GRS, after 1988 November. Some SEBs spots had been photographed within the Red Spot Hollow (Plates P21.6 and P23.4), but the last two SEBs spots became very large and almost halted as they staggered up to the GRS.

Locations of the outbreaks

The sources of mid-SEB outbreaks closely resemble those of classical Revivals in their shape, position, and motion (roughly stationary in System II during each outbreak). So it is worth looking for triggering factors as we did for SEB Revivals.

System II: The positions of these nine mid-SEB outbreaks all fall in a limited range of λ_2 with respect to the GRS. Within less than $100°$ f. the GRS, any individual outbreaks would not be readily distinguished from the continuous activity. Within less than $100°$ p. the GRS, mid-SEB outbreaks have not been seen, except in 1966–68. If more specific triggering agents are to be sought among the visible System II features, it is intriguing that the 1930 and 1931/32 outbreaks both began adjacent to a slow-moving white spot at the f. end of the South Tropical Disturbance as did the classical Revival of 1928. Similarly the first 1985 outbreak seemed to diverge from a stationary white spot in the STropZ, and the 1986 outbreak appeared adjacent to a Little Red Spot.

System I: The first 1985 outbreak may have been triggered by the passage of the long-lived SEBn white spot in System I, but this may have been mere coincidence.

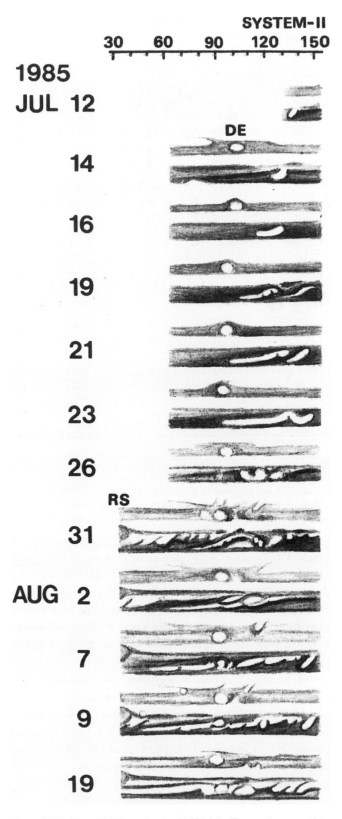

Figure 10.26. The mid-SEB outbreak of 1985 July. These strip-maps of the STB and SEB are by I. Miyazaki and other members of the Oriental Astronomical Association (T. Asada, I. Hirabayashi).

System III: Are the hypothetical Reese sources involved? Rogers (*1989a*) compared the nine mid-SEB outbreaks with the Reese predictions as shown in Table 10.5 and Figure 10.25. Two of them coincided with Reese lines B and C to within one degree. Three of

Table 10.5. *Mid-SEB outbreaks*

Date		Discoverer	λ_2	Reese source λ_2	(O–C)	GRS λ_2	
1930	Dec 14	BAA (UK)	(18)	C: 341	(+37)	257	Calculated from table in BAA report. First drawn 1930 Dec 21. $\Delta\lambda_2 = -25.5$.
1931	Nov 11*	Phillips (UK)	(34)	B: 1	(+33)	235	Seems to be shown alongside f. end STropD in Phillips'
1932	Feb 4	Schlumberger (France)	$\approx$339	B: 339	($\approx$0)	229	sketch in BAA Memoir. Possibly observed 1932 Jan 25 by Peek (transit small spot λ_2 311). First definite drawing by Schlumberger.
1966	Nov 24?	ALPO (USA)	297	B: 322	(−25)	28	Calculated from table in ALPO report; white spot, $\Delta\lambda_2 = -105$. Another 'very bright oval' on 1967 Jan 29 at λ_2 303, $\Delta\lambda_2 = -86$. Atypical, p. GRS. Similar activity reported p. GRS by NMSUO in 1967/68.
1979	Feb 19	Voyager 1					
1979	Feb 21	Néel (France)	152	C: 151	(+1)	55	
1979	May 31*	Voyager 2	179	B: 231	(−52)	55	
1979	Nov 1:	Rogers (USA)	170	B: 192	(−22)	55	
1980	Nov 18	Dragesco (Benin)	190	–	–	51	Started after Nov 6; well-developed by Nov 18
1985	Jul 12	Miyazaki (Japan), Andrews (UK)	130 (145)	–	–	25	Initial λ_2 130, but subsequent outbreaks at λ_2 145.
1985	Nov 3	Miyazaki (Japan), Rogers (UK)	193	A: 174	(+19)	25	
1986	Dec 11	Randall, Moseley (UK)	230	B: 246	(−16)	19	

This Table is adapted from Rogers (*1989a*), with the addition of observations marked *.
Columns 1, 2: Date and discoverer of first observation of white spot.
Column 3: Initial longitude of white spot (in brackets if the observation may have been some time after the actual outbreak)
Column 4: Predicted λ_2 positions of Reese sources A, B, and C are from Reese (*1972a*), calculated as in Table 10.4. (If the dark streak of a SEB Revival source corresponds to the Nf. edge of an expanding, prograding white spot, as seems likely, the difference in longitude between the streak and the white spot is negligible.)
Column 5: Difference between observed and calculated longitude.
Column 6: Longitude of GRS.

the others fell between 16° and 25° p. locus B, as did recent classical Revivals. These points defined a good line 20° p. the extrapolation of line B, suggesting that source B had shifted 20° p. between 1958 and 1966. The other four outbreaks were not consistent with the Reese loci.

Unfortunately, 3 of these 5 'successes' have been thrown into doubt by further research since the author published this analysis. The 1932 and 1979b outbreaks probably started several months earlier than listed then, and it is not clear that the 1966 event was the same type of outbreak (Table 10.5). As a result, one cannot say that the mid-SEB outbreaks show any tendency to arise at the Reese longitudes.

Spacecraft observations

The first region to describe is the perennial SEBZ activity f. the GRS. This region was spectacular in the Voyager movies (e.g. Figs. 10.2&9), especially where the SEBs jetstream is squeezed against the prograding mid-SEB and SEBn material, on the Nf. edge of the Red Spot Hollow. White spots were billowing out of the furious turbulence at this point. But the largest and strongest white spots were arising tens of degrees further f., so were not caused simply

by this local instability. The dark twisted filaments in this region were very reddish (T. Owen and R. Terrile, *1981*), although this colour did not apply on the large scale visually (section 10.2).

The SEBs edge was also very disturbed just f. the GRS, and some retrograding jetstream spots seemed to appear at the f. corner of the GRS, though they may not have persisted as stable features. The jetstream spots were shown to be anticyclonic rings, and a study of Voyager 1 images (Rogers, *1989a*) showed they had various origins. One emerged as a complete ring from the anticyclonic circulation around the GRS, although it was later disrupted while passing the rifted region. Others, such as number 5 in Figs. 10.27&28, emerged from the SEBZ rifted region itself. Number 5 originated from the tip of a white streak which had been roughly stationary in the turbulent SEBZ, but which was eventually propelled south right across the latitude of the jetstream (Fig. 10.27); this allowed the jetstream to break it off and twist it anticyclonically to form a retrograding eddy (number 5). However, this and other spots continued to interact, anticyclonically, with the turbulent SEBZ features, and did not acquire their complete ring shapes until they had retrograded away from the rifted region. The stabilisation of these jetstream spots typically occurred $\approx$90° f. the GRS

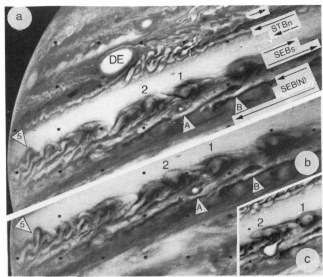

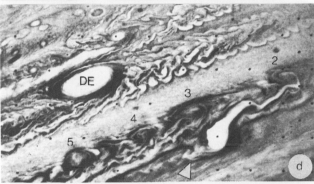

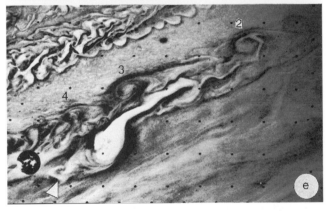

Figure 10.27. Voyager 1 photographs of the start of the 1979a mid-SEB outbreak. All are violet-filter images, spatially filtered. (NASA images; from Rogers, *1989a*.)

(a)   Image 15985.20: 1979 Feb. 20d 01h (UT at spacecraft). This was one rotation after the start of the outbreak within a mid-SEB dark spot (arrowhead A). Arrowhead B indicates a similar dark spot. Retrograding SEBs spots are numbered, including number 5 which was newly emerging. Major jetstreams are indicated by arrows.

(b)   Image 16010.09: Feb. 20d 21h.

(c)   Image 16047.15: Fed. 22d 21h (reduced scale).

(d)   Image 16157.45: Feb. 25d 19h. The SEB outbreak now has a long tail to Sf., entangled with SEBs spot 2, and is pushing out light veils of cloud Np. across SEB(N) (arrowhead). In (d) and (e), the broad dark rim around the outbreak is an artefact of the image processing.

(e)   Image 16207.36: Feb. 27d 10h. The tail to Sf. and veils to Np. have become more complex. The long-lived SEBn white spot is in the terminator shading at right, and Io is at bottom left.

(Fig. 10.27). Most SEBs jetstream spots came from rifted regions directly, or from SEBs turbulence which coalesced into rings further along the jetstream. (SEBs turbulence was more concentrated in the Voyager 2 maps; Fig. 10.64.)

In their mode of formation, their interactions with each other (M. MacLow and A. Ingersoll, *1986*), and their interactions with the subsequent mid-SEB outbreak (see below), these rings seemed to behave as surface eddies without any sign of deep-seated roots.

The mid-SEB outbreak of 1979 February began while the planet was being photographed by Voyager 1 only two weeks before the fly-by. It began within one of several small dark spots in the SEBZ. No features were detectable with the Reese drift ($\Delta\lambda_2 = -7.7$). Instead, the narrow, tapering SEBZ $\approx100°$ f. the GRS contained several long dark streaks aligned with the shear, sometimes carrying condensations which were almost stationary in λ_2. It was one of these condensations, with $\Delta\lambda_2 = +5$ (lat. 16.3°S), that produced the outbreak. The dark condensation became somewhat more distinct and oval around Feb.16, and acquired a complex halo of apparently cyclonic streaks on Feb.19; however, these developments were not unique as the neighbouring condensation (12° f.) had shown a similar aspect a few days earlier. Thus there was no clear warning of the outbreak, which began on Feb.19 as a tiny brilliant white spot precisely in the centre of the dark spot (Fig. 10.27).

The initial white spot expanded at 6.4×10^6 km²/day over Feb.20–22, then slowed (Fig. 10.28 inset). This rate is consistent with convection powered by water condensation below the visible clouds (Chapter 14.6).

The initial white spot had no evident influence on adjacent features until it actually spread across them; the retrograding rings, for example, were unaffected at first. The main white cloud pushed Np., breaking across SEB(N) on Feb.22. Meanwhile, a narrow white 'tail' emerged Sf. on Feb.21, and on Feb.22–24 its tip became caught up in the anticyclonic, retrograding motion of SEBs spot 2. By March 3 this connection had broken and a new tip was becoming caught up in SEBs spot 4 as it passed.

Another bright cloud erupted from the same source on March 3 (not shown).

The Voyager 2 movie showed the same process being repeated several times, both at the original site ($\lambda_2 \approx 140$) and 40° further f. ($\lambda_2 = 179$). The latter source first appeared on 1979 May 31, in a dark streak just like that where the earlier outbreak started. At each site, the movie showed that a white cloud erupted from a tiny point, and spread both Sf. forming a long sinuous tail, and Np. forming broad veils; then, as the first white spot dissipated, two more erupted in rapid succession from exactly the same site. Clearly the source was extremely localised.

Although one cannot be certain of relative altitudes from the Voyager photographs, they give the impression that the main white cloud was expanding above the surrounding dark features, while the Sf. tail, like so many other sinuous features in Voyager images, was getting tangled up in circulations at its own level but not readily mixing with them.

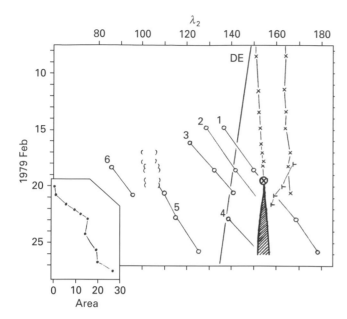

Figure 10.28. Drift chart for selected features around the 1979a outbreak, from Voyager 1 data. Longitudes were measured on Voyager 1 violet-filter strip-maps, to Feb 20, and thereafter on images like those in Fig. 10.27. The longitude of the GRS (λ_2 55) was used as the reference point, and the motion of oval DE measured from the Voyager strip-maps agreed well with the visual estimates. Symbols: o, SEBs spots (for clarity, only a few points are plotted); ( ), SEBZ white streak; x, SEBZ dark spots; ⊗, start of outbreak; ⊢, SEB(N)s feature. *Inset*: Area of the main white cloud of the mid-SEB outbreak, in millions of km². (From Rogers, *1989a*.)

10.5 THE GREAT RED SPOT[5]

Cassini's spot[5]

The dark oval spot tracked by J-D. Cassini and others from 1665 to 1713 (Chapter 1.2, Fig. 1.8) resembled the present GRS, and has often been identified with it, but this identification has never been certain. It had roughly the same latitude as the present GRS, lying in the STropZ and touching the STB. And like the GRS, it was conspicuous when the SEB was very faint (as in 1665 and 1672) but disappeared when the SEB revived. It was only visible in some years, viz.: 1665–66; 1672–74; 1677; 1685–87; 1690–94. The frequency of these intensifications was similar to those of the GRS nowadays. In fact, after the spot disappeared at the end of the 1674 apparition, Cassini may well have seen a classical SEB Revival, or perhaps an associated disturbance of the EZ. He wrote:

> That spot was invisible in 1675 and 1676, and during those two years there were some very important changes on Jupiter. A bright zone which lay between two dark belts [the SEBZ?] was broken into many little pieces resembling islands, as if the two dark belts were great rivers bordering on each other leaving these islands between them. Then the islands were

---

[5] For earlier reviews on the GRS, see: Antoniadi (*1926*), Peek (*1939*), Reese (*1953b*), Fox (*1967*), Sanchez-Lavega (*1985*), and Beebe (*1990*), as well as Peek's book.
    The 19th century observations were reviewed by Denning (*1898a, 1899*), and Kritzinger (*1912*).
    The 17th century observations of Cassini's spot were reviewed by Denning (*1899*), Antoniadi (*1926*), Falorni (*1987*), and Chapman (*1968*); also see the original descriptions by Cassini (*1666, 1672*).

entirely erased, and the whole region of the two belts and zone formed one very large dark belt. But after Jupiter emerged from conjunction with the sun in 1677, the belts resumed their previous form and situation, and the permanent spot reappeared. [Translation and note from Chapman, *1968*.]

Colour in the spot was not mentioned in the published reports, and would not have been shown clearly with the primitive refractors of the time. However, the spot was clearly shown as red – just like the Great Red Spot of the early 1970s – in a painting done around 1700 by the artist Donato Creti. (The painting is in the Vatican Museum, and was reproduced in a book by F. Hoyle, *1962*.) This is one of a series of paintings by Creti showing astronomers in romantic landscapes observing the planets. Perhaps one of the astronomers in the 1690s suspected the red colour, not confidently enough for a scientific report, but enough to tip off the artist that it would be worth including in his painting.

Cassini's spot differed from the present GRS in two respects: its shorter length and its slower motion. The length was ≈12°, though this cannot have been very precise with the primitive optics of the time. The rotation period was quoted by Cassini as 9h 56m or just under – much slower than the present GRS has ever shown. (Cassini's value did not take account of the speed of light, but this would make little difference to speeds deduced over more than a year.) Table 10.6 lists the rotation periods available. The only subsequent sightings (after a long gap) were in 1708 and 1713, by Cassini's nephew Maraldi, who gave a period of 9h 56m in the final year.

Because nothing resembling the GRS nor Hollow was seen for the next 118 years, it is uncertain whether Cassini's spot really was the GRS. This question will be discussed in Chapter 14.4.

Early history of the GRS

Although the Great Red Spot did not become generally known until 1878, the Spot or the Hollow had actually been observed many times in the nineteenth century.[6]

The earliest definite record of the GRS, or rather of the Red Spot Hollow, was made on 1831 Sep. 5 by S.H. Schwabe (Plate P1.1). He drew the Hollow on 20 occasions between then and 1856. The GRS itself was never seen during this period. Schwabe usually drew the Hollow ≈26–35° long with the range of shapes that are now well known, though sometimes it was longer and shallower. It was also drawn by J. Schmidt in 1845/46 and 1850, and by Bond in 1848 (cited by Kritzinger, *1912*). The next drawings were made by the Rev. W.R. Dawes on 1857 Nov. 27 and Dec. 5, showing the Hollow with a dark arch around the Sf. side (Plate P1.4).

In 1858/59, when the SEB was faint, William Huggins drew the Spot as a dark ring with a light interior (Plate P1.7); often it appeared as a well-defined ellipse with dark ends. By comparing the transit times given by Huggins for the ellipse with those by J. Baxendell for a dark spot which marked its f. end, its length can be

---

[6] The 19th century observations were collated by Denning (*1898a, 1899*) and Kritzinger (*1912*). Several are shown in Plates P1, P2 and P12. The present account also draws on unpublished observations in the archives of the Royal Astronomical Society, and numerous reports in contemporary issues of *The Astronomical Register, The Observatory*, and *MNRAS*.

Table 10.6. *Drift rate of Cassini's 'permanent spot'*

	P (9h+)	Δλ₂		P (9h+)	Δλ₂
Cassini			Denning (*1899*)		
1665–66	56m	+14	1666 Jan–1672 Mar	55m 54.2s	+10
			Chapman (*1968*)		
1672 Jan–	55m 53.5s	+9	1672 Jan–Mar	55m 57s	+12
1674 Oct?				(±3s)	
1685–87	55m 52s	+8	1690 Dec–1693 Feb	55m 49s	+6
			1693 Feb–1694 Feb	55m 39s	−1
Maraldi					
1708	55m 48s	+5			
1713	56m	+14			

This Table lists the rotation periods published by Cassini and Maraldi, and re-analysed by Denning (*1899*) and Chapman (*1968*). Chapman did not consider previous determinations to be reliable. Denning also calculated a period of 9h 55m 58.7s to link Hooke's spot in 1664 with Cassini's in 1665–66, but as it is not clear whether these were the same spot, this period must be discounted.

estimated as 33° in longitude. By 1860, the Hollow was again prominent and the Spot almost invisible (Plate P2.1–3). Drawings by Schmidt and others in 1859–1862, cited by Kritzinger (*1912*), showed the Hollow ≈42–46° long.

In 1869/70, as the SEB faded again, the GRS was rediscovered as a 'great southern ellipse'. According to J. Gledhill in Yorkshire, 'this fine object was easily seen every clear night' (Plate P2.7). But it must have been quite faint as Dawes could see nothing more than a suspected wisp here. Professor A.M. Mayer in Pennsylvania, using a 15-cm refractor, described it thus:

> Especially was my attention riveted on a ruddy elliptical line lying just below [S. of? – *ed.*] the SEB. This form was so remarkable that I was at first distrustful of my observation; but keeping my eye steadily upon it I perceived that the ellipse became more and more distinct as it advanced towards the centre of the disc. I followed this remarkable form until it was about bisected on the W. limb; when it had gradually faded from view.... Its major axis to its minor axis is, in its present projection, as 1 to 1.51. Can we be so bold as to regard it as a great gaseous mass, having its origin in the equatorial regions, and sweeping S. (as with the terrestrial cyclones of the S. hemisphere), and flattened by the rapid rotation of the planet? (Mayer, *1870*.)

Gledhill's and Mayer's data, though very disparate, give a mean length of 24° (±7°).

In early 1872, as the SEB and Hollow revived, the ellipse had become a dusky oval mass (Plates P2.9 and P12.2). No red colour was evident. Then a year later the red colour was discovered, thanks to one of the largest telescopes ever used for visual observations of Jupiter: the Earl of Rosse's 72–inch (1.8 m) reflector. Lord Rosse and Dr Copeland recorded reddish colour in the Spot for the first time on 1873 Jan.22 and on subsequent dates (Plate P12.3&4): 'Of all the features presented to our view by Jupiter during the opposition of 1873, probably the most remarkable is the great

break in the southern side of the equatoreal belt in long. 260°.... Following and filling up this break is a brick-red area that was seen most fully on February 6.... The red region may extend some 30° of longitude.' (Rosse and Copeland, *1874*.)

In 1876, the Spot was drawn as a distinct dark oval by T. Bredichin at the Moscow Observatory; but the planet was then approaching its furthest south declination, and the best views were had by Australian observers (Plates P3.1 and P12.6). They regarded the GRS as a familiar object, and in 1877 they nicknamed it 'the pink fish' from its shape and colour. Then in summer, 1878, observers in Europe and America also noticed the Spot, which had become an astonishing sight. It had a remarkable brick-red colour, and was bordered by a bright white halo. Its ends were slightly pointed and the f. cusp was a small dark spot. From 1879 to 1882, it remained strikingly prominent, 34° long and strongly reddish, floating in isolation as the SEB(S) disappeared (Plates P3.2–4 and P12.7–9).

The Spot faded rapidly in 1882, and for the next 35 years it underwent only moderate fluctuations in appearance. Some observers still saw a pink colour while others, with equally good instruments, regarded it as grey.[7] The range of aspect and colour was similar to that in the more recent era of SEB stability, from 1976 to 1989.

By 1900, the GRS was recognised as a permanent feature of the planet. Its subsequent history has been tied in with that of SEB Revivals and S. Tropical Disturbances, as reviewed in sections 10.3 and 10.7.

Shape and colour of the GRS

The essential property of the GRS is its oval outline. How this oval is displayed is variable, but even when the Spot itself is completely invisible, the oval is still represented by the curve of the SEBs edge round the Red Spot Hollow. According to modern observations, the oval marks the perimeter of the Spot's circulation, and the Hollow marks the course of the SEBs jetstream as it is deflected around the Spot.

The STBn edge usually runs precisely tangential to the GRS, but sometimes it too appears to be slightly deflected to avoid the Spot; sometimes this appearance is due to partial fading of the STB in these longitudes.

The main variations in the appearance of the GRS have been described already, but we should now summarise the range of aspects observed (Fig. 10.29). These aspects are by no means distinct, and can change rapidly according to the darkness of the SEB, or the presence of a dark rim around the Spot, or the amount of reddish material in the Spot itself. (Some confusion can arise when observers describe mere fragments of dark or reddish material as 'the Spot'. The only consistent rule is to define 'the Spot' as the whole oval, within which parts may be light or dark.)

[7] Readers of the original BAA Memoirs should be warned that A.S. Williams recorded reddish colour in the Spot even when most observers could not see it at all. Williams did discover many important features on the planet, but he used only a 6½-inch (16-cm) reflector, and successive Jupiter Section Directors have regarded his detailed representations with some caution (Chapter 1.3). To Williams, a tiny black spot was still often discernible at the f. tip and sometimes at the p. tip of the GRS, up to 1891.

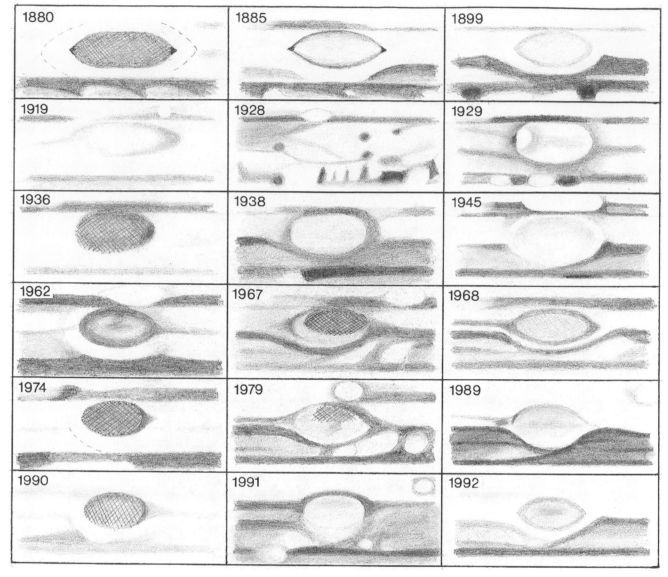

Figure 10.29: Various aspects of the Great Red Spot. Left column: while SEB is faint; 'red spot' or 'doughnut' aspects. Middle column: while SEB is active, viz. during a SEB Revival (1928), or during continuous SEB activity (1967, 1979), or just after a Revival (1938, 1991). Right column: while SEB is quiet; aspects such as 'pale spot in hollow' (1899, 1945, 1992), 'ring' (1929, within STropD), 'doughnut' (1968), and 'arch' (1989). Most views were taken from drawings and/or photographs made with fair-sized telescopes. Each panel is 70° wide. Cross-hatching indicates reddish colour. This series also shows the evolution of the STB, with the white ovals visible in some panels from 1945 onwards, gradually shrinking.

(i) The most well-known, though not the most frequent, aspect is a dark oval with a faint or absent Hollow. This occurs when the SEB is faint. It may be a 'red oval', a 'dark oval', or a 'doughnut', according to the colour of the interior. The 'doughnut' aspect, with pale interior, was common in early years but has been rare recently. (ii) The reverse is the 'hollow' aspect, when the SEB is dark and the GRS is invisible, its position only betrayed by the Hollow. More often the S. arch of the oval is still visible, supplemented with a greater or lesser amount of dark grey material that is continuous with the SEB at the f. shoulder, giving an 'arch' aspect. (iii) The 'ring' aspect is shown when the 'hollow' or 'arch' aspect is augmented by even more dark material surrounding the GRS. Sometimes this is a narrow dark rim, which may be difficult to distinguish from the outer part of the Spot itself. Sometimes it is a more extensive darkening, in which the GRS is silhouetted as a light oval; this often occurs following SEB Revivals and used to occur when the great STropD was passing. A 'ring' aspect was also familiar in the years when the great STropD was passing the GRS; the GRS would stand out as a light oval surrounded by the dusky shading of the STropD. This has been felicitously called the 'chrysalis' aspect (Slipher, 1964) or 'cocoon' aspect (Sato, 1970). In this situation, the GRS is sometimes a fully white oval.

During the era of triennial SEB Revivals, the GRS cycled continuously between aspects (i) and (iii).
(iv) Occasionally the GRS fails to fade even though the SEB revives, so a distinct dark oval persists within a well-formed Hollow. (This was the aspect in 1964 and 1968; Plate P13.)
(v) In the most common aspect, the Hollow contains only a faint or incomplete dusky oval, with a more or less dark rim. One can only give complex terms for such a complex appearance, such as 'pale

oval in hollow' or 'red smudge in ring'. Sometimes the complete GRS can be made out, as a ring or as a dusky oval; this was common between about 1892 and 1917, when the Hollow was significantly larger than the Spot, and the Spot was not connected to the SEB. This aspect has been more difficult to detect in recent years, as the Hollow has been wrapped more tightly around the Spot and they are often continuous at the f. end. But the oval form is still often evident, both at high resolution (thanks largely to its dark p. cusp being separated from the SEB), and at low resolution (thanks to an 'arch' to the south and the Hollow to the north). Inside the oval outline, sometimes only a part of the GRS oval is shaded, either grey, buff, pinkish, or frankly orange. Distinct orange patches are more commonly visible with large telescopes, especially in the S. half. The N. half tends to be lighter and sometimes even contains white spots.

This was the aspect during the Voyager 1 encounter (Plate P17); it could be described as 'orange patch in arch'. The dark ring or arch, as recorded in 1976–1989, always waxed and waned along with SEBs jetstream activity – even when the jetstream spots were not reaching the GRS due to a STropD. This ring seems to result from SEB material flowing around the rim of the Spot, and sometimes leads to the emission of a S. Tropical Band from the p. end, as in the Voyager 1 pictures. However, the fluctuations in the Spot's colour seemed to be capricious and are unexplained.

Although the red colour had attracted such attention in 1878, it seems that such an intense colour did not recur again for many decades. Peek, who gave careful attention to colours on the planet, wrote in his book:

> The author would willingly forgo his duty to discuss the actual redness of the Great Red Spot, especially as he himself has never noticed anything more striking in its coloration than a faint tendency to pinkness. He is quite incapable of imagining what its colour must have resembled during the notable years of 1879 to 1881 and can only state his belief that the lavish application of bright red pigments to the numerous tinted drawings that are extant conveys a very exaggerated impression of its true aspect. (Peek, *1958*.)

Peek would undoubtedly have revised his opinion had he lived to see the GRS in the early 1970s. Indeed, he noted in 1958 that the Spot had a 'brick-red lustre that he had not seen before the apparition'. But it was from 1961 onwards, during the years of inconclusive SEB outbreaks, that the Spot was intensely orange-red. It was partially eroded or dimmed in 1966–1968, but when the SEB was whitened again from 1969 to 1975, the Spot was again uniformly dark and brick-red – an awesome sight, whose colour is well represented in the published Pioneer photographs (Plate P16.1).

The nature of the reddish colour is unknown; it will be discussed in Chapter 16. Even when it does not appear reddish visually, photographs usually reveal the GRS as dark in the ultraviolet and violet, and bright in the infrared, indicating either small patches of red material, or material with an absorption mainly outside the visible spectrum.

Altitude of the GRS

The reddish clouds of the GRS are among the highest clouds on the planet. This has been known for several decades, mainly due to professional studies at infrared wavelengths (Chapter 4).

First, the GRS is always bright in the 0.89 µm methane absorption band, indicating that it is above much of the atmosphere. From 1968 to 1972 it was always the brightest feature at this wavelength (Figs. 4.9&10).[8] Second, it emits very little at 5 µm and 10 µm, indicating that it is cold. This was still true of the interior in 1979, although the dark rim was then warm like the belts (Fig. 4.15). Third, measurements of polarisation of the infrared and visible light in 1971 showed that the GRS was about 5 km above the STropZ.

These results were confirmed by the Pioneer spacecraft, which viewed the planet at much greater angles as they flew past. The lower polarisation of light from the Spot implied that it was 5 km above other zones and 2 km above the whitened SEBZ (Fig. 4.8). The infrared spectrometer on Voyager confirmed the coldness of the GRS and the atmosphere just above it (Fig. 4.19).

Dimensions of the GRS

In the years of its glory from 1879 to 1882, according to observations analysed by Denning (*1885*), the GRS had a length of 34° or 39 000 km, and a width of 10° or 12 000 km. It remained over 30° long until 1920, but since then it has never been as long again. Its average dimensions in the 1970s and 1980s were 21° (24 000 km) × 12° (14 500 km). The length is charted in Fig. 10.30. For reliable length measurements, of course, the elliptical outline of the Spot must be visible, and in Fig. 10.30 we have selected such measurements – preferably either the external diameter of the 'oval' aspect, or the internal diameter of the 'ring' aspect. There are variations in the size of the Spot on shorter timescales, which bear no relation to its aspect or colour. However, the chart as a whole suggests that the GRS may have a tendency to shrink as time goes on; the trend has been quite steady since 1920, at 0.14°/year.

It would therefore be very interesting to know how big the GRS was early in the nineteenth century. Unfortunately, before 1879 there were no accurate measurements, and the author has only been able to find two direct quotations of the length: 33° in 1858/59, and 30° in 1872/73 (see above). Other points on Fig. 10.30 are measurements from drawings. There is no sign of any overall shrinkage of the GRS during the nineteenth century. But the length evidently fluctuates, and there could still be a long-term shrinking trend. Such a trend was shown by the STB white ovals (Chapter 11.3) and by the circulating current of the great STropD (section 10.7). The possible similarities between these features and the GRS will be discussed in Chapter 14.4.

Latitude of the GRS

The latitude of the GRS centre since 1950 has been 22.4°S (Appendix 2).

This value relates to times when the oval outline was clearly evident on photos – mostly the 'red oval' aspect. At other times, the

---

[8] Methane imagery was described by Owen (*1969*), Kuiper (*1972a,b*), and Minton (*1972a*). For other data, see Chapter 4.

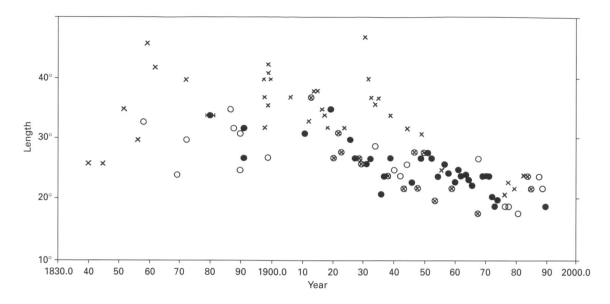

Figure 10.30: Changing length of the GRS. Symbols: ●, external dimension of dark oval; ⊗, internal dimension of ring; o, less definite dimension of dusky GRS or (before 1878) estimates from drawings; X, Red Spot Hollow. Before 1878, sources for the GRS itself are described in the text. Early sources for the Hollow (all imprecise) are drawings by Schwabe (lengths <40°) and by J. Schmidt (lengths ≥40°; cited by Kritzinger, *1912*).

The drawings by Schwabe are in the RAS archives, listed by Denning (*1899*), and the originals have been checked by the author; they are of fairly primitive quality but Schwabe does seem to have taken care over the scale. His earliest drawings in 1831 and 1832 show only a long, very shallow bay, which may or may not be the true Red Spot Hollow. After 1878, measurements are from the usual reports.

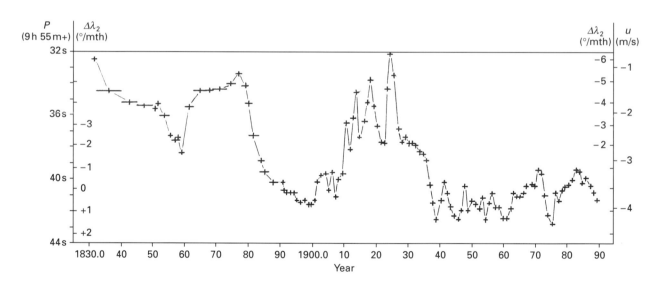

Figure 10.31. Changing drift rate of the GRS in longitude. These are mean rates from one opposition to another. Data up to 1889 are from Denning (*1898, 1899*; reproduced in Peek's book); some span several apparitions.

Subsequent values are from the usual reports. (For values within apparitions, see Fig. 10.6.)

northern outline is incomplete; however, the same latitude is then found for the centre of the GRS circulation (measured by Reese and Smith (*1968*) and by Voyager). But one must be cautious even with the 'red oval' aspect, as the N. edge can still be anomalously far south, as in 1928 and 1972.

It would be most interesting to know whether the GRS latitude has changed. Unfortunately, there were no accurate measurements

before 1950; the measurements in Appendix 2 from micrometry and from drawings seem to be scattered, and do not show a definite trend. Since 1950, the average latitude has been 22.4°S throughout, with no definite evidence for variation.

Motion of the GRS in longitude

Although the GRS is the most stable spot on the planet, it does not

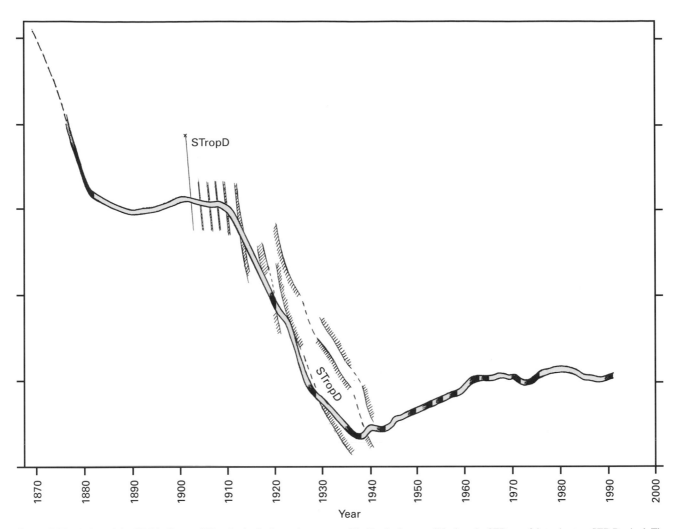

Figure 10.32. Motion of the GRS in System II longitude. Scale marks are at $\lambda_2 = 0$. The GRS has run five times round the planet relative to System II. The line is shown solid when the SEB was faint prior to a SEB Revival. The motion of the great STropD is also shown (shaded track).

remain fixed relative to System II (which was originally chosen to be as close to the mean rotation period of the GRS as possible) nor to any other longitude system. Its changing speed is shown in Fig. 10.6&31. Fig. 10.32 shows that it has drifted at least ten times around the planet relative to System II, and at least three times back and forth around the planet in any other longitude system that could be defined, during the past 160 years. Thus it cannot be attached to a solid core of the planet.

Two opposing influences on the drift of the GRS can be recognised. First, it tends to decelerate substantially when a major fading of the SEB takes place. But it is accelerated when a STropD passes it.

Peek noted that decelerations had occurred in 1880, 1919, 1926, and 1936, these being the years when the GRS became intense during the buildup to SEB Revivals. It also maintained an unprecedentedly slow rotation period from 1942 to 1961, the years of triennial SEB cycles, and again showed major decelerations in 1972–75 and 1988–90, during the two most recent SEB fadings. The linkage with SEB fadings is thus beyond doubt.

An important point not previously made is that when the SEB is faint for several years, the GRS continues to decelerate throughout this period, by $\approx 10^{-8}$ m/s^2 ($\approx 0.07°$/month^2) (1879–82; 1926–28; 1936–38; 1972–75; see Fig. 10.31.) The slowest speeds ever recorded ($\Delta\lambda_2 = +2$, period 9h 55m 41s) were in 1939 and in 1975, the latter being at the end of a prolonged fading of the SEB. The detailed observations of the two most recent events showed that the GRS began to decelerate while the SEB was fading (1972) or even earlier (1988). So the deceleration is not a consequence of the visible changes, but an early and independent effect of the same underlying cause. The continuing deceleration confirms that this state is not stable. The deceleration ends when the SEB revives.

The acceleration of the GRS, in conjunctions with the great STropD, was notable from 1902 to 1925. This was mainly because the GRS always moved faster (by 1–2°/mth) when it was within the Disturbance (Figs. 10.31&33). The GRS reached its fastest speed ($\Delta\lambda_2 = -6$, period 9h 55m 32–33s) in 1924 and 1925, then decelerated again. Most subsequent STropDs have not survived to reach the GRS, but one in 1970 did, and that may be why the GRS accelerated in that year even though the SEB was fading.

The interval covered by these correlations does not include the most extreme speeds which the GRS can sustain. From 1831 to

1880, the GRS moved faster than in any subsequent long interval (e.g. $\Delta\lambda_2 = -6$ in 1831–32, and –5 throughout 1869–1880). We do not know whether there were any immediate reasons for this fast drift. It may be a sign that the GRS has an intrinsic long-term deceleration, evolving in the same way as the STBs white ovals (Chapters 11.3 and 14.4).

A remarkable modulation of the Red Spot's longitude was discovered by the NMSUO photographic measurements from 1963 to 1974. This was an oscillation with a period of exactly 90 days and an amplitude of about 1.0° (Fig. 10.34a), which persisted throughout all these years. Thus it was still evident in 1971 and 1972, its phase undisturbed in spite of the 1971 acceleration and SEB Revival; and BAA observations (Fig. 10.34b) showed that the same oscillation persisted even through the great 1975 SEB Revival, although it was not detected in 1976/77. The oscillation had also been detected visually by the ALPO and BAA in 1962/63, although the largest deceleration then (as in 1975) could also have been due to the impact of the reviving SEB(S). The oscillation could only be detected because the GRS remained a well-defined dark oval throughout most of these years. Since 1976, the more irregular aspect of the GRS has prevented longitude measurements being made precisely enough to show whether any oscillation still persists.

But as long ago as 1905, Molesworth wrote in *MNRAS*: 'A few years ago, from the movements of the two shoulders of the Red Spot Bay, I was led to strongly suspect a period of [about] 90 days between maximum variations, but I have been unable to confirm this since.'

Note that 90 days is the period in which material on the SEBs jetstream, moving at $\Delta\lambda_2 = +120$, would encircle the planet; so the GRS oscillation might be coupled to some largescale feature of the jetstream. But no such feature has ever been detected. Anyway, the 90-day period is more constant than the speed of the jetstream.

There can be even shorter-term shifts in longitude. Visual observations sometimes show shifts of a few degrees within a week or so, and G. Solberg's NMSUO measurements proved that such shifts occur – especially when the GRS is interacting with SEBs jetstream spots.

Internal circulation of the GRS

The pattern of currents bordering the GRS has been suspected since the STropD flowed round its S. edge in 1902, and since retrograding SEBs spots curved round its N. edge in 1928, but the internal circulation was not revealed until E.J. Reese and B.A. Smith of NMSUO photographed it in 1966 and 1967 (Fig. 10.35; Reese and Smith, *1968*). The GRS at that time still contained a red oval, but it was reduced to the southern 2/3 of the usual outline, as its edges had been eroded by SEBs jetstream activity. In 1966 Jan. and Feb., Reese and Smith recorded a spot on the SEBs jetstream which split at the p. edge of the Red Spot Hollow. One half ran within the N. edge of the GRS oval, distorting it as the spot was torn apart; the other half skirted round the Hollow and lost most of its retrograding speed as it re-emerged onto the SEBs f. it. Also, they observed a spot on the STBn jetstream which made a complete anticlock-

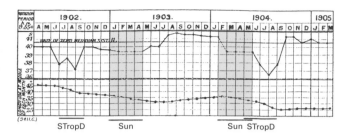

Figure 10.33. Changing drift rate and longitude of the GRS, 1902–04, showing acceleration when it was within the STropD. Conjunctions with the STropD and with the sun are marked below. (From BAA Memoir.)

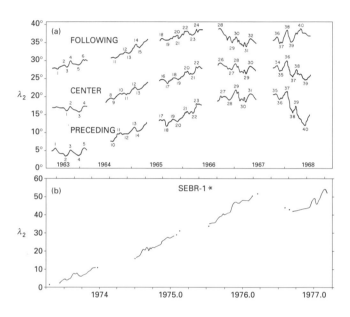

Figure 10.34. Changing longitude of the GRS in recent decades when it was a dark red oval, showing 90-day oscillation.
(a) 1963–1968; 90-day cycles are numbered. (NMSUO data; from Solberg, *1969*, with kind permission from Pergamon Press Ltd.).
(b) 1973–1977; 90-day cycles are marked at top (BAA data; from BAA Memoir). (Similar results were obtained by the SAF.)

wise loop around the GRS before resuming its original course. It attained a speed of $-8°/day$ along the GRS S. edge, and completed the circuit in 9 days. Finally, in 1966/67, they observed two dark spots on opposite sides of the GRS, which orbited around it with a period of 12 days for at least 2 months! All these spots moved around the original perimeter of the GRS, not the reduced red oval, with speeds of 71 m/s along the N. and S. edges and 20 m/s along the p. and f. edges.

Spacecraft observations of the GRS

The Pioneer spacecraft viewed the planet with the SEB whitened and the 'red spot' aspect of the GRS at its most extreme. The Pioneer 11 closeup (Fig. 12.2) showed the GRS to have a mottled though otherwise uniform interior, with only hints of flow lines.

The Voyagers saw a fairly typical 'arch' aspect of the GRS,

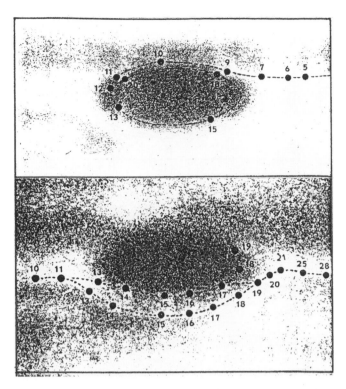

Figure 10.35. Circulation of spots around the GRS. (Upper:) Spot A from STBn; dates are for 1966 Jan. (Lower:) Spot B from SEBs; dates are for 1966 Feb. (From Reese and Smith, *1968*.)

though it contained an oblong orange patch, and the SEBs activity and the incipient S. Tropical Dislocation produced an encircling dark ring during the Voyager 1 encounter. The images confirmed the circulation of the GRS. This was particularly striking in the Voyager 1 movies (Fig. 10.36&37), which showed SEBs jetstream spots being caught up in the GRS circulation. When the circulation carried these spots to the f. or p. end of the GRS, they tended to hesitate, being visibly pulled in two directions by the conflicting currents, and sometimes splitting. During the Voyager 2 encounter, due to a new STropD, SEBs spots were not reaching the GRS, but the circulation continued unchanged (Fig. 10.38).

The internal circulation was divided into an inner oval, with puffy clouds that showed only lazy slopping motions, and an outer zone with rapid circulation.[9] In the inner zone, the puffy clouds included possible groups of convective cumulus cells. The wind speeds were highest just inside the outer edge, at 110–120 m/s; the circulation time was 6–8 days, and the vorticity peaked at about 6×10^{-5} s^{-1} (0.2 km/hr/km) across the north-south axis – about four times as great as the shear across the STropZ at other longitudes. The spiral lines are streak-lines, not stream-lines. No net inflow or outflow could be detected, implying that material would take months or years to move in or out of the GRS.

In the Voyager colour pictures, the large central flocculent oval was orange, and orange colour also extended over the highly streaked flow region in the southern half of the Spot (Plate P17). J. Mitchell and colleagues (*1981*), noting that the calm centre had no

[9] Details are from Mitchell *et al.* (*1979, 1981*). See Figs. 10.37&38, Fig. 3.6, and Plate P17.4. For measurements of windspeed, see Fig. 14.1.

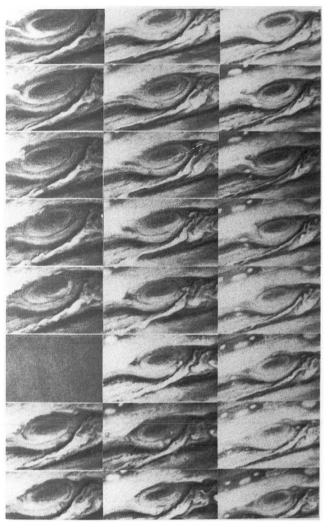

Figure 10.36. Circulation of the GRS viewed by Voyager 1. This is a series of blue-light images on every other jovian rotation, starting at lower right (1979 Jan. 15), reading upwards, and ending at upper left (Feb. 3). (NASA images, from Smith et al.(*1979a*) and Morrison and Samz (*1980*).)

equivalent in the STB white ovals, suggested that this is where convection brings up the orange material. In contrast the northern part contained large whitish areas, streaked by the vigorous circulating winds.

The movies from Voyager 1 and Voyager 2 both showed that there was a single whitish area orbiting around inside the Spot, which performed four complete circuits within the period of each movie. In fact it was followed for 10 circuits during the 60 days of the Voyager 1 approach (Figs. 10.36&39A). It was always reduced to a long streak when in the southern half, but broadened out and brightened irregularly each time it reached the northern half, distorting the regular flow pattern. On at least one circuit in each movie, when it was passing through the Sp. quadrant, it captured more white material from an external jetstream spot that happened to be circulating around the southern edge of the GRS. One such capture gave rise to the long spiral filament connecting it to white STropZ material that was still visible in Plate P17.4 and Fig. 10.39B.

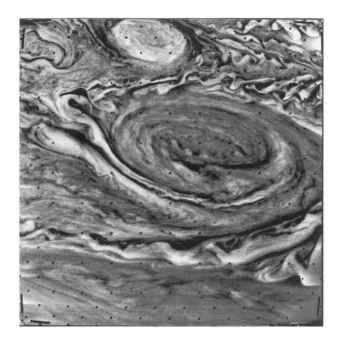

Figure 10.37. Voyager 1 view of GRS and vicinity on day of encounter, 1979 March 5, 04h (including rifted region f. it and STB Fade). (Also see views 2–3 rotations previously in Fig. 10.2, Fig. 12.4, and Plate P17.4. Image 16379.40, violet, filtered, wide-angle. NASA.)

This behaviour suggested that the whiter part of the GRS was being maintained, against the reddish material, by repeated influx of white clouds from the SEBs jetstream spots (as suggested by R. Beebe, *1990*). Likewise, the breakup of the Red Spot during SEB Revivals could simply be due to SEBs spot material displacing the red material. But this theory does not explain how the white clouds (and not the dark clouds) of SEBs jetstream spots could persist for so long after being dragged into a circulation very different from the one that produced them. Also, a different explanation is needed for periods when the GRS is pale and yet the SEBs is quiet.

It is interesting that the boundary between red and white portions during the Voyager encounter was approximately aligned with the SEBs, as is also frequently the case in visual observations. Could it be that, at altitudes above the visible clouds, the retrograding jetstream blows straight across the GRS, thus restricting red material to its southern half? No such behaviour is predicted by the thermal wind equation, but it would accord with the behaviour of the renascent SEB(S) during some Revivals. Also, the famous colour-enhanced closeup of the GRS from Voyager 1 (Plate P17.4 and Fig. 10.39B) actually shows cloud streaks of different colours crossing each other, presumably at different altitudes; one pink or whitish streak, crossing darker streaks that are presumably at lower altitude, is again aligned with the SEBs edge. However, this alignment may be accidental.

Under the GRS?

The physical aspects of the GRS will be discussed in Chapter 14, after we have considered observations of other circulations that may be analogous: the STropD (section 10.7) and the STB white ovals (Chapter 11.3). But an ingenious analysis of the Voyager data deserves mention here.

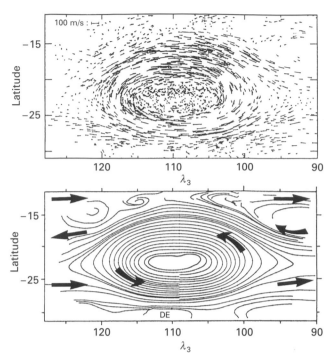

Figure 10.38. Velocity field in the GRS, from Voyager 2 images. North is up, in contrast to other figures. (Top:) Measured velocity vectors; position is marked with a dot, and the line points downwind, with scale shown at top left. (Bottom:) Deduced wind trajectories. Small dots mark intervals of 10 hours. (From Dowling and Ingersoll, *1988*.)

The Voyager 1 velocity field was essentially the same (Mitchell *et al.,1981*). The white oval on the S. edge was BC for Voyager 1 but DE for Voyager 2.

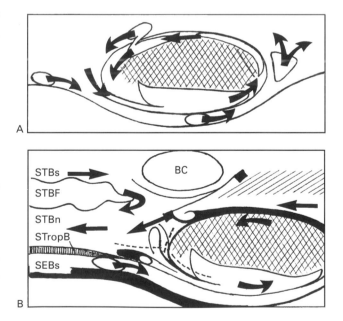

Figure 10.39. Diagrams showing features moving around and within GRS from Voyager 1 images. (A) Circulation of jetstream spots outside the GRS, and of the white patch inside the GRS, including transfer of white material into the GRS. As in Fig. 10.36. (B) Spots and streaks in Plate P17.4, the false-colour closeup. This shows a quadrangle bounded by the GRS, SEBs, STB Fade (STBF), and oval BC, within which a S. Tropical Band (STropB) was developing. Dashed lines indicate streaks of redder material which cross other streaks. This image was taken on 1979 March 4d 7h, 9 hrs after Fig. 10.2, coincident with Fig. 12.4, and 21 hrs before Fig. 10.37.

If one *assumes* that the GRS flow is shallow and isolated, as in most of the modern models (Chapter 14.2), the general shape of the atmosphere below the GRS can be inferred. In the flow of a river, variations in speed reveal variations in depth; and in a similar way, one can calculate the topography under the GRS, as was done by T. Dowling and A. Ingersoll (*1988, 1989*) (Fig. 10.40). What they tracked was the variation in vorticity (Chapter 14.2) around the GRS. As angular momentum is conserved, when the flow passes over a shallower region, its area must increase and its vorticity must decrease. Thus they deduced that the bottom of the GRS is flat and shallow to the north, but slopes down deeper towards the south. Adding a similar analysis for the STB oval BC, which was adjacent, gave a continuous profile from mid-SEB to the SSTB (Fig. 10.40).

In fact, the deduced thickness is an abstraction becase the winds in the real atmosphere will not have an abrupt lower boundary. So a lesser 'thickness' for part of the GRS really means that the winds extend less deeply or, in other words, that the underlying motions at a given depth are slower. Thus the map of 'thickness' is also a map of the zonal speed under the GRS against latitude. It shows that the SEBs jetstream does not extend as deep as the STBn, STBs, and SSTBn jetstreams – at least below the GRS and white oval. As the circulation also diminishes with altitude above the GRS (Chapter 4.5), this model implies that the wind speeds are strongest at the level of the visible clouds.

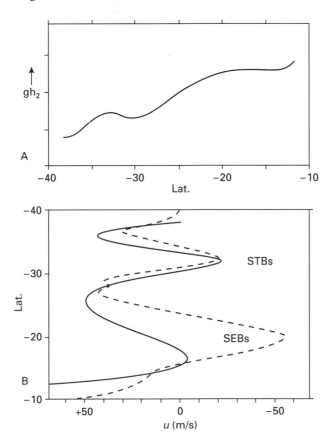

Figure 10.40. Profile of 'depth' of the GRS against latitude, according to one of the models of T. Dowling and A. Ingersoll. (A) A parameter representing topography, indicating that the flow extends deeper further south. (B) East-west wind speeds, at surface (dashed line; Limaye, *1986*) and deeper (according to the model). (Adapted from Dowling & Ingersoll, *1989*.)

10.6 THE SOUTH TROPICAL ZONE

The major phenomena of the STropZ are described in the following sections of this chapter: Disturbances and Dislocations (section 10.7), Bands and other rapidly-moving shadings (section 10.8), and STBn jetstream spots (section 10.9). Here, we note the general appearance of the STropZ and some of its minor phenomena.

The colour has almost always been white. There have only been a few episodes of general ochre coloration, always in association with reddish SEB coloration. These were following the SEB Revivals of 1928/29, 1938 (STropD only), 1952/53, and 1958/59 (ALPO); and also, less prominently, in 1906/7 (yellow), 1914 (pinkish within the STropD), 1933, and 1948. Orange colour also occurred along the S. edge of the zone in 1927–29 and 1970, in association with STBn jetstream activity. Grey shading is more common during SEB Revivals.

At times when the SEBs is quiet and there is no STropD, the STropZ is often featureless. But slow-moving spots sometimes develop. Some seem to be small SEBs spots that stray from the adjacent jetstreams (section 10.9). Others are large white spots or bays or shadings. White spots indenting the SEBs are presumably anticyclonic, and can be quite long-lived. One existed from 1983 to 1985, and another from 1987 to 1992, both remaining almost stationary in λ_2 (Fig. 10.5). The latter spot changed its aspect in parallel with the GRS during the SEB fading and revival, according to high-resolution photographs (Fig. 10.41). It developed a red lozenge in its southern part after the SEB faded in 1989/90, along with dusky bluish shading around its f. rim; the aspect thus resembled the GRS in miniature, and perhaps the SEB fading caused this development as it had also caused an intensification of the GRS. After the SEB Revival in 1990, this spot (or a similar one) reappeared as an oval ring at 22.2°S, which became a white oval enclosed by an advancing S.Trop.Band. By 1991/92 it reverted to its original aspect as a bright white spot on SEBs.

This was only the second 'little red spot' to be recorded in the STropZ; the first was seen in 1986 as part of a S. Tropical Dislocation (section 10.7). Both were smaller than their North Tropical equivalents, and their colour was only detectable photographically.

10.7 SOUTH TROPICAL DISTURBANCES AND DISLOCATIONS[10]

Some of the shadings that appear in the STropZ have a distinctive structure and motion that sets them apart as a special type of phenomenon. These are the South Tropical Disturbances. One that lasted from 1901 to 1939 was famous as *the* great STropD, but at least six others have occurred over the last 140 years. They share the following characteristics.

(a) They start small (<20° long) and expand longitudinally.

(b) They often begin near, or can be traced back to, the p. end of the GRS, though sometimes they arise elsewhere.

[10] Adapted from Rogers (*1980*).

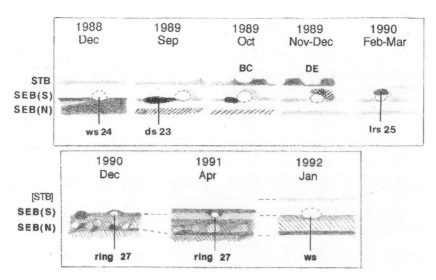

Figure 10.41. The STropZ white oval, 1987–1992. Feature numbers are from BAA reports. ////, blue shading; \\\\, red shading. It was a white bay until 1989 Nov, then became a Little Red Spot (no.25) after the SEB faded, and a ring (no.27) as the SEB revived, in parallel with the GRS – finally reverting to the original white bay aspect.

(c) The Disturbance is initially dark, and the STB and SEB(S) are drawn in towards it. The STB may be fainter and the SEB disturbed, and one or both belts may actually be dislocated to form a 'S. Tropical Streak' (Fig.10.42).

(d) The p. and f. ends are concave, usually with a white spot adjacent to each. The p. end is often the better-formed.

(e) The drift ($\Delta\lambda_2$) is initially between $-8°$ and $-16°$/mth, similar to the South Temperate Current.

(f) When the SEB(S) fades in preparation for a Revival, the STropD also fades.

(g) The Disturbance lasts for at least 11 months.

Apparently the most fundamental element, but the most difficult to observe, is the 'Circulating Current' – the reconnection of the SEBs and STBn jetstreams at the p. end of the Disturbance, so that small spots retrograding on the SEBs are bounced onto the STBn and run back the way they came. This behaviour has only been observed for the great STropD of 1901–1939, and for the closely-studied Disturbance of 1979–1981. However, the concave p. edge of a typical Disturbance and the white spot there (presumably anticyclonic) suggest that the Circulating Current is an integral part of all STropDs. One should perhaps define a STropD as 'a structure which reconnects the Circulating Current', but it is rare for this criterion to be fulfilled, as suitable jetstream spots are often unavailable or, in less than favourable conditions, unobservable. One should also note that recirculation has only been observed at the p. edge of a STropD, from SEBs to STBn; the completion of the circuit at the other end of a Disturbance was never proven for the great STropD, and was definitely not occurring at the Disturbance observed by Voyager.

In summary, a STropD appears to be the exact opposite of the GRS, in its morphology (all sides concave instead of convex), in its effect on the neighbouring belts (attractive instead of repulsive), in the circulation at its ends (away from it rather than into it), and in its behaviour prior to SEB revivals (the STropD fades, the GRS intensifies). These features suggest that whereas the GRS has the charac-

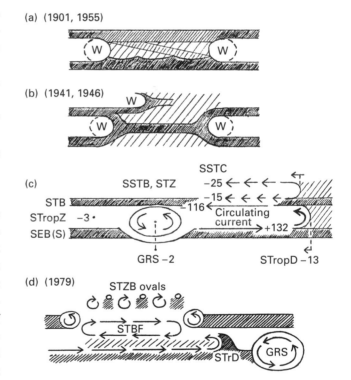

Figure 10.42. Aspects of S. Tropical Disturbances. (a,b) Classical appearances of STropDs. 'W' marks white spots. (c) Average drifts ($\Delta\lambda_2$) as given by Peek. (d) The situation viewed by Voyager 2. (From Rogers, *1980 & 1986*.)

ter of a 'super-zone' and is higher than the surrounding zone, a STropD is more belt-like in character. This may be described as a partial dislocation in the belt-and-zone pattern; in the most extreme cases the dislocation becomes complete as a belt segment replaces the zone, and neighbouring belts fade or are drawn into it.

Seven long-lived disturbances are on record which, although they have many differences, seem to have enough in common to qualify as classical STropDs. In addition, there have been several

Table 10.7. *South Tropical Disturbance(s) of 1857–1862 (Der Alter Schleier)*

	Mid. RSH	p. end Schleier	f. end Schleier	Length of Schleier
1857 Nov 28	240	14	66	52°
1860 Feb 3	181	202?	(var.)	≈40°–140°
[1861 Mar 13	134	332?	≈34?	≈62°?]
1862 May 15	90	192	245	53°

λ_2 is given, from Kritzinger (*1914*) plus the author's measurements (see text). The date is that to which longitudes are referred, not necessarily opposition. The 1861 entry is doubtful.

similar phenomena (particularly in association with S. Tropical Dislocations) that have some of the same properties and may have been incomplete or shorter-lived versions. All these phenomena are listed in Table 10.8 and described individually below.

South Tropical Disturbances in the nineteenth century

1857–1862? (Table 10.7): Drawings showed a large and unmistakable STropD in 1857 and again in 1862, and possibly related structures in between. These may have been a single Disturbance, or they may have belonged to a series of shorter-lived events as in the 1980s. No overall description of these phenomena has yet been given. H. Kritzinger (*1912, 1914*) linked up observations from 1860 to 1869 (and even, very speculatively, earlier and later, though he did not cite 1857–1859), proposing a single STropD which he called the Alter Schleier ('old veil'). However, the evidence was not persuasive except for 1857–1862. The data cited below for 1859–1862 are mostly measurements from unpublished drawings, by J. Schmidt and S. Gorton (cited by Kritzinger, *1912*), and by S. Schwabe (studied by the author in the archives of the Royal Astronomical Society).

In 1857 Sep.–Dec., an impressive STropD 50° long was drawn by W.R. Dawes (Plate P1.5&6), on the opposite side of the planet from the Red Spot Hollow. Its origin and motion were uncertain, as there was some uncertainty in the times of the drawings. The edges of the Disturbance merged with the STB and SEB(S), which were both highly disturbed with many internal white spots. It seems to have disappeared in 1858/59 when SEB was faint.

After the SEB revived in 1859, Schwabe drew the STropZ and SEB as a single massive dark belt at almost all longitudes. His drawing in Plate P1.8, which seems to show the p. end of a STropD, actually shows the f. end of the Red Spot Hollow. As this dark material cleared away in early 1860, some remained f. the Red Spot Hollow, but the longitudes of the p. or f. ends drawn by Schwabe (and also recorded by Gorton and by Schmidt) did not form a consistent pattern. This may have been rapidly-changing STropZ shading in the aftermath of the SEB Revival. But for ≈140° f. the Red Spot Hollow, the zone was consistently dark and this may have been the revived STropD (Plate P1.9).

In 1861, the only possible records of a STropD were doubtful ones by Gorton. As he used only a 3½-inch (9–cm) reflector through the window of his house, his work must be of little value!

In 1862, there was again a definite STropD. Most of the records were by Schmidt, including transits of both ends, but the STropD was also drawn several times by Schwabe (e.g. Plate P2.5&6). The position of the p. end was initially variable, but both ends stabilised to give $\Delta\lambda_2 \approx -20\ (\pm7)$ from 1862 April-June. According to Kritzinger:

'The similarity between the 'Schleier' [STropD] frequently observed since 1903 and that noted by Schmidt in 1862 is extraordinarily great. Not merely in the whole character, but also in the uniqueness. Thus the pair of bright spots, which to some extent mark the beginning and end of the

Table 10.8. *South Tropical Disturbances*

Date first seen	Drift (Δλ2): p.end	f.end	Distance p.GRS	Max. length	Date last seen: definite	possible
(1857 Sep)	?	?	≤140°	≥50°	(1857 Dec)	(1862 Jun)
1889 Jun 1*	−13	−13	200°†	≈10°	1890 Jul	1890 Aug
1901 Feb 28	(See Fig.10.6)		80°	230°	(1939 Dec)	
1941 Oct 22	−10.1	−11.3	0	33°		
[1942/43:	−10.0	−14.2			1942 Oct	
(1946 Jan)	−14.5	−10.2	≤60°	≤40°		
[1947:	−8.9	−13.7			(1947 Jul)	
(1955 Sep)	−8.2	−6.7	≤40°	42°		
[1956/57:	−14.3	−16.5			(1957 Jul)	
?(1967 Jan 31)	−8.6	(−4.1)	120°	41°	(1967 May)	1968 Feb
1970 Jul 19	−12.3	(−7.3)	250°†	30°†	1971 Jun	
[1971:	−25.0	(−21.4				
	to −5.6	to −11.6)				
?1975 Dec 20	−18		10°	60°	(1976 Mar)	
(First S.Trop.Disl'n.)						
1979 Jan 8	−12.8	(+1.2)	0			
[1979/80:	−12.6	−2.6		150°		
[1980/81:	−4.1	−2.8			1981 Jun	(1982 Jan)
(Second S.Trop.Disl'n. The dynamical STropD was a short complex at the f. end, which arose in 1979 May.)						
1984 Feb–Apr	−5.6	(0.0)	0	40°	1984 Sep	1985 Aug
		to −2.6				
(STropD at f. end of third S.Trop.Disl'n.)						
1985 Sep*		0				
[1986/87:	−11.4	−11.4		5°	(1987 Jan)	
(LRS at f. end of fourth S.Trop.Disl'n.)						

Adapted from Rogers (*1980*), with the addition of recent S. Tropical Dislocations. The table lists all known STropDs plus (in italics) some atypical, shorter-lived phenomena.

Columns 1, 6: Date is given in brackets if limited by solar conjunction.

Columns 2, 3: $\Delta\lambda_2$ for each end. Where more than one value is given, they are the drifts reported in successive apparitions (not corrected for phase effect). Drifts are from ALPO for 1955–57, 1967, and f. end in 1970; from NMSUO for other parts in 1970–71; from BAA for others.

Column 4: Initial distance from f. end of Disturbance to p. end of GRS.

* The 1889 and 1986 features were little dark ovals.

† The 1889 and 1970 STropDs probably arose from STBn spots which originated closer to GRS. In 1970, secondary f. ends extended the length to 100° according to Reese (NMSUO report).

Schleier, are very clearly recognisable and make possible a fairly accurate measurement of the drawings. Also the bright spots S. of Belt V [STB] are not wanting. The p. end of the Schleier is bluntly pointed and shows (on a copy) two thickenings, while the f. end is fluted.' (Kritzinger, *1912*; author's translation.)

No STropD was recorded in 1863.

It is not clear whether these were all a single STropD; the data are not adequate to establish accurate drift rates. In 1857, 1859, 1861, and 1862, the observations suggest drifts in the range $\Delta\lambda_2 \approx -15$ to -20. Motion of the ends at these speeds could connect up a single f. end in all these years except 1861. The motion of the p. end was even less secure.

1889–90 (Figs. 10.43 and 11.4): In spring 1888, a black spot emerged ($\Delta\lambda_2$ = –24) from the Sp. edge of the GRS, drawing a narrow S. Tropical Band behind it (A.S. Williams). The adjacent STB was faint and narrow so this resembled a S. Tropical Dislocation. Quite how the subsequent features developed is unclear, but Table 10.8 is compiled on the assumption that the SEBs projection accompanying a curious 'eye' in the STB in 1889 (Sells & Cooke, *1913*) was the same as the striking large dark STropZ spot accompanied by a small dark spot on the STB the following year, which Williams stated to have arisen in 1889 June. There are no other candidates on the 1889 maps of Sells and Cooke; however the implied motion during solar conjunction is only $\Delta\lambda_2 = -7$, slower than the more typical South Temperate Current drifts of –13 shown by both the 1889 'eye' and the 1890 dark spot. Remarkably, either of the latter drifts would extrapolate back to the GRS in spring 1888 when the S.Trop.Band was emerging (Fig. 11.4).

This large dark oval was a striking object in 1890. But it was not a typical STropD, and until 1985–86 it was unique in having the well-defined companion structures on the STB. It remained unexplained until 1985–86, when the same events were repeated as part of a S. Tropical Dislocation.

The dark spot encountered the f. end of the GRS about 1890 July 24, and apparently diffused out. However, on August 28 some complex faint shading was drawn p. the GRS (Fig. 10.43c), so it is possible that the Disturbance jumped across the length of the Red Spot. Meanwhile the small dark spot on the STB which had travelled along with the Disturbance, flanked by a white spot on each side, became more intense as it passed the GRS: within a year it was so intensely dark, and red, as to be the most striking feature on the planet, and it persisted until 1892 November (Chapter 11.1).

The great S. Tropical Disturbance of 1901–1939 [11]

(i) First appearance.

In 1901, one of several stationary SEBs projections developed into a very dark spot some 20° long and spanning the STropZ, about 90° p. the GRS. When first observed it was described as a black spot in the STropZ moving with the S. Temperate Current, very like the spot of 1889–90 (Fig. 10.44). It was connected to a dark streak on the STB by a grey wisp, and was preceded and followed by white spots. By the 1902 apparition it had become a long Disturbance, and in time it became the largest and longest-lived STropD ever recorded.

As the Disturbance expanded, its interior broke up into irregular clouds. In 1902 and 1904 it appeared striated, described by H. Thomson as like 'a rainstorm in the distance', whereas in 1903 the internal shading was diffuse with occasional lighter patches which moved with the Disturbance as a whole (Fig. 10.44). The bright spots at the p. and f. ends were constant features in these early years.

As with all subsequent STropDs, the STB was drawn slightly northward into the Disturbance, and was also slightly fainter in this region. Despite this it always remained intact. The SEBs was drawn southwards and disturbed within the Disturbance. Sometimes there were also large light spots in the SEB alongside it. Meanwhile, outside it, there were still almost-stationary projections on the SEBs. In 1904/5, according to Molesworth, they tended to be attracted to both ends of the Disturbance, decelerating as it approached them and following it as it receded.

---

[11] The early history was described not only in the BAA Memoirs but also, in more detail, by Denning (*1902, 1904*). The Disturbance was reviewed in detail in the BAA Memoir for 1932/33, and by Peek (*1939, 1958*). Also see Plates P4.9–P6.2, and P7.5-9.

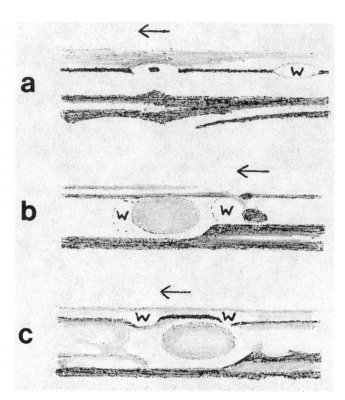

Figure 10.43. The STropD of 1889–90. *(a)* SEBs projection and STB 'eye' in 1889 June (after Sells and Cooke, *1913*). *(b)* STropD in 1890 May, approaching the GRS (after A.S. Williams). *(c)* Region of GRS on 1890 August 28, after its encounter with the STropD. Note wisps p. the GRS, and the dark bar south of it which later became extremely prominent. (After J.E. Keeler.) 'W' marks white spots. (From Rogers, *1980*.)

The great STropD: (ii) Growth and drift rate

The Disturbance started with a speed matching that of the S. *Temperate* Current, $\Delta\lambda_2 = -15$. Over the first five apparitions, the f. end decelerated somewhat (Fig. 10.6), so the Disturbance gradually but unevenly grew longer. It continued to grow throughout its life, as shown in Fig. 10.45. It gradually decelerated (Fig. 10.6). To reveal these long-term trends, it was necessary to exclude the times when the ends of the Disturbance were passing the GRS, when there were dramatic changes in the motion.

The great STropD: (iii) Conjunctions with GRS.

The initial prograding motion of the Disturbance brought it into conjunction with the GRS, and the remarkable outcome was that the Disturbance passed the Red Spot very rapidly and with only minimal interaction (Fig. 10.44). This first conjunction was described as follows in the BAA Memoir for 1902:

> The p. portion of the dark material overtook the GRS in June, and careful observations showed that, instead of passing over it, the dusky stream passed southwards round the Spot via the STB. On reaching the p. side of the Red Spot it struck across the STropZ, thus uniting the S. Temperate and S. Equatorial Belts. During the time that the long stream of dark material was sweeping past the Red Spot this region of the planet presented a remarkable appearance. The S. Temperate

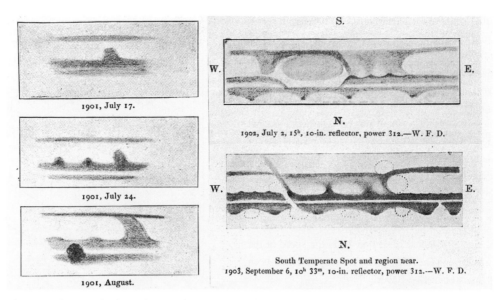

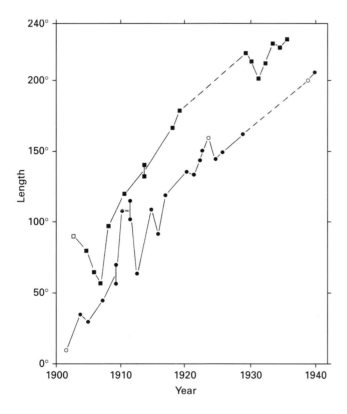

Figure 10.44. Origin of the great STropD. Each panel covers STB (top) to SEB (bottom). *(Left:)* 1901 (by Dr Kibbler). *(Right:)* 1902, 1903; its expansion and first passage past the GRS (by W.F. Denning). (From Denning, *1902, 1904.*)

Figure 10.45. Changing length of the great STropD. The lower line (circles) shows the length when the GRS was not included. The upper line (squares) shows the length when the GRS was within the Disturbance. Open symbols are imprecise. Growth was at ≈6.6°/yr (1901–1916) then 3.9°/yr (1917–1939).

and S. Equatorial Belts seemed merged together by the flooding of the intermediate STropZ with dark matter, in which the 'hollow' and place of the GRS stood out by contrast in small apertures as a bright oval spot.

The passage of the p. end was so rapid that the length of the GRS was added to the previous length of the Disturbance, making a total

of 90°. Three months later the f. end of the Disturbance also passed the GRS, and the length of the Disturbance was reduced again. Meanwhile the GRS, which had been stationary in 1901, developed a $\Delta\lambda_2$ of –0.5°/month in spring, 1902, and –2.0°/month during the three months of the conjunction, reverting to –0.5°/month in the autumn and even slower thereafter (Fig. 10.33).

The same process was repeated many times over the subsequent decades. In all there were 8 complete passages and one incomplete one (Fig. 10.32). Each time, the Disturbance avoided encroaching on the oval outline of the GRS, but each end in turn flowed very rapidly past it and re-formed p. the Red Spot. This time for passage seems to have been variable, being only a few days in some cases, but as long as six weeks for the sixth conjunction of the f. end of the Disturbance in 1914. The regathering of the p. end after the passage was often ill-defined for several weeks, so that the duration of passage could only be worked out in retrospect by extrapolation.

It was pointed out in 1914 that these rapid passages could be explained if the GRS was a vortex (Chapter 14.1), and with the discovery of the circulation of the GRS, it now seems natural that any coherent feature approaching it in this manner would be channeled round its south edge via the STBn jetstream. However, it is not so obvious that this would apply to the p. end of the STropD if it was in essence the end of the Circulating Current, in which the GRS was actually enclosed (Fig. 10.42c). If the STropD was indeed determined by this circulation, one can hypothesise that the white spots at the p. and f. ends of the Disturbance were themselves anticyclonic ovals, which could squeeze past the GRS without disruption, and act as nuclei for the re-formation of the Circulating Current on reaching the p. side of the Red Spot. But this would not explain the diffuse nature of the p. end after each passage. Alternatively, if the primary determinant of the STropD is the visible dark material, the circulation may really break down at each conjunction, but be re-induced when the dark material has streamed around to the p. side of the GRS.

During the earlier years of the Disturbance, both ends showed a clear tendency to accelerate while approaching the GRS and to move more slowly than usual while moving away again. This was well exemplified in 1912 and 1913.

During these years, overall, the Disturbance gradually decelerated while the GRS gradually accelerated, bringing their speeds closer together. The deceleration of the STropD may have been intrinsic, as will be discussed in Chapter 14, but there seems to be no explanation for the exceptionally rapid rotation period of the GRS between 1909 and 1935 other than the effect of the Disturbance upon it. The main effect was that the GRS moved faster while it lay within the STropD, the average difference being 1.5°/mth (Fig. 10.33). So the acceleration of the GRS was probably not due to friction, but due to its moving into a region where (as noted in section 10.1) the speed of the SEBs jetstream was lower. (The STBn jetstream may also have been slower within the STropD; section 10.9.)

Then in 1923, when the p. end of the Disturbance reached the GRS for the eighth time, the GRS began to move even faster than the Disturbance and the conjunction was aborted; by 1925 a gap of 40° longitude had opened up between them. Thereafter, the GRS decelerated to a more normal speed and the STropD did overtake it completely in the ninth conjunction that lasted from 1928 to 1938.

The great STropD: (iv) Later evolution
In 1911, the f. two-thirds of the Disturbance faded, leaving the p. section alone as a striking dark Disturbance 35° long, probably with a small white spot at its f. end as well as the big one at the p. end. This p. section remained fairly distinct for several years. Then in 1917/18, during a conjunction with the GRS, a white oval f. the GRS expanded to isolate the f. section as well. So in 1918–19 the f. section also temporarily took on the appearance of a separate Disturbance, 34° long with a white spot at each end.

This was just before the whole Disturbance faded away, 'as if veiled by mist', in the buildup to the SEB Revival. Thereafter, the body of the great STropD was generally faint, although the SEBs was still disturbed within it.

There are few published notes on the colour of the Disturbance; presumably it was not distinctive. Some notes indicate a cold grey colour in contrast to the reddish SEB, while others report the same 'warm' colour as the SEB. The whole Disturbance became ruddy following the SEB Revivals of 1928 and 1938; it was also yellowish-brown in 1933. In general, it is likely that the STropD derived both its dark material and its occasional colour from the SEB.

The great STropD: (v) The Circulating Current
There were only three SEB Revivals during the existence of the great STropD, all in the second half of its life. On all three occasions the STropD disappeared completely before the SEB Revival, and reappeared during it, along with the SEB(S). Only two of the Revivals were well observed. Each one began within the longitudes of the STropD.

The Circulating Current was the most remarkable aspect of the STropD to observers of the time, and it were described in detail in Peek's book. It was first observed during the great SEB Revival in

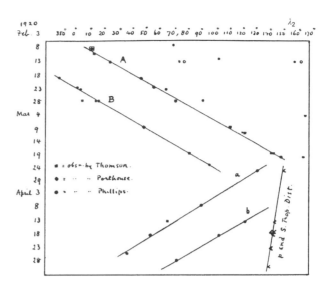

Figure 10.46. Discovery of the Circulating Current: chart of dark spots in 1920, on SEBs (A, B) and then on STBn (a, b). (Also see Plate P5.8&9.) (From the BAA Memoir.)

1919/20. Two large dark spots retrograded along the SEBs to the p. end of the Disturbance; then they were replaced by two spots prograding on the STBn, smaller than the previous SEBs spots but still dark and distinct (Fig. 10.46 and Plate P5.8&9). The idea that these were the same two spots, bounced off the STropD with a change in speed of 195°/month, must have seemed so ridiculous that it was not raised explicitly in the BAA report, but it was clearly implied by the published drift chart (Fig. 10.46), and was committed to print by Peek in 1926.

During the 1928 SEB Revival, retrograding SEB(S) spots passed out of the f. end of the Disturbance. The Circulating Current was not observed, but the p. end of the STropD (when it was eventually recovered) was only a few degrees f. the GRS. Small, rapidly changing spots were certainly present in this space, including some on the STBn, but the crowding of currents there was such that any circulating motions could not be disentangled.

When retrograding SEBs spots reappeared in 1931, observers were keenly alert for a renewal of the Circulating Current, but on reaching the p. end of the STropD the first train of spots simply disappeared. Then in 1932, as a further train of SEBs spots ran up to it, they were indeed reflected into the STBn jetstream. They appeared as short grey streaks on a tenuous band just N. of the STB, quite unlike the former SEBs humps in appearance, but the times of their appearance matched those of the arrival of the SEBs humps at the STropD, given a lag of about two days. This presumably represented the time it took for each spot to swing around the concave p. end of the Disturbance onto the STBn jetstream; in this brief interval their speed changed from +144 to -134°/month (a change of 124 m/s in speed and 6^m 21^s in rotation period!). Many more spots were observed performing this circuit from 1932 to 1934 (Figs. 10.47&48; Plate P6.5&6). It was very difficult to catch a spot actually in transit from SEBs to STBn, and the drawings suggested that they became stretched into very narrow streaks during this time, recovering somewhat on the STBn.

During the activity of 1932–34, the STropD was so long that the

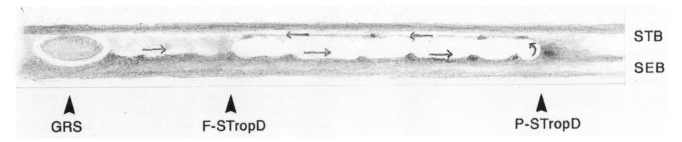

STB

SEB

▲   ▲   ▲

GRS   F-STropD   P-STropD

Figure 10.47. The Circulating Current in 1931/32. This diagram by T.E.R. Phillips shows it as a gigantic oval. (From the BAA Memoir.)

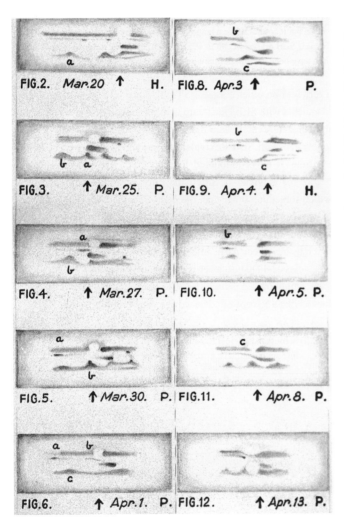

Figure 10.48. The Circulating Current in 1933. Drawings by Phillips (P.) and Hargreaves (H.) showing SEBs spots a, b, and c approaching the p. end of the STropD (a dark bar in STropZ, indicated by arrow) and being recirculated onto the STBn. The narrow inclined streak, formed by spot b on April 1 and c on April 8, was all that observers ever saw of such spots actually making the turn. (Also see Plate P7.5&6.) (From Peek, *1958*; reproduced by permission.)

sector of STropZ outside it, which contained the Circulating Current, was only 135° long and thus formed a giant anticyclonic oval (Fig. 10.47)! Its N. edge was at latitude 17.5°S, while the SEBs latitude elsewhere was 20.9°S (Table 10.3); its S. edge was formed by a narrow S. Tropical Band at 26.3°S, along which the circulating jetstream spots travelled, and which curved out of the p. end of the Disturbance in a structure nicknamed the 'smokestack'.

Whether the prograding STBn spots returned to the SEBs at the f. end of the STropD was never determined. This was not for want of trying, but most of the STBn spots were small streaks which faded away long before they reached the f. end. A few STBn spots could have recirculated in this manner, but only once was a STBn spot actually seen curving into the f. end of the STropD. The retrograding spots on the SEBs outside the STropD also might have arisen from the less rapidly retrograding spots that were arising *inside* the Disturbance, and which ran up to its f. end, but these tracks did not consistently connect up with those outside the Disturbance either.

In 1938, the SEB Revival occurred during solar conjunction, but when the planet reappeared, dusky spots were observed on the STBn jetstream p. the STropD, which might have represented a final display of the Circulating Current (Plate P7.7&8).

The great STropD: (vi) Disappearance

After many vicissitudes, the great STropD gradually faded away having reached the length of 230° in 1935. By this time the interior was little different from the remaining STropZ, and the white spots at the ends were only intermittently visible; only the two ends remained distinct, as faint arcs across the STropZ. In 1938, the p. end reappeared; the Disturbance was not clearly distinguished from the reddish shading accompanying the SEB Revival. Both ends were definitely re-identified the following year, and the last definite record of the Disturbance was by Phillips on 1939 December 27, just as the p. end reached the GRS for the tenth time.

South Tropical Disturbances from 1941 to 1971 [12]

In these years, at least four disturbances appeared which initially resembled the great STropD, but then contracted again, and disappeared after only 1–2 years. Their parameters are listed in Table 10.8. The first two were named 'S. Tropical Streaks' and were described in detail by Peek. As there were no evident triggering factors for them, it is tempting to speculate that these two STropDs, as well as various jetstream spots and oscillating spots at other longitudes (section 10.9), were relics of the breakup of the great STropD.

1941–42: The 'South Tropical Streak' of 1941 was first seen actually emerging from the p. end of the Red Spot Hollow. It finally emerged as a fully-formed STropD 30° long, with concave ends (Plates P8.1&2, P9.2). The SEB(S) was drawn up into each end of the Streak and absent alongside it. (The remainder of the SEB(S) was faint and quiet.)

[12] Reviewed and/or illustrated by Peek (*1958*), Sato (*1980*), Rogers (*1980*), and Sanchez-Lavega & Rodrigo (*1985*).

At this time the three great white ovals were forming in the STZ (Chapter 11.3), and the Streak lay within the longitudes of the future oval 'DE', which was then 94° long and separated from the Streak by the STB. However, in 1941 December the STB alongside the Streak began to fade, just as the f. end of 'DE' (the p. end of a segment of STZ belt) moved past it. Thus the region ended up completely dislocated, with the pattern:

STZB-Faded STB-Dark Streak-SEBZ

instead of the usual:

STZ-STB-STropZ-SEB(S).

This STZB was only ≈ 14° long, and the decline of the STropD began just as the STZB moved on away from it.

The Disturbance contracted until only 5° long, due to acceleration of its f. end, and then disappeared in 1942 October. Seven weeks after and a few degrees f. the disappearance, a feature was seen resembling the p. end of the old great STropD, but it was stationary, and marked the site at which the next SEB Revival began! We suggested in section 10.3 that this might have been the rim of a light oval in the STropZ; perhaps it arose as a remnant of the S. Tropical Streak?

1946–47: The new 'South Tropical Streak' was an almost exact replica of the previous one (Plate P8.5). It was a fully-formed Disturbance 12° long when first observed after solar conjunction. The drifts of its ends would converge back to a point 15–20° p. the GRS during solar conjunction, and it could have emerged from the GRS like its predecessor. Its colour was dark grey according to Focas and Reese. The SEB was again rather faint and quiet in preparation for a Revival, which occurred in late 1946.

The Streak formed within the longitudes of the future STB oval 'BC'. By 1946 April, exactly as in 1941–42, the STZB f. 'BC' was moving past it and the STB alongside it was fading, being drawn down into the Streak in complex fashion. During the next solar conjunction, the next incipient oval 'DE' was alongside the Streak as it started to shrink.

Like its predecessor, the Streak shrank in longitude until it vanished, as it was approaching the f. end of the Red Spot Hollow.

1955–57 (Fig. 10.49 and Plate P10.2): The STropD was first seen after solar conjunction, as a dusky bridge only about 5° long. It was growing at such a rate that it probably did not arise from the GRS, although it was only about 40° p. it. The bridge rapidly, although irregularly, grew into a Disturbance 40° long. After 1955 December it moved steadily without growing longer.

Like its two predecessors, the Disturbance appeared alongside the f. end of a STBs white oval (FA), but now that the ovals were fully formed, the STB was unaffected by the Disturbance.

The SEB had just completed a Revival and was generally dark and quiet. (The STropD arose at about the same longitude as that Revival.) As in the great STropD, the SEB(S) was disturbed alongside the Disturbance. A large whitish cloud in mid-SEB remained alongside the f. end of the STropD in early 1956 (ALPO). Reese's strip-maps (Fig. 10.49) suggest that this may have been SEBZ material piling up at the f. end of the Disturbance.

In 1956/57, when the SEB faded in preparation for the next outbreak, the STropD at first remained dark and well-formed and SEB(S) was drawn S. into it, but it contracted to a length of 20°, and then it too faded away. When the next SEB revival began in 1958 March, the Disturbance – which would then have been in conjunction with the GRS – did not reappear.

There was no other classical STropD until 1970, but some events associated with SEBs jetstream activity in the 1960s are worth noting here. In several years there was dusky material in the STropZ associated with SEB Revivals, but in spite of hopes expressed at the time, it never amounted to a coherent STropD. In 1962/63, before the Revival, occurred the only well-documented case in which something other than the great STropD caused a spot to circulate in the Circulating Current; this was a STBn jetstream spot just f. the GRS (section 10.9). And in 1967 there was a dark feature that may have been a true intermediate between small scale activity and a classical Disturbance. It did not show any of the interactions with neighbouring structures that characterize STropDs, but it looked like a STropD, and lasted at least five months.

1967 (ALPO, and other drawings): At a time of intense retrograding SEB(S) activity, which incidentally darkened the STropZ p. the GRS again, many white spots and dark bridges appeared in the STropZ and developed

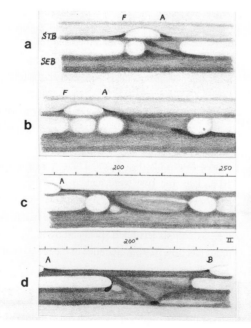
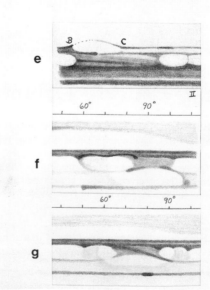

Figure 10.49. The STropD of 1955–57: strip-maps by E.J. Reese with a 15-cm reflector, covering the STB and SEB. STBs white ovals FA and BC are marked. λ₂ scale is given above maps. (a) 1955 Oct. 9; (b) 1955 Nov. 21; (c) 1956 Jan. 6; (d) 1956 Jan. 28; (e) 1956 May 4; (f) 1957 Mar. 17; (g) 1957 Apr. 11. (From the BAA archives.)

into the 24°-long 'Sectional Disturbance'. It had a white spot at the f. end only, and was connected to but did not distort the STB and SEB(S). According to the ALPO it had slightly negative drift, but this was unconfirmed.

1970–71 (NMSUO): A classical, albeit faint STropD was seen, 22° long. It was far from the GRS. However, it may have developed from a STBn spot seen a month earlier, in which case its initial motion was faster than shown in Table 10.8; and STBn spots of this type had been running p. from a South Tropical Band, which had emerged from the p. end of the GRS with $\Delta\lambda_2 = -38$ in 1969 March. Thus there may be tenuous similarities with the origin of the 1889 Disturbance. There were multiple apparent f. ends, giving a length of up to 100°. STBn and SEBs were drawn into the Disturbance as usual. Contrary to previous experience, the Disturbance arose after the SEB(S) had already faded to near-invisibility in preparation for the 1971 revival.

The p. end reached the GRS around 1970 Dec. 23; passage was complete by 1971 Feb. 14, implying a jump as with the great STropD (Reese, *1972a*). This interaction may have been responsible for the sudden acceleration of the GRS in 1970 December to a rotation period not seen since 1932. Conversely the STropD decelerated shortly afterwards.

The main STropD soon faded as the SEB had done, and also shrank until it was once again just a small STBn projection, which was last seen by Reese on 1971 June 29. This was the weakest and shortest-lived of the classical Disturbances.

South Tropical Dislocations, 1975–1991[13]

In the 1970s, a sector of the STB turned white, and since then this 'STB Fade' has repeatedly induced structures like STropDs p. the GRS, giving rise to a dislocation of the normal pattern of alternating belts and zones.

The STB Fade developed in 1975 between two of the great white ovals (FA and BC; Chapter 11.4), and this sector has remained faint, split, or absent ever since (up to 1992). Moving with the South Temperate Current, this STB Fade passes the GRS every two years or so. On some of these passages, it has induced a darkening of the STropZ on the p. side of the GRS, usually in the form of a S. Tropical Band although additional shading may be present. The term 'S. Tropical Dislocation' has been applied to this combination of the white STB and dark STropZ, and also to the remarkable suite of structures that develop from it. The dislocation of belts often extends into the S. S. Temperate domain (Chapters 12 and 13). The event is ended as a STB white oval passes the GRS, leading to the emergence of a STropD or similar structure p. the GRS.

Five S. Tropical Dislocations occurred from 1975 to 1991, each one ending with the appearance of a structure that in some way resembled a STropD. In 1975/76, the SEBs jetstream was interrupted at the p. end of the Dislocation. In 1979–81, the SEBs jetstream continued through the Dislocation to the f. end, where a classical STropD developed. In 1982–84, the third Dislocation was similar but much more prolonged. In 1985–86, the fourth Dislocation ended in the creation not of a STropD but of its opposite, a Little Red Spot. And in 1990/91, the fifth Dislocation extended completely around the planet.

[13] These have been described by Rogers (*1980, 1983, 1986*) and in BAA reports. A similar account with photos was given by Sanchez-Lavega and Rodrigo (*1985*).

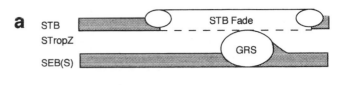

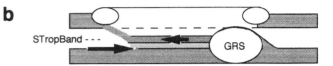

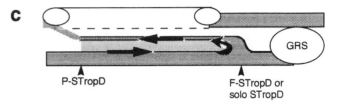

Figure 10.50. Typical progress of a S. Tropical Dislocation. (a) Before the Dislocation; (b) Dislocation begins with emergence of S. Tropical Band; (c) Dislocation ends with emergence of S. Tropical Disturbance or similar structure.

1975–76: (Fig. 10.51.) The STB from white oval FA to BC faded in 1975 September as it was passing the GRS. In October, the residual STB(N) drifted closer to the SEB(S), and the SEB(S) alongside the faded section darkened, having a distinct p. end with negative drift ($\Delta\lambda_2 = -18$). Retrograding SEB(S) spots in the SEB Revival then piled into each other and halted 28° p. the incipient Dislocation, merging to form a very dark little oval precisely on the SEBs edge, which then suddenly vanished. Then in December, the STropZ here became dark to create a Dislocation from FA to BC and the GRS (Plate P15.8). The Dislocation however differed from classical STropDs in that it was short-lived and, apparently, entirely governed by the STB. The f. end was last recorded in 1976 March before solar conjunction.

1979–82: (Figs. 10.42d,52,53. Also see Figs. 10.2 and similar Voyager pictures, and Plates P17, P18.3–6, and P21.1.) The STB Fade passed the GRS again in 1977, unobserved but with no lasting effect. It passed the GRS for the third time in 1979 January, and the second S. Tropical Dislocation began; the STropZ between ovals FA and BC was darkened by shading including a S. Tropical Band. Then, as oval BC pulled away from the GRS in 1979 May, the Dislocation formed a distinct f. end consisting of a large dark mass on SEBs, which lagged behind oval BC as it prograded more slowly. In contrast to the first Dislocation, Voyager photographs showed that the p. end had no effect on the SEBs jetstream spots. This was confirmed visually in 1979/80; SEBs spots retrograded along the length of the Dislocation until they vanished at its f. end (Plate P18.5). According to Voyager photographs (see next section), this 'f. end complex' was the site of the Circulating Current.

By late 1979, the Dislocation had turned into a fully-fledged STropD, the greatest for over 40 years (Plate P18). It showed all the classical characteristics: darkened STropZ, bright areas p. and f. it, longevity, slow motion now independent of the STB, and circulation of the Circulating Current. However, the Disturbance as defined by the Circulating Current was different from that defined by its appearance.

The visible Disturbance was some 140° long, and consisted of a narrow S. Tropical Band in a dusky, disturbed STropZ. Its p. end, having no effect on the passing SEBs spots, was difficult to locate, but was usually marked by a dusky column across the STropZ from 1979 October to 1980 April ($\Delta\lambda_2 = -12.6$). In 1979/80, there was also a small white oval interrupting the

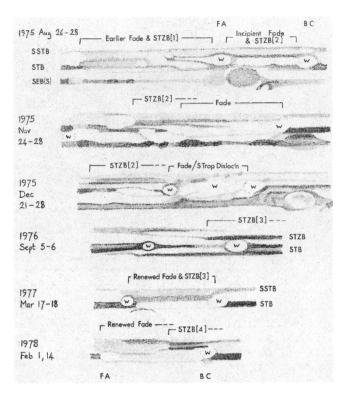

Figure 10.51. The first S. Tropical Dislocation, in 1975, with evolution of the STB Fade over the next few years. Strip-maps are from observations by the author except for 1975 Dec., which is from photographs by J. Dragesco and a map by P.J. Young. The 1975 maps incidentally show the Great Red Spot being obliterated by the SEB revival. Bright white spots are marked 'w', including STBs ovals FA and BC. 'STZB' indicates what is now known to be true SSTB. (From Rogers, *1980*)

S. Tropical Band 50° f. the p. end, which moved at similar speed ($\Delta\lambda_2 = -7$). In 1980/81 the Disturbance was only 86° long; the p. end of the shading was then close to the track of the white oval just mentioned, so it may have shifted from one small feature to another.

In contrast, the Disturbance as defined dynamically was restricted to the great dark complex derived from the f. end of the Dislocation, where the Circulating Current was observed by Voyager (see below), and retrograding spots disappeared according to visual observers. In 1980/81, SEBs activity declined rapidly, and for the first time dark and bright spots were detected by visual observers on the STBn jetstream (Fig. 10.63 and Plate P18.6). They were prograding along the broken-up S.Trop.Band p. the mass at the f. end, presumably either emitted from it or reflected from it in the Circulating Current.

In spring, 1981, the spot activity on the SEBs and STBn jetstreams virtually ceased, the S. Tropical Band faded away, and the p. and f. ends of the Disturbance collapsed. The p. end, formerly a dusky column, was replaced by a large dusky oval in mid-STropZ in 1981 May–July, which then vanished. At the f. end, the SAF reported several white spots p. and f. the f. end complex, with average $\Delta\lambda_2 = -28$ (range -21 to -34). If these identifications were correct, these could have been typical STropZ spots associated with the breakup of the dynamical STropD. (Also in early 1981, the STB Fade, which had just passed the GRS again without incident, began to overtake the STropD which it had induced two years earlier.) The f. end complex may still have been visible in spring 1982, but was not present thereafter.

1982–84: (Figs. 10.53&54.) In 1982 June–July, the STB Fade reached another conjunction with the GRS, and a third S. Tropical Dislocation began. As before, the STB(N) faded and was replaced by a S.Trop.Band stretching out between oval FA and the GRS. This time, developments were more prolonged and more complex than on the previous occasions.

By spring, 1983, ovals FA and BC were very far apart (140°), and the STB and STropZ had recovered in the p. half of this sector. In the STropZ there were only vague shadings and humps, some if not all being roughly stationary, but changing on a timescale of weeks. As oval BC crept away from the GRS, one prominent 'bridge' just Nf. BC resembled a STropD, but it did not last. By spring, 1984, oval DE too had passed the GRS, and there was still much disturbance in the STropZ in this sector, including a fragmentary S.Trop.Band and renewed SEBs jetstream activity (Plate P19.3&4). Thus the Dislocation had not yet developed a f. end; instead, the STB Fade was extending from oval BC to DE.

A STropD at last began to emerge from the p. edge of the GRS in 1984 Feb–Mar. Its relationship to the preceding events was unclear, as oval DE was already more than 40° further p., and the S.Trop.Band terminated near oval DE in a chaotically shaded region. However, the structure was identified as a STropD by its morphology (albeit variable), its motion (albeit not as fast as most Disturbances), and the fact that retrograding SEBs spots ran up to the white spot at its p. end and could not be traced further. A f. end to the STropD emerged in 1984 August, making it 30° long. (Remarkably, this was exactly alongside oval FA, which had completely circumnavigated the planet since it induced the p. end of the Dislocation two years earlier.) However, this STropD soon broke up, and it faded away in 1984 September. Meanwhile the STB recovered in all sectors outside the original FA–BC Fade, and the third S. Tropical Dislocation came to an end.

Thus the third S. Tropical Dislocation eventually followed the same pattern as the second, but more slowly; perhaps oval DE induced the formation of the STropD as oval BC had failed to do.

1985–86: (Fig. 10.55.) By summer, 1985, there was again a dark S.Trop.Band along the length of the STB Fade, starting where the STB stopped $\approx$30° f. oval FA, and extending 140° to oval BC and the GRS. The narrow STropZ between the Band and SEBs was cluttered with extensive shadings separating white ovals. In the p. half of the Dislocation, these were tracked as SEBs jetstream spots, but in the f. half, the residual STropZ was obstructed or constricted by dark shading.

After oval BC passed the GRS, the STBn also faded f. it (as in 1983), while a prominent shading developed alongside BC, occluding both the STB and STropZ latitudes (Fig. 10.55). This was the picture, suggestive of an incipient STropD, as the 1985 apparition ended.

By 1986 June and July, the only vestige of the Dislocation was a Little Red Spot, lying just Nf. oval BC and moving strictly with it! A tenuous S.Trop.Band still ran f. from it to the GRS, but this soon disappeared. The Little Red Spot was seen as a dusky 'bridge' or oval spanning the STropZ, associated with a stable complex of STB ovals and streaks, which now comprised the BC–DE sector of STB. The whole phenomenon was thus very like that of 1889 (see above). It was identified as a red oval in Jean Dragesco's photographs from the Pic du Midi Observatory showed that the 'dusky bridge' was a Little Red Spot (Plates P19.6&7, P21.5, and P22.4). Presumably it had formed in the 'dusky shading' alongside oval BC in 1985. It may have been actually pulled out of the Great Red Spot, or it may have arisen as an anticyclonic spot equivalent to one at the p. side of a STropD. So the fourth S. Tropical Dislocation created not a 'belt–like' STropD, but a 'zone–like' red oval. It was last seen in 1987 January, but disappeared before reaching the GRS again.

After 1986, almost the whole STB faded away and the STB white ovals became small and obscure. Two S.Trop.Band segments emerged p. the GRS (see below) but did not form coherent Dislocations. However, in the closing stages of the 1990 SEB Revival, a great S. Tropical Dislocation – the fifth – appeared in exactly the same way as in 1975/76.

1990–91: (Fig. 10.56.) In 1990 December, as the SEB Revival was quietening down, a massive S.Trop.Band developed p. the GRS. Its p. end began moving with $\Delta\lambda_2 = -17$ but then suddenly accelerated to $\Delta\lambda_2 = -88$. This p.

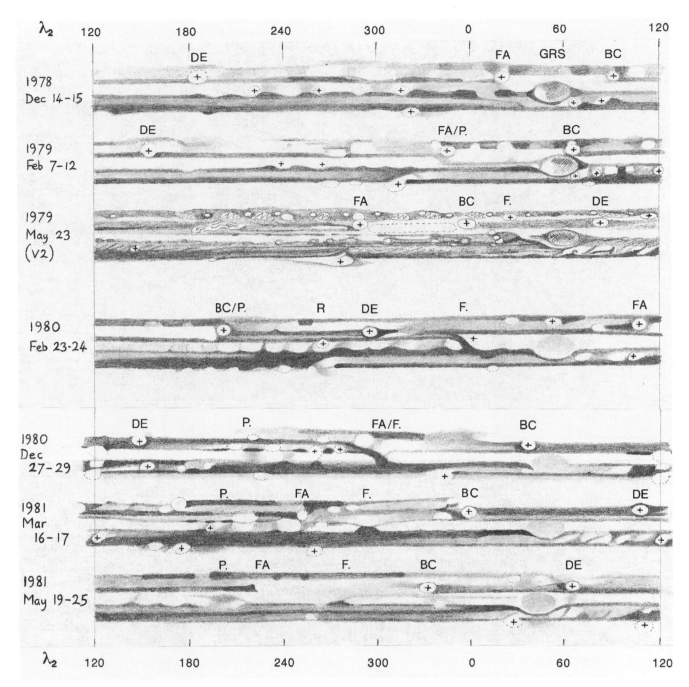

Figure 10.52. The second S. Tropical Dislocation, 1979–1981. Strip-maps are from observations by the author except 1979 May (Voyager 2), and 1981 May (photographs by I. Miyazaki and J. Dragesco, and some visual drawings). In 1981 May, the f. end complex was poorly observed but this 'festoon' aspect was drawn in April and June. The p. and f. ends of the visible STropD are marked. The p. end was variable and sometimes more distinct than shown here; the f. end was the dynamical STropD. 'R' was a ring moving with the STropD. Bright white spots are marked '+', including STBs ovals FA and BC, and the long-lived SEBn/EZs white spot. (The latter happens to be in the sector λ_2 250–350 in all maps except 1981 March, where it is at λ_2 175, and 1981 May, where it is just passing the GRS.)

end was associated with haze or 'rifts' interrupting the SEB(S). Althought temporarily held up on some stationary spots near oval FA, it eventually extended all round the planet. The f. end began to detach from the GRS alongside ovals BC and DE, and for two months there were rapidly–changing spots there moving with $\Delta\lambda_2 = -13$ to -89. But then the S.Trop.Band reformed, so by 1991 April this massive, 'cold grey' belt occupied the whole STropZ and enveloped the GRS (Plate P24.8&9). But during solar conjunction it faded away.

Bearing in mind the more localised circulations described in the NTropZ, it is possible that STropDs are only the largest of a range of circulations that can occur in the STropZ. For example, in 1990/91, while the full S. Tropical Dislocation was developing with ovals BC and DE on one side of the planet, a miniature version developed on the other side where oval FA was passing by. This feature consisted of a conspicuous white spot followed by a dark projection on the SEBs edge (no.25 in Fig. 10.56). It was stationary and short-lived and superficially unremarkable; only close

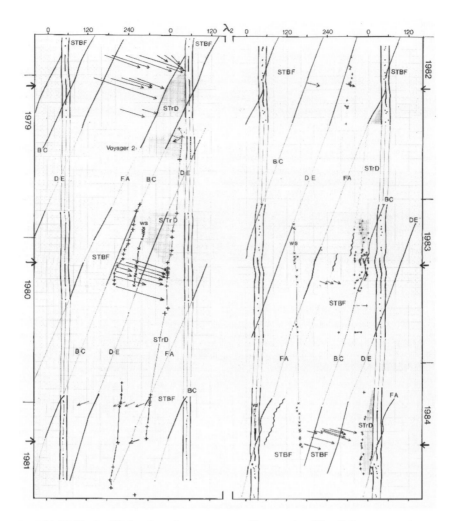

Figure 10.53. Charts of the second and third S. Tropical Dislocations, from BAA data, plus Voyager 2 in mid-1979. Only long-lived features are shown. STBF, STB faded sector. Wavy lines denote 'step-up' of STBn at p. end of STB Fade. Oblique arrows denote spots on SEBs and STBn jet-streams. Slow-moving STropZ features are shown as + (dark columns, including p. and f. ends of STropD) and x (white spots; in STropZs in 1980, in STropZn in 1983–84). Arrowheads at margins indicate dates of opposition. (From Rogers, *1986*, with revisions.)

scrutiny implied that this was another recirculation event, like those which have induced STropDs (see Chapter 13.1). Other minor spots and projections in the past may have had similar inter-esting dynamics.

Spacecraft observations: Origin of the 1979 STropD

The Voyager spacecraft observed the second S. Tropical Dislocation at close quarters, and the dynamical STropD arose from it in the early part of the Voyager 2 encounter. We can there-fore describe in detail how this Disturbance developed. Although it has been described in general terms (B.A. Smith and colleagues, *1979b*), it has not previously been analysed in detail.

First, the Voyager films showed that the STB Fade was a closed clockwise (cyclonic) circulation, in which the STBn and STBs jet-streams were reconnected at each end alongside ovals FA and BC (Fig. 10.57). The initial Dislocation (S.Trop.Band) did not inter-rupt the motion of the retrograding SEBs jetstream spots. Then, when the dynamical STropD appeared forming the 'f. end com-plex', Voyager 2 showed the anticlockwise (anticyclonic) Circulating Current reconnecting the SEBs and STBn jetstreams on its p. side. So the visible interruptions in the belt and the zone

turned out both to correspond to circulating currents, and the induction of the STropD by the STB Fade seems to have been due to a dynamical interaction between the jetstreams. However, the initial S. Tropical Dislocation (Band) did not show any distinctive dynamics, and it is still uncertain whether the dynamical changes or the visible albedo changes were primary. These questions will be discussed in Chapter 13.1.

The S. Tropical Dislocation was shown in the Voyager 1 images (Fig. 10.57). The darkening consisted mainly of a diffuse S.Trop.Band lying in the mid-latitudes of the STropZ (24.5°S), separate from the tenuous line of the STB(N). It had no definite p. end. What appeared visually as the p. end was made up of the very gradual darkening of the S.Trop.Band, plus a small slow-moving ring on it ($\Delta\lambda_2 = +2$) that happened to be near oval FA, and the occasional interactions of this spot with jetstream spots on the SEBs. When Voyager 2 viewed the S.Trop.Band, it extended much further p. like a puffy dark contrail. The relationship of the dark S.Trop.Band to the ensuing dynamical changes is not clear. The only distinct slow-moving features on the Band were a few small anticyclonic rings. Perhaps the p. ends and the small white oval that were observed in 1979/80 and 1980/81 were formed by one or

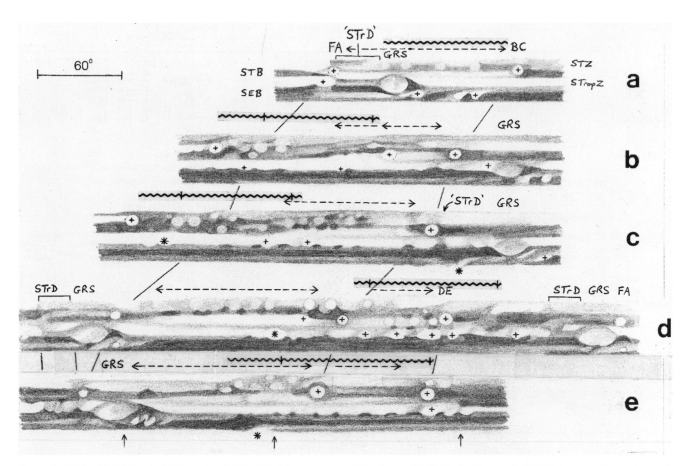

Figure 10.54. The third S. Tropical Dislocation, 1982–1984. Strip-maps are mostly from photographs by J. Dragesco and D. Parker. (a) 1982 July 8–17; (b) 1983 April 15–29; (c) 1983 June 26 – July 6; (d) 1984 May 20–26; (e) 1984 August 13–19. Dashed line marks STB Fade; wavy line marks rough extent of true SSTB. Bright white spots are marked +, including STBs ovals FA, BC, and DE; long-lived white spots in STropZ and EZs are marked *.

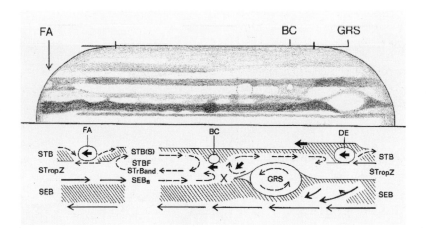

Figure 10.55. The fourth S. Tropical Dislocation, in 1985. Map by the author on 1985 Oct. 13, with a diagram of the hypothetical circulations. X marks a possible 'chiasma' where recirculation might have been occurring. In the next apparition, this had evolved into a Little Red Spot moving with oval BC (Plates P19.6&7, P21.5, P22.4). (From the 1985 BAA report.).

more such rings. If so, this STropD, like the great historical one, had a p. end defined by an anticyclonic circulation, but one which was too small to intercept SEBs jetstream spots.

The dynamical STropD developed immediately p. the GRS just after oval BC had passed it. The scene was set at the time of the Voyager 1 flyby in early March (Figs. 10.2&39B; Plate P17.4). The place was a calm, roughly square area bordered by four great circu-lations: the GRS, oval BC, the STB Fade, and the SEBs jetstream as it swung into the Red Spot Hollow. The last Voyager 1 photos show the SEBs spots continuing to retrograde into the Hollow, while the STBn jetstream between oval BC and the GRS was pro-ducing amorphous flecks of very dark cloud which were splashing p. the GRS. One has the impression that this square space was waiting for further development.

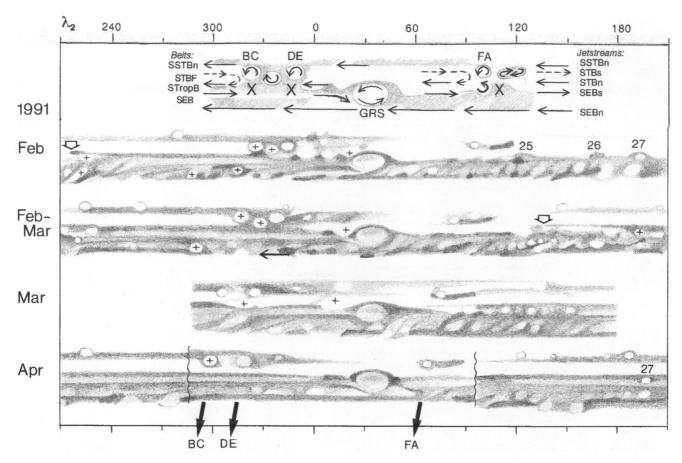

Figure 10.56. The fifth S. Tropical Dislocation, in 1990/91. These strip-maps carry on from those of the SEB Revival in Fig. 10.23. They are mainly from photographs by Miyazaki, the last two being supplemented with other observations. From top to bottom: 1991 Feb.2–6; Feb.27 – Mar.1; Mar.23–24; Apr.5–8 (λ_2 100–290) and Apr.13–16 (remainder). At top is a diagram of the hypothetical circulations. X marks a possible 'chiasma' where recirculation might have been occurring. Bright white spots are marked +, including STB oval BC. Slow-moving SEBs features are numbered 25–27 (cf. Fig. 10.41), and the rapidly prograding p. end of the S. Tropical Band is marked with an open arrowhead. (From the BAA report.)

Figure 10.57. The S. Tropical Dislocation from Voyager 1, 1979 Feb.17, showing the STB Fade as a closed circulation between ovals FA and BC.   (Image 15918.21, J1–16d, violet, filtered. NASA.)

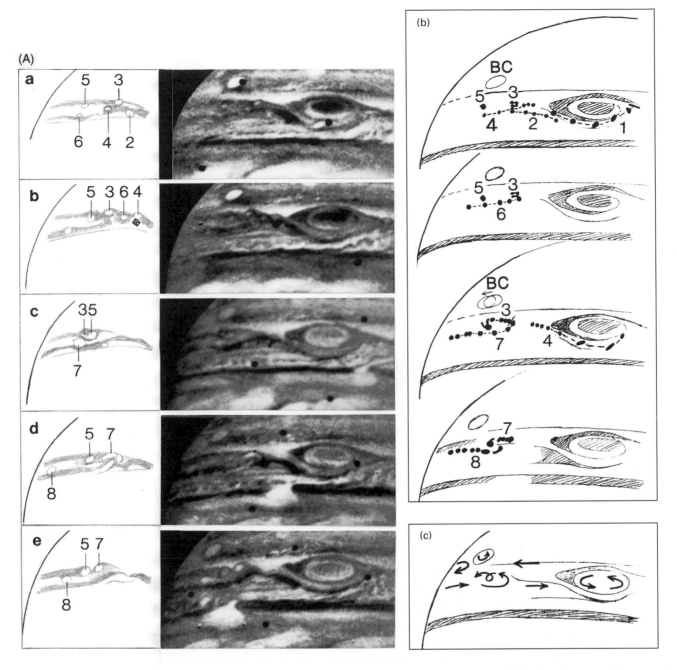

Figure 10.58. Origin of the STropD, from Voyager 2: 1979 April-May.

(A) Images at 4-day intervals. (a) 1979 April 28 (rotation 272, in the NASA numbering starting on 1979 Jan. 6). (b) May 2 (rotation 282). (c) May 6 (rotation 292). (d) May 10 (rotation 302). (e) May 13 (rotation 309). First two are violet images, filtered; remainder are green images, filtered. A grid of black dots is superimposed. Tracings at left identify the relevant spots. (NASA.)

More extensive sets covering this region were published in Hunt and Moore (*1981*) (all of rotations 269–302, April 27 – May 10) and in Smith *et al.*(*1979b*) (alternate rotations 311–325, May 13–19).

The images also show the long-lived SEBn/EZs white spot brightening as it passes the GRS.

(B) Motions of eight spots in the series from which these images were taken. Tracings show positions every 2 or 3 rotations, split between four panels for clarity. They cover the following rotations: (spot no.1) 267–281; (no.2) 267–278, then disappeared; (no.3) 267–281–292, then merged with no.5; (no.4) 267–281–300; (no.5) stationary; (no.6) 270–281, then disappeared; (no.7) 283–300–317, then merged with no.5; (no.8) 302–317. The dynamical STropD began when spots 7 and 8 orbited round no.5 at the end of this series (May 15–16, rotations 315–318).

(C) Diagram of the resulting circulation pattern.

By late April, when Voyager 2 photography began, the circulating current had still not been established, although oval BC had drawn well p. the GRS and the STropZ was heavily darkened. While SEBs jetstream spots (seen as white ovals against the surrounding shading) were still retrograding into the Red Spot

Hollow, there were also two similar anticyclonic spots slightly further south which were roughly stationary in λ_2: no.5 and no.3 in Fig. 10.58. It seems that the appearance of these two spots was the crucial step. No. 5 was probably the same as the one near oval FA in Voyager 1 coverage, and one may speculate that no.3 was a typi-

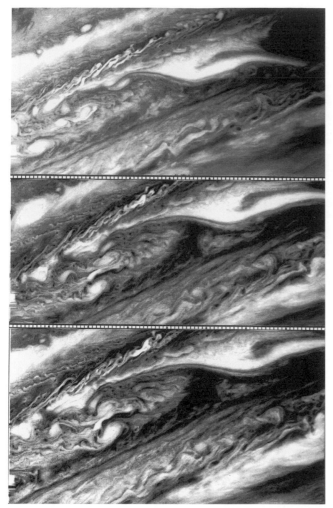

Figure 10.59. Circulation in the dynamical STropD, from Voyager 2, 1979 May 31 – June 2. From top to bottom: rotations 430, 434, 435. The GRS is at right in the top picture. Also see Plate P20.1. (Images 20378.56, J2–10d; 20428.54, J2–8d; 20440.59, J2–8d. All violet, unfiltered. NASA.)

cal SEBs jetstream spot that veered into the lacuna p. the GRS and thus dropped out of the jetstream.

The dynamical STropD developed early in May (Fig. 10.58). In these first Voyager 2 photos, SEBs retrograding spots were still swerving past spots 5 and 3 and twisting into shapes like the Chinese yin-yang symbol as they did so before continuing on their way. In one such interaction (with spot no.4), no.3 was propelled into prograding motion along the S.Trop.Band, and on May 5–7 (rotations 289–295) it orbited round no.5 and merged with it. Spot no.6 did still retrograde past this point, but the next (no.7) failed to do so; on May 9 (rotation 300) it swung south and halted just f. no.5, then gradually prograded towards it as no.3 had done. Both the prograding spot no.7 and the next retrograding spot, no.8, converged on no.5 from opposite sides and swung around it forming a beautiful vortex (May 15–16, rotations 315–318); thus the Circulating Current was started. From then on, the circulation built the dark hump into the great dark 'f. end complex', which by June 30 was an obvious barrier to all retrograding spots (Fig. 10.59). Each retrograding spot was recirculated at this point, swinging onto the STBn jetstream and eddying once or twice before being torn apart within a few days. No spots survived to emerge p. the Disturbance.

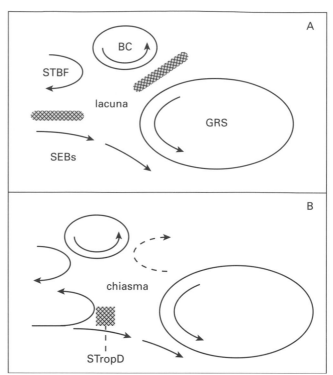

Figure 10.60. 'Chiasma' model of the origin of a STropD in a S. Tropical Dislocation. (A) As oval BC passes the GRS, a square lacuna opens up bordered by opposing currents. STBF = STB Fade. For actual view from Voyager 1, see Plate P17.4 and Fig. 10.39. (B) Rapid formation of a STropD could complete a 'chiasma', i.e. a pattern of meshing and diverging circulations. For actual view from Voyager 2, see Fig. 10.58; the dynamical STropD developed too late to produce this six-sided chiasma involving the STB circulations.

However, the retrograding flow of the SEBs jetstream still continued, carrying smallscale cloud textures and larger waves; it was only the SEBs anticyclonic spots that were skimmed off by the STropD.

Discussion: Origins of South Tropical Disturbances

According to this account, the origin of the STropD in 1979, which seemed to be such a deterministic process on the large scale, was attributable to random local interactions on the small scale. The opening of the 'square space' did not suddenly suck all retrograding spots into a new circulation; it may have just increased the chance that one passing spot would stray in that direction. This may have been the origin of spot no.3, which merged with no.5 to create the nucleus of the Disturbance. And spot no.5 did not immediately divert the other retrograding spots; it only caught one after several had swung past it and escaped. Having caught one, it rapidly established the Disturbance which devoured all subsequent retrograding spots.

Apparently the vorticity of the individual retrograding spots was fed into the Circulating Current, which thus grew with every mouthful. We will look at this in Chapter 14.4. It seems to be a dramatic instance of a theoretical process whereby the flows of the jetstreams and great ovals are attributed to chaotic mergings of smaller vortices.

But important questions remain. First, what is the role of the dark material in the STropZ? The formation of the S.Trop.Band

and the evanescent dark STBn material involved no dynamical rearrangement. Why did these features develop, and did they have any material effect on their surroundings, perhaps tending to entrap anticyclonic rings such as nos.5 and 3? The answer to these questions is unknown, but there will be more to say about them in section 10.8 and Chapter 13.1.

Second, as S. Tropical Dislocations unfold so regularly, is there a broader teleology at work? It may be instructive to look at the overall pattern of currents. The initial situation (Fig. 10.60A) had squeezed together the STBn jetstream with the oppositely directed flows of the STB white oval and the SEBs jetstream. The author suggested (1986) that formation of the STropD created a more coherent pattern, which we may call a *chiasma*, that is, an area bordered by diverging circulations (Fig. 10.60B). (The possibility that chiasmata are more generally involved in creating dislocation-type phenomena will be discussed in Chapter 13.1.) It may be that in the initial situation, the strong shear of the currents and the vacancy of the square space that opened up were unstable, whereas a STropD completing a chiasma is more stable, making it likely that random smallscale motions will tend to this outcome. Unfortunately for this theory, the Voyager 2 images as described above show that the dynamical STropD did not really develop until oval BC had moved on, so that a complete chiasma was not actually formed (Fig. 10.58). Nevertheless, the recurrent generation of STropDs in relation to the STB Fade and the GRS suggests that some such influence may exist.

However, the STropD once formed moved at its own characteristic speed, separate from the STB structures which had induced it.

The preceding description suggests that all the dynamical events may be understood (at least qualitatively) in terms of the visible surface features; there is no need to invoke deeper forces. This is consistent with the fact that the STropD did not completely divert the SEBs jetstream, although it did affect its speed, both inside and outside the Circulating Current (Chapters 10.1 and 14.4). One can therefore propose that, in spite of its impressive structure and stability, a STropD is mainly a surface pattern, below which the jetstreams continue on their courses.

Are small vortices like these always necessary to form a STropD? The observational record does not give full support to this idea. Several STropDs did arise during SEBs jetstream activity (1975, 1979, 1984) or immediately after a SEB Revival (in 1955), while the one in 1970 may have developed from a STBn jetstream spot instead. However, the great STropD appeared in 1901 when spots on SEBs were apparently stationary, and the STropDs of 1941 and 1946 appeared while the SEBs was entirely quiet. The birth of the great STropD was also not related to the STB nor to the GRS.

One would also like to know why the five S. Tropical Dislocations all ended in different ways. This may have been determined by the evolution of the STB Fade, which was gradually growing longer, and of the STB ovals, which were gradually growing smaller. Perhaps these trends altered the strength of the jetstreams and their behaviour at the chiasma with the GRS. Indeed, the fact that the Dislocations only appeared during these years, in

association with the FA–BC STB Fade and not with earlier such Fades, could have been a consequence of the gradual evolution of the STB white ovals in size and latitude (see next chapter). Alternatively, the differences between the five Dislocations may just reflect the random factors in the events that give rise to them.

10.8 SOUTH TROPICAL BANDS AND SHADINGS

We have mentioned various types of shading in previous sections, and it is time to bring these references together. STropDs are only the most distinct of several types of dark feature in the STropZ. These also include:

> isolated segments of S.Trop.Band;
> > the darkening of the zone in a S. Tropical Dislocation (which often comprises a fairly distinct S.Trop.Band);
> > the diffuse patchy shading that sometimes develops alongside a SEB Revival.

Like many STropDs, all of these features have prograding drifts; they arise just p. the GRS; and they tend to arise in the closing stages of SEBs activity.

When these features are distinct enough for speeds to be measured, they are typically in the $\Delta\lambda_2$ range -24 to -39 (mean -31.3) (Table 10.9, Fig. 10.61), forming a distinct group separate from the drifts of STropDs and of the STBn jetstream. (However, the three most recent S.Trop.Bands have moved at STBn jetstream speeds for at least some of the time.)

The isolated Bands listed in the Table, except the 1968 one, developed alongside STB Fade sectors as do S. Tropical Dislocations; the difference was that these Bands were just narrow lines that did not persist in a regional structure. Also like S. Tropical Dislocations, most of these Bands appeared during the declining stages of SEBs jetstream outbreaks (1968, 1987, 1989), although they emerged from the p. end of the GRS with no sign of a circulating current nor any physical connection to the jetstream spots.

A possible interpretation is that dark SEBs jetstream material is trapped around the GRS rim until jetstream activity dies down, when it is released into the STropZ, flowing to a variable degree under the influence of the STBn jetstream.

The irregular dark shading that sometimes appears in the late stages of SEBs activity (speeds also listed in Table 10.9) differs in that it arises diffusely alongside the SEBs, though also mainly p. the GRS.

These types of darkening clearly have much in common and they perhaps form a continuum from the most to the least structured. In the course of this chapter we have offered different *ad hoc* suggestions to explain each type: turbulent interactions for SEBs-associated darkening, release of trapped material for Bands, instability or compensatory darkening for Dislocations, and recirculation for Disturbances. Whether all of these explanations are true, or whether there is really a single cause for all these phenomena, remains to be seen.

Table 10.9. *South Tropical Bands and shadings*

Date	$\Delta\lambda_2$	(n)	Notes	Source	Lat.of Band (β'')
1888	−24		Dark spot on Band	Williams	−
1947	−31	(7)	Shadings	Reese*	
1948	−27	(7)	Shadings	Reese*	
1952/53	−34.6		P.end shading	ALPO	
1959	−29.2	(3)	Shadings	ALPO	
1962/63	−39	(15)	Shadings in Revival	ALPO	
1968	−26		Two successive Bands p.GRS, but p.end indistinct.	ALPO, LPL	−24.4 (NMSUO) (−24.9) (SAF)
1969	−38	(5)	Bands at p. end GRS	NMSUO	−24.6 (NMSUO)
1971	−33.5	(4)	Shadings before SEB Rev. arrived	NMSUO	
1975	−18		P.end 1st S.Trop.Disl'n	BAA	
1978/79	−13		Band in 2nd S.Trop.Disl'n	BAA	−24.5 (Voyager)†
1980/81	−28	(5)	White spots around f–STropD as it broke up	SAF	
1985	−30	(1)	Band in the 4th S.Trop.Disl'n	SAF	(−25) †
1987/88	−77	(1)	Band ($\Delta\lambda_2$ range −62 to −101 from different observers)	BAA	−25.3 (BAA)
1988/89	(≈−88)		Band	BAA	
1990/91	−17→ −88		Broad Band in 5th S.Trop.Disl'n	BAA	−25.1 (BAA)
Average:**					
−selected:	−31.3				−24.8 (±0.4)
−all:	−34.5				−25.2 (±0.7)

This Table lists fast-moving shadings and isolated S. Tropical Bands for which drifts were determined. (This table does not include STropDs (see Table 10.8), nor 'smokestack' S.Trop.Bands associated with STBn jetstream spots (see Table 10.10).) All the S. Tropical Bands were emerging from the p. end of the GRS and drifts are for the p. end. Notes:

* Reese, cited in ALPO 1959 report.

† Latitude measured by author from photos.

** Average, 'selected': $\Delta\lambda_2$ only up to 1971, and latitude for isolated S.Trop.Bands in this Table. Average, 'all': $\Delta\lambda_2$ including the very diverse values since 1971 (ranging from STC to STBn jetstream); latitudes of S.Trop.Bands including those from Table 10.10.

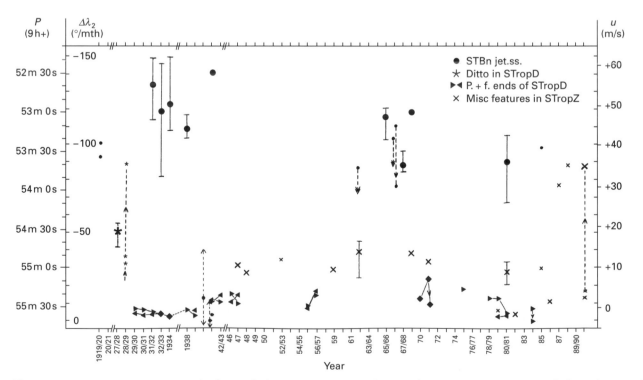

Figure 10.61. Speeds recorded for fast-moving features in the STropZ and in the STBn jetstream. (See Fig. 10.5 for slower-moving STropZ features.) From sources listed in Appendices 1 and 3. Speed is shown as *P* (rotation period), as $\Delta\lambda_2$ (degrees per month), and as *u* (m/s in System III). Symbols: ● STBn jetstream spots; ★ ditto within STropD; ▶ ◀ p. and f. ends of STropDs; × various STropZ features including dusky patches, p. end of S. Tropical Band, etc. Arrows show changing speeds of individual spots (including oscillating spots in 1940–42). Large symbols are apparition means of ≥3 features, and vertical bars indicate the range of speeds.

Table 10.10. *Records of the STBn jetstream.*

Date	$\Delta\lambda_2$: (n) (abnormal)	(normal)	Lat.(β'') of jet.spots or 'smokestack' Band*
1919/20		−92,−100	
1927/28	−50.2 (4) (in STropD)		
1928/29	−32, −35.5 [→ −88] (both in STropD)		
1931/32		−134 (11)	Band: −26.2*
1932/33		−118 (14)	Band: −26.5*
1934		−122 (9)	Band: −25.3*
1938		−108 (5)	
1940/41	−11.8± [oscillating −40 to +4]		
1941/42	+0.8± [oscillating −11 to +21]	−140 (2)	
		Weighted av. (Band): −26.3	
1962/63		−86 (1, →SEBs)	−
1965/66		−115 (4)	Spots: −25.6 NMSUO
1966/67		−103,−110 (→decel.)	Spots: −25.2 NMSUO
1967/68		−88 (3)	Spots: −26.1 NMSUO
1969		−118 (5)	Spots: −25.0 NMSUO
		Average (spots):−25.5 (±0.5)	
1979/80		−	Band: (−24.8) (SAF)
1980/81		−90 (6))	Band: (−24.6) (SAF)
		Average (Band): −24.7	
1985		−98 (1)†	
1990/91**	−13 (1) [spots p. GRS]	−84 (3)	
Average:		−109.3 (±17.0)	−25.5
Voyager (global; Limaye, *1986*):		−110	−27 to −29

This Table lists all records of spots on the STBn jetstream, both 'normal' spot outbreaks, and in the 'abnormal' column, other small dark spots like STBn jetstream spots which showed slower drifts. Speeds are from Reese and NMSUO (1962–1969) and from BAA (others).
* The 'smokestack' was the S. Tropical Band on which jetstream spots ran away from the great STropD. Latitudes for it were measured by the author from drawings by Peek and Hargeaves, normalised to micrometer measurements of neighbouring belts. (1932/33 value has greater weight.) Latitudes for 1965–1969 were measured from photos by NMSUO and are of high precision. Latitudes for 'smokestack' band 1979–81 were measured from photos by SAF and are of medium precision.
† The 1985 feature was a tiny notch, not a spot outbreak.
** See also Table 10.9 for other streaks, etc., including recent Bands that moved with jetstream speed.

10.9 THE STBN PROGRADING JETSTREAM

Finally, we reach the jetstream that defines the southern edge of the S. Tropical region. All the records of this jetstream are listed in Table 10.10.

Most of the phenomena of this jetstream have been described in the previous two sections. It was first observed in 1920 (Fig. 10.46), and since then it has most commonly been observed in similar circumstances, when it carries jetstream spots that are observed or inferred to have been transported from the SEBs jetstream via the Circulating Current. However, on three occasions there has been a small outbreak on STBn *before* a SEB Revival, associated with other curious phenomena, as follows.

1927–1928 (BAA): In 1927 September, while the SEB was faint, tiny dark spots on STBn had $\Delta\lambda_2 \approx -50$ (Plate P6.4). Two more such spots appeared in 1928 during the SEB Revival (one accelerating to $\Delta\lambda_2 \approx -88$). All these were within the boundaries of the faded STropD. This may have been a spontaneous jetstream outbreak, with this jetstream (like the SEBs jetstream) running slower inside the STropD. But such tiny spots could have been missed in many apparitions. The STBn was also marked in 1927 by the 'false Red Spot' (a faint loop resembling the GRS in oval shape, not colour, with $\Delta\lambda_2 = -6$.) The jetstream activity was associated with a reddish S.Trop.Band and was followed, in 1928, by major coloration spreading from the same latitude (section 10.3).

1940–1942 (BAA, and Peek's book): Oscillating spots. Two small but sometimes very dark spots appeared in the southern STropZ, one in 1940 and one in 1941/42, and both showed oscillating motion which was analysed in detail by Peek. The first had $\Delta\lambda_2$ smoothly varying between ≈ -40 and $+4$; between 1940 July and December, it performed two cycles of a damped oscillation with a mean drift of $\Delta\lambda_2 = -11.8°/\text{mth}$, and an oscillation period of 62 days. It gradually faded. Its latitude was constant at about 26°S.

The second spot, observed from 1941 November to 1942 March, initially seemed to perform one similar oscillation, but then continued to decelerate over the last 6 weeks of its life. This spot did vary in latitude, drifting northwards when it was retrograding. It started with $\Delta\lambda_2 \approx -11$, near the STBn edge, and after 1½ 'oscillations' ended up with $\Delta\lambda_2 = +21$, near the SEBs edge. (These were its extreme speeds; its average speed was +0.8.) Thus it recirculated across the STropZ, about 20° f. the edge of the GRS (Plate P8.1).

Earlier in 1941/42 there was also a group of four STBn jetstream spots with $\Delta\lambda_2 = -140$, and the oscillating (circulating) spot may have come from this group as it approached the GRS.

1962 (ALPO): A STBn dark spot in the STBn jetstream was reflected off a dusky column at the f. edge of the Red Spot Hollow in 1962 June 14, moving into the retrograding SEBs jetstream, while the SEB was faint (Fig. 10.62).

When many STBn jetstream spots are present, they usually lie on a narrow S.Trop.Band separate from the STBn proper (Fig. 10.63). This is the S.Trop.Band that curves out of the p. end of a STropD, called the 'smokestack' in the 1930s. Its average latitude is 25–26°S (Table 10.10). This is on the north edge of the broad jetstream observed by Voyager.

Otherwise, spontaneous STBn jetstream activity (that is, not derived from SEBs via the Circulating Current) has only been reported in the 1960s by NMSUO. The spots appeared singly or in small groups, during SEBs activity though not visibly connected to it. On approaching the GRS, some of the STBn spots were unperturbed, some were decelerated, and some were strikingly accelerated either past it or into it (Figs. 10.35&62). For example, one STBn spot in 1966/67 decelerated smoothly and halted 60° f. the GRS, meanwhile moving from 25°S to 23°S; then it disappeared. Of five spots recorded in 1969, two slowed down on approaching the GRS but leapt rapidly past it, and one showed the most rapid acceleration ever reported on the planet (from $\Delta\lambda_2 = -33$ to -168 within ≤ 1 day, a change of 87 m/s). A final spot in this series developed into the STropD of 1970 (section 10.7). The orange colour of the southern STropZ in 1970 may have been the sequel (Plate P13.4).

The speed of the jetstream was fastest in the 1930s, in the Circulating Current, while it was very slow inside the STropD (see above, 1927–1928); so the STBn jetstream may have behaved like the SEBs jetstream in this respect. But in later years its speed does not seem to have been affected by STropDs.

Spacecraft observations

According to the Voyager data, this is the broadest of the non-equatorial jetstreams. No outbreak was occurring during the Voyager missions (except for the turbulence at the STropD) but the jetstream showed various disturbances below the level of Earth-based resolution. There was intense rippling in the turbulent regions f. the STB ovals. There were the splashes of dark matter p. the GRS, noted in the description of the S. Tropical Dislocation. There were a few tiny amorphous dark spots in the STropZ, only $\approx 1000 \times 2000$

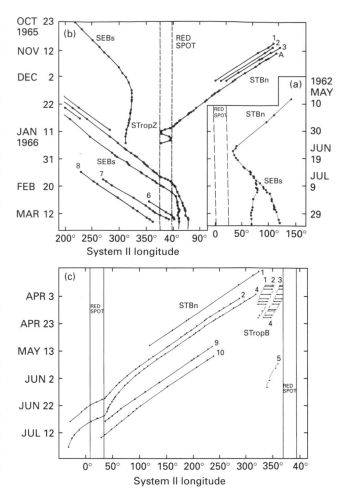

Figure 10.62. Motions of STBn jetstream spots near the GRS, from E.J. Reese and colleagues. (a) 1962, from ALPO report. (b) 1965/66, from NMSUO report by Solberg and Reese. Several retrograding SEBs spots are also shown, including one which shifted from 21.4°S to 23.1°S as it came to a halt in the STropZ. STBn spot A circulated around the GRS (Fig. 10.35). (c) 1969, from NMSUO report by Reese, including three dark streaks of S. Tropical Band (<—>).

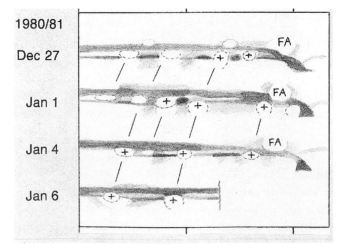

Figure 10.63. A possible display of the Circulating Current in 1980/81. The STB is shown with S. Tropical Band N. of it, carrying dark streaks and bright spots (+) in the STBn jetstream. They are moving away from the dynamical STropD at λ_2 305. Strip-maps by the author. (From the BAA Memoir.)

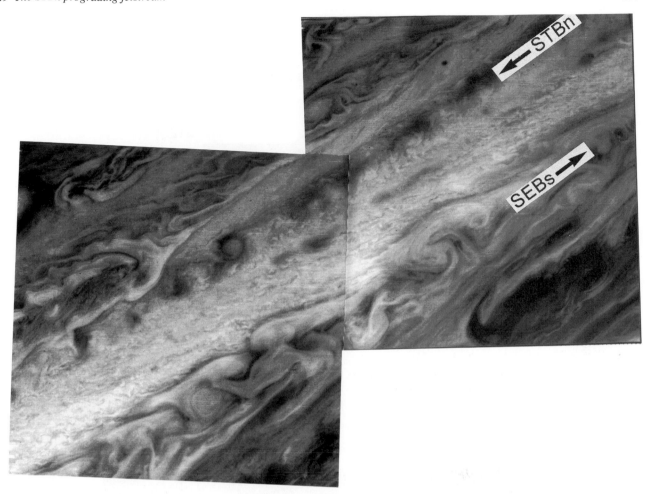

Fig. 10.64. Detailed texture of the STropZ and STBn, from Voyager 2. Compared to Voyager 1 images, there is more turbulence on SEBs and STBn, and the S. Tropical Band is advancing in the middle of the STropZ (represented here by small diffuse streaks). STB(N) consists of a chain of small dark patches without regular circulation. The STB 'orphan' filamentary region of cyclonic turbulence is ar left. (Images 20571.09&40, J2–3d, orange, unfiltered. NASA.)

km across (e.g. Fig.10.27), one of which had a drift measured as $\Delta\lambda_2 = -64$. Later, the leading streaks of S.Trop.Band ran along at the same latitude (section 10.7; Fig. 10.64). Also, there was a chain of tiny dark spots like amorphous waves on STBn p. oval FA, sepa- rate from the S.Trop.Band (Fig. 10.64). These were densest during the Voyager 2 encounter, and were superimposed on turbulent streaks but showed no evident circulation, as they lay within the broad jetstream.

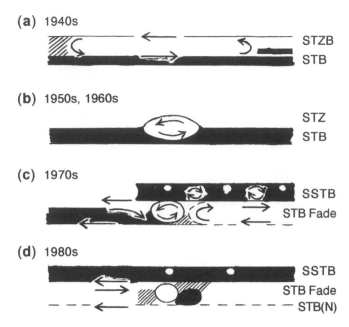

(a) 1940s

STZB

STB

(b) 1950s, 1960s

STZ

STB

(c) 1970s

SSTB

STB Fade

(d) 1980s

SSTB

STB Fade

STB(N)

Fig. 11.1. Typical appearances of the South Temperate region and the great white ovals that formed around 1940 and have been shrinking ever since. (a) 1940s ('proto-ovals'); (b) 1950s and 1960s (mature ovals); (c) 1970s (including Voyager data); (d) 1980s (with 'faded' STB). Arrows indicate motions observed by Voyager or inferred in other years. (From Rogers and Herbert, *1991*.)

Fig. 11.2. Voyager 1 close-up of South Temperate white oval 'BC', taken one day before encounter. See Fig. 10.2 for wide-angle view. South towards top right. (Image 16367.38, J1–1d, unfiltered, violet. NASA.)

11: South Temperate Region (27°S to 37°S)

11.1 OVERVIEW

Until the 1980s, there was always a distinct South Temperate Belt (STB) at most or all longitudes, more-or-less positioned between the latitudes that are now known to carry the STBn and STBs jet-streams. The STB is usually the darkest belt on the planet after the two Equatorial Belts, and has occasionally surpassed them in darkness, although it is narrower. There are often segments of the belt with differing darkness or width, as well as various light and dark spots in the belt or on the south edge. Usually, they all move with the South Temperate Current (STC). The most notable phenomena, to be described later in this chapter, have been first the appearance and evolution of three great white ovals on the STBs edge, and second, the occasional complete fading or whitening of discrete sectors of the STB (STB Fades). Such fadings culminated in the disappearance of the entire belt in the late 1980s.

Boundaries of the South Temperate Belt

The mean latitudes of the STB are listed in Table 11.1; the data are deferred to Figs. 12.7&8.

The STB north edge is usually between latitudes 26°S and 28°S, exactly tangential to the GRS. Sometimes the north edge is deflected or obscured to leave a narrow bright strip around the edge of the GRS, like the Red Spot Hollow in the SEB. Occasionally, instead, the STBn extends diffusely further north (e.g. 1920, 1958, 1963/64, 1970); then it is difficult to tell whether the GRS overlies the STB or vice versa. The STBn generally coincides with the STBn prograding jetstream, described in the previous chapter. (Activity on this jetstream seems to be unrelated to the state of the whole belt.)

The STB south edge, up to 1940, was typically at latitude 31–32°S. Thus the STB at most longitudes was a continuous, fairly narrow, dark belt. This was the case between 1938 and 1945, while the precursors of the three great ovals developed on the belt's south side. Since then, the STB between the three ovals has progressively broadened to the south over the decades, in several stages. The first broadening occurred in the 1940s as the sectors between the newly formed ovals extended around most of the longitudes of the STB. In the late 1970s, further broadening to the south took place,

Table 11.1. *Latitudes of the S. Temperate Belt*

	(1907–1937)	(1950–1979)
STBs	−31.3 (±1.3; 7)	−33.3 (±0.9; 16)
STB	−28.9 (±0.9; 27)	−30.2 (±0.7; 21)
STBn	−26.0 (±1.2; 7)	−26.8 (±0.7; 17)

The table lists zenographic latitudes, given as the mean (±s.d.; N) of apparition means. (S.d. = standard deviation; N = number of apparitions). Data are plotted in Fig. 12.7. Before 1938, the edges of the STB were measured in only 7 apparitions (1914–1930), by micrometer (4 apparitions) or on photos (3 apparitions; author's measurements). (Data for 1939, when the first signs of the 'proto-ovals' appeared, do not give internally consistent results and cannot be used, although the appearance was much as in the preceding few years.) From 1940 to 1979, data were summarised in Rogers and Herbert (*1991*) as plotted in Fig. 12.8. The STB gradually disappeared after 1979.

extending the STBs edge several degrees south of the corresponding jetstream.

The STBs jetstream, according to Voyager imagery, is at 32.6°S with a speed of $\Delta\lambda_2 = +42$. Like most other retrograding jetstreams, it has only been observed by Voyager. Nevertheless, we can infer whether it was in the same latitude 40 years earlier thanks to the unique circumstances of the origin of the STBs white ovals (section 11.3). As indicated in Fig. 11.1, it is presumed that the ovals developed by reconnection of the STBs and SSTBn jetstreams. Therefore, the STBs edge became the north edge of the white ovals, and the latitude of the STBs edge in 1938–1945 must have been the latitude of the STBs jetstream at that time. That latitude was 32.1°S in 1938, and averaged 31.3°S (±0.5°) in 1940–1945 (Fig. 12.8). Compared with the Voyager latitude of 32.6°S, the jetstream may have shifted by ≈1.3°, but this figure is so close to the margins of error that it may not have shifted at all.

In the Voyager images, much of the belt was disturbed (especially f. the three white ovals), and coarse turbulence near both edges allowed the STBn and STBs jetstreams to be tracked. The STBs jetstream coexists with many slow-moving spots and irregularities in the same latitude, which move with the South Temperate Current.

Features of the South Temperate Belt

The STB is usually a dark, solid belt, although it is sometimes double. It is usually devoid of colour; to the author, it is always cold grey, in contrast to the variably brown or warm grey tints of the other belts. Even in 1928/29, when an orange veil spread over the STropZ and STZ, the STB itself seems to have remained grey; Phillips and Ryves both recorded it as grey but 'immersed in strong reddish colour' or having an 'overlying red veil'. However, there have been occasional changes in the belt's colour. A large section of it was strongly brown in 1936, and in the following year this colour was restricted to the north half of the belt before it reverted to grey. The belt was also distinctly brown or even reddish from 1941 to 1943. (It was also reported as brown several times in the 1950s by ALPO observers, but these impressions seemed to vary with the observers). In other years, and from the 1960s onwards, the belt has always been described as grey.

The features most commonly seen on the STB are white ovals. Even before 1940, when the three great ones began to form, bright white ovals in the STZ or on the STBs edge were frequent. E.J. Reese (*1972b*) plotted the longitudes of these ovals as recorded by the BAA from 1914 to 1935 in a special system of longitude moving with the mean speed of the STC, and found that individual ovals could be traced over several apparitions – perhaps for as long as 20 years in one case (Fig. 11.3). Unlike the great white ovals of later years, these earlier ovals did not shrink over the years, but had an average length of about 14° throughout. While some of Reese's identifications over several years must be regarded as tentative, some of these ovals certainly persisted for several years at a time, in particular his oval A from 1929 to 1934, and his oval C from 1928 to 1933.

The development of 'STB Fades' (section 11.4) is usually controlled by the locations of the STBs white ovals. In the 1960s and 1970s, STB Fades were developing more frequently, usually bounded by the three great white ovals, and in the 1980s, the STB Fades extended to eliminate the STB at almost all longitudes, leaving the three white ovals flanked only by small dark or dusky patches, with only a very faint STB(N) remaining. At the same time, a major belt developed in the canonical latitudes of STZ (1983–85) or SSTB (1986–90), with the three white ovals on its north edge – a striking reversal of the picture that had been familiar in the early history of these ovals.

Outstanding dark features are less common but a few have attracted attention. The most notable was the 'Red Streak' of 1890–1892 (Fig. 11.4; Plate P3.9). It developed from a pre-existing dark spot between two white spots in the STB, which accompanied an unusual dark oval prograding in the STropZ (Chapter 10.7; Fig. 10.43). When this complex encountered the GRS in 1890, the STropZ spot disappeared while the STB dark feature became more intense. Within a year this STB feature was an intensely dark and intensely red sector of the belt, and was the most conspicuous feature on the planet. In the best seeing it always appeared to be

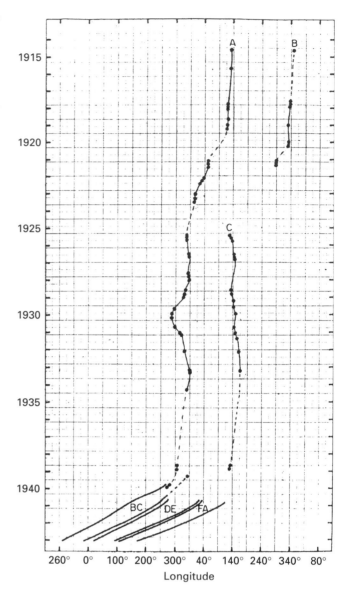

Fig. 11.3. Drift chart for an earlier generation of STBs white ovals, 1914–1935, plotted in a longitude system moving 0.5°/d faster than System II (λ_2 – 15°/mth). (From Reese, *1972b*).

divided into north and south components. It was still flanked by the two white spots; the preceding one was described as a 'curious little round, well-defined white spot of intense brilliancy' (BAA Memoir). The streak was still present in 1892, but it was then less dark and almost colourless, and was not seen thereafter.

The initial impression that the STropZ spot had actually been transformed into the STB Red Streak (Denning, *1902*) may seem implausible, although now we know that a very similar STropZ spot in 1986 was indeed a little red oval, and that the GRS has a circulation which could deflect such a spot into the STB, the possibility should not be totally dismissed. The more conventional interpretation would be that the Red Streak was a dark reddish-brown belt segment, like the 'barges' on the NEBn. In support of this, the red streak appeared less prominent relative to the STB when it was near the limb, which suggested that it was a low-altitude feature.

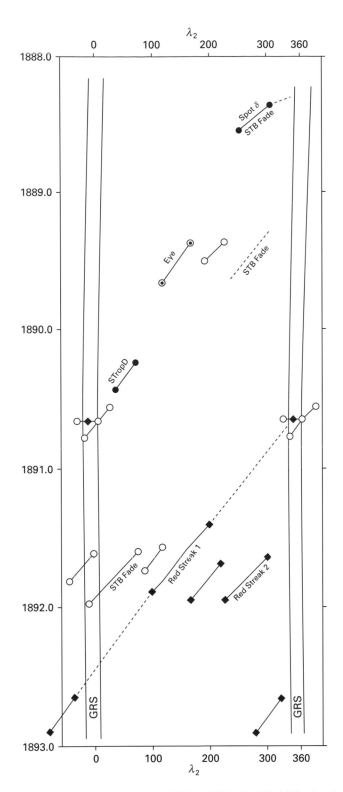

Fig. 11.4. Drift chart for features of STB and STropZ, 1888–1892, plotted in System II longitude. Data from Sells and Cooke (*1913*) and contemporary journals. Symbols: ● dark spot, STropZ; ◆ dark spot or streak, STB; ○ white oval, STB.

A very dark segment of STB immediately following a STB Fade attracted attention in 1935. Solitary dark bars were also conspicuous for a few months within STB Fades in 1926, 1967–68, and 1975. Finally, from 1986 to 1990 when almost the whole STB had disappeared, solitary dark grey streaks were again sometimes visi-

ble following one or other of the three great white ovals. In 1986, two such dark streaks were conspicuous on opposite sides of the planet: one was the centrepiece of a cluster of spots between ovals BC and DE, while the other was following oval FA, although the oval was temporarily invisible. In high-resolution photographs (Plates P19.6&7, P23.1), these two dark streaks looked very similar, tapering obliquely in the Sf. direction. The one following oval FA, named the STB Remnant, became much darker as it passed the GRS in early 1987, possibly like the dark streak in 1890; however, it shrank and faded away within less than a year.

Although none of these dark segments had been coloured with the exception of the 1891 Red Streak, recent photos by J. Bourgeois from the Pic du Midi have occasionally shown pale, reddish patches adjacent to the great white ovals.

South Temperate Zone

The South Temperate Zone (STZ) is very variable in appearance. Often there is a broad, fairly clear zone south of the STB, to which the name 'STZ' is applied. This zone is sometimes bright white (particularly so in 1924), sometimes more shaded, and sometimes occupied to a greater or lesser extent by spots and patches that appear indistinct or irregular. At high resolution, there are sometimes bright ovals in various latitudes. (An early display of such ovals was recorded by W.R. Dawes in 1857; Plate P1.5&6.) The colour of the STZ is usually the same white or yellowish as other zones; only in 1928–29 has a distinct reddish tint been observed, apparently extending from the STropZ.

Only as a result of the Voyager missions do we know that the typical broad STZ actually includes canonical SSTB territory, and a narrow 'STZ Band' (STZB) sometimes observed within it at 36°S or 37°S marks the SSTBn jetstream. Sometimes, as described in the next chapter, there are shadings which ought to be described as a true SSTB, and then the true STZ is very narrow. In some apparitions since 1978, while sectors of the STB were disappearing, some sectors contained a major belt in STZ latitudes (see section 11.4). Whatever the pattern of light and dark belts, features south of 36°S always belong to the territory of the South South Temperate Current (SSTC) to be described in the next chapter, while features north of 36°S, on the STBs edge, typically move with the South Temperate Current (STC).

South Temperate Current

The characteristic current of these latitudes is the South Temperate Current (STC), and with some interesting exceptions this has been one of the steadiest of the jovian currents. From 1880 to 1940 it had an average speed of $\Delta\lambda_2 = -15$, and the apparition averages all fell in the narrow range from −11 to −19, according to the tabulation by Peek (Fig. 11.5). However, these averages excluded a few spots which were moving distinctly faster than the general current for that latitude. These anomalously fast speeds were actually characteristic of the South South Temperate Current (SSTC), which normally occupies the domain further south, and have been described as 'SSTC invasions' of STC territory (section 11.2).

Around 1940, the history of the STC took a remarkable turn,

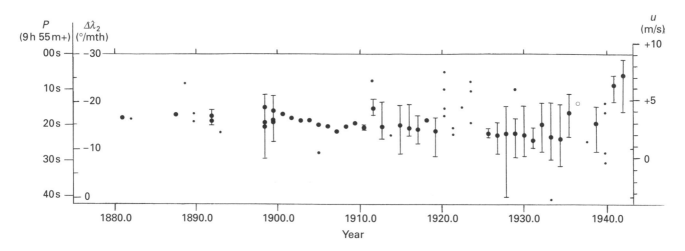

Fig. 11.5. History of the S. Temperate Current, 1880–1940. Earlier records were: 1787, 9h 55m 17.6s (Schroeter); 1862, 17.2s (Schmidt); 1872/73, 19.6s (Lohse). Later records, for the great white ovals, are in Fig. 11.10. Large symbols are the apparition mean and range (not error bars). Small points represent individual features. The chart includes some periods shorter than 9h 55m 11s which were definitely for S. Temperate features, which have been interpreted as 'invasions' of the S.S. Temperate Current.

when the entire STB and STZ accelerated to SSTC velocities ($\Delta\lambda_2$ ≈ −26). This change accompanied the subdivision of the STZ to form the three great white ovals. Normal STC velocities were not recorded after 1939; instead, the motion of the region was dominated by the motion of the three white ovals. As discussed in section 11.3 below, the speed gradually slackened (although with substantial fluctuations) and approached the previous STC value, reaching an average $\Delta\lambda_2$ of about −12 around 1990.

It is notable that all the features in the STB, whatever their size or appearance, have moved at the same speed as the three white ovals. These minor features are not tabulated here but included the dark streak in 1967/68, pale 'gaps' in the 1970s, small dark spots in 1987–89, and tiny features in Voyager images.

In retrospect, it is notable that the STC showed a slight deceleration from $\Delta\lambda_2$ ≈ −17 around 1880 to −12 around 1934 – an overall deceleration of 0.10°/month/year (Fig. 11.5), compared to the mean deceleration of the great ovals since 1940, 0.28°/month/year (Fig. 11.10). The pre-1940 chart even hints at a fluctuation with a period (≈12 years) and phase matching the oscillation shown by the post-1940 ovals, although it was too slight to be significant. So the current may have evolved before 1940 as it has done since 1940, although less obviously.

11.2 'SSTC INVASIONS' AND THE 'SOUTH TEMPERATE DISTURBANCE'

Before 1939, there were only rare instances of features on the STBs edge or in the STZ moving with speeds typical of the SSTC – that is, 'SSTC invasions' of STC territory. One or two spots with these speeds were recorded in six apparitions. In each of these apparitions, other spots moved with the normal STC.

The best-documented of these events were the series in 1919/20 and 1920/21 (see below). In 1923, the two features manifesting the 'SSTC invasion' were tracked for little more than a month so the speed may have been imprecise or transient. In 1911, 1928/29, and

1938, there were insufficient data on the spots to determine whether there were any unusual circumstances that might explain them.

The events from 1919 to 1921 were apparently connected with the so-called 'South Temperate Disturbance' of 1918/19 (Fig. 11.6). The S. Temperate Disturbance was a heavily shaded sector of STZ, linking the SSTB and STB, which moved with the speed of the SSTC. In appearance it resembled the S. Tropical Disturbance, which lay alongside it at the time. Like the STropD, it had a white spot at its p. end (and sometimes at its f. end), and complex structure internally, and it progressively grew longer, although this seemed to be by addition of new segments rather than by stretching. The latitude of the South Temperate Disturbance was unfortunately not measured, so we do not know whether it was a true homologue of the STropD in the STZ, or whether it was merely a segment of true SSTB (as described in the next chapter). It may well have included both, as in the next apparition, 1919/20, the 'SSTC invasion' was said to be manifested by two 'South Temperate' white spots (probably on STBs), each associated with a 'South South Temperate' dark feature. All four features were moving with SSTC speeds and could be traced back to the original S. Temperate Disturbance (Fig. 11.6). In 1920/21, the original S. Temperate Disturbance had disappeared, but a feature similar to its f. end was present at a different longitude, with a well-defined brilliant white oval just p. it on STBs (Plate P6.1). These features were all moving with the SSTC. They were shown on a photograph from the Lowell Observatory (Slipher, *1964*), from which latitudes have been measured (normalised to BAA micrometer measurements of the major belts; probably ±1.5°). The STBs white oval was at 32.4°S, clearly within normal STC territory, while the associated dark feature gave rise to a narrow band at 36.4°S (a typical STZB) and a dark belt centred at 43°S (a typical 'southerly SSTB'; see next chapter).

Thus the structures associated with the S. Temperate Disturbance, moving with the SSTC, did indeed extend over both

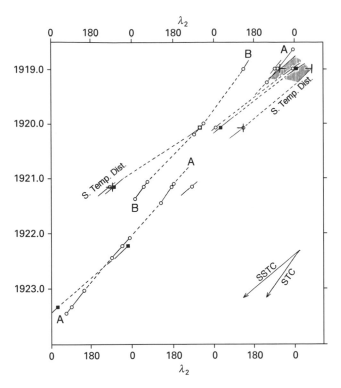

Fig. 11.6. Drift chart of S. and S.S. Temperate features including the 'S. Temperate Disturbance', 1918–1923, in System II longitude. Data from BAA Memoirs, with long-lived white ovals identified by Reese (*1972b*) (Fig. 11.3). Symbols: ○ white spot STBs (moving with STC or, unusually, with SSTC); ⊢---⊣ S. Temperate Disturbance (moving with SSTC); ■ dark spot in STZ/SSTB (moving with SSTC).

the S. Temperate and S.S. Temperate domains. The phenomena were similar to those attending the birth of the three great ovals, and theoretical discussion of them can be deferred until the three ovals have been described in the next section.

A unique interaction between white spots in the normal SSTC and STC latitudes was recorded in 1929/30. As the SSTC spot approached the STC spot, they both decelerated, and the SSTC spot then disappeared about the time of conjunction; the two spots 'may possibly have coalesced' although they were 'well separated in latitude' (BAA Memoir). The STC spot then resumed its original speed.

11.3 THE THREE WHITE OVALS[1]

From 1939 to 1941, changes in three widely separated portions of the STZ led to the formation of three great white ovals, called BC, DE, and FA, which dominated the structure and motion of the South Temperate region for the next 50 years. They began as long sectors of STZ ('proto-ovals') which have progressively contracted. They initially moved with SSTC velocities, and their speeds have gradually decreased, although with substantial fluctuations, which showed a 12-year periodicity until 1979. In latitude, the ovals have shifted north by 2°, with fluctuations that have been

[1] The origin of the STB white ovals was described by Peek (*1958*) and Reese (*1952*). Reviews of their development have been published from time to time: Reese (*1962a*); Cortesi (*1962*); Sato (*1969/1974*); Reese (*1971b*); Favero *et al.*(*1979*); Beebe & Youngblood (*1979*); and Rogers & Herbert (*1991*), from which this account is adapted.

correlated with the speed of the ovals. As the STBs edge has also moved southward, the three ovals have become more and more embedded in the STB. Fig. 11.1 shows schematically the typical appearances of the ovals, as well as the circulations inferred in and around them as described below.

These white ovals are the most long-lived features on Jupiter after the GRS. In fact, they are extremely similar to the GRS. This was initially apparent from their shape and their position on the south edge of a major belt, and it was demonstrated in detail by the Voyager spacecraft.

At the time of writing, the three ovals appear to be nearing the end of their life. They have all become quite small, and the smallest (oval FA) has seldom been seen for several years except in high-resolution photographs.

Origin of the three ovals

The three 'proto-ovals' appeared between 1939 and 1941 when the STZ became subdivided by the gradual appearance of three dark features, which initially appeared either as dusky sections of the zone or as segments of a south component of the STB. These dark features were unusual in that they moved with the SSTC. They were followed independently by the BAA and by E.J. Reese as they expanded longitudinally and confined the three intervening bright sectors of STZ into gradually contracting ovals (Figs. 11.7&8; Plate P9.2). Reese originally named the dark segments AB, CD, and EF, and the ovals formed between them therefore became known as FA, BC, and DE.

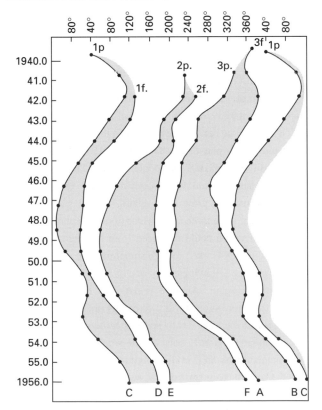

Fig. 11.7. Drift chart for STBs 'proto-ovals', 1939–1956, plotted in a longitude system moving 0.8°/d faster than System II ($\lambda_2 - 24°$/mth). Shaded areas denote the dark features that came to separate the three ovals. (From Peek, *1958*, reproduced by permission of Faber & Faber Ltd.) (See also Fig. 11.13.)

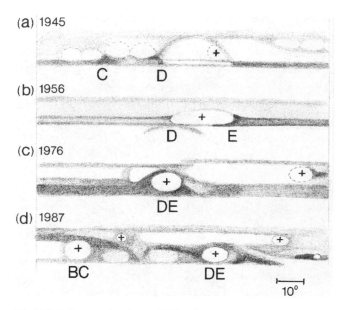

(a) 1945

C   D

(b) 1956

D   E

(c) 1976

DE

(d) 1987

BC   DE

10°

Fig. 11.8. Strip-maps showing oval DE, redrawn from the original observations. (a) 1945 May-June, from unpublished drawings by J. Focas (41-cm refractor, Athens); shows following end of BC and preceding end of DE; both proto-ovals extend out of the frame. (b) 1956 March, E.J. Reese (15–cm reflector, Pennsylvania). (c) 1976 November, J.H. Rogers (30-cm refractor, Cambridge). (d) 1987 October, I. Miyazaki (20-cm reflector, Okinawa) and J.H. Rogers (30-cm refractor); this was the only remaining segment of intact STB. The topmost belt in each map is (S)SSTB, and the last two maps also show segments of true SSTB towards the left-hand side. (From Rogers and Herbert, *1991*.)

The situation at the time of origin of the three ovals was as follows. The STB was fairly narrow, as it had been for several years. The darker features that initially split the STZ into the three 'proto-ovals' were sometimes dusky segments of STZ, and sometimes segments of a STB(S) belt component. They were described thus by Peek:

> The harbinger of the series appeared at the beginning of the apparition of 1939–40 as the p. end of a thicker portion of the STB. During the next two apparitions it took the form of the p. end of a dusky part of the STZ; but in 1942–43 it began to transform itself into a S. component of the STB and as such it remained until the end of 1947, when the zone to the south of it was beginning to grow dusky again...

> In all there were three main wider or double portions of the belt, all of which could be traced from the beginning of the apparition of 1941–42...though sometimes the preceding and sometimes the following end of one or other would fade, only to reappear again. Another remarkable transformation to which they were subject was a spreading out or diffusion, as it were, across the STZ; during some apparitions a section of the belt that had previously been double would not so appear, but the whole of the STZ in the longitudes it should have occupied would have grown dusky, with well-marked p. and f. ends that suggested miniature replicas of the p. and f. ends of the old S. Tropical Disturbance. Then the Zone would clear again and the double section of the belt be re-established. Such changes were actually seen to take place during the course of a few weeks. (Peek, *1958*.)

The early appearances were described in detail in the BAA Memoir for 1942/43. After 1944, these dark sectors separating the proto-ovals usually appeared as broad or double sectors of STB, without darkening the STZ.

The south edge of the proto-ovals was often delimited by a STZB or true SSTB segment. In 1939 there was a dark though narrow STZB, and in 1940, one photo showed this STZB to be at 37°S, continuous with the north edge of a segment of broad dark 'true SSTB'. Unfortunately, few data could be obtained for 1940, so we cannot say whether the true SSTB segment was physically related to the origin of the 'proto-ovals'. In 1941/42 this well-defined true SSTB was present at most longitudes, although it was often double and pock-marked by small light ovals. In 1942/43 the true SSTB had faded away again, and only faint traces of belts were present. These included a faint STZB at 37°S lying alongside at least one proto-oval, and a similar though incomplete STZB persisted until 1945 as the southern edge of proto-oval DE (Fig. 11.8, Plate P9.2).

The p. ends of the dark sectors, or f. ends of the proto-ovals, were more often well-defined than the opposite ends, and sometimes the STZB clearly curved round to outline the following edge of one of the proto-ovals. However, the proto-ovals were not always clearly defined, and there was considerable detail within the proto-ovals before they matured into distinct ovals with all-white interiors. In particular, a more distinct white oval was recorded within the preceding end of each of the proto-ovals in 1943–1945. A distinct white oval had also existed in 1939, which, in retrospect, probably occupied the preceding end of the newly forming proto-oval BC; Reese (*1972b*) suggested that it had existed since 1938 or earlier, and was gradually accelerating as the SSTC took hold of the STBs latitudes (Fig. 11.3).

If it is assumed that the ovals have always had the same anti-clockwise internal circulation, then their mode of origin in 1940 fits in well with the pattern of jetstreams as seen in 1979 (Fig. 11.1). The STBs and SSTBn jetstreams in 1940 would have evolved into the anticyclonic circulation of the three proto-ovals, and would have been in approximately the same latitudes as the equivalent jetstreams observed nearly four decades later by Voyager.

Lengths of the ovals

The 'proto-ovals' contracted very rapidly at first, and since 1950 the ovals have contracted more slowly. The lengths are plotted in Fig. 11.9. The contraction has clearly not been exponential; according to Beebe and Youngblood (*1979*), it was approximately linear from 1966 to 1979, and this behaviour has continued if the average of the three ovals is taken. Individually, however, ovals BC and DE remained about 8° to 9° long in the 1980s, while oval FA continued to shrink until only ≈5° long. During this decade, most of the STB disappeared, and oval FA was seldom visible after 1985. Its position in the whitened 'STB Zone' (STBZ) was most consistently marked by a faint dusky patch on its p. edge, which was reddish. Oval FA reappeared clearly after the 1990 SEB Revival, when each of the three ovals acquired a dark collar.

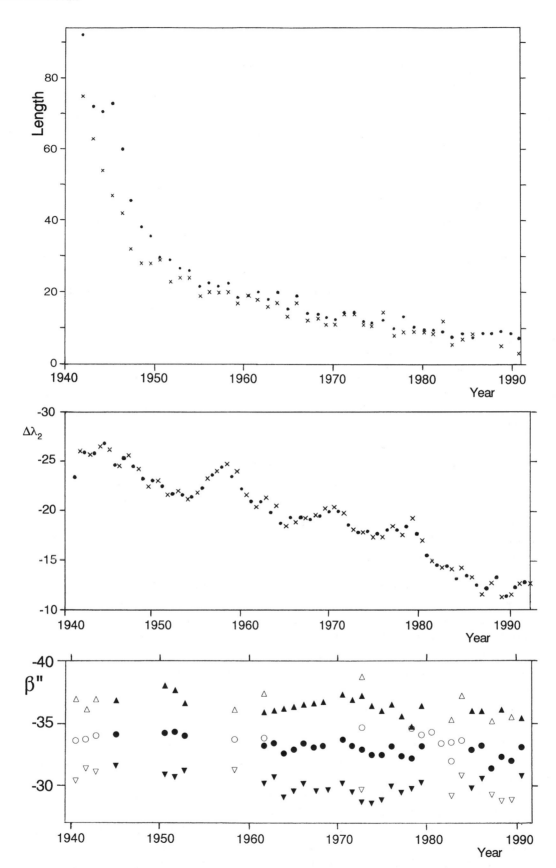

Fig. 11.9. Lengths of the three ovals plotted against time: ×, FA; ●, average of BC and DE. Fluctuations in later years may not be real. (From Rogers and Herbert, *1991*.)

Fig. 11.10. Speeds of the three ovals plotted against time: ×, within each apparition (corrected for phase effect); ●, between oppositions. (From Rogers and Herbert, *1991*.)

Fig. 11.11. Latitudes of the three ovals plotted against time: ●, centres; ▼▲, north and south edges. Solid symbols are high-precision measurements (±1° or better) and open symbols are medium-precision measurements. The least-squares fit to the high-precision measurements only is -0.041°/year with a correlation coefficient $r = 0.742$; with the first three 1940s points included, the fit is -0.036°/year with a correlation coefficient $r = 0.755$. (From Rogers and Herbert, *1991*; statistics by courtesy of A. Fitt.)

Oval FA had always tended to be the smallest, and was formed from the shortest sector of STZ. The difference in length amounted to 2–3 years of shrinkage in the 1940s and about 5 years of shrinkage in the 1970s, so it did not simply reflect a difference in age between the ovals. If the size was determined by the amount of energy initially incorporated from the jetstreams of the STZ, the persistent inferiority of oval FA could be because it formed from a shorter segment of STZ. Alternatively, if the size was determined by energy drawn continuously from outside, an initially smaller oval may have been less proficient in maintaining itself.

Motion of the ovals in longitude

The initial drift rates of the three 'proto-ovals', $\Delta\lambda_2 = -26°$/mth, were recognised as unusual at the time. They were still accelerating during their first few years, before dissociating from the SSTC and beginning a long-term deceleration towards the normal STC speeds. The average speeds of the ovals are plotted in Fig. 11.10.

Fig. 11.10 shows substantial fluctuations in speed, and G. Favero and colleagues (*1979*) pointed out that up to 1977 these comprised an oscillation with a period of about 12 years, superimposed on the long-term deceleration. The period corresponds to the jovian year, and the fastest speeds occurred around the time of aphelion, when Jupiter receives least warming from the sun. There was another maximum in speed in 1979; however, this was only 9 years after the previous maximum, and the ovals then slowed down abruptly. It will be interesting to see whether they show another maximum in speed before they finally disappear. (The next aphelion is in 1993.)

There are also interesting short-term trends in the speeds of the individual ovals. They tend to show a temporary acceleration when within about 50° of the Great Red Spot, travelling fastest when alongside it – although at any particular encounter there have often been irregular accelerations and decelerations. This behaviour was first reported by Reese (*1971b*) (Fig. 11.12). It has not been systematically analysed since but was confirmed by Italian observations in 1975/76 (UAI report) and by German observations in 1981–1992 (Fig. 11.13).

They also show individual variations in speed, and they have sometimes come within one diameter of each other, but they have never collided. Takeshi Sato (*1969/1974*) showed that the acceleration of the individual ovals was inversely related to their separa-

tion; that is, the ovals appeared to repel each other over distances of up to 180° (Fig. 11.14). But in 1988/89, ovals BC and DE approached until their centres were only 22° apart, and for three years they have retained much the same spacing – decreasing to 18° in 1990–92, and moving as a group, speeding up and slowing down together.

Motion of the ovals in latitude

Although the ovals began as sectors of STZ, they evolved into ovals nestling in the STBs edge, and then into ovals completely embedded in the STB. Have the ovals moved north or has the STBs edge moved south? Accurate latitude measurements showed that

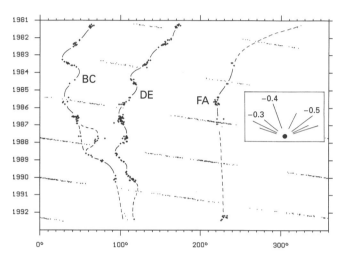

Fig. 11.13. Drift chart of the STB ovals, 1981–1992, plotted in a longitude system moving 0.42°/d faster than System II (λ_2 – 12.6°/mth). This shows that they tend to move faster while approaching the GRS and decelerate after passing it. The GRS is indicated by the nearly-horizontal rows of points. Chart by H-J. Mettig, from corrected visual data of the Arbeitskreis Planetenbeobachter. (Updated from Mettig, *1991*.)

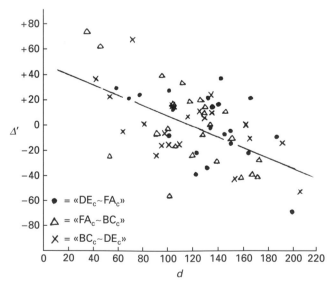

Fig. 11.14. Mutual repulsion of the STB white ovals, from Sato (*1969/1974*). d is the distance between the ovals in degrees longitude, and Δ' is the acceleration in degrees per (13 months)^2. The least-squares regression line is:

$$\Delta' = 48.5 - 0.406d,$$

with a correlation coefficient of 0.549.

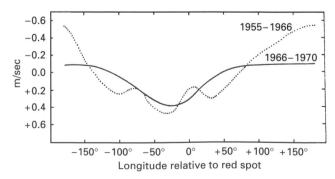

Fig. 11.12. Variation in drift rate of the three STB ovals according to their distance from the GRS. This shows that they tend to move faster while passing the GRS. Speed is given relative to the global STC average. (Speed positive in m/sec is negative in degrees/month.) (From Reese, *1971b*.)

both have happened, and implied that the the STBs jetstream must have remained at approximately constant latitude while the ovals burrowed into it.

The latitudes of the ovals were measured from high-resolution photographs and were plotted against time by Rogers and Herbert (*1991*) (Fig. 11.11). Small but real changes have clearly occurred. The changes are more clearly shown in Fig. 11.15 where only high-precision measurements, averaged over two or three apparitions, are plotted. The three ovals have drifted northwards by almost 2° during their lifetime. The latitudes of the edges of the ovals show more scatter than those of the centres, probably because the measurements cannot be as precise, but the overall northwards shift of the ovals seems to have been mainly due to expansion at their north edges, digging into the STB.

Much of the apparent scatter in the positions of the centres may be real. Remarkably, the maxima of latitude in Fig. 11.11 tend to coincide with the maxima of speed in Fig. 11.10. We therefore plotted latitude against speed in Fig. 11.16. A clear trend emerges. The latitudes correlate about as well with speed (Fig.11.16; correlation coefficient 0.741) as they do with time (Fig. 11.11: correlation coefficient 0.755).

It is unfortunate that no high-precision latitude measurements could be obtained for the late 1950s and the early 1980s, when the speed underwent some of its greatest fluctuations. Nor are there enough precise data to correlate latitude and speed for individual ovals within an apparition, except in one instance. In 1969/70, NMSUO measurements to ±0.1° showed that oval DE changed its latitude and speed together as plotted by crosses in Fig. 11.16, in very good agreement with the overall correlation.

The relationship between latitude and speed cannot be merely incidental to a decrease of both with time, as this would not explain the short-term fluctuations. Nor is it due to conservation of angular momentum, as the change in radius (cos β) (2%) is 40 times greater than the change in angular velocity (0.05%). Perhaps the most plausible cause of the correlation is that, as the ovals move north, they move more directly into the line of the retrograding STBs jetstream at 32.6°S. This might tend to slow them down. However, since the slower speeds were equivalent to the STC that had previously been typical of features on STBs as well as within the STB, this explanation begs the question of why the ovals ever moved faster in the first place.

In terms of this correlation, the latitudes of the ovals have shown exceptional scatter from the 1980s onwards. We have not found any good explanation for this, either in terms of systematic errors or in terms of jovian dynamics. It could be related either to the diminishing size of the ovals or to the changing pattern of the belts around them.

Spacecraft observations

Closeups of the white ovals are in Figs. 3.6, 11.2, 11.18, 12.2–4, and Plate P17.4. The Voyager imagery showed they roll between the SSTBn prograding jetstream tangential to the oval's south edge and the STBs retrograding jetstream which is deflected around the oval's north edge. They have the same type of anticlockwise, anti-

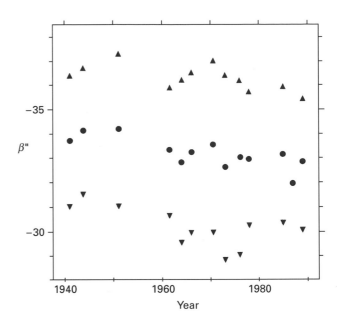

Fig. 11.15. Latitudes of the three ovals plotted against time: high-precision measurements only (plus the 1940s measurements), from Fig. 11.11, averaged over two or three apparitions. (From Rogers and Herbert, *1991*.)

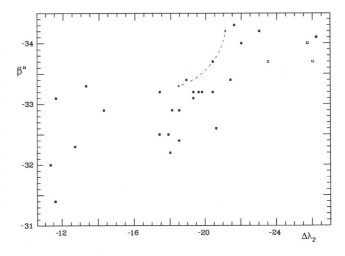

Fig. 11.16. Latitudes of the ovals plotted against speeds: the high-precision measurements plus the 1940s measurements (open symbols). The least-squares fit is:

$$-\beta'' = 30.8 + 0.125\,\Delta\lambda_2$$

with a correlation coefficient $r = 0.741$. Also shown are short-term measurements of oval DE in 1969/70 by NMSUO (crosses). (From Rogers and Herbert, *1991*; by courtesy of A. Fitt.)

cyclonic circulation as the GRS (Fig. 11.17). The puffy internal structure and opening spiral patterns indicate that rotation is fastest around the periphery, just as in the GRS, and the vorticity is virtually identical: 3×10^{-5} sec^{-1} overall, and 6×10^{-5} sec^{-1} near the outer rim.

For a long distance f. ovals BC and DE, the Voyager photos showed dramatic cyclonic turbulence in the belt, exactly like that following the GRS in the SEB. It consisted of 'folded filaments' aligning as chevrons (Figs. 3.6, 10.9, 12.3&4). Turbulence from these regions was propagating outwards on the STBn and STBs jetstreams. Also as in the SEB, such disturbed regions can exist

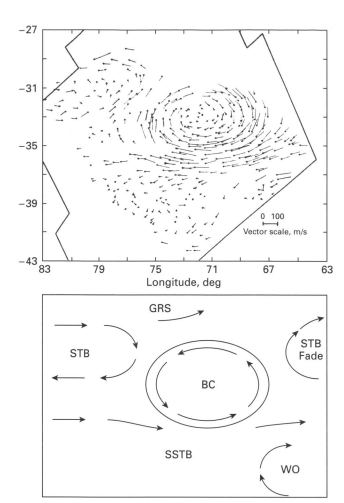

Fig. 11.17. Circulation in and around oval BC. (A) Winds measured from images 1.4 hours apart during the Voyager 1 encounter (including Fig. 11.2). North is up, and longitude is System III. Each dot marks the start of a vector, whose length is proportional to wind speed. (From Mitchell *et al.*, *1981*.) (B) Sketch of the circulations. (Also see Figs. 10.2,39,57.)

separately from the ovals. The Voyager 2 movies showed a vigorously disturbed sector of STB alongside the GRS, tens of degrees from ovals BC and DE on either side.

When two of the ovals drift a long way apart, a fourth light spot sometimes appears in the STB roughly midway between them, as if to maintain an even spacing of features. Both the Pioneer and Voyager spacecraft observed such a fourth feature, and it was found to be another region of cyclonic turbulence within the STB, like those which immediately follow the three long-lived ovals.[2] The appearance of this 'orphan' region appears to divide the STB into four sectors rather than three.

What produced the white ovals?

The three darker sectors that initially separated the proto-ovals were clearly in the true STZ, separate from the true SSTB segments which they superficially resembled. As noted by Peek, they most resembled the great South Tropical Disturbance which, coin-

cidentally or not, had disappeared in 1939 just as the darker sectors in STZ were beginning to appear. It is now clear that this resemblance was very close, in the following particulars:

> these features were long darkened sectors of a zone;
> they had concave p. ends abutting an anticyclonic circulation (often marked by a white spot), and less distinct f. ends;
> they progressively lengthened;
> they moved faster than normal for the domain, with a speed more typical of the next domain to the south;
> they progressively decelerated towards the normal speed for their domain;
> they lasted several decades.

Many of these similarities were also shared by the 'South Temperate Disturbance' of 1918–1921.

It is intriguing that the ovals arose by near-simultaneous division of the STZ at three roughly equidistant points. Possibly this was a manifestation of some sort of standing-wave pattern with wavenumber 3, that is, exactly three waves around the planet. The long-range mutual repulsion between the ovals could be another effect of such a planetary wave system, tending to maintain even spacing at wavenumber 3. Even so, the ovals do sometimes drift towards very unequal spacing, and the appearance of an 'orphan' fourth region of cyclonic disturbance at such times suggests that the STB can sometimes be divided by four waves rather than three. A possible basis for this regularity will be discussed in Chapter 14.5.

11.4 FADING OF SECTORS OF THE STB[3]

In 1961, the segment of STB 150° long between ovals FA and BC faded away, becoming replaced by white material like the adjacent STropZ. This was the first major STB Fade since the birth of the three ovals, and the first of a series of these events which became more and more extensive as the three ovals continued to shrink. In such events, a segment of STB usually bounded by two of the great white ovals all but disappears. The fading may begin patchily, and well-defined dark streaks sometimes linger for a few weeks or months (*1926, 1967/68, 1975*), but the Fade is usually complete within a month or two, leaving behind only a tenuous north component, STB(N). The whitened STB latitudes may be termed STB Zone (STBZ). Occasionally the STBZ may be cream-coloured (*1964, 1990*; Plate P13.1), but usually it is white, sometimes even brighter than the adjacent STropZ. Often, though, a new belt forms immediately to the south as if to provide a substitute for the STB.

The recorded STB Fades are listed in Table 11.2. In the 1960s, each of the three STB sectors faded in turn for one or two years. In 1975, a new STB Fade was limited at its p. end not by a great white oval but by the 'orphan' fourth region of cyclonic disturbance. This Fade only lasted for about 8 months, but as it recovered the adjacent sector faded, between ovals FA and BC, and this Fade became a permanent feature of the planet for at least the next 15 years (Figs. 10.51–57). Its further evolution will be described below.

---

[2] Rogers & Young (*1977*); Mitchell *et al.* (*1979*); Beebe *et al.* (*1989*) (their Figs. 132 and 139). See Figs. 3.7 and 10.64.

[3] This section is adapted from Rogers (*1980*) and subsequent BAA reports.)

Table 11.2. *Fadings of sectors of the South Temperate Belt*

No. (1)	Date first seen (2)	W.s.p. (3)	Length (4)	STZB or true SSTB (5)	W.s.f. (6)	Rate of growth (7)
1	(1888)	?	≥80°	–	–	
2	(1889)	–	≈90°	+	–	
3	1891 Aug	+	40°	+	+	
4	1926 Jul	+	160°	–	+	
*	1931 Jan	–	20°	+	+	
5	(1934)	–	90°	–	–	
*	(1951)	BC	27°	–	DE	
*	1959 May	FA	18°	–	BC	
6	1961 Aug	FA	150°	–	BC	+1.9, *+1.4*
7	(1964)	BC	120°	–	DE	+1.0, *+0.7*
8	(1967)	DE	120°	+	FA	*+1.4*, +1.4, +2.8, ≈0, *+0.9*
9	1975 Jan	+	90°	+	FA	[Short-lived]
10	1975 Sep	FA	60°	+	BC	(Fig.11.19)
11	1984 May	BC	80°	+	DE	[Short-lived]
12	1986 Jun	–	80°	+	FA	*–1.9*, (var.), +4.1, +0.9

Column 1: Number (*, miniature Fades).
Column 2: Date first seen; brackets indicate that it was at the beginning of the apparition.
Column 3: White oval spot on STB at p. end?
Column 4: Length of STB initially affected.
Column 5: Dark 'STZB' or true SSTB alongside Fade?
Column 6: White oval spot on STB at f. end?
Column 7: Rate of growth in degrees per month; successive figures give mean rates either within an apparition, or (in italics) from one opposition to the next, in temporal order from the start of the Fade.

These Fades were nothing new. As Table 11.2 indicates, several sections of STB faded earlier in the twentieth century. Nineteenth-century drawings also often showed breaks in the STB, for example in 1881, 1888, 1889, and 1891 (Fig. 11.4). In each of the last two Fades there was a faint STB(N) and a darker 'STZB' or true SSTB, and the short 1891 Fade was bounded at each end by a small white spot in the STZ with the typical STC drift of $\Delta\lambda_2 = -17$.

There were also miniature Fades in 1951 and 1959 when pairs of the great white ovals drifted very close together. In 1951, ovals BC and DE were only 27° apart edge-to-edge, and the STB between them was conspicuously double with a white rift between the components. In 1959, a similar bright rift developed temporarily between ovals FA and BC when they were only 18° apart edge-to-edge. (The long-lived ovals did not come so close again until 1988/89.)

The onset of STB Fades is probably influenced by the GRS. The Fades of 1889, 1961, and 1975 September, were first seen alongside the GRS; those of 1926, 1934, and possibly 1967 formed within 20° of it; and that of 1975 January, although first breaking 70° f. the GRS, spread much closer to it as the Fade reached its full length. The spread in relative longitudes is large, but so is the range of the dynamical influence of the GRS on the white ovals.

The conclusion of an STB Fade is a gradual affair, the STBZ becoming imperceptibly darker or narrower until the STB appears normal again. An exception occurred in 1928 when spots on the STBn jetstream seemed to produce the revival of the STB. But these spots were apparently derived from the SEB Revival via the Circulating Current; no other violent revival of the STB has ever been observed.

Spacecraft observations

At the time of the Pioneer 11 encounter in 1974 December, an STB Fade was just beginning between an internal light patch (an 'orphan' cyclonic disturbance) and oval FA (Rogers and Young, *1977*). The Pioneer 11 'snapshot' showed this STB section to consist of three narrow components: STB(N), STB(S) (a dark bluish component, which was the one that disappeared soon after), and STZB or SSTB(N) (which had been present for several months and soon developed into a dusky true SSTB alongside the Fade).

The Voyager spacecraft had a prolonged view of the fully-formed STB Fade between ovals FA and BC, and its findings were remarkable. The Fade section consisted of a bright white area of fluffy white clouds completely enclosed in a beautifully scalloped or braided border, around which the STBn and STBs jetstreams were connected to form a closed cyclonic (clockwise) circulation (Figs. 3.6, 10.57, 11.18, and Plate P17). Although these two jet-streams were also partially connected on the other side of oval BC and adjacent to oval DE, where there was no STB Fade, the connection seems to have been much better-organised around the STB Fade. One can speculate that the formation of this closed circulation was responsible for the white cloud cover that developed over it. Very similar features were observed by Voyager on several belts in higher southern and northern latitudes. If the formation of such a circulation tends to happen while passing the GRS, it seems to involve formation of a 'chiasma' of currents and to have some features in common with South Tropical Disturbances – see below and Chapter 13.

Spacecraft observations of the normal STB alongside the GRS may provide evidence for the instability of the belt at this point, which may be relevant to induction of STB Fades. The famous Pioneer 11 picture of the GRS and oval DE, which had just passed the GRS (Fig. 12.2) showed a rift through the STB between the oval and the STropZ. Also, Voyager 2 found extensive cyclonic disturbance alongside the GRS as described above.

Of course, we cannot be certain that the closed cyclonic circulation seen by Voyager was an essential feature of STB Fade segments.

STB Fades as dislocations

Most STB Fades have been associated with unusual segments of dark belt immediately to the south, in latitudes which more commonly appear to be STZ, so that the normal belt-zone pattern appears to be reversed. These belt segments have usually been described as 'STZB', but their latitudes indicate that they are actually segments of true SSTB, so they will be described in detail in the next chapter. As defined therein, true SSTB with a north edge at 36°S or 37°S is rarely present alongside normal STB; but during the induction or reinduction of STB Fades, this belt has often been present. Another curious feature of true SSTB segments alongside STB Fades is that they are often perforated by numerous light ovals

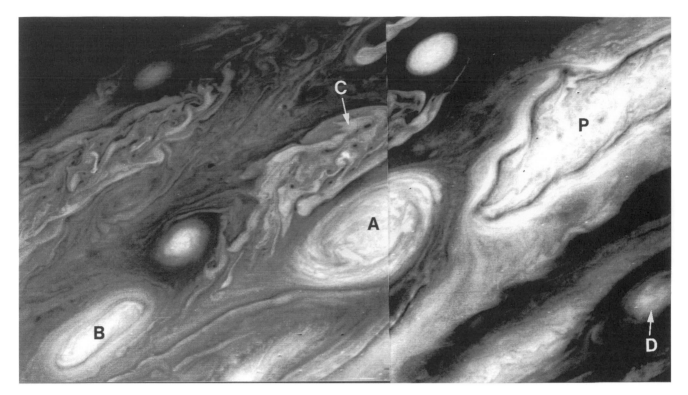

Fig. 11.18. White oval FA (marked A) and the p. end of the STB Fade (marked P), from Voyager 1. Also note cyclonic circulations in true SSTB, one being oval (B) and one filamentary (C), with small bright anticyclonic ovals on either side of them. A SEBs jetstream spot is also shown (D). The images are not perfectly aligned because of rotation during the 23 minutes between them. (Also see Figs. 3.6, 10.57, and 5.1 which overlaps this picture.) (Images 16313.32 and 16314.01, J1–3d, violet, unfiltered. NASA.)

at 38°S, which Voyager showed to be cyclonic circulations, resembling the STB Fade itself on a smaller scale. The archetypal dislocation (e.g. 1978/79 or 1982) thus includes a STB Fade alongside a 'true SSTB with ovals' – as well as further dislocations to north or south which are described in Chapters 10 and 12.

This picture of STB Fades as dislocations of the belt-zone pattern was suggested by the two 1975 cases, in both of which a dark true SSTB was present immediately south of the Fade, before drawing away in the p. direction in the current typical for its latitude (S.S. Temperate Current, SSTC). The 1975 FA–BC Fade, which had almost recovered during 1976, was re-initiated in 1977 January and 1978 January as two successive sections of pre-existing true SSTB drifted up to it in the SSTC (Fig. 10.51). The second of these was probably an extension (in the p. direction) of the same true SSTB segment that had originally formed alongside the STB Fade in 1975, which had now made a complete circuit of the planet relative to the STB! (However, the picture was complicated by the presence of yet another component alongside the Fade itself, a narrow 'STB(S)' between latitudes 32°S and 37°S, implying that the canonical STZ was darkened.) Since then, as almost the whole STB has disappeared, it has been replaced by a STB(S) or true SSTB as described below.

Although the true SSTB segments are clearly associated with STB Fades, they continue to move on their separate currents (SSTC and STC respectively) and so slide apart rapidly. It is not obvious what the physical relationship between them is. The 1967 event, in which the DE–FA segment faded in association with a

classic segment of true SSTB with ovals, is a challenging example. In early 1967 this segment of true SSTB with ovals was forming between ovals FA and BC, but the STB already seemed to be fading between ovals DE and FA (just p. the GRS). Perhaps the incipient fading of the STB would not have continued, nor been noticed, if the true SSTB segment had not drifted alongside it during the following months?

As described in the preceding chapter, from 1975 onwards, STB Fades in turn induced five successive South Tropical Dislocations p. the GRS – darkening of the STropZ culminating in the appearance of a feature resembling a STropD. There is little evidence for any influence in the reverse direction. In 1941–42 and 1946–47, the STropDs caused fading of the adjacent STB, fading which became complete as the sections of dark STZ that separated the proto-ovals moved up to the same longitudes. However, the STropDs and dark STZ segments had arisen independently, and by 1955, when the ovals were more mature, a STropD induced no sign of dislocation further south. There is no case of a STropD alone inducing a full STB Fade.

Before the Voyager encounters, the author had speculated that induction of STB Fades by segments of true SSTB was a 'dislocation' phenomenon similar to the induction of South Tropical Dislocations by STB Fades (Rogers, *1980*; see previous chapter). Both phenomena involve the reversal of the usual belt-zone pattern over a part of the planet, and both usually start adjacent to the GRS. Subsequent events have confirmed beyond doubt that these dislocations are real phenomena, but have weakened the analogy between them. The possible causal relationships will be discussed in Chapter 13.

One further general property of STB Fades is that they grow longer with time. Sanchez-Lavega and Rodrigo (*1985*) noted that the FA–BC STB Fade lengthened continuously from 1975 to 1983, and contrasted the gradual lengthening of this cyclonic circulation with the gradual shortening of the anticyclonic white ovals. Apart from a hiatus in 1983 (when a stretch 50° long f. oval FA recovered), the FA–BC Fade continued to lengthen thereafter until it extended more than half way round the planet (Fig. 11.19). Table 11.2 shows that all the other STB Fades since 1961 have likewise lengthened progressively. The mean rate of growth was 0.9°/mth (0.38 m/s) for the FA–BC fade (Fig. 11.19), and 1.2°/mth (0.51 m/s) for the others in Table 11.2.

Disappearance of the STB in the 1980s

The FA–BC Fade persisted and grew still longer during the 1980s, inducing the South Tropical Dislocations described in the previous chapter, and eventually covering almost half the circumference of the STB (Figs. 10.51–57). In 1983, ovals FA and BC were 149° apart, but the p. end of the Fade followed ≈50° behind oval FA at a stepup of the STBn edge. During 1983 the p. end of the Fade migrated back closer to oval FA, and remained 20°–40° f. FA through 1984 and 1985. Thus there was a short segment of 'normal STB' f. oval FA. In 1986, when the STB began to fade p. oval FA as well, this segment remained for a year as a distinct dark streak or 'STB Remnant'. It darkened substantially as it passed the GRS in 1986/87, but then rapidly faded and disappeared in 1987 Sep. From then onwards, oval FA was rarely detectable. The motion of oval FA, and of the p. ends of the STB Fade, was irregular in the late 1980s; this was illustrated by the STB Remnant in 1986 which may have oscillated with a period of about two months and an amplitude of about 6°.

The Fade between DE and FA began in mid-1986 as this sector was passing the GRS. For its first two years, only the f. half of the DE–FA sector faded, beginning at a gradual deflection of STB ≈50° f. DE. So there was a segment of normal STB f. oval DE, which shortened in 1988, just like the earlier one f. oval FA. Often, the f. end of this STB segment merged only gradually into the true SSTB Sf. it, so that the p. end of the STB Fade was imperceptible. This has been typical of such isolated STB segments since then. This short STB segment lasted until 1990 when it was reduced to a small dark spot.

The third sector, between BC and DE, had faded partially in 1984 May, and did so again in 1985 August while passing the GRS, but it recovered each time. However, it was becoming unusually short as ovals BC and DE drifted close together. In 1986, when their centres were only 57° apart, the intervening STB broke up into three dark wedges and two light ovals, which still persisted in 1987 although the central dark streak was less prominent. In 1988 January, as this complex of ovals began to pass the GRS, one of the light ovals seemed to open out into the STropZ and disappeared. Ovals BC and DE persisted, and by early 1989 their centres were only 22° apart – the closest they had ever been. They were still sep-

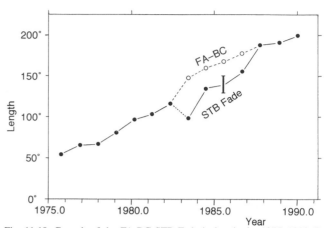

Fig. 11.19. Growth of the FA-BC STB Fade in longitude, 1975–1990. In 1983, the p. end of the Fade shifted tens of degrees f. from oval FA for several years; the open symbols indicate the FA-BC distance during these years.

arated by a single small dark blob. This blob, and the similar one lingering on the f. side of oval DE, were then the only remnants of the true STB.

At all other longitudes, the STB was absent except for a very tenuous STB(N), and the main belt in the 'south temperate' region was true SSTB – dark and continuous almost the whole way round the planet, for the first time in recorded history (Plates P22 and P23). In 1989, there were very faint shadings across the STBZ-STZ, and in 1990, sectors the STBZ were slightly yellowish, but nothing developed further.

The dark blobs between and following ovals BC and DE faded in spring 1990, and disappeared as they passed the GRS in 1990 September, so the STB was now completely absent. In late 1990 the SEB was reviving, and soon all three ovals reappeared with dark collars (Fig. 10.56); there was also an extra bright oval nestling between BC and DE, presumably cyclonic (at 30.5°S), and oval DE was reddish for a month or two! Dark streaks of STB re-formed f. ovals DE and FA, so the overall picture reverted to that of 1987. But it did not last; by 1991/92, the 'Remnant' f. FA was again the only dark material remaining.

In 1992/93, the STB is at last reviving fully. The remnant f. FA has been joined by two other long dark sectors, f. DE and f. another light oval. All three span STZ latitudes and trail in the Sf. direction. Although all features are still moving with the STC, some sectors have faded while others have darkened, and the net result in 1993 May is an almost complete STB.

11.5 THE SSTBn PROGRADING JETSTREAM

According to Voyager data, the next prograding jetstream to the south is at 36.5°S. This jetstream was previously unknown. If the jetstreams are taken as the canonical boundaries of the belts, it should mark the north edge of the SSTB. Sometimes it does, but only infrequently, as discussed in the next chapter. More commonly, this latitude is marked only by a narrow band ('STZB') or fragments thereof, or not at all. Fragments of STZB tend to drift with the SSTC. One unusual feature on this band in 1909/10 and

1911 was a smooth southward deflection where the belt crossed a dusky column in the STZ (the shape of the column would nowadays be likened to a cooling-tower). This complex moved with the SSTC, probably throughout these two apparitions.

In the Voyager imagery, the SSTBn jetstream was detected mainly by the circulations within the cyclonic features in the true SSTB (see next chapter), and by the motion of detailed cloud texture. There were no substantial spots moving with it.

Since the Voyager flyby, the SSTBn jetstream has been observed from Earth from 1988 December to 1990 March (BAA, and Rogers and Miyazaki, *1990*). With the STB absent, the main southern belt was true SSTB, except for a gap where there were only tenuous SSTB(S) and SSTB(N) (STZB). The jetstream was detected by the motion of several tiny dark spots on the SSTB(N) in the gap, as revealed by photographs by I. Miyazaki and others (Fig. 11.20). Later, several similar spots were tracked along the broad and dark sectors of SSTB. The spots had a minimum spacing of 12° to 15°. Their weighted mean speed was $\Delta\lambda_2 = -79$ in both apparitions (Table 11.3). There seems to have been a real scatter in the speeds of the spots, and the range included the speed of the SSTBn jetstream observed by Voyager in 1979.

It seems likely that these spots were similar in nature to the dark jetstream spots in concurrent outbreaks on the SEBs and NNTBs jetstreams, although the SSTBn spots were smaller and barely resolved. There was probably a distinct outbreak of spots on this jetstream in 1988/89. It is difficult to put limits on the frequency of such outbreaks because the planet is not often observed at high enough resolution to detect them, but no such outbreaks were detected by professional photography from 1965 to 1979, nor by the Voyager spacecraft.

There are some tantalising drawings in the BAA Memoir for 1932 and 1933 (Plate P7.5), showing a STZB carrying small dark spots much like those that defined the jetstream in 1988/89; but we will probably never know whether these spots represented an earlier outbreak on the jetstream.

Table 11.3. *Records of the SSTBn jetstream*

Appar'n	$\Delta\lambda_2$	(range) (n)	Lat.	Source
1988/89	−79	(−63 to −111)(3)	−35.8	BAA
1989/90	−79	(−75 to −90) (4)	−36.3	BAA
Voyager	−88	(global)	−36.5	Limaye(*1986*)

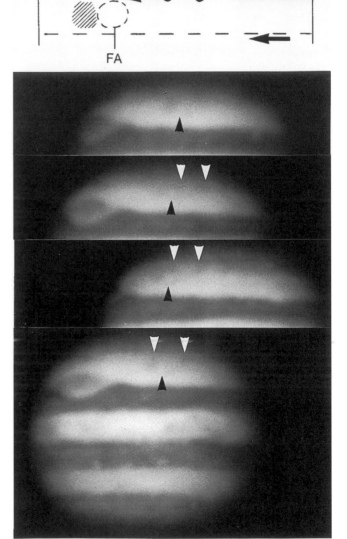

Fig. 11.20. Discovery of SSTBn jetstream spots in 1988/89. Top: sketch of two jetstream spots near oval FA. Bottom: Four photographs by I. Miyazaki showing motion of the spots. From top to bottom: Dec. 11, Dec. 18, Dec. 19, Dec. 21. Oval FA is arrowed from below and the two jetstream spots are arrowed from above. The GRS is near the p. limb. The first image shows the region before the jetstream spots appeared; the bright spot on the SEB near the f. limb is Io. (From Rogers and Miyazaki, *1990*.)

12: South South Temperate Regions (37°S to 53°S)

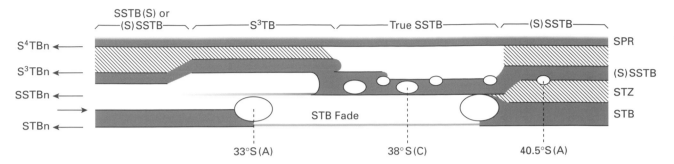

Fig. 12.1. Typical aspects of the S. and S.S. Temperate regions as viewed from Earth. Top and right: names of belts, etc. Bottom: classes of ovals, anticyclonic (A) and cyclonic (C). Left: jetstreams detected by Voyager.

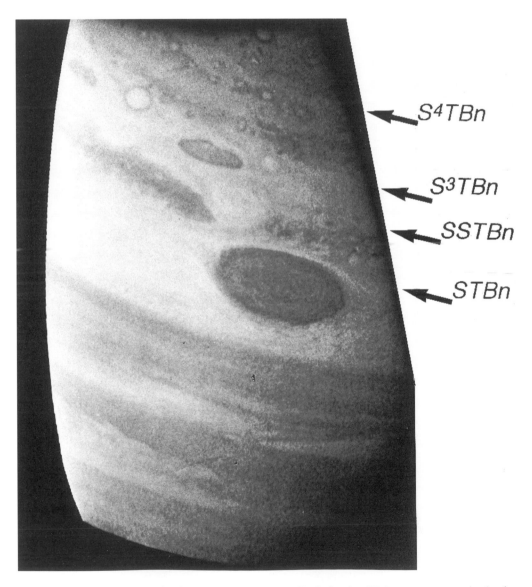

Fig. 12.2. Pioneer 11 close-up showing anticyclonic ovals in S. Tropical region (GRS), S. Temperate region (oval DE), S.S. Temperate region (two at 40–41°S, alternating with cyclonic dark SSTB 'barges'), and further south. (Image C3. NASA.)

12: South South Temperate Regions (37°S to 53°S)

12.1 OVERVIEW OF THE REGION

The latitudes to be considered in this chapter comprise two domains of the jetstream pattern, which would be expected to define the S.S. Temperate Belt and Zone (SSTB, SSTZ), and the S.S.S. Temperate Belt and Zone (S³TB, S³TZ). The reason for considering them together is not merely the observational difficulty of resolving features in this high-latitude region, but more importantly the fact that there is usually only a single major belt in the region, which does not fit into the jetstream framework. Thus there is a systematic mismatch between the belts and the jetstreams.

Unlike lower latitudes, the S.S. Temperate region cannot be covered satisfactorily by most amateur telescopes. The main features, such as small ovals, dusky patches, and tenuous belts, are generally smaller than those of lower latitudes, and perhaps of lower contrast, and can only be seen at all clearly with telescopes of ≥25 cm aperture. The best modern photographs surpass visual observations because they can record low-contrast detail accurately. But the Pioneer and Voyager spacecraft were needed to reveal the real, remarkable patterns in the region.

Most visual observers therefore see little detail in these latitudes. Often there is a distinct belt described as 'SSTB' bordering the south polar shading, but it is often irregular, and variable in width and latitude and intensity. Often, however, one sees just faint diffuse patterns which only doubtfully constitute a real belt. The STZ may be broad and clear, but it often contains shadings or markings which are difficult to define. When distinct belt segments can be defined, they can change on a timescale of months, making them difficult to track. With higher resolution, and particularly with photographs, more distinct belt segments become visible, sometimes accompanied by distinct small bright ovals.

No distinct colours have been recorded in the region.

The observations of a variable-latitude 'SSTB' contrast with the paradigm of the belts lying between the jetstreams. In an attempt to clarify the situation, the author has compiled precise measurements of belt latitudes over history, and compared them with the latitudes of the Voyager jetstreams. The resulting picture is paradoxical. Between the STropZ and the S. Polar Region, there are commonly two belts – STB and 'SSTB' – with a spacing of about 13°, but there are three sets of jetstreams, with a spacing of about 9°.

The Voyager spacecraft detected prograding jetstreams at 36.5°S, 43.6°S, and 53.4°S, which provide standard positions by which to define the SSTBn, S³TBn, and S⁴TBn edges. There is indeed a perennial boundary at 53°S, marking the edge of the S. Polar Region and sometimes of a narrow belt (S⁴TB); since 1950 it has been present in almost every apparition for which adequate observations are available (section 12.5). The dark shading of the S. Polar Region often appears to extend further north, to the 'SSTB' if present, or to the STZ, but the 53°S boundary is usually still present within it.

The paradox is that, between 36°S and 53°S, there is usually only one substantial belt, whose position often bridges the two units defined by the jetstreams. Although this belt has almost always been referred to as 'SSTB', it is usually further south than the canonical latitudes of the SSTB. Sometimes it approximately covers the canonical latitudes of SSTZ, when we will call it *'southerly SSTB'* or (S)SSTB, and sometimes it is a true S³TB. It is to be distinguished from the *true SSTB*, defined as lying between 36°S and ≈40°S, which is only infrequently a major belt. The true SSTB is most commonly seen in association with a STB Fade. Even then, it is often disrupted by numerous light ovals. Otherwise, true SSTB segments tend to be quite short, often with a sharp p. end marked by a white oval moving with the S.S. Temperate Current (SSTC). True SSTB features always move with the SSTC and thus drift in the p. direction relative to features on the STB.

The latitude data are summarised in Table 12.1; details are in Figs. 12.7&8.

The average position of the 'SSTB' has changed over the years. As noted by E.J. Reese (*1971a*), only the (S)SSTB and the STB have shown such longterm shifts, and the shifts of these two belts seem to have been coordinated. From 1918 to 1934, the belt was most commonly centred between 40°S and 43°S, and was thus a (S)SSTB in the canonical SSTZ latitudes. From about 1935 to 1980, it usually lay further south, bridging the canonical SSTZ and S³TB latitudes; the STB was then also centred further south because it was broader. And from about 1980 to 1992, the STB has

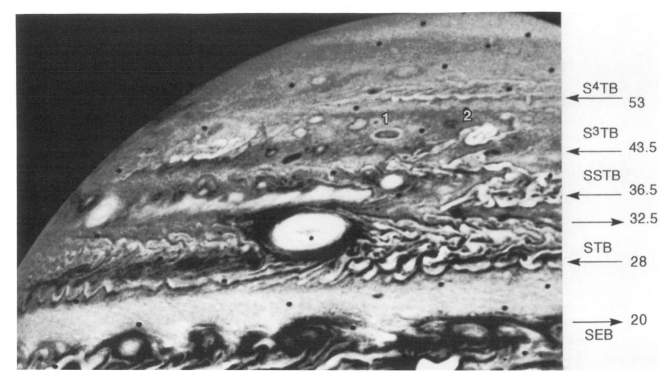

Fig. 12.3. Voyager 1 image of oval DE and regions further south, 1979 Feb. 22. Spatial frequency filtering has enhanced the dark collar around DE, but has not eliminated the true contrasts of the belts. Cyclonic SSTB latitudes include a whitened sector (at left, resembling the STB Fade) and a FFR (at right). Cyclonic S³TB latitudes include spots no.1 (dark oval) and no.2 (bright spiral) listed in Table 12.2B. The < shape connecting no.2 to the SSTB FFR was prograding on the S³TBn jetstream. The latitudes and directions of the jetstreams are marked. (Image 16071.51, violet, filtered. NASA.)

Table 12.1. *Average latitudes of belts, 1950–1989, and latitudes of jetstreams in 1979.*

Latitudes of normal belts (1950–1979)		Latitudes of STB Fade/True SSTB (1966–1989)		Latitudes of jetstreams (Voyager, 1979)	
SPRn	53.2 (±1.2) (15)	–		S⁴TBn	53.4 (–)
				S³TBs	49.3 (+)
(S)SSTBs	46.6 (±1.3) (14)	–		S³TBn	43.6 (–)
(S)SSTBn	42.1 (±1.1) (15)	SSTBs	41.9 (±1.8) (12)	SSTBs	(40) (0)
–		SSTBn	36.3 (±0.6) (9)	SSTBn	36.5 (–)
		STB(S)s	37.1 (±0.5)(3)†		
STBs	33.3 (±0.9) (16*)	STB(S)n	32.9 (±0.5) (3)†	STBs	32.6 (+)
STBn	26.8 (±0.7) (17)	–		STBn	27 to 29 (–)

These are averages of the high-precision measurements plotted on Fig. 12.8. Average zenographic latitudes are quoted as: latitude south (± standard deviation) (number of points).
*For STBs, the exceptionally high values for 1977–1979 were omitted; †STB(S) instead of true SSTB in 1977/78 and 1984–1985. The latitudes of jetstreams are from Voyager 2 (Limaye, *1986*), with a sign to indicate if the jetstream is prograding (–) or retrograding (+). Before 1950, the mean latitude of SSTB or (S)SSTB was 41.4°S (±2.3°; $N = 19$) (1907–1937).

disappeared and a complete true SSTB developed for the first time; there have also tended to be segments of a true S³TB. So it seems that, from the STropZ to the SPR, roughly coordinated changes occur to fit the typical number of broad belts onto the narrower grid of the jetstreams.

The 'SSTB' latitude varies not only with time but also, often, with longitude, and these variations also tend to be coordinated with the state of neighbouring belts. Thus in 1975, the general pattern of a normal STB and (S)SSTB was replaced in the limited longitudes of the STB Fade by a true SSTB and true S³TB; conversely

in 1988–1992, when most longitudes contained no STB but a true SSTB, the tiny remnants of STB flanking ovals BC, DE, or FA tended to be accompanied by a gap in the true SSTB and, sometimes, a streak of (S)SSTB or S³TB instead (e.g Plates P22.6 and P23.5). Other examples of such generally dislocated sectors are plotted in Fig. 12.8; see also Fig. 1.19. But rather than list more Earth-based examples, we can illustrate the point best by comparing the two Voyager photos of opposite sides of the planet in Figs. 12.3&4 – one alongside normal STB around oval DE, and the other alongside the STB Fade and South Tropical Dislocation. One pic-

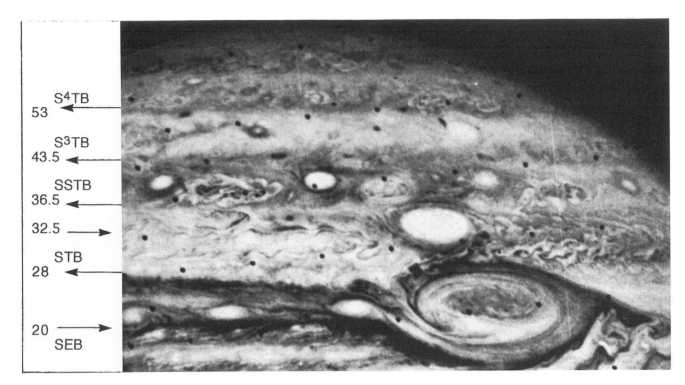

Fig. 12.4. Voyager 1 image of the GRS and oval BC and regions further south, 1979 March 4d 7h. This is on the oppposite side of the planet from Fig. 12.3 and is almost a negative of it. (The difference is not due to the colour filters used, as there are no largescale colour contrasts in the region, except for the blue south polar hood.) At left is the STB Fade and S. Tropical Dislocation. Note the chain of alternating cyclonic and anticyclonic spots in the SSTB. (Image 16353.31, green, wide-angle, filtered. NASA.)

ture is almost a negative of the other in terms of the pattern of belts and zones.

This coordination of dislocation through several belt-zone units is by no means precise or predictable; the belt segments are often irregular and they move on different characteristic currents. But there seems to be a strong tendency for them to occur. The reader can be reassured that these unexpected patterns are not the result of inaccurate measurements, nor of visual contrast effects. The same analysis produced latitudes for the tiny white ovals in the latitude of the canonical SSTBs edge, which turned out to be invariant to ±1° or so, and several observations show these ovals clearly on the *north* edge of the main belt in the region, when this is the (S)SSTB (e.g. Fig. 12.2).

Sectors of true SSTB wax and wane in albedo over a matter of months, and the 'dislocations' apparently arise because fadings of true SSTB sectors tend to occur while passing STB Remnants (BAA reports, 1987–1992). Conversely, when the STB was present, small streaks of true SSTB were sometimes seen alongside the south edges of the three STBs white ovals, and the streaks may perhaps darken as they drift past the STBs ovals. This was particularly suspected in 1985, and resembled the darkening of a STB streak while passing the GRS in 1986.

Apparently oblique belt segments are sometimes seen in the region. In general they may represent transitions between sectors with different belt patterns. At high resolution they often break up into irregular overlapping streaks and spots, but a few appear to be genuinely oblique. In view of the circulating currents revealed by Voyager in the region, some may represent true deflections of the local currents.

The SSTBs velocity minimum and the tiny white ovals at 40.5°S

Between the prograding jetstreams of SSTBn and S³TBn, one might expect Voyager to have seen a retrograding jetstream, but this was not the case: the minimum in velocity detected by Voyager at 40°S was simply the S.S. Temperate Current (SSTC). Retrograding velocities might have been missed because the region was so full of small ovals moving with the SSTC, but even in fairly undisturbed stretches of 'SSTB' (e.g. Fig. 12.3) there were no clear signs of a retrograding jetstream. However, the series of cyclonic 'intermediate ovals' at 38°S and anticyclonic tiny ovals at 40–41°S entailed locally retrograding velocities, and the latter ovals formed a well-defined line in place of a retrograding SSTBs jetstream.

These tiny ovals at the canonical latitude of SSTBs are clearly homologous, in structure, position, and dynamics, to the three white ovals on the STBs and the GRS on the SEBs. They are the one type of visible feature that does seem to be fixed in latitude in this region. Their average latitude from both Earth-based data is 40.6°S (±0.4°; Appendix 2), agreeing with the Voyager value of 40.5°S. They can only be resolved in the very best Earth-based observations, but they can be identified in the professional photos of 1950–1951, 1965–1968, and many apparitions from 1972 onwards including recent amateur photos (Plate P23), always at 40–41°S (Appendix 2). Since these ovals are so intimately related to the SSTBs velocity minimum, we can conclude that the latitude of this boundary has not changed. Before 1950, there are few photographs from which accurate latitudes for these tiny ovals can be obtained, but in 1914–1917 they were photographed and had an average latitude (albeit imprecise) of 40.6°S (±1.6°).

(a)

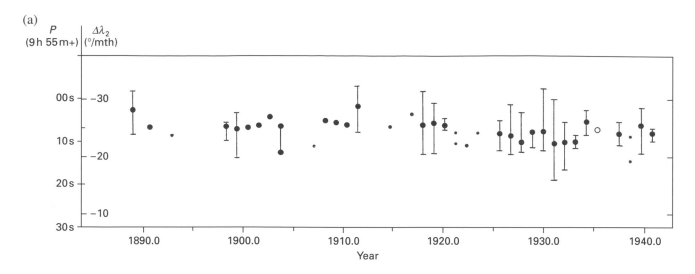

Fig. 12.5. History of the S.S. Temperate Current (SSTC). (From data listed in Appendices 1 and 3.) Speed is shown as P (rotation period), as $\Delta\lambda_2$ (degrees per month), and as u (m/s). Large symbols are apparition means of ≥3 features, and vertical bars indicate the range of speeds (not published before 1914). The chart does not include the 'proto-ovals' and other S. Temperate features which moved at similar speeds (Chapter 11). The paucity of records from 1947 to 1957 is partly because S.S. Temperate features were not readily distinguishable from the great S. Temperate ovals that were then forming.

Mean speeds from these data are as follows: 1887–1921, $\Delta\lambda_2$ = –25.3 (±1.5) (period 9h 55m 6.0s); 1922–1946, $\Delta\lambda_2$ = –23.5 (±1.2) (8.5s); 1949–1991, $\Delta\lambda_2$ = –25.9 (±2.0) (5.3s); overall, –25.1 (±1.9) (6.4s).

The Pioneer and Voyager spacecraft photographed impressive chains of these ovals, spaced 17°–30° apart around more than half the planet (section 12.2; and see Figs. 2.1&2); but this was probably not a permanent pattern. In 1985–1992, amateur photos showed these ovals clearly and there were always about seven of them, irregularly spaced around the planet, either on the true SSTB south edge or within the true SSTB where it was broader. Several of these ovals have been tracked for 5–6 years, which seems to be their approximate mean lifetime. One of them has been tracked over the whole period 1985–1992, and for most of that time it has marked the p. end of a gap in the true SSTB.

The S.S. Temperate Current and S.S.S. Temperate Current

Because we are dealing with two dynamical domains in this region, we would expect to find a slow current for each, and this is what has been observed.

The South South Temperate Current (SSTC) operates between 36°S and 43°S. It is observable in almost every apparition, and it was referred to in early reports as the 'Great Southern Current' on account of the constancy of its speed and the breadth of latitudes over which it appeared to operate. It governs features on the true SSTB and also, sometimes, on the (S)SSTB when this belt occupies the canonical latitudes of the SSTZ. It also governs spots in the 'STZ' when (as it commonly does) this bright zone extends south of 36°S. The SSTC also occasionally invades the South Temperate Region as discussed previously.

The SSTC is the steadiest current on the planet (Fig. 12.5). However there has been a slight, but statistically significant, shift in the mean rate: from 1922 to 1946 the mean $\Delta\lambda_2$ was –23.5, whereas it has been about –26 before and since. Short-term changes also occur: in 1988/89, features decelerated all round the planet, so the mean $\Delta\lambda_2$ reduced from –29.3 to –25.3 within a month or so.

The fastest speeds in the SSTC, in several apparitions, have been $\Delta\lambda_2$ = –31 to –34. Recently the BAA has found even faster speeds for a few exceptional spots, all small white ovals: $\Delta\lambda_2$ = –36.4 for one in 1973 (for three months), –42.8 for one in 1979/80 (for three weeks), and ≈–40 for one in 1986 (for just 5 days).

South of 43°S, features that can be tracked are very rare. In fact, they have only been recorded in 16 apparitions, as listed in Table 12.2A. In many of the best-covered years, no spots could be detected at all. Latitude measurements in this region have been rare and imprecise, but it appears that almost all the drifts for which latitudes were recorded were between 43°S and 50°S, and thus belong to the S.S.S. Temperate domain. (The handful of exceptions were all at 60°S, and demonstrate more diverse drifts in the South Polar Region proper, as discussed in Chapter 5.) The drifts between 43°S and 50°S are fairly scattered but can be regarded as representing the slow current of the region. Therefore they are here designated as the S.S.S. Temperate Current (S³TC).

In an attempt to clarify the behaviour of this current, the author has measured features on Voyager 1 strip-maps (Table 12.2B). Most of the measurable spots were at 47°S and their $\Delta\lambda_2$ ranged from –1.5 to –13, with an average of –8.5. These included cyclonic regions within a S³TB as well as tiny dark spots in a 'SSTZ'. Two other spots were at 49°S and had slightly positive drifts. The implied velocity gradient across the S.S.S. Temperate domain is consistent with the few measurements by Reese and also, remarkably, with the 'discovery' observations by Molesworth in 1900 (Table 12.2A). However, the data are not sufficient to demonstrate such a gradient as a permanent feature. In these latitudes, some of the measured spots may be so small that they trace out the underlying gradient of winds as revealed in the Voyager correlation analysis (Chapter 3), rather than adhering to a steady slow current. Such small spots may also be liable to peculiar motions. A final spot measured on the Voyager 1 maps, at approx. 50°S, initially moved

(b)

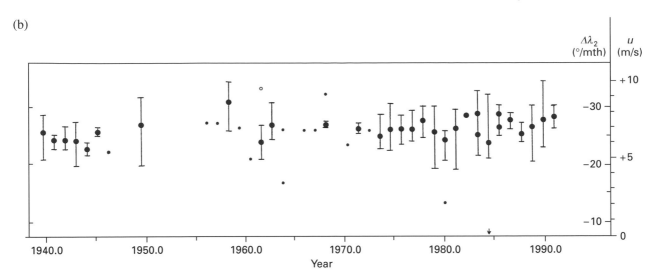

Fig. 12.5b. For legend see opposite.

with $\Delta\lambda_2 = +3$ but then accelerated to -39 for a week before decelerating again (see below). Perhaps it was partially entrained by the S^{4}TBn jetstream to the south. Further study of the Voyager images could reveal more details of the currents in this region.

Although dynamical domains have only been defined as far as 53°S, the Voyager images and Table 12.2 suggest that at least one more may exist further south, in latitudes covered in Chapter 5.

12.2 SPACECRAFT OBSERVATIONS

The true SSTB: a chain of alternating ovals

The most remarkable discovery of the Pioneer and Voyager spacecraft in these latitudes was the regular chain of ovals that each spacecraft saw in the canonical latitudes of the SSTB, consisting of alternating cyclonic 'intermediate ovals' at 38°S and anticyclonic tiny ovals at 40–40.5°S (Figs. 3.6&7, 11.18, 12.3&4, and Plate P20.1). In the Pioneer 10 map there were 8 pairs in 210°, and in the Voyager maps there were 9 pairs in 205°, spaced 20°–30° apart (Figs. 2.1&2). They all moved together with the SSTC.

The history of the ovals at 40.5°S was discussed above. In the spacecraft pictures, they typically spanned about 3000x2000 km. The bright oval was usually surrounded by a prominent dark ring or 'halo'. Some of them had a spiral arm structure. They had anticyclonic (anticlockwise) circulation, with vorticity very similar to that of the GRS and STB ovals (B.A. Smith and colleagues, *1979b*). In the sector which lacked these ovals, there were much smaller anticyclonic ovals at similar latitude.

The cyclonic structures, centred at 38°S to 39°S within the latitudes of true SSTB, were of two very different types. One ('intermediate oval') was a well-defined white oval with a double border, the inner border being a wavy line, exactly like the STB Fade. This type was clearly homologous to the STB Fade, in detailed structure and in being confined between pairs of anticyclonic ovals. The other type ('folded filamentary region', FFR) was an irregular turbulent structure that changed chaotically from day to day. In the

Voyager 1 maps, the two types occurred alternately, but during the Voyager 2 encounter, two of the intermediate ovals broke up into folded filaments. One case was described thus:

> A small bright core appeared in the oval's center. The bright core expanded longitudinally and was drawn into a bright S-shaped pattern. Within 5 rotations the oval ... was transformed into a region of folded filaments consistent with the prevailing pattern. (Smith *et al.*, *1979b*.)

The folded-filament regions were only partially closed circulations. They tended to emit turbulent streamers onto the adjacent jetstreams – surprisingly, onto both the SSTBn prograding jetstream and the S^{3}TBn prograding jetstream. Thus their activity crossed the boundary at 40°S. This behaviour can be clearly seen in the Voyager movies (see below).

The segment of canonical SSTB latitudes which did *not* contain true SSTB shading (Fig. 12.3 and Plate P21.1) was filled with fine-grained white clouds and had a wavy closed border just like that of the STB Fade, suggesting that this was a 'SSTB Fade' with the same closed cyclonic circulation, even though ripples on its south edge were not actually retrograding.

Yet another type of cyclonic feature was shown in the Pioneer 11 close up (Fig. 12.2): two dark brown 'barges'. These were the remains of a small streak of true SSTB that had been observed visually as it drifted past oval DE and broke into two parts, which became the two dark oblongs. Their appearance and their latitude (38°S; LPL) implied that they were cyclonic circulations like the NEBn barges. Like the white cyclonic ovals in the same latitude, these dark ovals were positioned between the tiny ovals at 40.5°S.

Spots and dislocations further south

Both Pioneer and Voyager revealed increasingly irregular patterns of streaks, ovals, and wavy lines towards higher latitudes. The Voyager imagery did however show that these structures were moving with the regular pattern of currents, even though there was no overall correlation between the currents and the belt/zone pat-

Table 12.2 *Drifts south of 43°S*

(A) Earth-based data

Years	P (9h 55m+)	Δλ₂ (N)	Lat. β″	Description	Observer
S.S.S. Temperate Current:					
1900 {	32.2s	−6.2 (8)	(−48)	SPRn	Molesworth
	22.2s	−13.5 (8)	(−43)	SSTZ	"
1901 {	22.8s	−13.1 (4)	(−49)	SPRn	"
	19.9s	−15.2 (5)	(−45)	SSTZ	"
1903	14.2s	−19.5 (4)	(−48)	SPRn	"
1907/08	24.3s	−11.9 (2)	–	SPR	Phillips
1909/10	25.4s	−11.1 (5)	–	SPR	Hawks
1928/29	30.4s	−7.5 (1)	(−45)	S³TB=SPRn	BAA
1934	34s	−4.6 (1)	–	SPR	BAA
1943/44	22s	−13.6 (1)	–	SSS	BAA
1945	27s	−10.2 (3)	–	SSS	BAA
1946	36s	−3.5 (3)	–	SSS	Peek
1965/66	41.1s	+0.4 (1)	−49.5	SSTBs	NMSUO
1967/68	40.0s	−0.4 (1)	−50.3	SSTZ	"
?*	33.0s	−5.7 (?)	−47.3	–	"
1978/79 {	33s	−6.0 (1)	(−48)	SSTZ	Rogers/BAA
	29s	−8.5 (7)	−46.8	*SSTZ,S³TB*	*Voyager†*
	48s	+5.5 (2)	−49.0	*SSTZ*	*Voyager†*
Average	29.4s (±7.8s)	−8.3 (14) (±5.7)			
South Polar Region					
1911	50.7s	+7.3 (2)	−60	Belt in SPR	Hawks
1967/68	11.0s (var)	−21.7 (1)	−60 (var)	S³TZ	NMSUO
1970	14.2s	−19.4 (1)	−60	S³TZ	"
1978/79	*30s*	*−8.0 (3)*	*−55*	*Belt on SPRn*	*Voyager*

Notes to Table 12.2A

See Appendices 1 and 3 for details of sources.

* This was some time between 1962 and 1970, as summarised by Reese (*1972a*).

†Voyager data from Table 12.2B, not included in averages.

Column 4: Zenographic latitudes (in brackets if imprecise). 1900–1903: estimated by Molesworth (may be β′?). 1928/29: measured from drawings. 1965–1970: directly measured from professional photographs.

Column 5: 'Description' is the location as described in the original report. In view of the analysis of the 'SSTB' in this chapter, 'SSTZ' is probably true S³TZ. 'SSS' (BAA description as 'S.S.S. Temp. spot' without further details), 'S³TB' and 'SPR' could indicate any latitude south of 43°S, although the agreement of these values with the others suggests that they were all in the same range of S.S.S. Temperate latitudes. All features were dark spots or streaks except for those recorded by NMSUO (Reese and Solberg), which were small bright spots.

tern. As one looks southwards from the STB, anticyclonic ovals get smaller, whereas cyclonic FFRs are still extensive and come to dominate some sectors (Figs. 3.7 and 5.1). Indeed, Figs. 11.18 and 12.6 show cyclonic FFRs tens of degrees long in the S³TB, while Fig. 12.3 shows a similar sector of S⁴TB, though this seems quieter and more like the enclosed bright sector of SSTB in the same picture.

The author has tracked the motions of various spots in Voyager 1 photos such as Fig. 12.3. At these high latitudes, tiny ovals that moved with the the broad S³TBn and S⁴TBn jetstreams looked just the same as the tiny anticyclonic white ovals which were moving in

Table 12.2 (*cont.*)

(B) Voyager data

No.	Δλ₂	Lat. β″	Description
S.S.S. Temperate Current:			
S³TC (N)			
1	−1.5 (±1.5)	−46.6	D. oval (C?)
2	−11 (±2)	−46.6	Spiral (C)
3	−5 (±3)	−47	p. end FFR (C)
4	−12 (var.)(±6)	−47	f. end FFR (C)
5	−9 (±3)	−46.8	D.s. 'SSTZ'
6	−13 (±3)	−46.8	D.s. 'SSTZ'
7	−8 (±1.5)	−46.8	Tiny ring
Av. (nos.1–7)	−8.5 (±4)	−46.8	
S³TC (S)			
8	+3,−39,−20 (±3)	−50	W.oval S³TBs
9	+4 (±2)	−49	W.oval 'SSTZ' (A?)
10	+7 (±3)	−49.5	D.s. 'SSTZ'
Av. (nos.9,10)	+5.5 (±1.5)	−49	
SPR			
11	−3 (±3)	−55	W. oval (FFR?) (C?)
12	−8 (±3)	−55	FFR (C)
13	−13 (±2)	−55	D. bar (C?)
Av. (nos.11–13)	−8.0 (±5)	−55	

Notes to Table 12.2B

Drifts were measured by the author from Voyager 1 images and maps, as described for Fig. 6.4, over time spans of 12–32 days. The features were mostly in the regions shown in Figs. 12.3&4. Most drifts were fairly steady but no.4 involved fusion of two features and no.8 varied wildly (see text). Descriptions include: FFR, folded filamentary region; (A), anticyclonic; (C), cyclonic. These circulations were observed directly or inferred from structure and latitude.

the SSTC and S³TC (Table 12.2B). The cyclonic FFRs were perturbing their surroundings, as the following examples show.

Spot 2: This S³TB FFR (marked in Fig. 12.3) was rolling up as a neat spiral cyclone. It was connected to a SSTB FFR by folded filaments across the S³TBn jetstream, which were being pulled out into a < shape by the jetstream.

Spot 8: A S³TB FFR Nf. this spot ejected a bright streamer S. and p. around spot 8 onto the S⁴TBn jetstream, which apparently caused its sudden acceleration within the S³TC (Table 12.2B).

Voyager 2 images were studied by M. MacLow and A. Ingersoll (*1986*), and many of the spot creations and interactions they described were in the S.S. Temperate region. Here, as elsewhere, new spots always arose out of the cyclonic FFRs, while the most common interactions were the orbiting and merging of anticyclonic spots. The following are some of their examples.

Appearance no.8: Nf. a S³TB FFR, a dark spot appeared on the S³TBn jetstream (Δλ₂ = −76).

Fig. 12.6. Pioneer 11 image showing white oval BC, with a S.S. Temperate FFR on its south edge, flanked by small bright ovals at 40–41°S. There is also an extensive FFR in the S³TB. (Image C5. NASA.)

Appearance no.9: From a SSTB FFR, an *anti*cyclonic light spot appeared, which merged with another anticyclonic spot, but the merging pair ejected a *cyclonic* small spot in the Nf. direction. (All were within the SSTC.)

Appearance no.11: From a SSTB FFR, a small bright spot appeared and was drawn Sp. into the S³TBn jetstream. It passed a white oval at 41°S, emitting a light filament which curved anticyclonically around the oval, and moved on.

12.3 HISTORY OF THE SSTB, 1900–1940

The next two sections present the first historical survey of the behaviour of the S.S. Temperate region.

Before 1940, the data were almost entirely visual, and conclusions cannot be absolutely certain. Nevertheless, there is good evidence for long-term shifts in the appearance of the region. Between 1918 and 1934, the appearance often differed from that before and since.

The problem with these early years is the paucity of accurate latitude measurements. The only original measurements are by micrometer, for the centre of the 'SSTB' (Fig. 12.7). In view of the difficulty of observing at these high latitudes, it is not surprising that there were usually no more than two measurements per apparition, that they showed a mean difference of 2°, and that belt edges were never measured. However, what one can do is to examine the reports on the 19 apparitions in which a micrometer measurement was given, to see whether the belts had any recognisable character

Table 12.3 *Selected descriptions of the SSTB before 1940*

(A) Mean 'SSTB' latitude 43°S to 45°S

	STZ, 'STZB'	'SSTB' = (S)SSTB
1911	Partly shaded, with STZB	Irregular
1914 1915	STZB at 37°S; in places, true SSTB with ovals	Double or complex or absent
[1920/21]	Bright white, but included 'S.Temp.Disturbance' with SSTC invasion	Distinct segment [ø]
1924	Bright white	Diffuse and broad
1925	Duskier, irregularly shaded, some white spots	Usually well-defined
1935	Clear	Distinct, quite narrow
1937	One segment of v.dark bluish true SSTB [ø]; elsewhere, shading and big white oval	Short v.dark section; elsewhere, 'SSTB' was further N.

(B) Mean 'SSTB' latitude <39°S

	STZB lat.		Description	Tiny ovals lat.
1914	−35.4	(BAA)	Narrow STZB; some places	
	−38.6	(ø)	have true SSTB with ovals	−43
[1915]	−36.5	(ø)		−40
1916/17	−38.7	(BAA)	Mostly broad clear STZ and 'SSTB' complex or faint, but one sector had complex shadings and STZB segments	
1917/18	−37.0	(BAA)	Again several narrow bands	
	−37.6	(ø)	inc. narrow true SSTB	−39.6

1915 and 1920/21 are added on the basis of measurements on Lowell Obs. photos which differed from the BAA values.
(ø) indicates data from Lowell Obs. photos only.

according to their latitude. The data have been supplemented by measurements on Lowell Observatory photos (in Slipher, *1964*), for 1914–1917 and 1920/21, normalised to the micrometer measurements as the limb was not definable on the photographs. The apparitions can be divided into the following three groups.

(i) Mean 'SSTB' latitude 43°S to 45°S, i.e. (S)SSTB:

These seven apparitions, with 1920/21 added on the basis of a photographic measurement, are listed in Table 12.3A. The last two entries (1935 and 1937) may represent the long-term transition to a high-latitude (S)SSTB. Before then, no distinct, continuous (S)SSTB was seen except in 1925. When an irregular or diffuse belt *was* seen in these latitudes, it was usually accompanied by at least fragments of a 'STZB' at ≈37°S and patchy shading in the true SSTB latitudes, often with visible white ovals which may have included both types observed by Voyager.

(ii) Mean 'SSTB' latitude <39°S, i.e. true SSTB:

There were 3 such apparitions, plus 1915 (photographic), as listed in Table 12.3B. Although the latitude suggested a true SSTB, in general the true SSTB was only seen as one of multiple belt segments; only in 1917/18 may it have been the major belt.

Tiny bright oval(s) were photographed on the true SSTB in most of these years. The rather imprecise latitudes are consis-

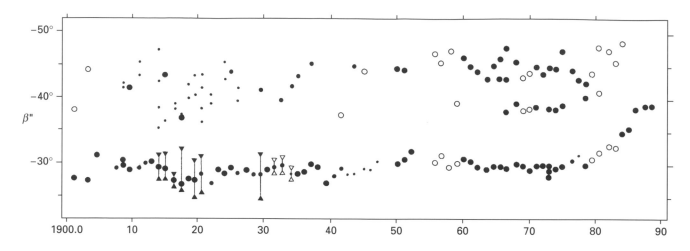

Fig. 12.7. Latitudes of the STB and SSTB, 1901–1989. This shows the zenographic latitude of the belt centre, from sources described in Appendix 2, plus the edges of the STB in a few apparitions. Large symbols are appari-tion averages; smaller symbols are one or two measurements, by micro-meter or from photos (by the author from photos in Slipher, *1964*). Open symbols have lower precision. Averages are listed in Tables 11.1 and 12.1.

tent with these being the 40.5°S ovals; 'intermediate ovals' slightly further north were also observed.

It may also be relevant that in 1900, Molesworth recorded SSTC for dark spots in a SSTB at 37°S, and his map from 1900 gives the impression of a 'true SSTB with intermediate ovals' as in the 1970s.

In categories (i) and (ii), the latitudes of the belt and the occasional sightings of tiny white ovals gave appearances not unlike those in the 1970s, and it may be significant that almost all these appari-tions were before 1918 or after 1934. In contrast, most of the years between 1918 and 1934 fall into category (iii).

(iii) Mean 'SSTB' latitude 40°S to 42°S:

These 10 apparitions were 1908/9, 1909/10, 1918–1922, 1926, 1929/30, 1932/33, and 1934. The 'SSTB' was often irregular or diffuse but sometimes distinct. It was in the lati-tude which, later, would be identified as the canonical SSTBs boundary marked by the tiny white ovals; but in these appari-tions tiny ovals were not visible so we have no clues as to whether the dynamical boundaries remained in the same lati-tudes while the belt shifted.

From 1918 to 1922, this 'SSTB' accompanied 'SSTC inva-sions' of the STZ to the north (see previous chapter). In sev-eral other apparitions, the 'STZ' contained a narrow but sometimes dark 'STZB' which must have been at the usual jetstream latitude of about 36°S or 37°S (only measured in 1920/21). In 1932/33, this band carried small spots whose motion was undetermined. Otherwise, the 'STZ' (including the north half of the canonical SSTB) was generally clear in most of these ten apparitions – even alongside the two long STB Fades of 1926 and 1934, which occurred under this régime without any accompanying true SSTB.

12.4 HISTORY OF THE SSTB, 1940–1990

The normal picture from 1940 to 1980 was of a broad dark STB and a fairly well-defined (S)SSTB as defined above, separated by a broad, fairly clear STZ; there was usually no true SSTB. But seg-ments of true SSTB appeared with increasing frequency, culminat-ing in the 1980s with the establishment of the true SSTB instead of STB as the major southern belt. The following chronicle is given in detail as no such survey has been reported before. The latitude measurements are shown in Fig. 12.8, taken from Rogers and Herbert (*1991*), who also summarised the main conclusions of this survey.

1940–1942 (including a true SSTB): In 1939, there was a massive dark (S)SSTB and only a narrow STZB. In 1940, separate narrow belts at 43°S and 37°S at some longitudes merged elsewhere to form a broad dark true SSTB between these latitudes; and in 1941/42 this SSTB was present at most longitudes, although with a variable south edge. It was often seen dou-ble, and was pockmarked by light ovals at 38°S, probably 'intermediate ovals'. In 1942/43, the true SSTB faded away again and only traces of belts remained, including a faint STZB at 37°S along one STBs proto-oval.

1943–1966: In 1943/44 and 1945, a dark (S)SSTB had revived at some lon-gitudes only; an incomplete STZB persisted along proto-oval DE; and there were many subtle shadings in the region. This pattern was still seen in 1948, and the dark (S)SSTB was still present in 1950–1952 (the intervening years being poorly observed), although the STZB had disappeared as the three white ovals matured. Thus in all well-observed apparitions from 1942 to 1952, the true SSTB was essentially absent, so there was a broad STZ bordered by a substantial dark (S)SSTB with its north edge at about 42°S. Further south there were no distinct belts; the SPR shading extended to a sharp border at 54°S (Mt. Palomar photos) or to a more-or-less diffuse (S)SSTBs edge (all other photos and drawings).

From 1953 to 1960, there are no precise data. The general picture remained much the same, except that the (S)SSTBn edge may have been further S. than before, at about 45°S, making this belt a true S³TB if so. An exception occurred in 1959 at some longitudes, where it extended north to about 39°S to encompass a small bright oval. Also, tenuous fragments of true SSTB were recorded in 1958 and 1959.

From 1961 to 1965/66, the picture still did not change much, except that the (S)SSTB as a whole moved northwards again. The professional photos sometimes revealed a very faint narrow STZB in the broad clear STZ; this was at 36°S, but the NMSUO values of 38°S probably refer to darker streaks on it alongside the STBs white ovals. The long STB Fades in 1961 and 1964/65 were not accompanied by any major new belt further south, although two drawings did suggest tenuous sections of STZB, and Reese

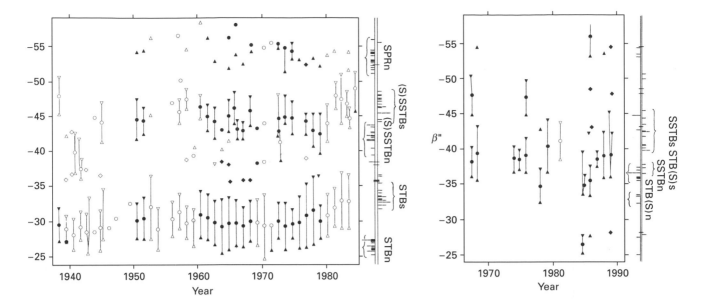

Fig. 12.8. Latitudes of the southern belts, 1938–1989. For clarity, different belt patterns are plotted in different panels: left panel, longitude sectors with normal STB; right panel, sectors with STB Fade and/or true SSTB from 1966 onwards. Circles, centres of belts; triangles, north and south edges (joined by bars for clarity; note that these are not error bars); diamonds, narrow bands. Solid symbols are high-precision measurements and open symbols are medium-precision measurements (including single micrometer measurements). Between the two panels is a histogram of high-precision measurements for the edges of belts: bars to left are north edges, bars to right are south edges, and bars in centre are narrow bands. The data are probably accurate to ±2° or better. For 1940–1945, the measurements are from a handful of professional photographs, normalised to Peek's STB micrometer measurements. For 1950–1979, published high-precision measurements were checked against original photos to ensure that the latitudes were not averages of several different belts, and supplemented with additional measurements from published professional photographs. For 1984–1989, the measurements are by the author from high-resolution amateur photographs. Details are given in Appendix 2 and in Rogers and Herbert (*1991*), from which this chart is reproduced.

noted dark and bright features in the 'STZ' in 1962/63 streaming round the S. edge of oval FA.

The improved coverage in 1965/66 showed the (S)SSTBn edge at 41°S or 44°S at different longitudes. There was also an isolated dark 'barge' centred at 40°S, as well as tiny white ovals at 41°S and 'intermediate ovals' at 38°S, all in the generally clear 'STZ'.

1967–1970 (including a true SSTB): In 1966/67 the longitudes between FA and BC, which contained the 41°S and 38°S ovals, developed extensive shading down to 36°S. Thus a substantial true SSTB, perforated by ovals, appeared for the first time in 24 years. Following oval BC, it coexisted for a while with a true S³TB, but between FA and BC the (S)SSTB had disappeared while the SSTB merged into the general south polar shading. In 1967/68, this 'true SSTB with ovals' lay alongside a new STB Fade from DE to FA, and thus presented the spectacle that was to recur during subsequent STB Fades. (It is however puzzling that the SSTB with ovals, and the STB Fade, had commenced separately before their drifts brought them into alignment.) In 1969, while the STB Fade persisted, the 'SSTB with ovals' was present but very irregular. In 1970, the STB had recovered, and SSTB ovals and dark blocks persisted but the shading was breaking up; it had more-or-less disappeard by 1971.

Meanwhile the (S)SSTB recovered, as short segments in 1969 and a well-defined belt in 1970; its N. edge contacted the tiny ovals at 41°S.

1971–1974 (including a true SSTB): The (S)SSTB or S³TB was present at all longitudes in 1971 and 1972; it varied greatly in latitude and intensity but major sectors of it had resumed its familiar position between 42°S and 48°S. It persisted in 1973 and 1974, and the chain of tiny ovals at 41°S on its N. edge was viewed by the Pioneers.

A new 'true SSTB with ovals' developed in 1973, around 2/3 of the planet, as the latitudes down to 36°S again developed patchy shading perforated by the 'intermediate ovals' with associated dark streaks and spots at 39°S. This largely cleared away by 1974, but belt segments remained at various latitudes between 36°S and 40°S (LPL; e.g. Fig. 12.2). One true SSTB segment remaining in 1974 lay alongside the STB p. oval FA which faded as it passed the GRS. This STB Fade recovered in 1975, as the SSTB segment drifted on past it, but this SSTB segment remained intact until 1977.

1975–1980 (including a true SSTB): In 1975, another STB Fade developed between FA and BC. This segment too was sandwiched between the GRS and a true SSTB segment at the time. This time, as the true SSTB replaced the STB, the belt further S. did not disappear; but it was unusually far S. in this sector, from 45°S to 50°S, so for once it occupied almost precisely the canonical latitudes of S³TB. In other sectors it was at different latitudes, and in 1976–1978 it was a (S)SSTB centred at 43°S to 44°S as usual.

The FA–BC Fade was the start of the progressive loss of the STB that continued throughout the 1980s (Chapter 11.4; Figs. 10.51–57). In the first few years this STB sector tended to recover, but the Fade was re-induced in 1976/77 and 1977/78 as dark segments of narrow true SSTB drifted past it. On the latter occasion, the SSTB segment was the same that had induced the original Fade 2½ years earlier, although it had extended in the p. direction. (Also during these years, the STBs edge was unusually far south, to about 36°S.)

In 1978/79, most longitudes contained a true SSTB (36–44°S) or more patchy disturbance; the true SSTB was particularly striking alongside the STB Fade. The chain of ovals at 39°S and 41°S, which looked so dramatic in the Voyager photos, also appeared exceptional in Earth-based photos and visual observations during that apparition. The (S)SSTB disappeared within that sector.

In 1979/80, the true SSTB was initially present at all longitudes, and alongside the STB Fade it was very irregular with ovals. But the pattern changed from month to month, and most of the N. half and the ovals faded away, leaving a (S)SSTB (41–47°S) with tiny ovals on its *north* edge. In 1980/81, the former SSTB continued to break up in one hemisphere (where a true S³TB had appeared); but in the other hemisphere, including the STB Fade, the SSTB (38–44°S) was again substantial.

1982–1985 (STB(S) and/or true SSTB): In 1982, the long STB Fade was again renewed while passing between the GRS and a long true SSTB with cyclonic ovals. In the following two years, this 'true SSTB with ovals' retained its identity (and some individual ovals were tracked from 1983 into 1984), and the latitudes to the south became generally shaded, but the picture was very complicated. Some longitudes contained a dark belt filling canonical STZ latitudes (33–38°S or 33–42°S), which bordered the STB Fade, with a separate SSTB(S). Meanwhile a true S³TB was present at all longitudes in 1982, and at most longitudes in 1983 and 1985, but rarely in 1984.

1986–1990 (complete true SSTB): In 1986, as the STB Fade extended around almost the whole planet, the major belt alongside it had reverted to a true SSTB (37–40°S), with 7 tiny white ovals on its S. edge at 41°S. This picture persisted with little change until 1990. (A true S³TB, centred at 49°S, was often still present but was narrow and sometimes inconspicuous.) So in these years, the true SSTB achieved an unprecedented status as the major southern belt.

Can we find any regularity in these very variable belt patterns?

The histograms alongside Fig.12.8 represent all the high-precision latitudes obtained for the edges of belts in this region; averages are in Table 12.1. The separation of 'STB Fade/True SSTB sectors' and 'normal sectors' is done only for clarity. If all the positions for belt south edges on both charts are taken together, there is a distinct cluster around 33.3°S marking the canonical STBs edge, but the positions of belt south edges further south seem to be rather uniformly distributed; in this they resemble the variable poleward edges of all the belts in the northern hemisphere. The north edge positions, on the other hand, are obviously clustered at the latitudes of the Voyager prograding jetstreams, i.e. at the canonical positions of STBn, SSTBn, and S⁴TBn. However, the S³TBn jetstream does not usually mark an albedo boundary; the (S)SSTB was unique among belts on the planet in that its edges habitually paid no regard to the jetstream pattern. Its north edge usually lay at about 42.1°S, 1.5° north of the Voyager position for the S³TBn jetstream. This cannot be due to variations with time because just such a (S)SSTB was observed in the apparitions just before and after the Voyager encounters.

Remarkably, there is also a minor cluster of north edge positions (for the 'STB(S)' in 1977/78 and 1983–85) which coincide with the latitudes of the STBs retrograding jetstream, suggesting that this jetstream can form an albedo boundary even when the albedo patterns are the opposite of the expected patterns.

The true SSTB is usually represented by nothing more than a tenuous band at 37°S (STZB or SSTB(N)). However, a true SSTB does sometimes develop, usually as part of a general dislocation in that sector. Then, the (S)SSTB always disappears or shifts south to become a true S³TB, and the STB usually disappears or undergoes some other major change. In 1940–42, this appearance at some

longitudes accompanied the global formation of the three STBs white ovals. And of the frequent manifestations of 'true SSTB with ovals' since 1966, all but one have been associated with the induction or reinduction of STB Fades. Conversely, all the STB Fades since 1966 had accompanying dislocation further south – usually the presence of a true SSTB segment, but STB(S) in 1976–78 and 1983–85 (Table 12.1). (Only the STB Fades of 1961 and 1964/65 lacked true SSTB.)

In conclusion, there have been three common patterns of belts in the region, which predominated at different eras (sketched in Fig.12.1):

> *(1)* (especially 1918–1934): Narrow STB; substantial SSTB(S); sometimes other narrow belts further south.
>
> *(2)* (especially 1935–1985): Broader STB, accompanying the development of the three STBs white ovals; usually a broad STZ with (S)SSTB covering SSTZ/S³TB latitudes, or a true S³TB.
>
> *(3)* (especially 1975–1990): STB faded, leaving only a tenuous STB(N); substantial true SSTB (often with cyclonic ovals); S³TB either in its canonical latitudes or dissolved in shading. This may be only part of a broader pattern that includes darkening of STropZ in the South Tropical Dislocation.

These are by no means the only patterns, and the real picture is often more confused, and there are often different patterns at different longitudes. How the transition from one pattern to another occurs, in longitude or in time, is an interesting question with no clear answer at present; some ideas will be discussed in Chapter 13.

12.5 THE S⁴TBn PROGRADING JETSTREAM AND THE EDGE OF THE POLAR REGION

We have now reached the latitude of 53°S, at which level the jovian atmosphere usually draws a sharp line around the planet, beyond which all is dark and obscure.

Voyager recorded a prograding jetstream at 53.4°S, which has never been detected from Earth.

However, evidence for the permanence of this jetstream is provided by the best Earth-based photographs in virtually every well-observed apparition since 1950. They have revealed a sharp edge to the shaded South Polar Region with a latitude of 53.2°S (±1.2°) (sometimes coinciding with a narrow S⁴TB) (Fig. 12.8). The scatter between apparitions is similar to that between individual measurements and is consistent with this boundary being constant at the jetstream latitude. This suggests that this jetstream too is fixed, and constrains dark material on its poleward side like lower-latitude prograding jetstreams.

IV. Physics and Chemistry of the Atmosphere

How the jovian atmosphere works is still largely uncertain, particularly where the global patterns of activity are concerned. So the reader may notice a distinct gap between the descriptions of global patterns in Chapter 13, which considers them from a phenomenological point of view, and Chapter 15, which gives the view of theoretical physics. (In Chapter 14, individual spots are looked at from both viewpoints.) In providing these accounts of current theories, the author makes no claim to be an expert on atmospheric physics, but hopes to offer both some explanations for visual observers and some ideas that may interest theoretical physicists. Perhaps the largescale patterns suggested in Chapter 13, as well as the 'check-lists' in Chapters 14.4 and 15.4, will provide useful challenges and tests for the various theories.

The main problem at present is that various theories of the atmosphere are incomplete and incompatible, because it is not known how deep the observed motions extend. Let us hope that physicists will soon be able to write a new account of the atmosphere, in which all the patterns that visual observers have documented will at last be explained.

First, in Chapter 13, we explore the hazardous topic of links between observed activity in different latitudinal domains. As there are rarely any visible connections between different domains, one can only seek for correlations between patterns of activity, and it is not difficult to find apparent correlations, in limited sets of data, which turn out to be merely due to chance. Some advertised regularities have indeed collapsed over longer spans of observation. However, some regularities do seem to be general, and these are the ones that will be described in this Chapter.

13: Possible large-scale and long-term patterns

13.1 DISLOCATIONS OF BELTS, ZONES, AND JETSTREAMS

South Tropical and Temperate Dislocations

We have seen that the usual paradigm of belts, zones, and jetstreams is sometimes violated in certain domains, producing 'dislocations' such that canonical belt latitudes are white and zone latitudes are dark (Fig. 13.1). Notable examples are the STB Fades (whitened STB; Chapter 11.4) and South Tropical Disturbances (darkened STropZ; Chapter 10.7). Both of these are associated with circulating currents between the relevant jetstreams. We have also seen that these phenomena can be linked in the South Tropical Dislocation (Chapter 10.7). Similarly, a STB Fade is usually accompanied by a dark segment of true SSTB, often studded with an array of cyclonic light ovals; otherwise, the true SSTB latitudes usually appear bright as part of the STZ (Chapters 11.4 and 12.1). The dislocation of the belt/zone pattern often extends to even higher latitudes. The problem to be addressed in this section is how these phenomena are coordinated across domain boundaries, and whether similar patterns occur across other sets of domains.

We will review evidence for two types of process. Some of the dislocations are now known to consist of circulating currents, and one can see proximate dynamical causes for the rearrangements that produce them. On the other hand, some dislocations consist of diffuse darkening or brightening with no dynamical changes.

It would be rash for the author to propose even one theory to account for these phenomena; let us throw caution to the winds and propose three. The present data cannot decide between them, and they are offered merely as a framework within which physicists may perhaps develop more coherent ideas. They may be labelled (1) the stochastic dynamics theory; (2) the chiasma dynamics theory; and (3) the albedo compensation theory.

These theories can best be described in terms of the South Tropical Dislocations, such as that observed by Voyager, which consisted of a true SSTB with ovals, a STB Fade, and a STropZ darkening which culminated in a S. Tropical Disturbance (STropD) as oval BC passed the GRS. (Note that causation in this case, as in most others, seemed to work from south to north – from SSTB to STB to STropZ. There are only a few weak instances of north-to-south influences, as when the 1941/42 and 1946 STropDs induced transient STB fading.)

(1) Stochastic dynamics theory. This theory would account for the dynamical changes purely in smallscale terms – that is, in terms of the unpredictable wanderings of the jetstream spots as seen by Voyager. It proposes that these wanderings lead to circulations which become self-perpetuating. The pros and cons of this theory for the origins of individual circulations are discussed in Chapter 10.7 and 14.4. To explain the more extensive, reproducible patterns of S. Tropical Dislocations, it proposes that passage of the STB Fade past the GRS causes subtle alterations in the spacing or speed

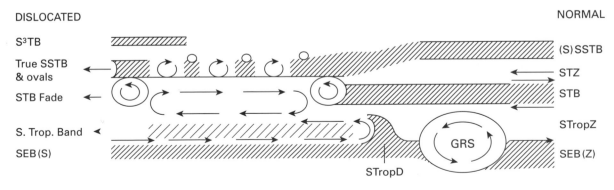

Fig. 13.1. Diagram of a South Tropical Dislocation, showing how the belt-zone pattern can be reversed from the S. Tropical to S.S.S. Temperate domains. (Also see Figs. 1.19, 10.50–58, and 12.3&4.) Arrows at left denote how the belt segments continue to move with the normal currents (SSTC, STC, STropC) in spite of their interactions.

of jetstreams, so that SEBs jetstream spots are more likely to drift off course and begin recirculating in the space that opens up there. This is what the Voyager pictures seem to show (Fig. 10.58). As no jetstream spots have yet been recorded at the onset of other types of dislocation, any stochastic dynamics theory of them would have to have a different basis.

(2) Chiasma dynamics theory. This proposes that the dynamical rearrangements take place in order that the largescale circulations can fit together more smoothly as they drift past each other. We define a chiasma as a place where jetstreams and circulations approach and recede in opposite directions, as sketched in Fig. 13.2.

In a South Tropical Dislocation, the initial situation has the STBn jetstream opposed to the nearby flows of the SEBs jetstream and the STB white oval; formation of the STropD can produce a chiasma which is a more smoothly interlocking pattern of circulations (Figs. 10.60 and 13.2b).

The author has suggested (Rogers, *1986*) that chiasmata in other latitudes may be involved in generating other circulating currents. In all these cases, the chiasmata are dark while the adjacent ovals or larger closed circulations are bright. The first such case is the South Tropical Dislocation. Secondly, the ends of the STB Fade, abutting white ovals FA and BC, constitute chiasmata (Fig. 13.2d). STB Fades tend to form in a sector of STB that is passing the GRS, and it seems plausible that this should occur by reconnection of the STBn and STBs jetstreams to form a chiasma when these are compressed in the passage of a STB white oval past the GRS (Fig. 13.2c,d). Thirdly, the spaces in the SSTB between cyclonic and anticyclonic ovals constitute chiasmata (Fig. 13.2e). STB Fades also seem to be induced by the passage of true SSTB segments (Chapter 11.4). However, whether this also involves chiasmata is unclear, as discussed below.

All the interactions involving chiasmata are invoked only to trigger the formation of dislocation structures. Thereafter, true SSTB segments, STB Fades, STropDs, and the GRS all have different characteristic speeds, and once any such structure has formed, it drifts away at this speed; its circulation then appears to be self-perpetuating. So the normal latitudinal dynamics appears to dominate even during a complete visible dislocation.

Unfortunately, there are difficulties in the chiasma explanation for each of these three interactions. The main difficulty in explaining a STropD is that the one observed by Voyager did not begin recirculating until the STB white oval had moved on; a chiasma was not really formed because the white oval was too far away (Figs. 10.58&60). In the case of a STB Fade beginning alongside the GRS, it is quite common to see transient partial fading of just the north side of the STB alongside the GRS, as if fading can begin by spreading of white clouds without any recirculation. And in the case of a STB Fade beginning alongside a segment of true SSTB with ovals, a direct linkage would require the formation of a circulating current in the STZ, for which there is no evidence. The best photographs have not revealed any physical connection across the STZ. Indeed, someone of a philosophical turn of mind might worry that, if the 'chiasma' mechanism for South Tropical Dislocations

looks like a world running by clockwork, the SSTB-STB interactions look more like action at a distance. This leads on to the third theory.

(3) Albedo compensation theory. This is essentially an admission of ignorance: it proposes that the changes are not primarily dynamical after all, but that the visible changes in brightness and darkness are primary: that there is an unexplained law requiring an alternating belt/zone pattern, even if this is temporarily at odds with the dynamical pattern. In the S. Tropical Dislocation, the S. Tropical Band arose before any evident dynamical change. In the case of the STB Fade and similar dislocations further south, the whitened belt latitudes may not always show recirculation; none was evident in Voyager pictures of the dislocated S³TB, nor in the Hubble Space Telescope picture of the faded STB, 12 years after Voyager (Plate P24.10).

Let us therefore survey the evidence for albedo compensation in regions other than the south tropical and temperate, before considering whether any possible mechanisms can be imagined.

Other examples of compensatory albedo changes

The SSTB-STB interactions seem to be an example of the general tendency in the south temperate latitudes for patterns of belts not to match the pattern of jetstreams, although the edges of some of the belts may be confined by the jetstreams. There are typically only two major belts across three dynamical domains (Chapter 12). Although the appearance of belt components south of the STB is often vague and variable, at least three characteristic patterns seem to recur:

Narrow STB – broad STZ – SSTB(S) – 'SSTZ' – SPR
(most common 1918–1934);

Broad STB – broad STZ – (S)SSTB – 'SSTZ' – SPR
(most common 1935–1980);

STB Fade – true SSTB – $\left\{ \begin{array}{l} \text{SSTZ – true S}^3\text{TB -SPR} \\ or \text{ shading into SPR} \end{array} \right.$
(most common 1981–1991)

(The last pattern is further extended towards the equator when a South Tropical Dislocation is present.) How the darkness of one latitude affects that of others, across several jetstreams, is entirely unknown. These patterns in Voyager photographs seem to be the result of multiple small circulations and indistinct shadings.

This sort of behaviour may also occur in the north temperate regions, although the patterns there are usually less distinct. An effect of features in one domain on the darkness of neighbouring belt segments could explain the 'serial behaviour' that is often suspected in those latitudes (Chapter 6).

Over the planet as a whole, the notion of constant proportions of 'dark material' and 'bright material', which interact in various ways, is one that has been mentioned over the years (e.g. N. Green, *1887*), but which has long been regarded with scepticism, not least by the author. However, our study has clearly shown some such tendency in the high southern latitudes. Is it possible that, in the better-observed lower latitudes, we have missed the wood for the trees?

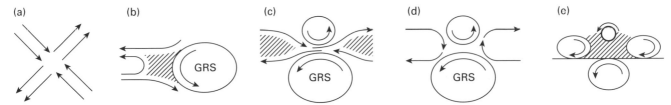

Fig. 13.2. Chiasma structures. (a) The basic flow pattern that defines a chiasma. (b) Chiasma formed as an incipient STropD emerges p. the GRS (but see Figs. 10.58&60). (c,d) Chiasma formed as a STB white oval passes the GRS, leading to reconnection of jetstreams to initiate a STB Fade. (e) Chiasma formed between SSTB white ovals (here shown with a STB white oval at bottom). (Adapted from Rogers, *1986*.)

Fadings of the SEB certainly appear to be different phenomena from the dislocations discussed so far, as they are not accompanied by darkening of the adjacent STropZ; indeed, if a dark STropD is present, it disappears along with the SEB (Chapter 10.3). SEB fadings are indeed accompanied by darkening and reddening of the GRS and sometimes of the EZ; but these are high-altitude red or orange colorations, apparently not analogous to dark belts. Moreover, SEB fadings are distinct events, linked to violent jetstream outbreaks locally and perhaps globally, whereas the dislocations further south have no evident connection with such events.

However, in the north tropical region there is evidence for largescale compensatory changes in albedo. There is a tendency for latitude shifts of the NEBn and NTB to occur in the same direction, maintaining a broad 'NTropZ' most of the time. As discussed in Chapter 7, the relationship is by no means absolute but it seems to be evident throughout recorded history. The 'displacements' of the NTB can be attributed to fading or revival of its S. component or of a neighbouring N. Tropical Band, in various colours. So here again, a variety of diffuse albedo changes result, overall, in complementary shifts on opposite sides of the NTropZ.

Perhaps these patterns may be governed by the rate at which heat escapes from the interior of the planet. Belts, being warmer and darker than zones, radiate more heat. If heat leaks from the interior at a constant rate, it may be necessary to maintain dark belt-like clouds over about half of each hemisphere in order to radiate the heat away. If dark cloud covers a larger area, increased cooling may allow high-altitude white clouds to spread. Or if white cloud covers a larger area, heat may build up below the cloud, leading either to its gradual erosion along the boundary of a belt, or else to a cataclysmic breakup.

A model along these lines for the grand-scale SEB Revivals was proposed by G. Kuiper (Chapter 14.6). A possible feedback mechanism on the micro-scale was proposed by R. West, D. Strobel and M. Tomasko (*1986*): dark cloud particles might act as nuclei on which white ammonia ice could condense or evaporate, according to the temperature.

A corollary of this theory is that the patterns of jetstreams and circulating currents serve only to organise the belts, not to create them. The belts and zones could exist without alternating jetstreams, as they do over most of Saturn. The theory also implies that red coloration in the EZ and GRS cannot be regarded as equivalent to belt-like darkening, as these reddish clouds are cold and elevated.

13.2 HEMISPHERIC ASYMMETRY

On the subject of largescale patterns, some comment seems necessary on the persistent asymmetries between the hemispheres. The most conspicuous asymmetries are:

(i) the perennial disturbance of the NEBs in contrast to the quiet SEBn (although this was different in the nineteenth century);

(ii) the presence of all the long-lived and well-organised features in the southern hemisphere (the SEB Revivals, GRS, STropDs, STB ovals, and even the SSTB ovals), in contrast to their much smaller and less organised equivalents in the north;

(iii) the exceptionally intense NTBs and SEBs jetstreams;

(iv) the displaced symmetry of the belts and currents of the South Temperate region with those of the North *North* Temperate region (Chapter 6).

Some of these asymmetries may well be contingent on others, and a hypothesis linking some of them will be offered in Chapter 14.4. But with only one example to study, and no theoretical explanation to guide us, we cannot be sure which asymmetries are causes and which are effects.

It is worth pointing out that several different latitudes showed anomalous behaviour from the 1880s to the 1910s. The unusual features were:

(i) perennial disturbances of SEBn (up to 1908) and *not* of NEBs (whose familiar activity began in 1913);

(ii) occurrence of triennial NEB Revivals (1893–1915) and absence of SEB Revivals (1883–1918);

(iii) slow drift of N. Equatorial Current (1887–1912);

(iv) slow drift of GRS (1883–1909).

13.3 GLOBAL UPHEAVALS[1]

The idea of global patterns of activity dates back many years. J. Focas (*1962*) and E.C. Slipher (*1964*), each reviewing a lifetime of observations, noted times when the entire planet appeared very disturbed (e.g. 1928, 1938, and more recently 1975), and other times when it appeared very subdued and quiet (e.g. 1940–41, and more recently 1989). The grand-scale changes in the S. Tropical region are the main contributors to these appearances; but is activity in other domains truly correlated?

[1] This section is adapted from Rogers (*1976, 1989b*). The association between EZ coloration and SEB Revivals was suggested by Peek (in a footnote to McIntosh, *1936*) and by Minton (in his LPL 1972 report).

In 1975, Wynn Wacker of the ALPO proposed a more specific scheme to describe these global upheavals. Most of the elements in his scheme are jetstream spot outbreaks, including SEB Revivals, NTBs outbreaks, and NNTBs outbreaks, plus episodes of coloration. We have seen that episodes of orange or yellow coloration tend to follow jetstream outbreaks and SEB and NEB Revivals, so it is reasonable to suppose that coloration episodes in the EZ (where jetstream spot outbreaks cannot be identified) may belong to the same category of disturbances.

Wacker (1975) proposed that all these disturbances tend to occur in a recurrent pattern of activity which he called a 'zenological disturbance'; the author prefers the term 'global upheaval'. The major elements in the pattern (modified from Wacker's listing) are as follows.

$T = -1$ to -2 yr	GRS rotation period begins to increase.
$T = -1$ yr	SEB(S) is very faint and GRS prominent.
	EZ becomes ochre with active white spots in the north.
$T = 0$	SEB Revival, causing GRS to fade and (in some years) spots on STBn jetstream.
$T = +1$ to 2 yr	(SEB becomes reddish.)
	NTBs and NNTBs jetstream outbreaks.
	(NTropZ becomes yellowish.)
	EZ quietens and becomes more bluish.
$T = +3$ to 4 yr	N.Temperate Current B appears in NTB.

This timetable links events in four domains: the SEB/STropZ (Revivals), the EZ (coloration), the NTBs jetstream, and the NNTBs jetstream. The observational record of these events is fairly complete and their dates of onset are easy to define, with the possible exception of some NNTBs outbreaks. All outbreaks of the relevant types are listed in Table 13.1 and plotted on Fig. 13.3.

When it was proposed in 1975, this scheme could indeed account for almost all the historical outbreaks of the specified types, but some doubt might have remained as to whether the coincidences were merely due to chance. Later that year, the coincidence of a vigorous SEB Revival with a great NTBs jetstream outbreak gave a boost to the hypothesis. From 1978 to 1980, the wave of activity that coincided with the Voyager encounters was also consistent with it, although the timings of outbreaks were not typical. What finally convinced the author of its validity was the global upheaval of 1990 (Plate P24), when SEB fading and Revival, EZ coloration, a great NTBs outbreak, and renewed NNTBs activity, all occurred within months of each other after 10 years of comparative inactivity.

The statistics of the associations are as follows. Notable EZ coloration, or a massive southerly EB thereafter, was present in 41 of the 104 apparitions from 1878 to 1991. These 41 apparitions witnessed 13 of the 15 SEB Revivals in Table 13.1 (the exceptions being 1952 and 1955). The 40 apparitions following them witnessed 7 of the 11 NTBs outbreaks. (Two or three other NTBs events began during the onset of EZ coloration [1942/43, 1989/90, and probably 1880], but that of 1891 was not associated with EZ coloration.)

The timetable listed above differs from Wacker's by the inclu-

Table 13.1. *List of jetstream outbreaks and equatorial colourings*

EZ/EB Colour	STBn Jetstream spots	SEB Revival or other outbreak	NTBs Jetstream spots	NNTBs Jetstream spots
1881–85	–	1882[e]	1880 Oct 17	–
–	–	–	1891 May 14	–
1897–99	–	–	–	–
1919–20	1919–20	1919 Dec 8	–	–
1925	–	–	1926 Jun 15[b]	(1926 Jul?)[e]
1927–29	–	1928 Aug 10	1929 Sep 27	1929 Oct
1934	1931–34	S: 1931–34 / Z: 1930–34	–	–
1937–39	1938	(1937–38)[c]	1939 Sep 23	1940 Oct
1943–44	1941	1943 Feb 7	1942 Oct 5[d]	(–>1943)[d]
1946–47	–	(1946–47)[a]	–	–
1949	–	1949 Jul 23	–	–
–	–	1952 Oct 20	–	–
–	–	1955 Feb 4	–	–
1958–59	–	1958 Mar 30	–	–
1961–65	1962[b]	1962 Sep 23	–	–
–	–	1964 Jun 14	1964 Jul 7[b,f]	1965 Dec
–	1965–68	S: 1965–68 / Z: 1966–68	–	1968 Jan
1968–71	1969	1971 Jun 21	1970 Aug 12[b,f]	(–>1970)
1972–76	–	1975 Jul 5	1975 Sep 23	1972 Apr
1977–80	1980–81	S: 1978–81 / Z: 1979–81	1980 May 11	1978 Nov
–	–	S: 1984–85 / Z: 1985–86	(1985 Jun)[f]	1984 May[c]
–	–	(S) 1988	–	1988 Aug[c]
1990–92	1991	1990 Jul[c]	1990 Feb 10	(–>1992)

This table lists all recorded episodes of activity since 1881, for equatorial colorations, SEB Revivals, episodes of long–running SEB activity, and other jetstream outbreaks – omitting minor activity recorded on NNTBs throughout recent years. For the SEB Revivals and transient jetstream outbreaks, the date of first observation is given. For long–running SEB activity, 'S' indicates SEBs jetstream spots and 'Z' indicates mid–SEB outbreaks of white rifts remote from the GRS. Details are in the preceding chapters.

[a] Belt revived between apparitions, no drifts recorded.
[b] Only one spot observed.
[c] Jetstream or Revival already active at start of apparition.
[d] Activity persisted from 1939–41.
[e] Disturbance seen but no rapid drifts recorded.
[f] Followed by revival of red NTB. (In 1985, no jetstream spots observed but NTropZ coloured.)
(Adapted from tables in Rogers, *1976, 1989b*.)

sion of NTropZ coloration, which seems to be associated with NTBs jetstream activity. On the other hand, Wacker also listed 'invasions' of the S. Temperate domain by the S.S. Temperate Current, around $T = 0$, and outbreaks in the NEB, around $T = +1$ to 2 yr, but it now appears that these are not correlated with the other phenomena. Nor are STropDs nor STB Fades.

One might have expected that NEB activity could be included, as it is analogous to SEB activity in many respects (Chapter 8). It is more difficult to put dates to NEB activity, because most manifestations of it (rifting, colour, and arrays of NEBn spots) are present to some extent in many years and only gradually wax and wane. The clearest indicator is the broadening of the belt. NEB activity is

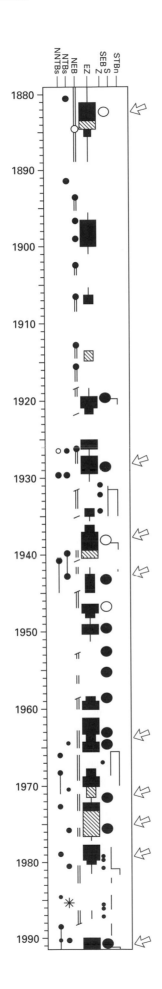

included in Fig. 13.3, but a systematic survey of all these indicators (Chapter 8) showed that NEB activity does not tend to coincide with the outbreaks that make up global upheavals; indeed, it may tend to avoid them. Given the prolonged and fluctuating nature of NEB activity, one cannot tell whether it is a very late sequel of global upheavals or whether it is an incompatible form of disturbance.

The character of global upheavals has varied during the 113 years since 1878, which can be divided into five eras.

In the *first era*, up to 1883, there was a great global upheaval in 1880–1882, with a major NTBs outbreak, major EZ coloration, and a SEB revival with GRS fading during solar conjunction. Previously, there were global upheavals in 1859–1860 and 1869–1872, with SEB revivals (1859 and late 1871), EZ coloration (1859–60 and 1869–72), and possibly a NTBs outbreak (1870/71). The SEB revivals may have been typical ones but no violent motions were actually observed. The present north-south asymmetry was different, with the SEB bearing equatorial projections and the NEB undergoing violent revivals; these aspects persisted in the second era.

The *second era* (1884 to 1918) showed several global anomalies as described in section 13.2. NEB Revivals occurred about every three years from 1893 to 1915 inclusive, but there were no SEB Revivals, and only one jetstream outbreak (NTBs, 1891), and only infrequent equatorial coloration. The great STropD appeared and the present equatorial asymmetry developed.

The *third era* (1919 to 1943), was that of the four great SEB Revivals. Each was part of a major global upheaval, with slowing of the GRS, major equatorial coloration, and outbreaks on other jetstreams. For a graphic account of one such global upheaval, one cannot do better than to read the BAA Memoirs for 1927–1929. However, the episode of continuous SEB activity from 1931–34 was not accompanied by other outbreaks. Cycles of NEB activity avoided the times of SEB Revivals, but did coincide with EZ and NTBs activity in 1925–27, and with continuous SEB activity in 1931/32, which may or may not be significant.

In the *fourth era* (1944 to 1959), there were SEB Revivals every three years, and most of these were also accompanied by EZ coloration, although some of the SEB Revivals and EZ coloration episodes were on a smaller scale than before. The GRS decelerations and other outbreaks stopped again until 1958, although minor jetstream outbreaks could have been missed due to incomplete coverage.

Fig. 13.3. Chart showing all episodes of activity of the types which have been implicated in global upheavals. For the EZ, blocks denote shading; solid if orange or yellow, hatched if grey or grey-brown, broad if intense, narrow if weak; a line denotes a trace of coloration or shading not included in statistics in the text. (The beginning and ending of EZ coloration is often indefinite.) For the NEB, dots denote Revivals, slashes denote expansions, and double lines denote other periods of activity. For other latitudes, dots denote jetstream spot outbreaks; open circles, outbreak inferred but no velocities measured; small circles, small outbreaks; lines, continuing jetstream activity. SEBZ and SEBs activity are shown separately when not linked in Revivals. (In some recent apparitions, smallscale jetstream activity has also been detected by photographs which go below the usual threshold of visual detection; these records are not included.) The major global upheavals are marked by arrows at right.

The *fifth era* (1960 to 1989), has seen frequent but irregular activity in all four domains. (Meanwhile, there were no typical cycles of NEB activity until 1988.) In the SEB, the 1962 Revival was a year late and abortive and was soon followed by the rather feeble Revival of 1964; since then, there have been great SEB Revivals interspersed with long episodes of continuous activity. NTBs outbreaks have had a different character to those in preceding eras (Chapter 7), being faster and sometimes producing a vivid orange belt, but not showing subsequent N. Temperate Current B drifts, and they have occurred every five years, except for 1985 when a unique coloration occurred instead. NNTBs activity has been detected in many years but this is partly because of improved coverage; Table 13.1 lists only the more substantial outbreaks. EZ coloration was frequently intense in the 1960s, coinciding with the SEB events of 1962 and 1964, but then becoming unsynchronised; and it condensed into a massive grey-brown EB from 1973 to 1976, but it was absent in the 1980s. Overall, it is difficult to define individual global upheavals in the 1960s, as episodes in different latitudes do not seem to have come in any particular order. This could be because they were really not correlated; or it could be because global upheavals had become prolonged and overlapped, with EZ coloration occurring several years before the related SEB outbreak, and NNTBs activity several years after. Anyway, the years 1975 and 1978–1980 produced clusters of outbreaks with many of the characteristics of classical global upheavals.

The complete global upheaval in 1990 (Plate P24) may have marked the end of the fifth era, and with another SEB Revival following only three years later in 1993, the pattern of activity may have changed again.

In summary, the first, third, and fifth eras have shown at least seven complete global upheavals, with SEB Revivals and impressive EZ coloration accompanied by one or usually more of the other typical outbreaks: 1880–82, 1927–29, 1937–39, 1961–64, 1969–72, 1972–75, and finally 1989–90. (There were also several less distinct global upheavals.) In these eras, EZ coloration almost always appeared months or years before the SEB Revival, accompanying the white obliteration of SEB(S) that preceded the Revival.

In contrast, the second and fourth eras showed approximately triennial Revivals of NEB and SEB respectively, with little activity in other latitudes. In these eras the EZ had only occasional episodes of activity. These often started in the same year as a SEB or NEB Revival, or the year after, but the EZ colour in these eras was seldom striking, and there was none at all during some NEB or SEB Revivals.

Although there seems little room for doubt about the association of the outbreaks in the four relevant domains, it is possible that the actual timetable is not determined. The order of the outbreaks does not always fit the timetable, and there may merely be a tendency for them all to occur within a year or two of each other.

What the various eruptions actually consist of is largely uncertain. They give a powerful impression of erupting bright, dark, or red material from below the visible clouds. If these are purely meteorological disturbances, they are remarkably localised and organised. White spots certainly do billow up from below, in mid-SEB and mid-NEB, as observed by Voyager; well-founded meteorological explanations for them will be described in Chapter 14.6. Whether dark spots are also brought up from below, or whether they are simply areas cleared of high cloud, is uncertain; but dark jetstream spots, although they affect high altitudes, do extend to a deep and warm level (Chapter 14.3). As for red material, we have not progressed much further than E.E. Barnard and A.S. Williams nearly a century ago. Barnard (*1892*) asserted that newly formed dark spots tended to become red over a year or so, and this is certainly the case with revived belts, the SEB, NEB, and sometimes NTB. Williams (*1920*), discussing possible variability in the latitudes covered by coloration, wrote:

> Assuming the tawny colour to be due to an effusion of steam into the upper regions of the planet's atmosphere, as was suggested in 1870, or as an alternative of dust, such as gave rise to our wonderful sunset effects after the Krakatoa explosion, then, on some disturbing influence arising in the southern hemisphere, the steam or dust-laden vapours, floating above the belts in the upper regions of the atmosphere, might be forced more north than usual, so as to overlie the NTropZ instead of the EZ.

The ochre colour is certainly high-altitude (Chapter 4), and can certainly cross domain boundaries (Chapter 3), but its origin and chemical nature are still unknown (Chapter 16).

13.4 POSSIBLE PERIODICITIES

Various periodicities have been proposed over the years for phenomena on Jupiter, but few have been borne out by subsequent events.

The one obvious periodicity has been the 3-year one that governed NEB Revivals from 1893 to 1915 and SEB Revivals from 1943 to 1958. It is possible, though, that this represented not an oscillation but a minimum interval between outbreaks. In support of this possibility, the SEB Revivals of 1955 and 1964, which occurred less than 2.5 years after the previous ones, were the weakest on record. The typical three-year interval in these eras may indicate that outbreaks were occurring as soon as conditions were ready after the previous one. In other eras, aspects of NEB and SEB activity have sometimes recurred at intervals of three years (R. Doel, *1977*), but major SEB Revivals and NEB expansions have occurred at intervals of three to 30 years with no sign of periodicity.

Otherwise, the most regular cycles, within limited eras, have been the 5- and 10-year periodicities of events on the NTB (Chapter 7). NTBs jetstream outbreaks occurred at intervals of 10 years from 1870 to 1891 and from 1929 to 1939, and have occurred strictly at intervals of 5 years from 1964/65 to 1990 (if one coloration episode without spots is included). Most persistent of all, though more variable in period, has been a 10-year periodicity in fadings and (quiet) revivals of the NTB, which is coupled to periodicity in its latitude, apparently because the NTB(S) component tends to fade first. Strangely, the NTBs jetstream outbreaks have not had any consistent relationship to the phase of the $\approx$10-year cycle of fadings.

A natural period to look for is that of the jovian year, 11.86 terrestrial years. The planet's tilt is too small to produce perceptible seasons, but the eccentricity of its orbit is such that it receives 21% more solar radiation at perihelion than at aphelion. This might be enough to affect atmospheric events.

Williams[2] reported strong twelve-year periodicities in the colours of the SEB, NEB and EZ in the nineteenth century. His proposed cycles of redness in the SEB and NEB were complementary, such that the SEB was reddest when the NEB was least red, coinciding with the time of aphelion. He also suggested a period of 12 years for the coloration of the EZ, again identifying peaks of tawny colour near the times of aphelion. Williams himself made numerical estimates of the colour for 57 years in an attempt to confirm the periodicities – but the irregular nature of the cycles meant that there was no consistent relation between the colour of the EZ and those of the belts, and these periodicities broke down in the early twentieth century. Reese (*1960*) brought the analysis up to date but found only a marginal tendency for EZ colour to occur near aphelion (correlation coefficient 0.36), and no correlation of colour in the NEB and SEB with the season of the jovian year. Subsequent EZ and SEB colorations have not tended to coincide with aphelion either.

It was also suggested that the NTBs jetstream outbreaks tended to occur near perihelion (see Chapter 7.3), but this has not been the case in recent decades.

Some of the regular currents and jetstreams have variable speeds, but there is only one persuasive case of periodicity: the speed of the STB white ovals fluctuated with an approximately 12-year period for most of their life, moving slowest near perihelion (Chapter 11.3). This may well have been linked to the jovian year.

The S. Temperate Current before 1940 had also shown hints of a much weaker 12-year fluctuation, in phase with the variation shown by the white ovals thereafter (Chapter 11.1).

The N. Equatorial Current may also be influenced by the jovian year, as it sometimes decelerates for a year or so, and five of the six instances have come 1–2 years after perihelion, when the NEB receives its greatest solar heating (Chapter 9.2). However, this is by no means a regular occurrence. When Reese and Beebe (*1976*) identified a pair of NEBs plumes that lasted for 12 years, they suggested that their variable speed was related to the jovian year, but of course it is impossible to establish this from a single cycle, and the plumes diverged considerably from the mean rate of the N. Equatorial Current.

As a cautionary tale, it is worth recalling that E. Graf and colleagues (*1968*) and D. Basu (*1969*) found a correlation of the prominence of the GRS with sunspot numbers – at least from 1894 to 1945. As it seems inconceivable that solar activity could influence the weather on Jupiter, one can only attribute the correlation to chance: the 11-year sunspot cycle happened to match the average interval between global upheavals over those years, and particularly those culminating in 1919, 1928, and 1938. After 1945, the correlation broke down. Thus one must remember that even a quite strong correlation between two sets of recurrent phenomena can be entirely spurious.

It is possible that the longer-term changes on the planet, such as the drift variations of the GRS or the shifts of EZ activity that occurred around 1908–1918, might be due to very long-term cycles, but several more centuries will be required to test such ideas.

---

[2] See Williams (*1899, 1920, 1930, 1936a,b*). Phillips (*1915*) and Reese (*1960*) derived numerical estimates of colour from the data in BAA and ALPO reports, to match those of Williams. (Reese (*1960*) only gave a brief summary, but his actual numbers up to 1954 were given in an unpublished letter to the BAA Jupiter Section.) The correlation between these different estimates of colour in the same years is far from perfect, indicating how difficult it is to make objective interpretations of colour.

Other suggested periodicities on the planet were reviewed by Favero *et al.* (*1979*).

14: The dynamics of individual spots

In Chapter 3, we looked at the basic dynamical properties of the various types of spots on Jupiter: bright and dark, cyclonic and anticyclonic. Now we can return to this topic, to see whether the detailed observations can be combined with theoretical approaches to produce anything like a complete model of the spots. Concerning cyclonic features, there is little more to say. But there have been plenty of ideas on the nature of anticyclonic features, above all the Great Red Spot.

14.1 EARLY THEORIES OF THE GREAT RED SPOT

To the earliest observers, it was the redness of the GRS that fired their imaginations, and there was an elision of perceptions of redness and incandescence which seems incomprehensible to the modern mind. For example, it was suggested to be 'a reflection of lurid light from the fires of an active volcano' (BAA Memoir for 1894/95), and the following year's report discussed 'whether it be.... gaseous emissions from some vast volcano, or a mere floating, incandescent cloud' (BAA Memoir for 1895/96).

As for its physical nature, the first hypothesis was that it was part of the underlying solid planet. Thus Nathaniel Green in 1887:

> The GRS may be fairly considered a portion of the more solid part of the planet, and therefore affords the best means for ascertaining the revolution period; all other markings being subject to frequent change or possessing special periods of revolution indicate their atmospheric character.... The extended white light surrounding the spot at its commencement, and continuing to the opposition of 1880, may have been due to the presence of a dense mass of clouds, amongst which the spot was seen, either in consequence of its elevation above them, or from the radiation of such an amount of heat from the spot, that they did not form over it.

Heat emanating from the spot was also suggested by some observers as the cause of the 'repulsion' of the SEB to form the Red Spot Hollow.

But the changes in its rotation period soon showed that it was not fixed to the planet's interior. A somewhat more advanced view, though lacking any physical theory, was expressed in the BAA Memoir for 1892:

...From the members of the Section, the current opinion is that the constitution of the Red Spot is not atmospheric in the same sense as the light or dark spots that appear irregularly on nearly every part of the disk...and that it is not strictly a protuberance from the more solid nucleus of the planet (if such there be), but that it is semi-atmospheric, consisting of a projection axially of a more dense stratum of the wonderful Jovian envelope, and probably of a higher temperature than its surroundings, hence its red-hot appearance during its prime.

Among physical scientists, though, the prevailing theory was that the GRS represented some sort of floating body – either a vast ellipsoid, or an elliptical plate. What is seen might be the body itself, or a cloud formation hanging above it. This theory was first proposed by G.W. Hough (*1905*, and earlier notes cited by Peek), who supposed that the GRS and other ovals were due to bodies floating on some sort of ocean which formed the visible surface of the planet. In 1939 R. Wildt, a pioneer of jovian chemistry, modernised the theory by postulating the 'ocean' to be of highly compressed gases rather than liquid. B.M. Peek (*1958*) added the ingenious rider that when the floating GRS changed its rotation period it would also change its depth, while conserving angular momentum; and this could account for the increased darkness when its rotation period was lengthened. The depth change would be accompanied by a density change, and Peek explained this in terms of a phase change in the material of the floating body. However, the discovery of the hydrogen/helium atmosphere presented this theory with a grave problem of how to make a solid body with suitable density and chemistry (as discussed by Carl Sagan, *1962*). In its final formulation, by W. Streett and colleagues (*1971*), this body was to be made mostly of solid hydrogen, whose density changed with pressure to keep it floating at a depth of several thousand kilometres – a 'Cartesian diver'.

Another oft-quoted theory, that of the 'Taylor column', was proposed by Raymond Hide (*1961*). In the 1920s, Sir Geoffrey Taylor had shown that winds blowing over a small cylindrical plateau could produce a matching circulation in the atmosphere above it, and Hide proposed that the GRS was just such a column above a plateau about a kilometre high on Jupiter. But experiments suggested that such a column could only extend a few times the height of the plateau. If so, because of its variable motion, the plateau would have to be a floating raft, and this theory amounted to a special version of the floating-body theory. Alternatively, Hide argued that if the solid surface was deep enough, it could vary its rotation period by swapping angular momentum with the massive atmosphere. The difficulties were discussed in detail by Gerard Kuiper (*1972b*).

But perhaps the nicest theory of this era appeared as a misprint in an English newspaper's account of the controversies: that the GRS is 'a whirlpool of emotion in the atmosphere of Jupiter'.[1]

Apart from their intrinsic difficulties, all these theories failed to explain the most important aspect of the GRS – its circulation. Although this was not revealed directly until 1966, it could be inferred from the sense of the jetstreams flanking it, which were recognised as recurrent features in 1927–1928; and the circulation was suspected even earlier, from the motion of the South Tropical Disturbance which seemed to stream round the southern edge of the Spot. H. Kritzinger (*1914*) maintained this theory, likening the GRS to a terrestrial cyclone and even estimating its period of circulation correctly as about ten days (although he thought the planet was hot like the sun). And Phillips, with his usual perspicacity, said in 1915 that the STropD-GRS interaction

> is very strong evidence, as sundry investigators have noticed, that the Red Spot is a vortex – analogous to a cyclone on Earth, though its prolonged existence shows that it must be of great strength and probably deep seated below the planet's visible surface. This theory is doubtless not free from objections. The pointed ends of the spot sometimes observed are difficult to explain, but it may fairly be said that on the whole the vortex theory at present fits the facts better than any other.

14.2 THE GREAT RED SPOT AND OTHER OVALS AS ANTICYCLONIC VORTICES[2]

Anticyclones and cyclones

As Phillips said, the GRS is a vortex. So are all the white ovals in homologous locations on the edges of other belts, as revealed by professional photography and by the Voyager spacecraft. Like the zones, all these ovals behave as high-pressure areas covered with white or red clouds. The cloud-tops are colder than their surroundings, but this may be because they are elevated cloud caps on top of columns of warm rising air. All these ovals have similar circulations to that of the GRS, with little if any net outflow, and the strongest winds blow around the periphery. The strength of the circulation is measured by the vorticity (strictly, its horizontal component, i.e. the rate at which the speed around the centre increases with radius). Both in the GRS and in the STB white ovals, the vorticity reaches about 6×10^{-5} s^{-1}, i.e. 60 m/s per 1000 km, which is four times the shear across the STropZ (Fig. 14.1). How are these

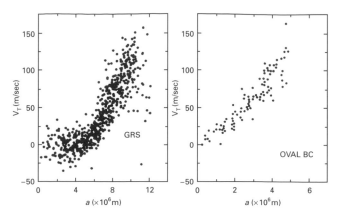

Fig. 14.1. How the speed of circulation varies with radius in the GRS (left) and oval BC (right). From Mitchell *et al.*(*1981*).

circulations organised on such a large scale, and how are they maintained for so long?

The circulation of the GRS is anticlockwise, which by definition is anticyclonic in the southern hemisphere. This implies that it must have higher pressure at the centre. The motion of air on a rotating planet is controlled by conservation of angular momentum which produces circulating motions perpendicular to an applied force, as in a gyroscope. This is the Coriolis effect. It requires that regions of high pressure are surrounded by winds blowing anticlockwise in the southern hemisphere or clockwise in the northern hemisphere; these are anticyclones. For regions of low pressure, the directions are reversed, and these are cyclones.

Nevertheless, the analogy of the (anticyclonic) GRS with a (cyclonic) terrestrial hurricane is not inappropriate. The fierce cyclonic winds around the low-pressure centre of a hurricane are at sea level, but at the higher altitudes to which the air rises, the storm is a high-pressure centre and has a weak anticylonic flow. Movies from weather satellites reveal this flow spiralling across the top of the oppositely directed low-level spiral. Similarly, we see the GRS from on top. Its vorticity is not much greater than in a terrestrial anticyclone. It reaches higher wind speeds because it is so large – as large as the Earth.

The first detailed model of the GRS as an isolated storm was by Gerard Kuiper, director of the LPL (*1972a,b*). This was closely modelled on terrestrial hurricanes, the basic principle being a convecting cloud system driven by the latent heat of condensation of water. Although this model does not fit the detailed structure of the GRS and other great ovals, it could be relevant to other spots on the planet.

Terrestrial hurricanes, like individual thunderstorms on a smaller scale, have a three-dimensional circulation in which moist air is sucked into a rising centre where water vapour condenses into cloud, releasing heat. The warm air rises, maintaining the circulation, so the centre is a convective column marked by cumulus or cumulonimbus clouds. At the top the cloud spreads out anticyclonically to form an 'anvil' much larger than the convective column itself. Storms with these dynamics occur on various scales on Earth, and the ones cited by Kuiper are the clusters of giant thunderstorms that occur over the tropical oceans where the trade winds converge. They have anvils covering up to 2000 km.

---

[1] Cited by R. Hide in Streett *et al.* (*1971*).
[2] Present theories of the GRS are reviewed by Ingersoll (*1981*); Williams (*1985*); Sommeria *et al.*(*1991*); Beebe (*1990*); and Read (*1992*).

On Jupiter, water condensation could occur in much the same way in the lowest cloud layers, as we will see in Chapter 16. Clusters of thunderstorms and their anvils could have diameters ten times those seen on Earth, and Kuiper proposed that the GRS is just such an anvil. Storms could be much more persistent than on Earth, because of their great size, the absence of an ocean boundary, and the persistence of internal heat flow by night as well as by day. His theory was consistent with several aspects of the motion of the GRS. However, it required the convection to be localised towards the f. end of the GRS, and predicted a fairly rapid outflow on spiral streamlines, such that the clouds would move outwards over 8–10 days. This was refuted by Voyager, which showed that the central regions are quite uniformly puffy; the surrounding spiral consists of streaklines, not streamlines; net outflow was undetectable over a couple of months, and small jetstream spots were actually being captured from outside.

Nevertheless, modified forms of this are not ruled out (cf. Flasar *et al., 1981*). And the model could perhaps apply to the smaller white spots in the NTropZ and NTZ, which have diffuse outlines. Jetstream spots have been seen to run across these, perhaps because they can disrupt an 'anvil' without hitting the more compact convective column (section 14.3). A similar theory also accounts for outbursts in equatorial plumes and in the equatorial belts (section 14.6).

Free eddies

In fact, theorists now realise that one can dispense with any sort of deep structure in theories of the GRS; convective columns, floating rafts, and hidden mountains are all unnecessary. The ovals can simply be regarded as huge, almost free-running eddies, and there is now a large range of theories which can produce more realistic models of them from a wide range of conditions.

The suggestion that the GRS is a free eddy was first made by G.S. Golitsyn in Moscow in 1970, and by Andrew Ingersoll at CalTech in 1973. The point is that eddies can persist for a very long time in the jovian atmosphere. This is partly because there is no solid surface to give friction, and partly because the cold cloud-tops allow little energy to escape by radiation. Some loss of energy and of vorticity must still occur by friction, but it can be made up from either of two sources: from merging of smaller eddies, or from shear-instabilities (waves) on the jetstreams. Both these processes arise naturally in many model atmospheres, and they are also the two processes by which new eddies can develop.

Merging of small eddies is a favoured mechanism, as has been shown both by Gareth Williams (*1975*) and by Andrew Ingersoll and P.G. Cuong (*1981*). The smallest eddies would arise by buoyancy; heat from the interior or from condensation causes hot air to rise and cold air to sink, and Coriolis forces convert this into horizontal eddying motion on the scale of tens of kilometres. Eddies can also arise on a larger scale due to horizontal temperature differences. In models, the eddies naturally merge either into jetstreams (Chapter 15), or into even larger eddies, ending up with a single great oval which has consumed all the others in the same latitude. The Voyager films seem to show this process at work when SEBs

jetstream spots merge into the GRS; they also showed jetstream spots merging with each other (often after several 'orbits' round each other).

One should remember, though, that large ovals on Jupiter do not tend to merge. The only observed merger of large anticyclonic ovals was of one pair on the NEBn. Normally when ovals come close (on NEBn, STBs, or SSTB), they settle into a fixed spacing, which is often stable for a year or more. (Indeed, S. Temperate ovals BC and DE have stayed separated by a distance similar to their diameters for the last five years.) This seems to imply that they are much more coherent and impenetrable than the looser spots seen on the jetstreams and in the models. A higher-resolution theoretical model might produce more realistic behaviour.

To understand the spots fully, we ought to understand the overall structure of the atmosphere, which is deferred to the next chapter; but in fact many different atmospheric models can produce surprisingly similar results. We do not know how deep the circulation goes, nor how much vertical motion there is, nor how the ovals are connected to the jetstreams; these are parameters that differ in different models.

Even the simplest model planet can generate large vortices, as shown by the work of Gareth Williams and colleagues at Princeton.[3a] They did computer simulations of a planet with a single thin layer of atmosphere and no jetstreams, just a global anticyclonic flow, unlike Jupiter but like Neptune.[3b] Vortices are easily generated even here, by merging of eddies or by shear-instability; they interact or dissipate in ways quite like ovals on Jupiter, and large anticyclonic ones are long-lived in temperate latitudes. They persist by drawing vorticity from the global flow pattern. The details of their behaviour depend on many parameters. For example, ovals of the size of the SSTB ovals and the jetstream spots are only just large enough to survive, but a weak shear-instability could grow into a large oval in about ten years, rather like the STB white ovals. In S. Tropical latitudes, a 'GRS' can form, though equatorial and SEBs jetstreams must be added to the model to prevent it from dissipating. All these ovals tend to drift westwards (retrograding), but applying the observed prograding jetstreams to the model can force drift values close to those observed for the GRS and the STB white ovals.

The eddies on the Williams planet illustrate two different ways in which these disturbances can interact. Under some conditions they merge during encounters, but under other conditions they pass through each other like waves; in fact, they behave as solitary waves or 'solitons'. A soliton is a wave that can propagate indefinitely even though it is not part of a train of waves.

Models of the GRS and other features as solitons were devised by T. Maxworthy and L.G. Redekopp (*1976a,b, 1978*). Applying

---

[3] *a.* Williams and Yamagata (*1984*); Williams and Wilson, (*1988*).
b. The Great Dark Spot on Neptune is an anticyclonic oval whose scale and latitude are very like those of the Great Red Spot, but it exists in a global anticyclonic flow rather than between narrow jetstreams. Saturn has no such giant oval, but that is because it is mostly covered by a vast equatorial jetstream. A Little Red Spot was seen by Voyager in Saturn's second anticyclonic zone to the south, at 55°S.

simple conditions to a planet with jetstreams, they could produce a solitary wave, which could be formed and sustained by shear-instabilities in the jetstreams. It could be either an elevated soliton, which resembled the GRS, or a depressed soliton, which resembled the S. Tropical Disturbance. The waves were shallow, implying that the GRS extends less than one scale height (25 km) below the white clouds of the STropZ. Solitons of different sizes moved at different speeds, and when a depressed soliton encountered an elevated soliton in the same latitude, they passed each other almost instantaneously just like the STropD used to pass the GRS (Fig. 14.2). The depressed soliton would have a limited length so they had to postulate that the great STropD consisted of two such solitons in tandem; this would be consistent with the observations in several years when the p. and f. sections did look like separate Disturbances. Although the model may be too simplified to be taken literally, it is so far the only coherent model of a STropD.

An alternative model of anticyclonic ovals, which has been popular recently, was developed by Ingersoll and Cuong (*1981*). They took the same case as Maxworthy and Redekopp, but with the jetstreams extending deeper into the planet. In this case, the eddies are not solitons which pass through each other, but 'modons' which merge, until the largest vortex in a zone sweeps up all the others

into an oval like the GRS (Fig. 14.3). These ovals are confined to a narrow upper layer (as in the soliton model), while the jetstreams extend deeper. These theoretical ovals are almost stationary relative to the east-west flow at the same latitude, so they are not yet perfect models of the real ovals, which nestle in the retrograding jetstreams. Another numerical model, by Williams and Wilson (*1988*), produces a range of ovals from combinations of different atmospheric processes (Fig. 14.4).

All the approaches described so far were theoretical. Recently, some experimental models have been attempted, using spinning tanks of water to imitate Jupiter. Anyone who enjoys watching vortices in the bath will appreciate the fascination of these experiments.

One was performed in Moscow by S.V. Antipov, M.V. Nezlin, and colleagues (*1986*), using a deep paraboloidal bowl 28 cm across which was spun until the water formed a layer 0.5–1 cm deep on the inner surface. To produce cyclonic or anticyclonic shear in the flow, the inner, middle, and outer segments of the wall could be spun at different speeds. In the anticyclonic flow, the experiment could produce an anticyclonic vortex like the GRS, with a westward (retrograding) drift (Fig. 14.5A). It had high pressure thanks to doming of the water over it, but its vorticity was highest at the middle, not near the edge as in the GRS. The Russian

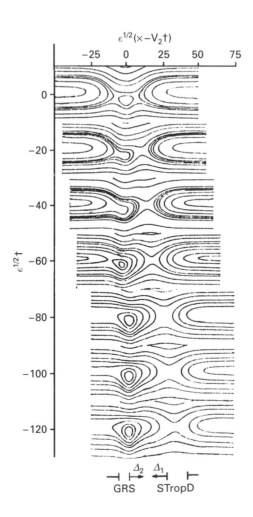

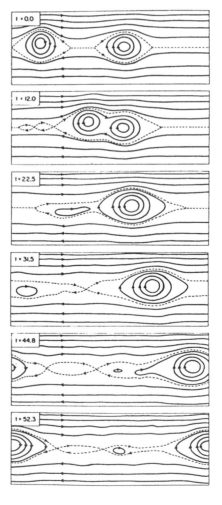

Fig. 14.2. Computer model of solitons: a depressed soliton ('STropD') passing an elevated soliton ('GRS'). Strip-maps are evolving from top to bottom. From Maxworthy, Redekopp, and Weidman (*1978*).

Fig. 14.3. Computer model of modons: anticyclonic eddies which merge. Each panel is a strip-map around the planet, evolving from top to bottom. From Ingersoll and Cuong (*1981*).

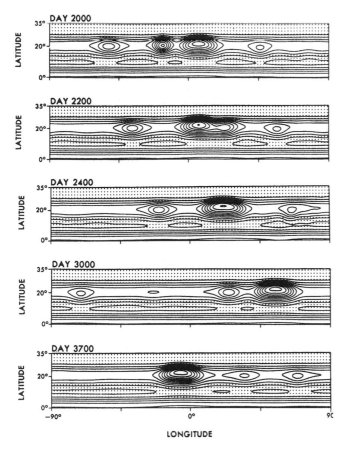

Fig. 14.4. Another computer model of anticyclonic eddies which merge. This string of anticyclonic vortices developed by shear-instability of a retrograding jetstream; with time, the largest vortex absorbs the smaller ones. From Williams and Wilson (*1988*), by courtesy of G. Williams.

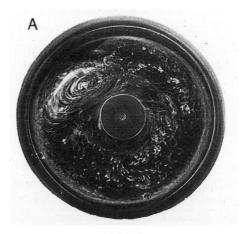

Fig. 14.5A. Experimental model in a spinning bowl with differential rotation. The flow is traced out by lines, and shows a single large anticyclonic oval. From Antipov *et al.*(*1986*).

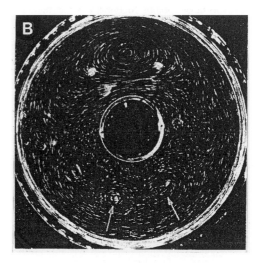

Fig. 14.5B. Experimental model in a spinning tank with radial flow. The flow is traced out by lines, and shows a single large anticyclonic oval, which consumes others that drift within its grasp. From Sommeria *et al.* (*1988*).

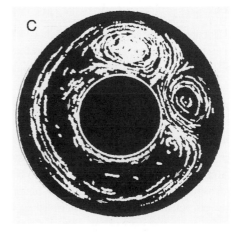

Fig. 14.5C. Experimental model in a spinning tank with differential heating and cooling. The flow is traced out by lines, and shows a single large anticyclonic oval, accompanied by a weaker cyclonic oval. The pair resemble the GRS and the disturbed region of SEB following it. Adapted from Read and Hide (*1984*), by courtesy of P.L. Read.

researchers also observed vortices merging in this system. In the cyclonic regime, they got no stable eddies except when the shear was very strong, and then they got a chain of cyclonic vortices with prograding drift which did not merge – very like the NEBn barges.

In Texas, H.L. Swinney's group[4] worked on a grander scale, with a spinning tank like a shallow jacuzzi. It looked nothing like a planet: it had a slightly convex bottom with a ring of inlets and a ring of plug-holes to produce a radial flow of water. The Coriolis effect produced a retrograding jetstream, and vortices developed over the ring of plug-holes. Anticyclonic flow formed nothing of interest, but cyclonic flow formed a wavy jetstream with cyclonic vortices evenly spaced around it; the faster the flow, the fewer the vortices. At maximum speed, they merged until only one giant vortex remained, consuming any small eddies that approached it (Fig.14.5B). Computer simulations by P. Marcus (*1988*) gave similar results to the rotating tank, and could also produce anticyclonic ovals. Thus even a very artificial model produces ovals or barges somewhat like those on Jupiter.

These models took no account of temperature variations; but ovals can also be produced due to horizontal temperature differences, as was done by Peter Read and Raymond Hide of the UK Meteorological Office.[5] Again they used a spinning annular tank,

[4] Sommeria *et al.* (*1988, 1991*); Meyers *et al.*(*1989*).
[5] Read and Hide (*1983, 1984*); see also Hide and Mason (*1975*).

but forces were produced by heating the fluid with an alternating electric current, and cooling it at the outer and inner walls. The temperature gradient set up sloping convection which set up growing eddies; this is a so-called 'baroclinic instability' similar in some ways to the origins of our weather in temperate latitudes on Earth. In the experiment, it produced a meandering retrograding jetstream enclosing a chain of anticyclonic vortices with warm rising centres. In conditions where the spinning tank had only slight baroclinic instability, it produced a single intense pair of vortices, resembling the GRS with the region of SEB disturbance following it (Fig.14.5C). At the bottom of the tank, the general flow was cyclonic like the SEB and the cyclonic vortex was the stronger; nearer the top, the flow was anticyclonic and the anticyclonic vortex was stronger, with the weaker cyclonic disturbance following it, just like the GRS and the SEB rifted region.

Whether this can be a realistic model of Jupiter remains to be seen; it had many arbitrary parameters, and there are no data on sources of heating or cooling on Jupiter. Nevertheless, the results looked like the jovian atmosphere in many respects. (Moreover, if the fluid were cooled instead of heated internally, as was done in a computer model, a similar meandering jetstream would be formed but with a chain of cyclonic vortices, like the NEBn barges.) In contrast to the previously discussed vortices which must consume vorticity from outside, these generate vorticity internally by convection, although this would not preclude them from sweeping up surrounding small spots as the GRS does.

The modelling has now advanced beyond arbitrary models to the stage where a large range of possibilities can be explored. To select the most realistic among them, the best strategy may be more precise modelling of the detailed evolution and interactions of the ovals, for comparison with the behaviour observed on Jupiter. In the hope of encouraging such progress, we will present a list of pertinent observational facts in section 14.4 after discussing the observed origins of the ovals.

14.3 JETSTREAM SPOTS AS ANTICYCLONIC VORTICES

The small dark spots on the jetstreams usually have a dark grey or sometimes bluish-grey colour, and strong 5–μm infrared emission which shows that they are warm, like the belts and cyclonic barges. This is consistent with the extremely dark appearance of the NNTBs spots when viewed on the crescent planet by Voyager; these aspects suggest a low-altitude spot which shows a greater degree of limb-darkening as long as there is no haze covering it.

However, a major discovery of the Voyager mission was that all jetstream spots (on both SEBs and NNTBs), are anticyclonic vortices, belonging dynamically to the zones. This implies high pressure and, in most other anticyclonic spots, is associated with thick white clouds. All these data suggest that the jetstream spots are anticyclonic systems whose circulation extends up to the altitude of the zone cloud-tops or higher, but that the area of the spot is kept free of white cloud and overlying haze, having a cloud structure like the belts. This could be because these spots (unlike large anticyclonic ovals) are warm

throughout the visible clouds, their vorticity being associated with upwelling of warm air. This should be testable by targeted infrared spectra. If it is true, the thermal wind equation would imply that the spots' circulation is most rapid at high altitudes (Chapter 4.6).

In judging the altitudes of the spots, there is particular interest in the rare observations of these spots encountering white ovals in the same latitude, when they appear to pass over or through the white ovals. A dark NTBs jetstream spot ran across a white NTropZ spot on 1939 Dec 27, and the red NTBs spot in 1964/65 twice did likewise; and a dark NNTBs jetstream spot ran across a white NTZ oval on 1987 Sep. 12. The white ovals concerned were rather diffuse, so the spots may only have penetrated a spreading 'anvil cloud'. Also, some SEBs jetstream spots ran across the northern half of the GRS in 1928 and in 1966, although they were greatly distorted in doing so. One should remember that these spots, although small on the scale of the planet, are several thousand kilometres across, whereas the visible cloud levels differ by only a few kilometres (Chapter 4). So the jetstream spots do not go over other spots, but through them. Perhaps, if a jetstream spot has sufficient momentum, it can carry on through a larger white spot disrupting its white cloud layer.

These dark grey spots thus contrast with other high-altitude anticyclonic systems which mostly appear as white ovals. But some of the jetstream spots are reddish and some do have white cores, so they can contain a mixture of clouds.

In examining the Voyager close-ups of the SEBs jetstream spots (Chapter 10.4), we noted that the details of their behaviour – their formation from the turbulent SEB region, their interactions with the white clouds in the SEB, and their coalescence to form the S. Tropical Disturbance – all gave the impression that they were shallow eddies with no sign of deep-seated roots. The same impression was given by the Voyager views of NNTBs jetstream spots merging with each other.

How are jetstream spots created? The relevant observations have been discussed in the chapters on the NNTB, NTB, and SEB. Sometimes the spots are all created at a single longitude, and the detailed observations show that they appear alongside regions of cyclonic disturbance. SEB Revival sources are the most obvious case, and on a smaller scale, Voyager showed examples of bright streamers in SEB and NNTB being broken off and twisted round by the adjacent jetstream to form the jetstream spots. In other outbreaks, the jetstream spots form at many longitudes simultaneously, apparently due to some instability in the jetstream flow.

The nature of this instability is not known. Theoretically, it has been argued that the latitude of the SEBs jetstream is the most prone to shear-instability because the velocity shear across it appears to be much greater than allowed by the 'barotropic stability criterion' (see next chapter). But jetstream spots are certainly not produced just because the jet is going too fast; there is no correlation of jetstream speed with outbreaks (Chapter 3). Moreover, this type of instability would not explain the outbreaks on the prograding jetstreams. Alternatively, the jetstream spots could be due to baroclinic instability (see next chapter).

Whatever the source of jetstream spots, their characteristic speed and size could be explained if physicists could give theoretical grounds for two conjectures: (1) that a shallow vortex would tend to avoid a shear zone and lodge in the nearest jet; and (2) that its size would be limited by the width of the jet.

14.4 THE FORMATION AND EVOLUTION OF ANTICYCLONIC CIRCULATIONS

How do large ovals come into being? As we saw in section 14.2, there are two main theoretical options, both supported by the Voyager movies. One is that the GRS and similar ovals were actually created by merging of smaller vortices. The other is that instability of the shear on the retrograding jetstreams causes them to meander, producing eddies within the curves. But these may not be the only possibilities, and the real processes might be more complicated. There is also the possibility that new circulations develop at 'chiasmata' (Chapter 13.1) so as to complete symmetrical patterns at the conjunctions of pre-existing circulations.

Let us therefore return to the observations, with a resumé (from the descriptions in the preceding chapters) of the origins of observed anticyclonic circulations. We do not know when or how the GRS arose, but there are detailed records of the origins of analogous circulations: the STB white ovals, South Tropical Disturbances, and several smaller anticyclonic ovals. It appears that such features form in a variety of ways, and are stable no matter how they arise.

Origins of Little Red Spots and NEBn white ovals

The arrays of NEBn barges and white ovals, which arise near-simultaneously in the aftermath of NEB expansions and/or rift activity, probably form as eddies within meanderings of the NEBn jetstream (Chapter 8.4). This process is probably a display of shear-instability triggered by the largescale disturbances in the belt, and we saw in section 14.2 how various theories and experiments can produce arrays of 'barges' by just such a process.

Whether the rare Little Red Spots and their littler brown kin are born in the same way is more doubtful. The best-observed example, a brown oval photographed by Voyager (Chapter 8.4), did arise by eddying where small dark patches on the NEBn retrograding jetstream were deflected by an adjacent barge, and swirled and merged until they formed an anticyclonic dusky oval. But the dark patches seemed to be an integral part of the process, and the question arises of whether they were tiny anticyclonic vortices which fed energy into the new oval. In fact they may have had underlying turbulent motions with a generally anticyclonic sense, but they were not typical jetstream spots; they looked more like amorphous waves. So some aspects of this story remain unclear. Other such brown ovals in the NTropZ, observed from Earth, were reported to form adjacent to areas of disturbance within the NEB. And one NTropZ white oval, in 1975, formed within a NTBs jetstream outbreak.

In the STropZ, one Little Red Spot seems to have been created at the f. end of a South Tropical Dislocation, perhaps in the same way as the STropD described below (Chapter 10.7). Conversely, another LRS evolved out of the southern half of a long-lived white oval, and may simply have been a colour change (Chapter 10.6). In the NTZ, a LRS formed during the breakup of a North Temperate Disturbance (Chapter 7.2).

Perhaps the most sensible conclusion is that, as anticyclonic circulations are such stable features, they can readily form in a variety of ways. Any of these processes can lead to much the same result.

Origin of the STB white ovals

The birth of the three great ovals was observed around 1940 (Chapter 11.3). They arose by the appearance in the STZ of three short dark features which resembled STropDs; they probably reconnected the jetstreams across the STZ. The dark barriers grew longer and confined the remaining zone to form the three ovals (Fig. 11.1).

There was no evidence for any process of merging of smaller ovals, nor of eddying within a planetary-scale wave, nor of chiasma formation. If any such structures existed at the time, they were invisible from Earth. One could only uphold the possibility of merging of smaller spots by invoking jetstream spots too small to track, on one or both of the jetstreams involved. Indeed, the SSTBn jetstream can carry jetstream spots which are so small as to be barely visible from Earth; while on the STBs, several tiny dark spots were actually seen as the great ovals were forming in 1940, but their motion was not established (and was suspected of being in the wrong direction). However, there was certainly no merging of intermediate-scale ovals. The possibility of a wave-like instability on the STBs jetstream was suggested by the near-simultaneous origin of the three widely separated divisions, but one would have to suppose that these were merely local effects of an invisible larger wave pattern, perhaps like the breaking of a wave (section 14.5).

Origin of a South Tropical Disturbance

Thus the origin of the STB ovals may have been equivalent to the origin of a South Tropical Disturbance – and that *has* been witnessed in great detail by Voyager (Chapters 10.7 and 13.1). Recurrent STropDs are triggered by a 'dislocation' of which the important feature may be a 'chiasma' formed next to the GRS, but Voyager showed that the effectual cause was the interaction and eventual merging of SEBs jetstream spots. They began to wander off course in a space formed next to the GRS, and ended up merging into a larger anticyclonic hemi-circulation that defined the STropD. So here was direct evidence for small vortices feeding into the larger circulation, although not quite in the manner of the models. They did not feed directly into a growing oval, nor into the normal jetstreams, but into a partial reconnection of the jetstreams that, if it had propagated long enough, could have evolved into a new great oval.

However, some questions remain. First there is the possibility that, in addition to the distinct circulations, amorphous dark and bright clouds might have some causative role. Second, some STropDs (as well as the STB white ovals) have appeared when no jetstream spots were in the vicinity. However, the characteristic jetstream spots are not essential to (and indeed are not explained by) the theoretical models, which do not require their merging vortices to have this particular size. It is conceivable that invisibly small eddies could do the job, though there would presumably have to be some external influence to make them recirculate in the first place.

Does a South Tropical Disturbance pump up the adjacent jetstream?

As seen in the Voyager movies, the new STropD was the most striking example of small vortices merging into a larger one. And visual observations showed that the SEBs jetstream speeded up over the next nine months. Indeed, in Chapter 10.1, we noted that the SEBs jetstream generally tends to retrograde faster when a STropD is present. Could this be a visible example of the cascade of energy from smallscale to largescale circulations?

We can make rough calculations to see if this is a reasonable hypothesis, using the example that coincided with the Voyager encounters, and taking the simplest possible assumptions. We work out angular momentum about an axis that is perpendicular to the axis of the planet, and ignore the rotation of the planet itself – a colossal omission, but one that may be justified because no change in latitude is involved. We can estimate the amount of angular momentum that was gained by the SEBs jetstream, and compare the amount of angular momentum in the SEBs jetstream spots that were destroyed. As the depths of these circulations are unknown, let us assume that they are the same.

The SEBs speed was constant at $\Delta\lambda_2$ = +119 (±2)°/month throughout 1979 January–June (including independent measurements by visual observers and by the two Voyagers), but had speeded up to +143°/month by 1980 February, a change of 24°/month or 10 m/s. The STropD formed in 1979 May, so the change occurred within nine months. We suppose that it affected the region from the GRS to the STropD, a distance of 280° longitude. (The STBn jetstream speed did not increase.) For the SEBs jetstream spots, we need to know the typical diameter (up to ≈4000 km) and the circulation period, which has not been measured accurately but from various Voyager picture pairs is probably ≈2.5 days. (These values are similar to those quoted for the little white ovals on the SSTB, which look very like the jetstream spots.) For simplicity, we assume that these spots circulate like rigid discs. With the frequencies seen in both 1979 and 1980, about 70 jetstream spots would have been consumed by the STropD over the nine-month interval.

From these numbers, we can calculate that the angular momentum change in the STropZ circulation was about 15 times the angular momentum delivered by the SEBs jetstream spots. Thus, if our assumptions are at all realistic, the spots made only a trivial contribution to the speed-up. If the jetstream goes deeper than the spots, the discrepancy is even greater.

There could be several ways round this result. Perhaps the calculation is hopelessly over-simplified, or some of the spots' energy cascades to different scales (as would be expected). A more advanced analysis could be worthwhile. Perhaps there is 15 times more angular momentum in even smaller, chaotic motions on SEBs than in the visible jetstream spots (though one must remember that the bulk flow of the jetstream was not recirculated). Or perhaps these spots are irrelevant, and power for the jetstreams comes

from other energy transfer processes, or baroclinic factors such as heating and cooling, or condensation and evaporation, of which almost nothing is known on Jupiter.

Why does the STBn jetstream not speed up along with the SEBs jetstream? A possible answer is that it extends much deeper, so extra momentum is diluted in a greater mass of air. But this raises awkward questions about the underlying jetstream dynamics which cannot be answered at present. In fact, the STBn jetstream was faster during the lifetime of the great STropD (Chapter 10.9), so perhaps it responds on a longer time-scale.

All this may shed some light on the problem of global asymmetry. Perhaps the north and south tropical regions represent alternative stable states. In the south, we have seen examples from Voyager images where the SEBs jetstream itself (in conjunction with the turbulent activity in the SEBZ) seemed to create jetstream spots; and where SEBs jetstream spots, in conjunction with adjacent circulations, formed a STropD; and where the STropD perhaps, somehow, pumped up the jetstream. Although none of the links in this triangle has been proved as a general process, it seems possible that some such circuit sustains the phenomena in the South Tropical region, and the absence of any of its components explains the absence of the others in the North Tropical region.

Origin and evolution of the Great Red Spot: Did the GRS arise within historical times?

Pushing to the limit the parallels between the South Temperate and Tropical regions, the author has proposed (*1980*) that the Great Red Spot arose in a way analogous to the STB white ovals, and also to the circulating current associated with the great STropD in the 1930s. (In 1931–1934, the ≈140°-long 'white oval' of STropZ excluded by the great STropD, containing the Circulating Current, could well have been an incipient GRS!) It seems possible that the GRS originally formed by the expansion of a great STropD. The region of the circulating current, outside the Disturbance, continued shrinking and decelerating until it became the GRS, while the expanded Disturbance eventually became the present SEB(S).

The possibility of this happening is documented in Fig. 14.6, which shows the historical changes in the length and speed of the GRS and of the STropD Circulating Current. The charts are re-scaled from Figs. 10.6,30,31, and are aligned as if the year 1901 for the STropD was equivalent to 1701 for the GRS. A third panel shows Cassini's spot of 1665–1713, which had smaller size and slower drift than the GRS has ever had. Clearly, despite great fluctuations, one could construct a single course of evolution from a Circulating Current through to a GRS and finally to a Cassini spot, shrinking and decelerating the whole time.

This evolution would be entirely parallel to that of the STB white ovals, as shown in Figs.11.9&10. It would be nice if we could just scale up the timecourse of the S. Temperate ovals to predict that for the S. Tropical features, but unfortunately, as the contraction and deceleration did not follow a simple law, this cannot be done.

The proposed evolution would also be consistent with the tendency of the SEBs jetstream to become stronger within the

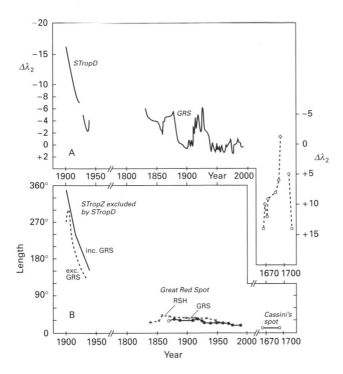

Fig. 14.6. Deceleration (A) and shrinkage (B) of anticyclonic circulations in the STropZ: STropD (left), GRS (middle), and Cassini's spot (right; values imprecise). The time scales are aligned as if the GRS originated as a STropD *circa* 1701. From Figs. 10.6,30,31. Compare with Figs. 11.9&10 for the S. Temperate white ovals.

Circulating Current, a trend which could be continued to create the even faster circulation of the GRS. In Chapter 15 I will suggest how this trend might be coupled to the deceleration of the eastward drift.

There might also be a trend in latitude. The STB ovals drifted gradually northwards, the N. edge showing the greatest shift, and it would be fascinating to know if the GRS has done likewise. Unfortunately, precise measurements of the GRS latitude do not go back far enough to detect any change (Appendix 2). However, the Circulating Current did erode deeper into the SEBs during its life, and still deeper intrusion would have brought it to the latitude of the Red Spot Hollow, so a northwards trend in the latitude of the N. edge is likely (Table 14.1). On the other hand, the S. edge. of the GRS is further south than the S. branch of the Circulating Current was; the centres of the two types of circulations have always been at the same latitude.

The relevant values of latitude, drift rate, and shrinkage rate, are summarised in Table 14.1.

Clearly the evolution of the great STropD did not exactly match that of a proto-GRS: the STropD died out. Perhaps this was because it was intrinsically too weak to survive; this seems possible, given that other STropDs have been much shorter-lived, and that one of the STB white ovals has always been smaller than the others. Or perhaps it was aborted because the GRS already existed.

One cannot exclude other possible histories for the GRS. For example, perhaps it really is the Cassini spot, whose size has fluctuated over the centuries; perhaps it became rejuvenated in the 1870s by merging with a STropD (the 'Alter Schleier'; Chapter

Table 14.1 *Evolution of the STropD and GRS.*

A. Length and drift rate

	Shrinkage rate (°/year)	Drift rate $\Delta\lambda_2$ (°/month)	u (m/s)
Circulating Current of STropD†			
1901–1916	6.6	$-16 \to -10$	$+3.6 \to +0.9$
1917–1939	3.9	$-10 \to -3$	$+0.9 \to -2.3$
Great Red Spot			
1858–1920	≈0.0	$-5 \to 0$†	$-1.4 \to -3.6$
1920–1990	0.14 (±0.02)	0†	-3.6
Cassini spot:	n.d.	+14 to +5	-6 to -10

B. Latitude (zenographic, °S)

	N.edge	Centre	S.edge
Circulating Current of STropD*:			
1908–1925	18.7 (±0.7)	22.4	26.2 (±1.1)
1929–1935	17.5 (±0.8)	21.9	26.3 (±1.4)
Great Red Spot:			
1879–1928	—	22.7 (±1.7)	—
1952–1990	16.5 (±0.4)	22.3 (±0.3)	28.3 (±0.5)

† Interactions between the STropD and GRS are omitted. Thus speeds for the STropD do not not include passages past GRS, and the speed for the GRS omits 1902–1938 when it was interacting with the STropD.
*N. edge of Circulating Current is latitude of SEBs; the same at all longitudes up to 1925, but lower outside the STropD from 1929 to 1935. During the latter years, SEBs inside STropD was at 20.9 (±1.1)°S. S. edge is latitude of STBn (1908–1925; Table 11.1), or of S. Trop. Band on which STBn jetstream spots ran (1932–1934; Table 10.9). (Errors quoted are standard deviations.)
The Cassini spot appeared to be restricted to the STropZ, but one cannot be precise about its latitude.

10.7). Williams and Wilson (*1988*) proposed an explicit scenario whereby varying baroclinic inputs could have produced the observed fluctuations. However, the proposal offered here seems to be the simplest.

The proposal is therefore that the GRS began life as a sector of zone excluded by a STropD in the early 1700s, and that the Cassini spot was the remnant of an earlier-generation spot (Fig. 14.7). The observations from the 18th and early 19th century are too scanty to confirm or refute this hypothesis, and in any case, the circulation need not have been visually conspicuous in its adolescent stages. It is possible that the present *Red* Spot really did appear for the first time in the 1870s. Presumably it will continue to shrink and decelerate (although no doubt with further fluctuations), until it reaches the condition of the Cassini spot, and perhaps gets absorbed into its successor. It looks as if this will take several centuries.

Summary: the observed dynamics of large ovals
It may now be helpful to summarise the observed facts about the dynamics of the GRS and of other anticyclonic ovals: the STB white ovals, the NTropZ Little Red Spots and their kin, and the Circulating

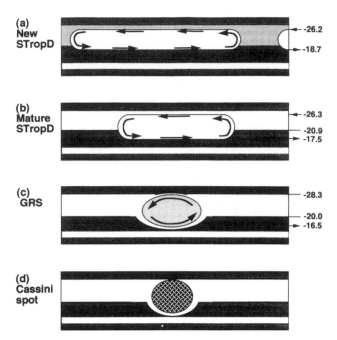

Fig. 14.7. Proposed evolution of the GRS. Compare with Fig. 11.1 for the S. Temperate white ovals. Zenographic latitudes are at right.

Current of the great STropD. This follows on from the generalisations about their appearance given in Chapter 3. Given the evident homologies between them, any theory of the GRS should also be capable of explaining all these ovals – as well as the smaller ones on the SSTB and NEBn, which do not show such elaborate dynamics.

1. All ovals of this type are anticyclonic.
2. The internal circulation has a maximum vorticity of 2–6 x 10^{-5} s^{-1}, greatest near the outer edge.
3. The GRS oscillates in longitude with a period of 90 days.
4. All such ovals drift eastwards (prograding) with respect to the mean slow current of the domain. Although the GRS now drifts westwards with respect to System III, there is no reason to think that System III is a relevant frame of reference. The GRS was almost stationary in System III in the mid-nineteenth century, and drifts slightly eastwards relative to smaller spots in the S. Tropical Current. The STropDs, STB white ovals, and NTropZ Little Red Spots all drift eastwards relative both to System III and to neighbouring spots. (This may be a function of their latitude.) Therefore, theoretical attempts to make the GRS drift westwards are misguided.
5. These ovals decelerate with time, until they move with the usual slow current for the domain.
6. They shrink in length with time. They are not stable for ever. (The contraction of these ovals contrasts with the progressive expansion of cyclonic features, both 'closed' STB Fade sectors and 'open' NEB rifted regions.)
7. The STB white ovals have moved slightly northwards with time. Thus they started off as a sector of the anticyclonic zone, but ended up in almost the same latitude as the retrograding jetstream, which is deflected around them. The GRS also deflects the retrograding jetstream.

8. They affect the adjacent belt for a long distance, setting up cyclonic turbulence to the west (f. side). The GRS also affects velocities in the STropZ and STZ for some 50° longitude on either side.
9. The GRS is unique, but the STB can contain three (or potentially four) such ovals, and the SSTB might contain up to 15.
10. They can originate in a variety of ways, not yet fully described by any simple theory.
11. They seldom if ever merge with each other. The STB ovals actually repel each other.
12. The interactions between the GRS and the great STropD were typically as follows:
—   The STropD accelerated towards the GRS and was retarded when moving away from it.
—   Each end of the STropD in turn moved very quickly past the GRS, but was indistinct for some time after the passage.
—   The GRS always moved faster when it lay within the STropD, which was a region where the SEBs jetstream was weaker.
—   It is not clear whether the lasting acceleration of the GRS and deceleration of the STropD was due to their encounters, or to other factors.
13. The GRS and STropD sometimes absorb smaller vortices (jetstream spots); but for most of their lives they do not, because there are none to absorb. They may of course be absorbing microscopic eddies, but most of the smallscale structure streams on past them. Therefore, more evidence is needed to determine whether or not they persist by feeding on smaller vortices.

14.5 WAVES AND OSCILLATIONS

Just as periodicity has long been a popular idea among amateurs, so waves and oscillations have recently been in fashion among space scientists, who will attempt to extract them from the most meagre data. The appeal of these patterns is not only that they would unite phenomena, but also that their exact periods (if detected) could powerfully constrain theories of the atmosphere. The visual and Voyager records of course form a rich database in which one can search for such regularities.

Planetary-scale waves: do standing waves control the patterns of ovals?

In Chapter 11 it was suggested that the origin of the three STB white ovals and their subsequent tendency to remain apart might have been influenced by a tendency for these latitudes to maintain a standing planetary wave with wavenumber 3 (that is, with a wavelength exactly 1/3 of the circumference of the planet), and that it could occasionally switch to wavenumber 4 to create an 'orphan' fourth region of cyclonic disturbance. However, there is no direct evidence on what kind of wave this might be.

Analogous models for other belts, inferred from the characteristic spacing of analogous ovals, suggest characteristic wavenumber 12–15 for the SSTB and wavenumber 1 for the SEB. (It appears that wavenumber 2 is not stable for the SEB, as the Great Red Spot and the SEB disturbance following it are the only stable features of their kind, whereas other comparable features that do sometimes arise – South Tropical Disturbances or mid-SEB eruptions – do not persist.)

However, the existence of any such wave pattern remains simply speculation, as the STB (for example) shows no visible sign of the supposed waves apart from the ovals themselves, and the spacing of the ovals in the STB and SSTB is often not at all regular. In most latitudes, the impression given is that ovals have a certain minimum spacing, which produces a regular array when they are crowded, but that they do not usually arise in a regular manner. This is usually the case on the NEBn, although in 1991 a set of 'barges' developed with an exact 'wavelength' of 24° longitude.

The NEBn ovals should be of particular theoretical interest because of the evidence that they arise by formation of waves in the NEBn jetstream, like the meandering of a river (Chapter 8.4). This could be an example of a theoretical mode of shear-instability in the retrograding jetstreams, called 'barotropic instability'. This will be explained in Chapter 15.2; essentially, the retrograding jetstreams seem to be going too fast, and would be expected to lose energy by becoming wavy. For instance, the models of Gareth Williams (Chapter 15.2) produce ovals via a shear-instability that is part barotropic and part baroclinic. This shear-instability may be what happens on the NEBn when an array of barges and ovals develops – typically just after a notable revival, expansion, or rift outbreak in the NEB. Such events probably involve a release of energy and/or a triggering of instability.

We do not know whether such events actually accelerate the NEBn jetstream. (SEB Revivals do not lead to acceleration of the SEBs jetstream, nor do they produce spots analogous to those on NEBn, except in rare instances.) If we adopt the conjecture that the speeds of the retrograding jetstreams are in fact limited by shear-instability, we can suppose that surplus energy is siphoned off into cyclonic and/or anticyclonic ovals. Once formed, these are apparently free to drift and evolve under their own dynamics.

The anticyclonic ovals of the southern hemisphere could also owe their existence to shear-instability, if the three 'S. Temperate Disturbances' that initiated the STBs ovals represented the crests of waves. The incipient STropD in the Voyager movies was also likened to a breaking wave, and this may be more than a metaphor; perhaps it did arise from a low-amplitude wave on the SEBs jetstream that built up and broke where the jetstream interacts with the GRS. According to our hypothesis, the GRS is the main sign of, and solution to, shear-instability on the SEBs jetstream.

In each case we conjecture that the large circulations develop to relieve shear-instability, acting as sinks for surplus energy and vorticity. They then evolve under their own dynamics as described in the previous section. Eventually, they become too small to drain the surplus energy that is generated in the domain, and the cycle begins again with new circulating currents. However, this hypothesis has not yet been fully formulated in terms of mathematical theory of the atmosphere.

Large-scale wave-like patterns

The main largescale periodic spacings seen are as follows.

(1) The NEBs plumes (typical interval 30°, 37 000 km). The periodic spacing of these is often obvious at a glance through the telescope, and the wavenumber of 11 to 13 in Voyager maps was typical.

Michael Allison (*1990*) interpreted the NEBs plumes as being due to trapped planetary-scale waves with this wavenumber, retrograding with respect to the material flow of the N. Equatorial Current, which was assumed to be faster than System I like the S. Equatorial Current. In his best-fitting model, these would be 'Rossby waves' with an equivalent depth of 2–4 km, trapped laterally by the NEB and vertically by a stable, deep atmospheric layer – perhaps the layer of convective water clouds. Such waves would indeed be favoured at a wavenumber of 11. The visible white plume cores would come at the crests of the waves where convective activity in the 'water layer' breaks through the visible cloudtops (see section 14.6).

This theory can also explain the great temperature fluctuations in the equatorial stratosphere, which were revealed by Voyager 1 radio occultations (Fig. 4.26). These are interpreted as being the same equatorial waves propagating vertically into the upper atmosphere.

(2) Trains of jetstream spots. During intense outbreaks, these spots are often fairly evenly spaced. Typical centre-to-centre spacings are: SEBs, 13°–20° (mean 17°, 20 000 km); NTBs, 13°–20° (16°, 18 000 km); NNTBs, 10°–20° (13°, 13 000 km). The spot diameters are 2000–4000 km.

(3) Ovals in slow currents. As mentioned above, spots may occur in regular arrays (e.g. NEBn, spacing 24°, 29 000 km; SSTB, spacing 24°, 24 000 km)) but probably do not represent periodic waves. They often occur singly or irregularly, as if from meanderings of unstable jetstreams. The regular arrays seem to be the most close-packed arrangement that is possible, with cyclonic and anticyclonic ovals fitting together like gearwheels.

On a smaller scale, the Voyager photographs revealed semi-periodic patterns of streaks or ripples in many latitudes.

(4) SEB: Voyager images showed particularly striking wave-like patterns in the SEB, with various latitudes and wavelengths (Fig. 10.11). More recently, images from the Pic du Midi showed regular chains of spots (wavelength 4°, 5000 km) at several latitudes in a quiet stretches of SEB (see BAA reports, 1988–1990).

(5) NEBn: Voyager images showed waves of diffuse darkening in the wake of the rifted region, apparently induced by the rift activity (wavelength 10°–12°, 13 000 km; Chapter 8.4, Figs. 8.7–10), and shorter-wavelength diffuse striations on the N. edges of barges (0.8°, 940 km; Fig. 8.6).

(6) NEBn (infrared) ('wavelength' ≈33°, ≈40 000 km). Recently, far-infrared scans of the planet have been made, and although the resolution is low (≈10°), there are suggestions of largescale waves at high altitudes in NEBn latitudes (≈15–20°N).[6] This pattern was detected in Voyager IRIS data, where IR spectra were converted into maps of the temperature at the 270-mbar level (above the visi-

---

[6] Magalhaes *et al.*, (*1989, 1990*); Deming *et al.*, (*1989*). See Chapter 4.4.

ble clouds), and in Earth-based scans in 1987 (8–13 μm emission). There were no corresponding visible features on NEBn. It would be valuable to have higher-resolution maps, to make sure that the pattern is not simply a spillover from the NEBs, but the indications are that it may be a separate, slow-moving phenomenon restricted to the upper atmosphere.

Mesoscale waves

Transverse wavetrains in the EZ (wavelength 0.24°, 300 km).[7] The closest-range Voyager photos of the EZ revealed the most regular and closely-spaced wavetrains of all, aligned roughly north–south and cutting perpendicularly across the cloud streaks (Fig. 9.14). These have been interpreted as gravity waves, which are caused by an air flow rising and cooling, then sinking and warming, in a regular oscillation (Chapter 9.2).

Oscillating spots

The best-documented oscillation is that of the longitude of the GRS, with a period of 90 days. It was followed through 48 cycles from 1963 to 1975 (Chapter 10.5).

Otherwise, true oscillations in longitude are very rare. Those that have been reported are as follows, with the period and number of cycles in brackets.

The first reported (by Peek) was the STropZ dark spot of 1940/41 (period 62 days; 2.5 cycles; see Chapter 10.9). A similar spot in 1941/42 showed similar but irregular behaviour.

NTBs and NNTBs jetstream spots have shown oscillations: NTBs red spot, 1964/65 (period ≈ months, irregular); NTBs white spot, 1990 (35 days; 1.2 cycles); NNTBs dark spot, 1968 (66 days; 1.2 cycles).

Other spots sometimes show uneven motion which may be oscillation or just irregular, e.g. the STB Remnant, 1986 (irregular, ≈60 days, 3 cycles), and a NEBn white oval, 1988/89 (75 days, 2 cycles). The long-lived NEBs plumes in 1973–1975 showed irregular 'oscillations' of periods 35–55 days, but these were apparently due to repeated collisions with other projections in the EZ.

Oscillations such as these are rarely detectable, and even more rarely are they truly regular. They could be merely accidental fluctuations in drift rates. The clustering of periods around 2–3 months may be an artefact of observational constraints, in that longer or shorter periods would be difficult to establish reliably within an apparition.

One other definite oscillation has been discovered, from a study of Voyager photographs of a NEBn barge (Chapter 8.4). The barge did not vary its drift rate, but oscillated in length and width with a period of 15 days, bulging and stretching while maintaining a constant area.

[7] Hunt and Müller (*1979*); Flasar and Gierasch (*1986*).

14.6 CYCLONIC DISTURBANCES AND BRIGHT PLUMES

So far we have mainly been discussing the anticyclonic features. What can we say about the cyclonic regions, in the belts, whose circulation indicates that they are low-pressure regions?

The chief stable spots of this type are the NEBn barges, and in section 14.2, we saw how several theoretical models and laboratory experiments have produced chains of stable cyclones like the NEBn barges. They can also produce a pair of ovals resembling the GRS and the cyclonic region following it (Read and Hide, *1984*; Williams and Wilson, *1988*). So far, not much attention has been given to making the cyclonic aspects of these models realistic, but if these models were developed further they might well be able to explain the cyclonic spots as well as they do the anticyclonic ones. We noted above that the NEBn barges may arise by shear-instability of the NEBn jetstream.

But most cyclonic regions are not stable. We are familiar with cyclonic instability on Earth, in the 'depressions' that provide most of our weather. On Jupiter too, although the physics must be different, it is the cyclonic belts that break up into 'folded filamentary regions' marked by turbulence, erupting white clouds, and sometimes thunderstorms.

The erupting white clouds within these filamentary regions, especially the 'rifts' in the NEB and SEB, resemble the active cores of plumes in the EZ. They all appear to be due to forceful convection, and the same theory has been applied to all of them. (Theoretical ideas about the overall structure of the EZ features have already been discussed in Chapter 9.) The prevailing theory is that these white clouds form atop convective columns of rising warm air, and that the heat that drives them comes from condensation of water vapour down below. This is exactly the same process that drives terrestrial thunderstorms, although the white 'anvil' clouds on Jupiter must include ammonia ice as well as water ice. Kuiper (*1972a,b*; section 14.2) applied this theory to all sorts of features on Jupiter, including steady-state structures in the anticyclonic zones and the GRS, but it can also produce short-lived storms in the cyclonic belts and the equatorial region.

Kuiper applied this theory to the entire cycle of the SEB Revival:

> Our meteorological discussion .... suggests an explanation of these roughly cyclic eruptions in the SEB. Condensation in the low, warm water-vapour layer should lead to a temperature inversion between it and the overlying NH_3 and H_2S condensation layers. The buildup of trapped heat would eventually become great enough to overcome the inversion, eventually causing an almost explosive growth of cumulus columns below the ammonia layer.... The amount of energy required for this breakthrough could take about 8 years to accumulate at Jupiter's rate of internal heat flux. This is about the order of magnitude for the observed time interval between major SEB outbreaks. (Kuiper, *1972a*.)

The observed minimum interval between outbreaks is 2–3 years, but this is acceptable given the uncertainties in these calculations.

The white cores of active equatorial plumes are typically 1000–2000 km across and consist of cloud clusters resembling the rising towers of terrestrial cumulus. A detailed analysis by Carol Stoker[8] (Fig. 14.8) confirmed the convective model: the plume cloud tops are ammonia ice at the 0.4–0.8 bar level (possibly mixed with water ice), but must be driven by convection from much deeper, probably in the water cloud layer at 4–6 bars. So the plumes have a vertical span of 50–60 km.

For the individual white spots, in plume cores or in rifted belts, one can measure the growth rate of the visible white cloud (the 'anvil') and thus, given a plausible estimate of the dimensions of the rising column, one can deduce the speed at which air rises and the rate at which energy is released.

During the Voyager encounters, about 1 in 4 of the NEBs projections had active plume cores, and measurements by Garry Hunt and colleagues[9] showed that they expanded at $3-6 \times 10^7$ km² per day, doubling their area in 10–20 hours. White spots erupting in the NEB rifted region showed similar or slower rates, and the author has measured a similar growth rate for the initial white cloud in the first two days of the mid-SEB outbreak (6.4×10^6 km² per day, doubling in 7 hours; Fig. 10.28). On a larger scale, as observed from Earth by NMSUO, white clouds in the 1971 SEB Revival

expanded from 6×10^7 km² to 5×10^8 km² in 12 days, implying a doubling time of 4 days. The doubling times of 10–20 hours for the equatorial plume cores imply vertical velocities around 0.2–0.4 m/s – well within the range found in tropical thunderstorms on Earth, although the jovian plumes are much larger.

A further parallel may be that the tropical thunderstorms on Earth propagate westwards, with new ones continually appearing at the leading edge of a cluster. It is possible that the jovian plumes likewise behave as a travelling wave, although the evidence is ambiguous. Thus mid-SEB sources drift west relative to System III (although, as argued for the GRS, their immobility in System II may be more relevant). The EZs white spot drifts west relative to the local flow, which is faster than System I. And it has been suggested that the EZn plumes may behave likewise, as part of Allison's wave model of the plumes (section 14.5) (although there is no evidence for such relative motion; Chapter 9.2).

Conversely the top of the plume gets caught by the prevailing high-altitude winds, so that mid-SEB clouds expand to the NE (Np.) from their source in System II, and the EZn plumes stream SW (Sf.) from their source in System I. So the towering cumulus anvil spreads out into a long plume at the higher altitudes.

---

[8] Stoker and Hord (*1985*); Stoker (*1986*).
[9] The theory described by Kuiper (*1972a,b*) was applied to the equatorial plumes by Conrath *et al.* (*1981a*) and Hunt *et al.* (*1981, 1982*). The measurements for outbreaks in the SEB were done by Sanchez-Lavega and Rogers (*1989a*).

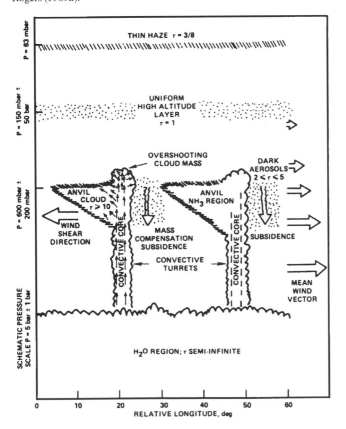

Fig. 14.8. Model of the equatorial plumes being generated by moist convection. Cloud towers initiated at the water cloud level rise to above 700 mbar where ammonia condenses and forms the white plume. The vertical dimension is vastly exaggerated. From Stoker (*1986*).

15: Theoretical models of the atmospheric dynamics

15.1 THE DEPTH OF THE ATMOSPHERE: THE EVIDENCE

In this chapter we consider the physics underlying the global pattern of winds on Jupiter. How deep do the jetstreams go, what determines their pattern, and what drives them? The banded structure is undoubtedly due to the planet's rapid rotation, which deflects winds to blow along lines of latitude due to the Coriolis effect. The physical factors involved are those which we met in the previous chapter: solar and internal heating, convection, vorticity, the Coriolis force, and baroclinic and barotropic instability. But now they must explain straight latitudinal winds in an alternating pattern rather than closed circulations.[1]

The physics of atmospheres is complicated, and there is no complete physical theory. This is partly because fluid motions are very difficult to analyse, and partly because there are interlocking processes over a vast range of scales, from individual clouds to the entire planet. So there is no prospect of accurately computing or simulating the atmosphere as a whole. Physicists must try to discover the laws that mainly govern the dynamics on each scale, and then link the hierarchy of scales together to represent the real planet. As everyone knows, even the atmosphere of Earth is notoriously unpredictable. On Jupiter, the problems are compounded by our ignorance of what goes on below the upper layers of clouds. While we can detect heat coming from below, we have no direct knowledge of the structure and motions of the deeper layers, which are all-important to defining the atmospheric dynamics.

Presumably the mantle of the planet rotates with the period of System III ($\Delta\lambda_2 = -8.0$), as this is where the magnetic field arises (Chapter 18). What we need to know is how the observed cloud-top winds meld into the deeper atmosphere.

We therefore start by summarising what little can be inferred about deep motions from direct observations.

[1] For reviews see: Williams (*1985*); Ingersoll (*1976a, 1981, 1990*); Ingersoll *et al.*(*1984*); Flasar (*1986*). Pre-spacecraft ideas were reviewed by Newburn and Gulkis (*1973*), and in a special issue of *Journal of the Atmospheric Sciences*, vol.26 (no.5.I)(*1969*). For the basic physics, see the book by Houghton (*1977, 1986*).

Can deep motions be observed?

Several candidates for visible deep-seated motions have been suggested historically.

(1)   The 'polar currents'. On Saturn and Neptune, polar features have periods identical to the radio period, but such features have not been measured on Jupiter. Zonal speeds do diminish and the cloud-tops are deeper towards the poles, but the observed currents actually belong to restricted domains of latitude (N^4TC and S^3TC). The motions in these currents are variable, mostly referring to small features of no obvious profundity (Chapters 6 and 11).

(2)   The Great Red Spot. But the variations in its speed show that it is not fixed, and as it is now known to be a circulation like other ovals, it seems no more likely to be deeply anchored than the others are (Chapters 10 and 14).

(3)   Sources of SEB Revivals. The longitudes of these 'eruptions' can be assigned to three hypothetical sources with constant rotation periods close to System III, as proposed by E.J. Reese in 1972 and strikingly confirmed in 1975 and 1993 (Chapter 10.3). However, there are some exceptions to the rule, and the eruptions themselves rarely show System III speeds.

(4)   The permanent slow currents. These currents control visible spots of all types (including the sources of SEB Revivals when they are active), and at various latitudes within a domain. They form a regular pattern across the planet with the highest speeds in high-latitude domains (Fig. 15.1). Do they perhaps betray an underlying flow pattern deeper than the jetstreams?

As we saw in the last chapter, there is no good theory to account for the speeds of the slow currents. From the structures of the ovals which move in them, one might suppose that the slow current in each domain could represent the characteristic speed of waves on the retrograding jetstream, or of circulations rolling between the jetstreams; but neither hypothesis seems attractive, as the speeds of the slow currents show no correlation with the speeds of the adjacent jetstreams. Therefore, the author ventures to suggest that the slow currents may represent a deep global flow pattern. The jetstreams would run over the surface of it and be shallower than the large ovals – except for the equatorial jetstream, which carries the apparently deep-seated plumes and interrupts the pattern of slow currents. This possibility has not yet been addressed mathematically.

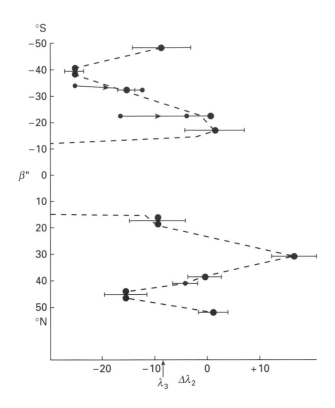

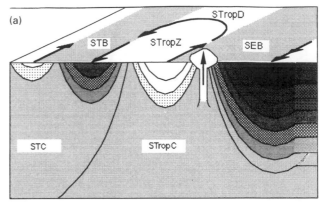

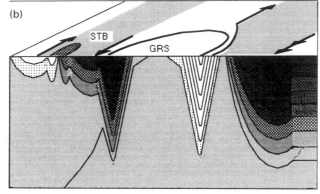

Fig. 15.1. Speeds of the slow currents, from Table 3.3, plus the faster moving features of STZ, STropZ, and NNTZ (smaller dots). The dashed line indicates the symmetrical pattern of these currents in each hemisphere.

Fig. 15.2. Diagram illustrating the author's hypothesis that all speeds diminish steadily with depth, down to deep-seated slow currents which govern the motion of large ovals. These north-south cross-sections are purely schematic, ignoring the curvature of the planet; vertical scale is not defined and is probably much exaggerated. The contours indicate east-west speed, with shading darker for eastward (prograding) winds and lighter for westward (retrograding) winds.

(a) Jetstream depths are shown as being correlated with their surface speeds. The underlying atmosphere is postulated to move at the rates of the slow currents. Convective white clouds erupt in mid-SEB from a location which is stationary in the STropC because there is no vertical shear at that latitude. A new STropD involves only surface recirculation, but may grow deeper with time.

(b) The GRS, having winds faster than the adjacent jetstreams, has become anchored in the underlying slow current. Smaller spots, such as a dark oblong in mid-STB, tend to develop at or near the latitudes with no vertical wind shear, perhaps because they involve vertical motions (e.g. a dark oblong in mid-STB indicated here).

At first it may seem awkward to suppose the large ovals to be deeper than the jetstreams. However, if we were to assume (in the absence of any evidence) that all wind speeds diminish with depth at a constant rate, then the large ovals would be deeper than the jetstreams as they have faster wind speeds; so they can perhaps be anchored in the underlying flow (Fig. 15.2). Small spots would not extend so deep, but they may form preferentially in latitudes where there is no vertical shear – that is, latitudes where the mean zonal speed equals that of the underlying flow. This could apply whether the spots are anticyclonic or cyclonic, as both may have vertical motions, and it could apply particularly to convective 'eruptions' in cyclonic regions, such as the sources of SEB outbreaks.

This hypothesis can also partially explain the behaviour of the STropD and similar circulating currents (STB ovals, GRS?), if we adopt the model proposed in Chapter 14.4, that such a circulation begins as a purely superficial eddying then evolves into an oval with accelerating wind speeds. Its initial propagation speed may be determined by the properties of a wave on the surface of the jetstream, or some other intrinsic dynamics; but as its circulation becomes faster, it grows deeper, and tends to become anchored in the underlying flow. This would explain why such features decelerate towards the normal slow-current speed.

Relation of belts and zones to jetstreams

The pattern of jetstreams indicates that zones are anticyclonic, high-pressure areas at the level of the visible clouds, while the belts are cyclonic, low-pressure areas (Chapter 3).[2] The belts also differ from the zones in having darker clouds, different cloud layering, and slightly higher observed temperatures (Chapter 4).

The higher temperatures in the belts, however, can be attributed to the lower albedo (more solar energy is absorbed in visible and near-infrared) and perhaps to the thinner clouds (more heat is radiated away in mid-infrared). There is no evidence that belts and zones differ in the upward flow of internal heat. The temperature difference is 2°–5° in the 0.2–0.5 bar range (i.e. in the visible

---

[2] First noted by Hess and Panofsky (*1951*); see also Ingersoll and Cuzzi (*1969*) and Williams (*1979*).

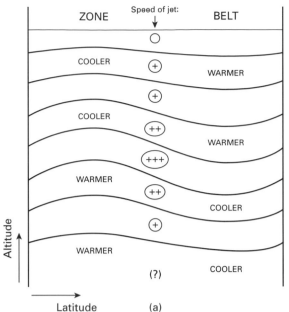

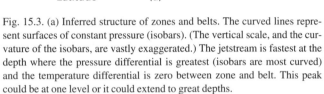

Fig. 15.3. (a) Inferred structure of zones and belts. The curved lines represent surfaces of constant pressure (isobars). (The vertical scale, and the curvature of the isobars, are vastly exaggerated.) The jetstream is fastest at the depth where the pressure differential is greatest (isobars are most curved) and the temperature differential is zero between zone and belt. This peak could be at one level or it could extend to great depths.
(b) Inferred structure of zones and belts, related to actual pressure levels

about one scale height apart (right) and including the cloud layers (Chapters 4 and 16). As infrared observations show the belts to be ≈3° warmer than the zones down to ≈0.5 bar and possibly deeper, the peak jetstream level has been put at 1–2 bars. This would also be expected if direct solar heating is a major driving force, as it takes place in the 0.3–2 bar range. But it is not known how fast (or even whether) the wind structure changes going to greater depths.

clouds), measured along surfaces of constant pressure. It appears to be similar down to ≈1 bar according to far-infrared maps, but these may be distorted by absorption in the clouds (Chapter 4).

The decline in wind speeds above the cloud tops, calculated from the thermal wind equation, was described in Chapter 4.6. The thermal wind equation can also make some predictions about the winds below the cloud tops, but there is uncertainty about its applicability. One view is that the jetstreams are mainly determined by temperature differences ('baroclinic' winds), in which case the thermal wind equation applies. A more recent view is that the jetstreams are largely powered by other factors, such as merging of eddies ('quasi-barotropic' winds), and the thermal wind equation is not a reliable guide to the deep motions.

The traditional explanation for the higher pressure in the zones, which omits any quasi-barotropic flow component, is that at deeper levels they are warmer than the belts. The reversal of the temperature differential occurs because the scale height is proportional to the temperature (Chapter 4.5) – that is, the decline in pressure and density with altitude is less in a warm air mass than in a cold one – as in a hurricane on Earth, where the top is anticyclonic while the bottom is cyclonic. The jetstreams will be fastest at a pressure level where there is no temperature difference between the belts and the zones (along a surface of constant pressure), and as we have just seen, this must be deeper than 0.5 bar. At the same level, the pressure difference will be a maximum (Fig. 15.3). Going down through deeper levels, the pressure difference and jetstream speed may decline rapidly or hardly at all. The temperature difference between belts and zones, multiplied by the depth of the motions,

must be about 240°K.km,[3] so the temperature difference is 2° if the motion extends 120 km below the jetstream peak, or 0.4° if it extends 600 km. This does not imply that temperature differentials power the winds; they are interrelated. If heat *is* being fed into the zones, it could power the winds (see below); or if vertical wind gradients are driven by some other mechanism, they will generate temperature differentials.

An early model held that these hypothetical temperature and pressure gradients entail convection – warm air rising in the zones and cool air sinking in the belts – so that a complete circulation is formed. This model fits comfortably with the thicker cloud layers in the zones, which would be expected in a convective updraft. It is also supported by the observation of lower vapour pressures of ammonia and water in the belts (Chapter 16). The net rising and sinking motions may be very slight and would not conflict with the observations of vigorous rising convection in turbulent regions in belts, as this can be balanced by sinking motions nearby. The convective model also implies slight net north-south motions, which the Voyager imagery was too brief to measure. However, it turns out that a rapidly-rotating planet cannot easily sustain such largescale convection, unless it can be achieved by deep global-scale cylinders (see next section).

The more modern view is that the east-west winds are balanced by north-south pressure differences, with only a minor contribution from temperature differences (Williams, 1985). Zones must still reach higher altitudes than belts for a given pressure, explaining

[3] Ingersoll & Cuzzi (*1969*). For simplicity, the temperature difference is assumed to be constant.

their thicker cloud cover, but nothing can be inferred about any deep structure.

Even if convection does not operate on the global belt-zone scale (≈10 000 km), it does appear to be happening on a medium scale (≈1000 km), forming bright cores in plumes and 'rifts' in belts, and on a small scale (up to 100 km), forming individual clouds as seen in the detailed texture of the zones. It is also assumed that the deep atmosphere must be in convection below the clouds; there is no other plausible way for the internal heat to escape. This implies that the temperature should increase with depth at a rate of 2°/km, the 'adiabatic lapse rate', which indeed it does (Chapter 4).

The giant planets compared

One might hope to gain insights into Jupiter's atmosphere by comparing it with those of the other three giant planets as revealed by Voyager (Fig. 15.4; Ingersoll, *1990*). There are surprising similarities and surprising differences.

Saturn is most like Jupiter, though it has a much faster and broader equatorial jetstream; the pattern of alternating jetstreams (and associated ovals) is squeezed into the temperate latitudes. However, the visible belts on Saturn are not confined by jetstreams like those on Jupiter. Neptune also looks remarkably like Jupiter, with its broad banding and its Great Dark Spot; but its equatorial wind is highly retrograde and this is the peak of an anticyclonic flow which extends over the entire planet. Uranus seems to have a global flow pattern similar to Neptune's, although only the equatorial region is actually retrograding, according to the very few features that could be tracked. (It may be significant that Uranus, the only planet without large visible spots, is also the only one without an internal heat source.)

So we can conclude that rapidly rotating gas giants always develop strong east-west wind patterns, even though their axial tilts are very diverse. But as the equatorial winds can range from +480 m/s on Saturn to –480 m/s on Neptune, they do not help directly in explaining the alternating pattern on Jupiter. The retrograde flows on Uranus and Neptune are simpler to understand, because they correspond to global anticyclonic vorticity, such that packets of air moving north or south tend to conserve angular momentum; thus the flow is stable. To produce a prograde equatorial jetstream, flanked by cyclonic belts, requires active pumping of angular momentum towards the equator by organised waves or eddies.

The equatorial jets on Saturn and Jupiter are probably comparable. As Saturn's atmosphere has twice the vertical scale of Jupiter's (twice the scale height, and probably twice the total depth), simple arguments of jetstream stability suggest that its jet would be twice as wide and four times as fast – which is almost exactly the case.

Even if the strong gradients of speed with latitude could be explained, this would not tell us why the winds on Jupiter are partitioned into multiple domains, rather than distributed in a smooth gradient over the entire planet as on Neptune. The reason could be that the maximum speeds are limited by some factor, perhaps concerning the depth of the atmosphere. (Because of Jupiter's strong gravity, it has the smallest scale height of all the giant planets; it also has the slowest maximum wind speeds.)

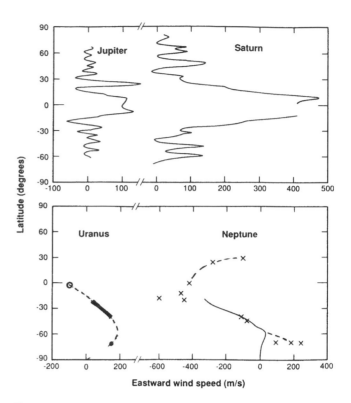

Fig. 15.4. Wind profiles of the four gas giants, from Voyager images. East-west wind speeds are measured relative to the radio rotation period which is assumed to represent the planetary interior. Adapted from Ingersoll (*1990*). (North is up, east is to the right. This is inverted relative to Fig. 15.1 and covers a much greater speed range.)

15.2 JETSTREAM DYNAMICS: WHAT DETERMINES THE GLOBAL PATTERN?

Dynamics of a shallow atmosphere

To develop theories of the atmosphere, physicists have turned to a more familiar planet for inspiration: Earth. It is obviously different from Jupiter in its size, spin, atmospheric depth, and chemistry. But the two planets have analogous sources of heating: from above (solar radiation), from below (internal heat on Jupiter; the sun-warmed ground or ocean on Earth), and from condensation of water clouds (plus ammonia and perhaps other clouds on Jupiter: Chapter 16). In the tropical atmosphere, Earth also has a 'zone and belt' structure, although the physics may be different from Jupiter's. There is an 'equatorial zone' marked by cumulus clouds and thunderstorms which represent a net updraft, driven by the latent heat of condensation of water into clouds. This zone is flanked by 'tropical belts', extending to about 30°N and S, in which there is a net downdraft and the skies are generally clear. Although the winds in the lower atmosphere are easterly (retrograding), there is often a prograding wind in the upper atmosphere.

At higher (temperate) latitudes, Earth's atmosphere behaves in a different way. The westerly (prograding) winds tend to be greatly distorted into planetary-scale waves (the simplest form being known as 'Rossby waves'). There are typically about four such waves spaced around the planet, obvious in satellite photos as they control weather systems. The same process might explain the waviness of two high-latitude prograding jetstreams on Saturn (called the 'ribbon' and 'hexagon'). But there is nothing of the sort on Jupiter's prograding jetstreams, which are always straight. Also, Earth's latitudinal gradient of temperature generates a vertical wind shear which causes 'baroclinic instability'; this results in vertical instability and horizontal eddying, leading to the cyclonic weather systems with which we are familiar in temperate latitudes. In fact, these eddies transfer their energy to larger scales and ultimately drive the westerly (prograding) winds themselves. On Earth, the scale of the baroclinic weather systems is ≈1000 km; on Jupiter, there may be similar systems with similar scale, but this is small compared to the jetstreams on the giant planet.

Can Jupiter be understood in terms of a shallow atmosphere like Earth's? There is now a robust set of theories which show that it can, developed by Gareth P. Williams of Princeton University and his associates.[4] The central idea, which only became generally accepted in the 1970s, is that smallscale eddies tend to merge so that kinetic energy naturally progresses from smaller to larger scales. On Earth, this is known to happen, as described above. Williams' model of Jupiter is a scaled-up and spun-up model of Earth, and he favours baroclinic instability as on Earth, driven by a pole-to-equator temperature gradient. On Jupiter, there is no evidence for such a gradient, but it does not much matter what generates the eddies; heat from the interior or from condensation could also generate eddies by buoyancy (Chapter 14.2 and section 15.3). Small eddies naturally merge into larger circulations until they reach the scale of Rossby waves, and thus evolve into east–west wind patterns with a scale very like that of the domains on Jupiter (Fig. 15.5). On this scale the dynamics becomes equivalent to barotropic (maintained by horizontal differences of pressure, not temperature), with no net vertical motion in belts and zones. Jetstream widths are limited by shear-instability (see below), which can generate large ovals as described in Chapter 14.5. The jetstreams could extend to depths of ≈200 km. This model closely resembles the jovian circulation (apart from the NTBs jetstream).

The main difficulty with all models has been to generate the great equatorial jetstream. It is not clear whether the EZ has the same dynamics as other domains, as the thermal wind equation does not directly apply at the equator, but the gradients and scales of the winds suggest that it is spun up by the same mechanism. Williams' (1978) theory can produce an equatorial jetstream but only by including special low-latitude forcing. As Raymond Hide has argued, the equatorial jetstream probably represents the final sink for energy and angular momentum originating at higher latitudes and drawn by waves or eddies towards the equator. But the exact process responsible is still unclear.

The scale and speeds of the jetstreams are presumably determined by the balance between processes feeding energy into them and processes taking energy out. These processes include instabilities (eddies) of various types. Whether a given type of instability feeds or drains the jetstream depends on the scale.

There are two types of instability which are important: baroclinic

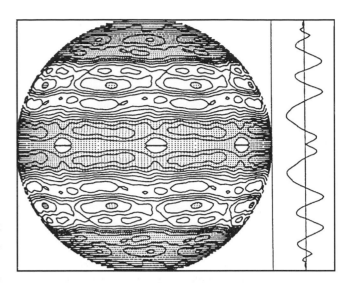

Fig. 15.5. A model of a shallow jovian atmosphere from computer numerical modelling by G. Williams. An initially turbulent atmosphere evolved to this pattern of alternating jetstreams. The profile of wind speed is shown at right; the peak speeds are 50 m/s east or west. No equatorial jetstream was produced in this model because it was computed for only one hemisphere, but similar models of the whole planet can produce one if eddy sources are strengthened in the equatorial region. (North is up.) From Williams (1978).

and barotropic. *Baroclinic instability* is caused by horizontal temperature differences which set up motions in three dimensions; it involves instabilities in vertical wind shear. Are the jetstreams stable against baroclinic instability? In practice, the straightness of the prograding jetstreams seems to show that they are stable, on the largest scale – presumably because the temperature differentials between belts and zones are small. In theory, the situation is not fully understood, but Barney J. Conrath and colleagues (1981b) conjectured that the width of the jetstreams evolves so as to match the typical scale of baroclinic instabilities (the deformation radius) – this being the optimal scale for baroclinic instability to transfer energy *into* the jets. Others think the deformation radius is likely to be much smaller. Williams (1979) suggested that some spots are due to baroclinic instability; perhaps jetstream spot outbreaks are particular candidates?

Barotropic instability is simpler to analyse as it involves only horizontal pressure differences, and motions which do not change with depth. It is a form of instability in horizontal wind shear. The barotropic stability criterion is derived from considerations of conservation of angular momentum, and puts a limit on how strongly the vorticity of the jetstream flow can vary with latitude.[5] There seems to be a problem: in essence, some of the jetstreams seem to be going too fast.

---

[4] Williams (*1975, 1978, 1979, 1985*). These models supercede earlier baroclinic theories proposed by Stone (*1967, 1972*), and a convective theory by Williams and Robinson (*1973*); see Williams (*1978*).

[5] The barotropic stability criterion is:

$$\frac{d^2 u}{dy^2} > \frac{2\Omega \cdot \cos\phi}{r}$$

where du/dy is the gradient of eastward wind speed with latitude (also called the vorticity for a purely horizontal, east-west flow), Ω is the planetary rotation rate, r is the planetary radius, ø is the latitude. The quantity $2\Omega(\cos\phi)/r$ is the 'gradient of planetary vorticity', usually denoted β (but not to be confused with the latitude). This formulation assumes a uniform thickness of flow, but incorporating variable thickness does not help the retrograding jetstreams. It was applied to Jupiter by Ingersoll and Cuzzi (*1969*), Stone (*1976*), and Limaye *et al.*(*1982*).

All the retrograding jetstreams appear to violate this criterion. These jetstreams would therefore be expected to break up or lose energy, most likely by becoming wavy and losing energy to large ovals.

As the jetstreams clearly *are* stable, in spite of the presence of large ovals, it seems likely that the stability criterion cited must be too naive. One possibility if the jetstreams are shallow (so the situation is not purely barotropic) is that their vertically-averaged speeds conform to the limit. Or perhaps the ovals themselves somehow stabilise the jetstreams, or perhaps there is some other factor, so that the correct limit is at the observed speeds. If so, as conjectured by physicists from Ingersoll and Cuzzi (*1969*) to Williams (*1978*) and Conrath and colleagues (*1981b*), the jetstreams are indeed pumped up to the limit of barotropic stability. In the real world, there may well be some combination of barotropic and baroclinic effects.

If these views are realistic, shear-instability really does limit the speeds of the jetstreams, and large ovals (like NEBn barges and the GRS) are produced in order to resolve the instability, as discussed in Chapter 14.5. They take up energy and vorticity that cannot be accommodated in the jetstreams.

Present speculations, then, are that the scales and speeds of the jetstreams may be set by the typical scale of baroclinic instability for feeding energy into them, and limited by barotropic instability that drains energy away. It also seems possible that these two instabilities may correspond to the two major types of spots that are commonly observed; jetstream spots on the prograding jetstreams, and large ovals deflecting the retrograding jetstreams, respectively. The soundness of these ideas will have to be tested by further theoretical analysis and more detailed spacecraft observations.

Dynamics of a deep atmosphere

Although the shallow-atmosphere theories can account for the main features quite well, it would be rash to assume that nothing happens in the deep atmosphere just because we cannot see it. Indeed, there is no evident reason why the observed motions should not have very deep roots.

As the zones and belts seem to involve convective motions, one might first ask whether they constitute single convective 'rolls', perhaps even as deep as they are broad. This seems to be impossible. Williams and Robinson (*1973*) studied models of this type, with giant convective cells driven by the internal heat. But they required more heat than the planet actually emits, and even then, they could only include a prograding equatorial jetstream if the depth of convection was less than 500 km, and could not produce near-polar jetstreams at all. Moreover, there are reasons why such large convective rolls cannot be stable; they should either break up longitudinally (perhaps to form ovals like the GRS) or rearrange into cylinders parallel to the axis of rotation.

A more stable deep circulation in a rapidly rotating planet might consist of rotating cylinders extending through the planet, as proposed by F.H. Busse (Fig. 15.6).[6] These might be parallel cylinders, which function as convective rolls at depth, and/or they might merge

[6] Busse (*1970, 1976, 1983*); Ingersoll and Pollard (*1982*); Ingersoll and Miller (*1986*).

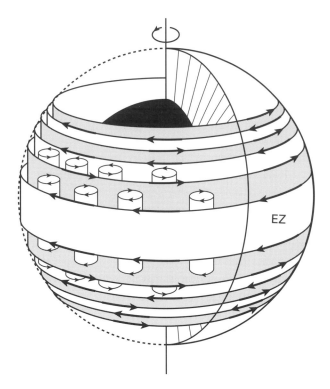

Fig. 15.6. How the deep motions might work. A pattern of convective cylinders might extend through the planet, according to the models of Busse (*1976*) and Ingersoll and Pollard (*1982*). The model has not yet been fully evaluated numerically, and how deep the cylinders could go is unknown. (South is up.)

to produce coaxial cylinders of differential rotation, whose surface projections would be the jetstreams. The cylinders could be powered by merging of turbulent eddies as in the shallow-atmosphere models. They might be much more stable than shallow jets. But they could only exist if there are no sharp density boundaries. There is some doubt as to whether they could remain coherent through the vast range of pressure and temperature in the interior of Jupiter.

Certainly they could not penetrate the planet's mantle, which is believed to consist of metallic hydrogen (Chapter 16.3), with a sharp outer edge somewhere between 65% and 80% of the radius of the planet. Therefore the cylinders should not extend to latitudes higher than 36°–48°, whereas the jetstream pattern extends to at least 61°N and S.

Another difficulty with this model is that the jetstreams, representing the surface traces of the cylinders, should be symmetrical about the equator. Obviously some of them are not: the SEBs and NTBs jetstreams interrupt the symmetry. In fact, at higher latitudes, the symmetry seems to be displaced by one domain (Table 6.1). Finally, the GRS is a problem: it obviously does not extend right through the planet, and analysis of its internal motions implied that the SEBs jetstream does indeed fall off below the visible clouds (Fig. 10.40).

So at present the shallow-atmosphere models seem more satisfactory in theory and in practice; but the issue is still open.

15.3 JETSTREAM DYNAMICS: WHAT PROVIDES THE ENERGY?

Several possible sources of energy for driving the jetstreams have been mentioned in passing, so now let us consider them in more detail. The ultimate energy sources must be the solar heating and the internal heat flow. But the heat may be transduced through latent heat of cloud formation, convection, and/or turbulent eddies, on its way to driving the jetstreams.

(1) Solar heating. First, could the equator-to-pole gradient of solar heating drive motions directly, by baroclinic instability? This seems unlikely. From solar heating alone, the poles should be ≈30° colder than the equator, and this difference would be more than enough to drive largescale motions. However, the actual difference in effective temperature is no more than 3° (Chapter 4.3). Admittedly, these measurements may not settle the matter as they refer to different levels in the atmosphere at poles and equator; there is a different cloud structure in the polar regions. Williams (*1985*) maintains that a 5° difference between equator and pole could exist and could be enough to drive the winds. But the Voyager data have been analysed up to 60°N and S and show no sign of such a temperature gradient (Fig. 4.20).

Why are the temperatures so uniform, if this analysis is correct? The reason could be that heat is transported from equator to poles – though there is no sign of any such transport across the prograding jetstreams. According to an attractive theory by Andrew Ingersoll, it could be because of local feedback between convection and solar heating.[7] The transport of heat from the interior by convection is represented by the adiabatic lapse rate of 2°/km, and as solar heating imposes slight deviations from this lapse rate, convection is suppressed at low latitudes and enhanced at high latitudes – a 'convective thermostat'. This may also be why there is little or no seasonal effect from the 21% change in solar heating between perihelion and aphelion.

However, within individual domains there could be a significant contribution from the gradient of solar heating – in addition to the greater heating of the belts because they are darker.

(2) Internal heating. The internal heat flow is of course likely to drive convection; but as we have seen, it is unlikely to do so on the scale of the zones directly. And there is no evidence that the internal heat flow is greater below zones than below belts. Instead, it may be transduced into smallscale convection and eddying by baroclinic mechanisms, and into latent heat through cloud formation.

(3) Latent heat of cloud formation.[8] The putative heating below the zones (and below the anticyclonic ovals) could be provided by release of latent heat when water or ammonia condense at the base of the clouds. This can trigger convection more effectively than the global heat flow alone can do. As the flow converges under the zones, more vapours are drawn in to amplify the local release of latent heat. The cycle would be completed, either by rain or snow falling from the clouds and evaporating deeper down, or by evaporation in the belts as they are warmer at the cloud tops.

(4) Merging of eddies. It is now realised that eddies tend to cascade into larger and larger circulations, and this may be the main source of energy for the jovian winds, whether they are shallow or deep (Chapter 14.2).[9] The eddies could arise from differential heating on any level and scale, but on a large scale are expected to merge, powering the jetstreams either via waves or directly.

Is there any direct evidence for merging of eddies into jetstreams? The most obvious eddies to look at are the jetstream spots:

(1) We have seen that jetstream spots merged into the South Tropical Disturbance (STropD), and their vorticity may have contributed a little to pumping up the SEBs jetstream in 1979–1980 (Chapter 14.4). But this was clearly a special case. Before the STropD developed, the SEBs jetstream spots tended to merge with the GRS.

(2) M. MacLow and A. Ingersoll (*1986*) studied the births and deaths of many spots in the Voyager 2 movies – mostly on the SEBs and NNTBs jetstreams. Of the spots which they saw disappear, most went by merger with another spot. Otherwise, 12 spots went into the STropD, nine were torn apart in cyclonic filamentary regions, and only two anticyclonic spots just disappeared. So, with the possible exception of those last two, the jetstream spots were not tending to dissolve into the jetstream flow.

The regions which did stand out in this study were the cyclonic filamentary regions, which were the main destroyers and creators of spots. The rapid turnover of turbulent clouds within them suggests that energy is being fed into convection in these regions. They disgorge both cyclonic and anticyclonic spots. So it is not clear whether the vorticity and energy of the jetstreams is generated here – or whether these regions are spoilers, tending to disrupt the flow of the jetstreams, and taking vorticity from them to create new jetstream spots. (This process may have been observed on both SEBs and NNTBs jetstreams: Chapters 6.4 and 10.9.)

(3) To test the eddy-merger theory globally, the Voyager imaging team measured the north-south and east-west motions within 1° latitude strips all over the Voyager maps, calculating the mean direction of eddying within them.[10] They found that most wind vectors pointed towards the jetstreams rather than away from them; to be precise, the 'eddy correlation coefficient' at each latitude was correlated with the latitudinal gradient of the jetstreams. This implied that the eddies on the ≈1000-km scale are indeed transferring their energy into the jetstreams.

However, this conclusion seems almost too good to be true, as the apparent rate of energy transfer is enormous – enough to replenish the kinetic energy of the jetstreams in only 2–4 months. If it occurs to a depth of 2.5 bars, it amounts to 10–15% of the total thermal energy flux from Jupiter. For Earth, only 0.1% of the thermal energy is converted into zonal winds in this way. It is uncertain whether the high efficiency implied for Jupiter is realistic. It is not in conflict with the long-term stability of the jetstreams if there is always surplus energy pumping them up to the limit of their stability, and indeed the SEBs jetstream does seem to respond to a

[7] Ingersoll (*1976*); Ingersoll and Porco (*1978*).
[8] Barcilon & Gierasch (*1970*); Gierasch & Conrath (*1983*).
[9] Rhines (*1975*); Williams (*1975, 1978*); Ingersoll and Cuong (*1981*).
[10] Beebe *et al.*(*1980*); Ingersoll *et al.*(*1981*).

STropD on a timescale of months. But L. Sromovsky and colleagues (*1982*) pointed out that the apparent observation of eddy transfer could be spurious, because of over-representation of parts of circulations where features are easiest to trace, such as the turbulent belt regions following anticyclonic ovals. With more uniform sampling, they concluded that the systematic errors were greater than any real transfer of energy. The evidence for the high efficiency cannot be dismissed but is perhaps not conclusive.

The Galileo probe may help to choose between the various energy sources, as it should measure the equatorial winds down to ≈20 bars. According to the Galileo team (J.B. Pollack and colleagues, *1992*), jetstream speeds should diminish with depth through the levels at which they are powered, so they might cease at a level as shallow as 2 bars if driven by solar heating, or 7 bars if driven by latent heat of condensation, or much deeper if driven directly by internal heat. However, as the equatorial jetstream may be the end-product of a cascade of processes, this single measurement may not be decisive.

15.4 TEN QUESTIONS FOR FUTURE RESEARCH

A theme throughout this book has been that the visual observations provide rich material for testing theories. In the last three chapters we have seen that many patterns remain unexplained – patterns in the behaviour of jetstreams, and ovals, and outbreaks of activity. Any set of theories intended to describe Jupiter's atmosphere must eventually meet the challenge of explaining them. Here we summarise ten of the outstanding questions.

1. What determines the speeds of the individual jetstreams? Why are they so constant in spite of being so different?
2. What determines the speeds of the slow currents?
3. What determines the stability and dynamics of S. Tropical Disturbances and large anticyclonic ovals (see checklist in Chapter 14.4)?
4. When the SEB fades, why does the GRS decelerate and darken?
5. Is there any single explanation for the various dark features that can emerge from the p. end of the GRS?
6. What determines the pattern of the belts where it is dissociated from the pattern of jetstreams, at high latitudes or in South Tropical Dislocations?
7. What is the nature of the instability that leads to a SEB Revival, or a NEB activity cycle, or outbreaks on the jetstreams?
8. What determines the periodicities of NTB fadings and NTBs jetstream outbreaks? (Also see the list of questions at the end of Chapter 7.)
9. How do NTBs jetstream outbreaks maintain two speeds simultaneously?
10. What connects the different outbreaks in a global upheaval?

16: The composition of the planet

16.1 THE CHEMICAL COMPOSITION OF THE ATMOSPHERE: OBSERVATIONS[1]

As Jupiter is so massive and has such a low density, it is expected to have retained the chemical elements in something like their cosmic proportions when it formed – that is, the proportions found in the sun and the interstellar gas (Table 16.1A). Thus hydrogen and helium, the two simplest and lightest elements, are expected to comprise 99% of its mass.

The dominance of hydrogen and helium was realised only in the mid-twentieth century. These two gases are difficult to detect, because they do not have absorption lines in easily accessible regions of the spectrum. Molecular hydrogen (H_2) was not directly detected until 1960,[2] by means of near-infrared spectral lines. Helium (He) has still not been directly detected in the troposphere, although it was detected by Pioneer 10 in the thermosphere (along with hydrogen atoms; Chapter 17.4) by its far-ultraviolet emission. The helium abundance of 10% by mole and 18% by mass has been determined indirectly (section 16.3).

The hydrogen-dominated atmosphere is chemically 'reducing' – that is, the other elements should exist mostly as hydrogen compounds – and indeed, most of the gases detected are the hydrides of other light elements: carbon, nitrogen, oxygen, and others.

The first molecules that were detected on Jupiter were methane and ammonia. Rupert Wildt (*1931, 1932*) identified them from several absorption bands that had long been known in the red-light spectrum. Most of the 12 molecules now known (Table 16.1B) have been identified in the same way: by absorption lines at various wavelengths from visible light to radio. Some of the best spectra are shown in Figures 16.1&2. Mid-infrared spectra are especially fruitful, but because of interference from the Earth's water vapour, they are best obtained from aircraft (notably NASA's Kuiper Airborne Observatory) – and from Voyager's IRIS instrument.

[1] The chemistry of Jupiter's atmosphere has been extensively studied, and this chapter is drawn mainly from the following major reports and reviews: Lewis (*1969*), Weidenschilling and Lewis (*1973*), Prinn and Owen (*1976*), Ridgway *et al.*(*1976*), Larson (*1980*), Atreya and Romani (*1983*), Bjoraker *et al.*(*1986a*), West *et al.*(*1986*), Carlson *et al.*(*1987*), Drossart *et al.*(*1989*). Otherwise, only a few individual research papers are cited – partly for brevity, and partly because most of the problems should soon be resolved if the Galileo probe works as intended.
[2] Kiess *et al.* (*1960*).

Table 16.1 *Molecules in Jupiter's atmosphere*

		(A)	(B)	
H_2	Hydrogen	0.88	0.90	(±2%)
He	Helium	0.12	0.10	(±20%)
CH_4	Methane	7.33×10^{-4}	2.5×10^{-3}	(±20%)
NH_3	Ammonia	1.52×10^{-4}	2.5×10^{-4}	(±20%) [≥2 bar]
H_2O	Water	1.22×10^{-3}	3.0×10^{-5}	(±60%) [6 bar]
H_2S	Hydrogen sulphide	3.30×10^{-5}	–	
PH_3	Phosphine	4.3×10^{-7}	5.0×10^{-7}	(±50%) [2 bar]
GeH_4	Germane	7.7×10^{-9}	7.0×10^{-10}	(±50%) [2 bar]
AsH_3	Arsine	4.1×10^{-10}	2.2×10^{-10}	(±50%)
Ne	Neon	1.7×10^{-4}	(Unobservable)	
Ar	Argon	7.0×10^{-6}	(Unobservable)	
Non-gaseous elements:				
Mg	Magnesium*	7.0×10^{-5}	–	
Si	Silicon*	6.6×10^{-5}	–	
Fe	Iron*	5.9×10^{-5}	–	
Non-equilibrium gases:				
CO	Carbon monoxide	–	1.5×10^{-9}	(±30%) [2 bar]
C_2H_6	Ethane	–	3×10^{-6}	(±100%)
C_2H_2	Acetylene (ethyne)	–	2×10^{-8}	(±100%) [≤0.1 bar]
HCN	Hydrogen cyanide	–	2×10^{-9}	(±100%) [≤0.1 bar]

(A) Abundances (by number and pressure) predicted from solar ratios of elements, with all fully hydrogenated. This list includes the 11 most abundant elements plus phosphorus (P), germanium (Ge), and arsenic (As). *Mg, Si, and Fe are expected to be buried in the core, probably combined with oxygen. From Cameron (*1982*).
(B) Abundances measured for molecules detected in Jupiter's atmosphere. [In square brackets, pressure level to which the measurement refers]. From Drossart *et al.*(*1989*). (Additional hydrocarbons are detected in trace amounts in the stratospheric polar hotspots; Kim *et al.*, *1985*.)

As different wavebands are emitted from different depths in Jupiter's atmosphere (Chapter 4), a spectral line reveals the concentration of the gas at and above the corresponding depth for that waveband. This ranges from 0.3 bars for visible light to 4–6 bars for 5–µm infrared radiation and ≈10 bars for 20-cm radio waves.

The following molecules have been detected.

Methane (CH_4), measured from the visible to the mid-infrared, is the most abundant gas after hydrogen and helium, comprising 2.5×10^{-3} of molecules at all levels in the atmosphere.

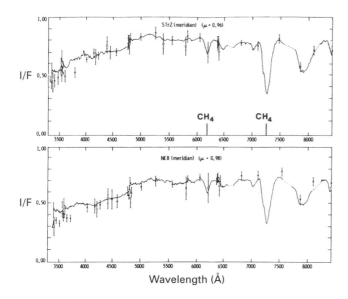

Fig. 16.1. High-resolution visible spectra of the STropZ (top) and NEB (bottom). Absorption bands of methane (CH_4) are indicated. From Orton (1975).

Ammonia (NH_3), measured from near-infrared to centimetre wavelengths, is the next most abundant: 2.5×10^{-4} of molecules, at 1 bar and deeper. But it falls off steeply above the 0.5-bar level due to condensation in the ammonia clouds at that level and photolysis above them (see next section).

Ammonia seems to be strongly affected by jovian weather. There is less ammonia in the belts than the zones.[3] Voyager infrared spectra (Fig. 16.3a) show ammonia depletion in the warmer areas of the NEB and SEB, with a molar ratio of 1.5×10^{-4} as against 3×10^{-4} in zones and cooler parts of the belts. This is confirmed by preliminary results in red-light ammonia lines. The clearest evidence is at radio wavelengths, which probe 0.5 to 10 bars. At these depths also, below the clouds, there is half as much ammonia in the equatorial belts as in the zones (Fig. 16.3b). This pattern is consistent with the theory that belts are regions of sinking gas, and the ammonia-poor gas drawn down from higher altitudes explains why the ammonia clouds are lower and thinner in the belts.

There is also less ammonia in the GRS, at cloudtop level – only one quarter of the amount in the STropZ, according to the Voyager IRIS. This clue to the dynamics or chemistry of the GRS is so far

[3] Infrared data: Gierasch *et al.* (1986),; Lellouch *et al.* (1989); Griffith *et al.* (1992). Red-light data: Baines *et al.* (1989). Radio data: de Pater (1986).

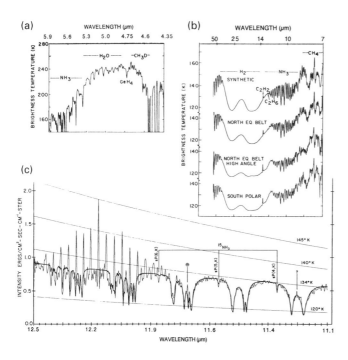

Fig. 16.2. High-resolution infrared spectra.
(a) From Voyager IRIS, 4–6 µm. NEB. Radiation in this waveband comes from deep below the clouds, permitting detection of trace gases that come from below such as germane (GeH_4). (From Hanel *et al.*, *1979a*).
(b) From Voyager IRIS, 7–50 µm. NEB and South Polar Region. The 'synthetic' spectrum at top reproduces the main sets of absorption bands due to H_2, NH_3, and CH_4, and small emission peaks due to C_2H_2 and C_2H_6. Note C_2H_2 (acetylene), which is in the stratosphere, is more prominent near the limb and in the polar region. These bands are identified not only by wavelength but also by multiplet structure, as shown in (C). (From Hanel *et al.*, *1979a*).
(c) Earth-based, 11–12.5 µm. Equatorial region. The main absorption bands repeating every 0.3 µm are due to ammonia (NH_3); a minor set is due to NH_3 with the ^{15}N isotope. The prominent set of emission lines centred on 12.2 µm is due to ethane (C_2H_6). Model curves are shown for black bodies and for a cloudy atmosphere with NH_3 absorption. (From Tokunaga *et al.*, *1979*).

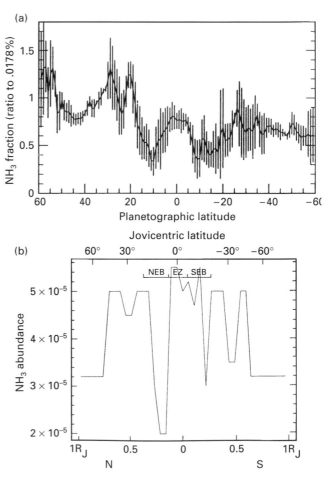

Fig. 16.3. Latitudinal profiles of NH_3 abundance at 0.5–1.0 bar, near the base of the ammonia clouds. (a) From Voyager IRIS spectra around 46 µm (Gierasch, Conrath and Magalhaes, *1986*). (b) From Earth-based radio maps (de Pater, *1986*).

uninterpretable. Conversely, in the stratosphere, ultraviolet absorption attributed to ammonia is stronger over the GRS than elsewhere.

Other compounds have been detected from the late 1960s onwards, from their infrared signatures.

Water (H₂O), measured at 2.7 and 5 μm, is, surprisingly, much less abundant than ammonia. There is none at 1 bar (estimated molar ratio $<2\times10^{-9}$), and only a little at ≈4 bars ($1-30\times10^{-6}$). The range of values is largely due to uncertainty in the depth and cloud-penetration of the emission at these wavelengths. The maximum depth probed may be within water clouds (see next section).

Water, like ammonia, is depleted in the warmer areas of the NEB and SEB, according to 5-μm spectra from aircraft and Voyager.[4] Water had a molar ratio of $1.5 - 6 \times 10^{-6}$ in these areas as against $5-50 \times 10^{-6}$ in the zones and the cooler parts of the belts – again attributable to downdraft in the hotspots.

Hydrogen sulphide (H₂S) has not been definitely detected. A report of its detection at millimetre wavelengths is ambiguous, and conflicts with a stringent upper limit at 2.7 μm ($<2\times10^{-8}$ around 1 bar). It is expected to be present deeper, below the clouds, but this is unconfirmed.

Phosphine (PH₃), *germane (GeH₄)* and recently *arsine (AsH₃)* (K. Noll and colleagues, *1990*), have been detected in proportions similar to those which are predicted from cosmic abundances – but not predicted from jovian chemistry. They should be unstable in the troposphere, phosphine (for example) reacting with water and becoming oxidised. Their presence may be due to updrafts from much deeper levels. However, so far there is no definite sign of variations in their abundances between belts and zones. There is a 60% excess of phosphine in the north polar region, according to P. Drossart and colleagues (*1990*).

Carbon monoxide (CO) is even more surprising as it is an oxidised compound, which must be unstable in a reducing atmosphere. But the observations show that it is well-mixed throughout the troposphere. So it too is probably brought up from deeper regions where it is thermochemically in equilibrium.

According to J.S. Lewis and M.B. Fegley (*1984*), the abundances of phosphine, germane, and carbon monoxide can all be explained by updrafts from a level at $\approx800-1300°K$ – well below any direct probing. Nitrogen (N₂) should also be brought up, to a molar ratio of $\approx6\times10^{-6}$, but does not offer spectral lines to permit detection. Recently, A. DelGenio and K. McGratton (*1990*) produced detailed models of convection governed by condensation and evaporation of water (like convection on Earth), which could account both for the enrichment of these unstable molecules from below, and for the local depletions of ammonia and water from above.

Hydrogen spectra provide direct evidence for such updrafts. In the 14–33 μm range, one can distinguish absorption by *ortho* and *para* forms of H₂, whose ratio depends on pressure. The Voyager IRIS spectra showed that the ratio at cloud-top level is not in thermodynamic equilibrium (especially at low latitudes), and that hydrogen must be rising from deeper, higher-pressure levels.[5] The

IRIS spectra seemed to show the disequilibrium strongest in the EZ, and possibly stronger in the tropical zones than in the belts. According to the most recent analysis, the difference lies in the rate at which the forms equilibrate. In the belts, conversion occurs throughout the width of the ammonia cloud layer. In the zones, it occurs only in the upper part of this layer – consistent with updraft being mainly in the zones. The latent heat of this conversion may contribute to heating the troposphere.

Hydrogen cyanide (HCN) was first definitely detected by A. Tokunaga and colleagues (*1981*), and is also not in chemical equilibrium. Its origin is uncertain. The most likely source seems to be photochemical (light-induced) reactions in the upper troposphere (Lewis and Fegley, *1984*).

Acetylene (C₂H₂) and *ethane (C₂H₆)* have also been detected, but in infrared emission (12–14 μm; Fig. 16.2C), which shows that they are in the stratosphere. Stratospheric methane also produces an emission band (7.8 μm). Acetylene and ethane are believed to be produced photochemically, by irradiation of methane in the stratosphere. A little ethane is predicted to descend to the cloud levels.

The ratio of ethane to acetylene emission is highest in the polar regions. In contrast, methane and acetylene emissions are enhanced in the variable polar 'hotspots' (Chapters 5 and 17), which also show traces of ethylene (C₂H₄) and other hydrocarbons, while ethane emission is sometimes reduced (reviewed by: S. Kim and colleagues, *1985*; T. Kostiuk and colleagues, *1989*). All this suggests that hydrocarbon chemistry, as well as temperature, is different in these regions – probably due to the extra energy provided by auroral particles shooting down from the magnetosphere.

The poles also show variable high-altitude blue 'hoods' (Chapters 4.7 and 5.2), whose composition is unknown. One speculation is that they may consist of hydrocarbon particles produced in the stratosphere due to auroral bombardment (R. West and colleagues, *1981*; G. McDonald and colleagues, *1992*).

16.2 THE CHEMICAL COMPOSITION OF THE ATMOSPHERE: THEORY

The nature of the clouds

Modelling of the atmosphere began with the working hypothesis that Jupiter contains the elements in their cosmic proportions (Table 16.1A). Indeed, the observed abundances are broadly consistent with a cosmic mixture of atoms, combined into the simplest molecules, from which some compounds have been removed by condensation into clouds or precipitation deep into the planet. Whether the ratios are exactly cosmic will be considered later.

Given the table of abundances, and the known profile of temperature and pressure with altitude (Chapter 4), it is possible to predict the levels at which clouds of various types should form.

The present cloud model was initially developed by John S. Lewis.[6] He predicted three main cloud layers (Figure 16.4A): ammonia ice uppermost (0.3–0.7 bars), ammonium hydrosulphide

---

[4] Bjoraker et al. (*1986b*); Lellouch et al. (*1989*); Carlson et al. (*1992a*).
[5] Reviewed by Gierasch and Conrath (*1983*). For recent spatial analyses see Gierasch *et al.* (*1986*) and Carlson *et al.* (*1992b*).

[6] Lewis (*1969*); Weidenschilling and Lewis (*1973*).

below it (NH$_4$SH; 2 bars), and water deeper still (5–6 bars). The water clouds could include an upper layer of ice crystals and a lower layer of water droplets, at temperatures of 250–280°K – surprisingly like the clouds of Earth. But the water will have a lot of ammonia dissolved in it – probably around 10%, so it is actually an ammonium hydroxide solution with pH greater than 10. The absence of water and hydrogen sulphide in the upper atmosphere is accounted for because they are sequestered in the water clouds and the ammonium hydrosulphide clouds respectively. These clouds only sequester a minor fraction of the ammonia.

These three cloud layers nicely match the physical cloud layers observed at infrared wavelengths (Chapter 4).

The Lewis model has survived through many refinements, and all the main features are still present in the recent version by Barbara Carlson and colleagues (*1987, 1992a*) (Fig. 16.4B). The best fit to the observations was achieved with twofold enrichment of carbon, nitrogen, oxygen, sulphur and phosphorus relative to hydrogen.

However, these models are only as good as the assumptions that go into them, and one must recognise that some of these are very weak. For example, the presence of sulphur (as H$_2$S) is purely hypothetical, and with it the existence of the ammonium hydrosulphide clouds. Moreover, the observed abundances of gases may not be in stable equilibrium; 'weather' can produce wide fluctuations, as it does for water vapour on Earth. We have already seen evidence for this on Jupiter in that water and ammonia are depleted in the belts. In the belts, in contrast to the zones, the clouds in the 'ammonia' region are thinner and a few kilometres lower, and may not be made of ammonia at all. R. West, D. Strobel and M. Tomasko (*1986*) proposed that the unidentified grey-brown particles of these clouds in the belts act as nuclei, on which white ammonia ice condenses in the lower temperatures and higher ammonia concentrations of the zones.

Is there rain or snow on Jupiter? According to Carlson and colleagues (*1988*), the ammonia clouds are like thick cirrus and are unlikely to precipitate much if anything. However, the thick water-ammonia clouds below will probably produce rain. Whether this is a drizzle or a downpour is still a matter for speculation. Perhaps it accompanies the violent thunderstorms at this level, whose lightning illuminates the cloud tops (Chapter 17).

What produces the colours?

To this important question, spectra have not given any answers. Most potential coloured substances (chromophores) do not produce distinct spectral lines, just very broad bands. Spectra of visible light (Fig. 16.1) show hardly any difference between the belts and the zones.

There are several colours that must be explained (Chapter 4.1). The origin of blue-grey and grey tints has hardly been addressed. (As both CO and CH$_4$ are present, is it conceivable that simple carbon could precipitate?) More attention has been given to brown colours, which are found both in the belts (though possibly above the grey colours) and also in the high-altitude, violet-absorbing haze. The colours from brown through yellow to reddish may form a continuum, and may be due to a single substance or group of sub-

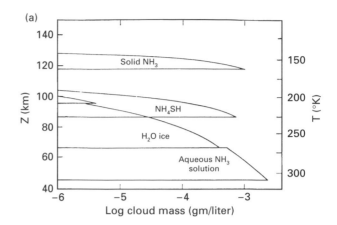

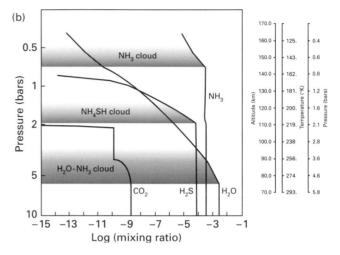

Fig. 16.4. Jupiter's cloud layers, as predicted from atmospheric chemistry. (a) Cloud densities versus altitude, from Lewis (*1969*). More recent models show more H$_2$O ice cloud and less NH$_3$–H$_2$O liquid cloud. (b) Gas composition versus altitude, with cloud levels shaded, from Carlson *et al.*(*1987*). This assumes twice solar abundances for elements other than hydrogen. The vertical scales at right are from Atreya and Romani (*1983*); note that the base of the altitude scale was defined differently in (a).

stances whose spectrum varies at different concentrations and temperatures (Chapter 9.4).

All the major compounds predicted to form clouds are chemically simple and would be white. The main candidates for coloured substances are as follows.[7]

Alkali metals such as sodium, dissolved in ammonia, were proposed by Wildt (*1939*) after he discovered the atmospheric ammonia. However, they do not seem plausible with present knowledge of the jovian chemistry and spectra.

Ammonium hydrosulphide (NH$_4$SH) is itself white, but can form yellow-brown polymers with an additional sulphide group. As ammonium hydrosulphide may be abundant, and polymer could be induced rapidly by solar ultraviolet light, this is a leading candidate.

Sulphur could be produced photochemically from H$_2$S and NH$_4$SH, along with the above polymers. There are various suitably coloured forms of elemental sulphur. But as there is no H$_2$S in the upper atmosphere, the process now seems unlikely.

[7] Reviewed in detail by West, Strobel and Tomasko, (*1986*). The photochemical reactions were calculated by Lewis and Prinn (*1970*).

Phosphorus likewise has coloured elemental forms, and could be produced photochemically from PH_3. The red form (once invoked for the GRS) does not have an appropriate spectrum, but there are other, more exotic forms that are yellowish. However, the GRS is not enriched in PH_3.

Organic polymers – that is, large carbon-rich molecules – are the most interesting and controversial candidates. One of the great surprises of the exploration of the outer solar system has been the abundance of organic molecules and of dark 'tarry' surfaces on satellites and comets. Everyone who does any cooking knows that brown polymers readily form when simpler organic molecules are energised – for example, in the oil in a frying pan. Stanley Miller began testing chemical reactions in various 'atmosphere' mixtures in the 1950s, and many experiments since then have produced organic molecules from Jupiter-like mixtures of gases, starting with those of Carl Sagan and Stanley Miller in 1960.[8] The most important energy source on Jupiter is probably solar ultraviolet light, which is likely to produce photochemical polymers of acetylene and other products. Other energy sources include lightning, thunder (shock waves), and magnetospheric particles. Experimentally, these sources are simulated by ultraviolet irradiation, electric discharges, or proton beams, in mixtures of CH_4, NH_3 and H_2S. When electric discharges are passed through methane-ammonia mixtures, the products include hydrogen cyanide, acetylene, acetonitriles, and a reddish polymer (for example, see C. Ponnamperuma, *1976*). HCN and acetylene can readily produce various coloured polymers, and acetonitriles can react with water to produce aminoacids.

The main problem with organics as chromophores is how to produce enough of them. They are unlikely to be stable, as downdrafts will carry them into hot areas where they are destroyed. The predicted rates of production are very low – thousands of years, against a few weeks for photochemical production of sulphur and phosphorus (Lewis and Prinn, *1970*). All calculations indicate that photochemical production of sulphur and phosphorus and NH_4SH polymers should be far greater than production of organics, in the troposphere. Even Carl Sagan and E.E. Salpeter (*1976*), who advocated organics most strongly, admitted this, and they suggested that the organics are actually produced by living organisms.

Is there life on Jupiter?

As physical motions on Jupiter have organised themselves into cycles in the form of great atmospheric circulations, could chemical reactions have organised themselves into cycles that constitute carbon- and water-based life?

It used to be said that the outer planets are too cold for life to exist. This may be true of their upper clouds, but deeper down, the problem is rather that they are too hot. In between is a region of Earth-like temperatures, probably decked with water-ice and water-droplet clouds. There is plenty that is prohibitive to life-as-we-know-it, most notably the alkaline solution of ammonia in the

water clouds, while other atmospheric gases are generally regarded as too poisonous to support life: methane, hydrogen cyanide, and carbon monoxide are highly inimical to humans. But these simple, reactive, carbon-containing molecules are the very ones from which life on Earth is supposed to have originated, and the experiments described above show that many complex molecules could arise from them on Jupiter. Jupiter may well have the materials and the conditions for its own forms of life to develop.

These considerations are not enough to imply that life exists; there are still problems that may be (literally) fatal. On Mars, nemesis comes in the form of the oxidising chemicals in the soil, which destroy organic compounds yet mimic the responses of life. On Jupiter, the villain may be the very vigour of the atmospheric circulation. With no solid or liquid surface, molecules would be cycled between the frigid heights and the scalding depths, where complex molecules would be destroyed. (The observed nonequilibrium molecules provide evidence for cycling down to a level at $\approx 1000°$K.) If so, on Jupiter there is no place to hide.

However, we do not know if this is really the case. In any case, living organisms might be able to protect themselves against these currents – small ones by reproducing fast enough so that a few progeny would always recirculate to the tolerable levels, and larger ones by floating or flying to maintain their levels. Sagan and Salpeter (*1976*) devised a complete jovian biology along these lines including balloon-like 'floaters' and winged 'hunters'.

An important principle, proposed by James Lovelock (*1965*), is that if life exists it will dominate the planet and its atmosphere. Life, by definition, multiplies; if it does not thrive, then sooner or later some accident will wipe it out. So a 'living' planet will have an atmosphere that is not in chemical equilibrium, as Earth's is not. In Jupiter's atmosphere, the divergences from chemical equilibrium are much smaller than those on Earth, and can apparently be attributed to inorganic agents such as ultraviolet light and lightning. The claim of Sagan and Salpeter is that the coloured substances actually constitute a massive, biological, violation of chemical equilibrium. So far, there are no data to rule out their ecology of 'floaters' and 'hunters', as so graphically described by Arthur C. Clarke in his novel *2010*. Even if these creatures do not exist, Jupiter is certainly a rich laboratory for the sort of reactions that could lead to life.

16.3 THE GLOBAL COMPOSITION

The data and models are now reaching sufficient precision to detect at least some deviations from cosmic abundances. In fact, carbon is definitely twofold enriched relative to hydrogen, and there may be a similar excess of other elements, while helium may be slightly depleted. These statements of course can only be made when these elements have been measured, or inferred, below the levels at which they condense in clouds.

The carbon excess is deduced from the spectroscopic measurements of methane, which does not condense at jovian temperatures and is therefore uniformly mixed throughout the atmosphere. An almost twofold excess of nitrogen (in ammonia) also seems likely

---

[8] These experiments are reviewed by Martin (*1972*), Sagan and Salpeter (*1976*), and West, Strobel and Tomasko (*1986*). Most recently: McDonald *et al.*(*1992*).

below the clouds. De Pater (*1986*) proposed that sulphur is several-fold enriched, in order to explain the depletion of ammonia within the ammonia clouds. And Carlson and colleagues (*1987, 1992a*) proposed that oxygen, sulphur and phosphorus are also twofold enriched – even though they are only detected in very small amounts.

One still cannot be sure that these models represent the true composition of the deep atmosphere, let alone the whole planet. Indeed, Bjoraker and colleagues (*1986b*) put the water clouds at a higher level in their model and proposed that oxygen is truly 50- to 100-fold *under*-abundant, as observed. Their model has been strongly rebutted by several groups, largely on the grounds that any oxygen depletion would be incompatible with the observed abundance of carbon monoxide. But as the origin of the carbon monoxide is itself unproven, this debate illustrates the great uncertainty in our knowledge of these abundances.

As for the bulk of the planet: it seems likely that the deep atmospheric abundances of CH_4, NH_3, H_2O, and H_2S will indeed reflect the global abundances of C, N, O, and S, because of the vast excess of hydrogen on the planet. Oxygen probably combines with silicon and metals to form a rocky core (as it does to form rocks on Earth), but this should only deplete about a third of the total oxygen if the ratios are cosmic. Whether sulphur and carbon are likewise sequestered is unknown.

If all these elements are indeed twofold overabundant, one should rather say that hydrogen and helium are twofold *depleted* relative to heavier elements. The missing hydrogen and helium could either have gone into the planet's mantle (see below), or have been lost into space when the planet formed. Both processes are quite plausible – and both could have also changed the helium:hydrogen ratio.

Loss of the lightest elements when the planet formed is plausible because the planets may well have accreted piecemeal from the nebula around the proto-sun. They must have condensed mainly from the nebular gas with essentially cosmic composition. But a dense ice-and-rock core depleted of light elements may have been necessary first to nucleate the condensation of the gas; and later, the planet may have been enriched by infall of many ice-and-rock planetesimals like Pluto. These would have lacked gaseous hydrogen and helium; they might have contained some hydrogen in water and other ices, but current opinion is that they would have been largely oxidised, like comets, consisting mainly of CO, CO_2, nitrogen, and hydrocarbons. So there is plenty of scope for deviations from the original cosmic abundances.[9]

Abundances have also been measured for minor isotopes of hydrogen (^2H, deuterium) and carbon (^{13}C), in the methane absorption lines. The deuterium abundance is comparable to that in the interstellar medium and the theoretical proto-sun.

The helium:hydrogen ratio

The helium abundance is very difficult to measure by remote sensing, in Jupiter or any other body, but much effort has been devoted

to determining it because it is very important not only for our understanding of Jupiter, but also for cosmology.

The first direct evidence for helium was from Pioneer 10 data, and the best estimate of the abundance is from Voyager data.[10] Two techniques were employed. One used IRIS spectra covering 14–33 µm, where H_2–H_2 and H_2–He collisions affect the absorption. The other used the radio occultation profile (Chapter 4.5): if the vertical temperature profile derived from infrared spectra is assumed correct, the occultation results can be fitted to it with helium abundance as a free parameter. Both methods agreed on a helium abundance of 10% (±2%) by mole and 18% (±4%) by mass.

How does this compare with solar and cosmic values? At that time, values for the sun and for nebulae in other galaxies clustered around 21% (±3%) by mass, compatible either with the jovian value, or with the 23–24% by mass that is thought to have been produced in the Big Bang (B. Pagel, *1989*). But recent estimates for the proto-sun are higher, 24–28% by mass, as the interstellar gas from which the solar system formed had been supplemented by supernovae (D. Gautier and T. Owen, *1989*). So the 18% in Jupiter's atmosphere is now thought to indicate a real depletion. (Helium is even more depleted in the atmosphere of Saturn.) This is bad news for cosmologists, who would have liked an independent check on the cosmic abundance.

This depletion could be due to selective loss during the planet's formation (see above), or to a 'helium rain' within the present-day mantle of the planet (see below).

16.4 BENEATH THE CLOUDS

A model planet[11]

To infer the structure of Jupiter far below the clouds, one must turn to theoretical physics, using what information exists on the behaviour of hydrogen and helium at extreme temperatures and pressures. The models that have been derived agree that the main features are as follows.

(1) A deep atmosphere/ocean of molecular hydrogen. The temperature increases with depth, and this atmosphere will have values above the critical point for hydrogen (13 bars, 33°K). So the gas merges imperceptibly into hot liquid, around a thousand kilometres below the cloud tops, at temperatures of 1000–2000°K and pressures of thousands of bars. Condensed silicates may float in a layer near this level. Other gases and liquids should occur roughly as predicted from cosmic abundances.

(2) A mantle ('envelope') of metallic hydrogen. This bizarre form of matter should begin when the hydrogen is sufficiently compressed – 15 000–25 000 km below the surface, at a pressure of 2–4 *million* bars. It will also contain helium in the metallic phase, but probably in less than the cosmic ratio (see below). It comprises most of the planet's mass.

(3) A core made of rock. Most of the planet's metals should have sunk into this core, oxidised with oxygen and perhaps other light

---

[9] Discussed by Gautier and Owen (*1983b,1989*), and Lunine (*1989*).
[10] Gautier *et al*. (*1981*), revised by Conrath *et al.* (*1984*).

[11] The first such model was by DeMarcus (*1959*). Later reviews are given by: Hubbard and Smoluchowski (*1973*); Smoluchowski (*1975*); Stevenson and Salpeter (*1976*); Pollack (*1984*); Hubbard (*1989*).

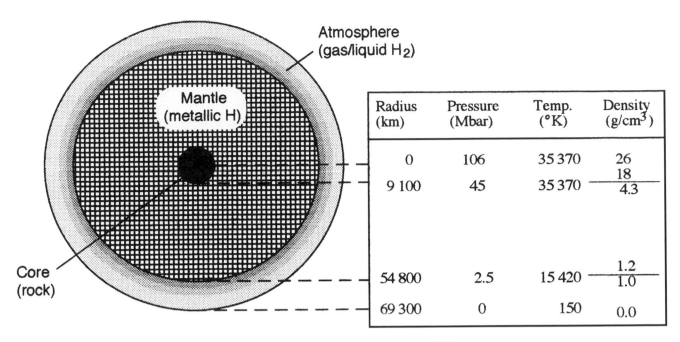

Fig. 16.5. Typical model of Jupiter's interior, as in Stevenson and Salpeter (*1976*). The core mass is 11.2 Earth masses.

elements. If the mantle consists of pure hydrogen and helium, and all the other elements (in cosmic abundances) have precipitated to the core, the core would have five times the mass of the Earth (5 M_E).

The proportions of these three parts depend on the assumed temperature and helium abundance. Typical models with helium abundances of 16.5% to 21.5% by mass give the following parameters for the rocky core: mass, 6–20 M_E(2–6% of Jupiter's mass); radius, 7 400–10 000 km; density, 26–34 g/cm³; pressure, ≈110 Mbar; temperature, 21 000–36 000°K. Figure 16.5 shows a representative model with a core of 11.2 M_E.

Does the metallic hydrogen mantle exist?

The prediction of metallic hydrogen is entirely theoretical. The required pressures are so extreme that the existence of this state has never been proved experimentally. Several teams of physicists are trying to achieve such pressures in the laboratory, with samples of hydrogen confined in small diamond cells through which they can make just one or two optical or electrical measurements of the material. Several times these teams have reported success, but given the difficult nature of the measurements, these claims have been disputed.

The metallic hydrogen mantle would be fluid and electrically conducting. Largescale convection in the outer part of this mantle probably provides the dynamo which generates the planet's magnetic field.

Does the rocky core exist?

One can deduce whether an oblate planet has a central mass concentration from the shape of its gravitational field, as revealed by subtle effects on the orbits of its satellites and on the Pioneer trajectories. The results can be summarised in the moment-of-inertia coefficient, which is 0.40 for a sphere of uniform density, 0.33 for

Earth, and 0.25 (±0.005) for Jupiter. This implies that Jupiter indeed contains heavy elements surpassing solar proportions. They could be uniformly distributed, but on other grounds are much more likely to form a very dense core, accounting for 3–10% of the planet's mass (10–30 M_E).[12]

Where does the heat come from?

Jupiter radiates heat that comes from its very hot interior (Chapter 4.3). The most likely source is primordial and gravitational heat: the planet has not finished cooling and contracting ever since it formed. It was probably so hot as to be incandescent when it formed, due to gravitational energy supplied as the components fell into the growing planet. In models by W.B. Hubbard (*1977*) and A.S. Grossman and colleagues (*1980*), with a planet of solar composition, the expected cooling time to the present state is almost exactly the age of the planet, 4.6×10^9 years. As the planet cools it contracts (to remain in adiabatic equilibrium) and this supplies further heat from gravitational energy. The present heat flux corresponds to cooling by 1°K per 1.4×10^8 years, with contraction at 0.7 mm/year.

Alternatively, the heat could entirely be supplied by contraction at a rate of 1 mm/year. This could be obtained without any core cooling if the liquid-to-metallic phase transition were creeping outwards.

Another possible source of heat is 'helium rain' in the mantle. Thermodynamic calculations indicate that there ought to be a lower proportion of helium there than in the atmosphere, and helium would be excluded altogether at temperatures below ≈10 000°K.[13] So as the planet cools down, helium should start to separate out and sink towards the core – the helium rain. Latent

---

[12] Stevenson and Salpeter (*1976*); Hubbard and Slattery (*1976*); Stevenson, (*1982*).
[13] Summarised by Lunine (*1989*) and Gautier and Owen (*1983a*).

heat would be released. This is thought to have been occurring in Saturn for billions of years, and may explain why helium is depleted in Saturn's atmosphere, and why Saturn emits even more internal heat than Jupiter. It is not certain whether it has started on Jupiter; the top of the mantle is predicted to be close to the critical temperature. If it has started, it could explain the slight helium depletion observed, and the internal heat flux – but the heat flux is more naturally accounted for by simple cooling.

Radioactivity, which keeps the Earth's core hot, cannot make much contribution to the heat in Jupiter. In Jupiter, the observed energy emission per unit mass is much higher than that of Earth, whereas the abundance of naturally radioactive elements is much smaller.

It is sometimes said that Jupiter is 'almost a star'. This is not really true. A star produces heat and light from nuclear fusion reactions, which require temperatures of millions of degrees – far higher than can occur in Jupiter. The smallest viable star is predicted to have about 80 times Jupiter's mass. Bodies smaller than this which radiate heat significantly are called 'brown dwarfs', so Jupiter lies at the borderline between a planet and a brown dwarf.

V. The Electromagnetic Environment of Jupiter

In the following chapters, all references are in Appendix 5, and all figures have north up unless otherwise stated.

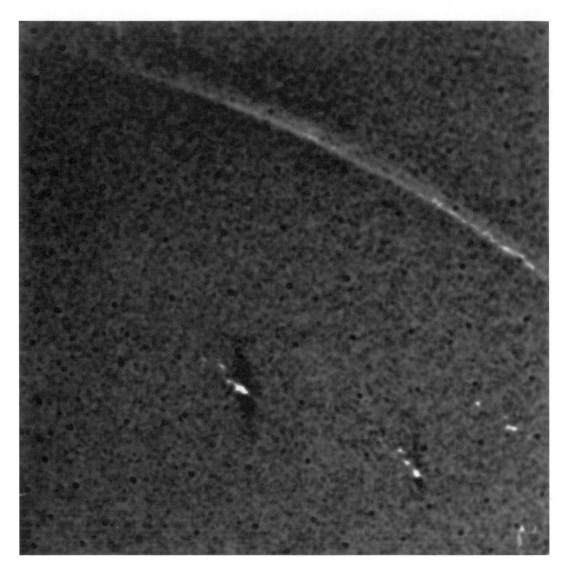

Figure 17.1. The night side of Jupiter from Voyager 1, showing lightning and aurora. The aurora marks the northern limb (an arc of about 30000 km), and most lightning flashes are in the N⁴TB. This was composed of three 192-second exposures, slightly offset, totalling 11 minutes. For zenocentric coordinates, see Cook *et al.*(1979) and footnote 1. (Image 16396.39. NASA.)

17: Lights in the jovian night

While Voyager 1 was flying through Jupiter's shadow, it took two long exposures of the planet's dark side, which revealed lightning bolts, a meteor, and aurorae flaring in the darkness (Fig. 17.1).

To understand what is known of these lights, we must also refer to one other glow in the jovian night sky, to be described more fully in the next chapter. This is the Io orbital cloud and torus. It is a luminous cloud of gas and plasma, scoured off Io by the magnetospheric radiation and distributed around its orbit. It focusses signals beamed up the planet's magnetic field lines from the lightning, and it beams particles down those same field lines to produce the aurorae.

These topics will lead on to a discussion of the ionosphere and the dayside airglow. As we saw in Chapter 16, all these luminescent phenomena may contribute to the strange chemistry of the upper atmosphere.

17.1 LIGHTNING

Lightning flashes are clearly visible in Fig. 17.1, and in other long exposures from Voyagers 1 and 2.[1] One can tell that they were brief flashes because the images are not triplicated, although this was a triple exposure.

They occurred in clusters, in four latitude ranges, all being in turbulent cyclonic belts (Table 17.1). Most (82%) were in the very 'filamentary' N^{4}TB latitudes, including several just north of anticyclonic N^{3}TZ white ovals (Chapter 6). Three in the NEB were all in the great rifted region (Chapter 8). There were none in the equatorial plumes.

These flashes had optical energies of 10^9 to 10^{10} joules – at least as great as the cloud-top 'superbolts' on Earth, which comprise only one in a thousand of our terrestrial lightning bolts. W. Borucki and M. Williams (1986) calculated that the lightning must be occurring far below the visible cloud tops, so that we see only the diffused light illuminating the clouds. They inferred that it occurs in water clouds near the 5-bar level, in giant thunderstorms with the same dynamics as those on Earth.

Lightning was not only seen but also heard by Voyager 1, as it

Table 17.1. *Zenographic latitudes of lightning flashes*

Latitude		Number
57–60°N	(N^{5}TB)	6
49–53°N	(N^{4}TB)	51
45°N	(N^{3}TB)	2
13–14°N	(NEB)	3

Total of 62 flashes, from Voyagers 1 and 2. (From Doyle and Borucki, *1989*, and Borucki and Magalhaes, *1992*.)

skimmed the orbit of Io. Voyager could 'hear' with its plasma wave experiment, which detected 'whistlers' attributed to lightning.[2] These are electromagnetic waves at audio frequencies, running outwards along the magnetic field lines, and sounding like a descending whistle. The falling tone is diagnostic of a sudden wide-frequency burst, propagated through a plasma in which the lower-frequency waves travel more slowly, and lightning is the only likely source of such bursts. Very similar whistlers are produced by lightning on Earth. Those detected by Voyager 1 are believed to arise from lightning strokes in the polar regions, at latitudes near 66°N or S, from where the magnetic field lines arch round to Io's orbit (Fig.17.2). The waves only produce the whistle when they enter the Io plasma torus.

17.2 METEORS

Meteoroids are expected to be concentrated in Jupiter's strong gravitational field – indeed, the Pioneer spacecraft were plentifully pummelled by them. So it was not unexpected that one Voyager darkside photograph showed a fireball plunging into the planet's atmosphere (A. Cook and T. Duxbury, *1981*).

17.3 AURORA[3]

The aurora seen in Fig.17.1 was as bright as a bright aurora on Earth, and had a very similar origin. As on Earth, the aurora forms

---

[1] This image was first presented by Smith *et al.* (*1979a*) and mapped approximately by Cook *et al.* (*1979*). The lightning locations have been repeatedly reanalysed and republished, most recently and accurately by Doyle and Borucki (*1989*) and Borucki and Magalhaes (*1992*).

[2] Gurnett *et al.*(*1979*); Menietti and Gurnett (*1980*); Kurth *et al.* (*1985*).
[3] For major reviews and previous references on the aurora, airglow, and ionosphere, see Broadfoot *et al.*(*1981*), Strobel and Atreya (*1986*), and Clarke *et al.*(*1989*).

an oval around each magnetic pole, where electrons or ions from the magnetosphere stream down into the ionosphere. In Jupiter's case, these particles come from the Io orbital plasma cloud and beyond, and the auroral oval roughly marks the ring of magnetic field lines that intersect Io's orbit (Figs. 17.2 and 18.11). This 'footprint' has a magnetic latitude of about 66°, which overlaps zenographic latitudes 55–80°N and S. It is fixed in System III longitude, as it is coupled to the magnetic field.

The aurora is best detected at far-ultraviolet wavelengths, where atoms and molecules of hydrogen radiate, and it is present on both dayside and nightside. These emissions can only be detected from outside the Earth's atmosphere, either by space probes such as Voyager, or by Earth-orbiting astronomical observatories such as Copernicus (which first detected the jovian aurora; S. Atreya and colleagues, *1977*) and the International Ultraviolet Explorer (IUE).

The Pioneer spacecraft did not detect any aurora. But in the Voyager observations, there was more plasma in Io's orbit (Chapter 18.3); and there was a bright aurora, spread around the whole orbital footprint (Figs. 17.2&3). There were 'hotspots' of brighter emission where the auroral oval reaches low zenographic latitudes, near $\lambda_3 = 180°$ in the northern oval, and drifting around $\lambda_3 = 300°$-90° in the southern oval. This pattern seen by Voyager has persisted throughout the 1980s, according to observations by IUE.[4] There are marked short-term fluctuations in brightness, but the average brightness has been much the same from year to year (in spite of the passing of a complete solar cycle).

The first fully-resolved image of the auroral oval – on the dayside, viewed from Earth orbit – was obtained by the Hubble Space

---

[4] Livengood *et al.*(*1992*); earlier references are therein and in Clarke *et al.*(*1989*).

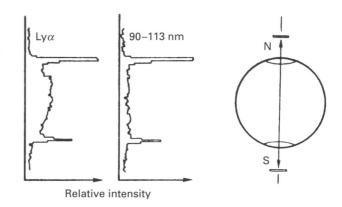

Relative intensity

Figure 17.3. Detection of the aurorae in two ultraviolet bands by the Voyager 1 UVS. The UVS slit was scanned from north to south as sketched at right. The recording of emission v. latitude is shown at left. (From Broadfoot *et al.*, *1979*.)

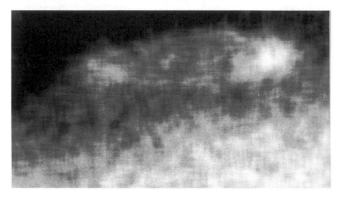

Figure 17.4. Image of the ultraviolet aurora from the Hubble Space Telescope: H_2 emission at 160 nm. The north pole is at top and the image shows the complete auroral oval with hotspots. (From Caldwell *et al.*,*1992*. By courtesy of John Caldwell, York University (Canada), and NASA/ESA.)

Telescope in 1992 (Fig. 17.4). This ultraviolet image confirms that the oval is indeed narrow with well-marked hotspots.

The aurora can also be detected from Earth using modern infrared technology (Figs. 4.12c and 17.5). One might not think it, but these images too show the dayside! At several wavelengths between 3.4 and 4.0 μm, where Jupiter is very dark due to absorption by methane, the aurora is bright due to emission by H_3^+ ions. Several research groups have made images such as these, all using the new electronic camera 'ProtoCAM' at the NASA Infrared Telescope Facility on Hawaii.[5] They clearly show the auroral arc and hotspots. The temperature of the H_3^+ ions is about 1000°K (±150°K). ProtoCAM recorded intensity fluctuations within only a few hours, as well as rapid movements of the southern hotspots. The latter may be due to interaction of one hotspot fixed in λ_3 with another fixed relative to the sun.

The energy powering these aurorae amounts to 0.3–1.0×10¹⁴ watts, compared to 10¹¹ watts for the aurorae of Earth. This energy comes from particles beamed down the magnetosphere, but it is not certain whether the most important particles are electrons, or protons, or ions of oxygen and sulphur. Electrons are the best candidates, as IUE spectra fail to reveal the heavy ions in the amounts required, nor is there massive proton streaming.

---

[5] Kim *et al.*(*1991, 1992*); Baron *et al.*(*1991*); Drossart *et al.*(*1992*).

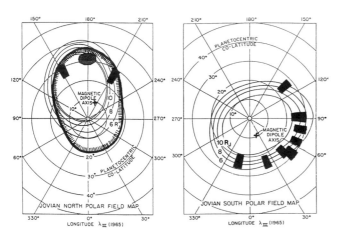

Figure 17.2. Location of the aurora in planetary polar coordinates. (Left:) North pole. The solid line with radial shading marks the area within which ultraviolet emission was detected by the Voyager UVS, from Broadfoot *et al.* (*1981*). The smooth curves are the predicted footprints of magnetic field lines that extend to 6, 8, and 10 times the radius of the planet; the '6 R$_J$' line maps to the Io torus and the innermost double line marks the limit of closed magnetic field (see next chapter). Black rectangles are locations of 'hotspots' observed in H_3^+ emission at 3.53 μm from Earth (Fig. 17.5). The small dark oval marks the hotspot of methane emission at 7.8 μm (Chapter 4.4 and 5.2). (Right:) South pole; locations of H_3^+ emission hotspots and predicted footprints of magnetic field lines. (Adapted from Drossart *et al.*, *1992*.)

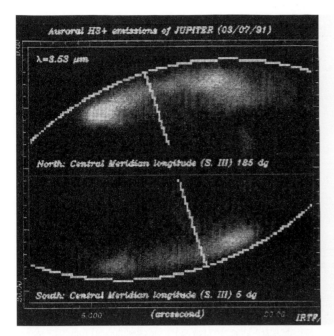

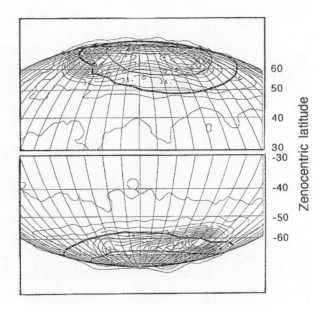

Figure 17.5. Images of the infrared aurora from Earth, in H_3^+ emission at 3.53 μm, using the ProtoCAM instrument on Mauna Kea, on 1991 March 7. (Left) Images of the north polar region (top, $\omega_3 = 185$) and south polar region (bottom, $\omega_3 = 5$). (From Drossart *et al.,1992*, by courtesy of P. Drossart.) (Right) Contours of images (top, north, $\omega_3 = 185$; bottom, south $\omega_3 = 91$), with zenocentric latitude and longitude. Bold lines are the calculated footprints of the Io torus (outer) and of the limit of closed magnetic field (inner). Small diamonds show auroral locations from Voyager UVS data. Fig. 17.2 maps the same data in polar coordinates. (From Kim *et al., 1992*.)

However, X-ray emission poses a puzzle. Soft X-rays were detected by the Earth-orbiting observatory Einstein, coming from Jupiter's auroral regions (A. Metzger and colleagues, *1983*). These emissions are most easily explicable by sulphur and oxygen ions; they seem to be too intense to come from the expected electron flux. So both ions and electrons may contribute to the aurora.

In the stratosphere under the auroral zones, and especially under the hotspots near λ_3 180, 60°N, there is strong infrared line-emission from methane and other hydrocarbons (Fig. 4.23 and Chapter 5.2). This shows that the stratosphere is heated and has unique chemistry below the aurorae.

17.4 AIRGLOW[3]

While the aurora shines brightly by day and by night, the same ultraviolet lines are emitted more faintly from all over the dayside of the planet. Although this emission is less spectacular than the aurora and other night lights, it betrays the effects of solar ultraviolet radiation, which is probably more important for atmospheric chemistry than all the other excitations put together.

The main airglow comes from the thermosphere which coincides with the ionosphere. The *ionosphere* is defined as the upper layer of the atmosphere where there are many free ions and electrons. Hydrogen molecules are dissociated and ionised here by solar ultraviolet radiation, and also by electrons and/or ions raining down from the magnetosphere. The main reaction comes from photons with wavelengths shorter than 80.4 nm, which can split an H_2 molecule into H^+, H, and an electron. The ionosphere extends several thousand kilometres above the molecular atmosphere (Figs. 4.18 and 17.6). It is detectable by occultations of spacecraft radio signals, which are refracted by the electrons. The Voyager occultations gave an electron temperature of 1000–1300°K. The *thermosphere* is defined as the upper layer of the atmosphere which is very hot. It was identified by Voyager's far-ultraviolet observations of the setting sun (Chapter 4.5), which revealed that the neutral gas some 1500–2000 km above the cloudtops was at 1100 (±200)°K. So the temperature of the neutral gas agrees with that of the electrons in the same altitude range, as would be expected theoretically. Since solar radiation alone is not a sufficient source of heat, the heat must come from elsewhere: from charged particles raining from above, or from atmospheric waves propagating up from below, or from both (Chapter 4.5).

The far-ultraviolet airglow comes from hydrogen atoms, helium atoms, and hydrogen molecules (Table 17.2). The atomic emissions (mainly the hydrogen line called Lyman-α) are scattered solar radiation, which is strongly re-radiated by resonance at these emission wavelengths. The H_2 emission may be produced likewise, at lower altitudes.[6] On the nightside, there is no global airglow except for a trace of H Lyman-α emission; this is probably due to resonant scattering of the interplanetary hydrogen glow.

The ionosphere varied dramatically between the Pioneer and the Voyager missions (Table 17.2). The Pioneer occultations showed a

---

[6] The H_2 emission was initially attributed to a rain of low-energy electrons ('electroglow'), but this seems very unlikely now that Voyager has discovered the same emission from the daysides of Saturn, Uranus, and Neptune (Fig. 17.7). On all four planets the emission only occurs in sunlight, falling off as the inverse square of the distance from the sun. On Saturn and Uranus, at least, it is brightest well below the ionosphere - at a level where there are 10^{12}–10^{13} hydrogen molecules per cubic centimetre, which on Jupiter is equivalent to around 10^{-7} bars. As proposed by Yelle *et al.(1987)*, the H_2 emission may be due to resonant fluorescence of sunlight after all.

(a)

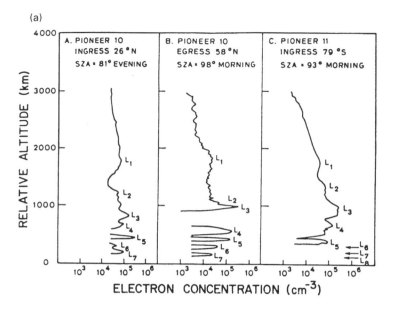

(b)

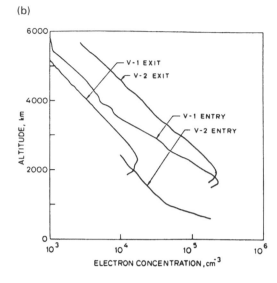

Figure 17.6. Profiles of the ionosphere from spacecraft occultations. (a) From Pioneer 10 and 11. SZA is solar zenith angle. Zero altitude is 50 km above the cloud tops. (From the profiles of Fjeldbo *et al.*, *1975*, *1976*). (b) From Voyager 1 and 2. V-1 entry, 12°S, late afternoon; V-1 exit, 1°N, pre-dawn; V-2 entry, 67°S, sunset (in auroral oval); V-2 exit, 50°S, sunrise. Zero altitude is at 1 mbar, 160 km above the cloud tops. (From Eshleman *et al.*, *1979a,b*.) (These versions of both figures reproduced from Dessler, *1983*.)

Table 17.2 *Properties of the aurora, ionosphere, and airglow, as observed by Pioneer and Voyager.*

	Pioneer (global)	Voyager (global)	Voyager (AURORA)**
IONOSPHERE			
Vertical extent	3500–4000 km	6000 km	≈ 4000 km
Daytime electron peak	10^5/cm³ at ≈ 1000 km altitude [above cloudtops]	2×10^5/cm³ at ≈ 2000 km altitude	12×10^5/cm³ at ≤ 800 km altitude
Scale height* of topside ionosphere	540–975 km	590–960 km	1040 km
Temperature (electrons)	850–1000°K	1100–1300°K	1600°K
Main mechanism of ionisation	Solar extreme-UV	Particle precipitation	Particle precipitation
References	Fjeldbo *et al.* (*1975, 1976*)	Eshleman *et al.* (*1979a, b*); Atreya *et al.*(*1979*)	Eshleman *et al.* (*1979b*)
ULTRAVIOLET EMISSIONS†			
H (Lyman α) (122 nm)	0.4 kR (day)	15–20 kR (day) 0.7–1.0 kR (night)	≈ 60 kR††
He (58.4 nm)	0.004 kR (day)	0.004 kR (day only)	?
H_2 ($\approx$90–170 nm)	–	3 kR (day only)	≈ 80 kR††
References	Carlson & Judge (*1974*)	Sandel *et al.*(*1979*) Broadfoot *et al.*(*1981*)	Sandel *et al.* (*1979*) Broadfoot *et al.* (*1979*)
SOLAR CYCLE	Minimum	Maximum	Maximum

Data from Strobel and Atreya (*1986*) and references given in the table.
*Scale height is the height over which the electron density falls by a factor of 2.72.
†kR = kilorayleighs.
††The brightness of the visible aurora in Voyager images, 14–22 kR (Cook et al.,*1981a*), was consistent with the 80 kR estimated in ultraviolet H_2 bands.
**V2 ingress, just before sunset

much lower, cooler ionosphere than the Voyagers; and in the ultraviolet, the Pioneers detected only one fiftieth as much H Lyman-α emission as the Voyagers did, indicating a much smaller thermosphere. Ultraviolet spectra from rockets and Earth-orbiting observatories confirmed that the overall H Lyman-α emission had

increased from 1–4 kilorayleighs in the early 1970s to 13 kilorayleighs in 1978 December, just before the Voyager encounters.[7]

Two factors may have made the difference. The solar cycle,

[7] Summarised in Broadfoot *et al.*(*1979*); also Cochran and Barker (*1979*).

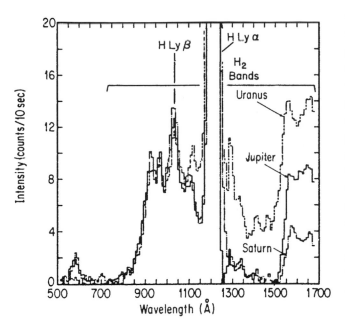

Figure 17.7. Voyager UVS spectrum of the dayside airglow from Jupiter, compared with spectra from Saturn and Uranus scaled according to the inverse square law of the distance from the sun. The match over the H and H$_2$ emissions shows that they are proportional to the brightness of sunlight. The unmatched hump at >1500 Å is reflected sunlight which depends on the albedo of the planet. (10 Å = 1 nm.) (From Yelle *et al.*, *1987*.)

which governs both the solar wind and solar ultraviolet radiation, was at minimum for Pioneer and maximum for Voyager. And the magnetosphere, especially the Io orbital cloud, was more energetic for Voyager – whether because of the solar cycle or because of variations in Io's volcanoes. It seems that whereas Pioneer's ionosphere was maintained mainly by solar ultraviolet, Voyager's was expanded and heated by electron precipitation from the more energetic magnetosphere. This extra energy may all have entered via the auroral zones, and the charged particles then diffused around the planet. One important difference was that Pioneer's ionosphere was similar by day and by night, whereas Voyager's was much reduced on the nightside – probably because higher concentrations of H and H$^+$ led to faster neutralisation of ions.

The dayside H Lyman-α airglow was ≈50% stronger in a limited range of longitude, called the Lyman-α bulge, according to the Voyager UVS (Broadfoot and colleagues, *1981*). The reason is unknown but it may be linked somehow to the magnetosphere.

After the Voyager encounters, as the solar cycle waned, IUE observations showed that H Lyman-α emission from Jupiter decreased in proportion to that from the sun, confirming that this airglow is mostly due to resonant scattering of sunlight (T. Skinner and colleagues, *1988*).

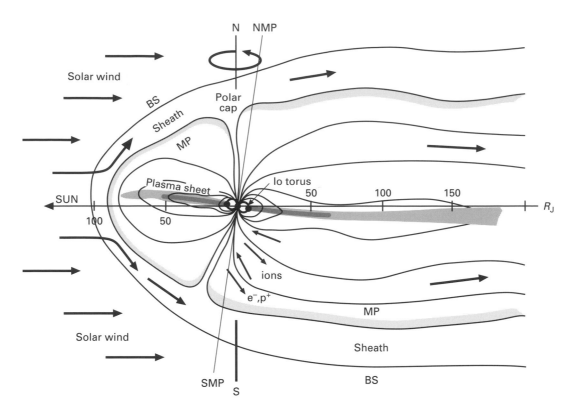

Figure 18.1. Overall structure of the magnetosphere – side view. The scale is in units of the radius of Jupiter (R_J = 71 400 km). N, S: north and south poles of rotation. NMP, SMP: north and south magnetic poles. BS, bow shock; MP, magnetopause; e⁻, p⁺, electron and proton streams detected by Ulysses. Magnetic field lines are sketched.

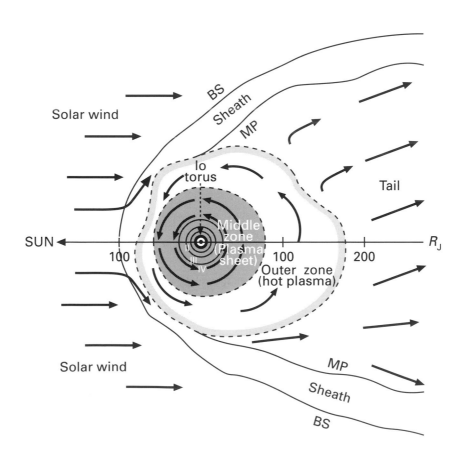

Figure 18.2. Overall structure of the magnetosphere – view from above the north pole. Arrows show plasma flow. It is largely corotating with the planet in the middle zone, but the extent of corotation in the outer zone is variable. The orbits of the galilean satellites are shown (Io torus, II, III, and IV).

18: The magnetosphere and radiation belts

18.1 OVERVIEW[1]

We now leave the planet behind and turn to the insubstantial realm of magnetic fields and electrical particles around it: the magnetosphere. Although it is a vacuum by our everyday standards, it does contain a rarefied plasma – that is, gas dissociated into positively charged ions and negatively charged electrons. The plasma is so energetic and emits such remarkable forms of radiation that it is one of the most intensively studied aspects of Jupiter.

The magnetosphere is also vast – in fact, it has been called the largest object in the solar system. With a diameter of ≈22–29 M km, it would appear four times as wide as the full moon if its outline could be seen from Earth.

The magnetosphere owes its existence to Jupiter's magnetic field. Jupiter's magnetosphere, like that of Earth, consists of plasma populations that are mostly confined to certain regions of the magnetic field and are mostly pulled round with it as the planet rotates. The magnetic field can be pictured as 'lines of force' like elastic cables running through space, anchored in the rotating planet. Charged particles such as electrons in a magnetic field are trapped because they can only move freely parallel to the lines of force. Motion across the lines of force is twisted into tight circles around them. (The radii and periods of these gyrations are variable but much smaller than those of the magnetosphere itself.) A typical particle will have a combination of motions, along and around the lines of force, so that it moves in a helical path, i.e. it 'spirals' around a magnetic field line.

As the magnetic field is stronger near the magnetic poles, the lines of force are closer together, and this produces a force that pushes the gyrating particles towards the weaker field near the magnetic equator. A trapped electron travels to and fro in a helical path around a magnetic field line; as the field lines converge towards the magnetic pole, the helix tightens, and at a critical field density, the particle is bounced back towards the opposite hemisphere. Consequently, the field can be described as a magnetic bottle, and this is why it can accumulate plasma in the form of high-energy 'radiation belts', both around Jupiter and around Earth.

In Jupiter's radiation belts, the trapped particles are some tenfold more energetic than in the equivalent belts of Earth, and several orders of magnitude more abundant. There are two other important differences between the two magnetospheres. First, Jupiter's contains a major internal source of material – the volcanic moon Io. Secondly, the faster rotation and larger size of Jupiter's magnetosphere mean that it is much more rotationally constrained than Earth's. Indeed, in some ways it resembles a low-energy model of a pulsar.

Coordinates and units

The magnetic field rotates with a period of 9h 55m 29.71s (±0.02s). Longitudes are measured in System III which is defined as rotating with this period.[2]

Latitudes are generally given in magnetic coordinates. The magnetic equator is tilted at 10° to the rotational equator, so the whole magnetosphere appears to wobble vigorously as it spins around with the planet's rotation (Fig. 18.1). Because of centrifugal force, the plasma structures are centred about an intermediate plane, the 'centrifugal equator' with a tilt of 7°; this is the 'equator' meant in this chapter unless otherwise stated.

Distances in the magnetosphere are usually expressed as multiples of the radius of Jupiter (R_J), measured from the centre of the planet; 1 R_J = 71 400 km. The orbital radius of Io is 5.9 R_J. As the magnetosphere is shaped by the magnetic field, the field lines are used as a coordinate system, labelled by their radial distance at the equator. Therefore, references to features at 5.9 R_J (for example) include the set of magnetic field lines that intersect the equator at that distance, even though they come closer to the planet at higher latitudes.

The magnetic field

The field itself arises deep within the planet – probably by 'dynamo' circulation within the outer shell of the metallic hydrogen mantle

---

[1] Since the spacecraft encounters, more papers have been published about the magnetosphere than about the planet itself. Full descriptions at the professional level are given in the books edited by Gehrels (*1976*) (post-Pioneer) and by Dessler (*1983*) (post-Voyager). The main compendia of the original data are in special issues of *Science* and *Journal of Geophysical Research* (Appendix 5.1).

[2] The exact definition is 870.536°/day (Riddle & Warwick, *1976*). Strictly, this is 'System III (1965)'; before 1976, an earlier system called 'System III (1957)' was used with period 9h 55m 29.37s, but this is now obsolete.

(Chapter 16.4). The field is not fully symmetrical, but beyond a few planetary radii it can be described quite well as a dipole, which is tilted at 9.6° to the rotation axis (towards $\lambda_3 = 202$) and offset by 0.12 R_J from the centre (towards $\lambda_3 = 149$) (Fig. 18.3). So the NMP is at $\lambda_3 = 202$ as viewed from a distance, but nearer $\lambda_3 = 180$ at the cloud-tops because of the offset (Fig. 17.2). The polarity is opposite to that of Earth. The overall tilted-dipole form was established from Earth-based radio observations, and its shape has been established in more detail by the spacecraft that have passed through it – most usefully, Pioneer 11, as it passed closest to the planet (Fig.18.4). The field at the planet's cloud-tops is estimated to be 4.2 gauss at the equator, and 10 to 14 gauss at the magnetic poles. (For comparison, the equatorial surface field on Earth is 0.30 gauss = 30 microteslas.)

The overall structure of the magnetosphere

The existence, and scale of energy, of the magnetosphere was known from Earth-based radio observations. But these data could not reveal the exact intensities of the magnetic field and the various particle populations, nor their detailed structure and overall extent. These factors were not sorted out until spacecraft flew in, and this was the greatest achievement of Pioneers 10 and 11 (Fig. 18.5). The radio emission detectable from Earth was found to represent only the innermost part of the magnetosphere. Even then, the view was seriously limited in space and time. Any one flyby only gives a snapshot of conditions along a single trajectory (although the tilt of the magnetic field gives the bonus that the spacecraft's magnetic latitude oscillates up and down by ±10° on every rotation). The Voyagers and Ulysses provided a more complete understanding, by penetrating new regions of the magnetosphere, by revealing the central importance of Io as a source of material, and by showing how greatly the structure changes with time (Figs. 18.6&7).

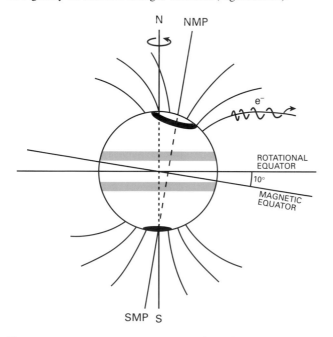

Figure 18.3. Jupiter's magnetic field: the basic model of the offset tilted dipole. N, S: north and south poles of rotation. NMP, SMP: north and south magnetic poles. The auroral regions are shown black. Also shown (not to scale) is the typical motion of an electron (e⁻) spiralling around a magnetic field line.

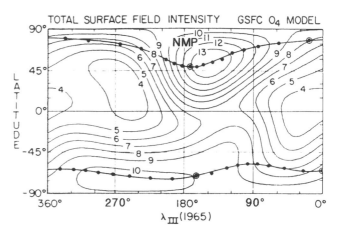

Figure 18.4. Magnetic field strength at the planet's surface, according to a model based on measurements by Pioneer 10 and 11. Magnetic field contours are in gauss, and reach 14 gauss at the north magnetic pole (NMP) and 10 gauss near the south magnetic pole (which is close to the south rotational pole). As the field is uneven, the magnetic poles are only approximately defined. The dark lines indicate the 'footprints' of the Io flux tube, which roughly coincide with the auroral regions (see Fig. 17.2). (From Acuña and Ness, *1976.*)

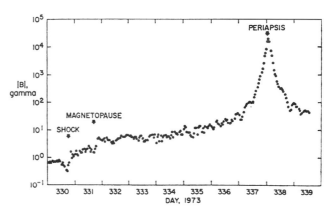

Figure 18.5. The first *in-situ* measurements of the jovian magnetic field: overview of the Pioneer 10 encounter. The vertical scale is magnetic field strength in units of 1 gamma = 1 nanotesla = 10^{-5} gauss. (From Smith *et al.*, *1974.*)

The outer boundary of the magnetosphere is formed by its interaction with the solar wind – the rarefied magnetic field and plasma that streams radially outward from the sun. Where the solar field is draped across the sunward side of Jupiter's field, it forms a sharp boundary, the 'magnetopause'. This lies at a distance from Jupiter that fluctuates between ≈110 R_J (8 M km) and ≈45 R_J (3.2 M km). Plasma in the solar wind is sharply deflected as it approaches the magnetopause, and the result is a 'bow shock', 10–30 R_J ahead of the magnetopause. The zone between the bow shock and the magnetopause is called the magnetosheath (Fig. 18.7).

On the side away from the sun, the magnetosphere is drawn out into a huge tail like a comet's. Its radius is ≈150–200 R_J (11–14 M km). Its length is estimated as typically half a billion kilometres – but pieces of it apparently break off, and were traversed by both Pioneer 10 and Voyager 2 near the orbit of Saturn (Chapter 2).

Around the planet, the magnetosphere is divided into three zones

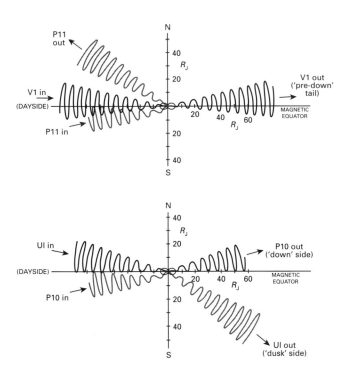

Figure 18.6. Trajectories of spacecraft – side view, relative to the magnetic equator. The dot at the centre is the planet, with the Io-related magnetic field lines marked. The position of the spacecraft is plotted in units of 1 R_J (=71 400 km). Because of the 10° tilt of the magnetic field to the rotation axis, the magnetic latitude of the spacecraft oscillates up and down with a period of 10 hours. **(Top)** Pioneer 11 and Voyager 1. The trajectory of Voyager 2 was almost identical to that of Voyager 1 in these coordinates. **(Bottom)** Pioneer 10 and Ulysses.

Table 18.1. *Plasma populations in the magnetosphere*

		Density (cm^{-3})	Temp. (eV)	Temp. (°K)
Solar wind:	Solar wind	10^{-1}	10 eV	10^5
Ubiquitous:	Energetic particles	–	≥ 1 MeV	–
Outer magnetosphere)	} Super–hot plasma	10^{-2} to 10^{-3}	30–40 keV	3–4×10^8
Middle magnetosphere (plasma sheet)	} Hot plasma	10^{-1} to 10^{-3}	1–3 keV	1–3×10^7
	Outer edge	10^{-2} to 1	300–800 eV	3–8×10^6
	Densest sheet	1–10	10–100 eV	1–10×10^5
Io torus	Warm outer torus	10^3		
	Electrons		5–10 eV	5–10×10^4
	Ions		30–50 eV	30–50×10^4
	Warm middle torus	3×10^3		
	Electrons		2–3 eV	2–3×10^4
	Ions		7–50 eV	7–50×10^4
	Cold inner torus	10^3		
	Electrons		1 eV	1×10^4
	Ions		1–5 eV	1–5×10^4

This is a summary of results from several sources, mainly Voyager experiments, and is intended only to indicate the order of magnitude of the plasma parameters. *Column 3:* Density of electrons or ions (cm^{-3}). *Column 4:* Typical energy of electrons and/or ions. The unit of energy is the electron volt (eV), which is the energy gained by an electron falling through a potential drop of 1 volt. Most of the populations have a thermal distribution and the table lists the characteristic energy which defines the temperature (*column 5*).

(Fig.18.2 and Table 18.1). The outer zone (beyond $\approx$30–70 R_J) is the most rarefied and the structure is variable; it was extensive during the Pioneer encounters but had mostly collapsed during the Voyager encounters. Field lines here are distorted, and as the planet rotates, they are stretched out far down the magnetotail on the anti-sun side, so plasma can escape down the tail. The middle zone is distinguished by an equatorial sheet of plasma that is largely corotating with the magnetic field. This also constitutes a circling sheet of electric current, which stretches and flattens the magnetic field here. Its inner boundary and its main source is a comparatively dense ring of plasma around the orbital distance of Io (5.9 R_J) – the Io plasma torus. Here, both heavy ions (largely of sulphur and oxygen), and clouds of neutral atoms (notably sodium), are supplied to the magnetosphere from Io (section 18.3). From 5.5 R_J to 1.5 R_J is the inner zone, dominated by the planet's intrinsic magnetic field, where total plasma density decreases but the highest-energy particles reach their peak intensities.

Towards the poles, some magnetic field lines may merge with the interplanetary field, leaving the magnetosphere open to the solar wind in a 'polar cap' or 'polar cusp' (Fig. 18.1). These configurations are a feature of Earth's magnetosphere. This region was first probed by Ulysses, which reached its highest magnetic latitudes inbound at 15 R_J, 34°N, and 9 R_J, 49°N. At each of these points, the ubiquitous hot plasma and energetic ions of the magnetosphere suddenly disappeared and conditions resembled the solar wind.

Even larger than the magnetosphere, though unconfined, is a rarefied nebula of sodium atoms that envelops the entire system (M. Mendillo and colleagues, *1990, 1992*). It can be imaged from Earth by its very faint orange sodium line emission , which extends to $\approx$400 R_J ($\approx$30 M km, 2.6° on sky) from the planet (Fig. 18.8). It is believed to consist of sodium atoms arising in the Io plasma torus from Na$^+$ ions or molecular ions which are neutralised by collisions (see section 18.3). Upon neutralisation, they are released from the torus at corotational speed ($\approx$74 km/s) which is much greater than local jovian escape velocity. The same process with the more abundant sulphur and oxygen species could be a major contributor to populating the outer magnetosphere.

Plasma composition, mass budget, and energy budget

Jupiter's magnetosphere is so enormous because it is inflated by hot plasma, and also (at $\approx$20–50 R_J) by the centrifugal force of the cooler corotating plasma sheet. The hot plasma keeps it expanded at high latitudes even though the density there is exceedingly low. There are also even more energetic particles which do not constitute a thermal population, such as electrons at energies greater than 30 million electron volts (>30 MeV), and ions at >7 MeV, speeding all over the magnetosphere (Table 18.1). Electrons with this energy travel at a substantial fraction of the speed of light.

The high-energy ions ($\geq$0.7 MeV) are mainly of hydrogen (H; protons), helium (He), oxygen (O), and sulphur (S). The apparent

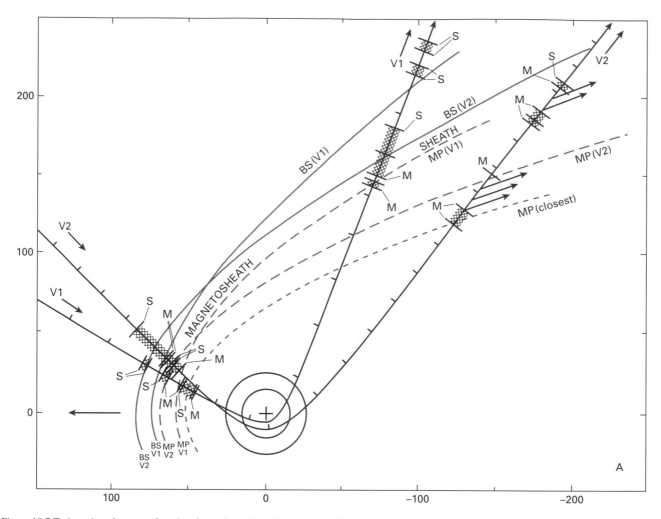

Figure 18.7 Trajectories of spacecraft – view from above the orbital plane, relative to the sun line, showing the fluctuating boundaries of the magnetosphere. Distances are in R_J relative to Jupiter (marked +). Circles are the orbits of Ganymede and Callisto. Parabolic or hyperbolic curves are models of the mean positions of the bow shock and magnetopause during each encounter; the innermost curve (dotted, same on both panels) models the magnetopause at its most compressed.
(A) Voyagers 1 and 2. S, bow shock crossing; M, magnetopause crossing; shading indicates periods in the magnetosheath; arrows on the Voyager 2 outbound track show the flow of ions down the tail. Voyager 2 may have re-entered the magnetosheath briefly even further downstream than shown here. The tracks are shown projected onto the orbital plane of Jupiter; they

did not stray far from this plane. Tick marks are put at 2-day intervals. Adapted from Bridge *et al.*(1979) with further details from Ness *et al.*(1979) and Krimigis *et al.*(1981).
(B) Pioneers 10 and 11 and Ulysses. During these encounters, the magnetosphere was greatly inflated. Open symbols show bow shock crossings; dark bars show magnetopause crossings; shading indicates periods in the magnetosheath. Each trajectory is shown in its own plane, aligned on the Jupiter-sun line. (If they were projected on the orbital plane of Jupiter instead, the only significant differences would be in the two lines shown in bold, outbound at high latitudes; for Pioneer 11, the track projected on the orbital plane would be the dashed line; for Ulysses, it would lie beneath the track shown.) Adapted from Mihalov *et al.*(1975) and Bame *et al.*(1992).

fluxes are dominated by protons, with much less oxygen and sulphur (Fig. 18.9), but this may be an artefact of the different energy distributions of ions with different masses and charges. It is estimated that the actual density of protons is only ≈ 15 times that of oxygen and sulphur ions, so the masses of each species are similar. This ratio is maintained throughout the outer and middle magnetosphere. The helium and carbon ions probably come from the solar wind, consistent with evidence that the carbon is totally ionised. They have greater relative abundance in the rarefied high-latitude regions probed by Ulysses. Conversely oxygen and sulphur come from Io; they are relatively more abundant closer in, and are only partially ionised (mainly O^{2+} and S^{3+}). Hydrogen ions are the most abundant in the solar wind, but here they most likely come from Jupiter's ionosphere, as do H_2^+ and H_3^+ ions which have also been identified.

The composition of the denser, cooler plasma sheet is even more dominated by sulphur and oxygen from Io (Fig. 18.10).

Thus, most of the particles of the magnetosphere come ultimately from Io. The total mass of the Io cloud and torus is estimated to be about a million tonnes. The rate of supply is at least one tonne per second, from Io (mostly sulphur and oxygen with some sodium). The magnetosphere also receives ≥ 20 kg/s from Jupiter's ionosphere (mostly hydrogen) and perhaps something from the solar wind (much less than 100 kg/s). Io and its neutral atomic cloud supply the Io plasma torus (section 18.3), which diffuses outward to form the plasma sheet (section 18.2). The composition of the hot and high-energy plasmas indicates that they too come from Io, though the route is uncertain. One source is neutral atoms that are ejected from the Io torus (see above); another is the

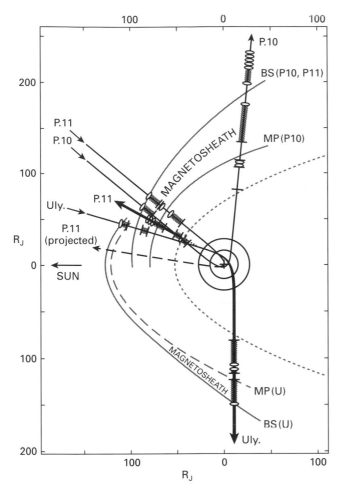

Fig. 18.7B. For legend see opposite.

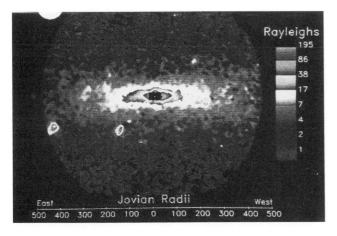

Figure 18.8. The extended sodium nebula of Jupiter: Earth-based image taken with a 589-nm filter, spanning 6° on the sky. (Image by M. Mendillo, Boston University; from Mendillo *et al.*, *1990*).

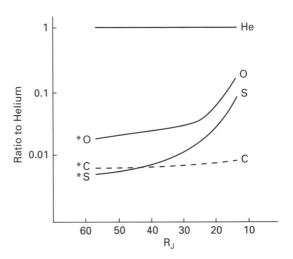

Figure 18.9. Ion composition versus distance from Jupiter: energetic (≈1 MeV) ions, from Voyager 2 (LECP experiment). This graph shows the apparent count rates, relative to helium. Hydrogen ions (not shown) were ≈200–6000 times more frequent than helium. However, the count rates are affected by the ion masses and energies, and the true abundances of O, S, He, and H ions are probably similar (see text). O and S come from Io while C and He come from the solar wind. Asterisks show solar wind levels during solar flares. From data in Krimigis *et al.*(*1979*); see also Hamilton *et al.*(*1981*).

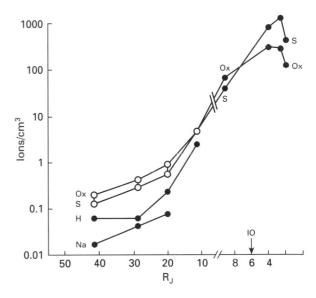

Figure 18.10. Ion composition versus distance from Jupiter: thermal ions, from Voyager 2 (plasma science experiment). This shows estimates of true densities of ions, based on a 'constant thermal speed' model at ≥20 R_J and an 'isothermal' model at ⩽12 R_J. Oxygen and sulphur were not clearly distinguished in the data at ⩾20 R_J, and helium was not measured. Oxygen is labelled Ox for clarity. (From data in Bagenal and Sullivan, *1981*.)

streams of protons and electrons detected at high latitudes by Ulysses (see section 18.2).

The energy of the magnetosphere is mainly supplied by Jupiter's rotation (Fig. 18.11). Ions and electrons are initially energised as they are spun up to corotation by the magnetic field, into the Io torus (section 18.3), and the field maintains them close to corotation speed as they diffuse out to the outer magnetosphere. Further

heating of the plasma torus comes from the very hot plasma that bathes the whole magnetosphere. How this vast reservoir gets its energy, in the outer magnetosphere, is still undetermined. One theory is that ions and electrons in the outer magnetosphere are energised by 'magnetic pumping', due to repeated cycling through the asymmetrically compressed magnetosphere on the day-side. Other mechanisms may also be active.

The magnetosphere is very dynamic. Firstly, being tilted, it wobbles wildly as it is whirled around by the planet's rotation. So it oscil-

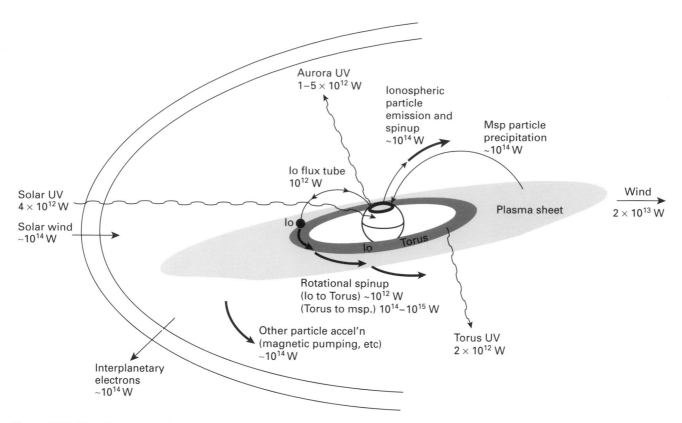

Figure 18.11. Towards an energy budget for the magnetosphere. Bold arrows are the major energy inputs from the rotation of the planet via its magnetic field. ('Msp', middle and outer magnetosphere.) Adapted from Fig.28 of Krimigis et al.(1981) and Table 10.2 of Dessler (1983).

lates and twists as viewed from any one point, be it the sun, or Io, or a spacecraft. Secondly, the tenuous outer fringe is buffeted by gusts in the solar wind; it pulsates and flutters on a timescale of hours. Longterm variations in the solar wind may affect the overall form of the magnetosphere. Thirdly, there are large changes in the Io plasma torus from year to year, perhaps due to activity in Io's volcanoes; these changes may affect the supply of plasma to all other regions.

There are still aspects of magnetosphere structure that have not been resolved. Some researchers believe that the surface asymmetries of Jupiter's magnetic field – not just its tilt – have effects all through the magnetosphere. Thus an 'active sector' covering λ_3 170–320 appears to have widespread effects, ranging from enhanced brightness of the Io torus, to enhanced emission of interplanetary electrons and a bulge in the bow shock. Indeed, Ulysses confirmed a Pioneer finding that the spectrum of energetic electrons has a synchronous 10–hour period all over the magnetosphere, and even outside it, being least energetic when λ_3 270 is facing the sun. These effects still cannot be fully distinguished from the effects of the magnetosphere tilt, and the matter remains controversial.

Radio emissions[3]

Jupiter is one of the brightest radio sources in the sky, especially at wavelengths between 10 and 500 metres (30 – 0.6 MHz). At these wavelengths, it emits clusters of intense bursts lasting a matter of

seconds to minutes, often grouped in 'storms' lasting 1–2 hours. This radiation was first recorded by C.A. Shain in Australia in 1950, although its origin from Jupiter was not recognised until the studies of B.F. Burke and F.L. Franklin (1955). Observers soon discovered that the radio bursts tended to occur when certain jovian longitudes were near the central meridian, but it took several years of study to show that the sources did not correspond to any visible features on the planet. The rotation period, now called System III, is in fact that of the magnetic field. But the origin of the radio bursts remained a mystery until E.K. Bigg, in 1964, realised that they were synchronised to particular orbital positions of Io. The bursts are just one aspect of the complicated interactions between Io and the magnetosphere that we will discuss in section 18.4.

But these 'decametric' emissions (defined thus by their wavelength) are only one of four quite distinct classes of radio emission from Jupiter. There are also decimetric, millimetric, and kilometric emissions, due to entirely different mechanisms (Figs. 18.12 & 13; Table 18.2).

Decimetric emission, discovered in 1958, is detected at wavelengths between 10 cm and several metres. It comes from the inner magnetosphere, between 1.3 and 3 R_J (Fig. 18.14). Early scans showed it peaking at about 2 R_J from the centre of the planet – that is, very close to the dust ring at 1.8 R_J (Chapter 19). But the best images, such as Fig. 18.14, show the strongest emission at 1.5 R_J with a sharp inner edge at only 1.3 R_J – that is, confined within the dust ring, in a region never explored by spacecraft. This is the highest-energy part of the radiation belts.

[3] Reviewed by Carr et al.(1983), and by de Pater (1990), and in several chapters in the book by Belton, West and Rahe (1989).

Table 18.2. *Radio emissions from Jupiter*

Class	Main wavelengths	Main frequencies	Timecourse	Av. total power	Notes
[Infrared] and millimetric	[4 μm]–10 cm	[10^5] – 3 GHz	Steady	[8×10^{17} W]	Thermal, from atmosphere
Decimetric	0.1 – 3 m	3 – 0.1 GHz	Almost steady	2×10^9 W	Synchrotron, from radiation belts
Decametric	10 – 500 m	30 – 0.6 MHz	Bursts	10^{10}–10^{11} W	Triggered by Io
Kilometric:					
(broad-band)	0.3 – 15 km	1000 – 20 kHz	Bursts	10^8–10^9 W	From auroral ionosphere
(narrow-band)	2 – 5 km	140 – 60 kHz	Intermittent	10^8 W	$P > 10$ hrs; from outer Io torus?

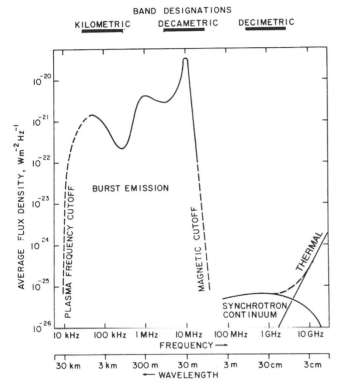

Figure 18.12. Radio spectrum of Jupiter, time-averaged. (Individual decametric bursts can be 1–2 orders of magnitude above the curve.) Adapted from Fig.7.1 of Carr *et al.*(*1983*).

Long before spacecraft visited Jupiter, the polarisation patterns showed that these radio waves are 'synchrotron radiation', emitted by electrons of 10–20 MeV as they spiral around the magnetic field lines. The emission radiates most strongly parallel to the magnetic equator. Therefore its apparent strength varies with the tilt of the magnetic field, both around the planet's 10-hour rotation and around its 12-year orbit. After correction for these effects, the intrinsic intensity has shown only a small variation on a timescale of ≈5 years (Fig. 18.15). The main fluctuation has been a 25% dip lasting from 1971 to 1974. This was accompanied by an apparent reduction in the tilt of the field from 10° to 9°, and a change in longitudinal asymmetry, and a slight contraction of the radiation belts. The event has been interpreted as a thinning of the outer part of the radiation belts, such that the remaining emission came from the more complex field region closer to the planet. It is also possible that the details of the magnetic field structure changed.

Millimetric emission is part of the thermal spectrum of Jupiter's atmosphere (Chapter 4.3; Figs. 4.13 and 4.21), detected at wavelengths up to about 10 cm.

At the other extreme is the *kilometric* emission, which can only be observed from space. Earth-orbiting satellites have observed similar kilometric radiation from Earth's magnetosphere, but Jupiter's was not discovered until Voyager flew its 'planetary radio astronomy' antennae into the system. Voyager revealed two components, which were reobserved by Ulysses.

The main emission, first detected several months before the Voyager encounter, was *'broad-band kilometric'* and occurred in bursts when the north magnetic pole was tilted towards the spacecraft. It is believed to originate at a spot just above the northern auroral ionosphere, fixed in local time on the dayside. There is also less frequent, much weaker emission from the south magnetic polar region.

Weaker emissions with wavelengths close to 3 km, *'narrowband kilometric'*, occurred in episodes lasting minutes to hours within which the intensity and wavelength were often quite steady, although sometimes the wavelength drifted. Remarkably, these episodes tended to recur with periods greater than 10 hours (10.25 hr for Voyager 1, 10.47 hr for Voyager 2). The most likely origin is in the outer part of the Io plasma torus at 8–9 R_J, where the plasma is indeed not quite corotating (see section 18.3). This radiation may arise from plasma oscillations of the same frequency, which were detected there as Voyager passed through, but the actual mechanism is still unknown. Ulysses observed similar emission, coming from five distinct sources spread around the Io torus at ≈8–10 R_J; they persisted over several jovian rotations and again had periods above 10 hours, lagging by 3–8% behind corotation speed.

At even lower frequencies, electromagnetic waves are coupled to plasma oscillations and comprise *'plasma waves'*. Many forms of these are locally confined. Voyager and Ulysses recorded rich polyphonies of plasma waves as they cruised through the magnetosphere.

18.2 COMPONENTS OF THE MAGNETOSPHERE

The outer zone and the tail

As spacecraft approach Jupiter, they encounter energetic particles from the planet long before crossing into the magnetosphere.

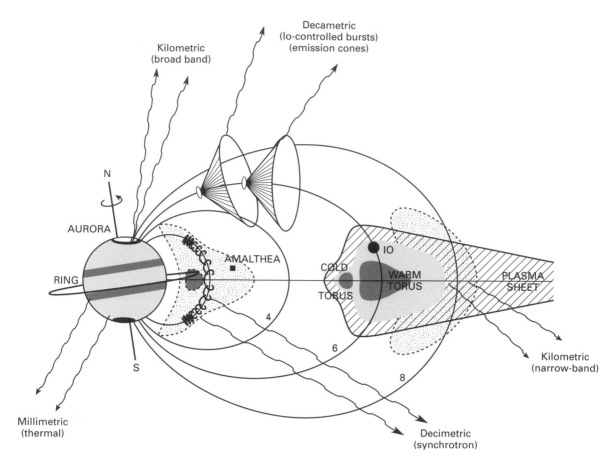

Figure 18.13. Scale diagram of Jupiter's inner magnetosphere, and origins of the radio emissions. Magnetic field lines are drawn at 2, 4, 6, and 8 R_J. Stippled areas are concentrations of energetic particles. The inner one (around 2 R_J) is the source of synchrotron radio emission. The outer one (around 7–8 R_J) is the peak of energetic ions around the plasma sheet; from here, ions and electrons plunge down the magnetic field lines into the auroral regions, and excite ionospheric electrons which soar up in the opposite direction.

Indeed, trains of such particles even reach the vicinity of Earth, where they have been recorded as 'anomalous cosmic ray events' – electrons detected by the Earth-orbiting Interplanetary Monitoring Platforms. Within 200 M km of Jupiter, jovian electrons dominate the interplanetary flux in the 3–6 MeV range. They are accompanied by ions, also streaming away from Jupiter, including heavy ions such as sulphur which are typical of Jupiter rather than the solar wind. The bursts last from a few minutes to 1–2 days, and the energies range from 30 keV to 30 MeV. They come from the outer zone of the magnetosphere, somehow being released across the magnetopause and bow shock.

Pioneers 10 and 11 both crossed the bow shock at 109–110 R_J and the magnetopause at 96–97 R_J, but then re-crossed one or both boundaries several times. Evidently the magnetosphere was pulsating over a matter of hours (Fig. 18.7). The Voyagers found the magnetosphere rather more compact, first crossing the bow shock at 86–99 R_J and the magnetopause at 67–72 R_J; but again there were multiple re-crossings and Voyager 1's final magnetopause crossing, like Pioneer 10's, was at 47 R_J. As Voyager 1 flew in, Voyager 2 was measuring the solar wind pressure upstream, and the major solar-wind gusts coincided with the transient compressions of the magnetosphere that twice caused Voyager 1 to re-enter the solar wind. Ulysses entered during a prolonged lull in the solar

wind, and its one traversal of the bowshock was further out than ever before, at 113 R_J; even then the bowshock was moving outwards, and the magnetopause was crossed at 110 R_J.

The field in the outer magnetosphere is weak, complex and variable on timescales less than an hour, especially near the magnetopause. On the sunward side, it is still closed, but when it is largest and weakest (about 5 nanoteslas or 5×10^{-5} gauss), this field is largely induced by fluctuating magnetospheric currents.

The plasma here is very rarefied and hot: just within the magnetopause, it is the hottest thermal population known anywhere in the solar system (including the sun), at 300–400 *million* °K! However, these particles are extremely sparse, only 10^{-2}–10^{-3} particle cm^{-3} (Table 18.1). It is the pressure of this super-hot plasma, not the jovian magnetic field itself, that holds off the solar wind. This plasma may be the source of the particles detected far outside the magnetosphere; one possibility is that 'splashing' of the boundary can release large packets of super-hot plasma and allow the outer magnetosphere to collapse like a punctured tyre, until it is refilled by more hot plasma from within.

This super-hot plasma has a characteristic particle energy of 30–40 keV, but this is low-energy in particle physics terms! There are also the sparser but much more energetic particles, from 1 to 30 MeV, which permeate the entire magnetosphere. The plasma is

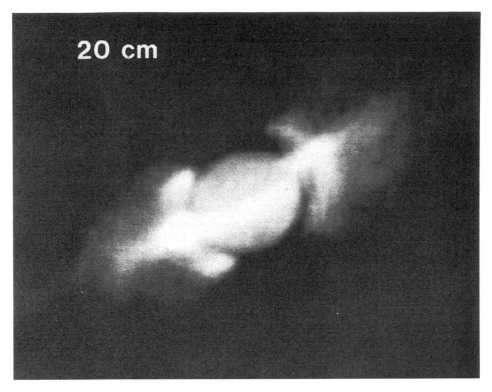

Figure 18.14. Image of Jupiter at 20.5 cm wavelength, from the Very Large Array, showing synchrotron emission from the inner radiation belts. From de Pater and Dickel (*1986*) and de Pater (*1986*).

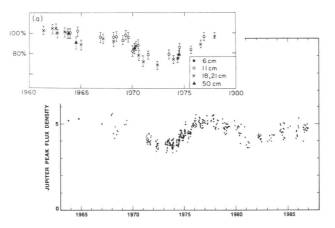

Figure 18.15. Variations in the decimetric radio emission. The intensity has been corrected for varying distance and tilt. (**Top**) At 6–50 cm, relative to the intensity in 1964. (From de Pater, *1981*.) (**Bottom**) At 11–13 cm; intensity in Janskys, reduced to 4.04 astronomical units. (By courtesy of M.J. Klein; adapted from Klein *et al.*(*1989*).

not always uniformly mixed: thus Voyager 1 passed through a zone especially rich in sulphur and oxygen at 48 R_J, which was not present in Voyager 2's data.

The outer boundary of the middle magnetosphere on the sunward side is complex and variable. Both Pioneer and Ulysses found that the outer magnetosphere, which was inflated at those times, was far from fully corotating. But the super-hot plasma observed by Voyager was close to corotating, right to the edge of the magnetosphere; the inflated outer zone had collapsed.

As Ulysses flew outwards at high latitudes, it became the first spacecraft to sample the dusk-side magnetosphere. Here, it detected narrow streams of energetic ions and electrons, travelling along the magnetic field lines (Fig. 18.1). The electrons were streaming away from the planet and the ions towards it; these ions may contribute to high-latitude aurorae. However, there were also streams of energetic protons streaming away from the planet – probably from the ionosphere in the auroral regions, accelerated upwards by an electric potential. They may be the long-sought source of the hot plasma of the outer magnetosphere. The electron streams likewise may be a source of the relativistic electrons observed in interplanetary space.

On their outbound trajectories, the spacecraft recorded even

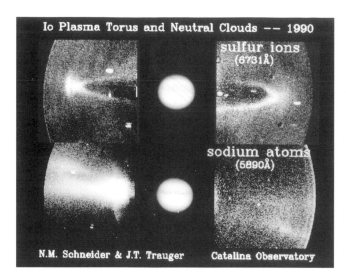

Figure 18.21. Images of the Io plasma torus (top) and sodium cloud (bottom), on 1990 Jan.11. They were taken through narrow-band filters at 673 nm (for S⁺ ions) and at 589 nm (for sodium atoms). Jupiter's image is recorded on the same exposure through a strip of filter with 0.01% transmission. Note that the plasma torus is a ring tilted to the rotational equator, while the sodium cloud is concentrated around Io but has faint jets. These images (and Figs. 18.22,24,25) were taken by Nick Schneider (LASP, University of Colorado) and John Trauger (JPL) with a CCD on the 1.5 metre Catalina Observatory telescope of the University of Arizona. (By courtesy of Nick Schneider.)

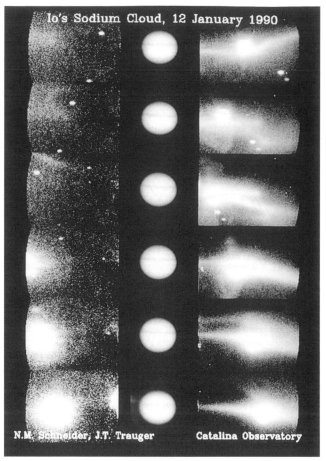

Figure 18.22. Images of the Io sodium cloud, covering about a quarter of Io's orbit on 1990 Jan.12. Io is the brilliant spot on the right; another satellite intrudes on the left in the lower images. The images show the stable slow-sodium 'banana' which always lies in the orbital plane near and preceding Io, and a rapidly changing fast-sodium jet which is directed out of the orbital plane. (Credits as in Fig. 18.21.)

The innermost part of the sodium cloud is normally masked by Io's brilliance, but it was recorded as an absorption band in spectra of Europa when Io eclipsed it in 1985 (Nicholas Schneider and colleagues, *1987, 1991b*). Measurements down to 700 km above Io indicated a very thin sodium 'atmosphere' with the surprisingly high temperature of ≈2000°K, and would extrapolate to a surface density of 8000 sodium atoms/cm³ (although the actual surface conditions were not determined).

The inner part of the sodium cloud, for a radius of 10⁴ km around Io, had an almost constant brightness from 1974 to 1985. This argues against the eruptive plumes as the immediate source of the sodium. (This constancy, to within a factor of two, contrasts with tenfold variation in S⁺ concentration in the plasma torus over the same time interval; see below.)

The sodium cloud is limited to the vicinity of Io, and is banana-shaped (Fig. 18.20). This is because the atoms are orbiting; those that diffuse inwards orbit faster, and those that diffuse outwards orbit slower. The cloud is limited by ionisation, the atoms having a lifetime of up to 20 hours. Impacts by the corotating plasma, overtaking the cloud at 57 km/s, split the atoms into ions and electrons, which are thus recruited to that same plasma torus. The sodium cloud is most extended on the leading (inner) side, where the corotating plasma is cool. On the trailing (outer) side it is ionised more quickly as the plasma is hotter. As viewed from Earth, the north and south sides are alternately brighter, in synchrony with Jupiter's rotation, because the tilted plasma torus bears down on each side in turn.

The outer fringes of the sodium cloud are highly variable and show long 'jets' and 'sprays', pointing forward along Io's orbit and out of plane. Sodium atoms are ejected at high speeds (10–100

km/s), mainly when the plasma torus intersects the cloud (twice per rotation). These fast atoms may be the visible jets. The jets and sprays were best shown in high-resolution images of the sodium cloud by Schneider and colleagues (*1991a*, and Fig. 18.22). They interpreted them as the tracks of sodium-containing ionised molecules (e.g. NaS⁺), arising by unknown means from Io, accelerating and oscillating in the magnetic field, and releasing neutral sodium at corotational speeds after ≈8–13 hours. This process is proposed as the origin of the extended sodium nebula of Jupiter (section 18.1).

Although the sodium emission is the brightest, this does not imply that sodium is the most abundant element in the cloud; it is just the most efficient emitter. Emission has also been detected from neutral atoms of potassium (767 + 770 nm), oxygen (630 nm), and sulphur (130 + 143 nm). It is not known how far they extend. Estimated densities are ≈6 atoms cm⁻³ for sulphur and ≈30–40 atoms cm⁻³ for oxygen,[4] consistent with an origin from SO_2.

---

[4] Smith & Strobel (*1985*); Shemansky (*1987*).

Ionisation and acceleration

The likely conclusion is that the major ions of the plasma torus (S and O) arise from the neutral cloud like sodium, by impact of electrons or ions already in the torus, although some may be formed directly by impact of electrons on Io's surface or atmosphere.

The new ions and electrons, being electrically charged, are immediately subject to Jupiter's corotating magnetic field. It does not simply drag them along with it. Rather, the electric field that it induces gives them a radial acceleration, so they begin circling around the magnetic field lines with a speed comparable to the corotation speed (57 km/s relative to Io). This gyrational energy, which comes from Jupiter's rotation, is the first source of heating in the torus. It amounts to ≈ 0.25–0.6 eV for electrons and several hundred eV for ions. (The actual ion energy in the torus is ≈ 40 eV, so the ions either fail to attain, or rapidly lose, their full gyrational energy.)

Later, collisions allow the ions and electrons to diffuse, mainly outwards due to centrifugal force. As they do so, the magnetic field continues to accelerate them towards maintaining corotation. Further heating of the plasma torus comes from the surrounding hot plasma.

The Io plasma torus[5]

The plasma torus encircles Jupiter along the centrifugal equator between 5 and 8 R_J, straddling the orbit of Io, and corotating with the magnetic field (Figs. 18.13,20,21).

It was discovered by means of its ultraviolet emission by Pioneer 10 in 1973, but was wrongly assumed (and announced) to be a ring of hydrogen. In fact, the emission must have been from heavier ions, which became apparent when optical emission from S^+ ions was observed from Earth in 1976, and when the full extent of the torus was revealed by Voyager in 1979. In any case, the torus viewed by the Pioneers was very weak by later standards – only 4% as bright as the torus viewed by Voyager 1. It had an average electron density of only ≈ 500 cm^{-3}. At some longitudes the emission was absent altogether. At the time of the Pioneer encounters, the inner radiation belts were smaller than in the Voyager years (section 18.1), and this may have been a coordinated change.

Thus the discovery of the Io torus was one of the major results of the Voyager 1 encounter. The UVS found emission from S^{2+}, S^{3+}, and O^{2+} ions (Fig. 18.23). During the Voyager 1 encounter, the average electron density was 2000 cm^{-3}, and the temperature about 10^5 °K. During the Voyager 2 encounter 4 months later, the UV emission was twice as bright, indicating that the average electron density had increased to 3000 cm^{-3} and the temperature had dropped by 30%. This enhancement was confirmed in 1981 by ground-based observations of S^+ and O^+ optical emissions.

The ultraviolet emissions have been observed not only by Pioneer and Voyager but also, over the subsequent years, by the International Ultraviolet Explorer satellite in Earth orbit. It has seen only minor variations since Voyager. There are also emissions in the red and near-infrared which can be detected by sensitive Earth-based imaging (Figs. 18.24&25).

[5] In addition to the references at the start of this chapter, there are recent reviews on the torus by Morgan (*1985*), and Strobel (*1989*) and Bagenal (*1989*).

The torus has been penetrated by three spacecraft: Pioneer 10, Voyager 1, and Ulysses. These sampled the electrons and ions of the torus directly. The Pioneer instruments gave ambiguous results for these 'low-energy' ions, which were misinterpreted as protons, but the results were actually consistent with corotating sulphur and oxygen ions as later sampled by Voyager – peaking in the torus near 6 R_J, but still present as far in as 2.9 R_J. The overall profile of the torus plasma, as measured by Voyager 1, is in Fig. 18.26. On this profile, the ion charge density is compared with the electron densities deduced from radio observations; the close agreement indicates that the torus consists mainly of the observed heavy ions, with few hydrogen or helium ions.

The conclusion of all these studies is that there are three concentric tori, as follows.

(i) Cold inner torus (5.0–5.6 R_J, inside Io's orbit)

This has been called a backwater, as it only receives about 2% of the plasma from Io, and did not produce UV emission detectable by Voyager, but it produces red-light emission detectable from Earth (Figs. 18.24&25). It consists of ions diffusing inwards against centrifugal force. They diffuse slowly, having time to radiate away much of their energy. As the plasma cools it collapses towards the equator, which enhances the efficiency of radiation and thus of cooling. This positive feedback makes the cold inner torus sharply defined.

It contains mainly S^+ and O^+, with much less S^{2+} and O^{2+}, and also includes SO_2^+ ions (Table 18.3, columns 1&2). The electron temperature is estimated at ≈ 1 eV (10^4 °K) (Table 18.1).

(ii) The torus peak (5.7 R_J, just inside Io's orbit)

The watershed between the cold inner torus and the warm outer

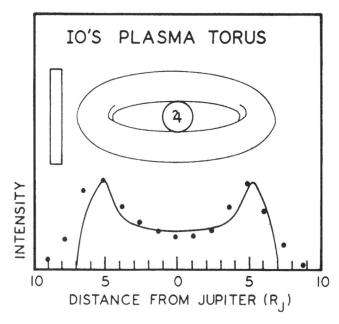

Figure 18.23. The torus UV emission as viewed by Voyager 1. The data points are a UVS scan at 68.5 nm, the wavelength of the brightest emission line, a mixture of S^{2+} and O^{2+} emission. The smooth line indicates the emission that would be seen from the model torus drawn to scale above. The vertical box represents the UVS field of view. (From Broadfoot *et al.,1979*.)

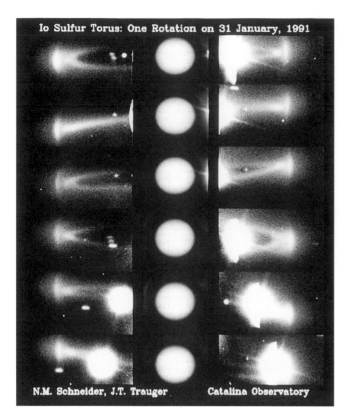

Io Sulfur Torus: One Rotation on 31 January, 1991

N.M. Schneider, J.T. Trauger   Catalina Observatory

Table 18.3. *Ion composition of the Io plasma torus*

Experiment:	PLS	PLS	PLS*	UVS(1)	UVS(2)
Distance:	$5.0\,R_J$	$5.3\,R_J$	$6.0\,R_J$	$6\,R_J$	$6\,R_J$
O^{2+}	48	26	160	40–80	14
S^{3+}	<6	<4	27	10–20	42
S^{2+}	14	39	430	420	370
O^+	250	350	130	660	850
S^+	91	1100	430	<350	260
Na^+	<21	<72			
SO_2^+	3.5	13	73		
Total ions	431	1630	1300		
Electrons	500	1700	1900	2000	2000

Densities are given in ions cm^{-3}, from the Voyager 1 plasma science experiment (PLS) and ultraviolet spectrometer (UVS). (O, oxygen; S, sulphur; Na, sodium.)

The PLS values are from Bagenal and Sullivan (*1981*). S^{2+} and O^+ could not be distinguished by this experiment. The values quoted are for an isothermal model. *The PLS values at 6.0 R_J (in the warm torus) are very uncertain because of thermal broadening of the spectral peaks.

The UVS estimates are greatly revised from the values given in early Voyager publications: (1) analysed by Smith and Strobel (*1985*); (2) analysed by Shemansky (*1987*). From the UVS, ion composition seems to be constant throughout the warm outer torus from 5.9 to 7.5 R_J.

Figure 18.24. Images of the Io plasma torus, in the light of S^+ (673 nm), taken as in Fig. 18.21, over one rotation of Jupiter on 1991 Jan. 31. Brilliant spots are satellites crossing the field. The tilt of the plasma torus relative to the rotational equator is obvious. The plasma torus is also slightly warped (e.g. first two panels). The vertical line or 'ribbon' visible at the ansae represents the warm middle torus and clearly varies with longitude. (Credits as in Fig. 18.21.)

torus, at 5.7 R_J, was very sharp in the Voyager 1 PLS data. It coincided with the peak ion density (peak charge density 3000–3500 cm^{-3}). It is slightly inside Io's orbit because of the orbit of the neutral clouds which are its source.

Earth-based imaging since Voyager has revealed a distinct 'warm middle torus' at the boundary, at 5.8 R_J (J. Trauger, *1984*)(the vertical ribbon in Fig. 18.24). This was slightly denser but cooler than the peak observed by Voyager 1, with an electron density 3000–4000 cm^{-3} (estimated energy 2–3 eV), and S^{2+} and S^{3+} at 500–800 cm^{-3} (7–50 eV). It fluctuates from year to year: Fig. 18.25 shows it present in 1988 and 1991 but absent in 1990 in the same longitude range.

(iii) Warm outer torus (5.9–7.5 R_J, at and outside Io's orbit)
This is the main pool of plasma, gradually diffusing outwards and gaining energy from the magnetic field and from the hot plasma that surrounds it. (This torus is often called 'hot', but the surrounding plasma is much hotter.)

The electrons mostly have a temperature of 5–10 eV (5–10×10⁴ °K). They are mixed with the ubiquitous hot plasma which has less than one hundredth of the density but a hundred times the temperature. The torus ions are at 10–100 eV with a mean of about 40 eV.

The warm outer torus typically emits about $3×10^{12}$ watts of radiation, mostly in the far-UV lines of S^{2+} and O^{2+} and/or O^+. This energy must be re-supplied with a timescale of only a few days. As

with the neutral sodium cloud, it is remarkable that the emission is so steady given such a short timescale for energy supply. The whole Io cloud-and-torus system is maintained in a fairly steady state, over months or years, by factors that are not yet understood.

The ion composition and densities are given in Table 18.3. Ion composition seems to be constant throughout the warm outer torus from 5.9 to 7.5 R_J. The Voyager 1 UVS estimates (columns 4 and 5) agree with the maps in Fig. 18.27. These densities are also consistent with the Voyager 1 *in-situ* plasma measurements (column 3 of Table 18.3).

The S^+ and S^{2+} emission, observed from Earth and from the Voyager UVS, shows longitudinal structure. The S^+ emission of the *inner* torus is enhanced in the 'active sector' ($\lambda_3 \approx 175–320$) where the magnetic field is weaker. This effect does not extend to the S^{2+} emission of the warm outer torus, but this does have some structure which is apparently not fixed in System III. The pattern is not completely understood; to complicate matters, the outer torus also varies severalfold in brightness from year to year. Patchiness in the outer torus was also revealed in the Ulysses flyby, both by the kilometric radio emissions from the torus (section 18.1) and by the distortions of the spacecraft's radio signal when passing through the torus. B.R. Sandel & A.J. Dessler (*1988*) showed that the S^{2+} emission could be drifting at the same rate as the narrow-band kilometric radio emissions and the sub-corotating plasma at 8–9 R_J; a rotation period of about 10h 13m could fit all these data. They designated this as 'System IV' (defined with a rotation of 845.05°/day) and suggested that it might represent differential rotation within the magnetised interior of Jupiter. However, it is not yet clear if there really is a single constant rotation period. Possibly the outer fringe of the torus is lagging simply because of the drag of outwardly flowing ions.

There is a distinct outer boundary to the plasma torus, where it

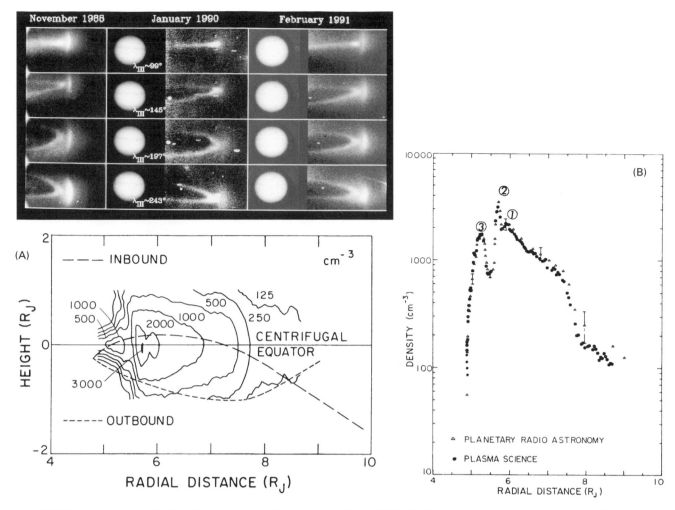

Figure 18.25 (top). Long-term variability of the plasma torus. These images in the light of S⁺ (673 nm) were taken as in Fig. 18.21 in three successive apparitions. (They are not intensity-calibrated, so only morphological comparisons can be made.) In the limited longitude range shown, the 'ribbon' of the warm middle torus was present in 1988 and 1991 but absent in 1990. (Credits as in Fig. 18.21.)

Figure 18.26. The Io torus: total plasma density (in units of proton charge per cm³) from Voyager 1's plasma science experiment. (A) Cross-sectional map. The spacecraft track is the dashed line, and contours elsewhere were extrapolated using a model of the magnetic field. (B) Measurements of ion charge density along the inbound trajectory, compared with estimates of electron charge density from Voyager's radio-astronomy experiment. (1) warm outer torus; (2) peak density; (3) cold inner torus. Io's orbit is at 5.9 R_J. (From Bagenal and Sullivan, *1981*.)

abuts the plasma sheet of the middle magnetosphere. Going from 7.5 to 9 R_J, the plasma density falls off steeply (Fig. 18.26), the temperature increases five-fold, and the ions lag behind corotation. Further out, they are spun up again closer to corotation speed.

18.4 THE IO FLUX TUBE AND RADIO BURSTS

The Io flux tube

As the plasma-filled magnetic field sweeps past Io, it induces an electric potential (about 400 kilovolts) across Io which drives an electric current, either through its body or (more likely) through its ionosphere. Plasma sweeping past Io tends to be entrained to the orbital speed of Io and accelerated north or south into this current, which becomes a so-called 'flux tube' of particles running along the magnetic field lines that intersect Io (Fig. 18.28). The flux tube is actually guided by a wave that propagates from Io along the field

lines, called an 'Alfvén wave'. On the side of the flux tube towards Jupiter, electrons stream away from Io towards the ionosphere, and ions stream in the opposite direction. On the side away from Jupiter, the directions are reversed.

Voyager 1 was targeted to fly through the flux tube as it passed 20 500 km from Io's south pole. Although the spacecraft failed to enter it (the tube apparently curved to one side), its presence was confirmed by an obvious magnetic field perturbation (Fig.18.29). This implied that the flux tube carried a current of 2.8 million amps.

The decametric radio bursts[6]

Finally, we can return to the decametric radio emissions, the amazingly intense bursts that first revealed the existence of Jupiter's magnetosphere.

[6] Recent major reviews are by Staelin (*1981*); Carr *et al.*(*1983*); Belcher (*1987*); Genova *et al.*(*1989*).

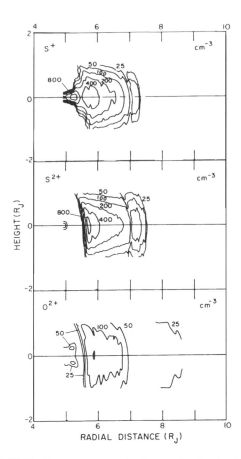

Figure 18.27. The Io torus: cross-sectional maps showing densities of S^+, S^{2+}, and O^{2+} ions, measured as in the previous figure. (From Bagenal and Sullivan, *1981*.)

These emissions span frequencies from 0.6 to 39 MHz. The bursts as recorded on Earth last only seconds, but this is due to interplanetary twinkling in the solar wind. The Voyager radio-astronomy detectors showed that the bursts actually last for 3–6 minutes and are separated by similar intervals. They tend to appear in storms often lasting an hour or two.

The bursts are controlled by the tilted magnetic field and by Io, and originate quite low over the jovian ionosphere. They mostly occur when the north magnetic pole ($\lambda_3 \approx 180$ at surface) is 30–80° preceding or following the central meridian. There are much weaker signals related to the south magnetic pole; given the asymmetry of the magnetic field (Fig. 18.4), the south magnetic pole is both weaker and less offset from the rotational axis. Within these intervals most of the bursts, especially at the higher frequencies, are synchronised with the orbital position of Io. When the frequency of bursts is plotted against the longitudes of System III and of Io (Fig. 18.30), these patterns show up as clusters which are usually (though inaccurately) called 'sources' (A, B, C, D, and A').

Clearly these are not true 'storms' nor 'sources'. Presumably the emission is liable to occur every time Io is passing through the appropriate System III longitudes, but it is beamed so that we only detect it from a certain angle. This was confirmed by Voyager: from whatever angle it was viewing, Voyager saw the same dependence on λ_3 and λ_{Io} as from Earth.

Voyager revolutionised our knowledge of the decametric bursts,

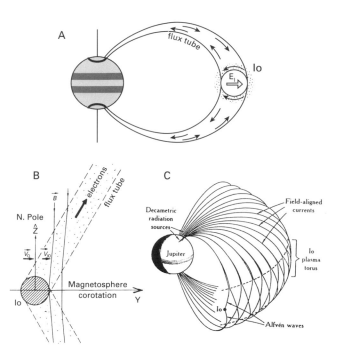

Figure 18.28. Theory of the Io flux tube. As the magnetosphere overtakes Io, it generates an electric potential of 400 kilovolts across the moon, which drives an electric current. Solid lines indicate magnetic field lines. (**A**) View along Io's orbit (not to scale). E_I indicates the induced electric potential. Arrows indicate the flow of electrons; ions flow the other way. (**B**) View towards the planet; Y points in the direction of magnetosphere rotation and Z points to the north pole. V_o represents the average velocity of plasma and V_{Io} its reduced velocity as it passes Io. Magnetic field lines (*B*) are bent as they pass Io. The propagating locus of this bend (the Alfvén wave) fills with accelerated electrons, comprising the flux tube. Adapted from Fig.9.6 in Dessler (*1983*). (**C**) Perspective view showing how the Alfvén waves are predicted to be reflected by the ionosphere to produce a standing-wave pattern downwind of Io. (From Gurnett and Goertz, *1981*.)

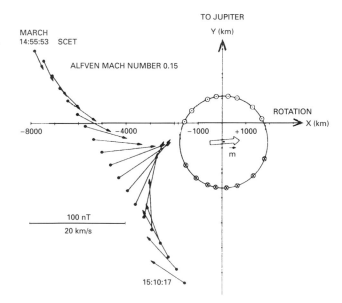

Figure 18.29. Observation of the Io flux tube. Voyager 1 trajectory is shown by dots; magnetic field vector measurements are shown by arrows. The field indicated the location of the flux tube shown by the circle, centred 5000 km off Voyager's track at 15h 06m. Plasma flow is from left to right. Io was 20 500 km above the plane of the diagram. The magnetic field perturbation was accompanied by sharp peaks in ion and electron fluxes, and a sharp dip in high-energy electrons (not shown). (From Acuña *et al.,1981*.)

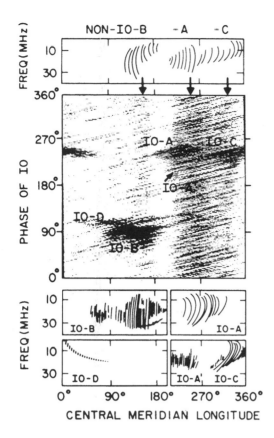

Figure 18.30. Relationship of decametric bursts to the magnetic field and to Io. Times of bursts at about 20 MHz are plotted against λ_3 (x-axis) and λ_{Io} (y-axis). Dense areas define the Io-dependent sources; vertical bands define the Io-independent sources. Panels above and below show typical spectral arc shapes for each source. (Diagram by J.R. Thieman published by Carr *et al.*,1983.)

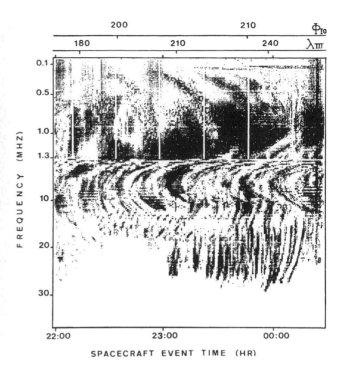

Figure 18.31. Typical spectrograms of decametric radio emissions from Voyager 1, near closest approach. In this frequency–versus–time plot, all the emission is resolved into arcs. At middle frequencies the bursts occur continuously ('lesser arcs'), but at high and low frequencies they occur in 'storms' ('greater arcs'). A storm from Io-dependent source A' occurred from 23h 05m to 00h 10m. The experiment's resolution was reduced below 1.3 MHz (upper section). (From Boischot *et al.,1981*.)

as it could view from different longitudes and latitudes, it was close enough to the source to receive a very strong signal, it was unperturbed by interplanetary or ionospheric distortions, and it covered low frequencies unobtainable from Earth. The major discovery was that all the bursts are in fact 'spectral arcs' (Fig. 18.31). That is, when frequency is plotted against time, each 'burst' either starts as a broad-band emission which separates into higher- and lower-frequency whistles, or vice versa. There are two classes of arcs, broad-band ('greater arcs') and narrow-band ('lesser arcs'). (Part of the arc structure was also resolved from Earth about the same time, by a prototype of the Voyager receiver on the French radio-antenna array at Nançay.) Each of the 'sources' (A,B,C,D,A') has a characteristic pattern of arcs, distinguished by their spectral shape (Fig. 18.30) and also by their polarisation. These properties, combined with the longitude patterns, show clearly where the emission of each 'source' originates.

The arcs must represent the true sources which are detected only when suitably pointed relative to the observer. According to a simple model proposed by the Voyager team, these sources are packets of electrons spiralling around particular magnetic field lines – probably in or close to the Io flux tube – emitting radiation in a series of conical shells centred on the field line (Fig. 18.13). The radiation could be synchrotron emission (from relativistic electrons moving in helical paths) or cyclotron emission (from lower-energy electrons moving in circular paths). At each point on the field line, the emission cone has a certain angle and a certain frequency, and the 'arc' is detected as the series of cones sweeps across the observer. Typical opening half-angles for the cones are 65–80°, corresponding to the preferred distance of the sources from the central meridian of the observer.

Cyclotron emission is the most likely mechanism, and the sharp upper-frequency cutoff at 39.5 MHz represents the cyclotron frequency in the strongest field encountered by these electrons: 14 gauss. This is identical to the strongest field present around Jupiter – at the north magnetic pole near the surface of the planet. A corollary is that these electrons cannot be stably trapped, as they would plunge right into the atmosphere if they were to travel to the weaker south magnetic pole.

The strongest two 'sources', A and B, are clearly in the northern hemisphere (Fig. 18.32). This is indicated by their high-frequency cutoff just mentioned; by their polarisation; by their relation to the λ_3 of the NMP; and by the fact that they are more frequently seen when the north rotational pole is tilted towards the observer. Bursts are seen when Io is approximately aligned with the north magnetic pole, close to the preceding or following limb – perhaps most pertinently, when the footprint of the Io flux tube on the auroral oval is near its lowest zenographic latitude. A and B clearly represent the same source region seen from opposite sides. In fact, for the best symmetry of A and B, the relevant field lines would be ≈15° ahead of Io in its orbit, that is, ≈15° downwind of Io in relation to the

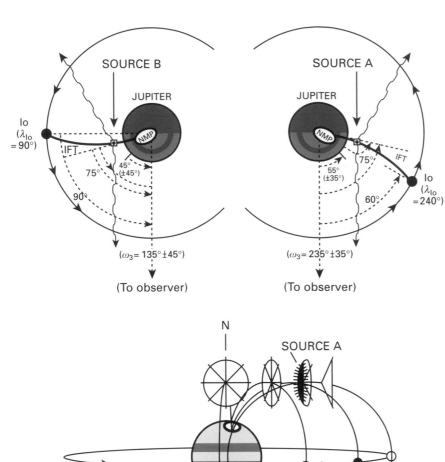

Figure 18.32. Configurations of the magnetic field and Io that produce decametric signals. (Radius of Jupiter exaggerated.) **(Top)** Views from above the north pole, in the configurations required for 'source B' and 'source A'. NMP, north magnetic pole; IFT, Io flux tube. The auroral region is shown black, and the IFT is postulated to end 15° of longitude ahead of Io, such that the emitting region for both 'sources' is ≈75° from the central meridian. Emission arises from conjunction of the IFT with the low-latitude auroral region. (Alternatively, emission may arise from field lines some way downwind of the IFT.) Wavy lines indicate decametric emission arising in a cone with opening half-angle 75°. **(Bottom)** View from Earth, showing the IFT at various longitudes. Emission cones with opening half-angles ≈65–80° intersect the observer only when they occur ≈65–80° preceding or following the central meridian. Each spectral arc arises along one field line, along which the emission varies in frequency and in cone angle.

corotating magnetosphere whirling past it. This is indeed very close to the footprint of the flux tube on the ionosphere, according to F. Genova and M.G. Aubier (*1985*).

Conversely the weaker source D and part of source C form a symmetrical pair arising from the southern hemisphere. This is indicated by their high-frequency cutoff ($\leq$32 MHz, consistent with the 10-gauss field of the SMP); by their polarisation; by their λ_3; and by the fact that they are more frequently seen when the south rotational pole is tilted towards the observer.

With this model as the starting point, various theories of the emission mechanism have been developed, but there is still no comprehensive understanding of it.

A complication is that the bursts may not arise from the Io flux tube itself, but from flux tubes some way downwind of it. Genova and Aubier (*1985*) calculated the likely twisting of the Io flux tube in the nonuniform planetary magnetic field, and concluded that although some bursts come from the north or south foot of the Io flux tube, most Io-dependent bursts may come from field lines ≈40° preceding (downwind of) the flux tube. It has been calculated that the Alfvén wave guiding the Io flux tube could be reflected multiple times from the ionosphere, producing a standing-wave pattern downwind of Io (Fig. 18.28C). This might provide a disturbance at the appropriate longitude to generate the bursts. Another unsolved

problem is the periodic pattern of bursts, which repeat at intervals of 3–6 minutes (2–4° longitude). Among various hypotheses, this behaviour has been attributed to the proposed standing Alfvén waves downstream of Io, or to diffraction of single radio bursts by plasma condensations.

The Io-dependent 'sources' also produce bursts that last only for milliseconds ('S-bursts'), with very narrow bandwidths. They typically sweep to lower frequencies, which shows that they come from electrons of 3 keV ascending on field lines from Jupiter's ionosphere.

The Io-independent A and B events often appear as less-distinct versions of the corresponding Io-dependent bursts, but with all the same properties. Apparently these are similar bursts from the northern auroral zone that are triggered by something other than Io. As the relative intensity of the Io-independent 'sources' changed when Voyager viewed them from the dark side, they may be controlled by the sun in some way – possibly by the action of ultraviolet radiation on the ionosphere.

In conclusion, as the magnetosphere rotates past Io, it is kinked by the moon and sets up a wake of waves. These channel plasma to and from the ionosphere. Somewhere in this complicated system, as bunches of electrons spiral in the field lines above the auroral zones, the radio bursts arise.

VI. The Satellites

Jupiter has 16 known satellites. They fall into four groups of four (Table 19.1). The largest four – the galilean moons – are well-known to visual observers and are all fascinating worlds in their own right. The other moons are all very small. One group of four orbits very close to the planet and is accompanied by a tenuous ring. The other groups orbit very far from the planet and are probably fragments of captured asteroids.

Since the Voyager encounters, more has been published about the satellites than about Jupiter itself! Is this because people have been more excited by these totally new worlds, or is it because people feel more at home with worlds that have solid surfaces? Whichever is the case, all the Voyager data have been thoroughly reviewed: in the special issues of Science, Journal of Geophysical Research, and Icarus (Appendix 5.1); in Morrison (1982); in excellent textbook-style chapters in four books (Morrison and Samz, 1980; Beatty and Chaikin (eds.), 1981, 1990; Hamblin and Christiansen, 1990; Rothery, 1992); and in two whole volumes of the Arizona Space Science series (edited by Morrison, 1982, and Burns and Matthews, 1986). There are even more publications on Io alone (Chapter 21). So the following chapters will be derived mainly from these sources, and only a few original references will be cited. All references are in Appendix 5.

As most of the images and maps of the satellites are from Voyager, they will be shown with north up unless otherwise specified, and the terms 'east' and 'west' will be used in the same sense as on Earth. The coordinates of longitude depend on the fact that the eight inner satellites always keep one side facing Jupiter. Thus longitude 0° is the centre of that face (the subjovian point), 90° is the centre of the leading hemisphere (the apex), 180° is the centre of the face away from Jupiter (the antijovian point), and 270° is the centre of the trailing hemisphere (the antapex).

Table 19.1 *Satellite orbits*

Satellite		Orbital Semimajor Axis ×10⁶ m (Planetary Radii)		Orbital Period (days)	Eccentricity	Inclination
Ring:	inner edge	122.0	(1.71)	0.274	–	–
	outer edge	129.1	(1.81)	0.298	–	–
JXVI	Metis	127.96	(1.7922)	0.2948	<0.004	0.0°
JXV	Adrastea	128.98	(1.8065)	0.2983	<0.005	0.2°
JV	Amalthea	181.3	(2.539)	0.4981	0.003	0.4°
JXIV	Thebe	221.9	(3.108)	0.6745	0.015	0.8°
JI	Io	421.6	(5.905)	1.769	0.0041	0.04°
JII	Europa	670.9	(9.397)	3.551	0.0101	0.47°
JIII	Ganymede	1 070	(14.99)	7.155	0.0015	0.19°
JIV	Callisto	1 883	(26.37)	16.689	0.007	0.28°
JXIII	Leda	11 094	(155.4)	238.72	0.148	27°
JVI	Himalia	11 480	(160.8)	250.57	0.158	28°
JX	Lysithea	11 720	(164.2)	259.22	0.107	29°
JVII	Elara	11 737	(164.4)	259.65	0.207	28°
JXII	Ananke	21 200	(296.9)	631	0.169	147°
JXI	Carme	22 600	(316.5)	692	0.207	163°
JVIII	Pasiphae	23 500	(329.1)	735	0.378	148°
JIX	Sinope	23 700	(331.9)	758	0.275	153°

Adapted from data in Burns and Matthews (eds., *1986*).

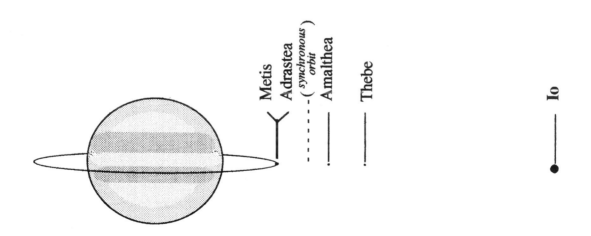

Fig. 19.1. Orbital positions of Jupiter's ring, inner satellites, and Io.

19: The inner satellites and the ring

Orbiting in a blitzkrieg

An astronaut hovering above Jupiter's clouds shortly after sunset would see a remarkable sight: a thin bright line arching up in the twilight glow. This is Jupiter's ring. Higher up in the darkening sky, it fades from sight as a continuous ring, but may still be traced as a speckled line of faint 'stars'. Within this line are several much brighter 'stars' – the four inner moons, orbiting in or beyond the ring, shining as brightly as do the planets in the sky of Earth. Only the brightest of them, Amalthea, would show a disc to the human eye. At 5 arc-minutes across, it appears almost as large as the furthest galilean moon, Callisto (9 arc-minutes across) – though Callisto and the other galilean moons appear far brighter than Amalthea. The motions of these little moons are striking, too. While the galilean moons set in the west along with the stars, Amalthea creeps only sluggishly towards the west as its orbit almost keeps pace with Jupiter's rotation, and the satellites within the ring actually rise in the west and arch towards eclipse in the east.

The two satellites within the ring are called Metis and Adrastea. Further out is Amalthea and then Thebe. They belong to the category of tiny, close-in, non-spherical satellites that are found around all the giant planets. These are sometimes dismissed as 'the rocks'. The largest, Amalthea, is only 270 km long (Table 19.2). Little is known about their composition, but they are believed to be made of rock, not ice, for three reasons. First, their surfaces are all very dark (albedo 5–10%) though this may be because they are drenched in radiation and dust. Second, it is likely that ice never formed so close to Jupiter (see next chapter). Third, Amalthea must be strong and dense in order to maintain its very elongated shape (see below).

The rigours of the environment so close to Jupiter determine the properties of these satellites. They are wrenched by tides, battered by meteoroids, and seared by radiation.

Tides are the effects of orbiting in a non-uniform gravitational field. For a satellite so close to Jupiter, the side towards the planet feels a stronger pull than the side away from it. The result is that the satellite ends up with its long axis radial to the planet, turning exactly once per orbit to keep it that way – 'synchronous rotation'. All the inner satellites must have achieved this state long ago. If such a satellite were made of loose rubble, it would adopt an equilibrium ellipsoidal shape, the more elongated the closer it was to the planet.

But if a satellite were too close to the planet, there would be no stable ellipsoid; it would just be pulled apart by the tides. The critical distance is called the Roche limit. Here, the planet's gravity competes with the satellite's own gravity, so the critical distance depends on the satellite's density. It is close to the orbit of Amalthea for an 'icy' density of 1.0 g/cm^3 and coincides with the orbit of Adrastea for a 'rocky' density of 3.5 g/cm^3. Small satellites can survive within the Roche limit thanks to their internal strength, but debris knocked off them by meteorites never falls back; instead it forms the planet's ring.

Conversely, for satellites outside the Roche limit such as Amalthea, debris does eventually return to the satellite even if it was flung free at first. The satellite sweeps the debris up out of its orbit. In fact, such a satellite can be completely smashed by a large impact, and then reassemble from the fragments. Such reincarnations have probably happened to the medium-sized inner satellites of Saturn and Uranus, according to the calculated frequencies of impacts and the observed frequencies of large craters in those systems.

Meteoroids (which are termed 'meteorites' only when they land) are the second major agent of damage. They are concentrated and accelerated by Jupiter's gravity, and so they impact with enhanced frequency and energy on the leading side of the satellite as it sweeps round in its orbit. The high concentration of meteoroids near Jupiter was verified by the impact detectors on the Pioneers and Ulysses – especially by Pioneer 11, which unknowingly scraped just inside the inner edge of the ring, and had at least one impact as it did so.

Radiation from the magnetosphere is the third assailant on these moons. They absorb so much of it that there are deep plasma 'shadows' at the orbital radii of Amalthea and of the ring (Chapter 18.2). It deposits ions on their surfaces – notably sulphur, which ultimately comes from Io, and is probably responsible for Amalthea's reddish colour.

The ring, Metis, and Adrastea[1]

Jupiter's ring, 57 000 km above the cloud-tops, is the most tenuous of all the giant planets' rings, and does not cast a perceptible

[1] Voyager images of the ring were analysed by: Owen *et al.* (*1979*); Jewitt and Danielson (*1981*). For recent reviews see: Jewitt (*1982*); Thomas and Veverka (*1982*); Showalter (*1989*). The most accurate orbits for the inner satellites are in: Synnott (*1984*); Nicholson & Matthews (*1991*).

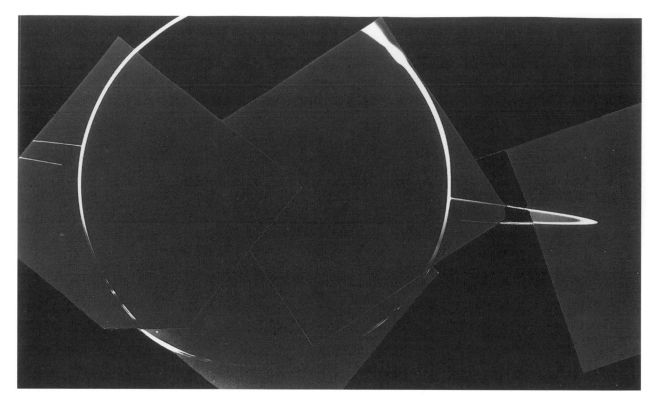

Fig.19.2. The ring viewed by Voyager 2 from within Jupiter's shadow, backlit by the sun. (Montage P-21774; NASA.)

shadow. It was discovered by Voyager 1 in a long exposure taken as the spacecraft crossed the equatorial plane (Fig.19.3), and was viewed in more detail by Voyager 2. It appeared particularly bright when viewed from within Jupiter's shadow, backlit by the sun (Fig.19.2). From Earth, it is too faint to be detected in visible light, but it can be imaged in the near-infrared – at wavelengths where Jupiter's glare is reduced by methane absorption. This was first

achieved soon after the Voyager encounters (E. Becklin and C. Wynn-Williams, *1979*), and the best images are those made at 2.2 μm with the Mt. Palomar telescope (P.D. Nicholson and K. Matthews, *1991*) (Fig.19.4).

The Voyager images reveal a narrow ring, 6000 km wide, with a brighter strip in the middle. It is less than 30 km thick. The main ring seen dimly in back-scattered light must consist of fairly large pieces (centimetres, metres, or even kilometres in scale), which are rough, dark, and red, much like the surface of Amalthea. The total mass of the ring must be no more than 10^{17} kg, which is similar to the estimated mass of Adrastea; it may be much less. The ring particles consist of rock rather than ice, according to an Earth-based near-infrared spectrum, which again resembles Amalthea's (G. Neugebauer and colleagues, *1981*).

But the bright particles seen in the main ring when illuminated from behind must consist of fine dust, only a few microns across. Voyager's radio signals were undimmed by passing through the ring, so there is probably a deficiency of particles larger than a few microns.

There are three extra components to the ring, all extremely faint, according to the Voyager 2 pictures (Fig. 19.5).

(i)   The inner sheet. This extends from the main ring down almost to the cloud-tops.
(ii) The halo. This is an inflated ring with the same radius as the main one, but extending ≈10 000 km north and south of the ring plane. It probably consists of dust smaller than 0.5 μm, which has been given an electric charge and then lofted out of plane by electromagnetic forces.

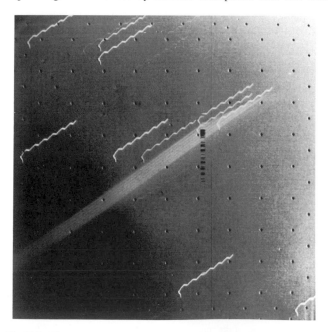

Fig.19.3. The Voyager 1 discovery image of the ring, viewed against the star cluster Praesepe, which fills the field. (Compare with Fig. 1.1.) (Picture P-21258; NASA.)

(iii) The gossamer ring. This is an even fainter diffuse glow outside the main ring, extending out almost to the orbit of Thebe (M. Showalter and colleagues, *1985*; Fig.19.5). Its nature is unknown.

The jovian ring cannot be stable. Within a few thousand years, the dust particles must be ground ever finer by collisions with each other and with meteoroids, and eroded by plasma impact, until they are wafted away by the effects of solar radiation and plasma drag. Indeed, the inner sheet and the halo probably consist of dust being lost in these ways. So the dust has to be re-supplied, and it probably comes from meteoroid impacts on the large ring particles, including the moons Metis and Adrastea.

Both these moons were discovered by Voyager (Table 19.2; Fig. 19.6) but were barely resolved by its camera. Metis was only recorded in a few images which caught it and its shadow in transit against the clouds. But these moons have now been recorded more clearly from Earth in the remarkable infrared images of Nicholson and Matthews (*1991*) (Fig.19.4). These images have established precise orbits for them. Metis orbits within the ring, and Adrastea precisely at its outer edge – to within the resolution of the Voyager ring pictures, which is 200 km. (The Voyager resolution was limited by smear as the spacecraft was moving so fast.) Both tiny moons are thought to be not only sources of ring particles, but also 'shepherds' for confining the ring – Metis perhaps by trapping particles in its own orbit, and Adrastea by preventing them from diffusing further out from the planet. Adrastea's presumed shepherding role is paralleled by small moons in the ring systems of Saturn and Uranus. However, the precise mechanisms involved will not be known until the ring is imaged at higher resolution by Galileo. The orbits of Metis and Adrastea appear to be stable. They are not involved in any orbital resonances, and they are too

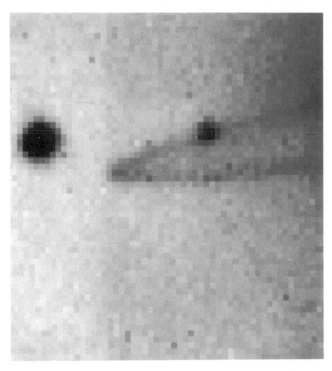 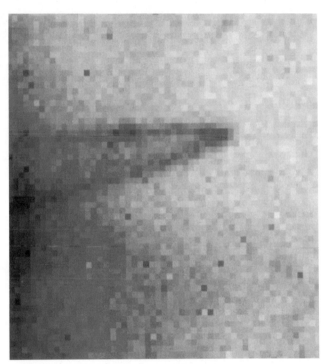

Fig.19.4. The ring and inner satellites imaged from Earth, using the 5-metre Mt. Palomar telescope in 1988. The (negative) images are at 2.2 μm wavelength, where Jupiter itself is very dark due to methane and ammonia absorption. Even so, there is some scattered illumination from Jupiter at the inner edge of each image. (Left:) Following side, with Metis as a dark spot in the ring, and Amalthea very intense at left. (Right:) Preceding side, with Adrastea just visible mid-way along the upper part of the ring image. (From P.D. Nicholson and K. Matthews, *1991*; by courtesy of P.D. Nicholson.)

Table 19.2 *Physical properties of the inner moons*

	Prelim. name	Diameters (km)	Geometric albedo†	Mean opposition m_v†	Discovery ref.
XVI Metis	(1979J3)	40×40*	0.05–0.1	17.4	Synnott (*1981*)
XV Adrastea	(1979J1)	25×20×15	0.05–0.1	18.9	Jewitt *et al.* (*1979*)
V Amalthea	–	270×170×150	0.06	14.1	Barnard (*1892*)
XIV Thebe	(1979J2)	110×90*	0.05–0.1	15.5	Synnott (*1980*)

*Longest dimension, radial to Jupiter, is not known.
Sizes are very approximate for all except Amalthea.
†See Table 20.1 for definitions

Fig. 19.5. Enhanced version of the backlit image from Voyager 2, with illumination as in Fig. 19.2, showing the faint 'halo' above and below the main ring, and the 'gossamer ring' extending in-plane to lower right. (As in Showalter *et al.*(*1985*). Image 20693.02; NASA.)

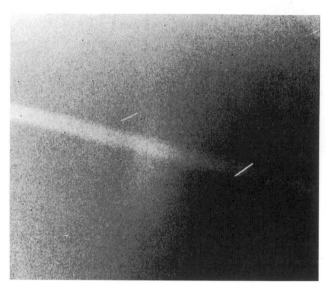

Fig. 19.6. Discovery image of Adrastea (1979J1), in a view of the edge-on ring from Voyager 2. Movement during the long exposure has elongated the images. Note that the satellite image (the brightest streak, in the ring) is not parallel to a background star trail (left and above), indicating movement of the satellite. (As in Jewitt *et al.*(*1979*). P-22172; NASA.)

small to raise tides on Jupiter which might affect their orbits significantly, so they are probably orbiting in their original positions. But being at the Roche limit, they are continually ground down by impacts.

Amalthea and Thebe

Amalthea was the only inner moon known before Voyager, having been the last moon of any planet to be discovered by a visual observer – E.E. Barnard in 1892, who was using the 90-cm refractor at Lick Observatory.[2] At magnitude 14 in the glare of Jupiter, Amalthea is very hard to see at all.

Voyager showed that it is very elongated (270×170×150 km) and very irregular (Fig. 19.7). There are two craters comparable in size to its radius, named Gaea (75 km across, covering the south pole), and Pan (90 km across, in the northern 'hemisphere'). Each is at least 10 km deep. There are also many other hollows and ridges on its leading side (Fig. 19.7A). Whether there are many other craters is unclear, given the limited resolution of the Voyager pictures, and the general unevenness of the surface. So it has clearly been bombarded enough that it may well have suffered an impact big enough to shatter it, as would also be predicted theoretically.

However, its shape is more elongated than an equilibrium ellipsoid would be, implying that it is made of strong material and is not a re-accreted rubble-pile. Its gravity is estimated to vary from 0.55% to 0.85% of Earth's gravity across its surface. Perhaps it is so strong that it resisted shattering, or large fragments were chipped off but were then ground down and lost through non-gravitational processes.

Its surface is mostly very dark and reddish. With a colour index (difference between B and V magnitudes) of 1.5, compared to 1.4 for Mars, it is the reddest object in the solar system (R. Millis, *1978*). The colour (though not the darkness) can be attributed to sulphur implanted from the magnetosphere. However, it has bright patches on some major slopes – of ridges and of crater walls – and these are green (Fig.19.7D)! Their nature is unknown. For example, they could be bedrock, exposed where slopes cannot support dust, or glass, formed by meteoritic melting and stained by sulphur.

About Thebe, there is little more to say. It was discovered in Voyager images when it and its shadow were spotted in transit over the planet (Fig. 19.8). It is ≈100 km wide. Its length radial to Jupiter is unknown. It has also been detected from Earth by infrared imaging, and like Amalthea it is reddish.

[2] Discovery references: Barnard (*1892*); Baum (*1992*). Voyager references: Veverka *et al.* (*1981*); Thomas and Veverka (*1982*).

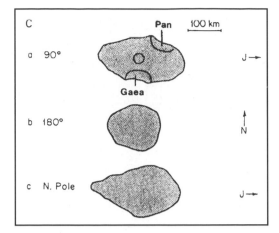

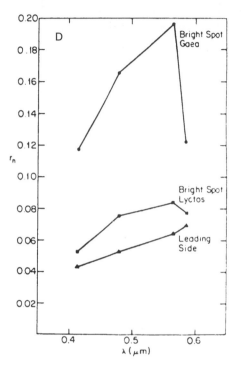

Fig.19.7. Amalthea. **(A)** Voyager 1 view of the leading side, showing bright patches and the two large craters. Compare with the diagram a in (C). (NASA image 16377.34, processed by Simon Mentha.) **(B)** Voyager 2 end-on view with the satellite in transit over Jupiter. (NASA image 20655.32, processed by Simon Mentha.) **(C)** Diagram of profiles along three axes. (Adapted from Veverka et al., 1981.) **(D)** Spectra of Amalthea and of bright spots on it, from Voyager images. (From Veverka *et al., 1981.*)

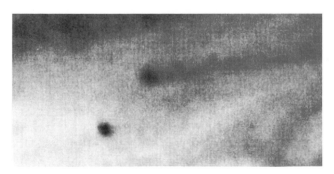

Fig. 19.8. Discovery image of Thebe (1979J2), taken by Voyager 1 on the day of its close encounter, showing Thebe (darkest spot) and its shadow in transit. (P-22580; NASA.)

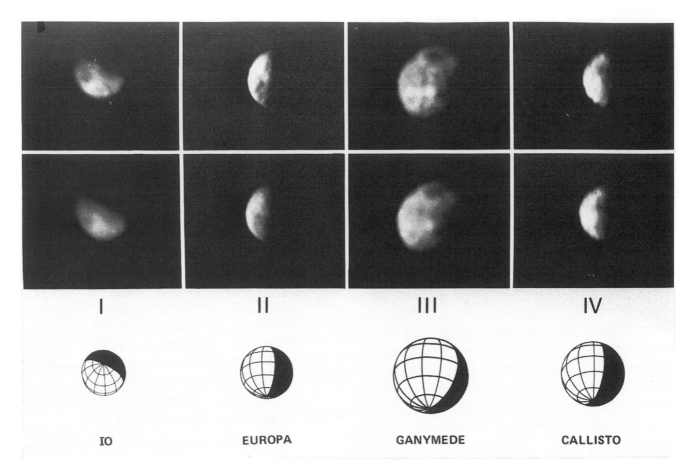

Figure 20.1. The galilean moons as viewed by Pioneers 10 and 11. Top, blue image; middle, red image; bottom, orientations (north is up). (NASA images, from Fimmel *et al.*, *1977*.)

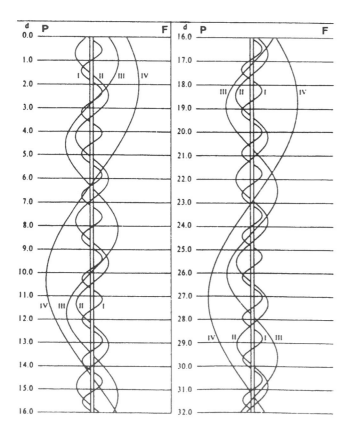

Figure 20.2. How the positions of the satellites change during one month. Diagram from U.S. Nautical Office.

20: The galilean satellites

atellite was named after Callisto, an
for her love, she was changed into a
s named after Io, princess of Argos;
to a white heifer and pursued all over
by Zeus' wife Hera. In the case of
a, Zeus himself took on the shape of a
other of King Minos of Crete. Finally,
med after the god's even greater pas-
ede, son of the king of Troy; Zeus in
m away to serve as his cupbearer.
not recognised before the invention
ht enough (apparent magnitudes 5.6
ble to the naked eye, were it not for
is probable that a few people with
em with the naked eye – or at least
as reviewed by D. Dutton (*1976*),
cussions of it can be found in the
, *1901, 1954*).
n moons are of historical interest,

	Bond albedo	Radar albedo	Ice content
			0%
	0.50	0.06	0%
	0.62	0.7	10–15%
	0.35	0.4	51–57%
	0.13	0.2	52–58%

dent light, under near-vertical
values are from Morrison and Cruikshank
transit, by Dollfus and Murray
individual features from

and angles. From Burns and Matthews (eds.,

Orbital data are in

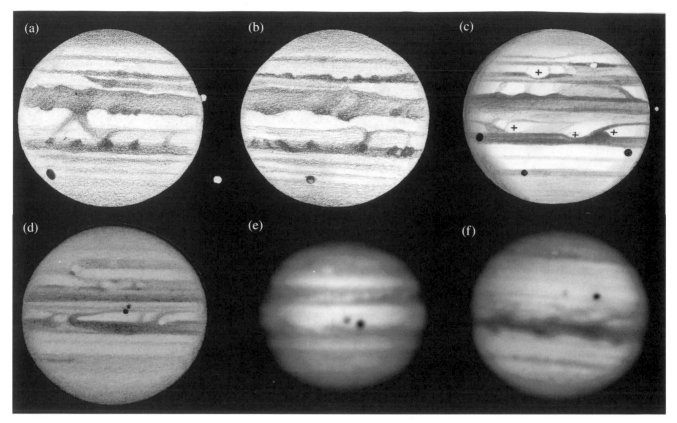

Figure 20.3. Drawings of Jupiter with satellites and their shadows in transit. South is up; transits from right to left. (From BAA archives.) **(a)** 1921 Feb 14, 21.27 UT, ω_2 = 96; C.F. du Martheray (13.5-cm refractor). I reappearing from occultation; shadow of IV in transit; IV approaching transit (lower right). On the planet, p. end of STropD is visible. **(b)** 1921 Feb 14, 23.55 UT, ω_2 = 185; du Martheray. IV in transit showing uneven contrast against the N.N. Temperate region. F. end of STropD is visible. **(c)** A similar configuration of I, IV, and a STropD: 1980 Jan 25, 11.58 UT, ω_2 = 323; J.H. Rogers (32-cm reflector). I reappearing from occultation; IV dark in transit on NNTB; also III dark in transit on NTropZ (right side) preceded by its shadow on NEB (left side). **(d)** 1979 Jun 21, 10.46 UT, ω_2 = 303; I. Miyazaki (20-cm reflector). Shadows of II and IV near centre of disc; shadow of IV appears smaller and more diffuse. **(e)** 1985 Aug 7, 04.08 UT, ω_2 = 104; D.C. Parker (32-cm reflector), photograph. III dark in centre of disc, followed by its shadow. **(f)** 1990 Jan 13, 21.23 UT; G. Viscardy (520-mm reflector); photograph with red filter. Io in transit on whitened SEBZ, followed by its shadow. (For enlargement see Fig. 20.15g.)

as they provided an early celestial timepiece; their orbits could be presumed to have constant periods. The first measurement of the finite speed of light, by Ole Römer of Denmark in 1675, was achieved by timing the eclipses of the satellites by Jupiter's shadow. Such timings were also proposed as a solution to the vital problem of determining time, and hence longitude, for ships at sea. But given the unlikelihood of being able to make accurate telescopic timings from on board a rolling ship in a storm, this method did not recommend itself to the British Admiralty.

20.2 VISUAL AND PHOTOGRAPHIC OBSERVATIONS

Transits, occultations, and eclipses

The four moons can be seen easily with binoculars, which reveal them moving from night to night or even from hour to hour. As we see the system almost edge-on, never tilted by more than 3.4°, the moons usually appear to lie close to a straight line. Their positions preceding or following the planet can be plotted with time, as in Fig. 20.2. Such diagrams are published as predictions to help observers identify the satellites, and it is fun for a novice to make one himself from observations with binoculars.

The moons pass in front of and behind the planet, once in every orbit (except for Callisto in years when the system is tilted more than 2.7°). These events are always enjoyable to watch through a telescope. The passage lasts a matter of hours and the actual crossing of the planet's limb takes only a few minutes, so one has a real-time view of orbital motions.

When the moon passes behind the planet, it is in occultation. When it disappears in the planet's shadow, it is in eclipse. Before opposition, the satellite will disappear into eclipse on the preceding side of the planet then reappear from occultation on the following side. After opposition, the order is reversed. For Callisto, Ganymede, and occasionally Europa, there are times near quadrature when the satellite can reappear from eclipse before entering occultation, or vice versa.

Of all the phenomena, eclipses can be timed with the greatest accuracy, and they have always been used for refining the orbital parameters of the satellites. This work has continued into the 1980s as the ALPO pursued a programme of eclipse timings. These confirm that the orbital predictions developed for NASA by J.H. Lieske (*1981*) are much more accurate than previous ones, but suggest that even these can be out by up to 200 km (J.E. Westfall, *1983–84, 1992*).

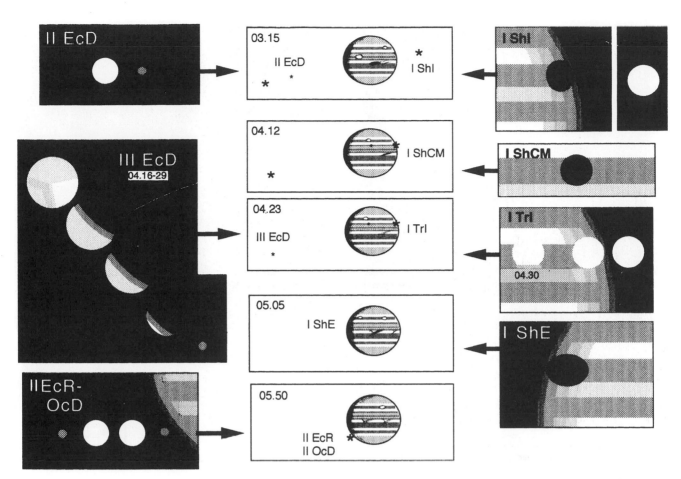

Figure 20.4. Phenomena of the satellites observed by the author with a 30-cm refractor, 1976 Sep.5. South is up. When II moved into and out of eclipse (EcD, EcR), the crescent aspect was not observed because of the small disc size, in contrast to the eclipse of III. II occultation (OcD) occurred at the dark limb of Jupiter. The transits of Io's shadow (ingress, ShI; transit, ShCM; egress, ShE) and of Io itself (ingress, TrI) were along the SEB(S).

When the moon passes in front of the planet, it is in transit, and it will be preceded or followed by its shadow, in shadow transit. The appearances of the four satellites during transit are strikingly different, because they differ in surface brightness (albedo) (Table 20.1; Fig. 20.3). All of them appear bright at the start of transit, as they are projected against the planet's limb-darkening, but then they change. Europa, the one with highest albedo but also the smallest, is normally invisible. (However, it has occasionally been seen as a tiny bright spot against a belt, and in 1989/90 BAA observations it was once seen bright against the dark GRS, and was photographed as a faint dusky spot against the STropZ.) Io is often invisible too, but against a bright zone it appears as a grey spot. Ganymede is much darker, and larger, and appears as a very dark spot in transit. Callisto, the darkest, appears almost as black as a satellite shadow. When Callisto begins a transit, it typically 'turns dark' just after second contact, as a striking reminder of how strongly our perceptions of surface brightness depend on contrast. However, the apparent darkness of Ganymede and Callisto in transit is variable, presumably according to the darkness of the atmosphere below.

The shadows during transit normally appear round and black. That of Callisto may appear diffuse, as its visible shadow is largely penumbra; the satellite is far enough out to appear only 1.5 times the sun's diameter as seen from Jupiter's cloud-tops. Near quadra-ture, if viewed at high resolution, the shadows can appear noticeably oval (Fig.20.4). Of course they are oval when projected on the planet, but this is only apparent from Earth when there is a substantial angle between Earth and sun.

Within a day or two of opposition, a satellite can partially hide its own shadow during transit. This can occur within about 56 hrs of opposition for Io and within about 17 hrs for Callisto. This phenomenon was reported as early as 1885 by R. Copeland, and has since been reported at more than a dozen oppositions from 1893 (Fig. 20.14) to 1992 (Fig. 20.5).

Mutual phenomena

Every six years, we see the jovian system exactly edge-on, as the plane of the satellites' orbits intersects the Earth and sun. Then the satellites can be seen to occult and eclipse each other. These 'mutual phenomena' happen for several months, during which different pairs of satellites can be involved successively, as the Earth is moving round the sun and as the four satellite orbits are not quite in a single plane. The events can last just a few minutes, or as long as an hour, depending on the relative positions of the moons in their orbits.

A mutual occultation is not very remarkable, and good seeing is required to be sure whether the satellite discs are indeed overlapping. Poor seeing can broaden the apparent discs. A mutual eclipse

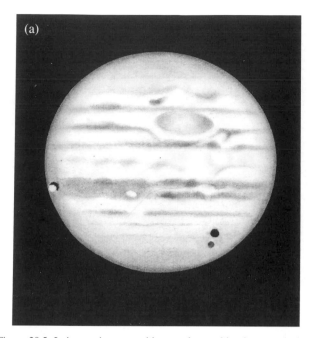

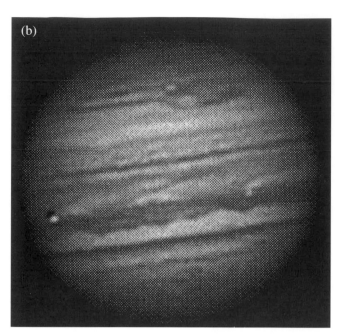

Figure 20.5. Io in transit at opposition, partly occulting its own shadow. (See also Fig. 20.14.) South is up. (From BAA archives.)
(a) 1933 Mar 7, 22.35 UT; T.E.R. Phillips (46-cm reflector). Io and shadow on preceding limb; also IV and its shadow on lower right part of disc.
(b) 1992 Feb 28, 03.22 UT; D.C. Parker (41-cm reflector), CCD image.

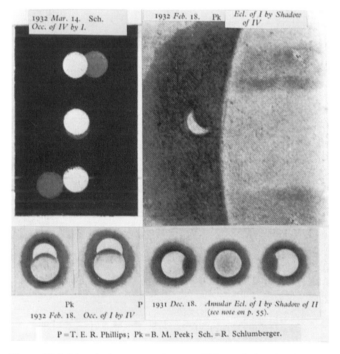

Figure 20.6. Mutual phenomena, drawn at high resolution. South is up. (From the BAA Memoir for 1931/32.)

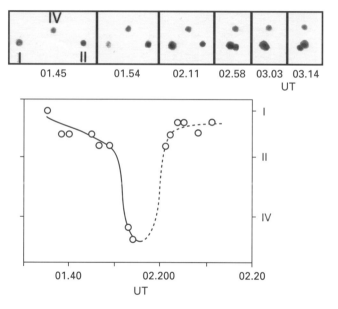

Figure 20.7. Visual observation of mutual eclipse and occultation of I by II: 1991 Jan 12, J.H. Rogers, in mediocre seeing. Above: sketches of the satellites in negative. Below: estimated light-curve; y-axis marked at intervals of 0.5 magnitudes.

is more striking, as the satellites need not appear close to each other in the field of view. What happens is that one satellite suddenly starts fading, and may even disappear within a few minutes, depending on the magnitude of the eclipse (Figs.20.6–8). Mutual occultations have been observed by the BAA since 1907/8, and mutual eclipses since 1920. For 1926 and 1931/32, detailed predictions were provided and many mutual phenomena were observed. Since then, many mutual phenomena have been observed but only a few reported in detail. Examples are in the ALPO reports for 1954/55 and 1955/56, and BAA reports for 1973, 1979/80, 1985

(W.E. Fox, *1986*), and 1990/91. Professional photometry of these events produced the most accurate pre-Voyager values of the satellite diameters. The events are also very sensitive tests of the accuracy of the orbital elements. Even in 1991, amateur CCD photometry of mutual events revealed a 308-km error in the best current ephemeris (A. Mallama, *1992*; Fig. 20.8).

One of the commonest of these events is an eclipse of Io by Europa, and as Europa is the smaller of the two, these eclipses are annular. As described in Peek's book, these eclipses have an unexpected visual appearance which has important implications for

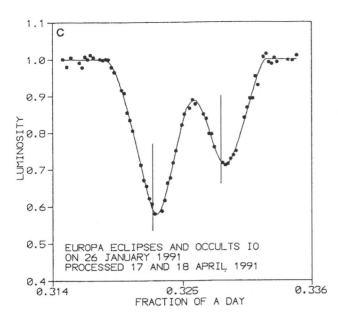

Figure 20.8 Photometric record of mutual eclipse and occultation of I by II: 1991 Jan 26. Vertical lines mark the predicted mid-times of the phenomena. (From Mallama, *1992*).

observations of surface markings (see below). Although the encroaching shadow is resolvable in the partial phase with a large enough telescope, the annular phase is not perceived as such; Io simply appears grey. These observations were with 20-cm or 31-cm telescopes. Peek described experiments with artificial images which confirmed that details on this scale cannot be resolved optically. This illustrates the limitations of resolving detail on these tiny discs. Their apparent diameters at opposition range from 1.0″ arc to 1.7″ arc (Table 20.1), while the best obtainable resolution (the Dawes limit) is 0.6″ for a 20-cm telescope. All drawings of satellite phenomena that show large discs with sharp edges must have involved some visual interpretation of the actual image.

The most engaging event of all is one that Peek, in his book, said that he 'would deem it a great privilege to behold': the eclipse of one satellite by another when in transit across the planet, so that 'the two shadows will coalesce and the satellite will turn black – surely a delightful spectacle!' This has now been observed twice: by several BAA observers on 1980 April 7, and by the author on 1991 January 15, each occasion being an eclipse of Io by Europa. In the latter observation, although the seeing was too poor for the shadow falling on Io to appear black, the eclipsed satellite clearly appeared on schedule then faded again as it slipped in and out of the shadow (Figs. 20.9&10).

Colours

Judging the colours of the moons visually is a different matter from judging the extended colours on Jupiter, but no less liable to subjective or systematic errors. The author can offer no opinion on this, since he has little sensitivity for colour in stellar images such as those of the moons. Some of the more systematic colour notes by visual observers are summarised in Table 20.2.

The colour differences appear most vivid at mutual occultations; for example on 1932 March 14, R. Schlumberger reported Io

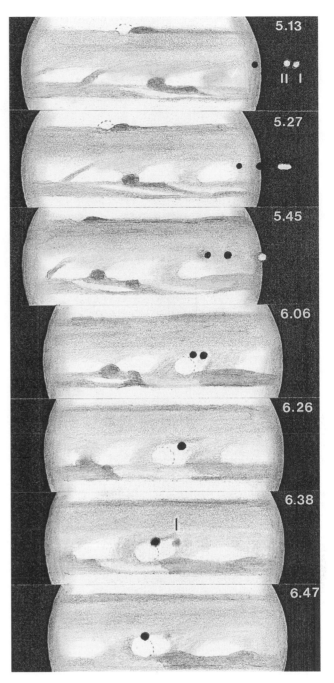

Figure 20.9. Occultation and eclipse of Io by Europa while both were in transit: 1991 Jan 15, J.H. Rogers. South is up.

slightly orange-yellow as against Callisto much duller and bluish-grey (BAA; Fig. 20.6); on 1985 August 5, W.E. Fox found Ganymede golden yellow with Callisto quite blue (Fox, *1986*); and on 1991 January 16/17, R.J. McKim and M. Foulkes independently recorded Europa as blue versus Ganymede as yellowish (BAA).

The most notable thing about Table 20.2 is the absence of strong colour from Io, in contrast to the impression given by the Voyager pictures. Most observers agree that Io and Ganymede are similarly yellowish, and Patrick Moore (*1965*), reviewing previous estimates, commented that he normally sees Io as the whiter of the two.

Actual spectra (Fig. 20.11) show that Io is indeed redder than the other three, but only if one includes its steep absorption in the

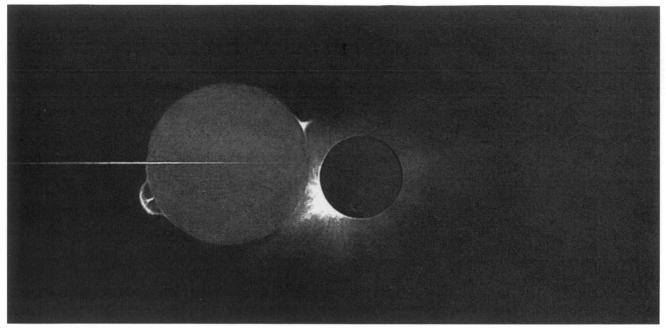

Figure 20.10. Artist's impression of the event of Fig. 20.9 seen from the cloud-tops of Jupiter. With the sun eclipsed by Europa, but Io not yet in Europa's shadow, the sun back-lights the volcanic plumes and a part of the jovian ring. The moons are dimly lit by light reflected from Jupiter. (By the author.)

Table 20.2 *Visual colours of the galilean moons*

P.B. Molesworth (*1891* and *1898*):
I   Pale yellow with slight tinge of rose
II Very pale yellow [but in 1903/4 BAA Memoir,
    he reported it as the yellowest]
III Pale yellow ('primrose yellow') (ruddier
    when close to planet's disc)
IV Dull pale blue [in 1903/4 Memoir, bluish
    or purplish or greyish white]

E.J. Reese (in BAA reports for 1951/52 and 1952/53, cf. Appendix 3):
I   Yellow or cream
II White
III Yellow
IV Pale bluish-grey

P. Moore (*ibid.*):
I   Usually white
II White
III Yellow
IV Variable, usually faint reddish
    violet

E.M. Antoniadi (*1939*):
I   Fiery yellow
II Bright yellow
III Strong yellow
IV Dull yellow or ruddy
    yellow

D.P. Avigliano (*1954*):
I   White
II Yellow-white
III White-yellow
IV Dull greyish-white, but variable

B. Lyot (*1943, 1953*) and
  colleagues:
I   Yellowish
II Brilliant white
III Yellowish
IV Dark maroon

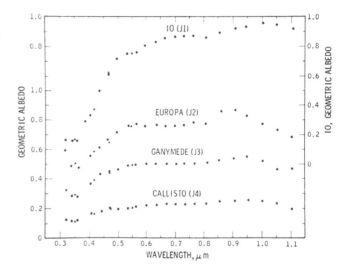

Figure 20.11. Spectra of the moons from near-UV to near-IR. (From Morrison and Burns, *1976*).

violet; it also has a unique brightening in the near-infrared. These ends of the spectrum are better detected photographically than visually. The published Voyager pictures of Io, made up from three images in separate colours, were artificially red because violet images were used for the blue component. From blue to red, Io's spectrum is unremarkable, and according to the standard colour index, Io (+1.2) is less red than Mars (+1.4).[3] Spectra also show no

[3] Young (*1984*); McEwen (*1988*). The colour index is B minus V magnitude.

real difference in colour between Europa, Ganymede, and Callisto. The apparently different colour of Callisto may perhaps be an optical illusion due to its being fainter than the others, but the physiological basis of such an effect is not known.

Surface markings

Surface markings can be seen on all the moons, given a large enough telescope, but they are so difficult that it is hard to distinguish fact from fantasy in the published record. Even the best observations cannot be matched to the Voyager maps except in the broadest terms.

Sad to say, the search for detail on these tiny discs tends to bring out the worst in observers. There is always a risk when the observer can be given prestige for seeing something fainter or smaller than anyone else. Under such an influence, suspicions can evolve into certainties, and thus all observations become suspect. This is not a new problem. As long ago as 1898, A.E. Douglass of the Lowell Observatory reported seeing a pattern of 'sharp distinct black lines' on Ganymede and Callisto (as well as on Mars and Venus). And the publications of one national amateur association in the 1950s contain drawings of surface detail on the moons made with apertures as small as 10 cm! The only value of such material is to demonstrate that some observers can delude themselves into seeing 'features' which are optically impossible. (Compare the satellite diameters quoted above with the Dawes limit of 1.2″ for 10 cm aperture, and consider the effects of limb contrast on these small discs.) Some observations made with larger instruments may be no more reliable.

Indeed, there have been several reports of the satellites' shadows appearing double in transit; even Peek, in his book, reported such an impression. As he realised, this can only have been due to some atmospheric or instrumental effect, and it emphasises the need for confirmation, under different viewing conditions, of anything seen near the limit of visibility.

For many years, even the rotation periods of the satellites were uncertain.[4] It was expected that tidal forces would constrain them to one rotation per orbit, so that they would always keep the same face facing the planet, as our own Moon does. Confirmation of this by observation of surface features on Ganymede was first claimed by J.H. Schroeter in 1801 – and with more credibility by J. Schaeberle and W. Campbell (*1891*), who published many drawings made with the 36-inch (91-cm) refractor on Mt. Hamilton. However, some amateurs including P.B. Molesworth (*1898*) and W.H. Steavenson (*1915*) instead proposed non-synchronous periods. The question was settled by T.E.R. Phillips (*1921*; also 1927/28 BAA Memoir) using an admirably simple method. Instead of searching for minute markings, he simply recorded whether Ganymede's disc appeared elongated and what its position angle was – factors which presumably depend on unresolved surface detail. He found that the same appearances recurred at the same points in the satellite's orbit over six years of observation, and thus the rotation was indeed synchronous.

For the other satellites, synchronous rotation was proved by the photometric light curves from the Lick Observatory in the 1920s (Stebbins & Jacobsen, *1928*), which showed that all the moons varied in brightness strictly according to their orbital longitude. The light curves have not changed since then (Fig. 20.12). (Actually, Sir William Herschel had deduced synchronous rotation from his visual brightness estimates in 1794–1796; but such visual estimates are bedevilled by the glare of the planet, and the true light variations are too slight to be detected visually, except perhaps in the case of Europa.) The synchronous rotation periods were confirmed by observations of surface features at the Pic du Midi in the 1940s (B. Lyot, *1943, 1953*).

Surface markings are most readily seen when the moons are in

[4] The early observations were reviewed by Antoniadi (*1939*) and Lyot (*1943*).

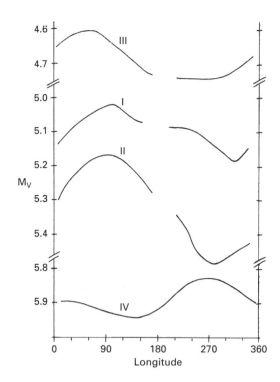

Figure 20.12. Orbital light-curves of the moons. (Adapted from Morrison and Burns, *1976*.)

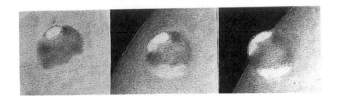

Figure 20.13. Ganymede in transit: high-resolution view, 1962 Sep 13/14, by Audouin Dollfus (107-cm reflector at Pic du Midi). South is up. (From Dollfus and Murray (*1974*), by courtesy of A. Dollfus.)

transit, by contrast with the sunlit clouds below – at least on Ganymede and Io. (Bright Europa and dusky Callisto usually appear featureless even then.) In transit, of course, we see that side of the moon that always faces away from Jupiter. Ganymede's main feature is an oblique dark band, while Io's is a bright equatorial zone. Given ideal seeing, these markings on Io and Ganymede are visible in a good 20-cm telescope (e.g. Avigliano, *1954*). They have even been drawn by E.J. Reese with a 15-cm reflector (ALPO 1951/52 report) – which is just within the bounds of possibility.

Ganymede's 'oblique band' is resolvable into two large dark patches, which were first seen by 'the eagle-eyed Dawes' in 1849 with a 16-cm refractor, if the features in his drawing were not a fortuitous illusion (W.R. Dawes, *1860*). Being the largest moon, Ganymede covers a significant span of the planet's limb-darkening, and so it may transform into various 'shapes' during transit as the contrast between its markings and the underlying atmosphere changes. A low-resolution example is in Plate P15.5. More detail has been shown by many acute observers including Molesworth (*1898*), Phillips and Steavenson (*1917*), Lyot (*1953*), Dollfus and

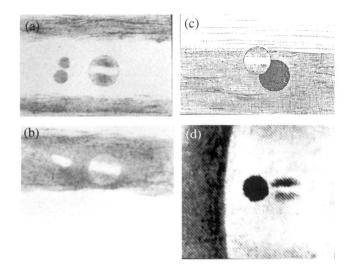

Figure 20.14. Io in transit showing the bright equatorial zone, as observed by E.E. Barnard. South is up. (a) 1890 Sep 8; (b) 1891 Aug 3. At left is the observation, at right his interpretation in terms of a bright equatorial band. From Barnard (*1891*). (c) 1893 Nov 19; his 'perfectly satisfactory' observation with 36-inch refractor at Lick. From Barnard (*1894*). (d) A similar observation by T.E.R. Phillips on 1932 Feb 3. From BAA Memoir for 1931/32.

Murray (*1974*) (Fig. 20.13), and Murray (*1975*). Their drawings all agree well. The darkest spot, in the Np. quadrant of the satellite, is the 'mare' Galileo Regio in the Voyager images, although visual observers have tended to draw a small spot rather than the true large circular area (Fig. 20.29).

Io's bright equatorial zone can also produce strange aspects in transit. When Io is near the limb, or projected on a belt, this bright zone can appear as an isolated bright streak, and when it is projected on a zone, the darker polar regions can appear as two separate dark spots. These aspects of Io were first recorded by Holden at the Lick Observatory in 1888, and re-observed and explained by E.E. Barnard (Fig. 20.14). The bright equatorial zone seems to be a permanent feature of Io. It was recorded in 1898 and 1903–5 by Molesworth (BAA Memoirs), in 1926–27 by Antoniadi (Antoniadi, *1939*), in 1929 and 1931/32 by Phillips (BAA Memoirs), and by later observers with large telescopes as described below.

Given what we now know about Io, one would dearly like to know if its visible markings have changed through history. The aspect in transit (i.e., with longitude 180° on the central meridian) is the only one for which one can make reasonably confident comparisons between different observers. In addition to the permanent equatorial zone, two smaller markings have been seen: a north–south 'band' just p. the central meridian, at λ ≈ 150°, which may be permanent; and a possible darkening of the f. limb (towards λ = 270°), recorded only in the last few years. Fig. 20.15 shows a panel of views redrawn from the following reports.

The first drawings showing the 'N–S band' at 150° were made with the Pic du Midi 38-cm refractor in 1941 (Lyot, *1943*). With the 60-cm refractor in 1943–45 (Lyot, *1953*), the same observers placed the N–S band on the 180° meridian instead, but they attributed the difference to the increased resolution which revealed an extra dark

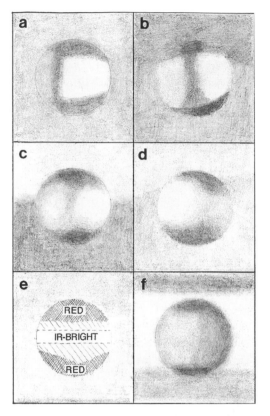

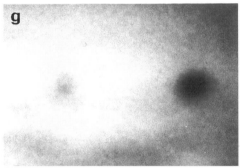

Figure 20.15. Io in transit: high-resolution views, 1941–1990, redrawn by the author. South is up. (**a**) 1941: Camichel, Gentili and Lyot (38-cm refractor), from Lyot (*1943*). (**b**) 1943–45: Camichel, Gentili and Lyot (60-cm refractor), from Lyot (*1953*). (**c**) 1966: A. Dollfus (83-cm refractor), from Dollfus and Murray (*1974*). (**d**) 1973: J. Murray (108-cm reflector), from Murray (*1975*). (**e**) 1973: R.B. Minton, from multicolour photos (154-cm reflector), from Minton (*1973*). (**f**) 1989: D. Gray, single drawing (41.5-cm Dall-Kirkham), from Rogers (*1990*). (**g**) 1990: G. Viscardy, photograph enlarged from Fig.20.3 (52-cm reflector); Io at left, shadow at right; from Rogers (*1990*).

spot. In 1966, at the Pic du Midi and at Meudon (Paris), Audouin Dollfus again drew the band at 150° (Dollfus and Murray, *1974*). In 1973, at the Pic du Midi, John Murray drew this band at 150° when the satellite was off the disc, but in transit he showed only a hint of it (Murray, *1975*). Also in 1973, photographs by R.B. Minton (*1973*) clearly showed the bright equatorial zone (and dark red colour of the polar regions), but did not show the 'N–S band'. The 1979 Voyager photos (Figs. 20.16 and 20.29) showed that the 'N–S band' is a real though very irregular marking, studded with several volcanoes, including on the equator one of the most active ones, Prometheus. So it seems likely that this band has been present for at least 38 years.

Figure 20.16. Io at the end of a transit, viewed from Voyager 1. South is up. This was taken from close range, at the start of the flyby of Io, and central meridian longitude on the moon is 190°; for the normal mid-transit longitude of 180°, the dark spot just p. the central meridian should be central. (More distant Voyager images of Io in transit all have the trailing limb in shadow.). In the background is Jupiter's NEB with a dark barge and bright plume. (NASA image 16372.37, processed by Simon Mentha.)

All these observers found the equatorial zone bright up to the p. and f. limbs. Murray in 1973 drew the f. half slightly less bright than the p. half, which may represent a change. This contrast was confirmed by pre-Voyager lightcurves and by the Voyager maps. Voyager's best view in transit is Fig. 20.16. The next high-resolution views were by BAA observers in 1988–90, and led to the suggestion that one or both limbs might have become dark (Rogers, *1990*). The author first suspected this in 1988/89 when he saw Io appear remarkably dark against the pale GRS, and one photograph by I. Miyazaki showed it in transit over the dark SEB as a tiny bright spot, not the usual streak. In a photo by G. Viscardy in 1989/90, when transiting against a whitened SEB, Io seemed to show dark p. and f. limbs (Fig. 20.15g). A drawing by D. Gray (Fig. 20.15f) suggested that the f. limb in particular was dark, as it showed the moon as a dark crescent. While these data are not conclusive, the trailing limb at least may have darkened since 1973.

It is interesting that this region was reported to be possibly slightly darker and redder in the Voyager data of 1979 than in rotation curves of 1973 (approx. longitudes 210°–270°) (P. Simonelli and J. Veverka, *1984*). In 1979, this region included the dark deposits of the greatest of the active volcanoes, Pele, centred at 255° and changing over a few months (Chapter 21). Fig. 20.16 shows this area as only a narrow dark strip along the trailing limb, and if rotated to put 180° on the central meridian, this strip would be too narrow to see from Earth. So any change may have occurred since 1979. Perhaps we will find out what changes have really occurred after the Hubble Space Telescope optics are fixed.

Markings at other longitudes, which are viewed when the satellite's disc glares against the dark sky, can only be reliably recorded with very large telescopes. Some of the best drawings have been made from the Pic du Midi in France (e.g. Fig. 20.17), and in the 1940s Bernard Lyot and colleagues made complete maps of the four moons with the 38-cm and 60-cm refractors there (Fig. 20.18). From their drawings in the years 1958 to 1972, Dollfus and Murray (*1974*) also made maps, which agree well with Lyot's except for that of Io; here it seems that some features may have changed in intensity, although the differences may be within the range of uncertainty. Murray described the difficulties well:

In the first place the contrast of the bright disk against the dark sky makes the limb appear bright, and tends to hide markings at the edges. Since the disk is so small, this edge effect affects a large fraction of the disk, hiding or softening markings up to 45° or more from the limb. Second, atmospheric turbulence ensures that the disk is never stationary, even under the best seeing, so that markings constantly move, change shape, vary in size, and change in intensity. When the seeing is good, surface markings are clearly visible, indeed much more clearly visible than is generally supposed, yet...measurements of

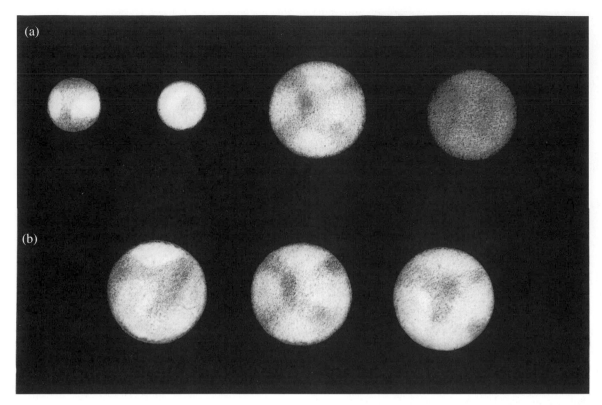

Figure 20.17. Drawings of the moons by A. Dollfus in 1982 April (2-metre reflector at Pic du Midi). South is up. **(a)** Central meridian longitudes: I, 237; II, 87; III, 283; IV, 213. **(b)** Ganymede on three successive nights; longitudes: 225, 284, 334. (By courtesy of A. Dollfus.)

position can yield differences of 20° for markings near the center of the disk. (Murray, *1975*.)

The maps by Lyot (*1953*) remain the best visual maps of the satellites, and can be compared with the Voyager maps (Chapters 21–24).

Lyot's map of *Callisto* shows low-contrast dark mottlings, and several light spots in high latitudes. The Voyager map does show bright patches around fresh craters, and one great ray crater (called Adlinda) is probably Lyot's south polar bright spot. Otherwise, individual features cannot be identified with those on Lyot's map.

Lyot's map of *Ganymede* is the most detailed; he commented that this moon resembles Mars as seen with an equivalent disc diameter in a very small telescope. There are dark bands and patches, which avoid the poles. Around the poles there are light areas ('polar caps'); the north polar region is particularly bright and has been noticed by other observers.[5] Lyot also noted brightening of the following (morning) limb at some longitudes; this possible indication of an atmosphere will be discussed below. When compared with the Voyager map of Ganymede, the match is surprisingly poor. This is partly due to uneven portrayal of albedo in the Voyager map; but of the most distinct dark regions seen by Voyager, Galileo Regio was drawn too small by Lyot, and Nicholson Regio was not drawn at all. The light bands recorded by Voyager were not clearly resolved by Lyot. The polar caps, though, are genuine.

Lyot's map of *Europa* consists of a dark band or dark patches along the equator. This is also the aspect in the Voyager images.

[5] E.g., Schaeberle and Campbell (*1891*); Molesworth (*1898*); Dollfus and Murray (*1974*).

Lyot's map of *Io* is dominated by the bright equatorial zone, crossed by several north-south bands, which connect dark spots lying in the mid-latitudes. This agrees quite well with the Voyager map; at least half the dark spots can be tentatively identified with complex dark areas (not individual volcanoes). The major difference is that the side from λ 260–360°, which was all dusky and reddish in the Voyager images, contained a large light area in Lyot's map. This is the side dominated by the most flamboyant volcanoes including Pele and Loki, so it may well have changed; but the example of Ganymede shows that one cannot be certain of this from the records available.

It is conceivable that varying volcanic activity could make a detectable change in the total brightness of Io. According to Morrison and colleagues (*1979*), the story on this is the same as for surface patterns: any change in apparent brightness is within the possible range of error. The difference between Io and Ganymede was recorded as only 0.24 magnitudes in the 1920s, 0.33 magnitudes in the 1950s, and 0.4 magnitudes in the 1970s. However, the apparent darkening might be due to systematic errors, due to the reddish colour (different wavebands have been used) or to scattered light from Jupiter.

Images from Pioneers 10 and 11 were somewhat sharper than Earth-based drawings (Fig. 20.1), but were taken from angles that made them difficult to compare directly with Earth-based or Voyager views. They still could not reveal the nature of the markings.

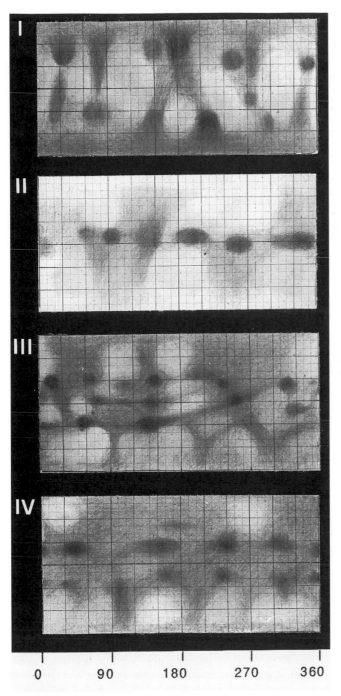

Figure 20.18. Maps of the moons by B. Lyot (*1953*). From top to bottom: Io, Europa, Ganymede, Callisto. South is up; major grid-lines are at 30° intervals. (By courtesy of A. Dollfus.)

20.3 THE SATELLITE SURFACES[6]

Even without resolving surface details, astronomers could find out some basic facts about the moons' surfaces from Earth-based measurements of their light and infrared radiation. The results can be summarised as follows. *Callisto's* surface is dark and rather rough, but nevertheless consists largely of water-ice; it may be mixed in

[6] Pre-Voyager observations were thoroughly reviewed by Morrison and Cruikshank (*1974*) and Morrison and Burns (*1976*); post-Voyager data in the book edited by Burns and Matthews (*1986*). The next three sections are largely adapted from their accounts, with more recent data as referenced.

with jumbled rocks or a meteoritic 'soil'. *Ganymede* has dark patches resembling Callisto, but its surface is largely covered with water-ice with a frosty texture, which presumably corresponds to the brighter areas. *Europa* is entirely covered in water-ice, which is a very bright and fine-textured frost that must be quite fresh in geological time. It has the highest albedo, almost the same as that of pure white snow. *Io* was the greatest puzzle of all before the Voyager encounters, with a bright but patchy surface, a texture like frost, but no water-ice at all. Now we know it is largely covered by sulphur and sulphur dioxide.

Photometry and polarimetry

The first accurate photometry was done by Stebbins and Jacobsen (*1928*) of the Lick Observatory. Their work was extensively confirmed from Earth in the 1970s, and by Voyager in 1979.[7] Each satellite has one hemisphere that is redder and 10–30% darker than the other. For the inner three, this is the trailing hemisphere, but for Callisto it is the leading hemisphere. The inner and outer satellites of Saturn show the same pattern.

This pattern could well be due to an external agent, darkening or brightening the surfaces. There are two obvious candidates, both of which should be most intense for the inner satellites. One is magnetospheric radiation; charged particles will mainly impact the trailing hemisphere, as the plasma is swept round by the corotating magnetic field. Charged particles, including sulphur ions, would darken and redden an icy surface, and this may be the explanation for the darker trailing hemispheres of the inner three moons (R. Wolff and D. Mendis, *1983*; see below). The second candidate is meteoroids, which will mostly hit the leading hemisphere. Large meteorites would brighten a surface of 'dirty ice' by excavating fresh ice, but the rain of smaller meteoroids would dirty a bright fresh surface, so they are probably not the dominant influence on the inner three moons. They may however be responsible for Callisto's asymmetry. Callisto is uniquely affected by electrically charged micrometeoroids in orbit around Jupiter, which impact its leading hemisphere.

However, for Io and possibly others, the Voyager images have shown that internal processes largely shape the surface patterns.

Surface texture can be inferred from photometry at different phase angles (Fig. 20.19). A rough, dark surface, such as a loose soil, darkens rapidly as the phase angle increases, due to shadowing between the particles, and shows a particularly steep 'spike' of brightness within a few degrees of opposition. This is the case with Callisto, whose phase curve resembles that of the Moon or Mercury. Its leading hemisphere shows the steepest curve, and becomes as bright as the trailing hemisphere at opposition. The trailing hemisphere must be more compacted. These properties are confirmed by studies of Callisto's polarisation, which is strongly negative unlike that of the other three moons, especially on the leading hemisphere. This too implies that the leading hemisphere is very rough and dark, while the trailing hemisphere is smoother. Dollfus' (*1975*) interpretation was that Callisto's crust was for-

[7] Johnson *et al.*(*1983*); McEwen (*1988*).

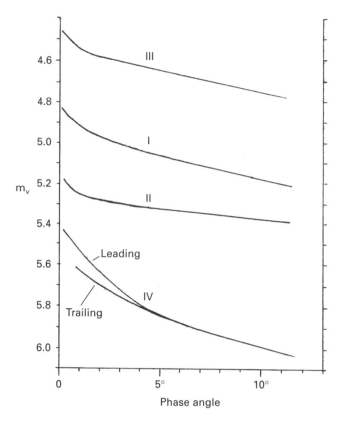

Figure 20.19. How the brightness depends on phase angle: Earth-based photometry. The phase angle is the angle between sun, satellite, and observer. Adapted from Morrison and Burns (*1976*). More complete curves including Voyager data out to phase angles of at least 120° are given by Buratti (*1991*) and Domingue *et al.* (*1991*).

merly of ice with embedded rocks, but that the surface layer of ice has evaporated leaving bare rocks lying in heaps; on the leading side, the rocks have been largely pulverised by meteoroids.[8]

The inner three moons show much shallower variations in brightness and polarisation with phase angle, especially Europa whose curve is almost flat (Fig. 20.19). These curves suggest bright, fragmented material that scatters light well, such as frost or other fine crystals. (Ganymede's curve may well be an average over regions resembling both Callisto and Europa.)

Radar reflections

Radar, operated from the great Arecibo and Goldstone radiotelescopes at a wavelength of 13 cm, can probe the the surface and several wavelengths below it. Europa, Ganymede, and Callisto all give amazingly strong reflections which are still not definitely explained (reviewed by S.J. Ostro, *1982*). The average geometric albedoes at 13 cm (Europa 0.7, Ganymede 0.4, Callisto 0.2) dwarf those for the terrestrial planets, Moon, and asteroids (0.01–0.08). (For Ganymede, the value is lower for a large dark area.) The echoes are also bizarre in that circular polarisation is preserved rather than reversed.

Various implausible theories were advanced to explain these echoes: for example, buried craters, which are surely absent from

Europa (Chapter 22). The most likely theories invoke either a surface of ice, fractured and pulverised by meteoroids, or else deep snow containing chunks of ice or rocks.[9] Scattering in the upper few metres of this ice-field or snow-field might produce the echoes, although the exact mechanism is uncertain.

Io also reflects radar, but more weakly (Ostro and colleagues, *1990*). It is more similar to rocky planets in its radar albedo (0.06) and polarisation properties (most of the signal being reversed).

Infrared and ultraviolet spectra

Water-ice can be recognised by several infrared absorption bands, notably at 1.5 and 2.0 μm. These major bands are obvious on Ganymede and Europa, and much weaker on Callisto[10] (Fig. 20.20). Water-ice absorbs even more strongly at 3 μm and above, and Europa and Ganymede are essentially black at these wavelengths. However, Callisto has an albedo of 0.04 at 3 μm, so some of its surface must be ice-free.

On Callisto, the weakness of the major bands implies that clean ice only covers about 25% of the surface. This may represent small patches of pure ice in a generally rocky surface, or a general crust of ice thoroughly dirtied by fragments of rock or meteoroids. On Ganymede, the clean ice covers about 65%; presumably the largescale brighter areas are mostly ice. But the minor bands seen in the best spectra of Callisto and Ganymede indicate that the darker areas also consist largely of ice, albeit very dirty, so about 90% of each surface is covered in ice altogether.

On Europa, almost-clean water-ice covers 100% of the surface or not far short of it.

What causes the brown tint of these three moons? They have very similar visible-light spectra (Fig. 20.11), consistent with the reddish component being ferric minerals (typical of meteoroids or of rocks). However, it may alternatively be due to sulphur and other aspects of radiation-darkening (Wolff and Mendis, *1983*). This seems to be the explanation for the comparatively weak colouring of Europa, which darkens smoothly from the leading to the trailing hemisphere according to Voyager photometry, following a cosine law as would be expected for darkening by an external agent. Spectra from the International Ultraviolet Explorer satellite in Earth orbit revealed a trace of sulphur dioxide (SO_2) embedded in the ice of the darker hemisphere (A. Lane and colleagues, *1981*). This is not a pure SO_2 frost, and is attributed to implantation of sulphur ions from the magnetosphere. The observed amount of sulphur atoms could accumulate in only seven years! Elemental sulphur may also be present and explain the colour. But the radiation hitting the trailing hemisphere not only implants ions, but also sputters atoms and molecules out of the surface ice. There must be a complex balance of ion implantation, sputtering and dissociation of water and SO_2 molecules, and re-condensation of these molecules all over the satellite.[11] As SO_2 is only observed on the trailing side, it is calculated that there must be an extra source of water-

[8] Polarisation curves were presented by Dollfus (*1975*) and Veverka (*1975b*).

[9] Goldstein and Green (*1980*); Ostro and Shoemaker (*1990*); Hapke (*1990*).
[10] Pilcher *et al.*(*1972*); Fink *et al.*(*1973*); Clark and McCord (*1980*); Clark (*1980*); Sill and Clark (*1982*).
[11] Sieveka and Johnson (*1982*); Squyres *et al.* (*1983*); Eviatar *et al.* (*1983*).

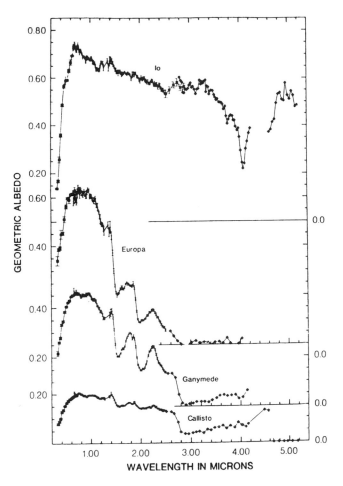

Figure 20.20. Infrared spectra, from Clark and McCord (*1980*).

frost being deposited at about 0.04 – 0.1 μm thickness per year. This conclusion is tentative because it depends on the concentration of sulphur ions in the magnetosphere, which is certainly variable; but it fits in well with other evidence for fresh water-frost on Europa.

There is one remaining mystery about the spectra of Europa: a pair of weak bands at 2.2 and 2.3 μm, recorded in 1980 and 1985, but absent in higher-quality spectra in 1986 (R.H. Brown and colleagues, *1988*). The authors could not identify the bands nor account for their disappearance.

Io has a spectrum quite different from the others, showing no water-ice at all (Fig. 20.20). Before Voyager, Io was thought likely to be covered in salts, perhaps left behind by the evaporation of water. The only distinct absorption band is a sharp one at 4.1 μm, discovered in 1978 from Mauna Kea and from the NASA Kuiper Airborne Observatory, but not identified until after Voyager. Then astronomers realised that this band indicates frozen SO_2.[12] Likewise, strong ultraviolet absorption below 0.32 μm, discovered by the International Ultraviolet Explorer, was identified as due to SO_2. The SO_2 frost on Io, in both infrared and ultraviolet absorption, is mostly on the brighter side, and is believed to correspond to the white equatorial zone. The infrared spectrum at 3–4 μm also

shows minor features due to H_2S and possibly water (as H_2SO_3), mixed into the frozen SO_2 at ≈3% and ≈0.1% respectively.[13] H_2S varies with time (see below).

Sulphur is also believed to be a major constituent of Io's surface, but the evidence is not yet conclusive. Much of the satellite's ultraviolet absorption may be due to sulphur, which absorbs below 0.4–0.5 μm depending on temperature, whereas SO_2 only absorbs below 0.32 μm. Even before Voyager, Kuiper (*1973*) proposed that the dark polar caps might be sulphur produced from H_2S, and Nelson and Hapke (*1978*) proposed that the ultraviolet absorption might be caused by sulphur near volcanic fumaroles. But as the evidence for sulphur and its origin is indirect and largely due to the Voyager mission, further discussion of sulphur is deferred to the next chapter.

Surface temperatures

The average temperature of the sunlit disc, if the satellite were a black body rotating only slowly, would be 145°K at the distance of Jupiter. (The noon temperature at the equator would be 170°K.) The actual temperature of the disc can be deduced from the infrared emission. Brightness temperatures at 8–28 μm, from several studies listed by Morrison and Cruikshank (*1974*), are as follows: Callisto, 142–157°K; Ganymede, 132–143°K; Europa, 120–131°K; Io, 128–139°K. Although brightness temperature is not identical to true temperature, and these studies used satellite diameters slightly different from the true ones, they are a fair measure of the actual disc-average temperatures.

More accurate values were measured by the Voyager IRIS (R. Hanel and colleagues, *1979b*). The peak temperatures around the subsolar point were 155°K for Callisto and 145°K for Ganymede. Night-time temperatures were 85°K on Callisto, Ganymede, and Europa, and the lowest temperature just before dawn (measured on Callisto, but probably applicable to all three) was 80°K.

For all except Io, these values are just what would be expected given the satellites' reflectivities. Ganymede and Europa are colder than 145°K because they reflect much of the sun's light. But Io, again, is odd; it is distinctly warmer than Europa in spite of its high albedo, and Earth-based observers noted anomalously high emission in the 7–10 μm region. The origin of this heat was made clear by Voyager, and this and subsequent observations of the thermal emission will be described in the next chapter.

20.4 THE SEARCH FOR ATMOSPHERES

There are no substantial atmospheres on the galilean moons, as shown by the speed with which they cool down in eclipse and heat up again on emergence. Before Voyager, there was possible but inconclusive evidence for very thin atmospheres on Ganymede and Io – and also evidence to the contrary. Voyager has ruled out the atmosphere considered most likely on Ganymede, and has shown that Io's atmosphere is bizarre and inconstant.

---

[12] Fanale *et al.* (*1979*); Nash and Nelson (*1979*). The IUE result and equatorial distribution were reported by Nelson et al. (*1980*) and Howell *et al.* (*1984*).

[13] Nash and Nelson (*1979*); Salama *et al.* (*1990*). Solid H_2S was also identified from a 3.9 μm band, by Matson and Johnson (*1988*) and Nash and Howell (*1989*).

First of all, what sort of atmospheres could the moons have? Methane or ammonia would be unlikely as the satellites are too warm to hold their ices, and too light to hold the gases without any frozen reservoir. In any case, they are excluded by observations: infrared spectra show no methane nor ammonia to an upper limit of 10^{-7} bar on all four moons.[14] Water is more interesting, as the outer three moons carry large amounts of ice. At their mean surface temperatures, the ice would not evaporate – less than a metre would be lost over the age of the solar system. But warming above 150°K, which may occur in dark regions near the equator, could evaporate some water. If so, the water vapour could dissociate to form a mixed oxygen-plus-water atmosphere on the three icy moons.[15] Water vapour would exist only on the day-side, peaking at 10^{-10} to 10^{-9} bar; oxygen (being non-condensible at the temperatures of these moons) would exist all round, at between 10^{-12} and 10^{-6} bar. Presumably the water would tend to freeze out nearer the poles.

Another possible atmospheric gas is nitrogen. It is the main constituent of the atmospheres of Earth and Triton, and cannot be detected spectroscopically. Of course, there could also be more exotic volatiles that we have not thought of – as turned out to be the case on Io.

What of the observations? If there is a condensible atmosphere in equilibrium with its solid, it could form bright polar caps of ice. The bright polar regions of Ganymede and Europa would fit the bill nicely, and water ice would be the prime candidate. This is quite consistent with the Voyager images. On Europa, the whole surface is white and the polar regions are just not so disfigured as the rest; they may acquire fresh ice from water vapour released very slowly or very rarely elsewhere. On Ganymede, Voyager showed unmistakeable polar caps, pure white in contrast to the yellowish tint of the other bright regions, lying diffusely over the more distinct markings (Fig. 20.29). However, there is no indication as to how recently these caps were deposited; the surfaces of both satellites may date back a billion years (Chapters 22 and 23).

Lyot's visual observations from the Pic du Midi (Lyot, *1943, 1953*) also suggested an atmosphere on Ganymede. He repeatedly saw a bright following (morning) limb, only on a restricted range of longitudes on Ganymede, which did not rotate onto the disc nor reappear at the evening limb. Murray (*1975*) also reported limb-brightening on Ganymede. Lyot suggested that this could be some sort of frost or fog that condenses during the night and evaporates after sunrise – a phenomenon well known to observers of Mars. However, these observations were at the limit of visual abilities, and no such thing has been reported in the Voyager data.

Atmospheric vapour might also condense out during the much shorter time – hours rather than days – while a satellite is eclipsed by Jupiter. If so, the satellite might appear brighter immediately after eclipse, before the frost evaporates. With this in mind, A. Binder and D. Cruikshank (*1964, 1966*) searched for such 'post-eclipse brightening' on the three inner satellites, and found it on one: not Ganymede, but Io. The brightening was of 15% or 0.1

magnitudes in blue light, and lasted ≈15 minutes. This post-eclipse brightening has been recorded several times subsequently – but more often it has failed to occur.[16] It was not detected by Voyager (despite careful photometric monitoring; Fig. 20.21), and has not been seen since then. R. Nelson and colleagues (*1993*) monitored 14 eclipse reappearances during the 1980s, and found a possible post-eclipse brightening of 2% in many cases, but never anything approaching 15%. As Io's reappearances are always quite close to the limb of Jupiter, it is conceivable that earlier observers were tricked by stray light from the planet.

However, the brightening may well be real but intermittent. It could well be due to freezing-out of Io's transient SO_2 atmosphere, which varies with volcanic activity. H_2S may also be involved, as Nash and Howell (*1989*) found H_2S absorption in an infrared spectrum taken 35 minutes after the end of an eclipse, which had disappeared 8 minutes later. After two other eclipses, no such absorption was present. Recently, an amateur astronomer was given time on the Hubble Space Telescope to search for Io's post-eclipse brightening in the ultraviolet (Fig. 20.22).

Occultations provide another way of searching for atmospheres. For Ganymede, there has been just one event seen from Earth, with inconclusive results. This was an occultation of a magnitude-8 star on 1972 June 7, recorded photometrically by R.W. Carlson and colleagues (*1973*). They reported a gradual dimming, indicating an atmosphere of more than 10^{-6} bar, but this was uncertain in view of the poor signal-to-noise ratio. Voyager 1's UV spectrometer observed an occultation of the star κ Centauri by Ganymede, looking for absorption lines at 0.9–1.7 nm, and found no absorption at all. This proved that there is no atmosphere (of oxygen, water vapour, carbon dioxide, nor methane) down to a limit of 10^{-11} bar (A. Broadfoot and colleagues, *1979*). (Voyager's UVS also excluded an atmosphere on Callisto, by showing that there was no day-side airglow.)

Io occulted a magnitude-5 star, ß Scorpii C, on 1971 May 13. The occultation was not sharp, because Io generated diffraction fringes and because the star turned out to be a close double! But this event showed that there was no atmosphere down to an upper limit of 10^{-7} bar. Then the Pioneer 10 spacecraft was aimed so as to be occulted by Io, and the sharp cutoff of its radio signals gave an upper limit of 10^{-8} bar for any atmosphere near the terminator. However, the Pioneer occultation did reveal an ionosphere on Io (A. Kliore and colleagues, *1975*). On the day side, this had a peak electron density of 6×10^4 cm^{-3} at an altitude of 100 km, and it ranged from the surface to at least 700 km altitude; this implies that a thin atmosphere should be present below it, perhaps ≈10^{-9} bar. On the night side, the electron density was an order of magnitude smaller.

The idea of an atmosphere on Io was thrown into confusion with the discovery, before and during the Pioneer missions, of the emission from sodium and various ions in its vicinity. But it was soon shown that these are actually spread around its orbit, not bound to the satellite (Chapter 18). The true atmosphere of Io was detected by Voyager, and will be described in the next chapter.

[14] Fink *et al.* (*1973*). The atmospheric pressure on Earth is roughly 1 bar.
[15] Yung and McElroy (*1975*); Kumar and Hunten (*1982*); Wolff and Mendis (*1983*).

[16] Reviewed by Frey (*1975*); Fanale *et al.* (*1981*); Howell and Sinton (*1989*); Nelson *et al.* (*1993*).

Figure 20.21. Io emerging from eclipse, viewed by Voyager 2. South towards top left. Central meridian longitude ≈30°. (NASA images 20591.49, 51, 53, taken at 100-second intervals; processed by Simon Mentha.)

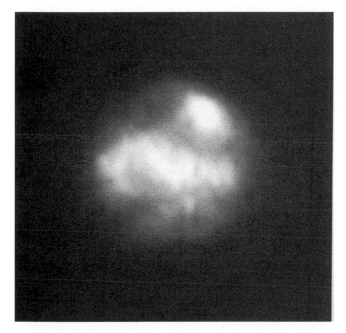

Figure 20.22. Io shortly after an eclipse, imaged by the Hubble Space Telescope, 1992 April 20, as part of an amateur project by James Secosky. South is up. This image is in the near-ultraviolet at 360 nm (similar to Voyager's UV waveband), so the bright areas probably represent SO_2 frost. Within the limits of resolution, the image appears identical to the Voyager one in the previous figure. Taken with the Wide-Field/Planetary Camera and corrected for mirror aberration. (By courtesy of J. Secosky.)

20.5 THE SATELLITE INTERIORS

We have only two kinds of direct data on the satellites' interiors: the surface chemistries, and their bulk densities. Much more has been inferred from their surface geographies revealed by Voyager, but these inferences are all controversial.

The densities were only established accurately by spacecraft. The masses had been worked out to within ±20% in the 1920s by De Sitter and by Sampson, from the mutual perturbations of the satellites' orbits. Better values were given by the Pioneer and Voyager encounters, from the accurate radio tracking of their trajectories as they swung near the moons. The diameters of the moons were established roughly from visual observations, then more precisely from the lightcurves of their mutual occultations. The diameters of Io and Ganymede were established even better by their occultations of stars described above. These values were confirmed by the Voyager images, which also provided accurate values for Europa and Callisto. The final values, and the deduced densities, are listed in Table 20.1.

The striking result is that the densities of the satellites decrease steadily with distance from Jupiter, just as the densities of the planets decrease with distance from the sun. The four moons thus seem more than ever like a miniature solar system. Io and Europa have sizes, masses, and densities very similar to the Earth's Moon. Ganymede and Callisto are larger but less dense (Fig. 20.23).

What are they made of? The likely composition was worked out from basic chemical principles by John S. Lewis (*1972, 1973*) and his main conclusions still hold; a recent review is by G. Schubert and colleagues (*1986*). Suppose that planetesimals, including moons, condensed from a hot primordial nebula of cosmic composition. When the temperature dropped below 400°K, all the heavy elements would have condensed out into dust or rocks, giving bodies with a density of about 2.8 g/cm^3. Io and the Moon actually have somewhat higher densities, indicating more fractionation at some stage; and as Europa's surface is evidently the product of a substantial crust of water-ice (Chapter 22), it too must have a higher density underneath. (It is likely to be 85% rock and 15% water by volume.) As for Ganymede and Callisto, their densities of about 1.9 g/cm^3 are closer to what would result from the next step of condensation. This is the condensation of water into ice, which occurs below 170°K, and should lead to bodies of 1.5 g/cm^3. So Ganymede and Callisto must consist largely of water-ice (40–50% by mass), consistent with the observations of Ganymede's surface. Models for the satellites' structures are shown in Fig. 20.24.

All four satellites should have melted during their formation, due to the heat of collapse and of radioactivity as well as any heat in the original nebula. (Callisto could just have escaped melting and remained undifferentiated, but this is thought unlikely by most researchers.) So Callisto, Ganymede, and Europa are expected to

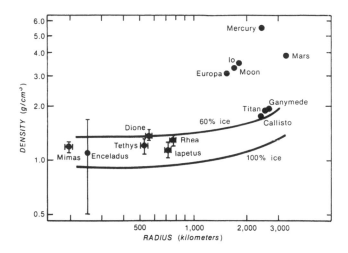

Figure 20.23. Densities of the galilean moons, compared with other moons and planets. The curves show densities predicted for moons consisting of 100% water-ice and for 60% ice:40% rock. (From Morrison, *1982*.)

have largely fractionated into a rocky (silicate) core and a water mantle which is now at least partially frozen. Two aspects of this are uncertain. First, some water may have remained bound to silicates in the outer part of the core for a while after formation, and perhaps even today; progressive exudation of water from the hydrated silicate layer could have increased the volumes of the moons by several per cent. Second, it is uncertain whether the water mantles should be fully frozen or not. They could be kept partly melted (slushy) by radioactive heating from the core (plus tidal heating for Europa; see below). High salt concentrations could also keep the mantles liquid. Or they could have frozen solid, but only if the ice can flow well enough to convey the core's heat outwards by glacial convection; the likelihood of this is controversial. This frozen but geologically plastic mantle is called an 'asthenosphere'. Progressive freezing of the mantles would also have contributed to global expansion; but this would be offset by conversion of ordinary ice into denser forms under pressure.

The surfaces are now so cold that they are very solidly frozen. Even so, the ice below is not as strong as rock, so if these icy moons originally had any substantial mountains or craters, they should largely have subsided by now. This is indeed the case over most of their surfaces (Chapters 22–24).

The scenario described above implies that the progression of satellite densities is due to heat that Jupiter released as it formed, that prevented water from condensing in the inner part of the 'proto-jovian nebula', 4.6 billion years ago. This may indeed be true. But now we know of more immediate reasons for loss of water from Europa and, especially, from Io. These reasons arise from the satellites' orbits.

20.6 THE SATELLITE ORBITS

A regular visual observer soon discovers that the inner three moons repeat their cycles with a period of almost exactly a week. So every Saturday evening, these satellites show more-or-less the same phenomena as the week before. (As it happens, roughly the same face of the planet is presented at weekly intervals as well.) Though the

approximate matches to the rotation periods of Earth and Jupiter are merely coincidence, the coupling between the satellite positions is of crucial importance. This is what has condemned Io to its unique fate.

The orbital period of satellite I (Io) is almost exactly half that of satellite II (Europa), which in turn is close to half that of satellite III (Ganymede). Thus:

$$4P_{\rm I} \approx 2P_{\rm II} \approx P_{\rm III}.$$

The ratios are not exactly one half; but an exact relationship does exist. This is:

$$\frac{1}{P_{\rm I}} - \frac{3}{P_{\rm II}} + \frac{2}{P_{\rm III}} = 0.$$

A corollary of this is a relationship between the orbital longitudes of the satellites (strictly, the mean longitudes, measured from the centre of Jupiter):

$$\lambda_{\rm I} - 3\lambda_{\rm II} + 2\lambda_{\rm III} = {\rm constant}.$$

In fact, this constant is exactly 180°.

This means that the three moons cannot all line up on the same side of Jupiter. As seen from Earth, if two are in transit, the third must be in or near occultation. From time to time these triple events are noticed by observers (and can then be seen again in several successive weeks); some neat examples are shown in Figs. 20.4 and 20.25. Note that the motion of Callisto is not related to that of the inner three, so it is possible for three moons (or shadows) to be in transit simultaneously if one of them is Callisto, while the remaining moon is in occultation or eclipse. This rare and impressive spectacle was first observed in 1867 (S. Gorton, *1867*; W. Denning, *1923*). More recent examples were described by W.E. Fox (*1978*), and one is shown in Fig. 20.25.

Why is the constant exactly 180°? The reason was shown by P. Laplace (*1805*), in the heyday of Newtonian mechanics. He proved that this is a stable resonance. If the formula above were to deviate from 180°, gravitational perturbations between the satellites would bring them back to that value.

The dramatic consequences of this resonance were not realised until the space age, even though the physical principles involved were known a century ago. The breakthrough was made in 1979 by three physicists – S.J. Peale, P. Cassen, and R.T. Reynolds, of NASA and the University of California. Just three days before the Voyager 1 flyby, they published a paper entitled 'Melting of Io by tidal dissipation' which ranks with the discovery of Neptune as one of the greatest successful predictions in the history of science. They argued that the Laplace resonance must impose enormous tidal strains on Io, and predicted that 'widespread and recurrent surface volcanism would occur', and that 'consequences of a largely molten interior may be evident in pictures of Io's surface returned by Voyager 1.' They were right.

Their point was that the satellites pull on each others' orbits in a regular way, dictated by the Laplace resonance, which forces the orbits to be non-circular. Conjunctions of I and II occur when they

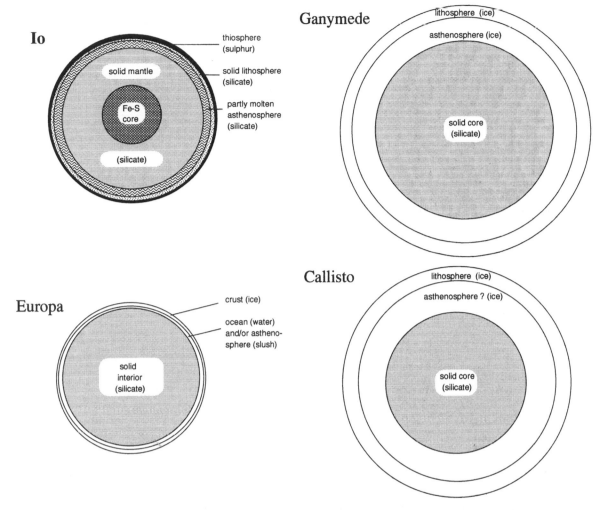

Figure 20.24. Models of the interiors of the moons. For Io, the outermost few kilometres is the 'thiosphere', made of molten sulphur with a solid sulphur and SO$_2$ crust. For Ganymede and Callisto, the state of the water-ice below ≈100 km is uncertain: it may be partly melted (slush), or frozen but plastic ('asthenosphere'), or there may be an undifferentiated ice/rock mantle. (Adapted from Schubert *et al.*,1986.)

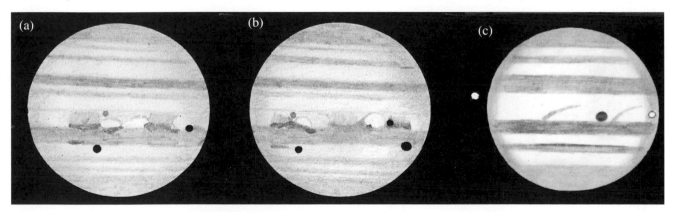

Figure 20.25. Simultaneous phenomena of the galilean moons, illustrating the coupling between the orbits. South is up; transits from right to left. (From BAA archives.) **(a,b)** 1956 April 21, 08.50 and 09.15 UT, drawn by Ben Burrell (in daylight, 25-cm reflector). In transit are I (grey spot on equator), shadow of I (on right side), shadow of IV (on NTropZ), and later shadow of III. Meanwhile II was eclipsed and occulted. **(c)** 1985 Sep 11, drawn by Michael Foulkes (20-cm reflector; satellites enlarged). I approaching occultation, III in transit (dark), and II entering transit.

are near perigee and apogee respectively; conjunctions of II and III occur when II is near perigee. (III can be anywhere in its orbit; it has an unforced eccentricity which requires a separate explanation.) These forced eccentricities mean that the three moons, especially Io, are subject to tidal effects from Jupiter (Fig. 20.26). A satellite in an eccentric orbit moves faster near perigee and slower near apogee according to Kepler's laws. So, as seen from Jupiter, it has a varying distance (leading to variable tidal strain), and a varying rotation (as with the 'libration' of Earth's Moon, leading to tidal torques). If Io was the only satellite, it would long ago have adopted a circular orbit, keeping one face towards Jupiter. As it is, II and III keep Io's orbit eccentric, so the tidal strains and torques heat up the interior. The energy comes from Io's orbit, as if by friction between the different parts of Io.[17]

Thus the satellites deform each others' orbits, in conflict with Jupiter's tides which tend to circularise them. Io bears the brunt of this struggle because it is closest to the planet. The actual amount of heating is difficult to predict because it depends on Io's internal structure. It will be much greater if the interior is partly molten, so melting should be a 'runaway' process. But the most efficient heating occurs if just a thin shell near the outside is molten; therefore, many researchers believe this is the most likely state. In this state, tidal heating can produce the observed heat output of Io's volcanoes (Table 20.3). Europa also suffers significant heating which probably explains its fresh-looking surface, although the actual rate is very uncertain for similar reasons; tidal heating may be enough to maintain a liquid water shell, although it should not have been enough to melt the mantle if it was initially frozen (Chapter 22). However, one must note that all tidally heated moons – Io, Europa, and Saturn's Enceladus – display so much geological activity that one suspects they do have molten mantles, and the best predictions of tidal heating are probably underestimates. Ganymede, however, gets negligible tidal heating as it is too far from the planet.

[17] In mass and in distance from its planet, Io closely matches Earth's Moon, so its tidal influence on Jupiter is similar to the lunar tides on Earth. However, Earth's effect on the Moon is of course much weaker than Jupiter's on Io. In the Earth-Moon system, the tides still trigger earthquakes in both bodies, but these factors have not been enough to circularise the Moon's orbit.

The jovian tide dictates that Io is prolate (stretched along the axis radial to the planet). If it has uniform composition, its diameter is predicted to be 22 km larger than average along a line radial to the planet, and 22 km shorter along its orbit. Measurements of the Voyager images gave a value of 15.3 (±0.2) km (Gaskell et al., 1988), consistent with tidal theory if there is a denser core. For Europa, the corresponding tidal bulge is predicted to be 2.3 km.

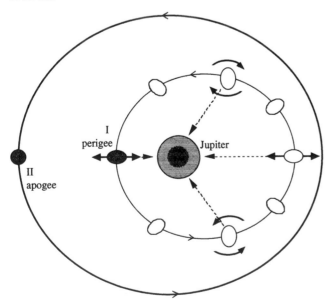

Figure 20.26. Forced eccentricity and tidal heating of Io. Io's orbit is forced to be elliptical by the repeated conjunctions with Europa. Therefore Jupiter's gravity (dashed arrows) imposes variable stretching as Io's distance from the planet varies, and variable torque as Io's orientation relative to the planet varies (heavy arrows).

Table 20.3. Estimates of the heat budgets

Input:	I	II	III	IV
Tidal heating	17–900	0.7–14	–	–
Radioactivity	4.9	2.6	4.7–5.9	3.0–3.9
Primordial heat	(< 5)	(< 2)	(< 10)	(< 5)
Electrical heating	< 2	–	–	–

Output from Io (see Chapter 21):

Heat from volcanoes (hotspots measured from Earth)	600 (±200)
Heat through crust	≤200
Kinetic energy of volcanoes	<200

Energy fluxes are given in units of 10^{11} watts, as reviewed by McEwen et al.(1989), Greenberg (1989), and Schubert et al.(1986).
Tidal heating: lower estimates are for a completely solid moon and 'acceptable' Q_J. Upper values are for a molten mantle and equilibrium orbit.
Radioactivity: Decay in the core, assuming original solar composition.
Primordial heat: Accretional heat, assuming steady loss over lifetime of the system. In fact, it would all have been lost in early times, and these figures are given only as an indication of the former geological importance of primordial heat.
Electrical heating: Power in the Io flux tube; but most is lost via Io's ionosphere and does not heat the moon.

This extraordinary behaviour compels us to ask: why? Why is the ratio of periods so close to 1:2:4, but not exactly there? And why did the system get this way?[18]

The reason that the 1:2:4 resonance is not exact is straightforward. The forced eccentricities are inversely proportional to the amount, v, by which they differ from exact resonance:

$$v = \frac{1}{P_I} - \frac{2}{P_{II}} = \frac{1}{P_{II}} - \frac{2}{P_{III}} = 0.74°/\text{day}$$

(v is the rate at which the longitude of conjunctions, and of perigees, precesses around Jupiter.) So if v was smaller, the forced eccentricities would be much larger. The orbits would become chaotic and the heat dissipation would be stupendous! (The dire effects of exact resonances on orbits can be seen in the asteroid belt, where there are gaps at resonances with Jupiter's orbital period.)

But why is v not larger? The heat generated in Io comes from the kinetic energy of its orbit, which should be slowly shrinking as a result. As the satellites are locked in the Laplace resonance, the energy loss must actually be shared between the three satellites, but the shrinkage should still be noticeable. Io's period should be decreasing by about 3 parts in 10^{10} per year, and it should be free to evolve away from the 1:2:4 resonance. Why has it not done so and thus saved itself from endless torment?

The only likely counterforce comes from the tides that Io raises on Jupiter. The satellite raises a tidal bulge on the planet, which is

[18] The following discussion is based on Greenberg (1989).

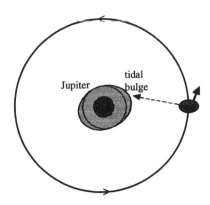

Figure 20.27. Tidal expansion of a moon's orbit. The slight bulge raised in the planet by the moon's gravity rotates ahead of the satellite, and thus pulls it forward in its orbit (dashed arrow), and thus increases the energy in the orbit and causes it to expand (heavy arrow).

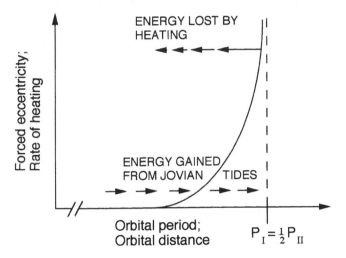

Figure 20.28. Sketch of the opposition between tidal forces on Io. Its orbit loses energy by heating, depending on how close it is to the resonance, but gains energy from jovian tides at a rate that is not so sensitive to this distance. No scales are given as the magnitudes of the effects are still uncertain.

carried ahead of the satellite by the planet's rotation (Fig. 20.27). So the bulge tends to pull the satellite forwards. This accelerates it and thus puts energy into the orbit from the planet's rotation; the orbit is gradually raised and the period increased. Perhaps this effect just balances the loss of orbital energy by heating of Io (Fig. 20.28). If so, the energy for Io's volcanoes comes ultimately from Jupiter's rotation.

However, the rate of this process depends on how readily Jupiter's interior deforms, which is expressed as the 'tidal dissipation parameter'. Most theorists agree that this should be so high that the outward force on Io's orbit would be negligible. No straightforward theory of Jupiter can supply enough energy to Io's orbit to balance that lost by the presently observed heating.

One possibility might be that Io's heating is cyclic.[19] When ν is small, Io suffers intense runaway heating and melting, so its orbit shrinks and ν increases; the heating drops, the interior solidifies, and it creeps outwards again until ν is again small and the heating resumes. The period could be around 10^8 years. Thus Io might

be particularly hot and its orbit shrinking today, but over cosmic time it would be in equilibrium. If the moons have been in such a deeply resonant state, it might also explain the fresh, cracked surface of Europa, for which present tidal heating is thought to be insufficient. It might also explain the present eccentricity of Ganymede's orbit, which should be damped out within 10^8 years by Jupiter's tides.

However, observations over the last 300 years show that Io's orbit is not changing. Shrinkage at the rate predicted from the heat output would have changed the times of eclipses by several minutes over this interval. But J.H. Lieske's studies of the recorded times of eclipses, going back to Roemer in 1668–1678, show that the orbital period is changing by less than 1 part in 10^{11} per year.[20]

This seems to leave only two possibilities. One is that Io's present heat flux is indeed balanced by large tidal dissipation in Jupiter itself, orders of magnitude above that allowed by theorists, implying that the internal structure of Jupiter is not understood. The other is that Io's heat flux over the years 1979–1991 is a sudden, unusual outburst, and that the average heating over the past 300 years has been at least ten times less. Either way, we still have a lot to learn.

If Io is indeed in equilibrium, we can understand how the resonance might have arisen (C.F. Yoder, *1979*). The 1:2:4 ratio is not a favoured resonance, but a barrier through which the orbits cannot pass. Suppose that P_I was initially well below $\frac{1}{2}P_{II}$. Jupiter's tidal influence pushed Io's orbit outwards most rapidly, but as it approached $\frac{1}{2}P_{II}$, the resonance and consequent heat dissipation prevented any further expansion of Io's orbit alone. Orbits I and II then expanded together, but the same effect prevented them ever quite reaching $P_{II} = \frac{1}{2}P_{III}$. So now all three orbits are linked, just short of exact resonance, and here the second-order effect worked out by Laplace keeps their week-by-week motions coupled. In principle they could all expand until $P_{III} = \frac{1}{2}P_{IV}$, but they will not do so within the lifetime of the solar system. P_{IV} is sufficiently large that there are no resonant effects involving Callisto.

So this is the mechanism that melts Io's interior, drives its volcanoes, raises temperatures on its surface, covers it with sulphur crystals and sulphur dioxide frost, creates its blotchy colours, spurts out transient atmospheres, and supplies the sodium and sulphur torus around its orbit. In the next chapter, we will look at these phenomena in detail.

---

[19] Greenberg (*1982, 1989*); Ojakangas and Stevenson (*1986*).

[20] Lieske (*1987*). This conclusion was confirmed by Greenberg (*1989*). Both authors refuted another study which reported a shrinkage at close to the predicted rate.

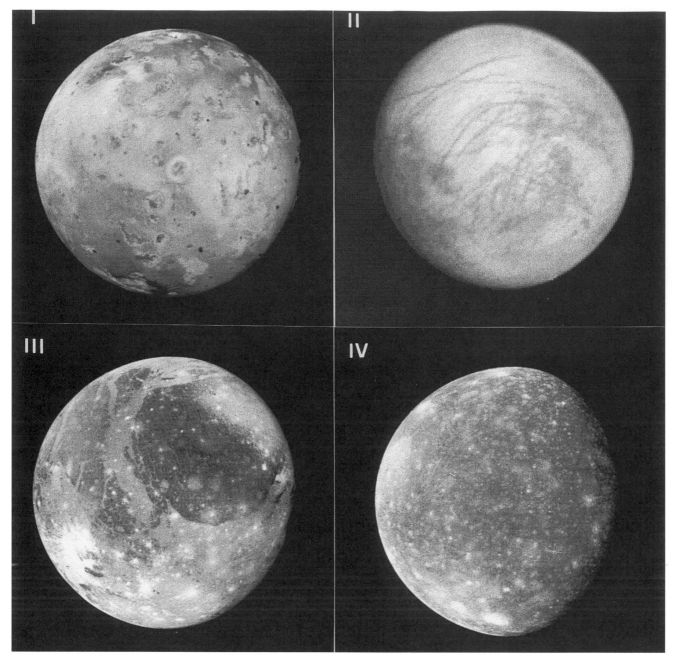

Figure 20.29. Photographs of the four galilean moons from Voyager. North is approximately up. (NASA images processed by Simon Mentha.) I, Io (image 16368.38, blue filter). Central meridian longitude (ω) $\approx$ 160. Volcano Prometheus in centre. II, Europa (image 16323.22), $\omega \approx$ 190. III, Ganymede (image 20608.11), $\omega \approx$ 140. Large dark area is Galileo Regio, and north polar cap overlies its northern part (upper right). IV, Callisto (image 16399.48), $\omega \approx$ 0. Major bright areas are Valhalla near left-hand limb, Adlinda at bottom.

21: Io

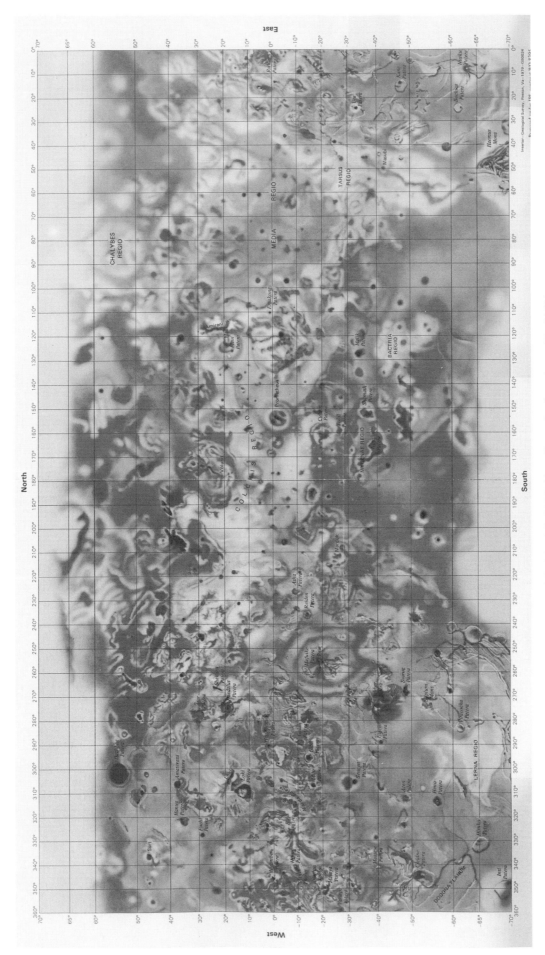

Figure 21.1. Map of Io from Voyager images. From the U.S. Geological Survey, map I–1240, airbrush, by P.M. Bridges. (For south polar region, see Fig. 21.22.)

21: Io

21.1 OVERVIEW[1]

The Voyager images solved the mysteries of Io. They revealed the most volcanic world in the solar system, with plumes from 8 active volcanoes spurting 70–300 km into space, and a surface completely covered in recent volcanic products. Voyager's IRIS confirmed that the active volcanoes were hot. Here is a world that surpasses science fiction.

No impact craters are visible on the surface, down to Voyager's resolution of 2 km, so the surface must be younger than a million years. In fact, the real-time observations imply that it is largely resurfaced within only tens of years.

The principal heat output, 6 $(\pm 2) \times 10^{13}$ W, is in the form of infrared emission from the active volcanoes. Heat lost by conduction through the crust is estimated as 2×10^{13} W or less, and kinetic energy of the volcanoes is also less than 2×10^{13} W. The estimated total of $\approx 7 \times 10^{13}$ W is close to the maximum that tidal energy can supply (previous chapter). It corresponds to a global average of 1.5 (± 0.5) W/m^2, twenty times the average for Earth, and about the same as in Earth's most active geothermal areas.

The surface probably consists largely of sulphur, with extensive deposits of frozen sulphur dioxide (SO_2). SO_2 is certainly vented from the volcanoes, and sometimes produces a local atmosphere before it freezes out. There appear to be two classes of active volcanoes, whose main emissions are SO_2 and sulphur respectively. Most of the active plumes belong to the first class and are essentially SO_2 geysers, spurting continuously, with temperatures observed by Voyager around 400°K. Larger plumes belong to the second class, erupting only transiently, with temperatures around 600°K, in the range of molten sulphur. There is also evidence for eruptions involving silicate lava, as in volcanoes on Earth; this would be even hotter, but no such eruptions were taking place during the Voyager encounters.

[1] Io has been exhaustively described in the scientific and popular literature, not only in the references given in the preface to Part VI, but also in the special Voyager issues of *Nature* and *Journal of Geophysical Research*; in the book edited by Belton, West and Rahe (*1989*); and in semi-popular articles by Johnson and Soderblom (*1983*) and by Morrison (*1985*). This chapter is drawn from these sources, and in particular from chapters by Schaber (*1982*); Nash *et al.*(*1986*); McEwen, Lunine and Carr (*1989*); and Howell and Sinton (*1989*).

The sources of the plumes are dark areas, and look like calderas or broad fissures. There are hundreds of other calderas, visible as dark spots in the Voyager pictures, that were not erupting during the Voyager encounters (although some of them were warm as viewed by IRIS). Many calderas are surrounded by lava flows; it is uncertain whether these are of sulphur or silicates.

Otherwise, most of the surface consists of flat plains, broken into irregular, layered plates in places. Relief in the plains is no higher than 1 km, except for the deepest calderas and one volcanic cone, which reach 2.5 km. But there are also isolated rugged mountains, sharply delineated from the surrounding plains, up to 9 km high; these are believed to be protrusions of the silicate rock layer below. Both the layered plates and the high mountains are most notable in the polar region.

There is a distinct longitudinal segregation of volcanic types. Longitudes $\approx 0°$ to 240° have the bright equatorial zone, probably consisting of SO_2 deposits, and it is in this zone (to 30°N or S) that almost all the SO_2-type plumes were observed. Longitudes ≈ 240–360° are darker and redder, with less SO_2 but probably more sulphur, and these longitudes are marked by the short-lived, larger eruptions. These longitudes also revealed more calderas with lava flows (Fig. 21.7), and more mountain massifs; Voyager's imagery was best in this region, but some such features would probably have been detected if they had existed at other longitudes. This is also the trailing hemisphere and some of the redness may be due to sulphur deposited from Jupiter's magnetosphere, but the large volcanic eruptions are clearly the main source. The near-constancy of the rotational light-curve as observed from Earth (Chapter 20) suggests that such eruptions have favoured this hemisphere in the long term, not just during the Voyager encounters.

Apart from these broad divisions, there is no global-scale pattern to the volcanism. This is consistent with the surface being a thin, largely sulphur crust, floating on a widespread sulphur ocean, and thus unable to support any global stress patterns. As we will see in section 21.4, the regions covered in white plains are likely to have such a structure: they show mainly Prometheus-type plumes that are attributed to SO_2 volcanism driven by liquid sulphur below. The darker trailing hemisphere also shows plenty of evidence for sulphur volcanism, driven by silicate magma below: the Pele-style

Figure 21.2. Figs.21.2–6 are put together as an imaginary circuit around Io, showing 5 of its 8 erupting plumes. This is a mosaic from Voyager 1. Pele is prominent in centre of 'hoof-print'; Loki is near limb at upper left. White plains cover equatorial region on right side. Prometheus plume is on limb at right. (NASA images 16377.50,52,54,56; white light; processed by Simon Mentha.)

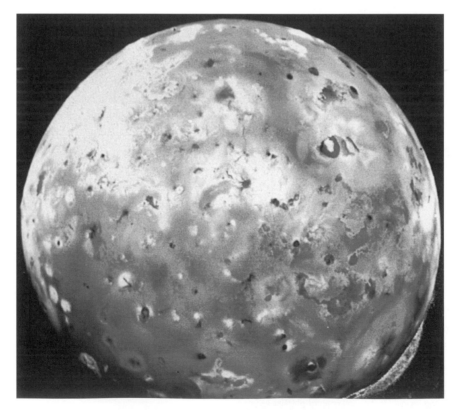

Figure 21.3. Pele plume is above limb at lower right (enhanced). Loki in upper right quadrant. (NASA image 16391.37, processed by Simon Mentha.)

Figure 21.4. The picture on which Io's volcanic eruptions were discovered, taken 3 days after the Voyager 1 encounter (Morabito *et al.*, *1979*). The Pele plume is above the limb at lower right, and the Loki plume shines brightly against the terminator. The dark side of Io is faintly lit by reflected light from Jupiter. (NASA image P-21306 = 16481.09.)

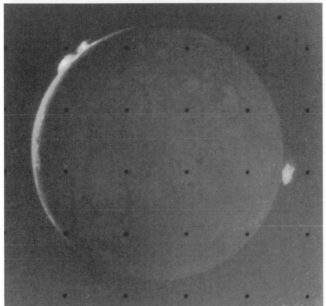

Figure 21.5. Three plumes backlit by the sun, in a view of Io's dark side taken by Voyager 2 during its 'volcano movie'. The dark side is faintly lit by light reflected from Jupiter. Loki is at right, and Maui and Amirani at top left. (NASA P-21773.)

Figure 21.6. A view from the opposite side, in ultraviolet light, also from Voyager 2. At this wavelength (360 nm), the plumes on the limb are bright even in direct sunlight:Maui and Amirani at right, Loki at left. SO_2 deposits are bright but the remainder very dark; the Pele ring is dimly visible on the left. Similar orientation to Fig. 21.2. (NASA image 20621.30, processed by Simon Mentha.)

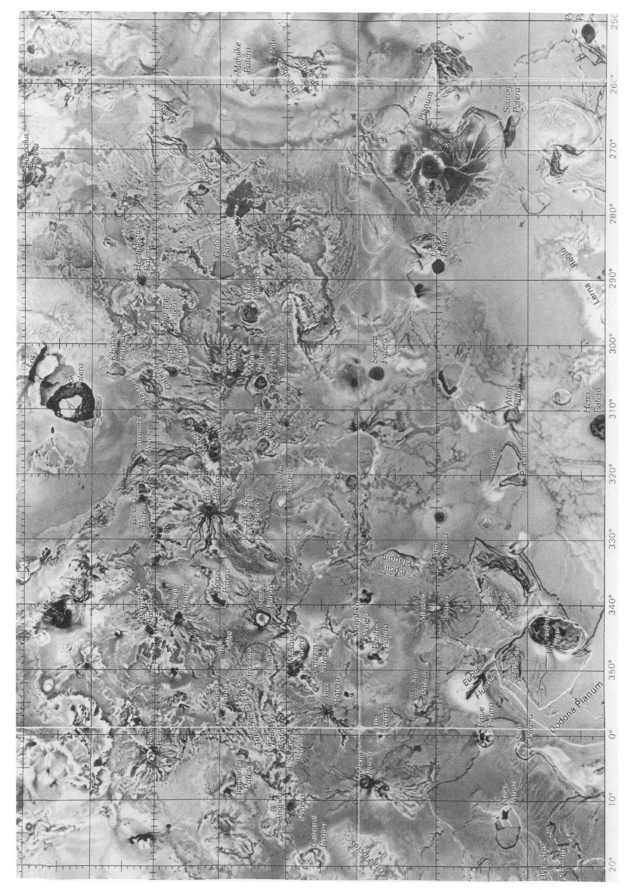

Figure 21.7. Map of the regions viewed most closely by Voyager 1, showing many volcanoes with diverse lava flows. From the U.S. Geological Survey, map I-1713, airbrush by P.M. Bridges, Mercator projection.

eruptions, the great heat flow from the Loki area, and the lava flows. There may also be direct silicate volcanism here, perhaps creating some of the lava flows; and the rugged mountains with which this surface is studded are evidence of larger and deeper upheavals in the silicate crust below, which is otherwise hidden from us.

Colours

The extraordinary patchwork of colours can be classified into four distinct types: white, 'reddish', grey, and dark grey. Most areas display one of these types or a mixture of them (A.S. McEwen, *1988*). *White* areas include the equatorial zone, and also many local diffuse patches around calderas, lava flows, scarps, and mountains. These are all believed to be SO_2 frost. *'Yellow'* and *'reddish'* areas comprise a large fraction of the surface; in the Voyager pictures they appear in various shades of yellow, orange, or red, but the true colours are subtle shades of yellow. These colours are found in plume deposits and lava flows, and may be identified with sulphur (see below). *Grey* areas cover much of the polar regions, though interrupted by large splodges of white or reddish colour; the grey material is not identified. *'Black'* (*dark grey*) deposits are found around a few calderas, and perhaps represent eruptions of volcanic ash.

The published Voyager pictures give a misleading impression of colour. They are strongly biassed towards the red, not only because of the usual Voyager colour exaggeration, but also because they were made from violet, blue, and orange images rather than blue, green and red. So the extreme violet absorption is portrayed as a red colour that is not evident to the human eye. Actual spectra of different regions are shown in Fig. 21.8. In fact the 'reddish' colours are shades of yellow. Beautiful composite images and maps with better-balanced colours have been produced by A.S.

McEwen of the U.S. Geological Survey.[2] Unfortunately, mass reproduction of the subtle tints is difficult, and some have appeared in a sickly green which is as bad as the 'Voyager red'.

One should remember that Io's surface is very bright. Even the warm calderas, which appear black in most Voyager pictures, actually range from near-black (albedo 0.05) to light brown (albedo 0.2–0.4).

For convenience, we will describe colours by their appearance in the published Voyager pictures, but their artificiality needs to be borne in mind.

Nomenclature

The Voyager maps of Io carry mythical names given by the International Astronomical Union (Figs. 21.1,7,22; M. Davies, *1982*). The active volcanoes are named after fire gods and sun gods from around the world. The calderas are named for similar gods and heroes, with the Latin suffix 'Patera'. Names from the classical myth of Io are given to coloured regions (Latin singular 'Regio'), plains ('Planum'), and mountains ('Mons').

21.2 THE ACTIVE VOLCANOES

The view from Voyager

Eight erupting plumes were visible in the Voyager 1 images (Table 21.1).[3] The plumes were first noticed as bright features projecting from the limb (Figs.21.3 & 4), where some of them have an umbrella shape; this is the sign of material ejected symmetrically at

[2] Published in: Johnson and Soderblom (*1983*); Morrison (*1985*); McEwen (*1988*).
[3] Originally described by Morabito *et al.*(*1979*), Strom *et al.*(*1979*) and Strom *et al.*(*1981*).

(a)   (b)

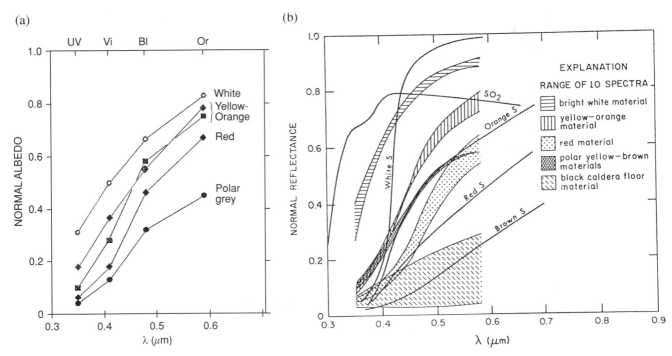

Figure 21.8. Low-resolution visible spectra of differently coloured regions of Io, from photometry of Voyager images in four wavebands. (a) From data of McEwen (*1988*); accurate to ±0.06 albedo units. (b) From Soderblom *et al.* (*1980*). Io spectra (ranges shown shaded) are compared with spectra of solid SO_2 and various forms of sulphur. The wavelength of the sulphur absorption edge depends on temperature. For additional spectra see Nash *et al.*(*1986*) and Moses and Nash (*1991*).

Table 21.1 *Active volcanoes seen by Voyager*

Plume	Name	Lat/long	Width (km)	Height (km)	Plume shape	Source shape	Active in: V1	V2
1	Pele	−19, 256	1200	300	Umbrella	Stripe beneath scarp	++	−
2	Loki (W)	+19, 305	560	200	Diffuse column & UV envelope	End of large stripe	+	++
3	Prometheus	−2, 152	260	75	Umbrella	Spot	+	+
4	Volund	+23, 177	150	80	Column topped by cloud	End of stripe	+	(nd)
5	Amirani	+28, 114	270	90	(Low res.)	End of long stripe	+	+
6	Maui	+19, 122	300	80	Broad, within umbrella	Short stripe	+	+
7	Marduk	−27, 210	200	80	Umbrella?	Irreg. spot	+	+
8	Masubi	−44, 54	(faint)	60	(Faint)	End of stripe	+	+
9	Loki (E)	+17, 301	–	30	Diffuse	End of stripe	(+)	+
10	Surt	+45, 338	(1400)	–	–	Caldera	−	−
11	Aten	−48, 311	(1400)	–	–	Caldera	−	−

Positions and heights are summarised from McEwen *et al.*(*1989*).
Width is from ultraviolet images and is roughly equal to diameter of surface deposits. For Surt and Aten, only surface deposits were observed.

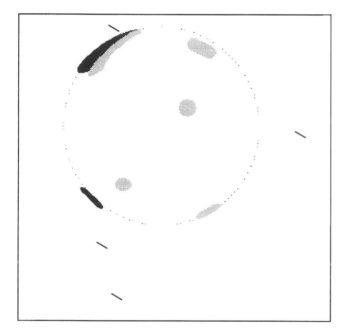

Figure 21.9. The dark side of Io in eclipse; the only light comes from aurorae above the poles and the plumes. This is a negative sketch from a Voyager 1 image, taken while both Io and the spacecraft were in the shadow of Jupiter. In the original 3-minute exposure (image 16395.39; Cook *et al., 1981b*), the glows were only just discernible and they formed multiple images because of spacecraft motions. They are aurorae above the north pole (top left), south pole (bottom right), and plumes 1, 7, 3, and 5+6 (from lower left to upper right). Streaks are star trails. Dotted line indicates the (invisible) limb of Io.

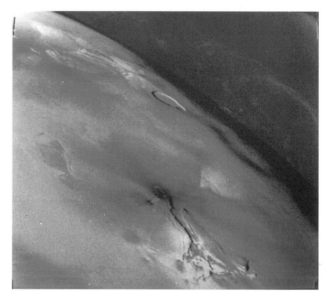

Figure 21.10. Close-up view of Pele, with the plume seen as filaments against the sky (enhanced). Dark streaks radiate from the vent at centre, which lies at the base of a cracked plateau (Danube Planum). (NASA image 16391.30, processed by Simon Mentha.)

a fixed velocity, and travelling in ballistic trajectories. Although the vents must be driven by gas, most of it will condense into 'snow' as it expands into the near–vacuum above, to fall parabolically back to the surface. Other plumes appear more like spikes or clouds, indicating a narrower or less uniform vent. The plumes appear bluish in scattered sunlight, probably because they consist of micron-sized 'snowflakes' and dust. Once identified, the plumes can also be recognised against the sunlit disc, where they appear dark with diffuse spider-like shapes. Obviously the gas entrains plenty of fine dark particles as well as white SO_2 snow.

In a view of Io's dark side in eclipse (Fig. 21.9), the plumes can be seen faintly glowing in the dark, due to aurorae over them. (The vents themselves are not glowing.)

By the time of the Voyager 2 encounter, 4 months after Voyager 1, there were changes. The largest and hottest plume, Pele, had ceased erupting, and the second largest, Loki, had changed its form. But at least 5 of the other 6 plumes, typified by Prometheus, were still active. Pele and Prometheus represent the two different types of volcanism, with Loki an intermediate case (Table 21.2).

The Voyager 1 IRIS scanned several of the erupting volcanoes and found them to be hotspots, coincident with the vent to within the resolution of the instrument. Of 10 hotspots detected, 3 were erupting plumes at temperatures from 400 to 650°K, and 7 were other very dark calderas at 220–380°K. These temperatures referred to the vent regions, 10–80 km across, but each source

Table 21.2 *The two classes of plume eruptions.*

	Pele Type	Prometheus Type
Eruptive height	250–300 km	50–120 km
Diameter of plume deposits	1000–1500 km	200–600 km
Eruptive velocity	1 km s^{-1}	0.5 km s^{-1}
Eruptive duration	Days–months	Months–years
Frequency of new eruptions	10 yr^{-1}	<3 yr^{-1}
Associated surface temperatures	600–650 K	<450 K
Plume optical densities	Optically thin	Portions optically thick
Spectra of plume deposits	Relatively dark red	Bright white
Compositional association	S-rich, SO_2-poor	SO_2-rich
Global distribution	Longitude 240–360°; possibly longitude 0–100°	Latitude +30° to −30°
Geologic interpretations	S vaporized by molten silicates; dominantly silicate crust	SO_2 vaporized at low T (≈ 400 K), perhaps by molten S; abundant SO_2 in crust

From McEwen, Lunine and Carr (*1989*).

also showed a larger, cooler region around it which may have been the plume gases and/or surface deposits (J. Pearl and W. Sinton, *1982*.)

(i) Pele

Pele was the first plume spotted, and the largest, reaching a height of 300 km and a width of 1000 km. A close-up of the plume processed for maximum contrast (Fig. 21.10) reveals wispy filaments. The giant 'hoofprint' that caught the eye in distant views of Io, 900 by 1400 km (Fig. 21.2), marked the fallout from the plume. Bright 'shadows' are cast by two low plateaux within the firing zone, suggesting that a surface wind blows away from the gas-rich vent region. The big indentation in the 'hoofprint' was apparently due to a distortion of the vent itself. North of the 'hoofprint', diffuse broad reddish streaks may be rays of ejecta reaching out to 1500 km from the vent. South-east of the 'hoofprint', there was a transient surface brightening lasting only a few hours during the Voyager 1 encounter, probably due to variations in the ejecta.

The dimensions indicate an ejection speed of 1 km/s, faster than other volcanoes on Io or on Earth. IRIS recorded a temperature of 650°K (±50°K) at the vent, well above the melting point of sulphur. The vent itself (Fig. 21.10) consists of a narrow dark 'fissure' along the edge of a complex set of plateaux. These plateaux may be fragments of a single plate (Danube Planum), split by rift valleys that have expanded to some 30 km in width. They lie on a broad raised area that could mark a deep convective hotspot (see Fig. 21.19).

By the time Voyager 2 arrived, Pele was inactive. The deposits then formed a circle rather than a hoofprint; the obstruction in the vent must have cleared before the eruption ceased.

(ii) Loki

The second-highest eruption, Loki, has the temperature, height, and longevity of the Prometheus type, but has a very large source complex, and sometimes enlarges to Pelean proportions. The plume seen by Voyager 1 was uneven and diffuse; the main part was ≈100 km high but a faint extension, better seen in ultraviolet,

reached 190 km (Fig. 21.11A). The UV-scattering 'envelope' probably consisted of very fine 'snow' crystals only nanometres in size.

The vent was one end of a large black stripe, 200×25 km. The plume was rising across the width of the stripe, and short eruption streaks were also emerging from the other end (Fig. 21.11). Adjacent to the stripe (or 'fissure') was a 'black' area 250 km across, which was quickly dubbed 'the lava lake'. IRIS detected a small hotspot at 450°K, presumably the erupting vent, plus a much larger area averaging 250°K, which seemed to coincide with the 'lava lake'. The vent temperature is suitable for liquid sulphur, and the 'lava lake' is suspected of being liquid sulphur under a thin solidifying scum (although its temperature, which is warm for Io but would be arctic for Earth, has also been attributed to diffusion of the warm plume gases). Although it looks black in the pictures, its albedo is actually about 0.4, much too high for liquid sulphur.

The 'lava lake' contains a large cracked bright patch which may be a raft of solid sulphur, and many tiny dots which could be sulphur 'icebergs'. The 'lake' and the stripe overlap other outlines that are not so dark; the whole complex looks like a huge lake covered by crusts of various ages (Fig. 21.11C). A diffuse annulus ≈800 km across marked the fallout zone from the Loki plume. A similar smaller annulus surrounded the 'lava lake', which may itself have been venting in the recent past.

Voyager 2 found Loki still active. Indeed, the plume soared to 350 km altitude two days before encounter, though it subsided to about 180 km again on the day of encounter. It was now erupting strongly from both ends of the stripe, and the fallout pattern had changed. Also, the northern half of the 'lava lake' had become lighter, presumably freezing over, and perhaps being powdered by fallout from the plume.

(iii) Other large eruptions

The whole region around Loki seemed to be moulded by similar eruptions (Fig. 21.11C). North of Loki was an irregular grey area with overlapping outlines, similar to it in size and colour, and with a 'black', still-warm caldera at one end (Amaterasu Patera). There seemed to be traces of a fallout pattern around this complex. There

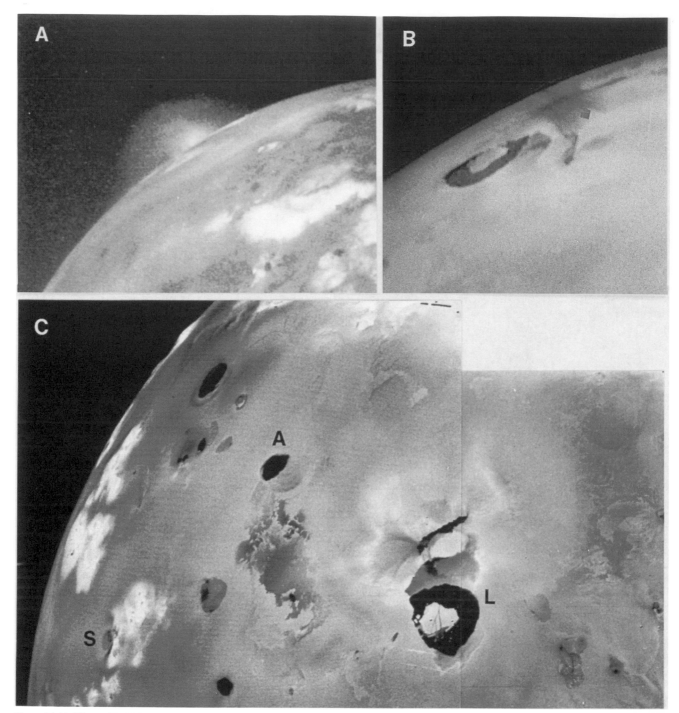

Figure 21.11. Views of Loki from Voyager 1. (NASA images processed by Simon Mentha.)

(A) Loki on the limb; ultraviolet image. (Image 16375.44.)

(B) Oblique view. The plume is visible as dark streaks rising vertically from the 'fissure'. (Image 16377.56, white light; a grid dot has been incompletely removed from the edge of the plume.)

(C) Near-vertical view of the Loki area, showing the 'lava lake' (L) and the 'fissure' with radiating dark streaks and halo. There is no relief in this area and the Loki deposits mask the edges of neighbouring lava flows. Neighbouring calderas include Amaterasu Patera (A), which was a hotspot, and Surt (S), which erupted soon after the Voyager 1 encounter. To the north (upper left), white patches are revealed as hills on the horizon. (Images 16388.58 and 16389.42.)

were plenty of other calderas too. One of them, Surt, was grey and unremarkable during the Voyager 1 encounter, but in the Voyager 2 images it was black and surrounded by a fallout annulus as large and prominent as Loki's (Fig. 21.12). Clearly there had been a major eruption here, though there was no noticeable plume when Voyager 2 arrived. Looking at the photographs of this area (Fig. 21.11C), one has the impression of a very thin crust overlying the 'black' magma, punctured by calderas and lava-lakes which open up, vent for a while, then freeze over and gradually revert to bright crust. The colour patterns on the crust here seem to represent a history of breakage and refreezing from below, and of painting by ejecta fallout from above. The more ragged borders may outline massive floods of some sort of lava. It may be premature to name these volcanoes as they may be unrecognisable after a few decades!

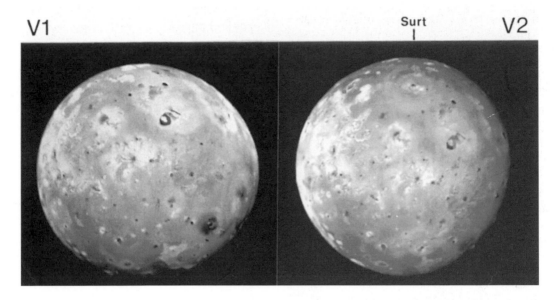

Figure 21.12. Changes in the trailing hemisphere between Voyagers 1 and 2. The deposits around Loki changed, and new haloes appeared around Surt and Aten. (NASA images 16390.49 and 20641.52, both white light; processed by Simon Mentha.)

One other eruption, very like Surt, occurred between the Voyager encounters. It was southwest of Pele, and was named Aten. Like Surt, it was a pre-existing caldera 50 km across which darkened and acquired a halo 1400 km across (Fig. 21.12).

(iv) The Promethean plumes

The other 6 plumes viewed by Voyager 1 were smaller (70–120 km high; Table 21.1). At least 5 of them were still active 4 months later, which implies that their mean lifetime is at least several years. Voyager 2 recorded an 8–hour 'volcano movie', looking at the dark side of Io. This showed plumes 5 and 6 (Amirani and Maui) lit by the evening sun on one limb, and on the other limb the Loki plume emerged as it rotated into sunlight (Fig. 21.5). The plumes did not change in form during the movie, but they brightened as they became more directly backlit, probably because of scattering by micron-sized particles.

Some of these plumes are umbrella-shaped, while others are columns or broad clouds. The vents are 'black' features ranging from small calderas to long stripes; as at Loki, the eruption is usually from one end of the stripe or 'fissure'. Each plume is surrounded by a diffuse halo of bright and/or dark deposits; a bright ring is usually prominent and is probably SO_2 snow.

The best-observed such plume, called Prometheus (Fig. 21.13), is seen erupting dark material from a fountain 15 km across into a dark ring 160 km across. The striking bright ring surrounding this was presumably produced earlier. The vent is a small dark spot, presumably a small caldera, within a more complex marking.

These plumes were not on the side viewed most closely by Voyager 1, but the Amirani/Maui pair did enter the IRIS field of view, and comprised an (unresolved) hotspot at about 400°K. Other erupting plumes were not detectable as hotspots, perhaps because their vents were too small.

Voyager may have missed some small plumes. Several circular

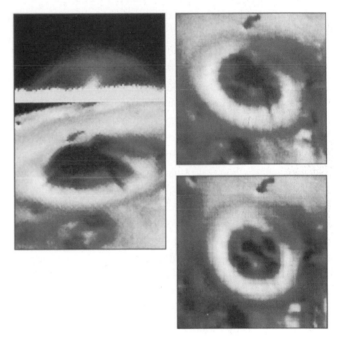

Figure 21.13. Prometheus, from different angles. (NASA images, as in McEwen, *1988*.)

features contained radial dark streaks as in the verified eruptions but were not photographed on the limb. But small eruptions seemed to be truly less common; Voyager could have identified plumes with heights of 30–70 km, and saw none. Some calderas with diffuse haloes were certainly quiet during the flybys.

(v) Other hotspots and calderas

The IRIS scanned 30% of Io and detected seven hotspots apart from the active plumes, and all coincided with dark calderas: Amaterasu, Babbar, Creidne, Svarog, Ulgen, and two unnamed ones.[4] The temperatures were 220–380°K, all cooler than the erupting volcanoes and below the melting point of sulphur (390°K). One of these – an unnamed spot at 81°S, 330°W, with a very mild temperature of 220°K – was a small hole in the layered deposits at the south pole (Fig. 21.22).

Babbar Patera, a large round caldera and hotspot SW of Pele, is unique in having a large, very dark grey halo around it. Perhaps this resulted from an explosive ash eruption?

One other sign of activity was photographed in Mazda Catena, a chain of overlapping calderas which mostly have light brown surfaces (see Fig. 21.17). In a U-shaped caldera at one end, the surface was 'black' in one image, but masked by bright bluish material six hours later – perhaps a smallscale venting of gas.

The view from Earth[5]

Although the volcanoes cannot be seen from Earth, they can be identified by their heat, thanks to recent advances in infrared technology.

Io's anomalous infrared emission was a puzzle before the Voyager encounters. There is excess emission around 7–10 μm, which is variable. And during eclipses, when the surface temperature plunges, a residual emission remains as though 1–2% of the surface consists of hotspots in the range 200–600°K (Fig. 21.14). These are the active volcanoes.

The approximate longitudes of the hotspots can be deduced by measuring the changes in emission as the satellite rotates. At 9 μm, where hotspot emission is the major component in the flux, it was found that more than half the emission came from a single source near longitude 310°, with a diameter of ≈200 km and a mean temperature of ≈350°K. This was Loki with its 'lava lake'. The identification was confirmed by the polarisation of the radiation, which rotates with the satellite; this gave the latitude as well, coincident with Loki.

More precise mapping was permitted by the series of mutual phenomena that occurred in 1985. Photometry of Io being occulted by Callisto, at 3.8 μm, confirmed that a large fraction of the flux came from Loki.

But now the hotspots can be directly resolved. Very large telescopes are needed, aided by modern imaging technology. The first results came from speckle interferometry, in 1984. More recently, Io has been resolved by a new electronic camera called ProtoCAM at the Mauna Kea observatory (J. Spencer and colleagues, *1990*). Both methods have again confirmed Loki as the major source, but some views also revealed newly active sources (see below).

(i) Overall variability

The overall heat flow, as judged by measurements at 8–10 μm, has not changed much from the early 1970s into the mid-1980s. The global average is 1.5 (±0.5) W/m² , or a total of 7×10¹³ W, of which Loki contributes almost a quarter – that is, almost half the emission from one hemisphere.

But at 5 μm wavelength, short enough to reveal just the hottest hotspots, local eruptions can be detected. They do not contribute

[4] Pearl and Sinton (*1982*); McEwen *et al.* (*1989*).
[5] Previously reviewed by Pearl and Sinton (*1982*), Morrison (*1985*), and Howell & Sinton (*1989*), which should be consulted for original references.

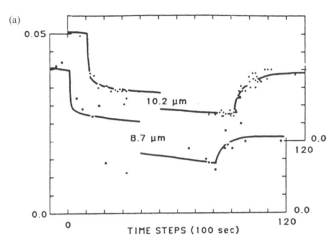

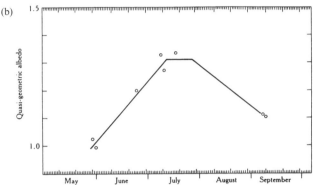

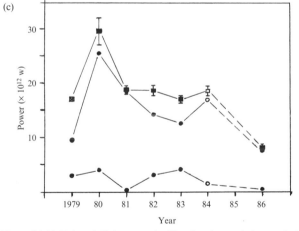

Figure 21.14. Infrared 'light-curves' of Io, showing variations at Loki, on timescales of minutes, weeks, and years.

(a) Infrared emission from Io during eclipse. 'Light-curves' at 8.7 μm (scale at left) and 10.2 μm (scale at right). Ingress into Jupiter's shadow at left, egress at right, recorded on dates before and after opposition respectively. The residual emission during eclipse is due to the volcanoes, and the intensity difference between ingress and egress is mostly due to rotation of Loki across the disc. From Sinton and Kaminski (*1988*).

(b) The brightness of Io during 1984, as measured at 4.8 μm, during an enhancement in the activity at Loki. From Howell and Sinton (*1989*).

(c) Tentative record of the power emitted from the Loki complex, 1979–1986. Brightness is from eclipse photometry as in (A), and could be inaccurate because it is derived from comparison of ingress and egress several months apart. The 1979 data are from Voyager; but as these were derived by a different technique, one must be cautious in comparing them with the Earth-based results. Squares, total power; circles, estimated fractions in large warm component (greater) and small hot component (lesser). Since 1986, Loki has continued to fluctuate between the two extremes shown here. Plotted from data in Sinton and Kaminski (*1988*) and Howell and Sinton (*1989*).

Table 21.3 *New eruptions detected from Earth*

Date	Technique	Lat, long	Diam.(km)	Temp.(°K)	Type	Refs.
1978	Total flux	(Leading)	50	≈600	S	
1979	Total flux	(Trailing) =Surt or Aten?	(nd)	–	–	
1984	Polarimetry	(Leading)	(?)	–	–	
1985 Jun	Rot'n curves	≈260°W =Pele?	small	≈600	S	
1985 Jul	Mut.occ.	–23, 79 (±6°) ='Poliahu'	≤20	≈550	S	Goguen et al. (1988)
1986 Aug	Total flux	(Leading)	≈30	≈900	SiO	Johnson et al. (1988)
1989 Dec	Image	–22, 37 (±3°)	28	460	SO₂	Spencer et al. (1990, 1992)
1990 Mar	Image	– 9, 50 (±3°)	(nd)			Miller (1990)
1992 Jan	Image	–11, 39 (±4°) ='Kanehekili'	20	465		McCleod et al. (1991)
1990 Mar	Image	–19, 255 =Pele	(nd)	–	–	

Weaker eruptions detected by Sinton's group and imaged by Spencer et al.(1992) are not listed.
'Type' indicates whether the eruption is thought to be driven by sulphur (S), silicate (SiO), or SO₂, according to its temperature.
Note the hot eruptions of 1978, 1984, 1985, and 1986 were all on the leading hemisphere, contrary to expectations from the Voyager coverage.

much to the time-averaged heat output, but probably represent dramatic eruptions such as Pele, or even rarer silicate-lava eruptions.

The first such outburst was detected in 1978 from the Kuiper Airborne Observatory; the 5-μm intensity increased five-fold within five hours! It implied a temperature of 600°K over 0.01% of the surface. The idea of a volcanic eruption was considered, but not taken seriously, until the Voyager encounter. Then William Sinton, at Mauna Kea, began monitoring for such events. He detected another five-fold flare a month before the Voyager 2 encounter, which ceased within a day or two; this may have been the Surt or Aten eruption. From then until 1986, only minor changes were recorded, less than two-fold. Most of the outbursts have been on the trailing (Pele-Loki) hemisphere; but recently the largest ones, surprisingly, have been on the leading hemisphere.

(ii) Loki

In 1984 July, Loki was brighter than in previous observations. It was resolved both by speckle interferometry and by infrared polarimetry, and contributed 30–40% of the flux from the trailing hemisphere. Two months later it had halved in brightness (Fig. 21.14b).

Sinton and colleagues have analysed infrared photometry of eclipses by Jupiter (Fig. 21.14), to extract the flux due to the Loki region throughout the 1980s. The deduced flux seems to have been high in 1980 and low in 1986. When the data are split into two components, as in the Voyager data, the small hot component (the vent?) seems to have fluctuated in size and cooled (from a nominal 565°K in 1980 to 475°K in 1986), while the large warm component (the plume or lava lake?) has remained at 250–360°K but shrunk in area. However, these results could have large errors, as they depend on comparing eclipse disappearances and reappearances several months apart – during which the activity of Loki can change markedly.

In recent resolved images, Loki has been shown to fluctuate dramatically over intervals of several months (Spencer and colleagues, *1990, 1992*). The maximum flux (accounting for 80% of Io's total emission) is about ten times the minimum flux; Voyager

observed an intermediate level. Thus, Loki or a nearby source was very bright in 1989 December, faint in 1990 March, very bright again in 1991 January, and faint again in winter, 1991/92.

(iii) New hotspots

These infrared techniques have also located several new hotspots which were not present during the Voyager 2 flyby (Table 21.3). They include one or two new eruptions of Pele, and the first detection of a silicate lava eruption.

1984: Infrared polarimetry detected a new source in the leading hemisphere.

1985: An occultation by Callisto on 1985 July 10 revealed a new hotspot (or the same one?) at 23°S, 79°W, at ≈550°K. (This region was not well observed by Voyager.) However, it was not present during similar mutual occultations on 1985 July 26 nor December 24. This was probably a Pele-type eruption, though on the leading hemisphere.

Meanwhile at about 260°W (the longitude of Pele), a small ≈600°K source was present in 1985 June and had switched off the following month, according to rotation curves at 4.8 μm.

1986: In 1986 August, a major flare was recorded in integrated flux, the first since 1979. Again it was on the *leading* hemisphere. The most exciting aspect was the temperature: ≈900°K, over an area ≈30 km across. This was well above the boiling point of sulphur in a vacuum (715°K), and so must have been due to silicate lava – the only such eruption that has been observed.

1989/90: The first resolved images in the infrared, in late 1989, showed not only Loki but also a new source at 460°K. Spencer and colleagues located it more precisely by timing its disappearance when occulted by Jupiter; it was at 22°S, 37°W (Fig. 21.15; Table 21.3). They proposed to name it Kanehekili, after a Hawaiian thunder god. It has remained present at least up to 1992 January, so this seems to be a new Prometheus-type eruption – although small discrepancies between positions suggest that more than one source may be active.

Meanwhile on the other side of Io, Pele erupted again. This was revealed by speckle-interferometry imaging in 1990 March (Fig. 21.16).

21.3 THE DORMANT VOLCANOES

The most prominent features in the landscape of Io are the calderas – at least 200 with diameters greater than 20 km. They arc distrib-

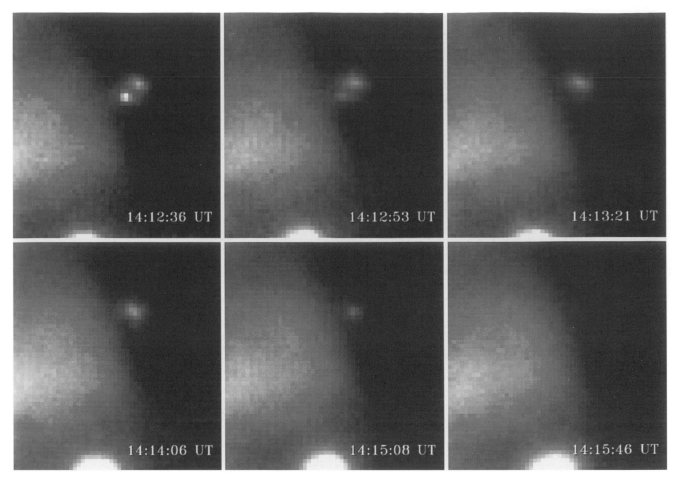

Figure 21.15. Earth-based images of Io at 3.8 μm wavelength, resolving Loki and a new source. Images were made on 1990 Oct. 12 with the ProtoCAM instrument at the NASA IR Telescope Facility in Hawaii. Io is in Jupiter's shadow so the only emission comes from the volcanoes, which disappear as they are occulted by Jupiter. The right-hand spot is Loki (much fainter than in 1979). The brighter, left-hand spot is a new source discovered at Christmas, 1989, named Kanehekili. The very bright spot at bottom is Ganymede in transit, shining by reflected sunlight. (Images taken by J. Spencer, D. Toomey and C. Kaminski; by courtesy of John Spencer.)

uted almost at random, up to latitude 80°S, and occupy about 5% of Io's surface area. The average diameter is about 40 km, ranging up to 250 km at Loki. Their outlines are sharp; most are nearly circular or oval, while some are irregular or fissure-like. They have steep walls and flat floors, ≈1–2 km deep.

The calderas are not perceptibly raised; even for those which have radiating lava flows, the external slopes are imperceptible. So in terrestrial terms, they do not resemble the shield volcanoes like Hawaii, but rather the larger calderas like Valles (New Mexico) or Yellowstone (Wyoming). These erupted mainly ash, with catastrophic collapse of the magma chamber, but the similarity may be misleading. If the ionian calderas are holes in a sulphur crust (a theory which is disputed), they probably opened by melting or foundering of the crust, and the crust may be too weak and the lifetime of the volcanoes too short to permit a mountain to be built.

About half the calderas have visible lava flows around them, implying that they must be slightly raised (or must have been when the lava flowed). Some flows are broad floods, perhaps reaching up to 700 km. More distinct lava flows seem to be concentrated in the southern hemisphere on the trailing side (Fig. 21.7 and 21.17). They look quite like basaltic flows in Hawaii, but are much longer (hun-

dreds of kilometres) and wider. The forms are diverse, but many of the flows are long, narrow, and sinuous, clearly the product of very fluid lava. Their colours in Voyager pictures range from yellow to red to brown to dark grey, although the true colours are subtle yellows and greys. Whether these flows are of silicate lavas (such as basalt), or of sulphur, is a matter of controversy (section 21.4). An important datum is the depth of the calderas, as a sulphur crust could not support relief of more than 1–2 km. The depth has only been accurately measured for one of the deeper ones, Maasaw Patera (Fig. 21.18), which is 2.2 km deep. It resembles a martian volcano with its multiple-collapse caldera and its densely radiating and branching lava flows. In contrast, Ra Patera has longer, sinuous flows in various colours (Fig. 21.17). Adjacent volcanoes can have surprisingly different patterns, ranging from the flowery patterns of Ra Patera to a rare pair of compact shields just to the west. This suggests that different vents tap magma chambers of different compositions.

Many lava flows have bright white puffs along their edges, which are believed to be deposits of SO_2, vented from the ground as the hot lava flowed over it. Similar bright frosts occur around some calderas (e.g. Fig. 21.18).

Only one other shield volcano was photographed by Voyager – a

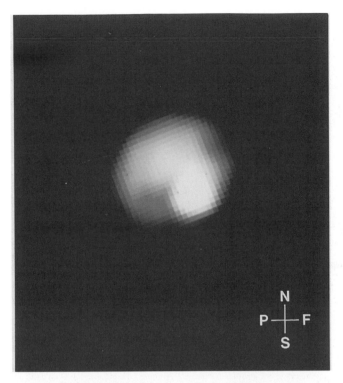

Figure 21.16. Earth-based image of Io at 3.8 μm wavelength, resolving several volcanoes including a new eruption of Pele (the very bright spot). Image made on 1990 March 8 by speckle interferometry using the Multiple Mirror Telescope in Arizona. Central meridian longitude 236°; image back-to-front. (Derived from McLeod *et al.*(*1991*); by courtesy of Brian McLeod.)

shallow cone 2.5 km high and 85 km across, with a 5-km crater at the summit, located in the Pele fallout zone (Fig. 21.25).

21.4 THE PHYSICS AND CHEMISTRY OF IO: INTERIOR, VOLCANOES, AND ATMOSPHERE

The interior of Io[6]

We saw in Chapter 20 that Io must consist mainly of silicates, which are predicted to be partially molten due to tidal heating. In the most extreme model, a silicate crust overlies a largely molten mantle. However, because tidal theory has to strain to match Io's observed heat output, theorists favour the model that gives most efficient tidal heating, in which most of the silicate layers are solid. In that case, as on Earth, there would be a rigid 'lithosphere' overlying a thick 'asthenosphere'[7] – that is, a plastic layer in which convection is possible, 8–100 km thick (Fig. 20.19). Only a very thin layer of silicate magma, below the asthenosphere, would be fully molten – though probably local melting in the asthenosphere would create magma chambers to fuel surface volcanism. This largely-solid model seems favoured by the existence of 9-km-high mountains, which require a lithosphere at least 30 km thick to sup-

[6] Carr (*1986*); Nash *et al.*(*1986*).
[7] The terms 'lithosphere' and 'asthenosphere' are also used for Earth. They are defined by mechanical properties. They are not synonymous with 'crust' and 'mantle', which are defined by chemical properties.

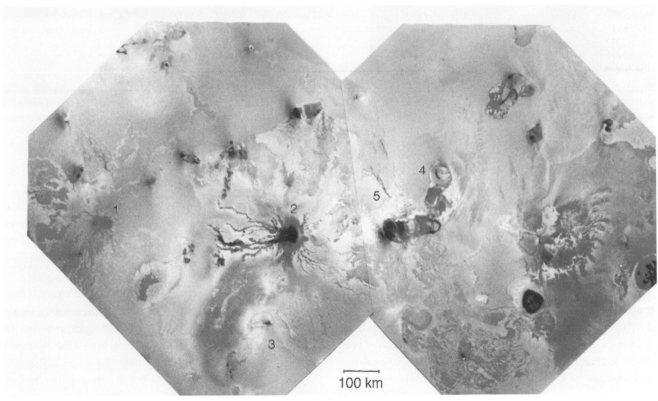

Figure 21.17. Volcanoes in the equatorial region. They include: (1) Dinjin Patera, with radiating flows. (2) Ra Patera, with many radiating lava flows of different colours. (3) Huo-Shen Patera, with a diffuse halo indicating a recent plume eruption, which seems to have buried the Ra lavas on that side. (4) Mazda Catena: a chain of calderas. The bright patches in the north-ernmost (topmost) bay of Mazda are gaseous veils which were not present 6 hours earlier. (5) Near Mazda is a ridge which looks like the product of a fissure eruption of silicate lava; flows down its flanks were evident at higher resolution. (NASA images 16390.06, 12; both blue; processed by Simon Mentha.)

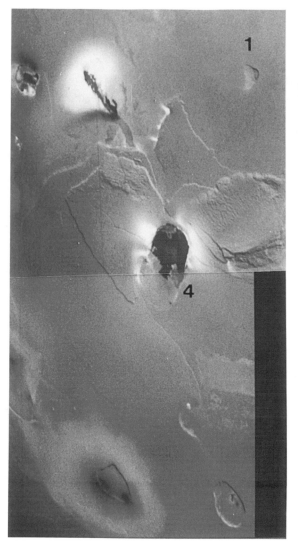
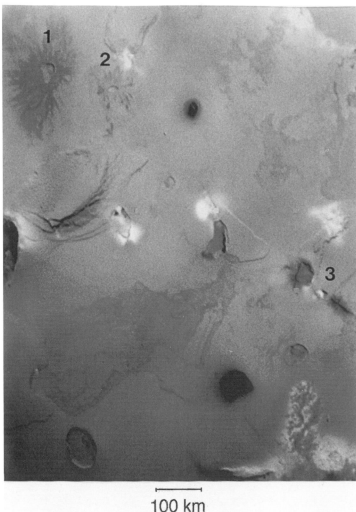

100 km

Figure 21.18. Volcanoes and plateaux in the southern hemisphere. They include: (1) Maasaw Patera, with radiating lava flows (only visible in blue image). (2) Agni Patera, similar but smaller. (3) Aten Patera, lying on a possible fault line; it erupted and became black soon after the Voyager 1 encounter. (4) Creidne Patera, a very large caldera like Loki, and a hotspot. Creidne is embedded in high, layered and tilted plateaux, called Euboia Montes. To top left is a long dark lava flow with a bright white flare; two other flows from the same irregular dark source have flowed to the foot of the plateau. Several other white flares are visible, probably from venting of SO_2, including some around the rim of Creidne Patera. *(Left:* Images 16390.38,40, in white light. *Right:* Image 16392.39, blue, taken later in the encounter with wide-angle camera, with lower sun angle. All NASA images processed by Simon Mentha.)

port them – at least locally. If the mantle were fully molten, tidal theory would predict a silicate crust only 8–18 km thick. But one must note that tidal theory has difficulty in providing enough tidal heating for *any* of the moons that are predicted to experience it – Io, Europa, and Saturn's Enceladus – so it is possible that the theory is deficient, and Io's interior is indeed 'largely molten' as Peale, Cassen and Reynolds predicted in 1979.

Whether it is molten now, or was only molten when it formed, Io's interior has probably separated out a core of iron and iron sulphide like that of Earth, 350–700 km in radius. (Iron oxides have also been suggested for the core.) This core is itself expected to be partly molten, and so to generate a magnetic field. Voyager did not detect such a field, but may not have passed close enough, and any signal may have been overwhelmed by the effects of the Io flux tube.

The shape of Io, determined by trigonometry of Voyager images, gives some insight into the interior. The ellipticity implies that there is a dense core (Chapter 20.6), and there are several broad swells and depressions of 1–2 km alternating around the moon (Fig. 21.19; R. Gaskell and colleagues, *1988*). One of the swells coincides with Pele but others are unrelated to major eruptive centres. They may indicate regions of hotter asthenosphere.

Near the outer surface, all theorists agree that there must be a solid silicate crust (Figs. 20.24 and 21.25). And there is certainly enough heating to generate silicate magma below it, which transports the heat to the surface, and must cause volcanism in the crust. We cannot see the top of this crust (except on the high mountains), but the magma can probably replace it at a rate of several millimetres per year, while the lowest part of the silicate crust is re-melted or dissolved. So this crust is envisaged as a stack of lava flows, being turned inside-out very rapidly, and there must have been plenty of chemical differentiation to produce shells of various compositions.

This theoretical picture of Io is not unlike the Earth, with the

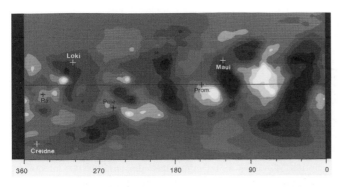

Figure 21.19. Map of global topography, from trigonometry of the Voyager images, in cylindrical projection. Contours are at intervals of 0.5 km; white is high, black is low. Major volcanoes are marked. (Adapted from Gaskell *et al.*(*1988*); by courtesy of Robert Gaskell.)

iron-rich core, the largely solid silicate mantle and crust, and silicate volcanism. But because Io has so much more heating, it has driven off all its water and carbon dioxide, which are important in driving terrestrial volcanic eruptions, as well as in our atmosphere and oceans. The most volatile substances remaining on Io are sulphur and its compounds.

The lava flows: silicates or sulphur?

The surface that we see is believed to be a crust made largely of sulphur, partly overlaid with sulphur dioxide. The evidence for surface sulphur is indirect but seems persuasive. It is:

(i) the low-resolution spectra of Io from the ultraviolet (very dark) to the infrared (very bright), which match powdered sulphur in the laboratory – particularly the colour of the 'yellow' component in Voyager photometry (Fig. 21.8). Most likely, the steep absorption below 0.32 μm is due to SO_2, but the absorption below 0.5 μm is due to sulphur;

(ii) the presence of sulphur ions in the plasma torus (Chapter 18.3);

(iii) the temperatures of the volcanoes, which cluster around the melting point of sulphur;

(iv) the successful theory of SO_2- and sulphur-driven volcanism (see below).

There may also be other sulphur compounds on the surface. Polysulphur oxide, disulphur monoxide (S_2O), and sodium and potassium sulphides have all been proposed and could exist.

Sulphur, which melts at 385–393°K, is presumably melted by the silicate volcanism below, generating sulphur volcanism at the surface. Whether there is also substantial silicate volcanism at the surface is a controversial issue.

The main issue is whether the visible lava flows, and a large proportion of the upper crust, are made of silicates (as on Earth) or of sulphur. According to the proponents of silicates, the 'lava lake' and warm calderas may be filled not with liquid sulphur but with recently solidified basalt lava, merely tinted by sulphur. (Liquid lava would be much hotter, over 1000°K.) The strongest argument for silicate lava is that a sulphur crust could not support the 2-km

relief measured in Maasaw Patera and on one volcanic shield. This does not seem to be conclusive, first because the largescale properties of sulphur under ionian conditions are not well known, and second because these two features may be extreme cases. On the contrary, it is impressive that the typical relief of the calderas and plateaux is generally limited to the 1–2 km that sulphur should be able to support. Another argument is that sulphur could not support the walls of Pele-style vents, which are far hotter than its melting point; but these could well be protected by a refractory coating of silicates (see below). Silicate glass, spattered from large eruptions, could strengthen a mainly sulphur crust.

The silicate lava theory leaves two questions unanswered. First, as there is so much evidence for sulphur, where *are* the sulphur flows? Second, why do these supposed silicate volcanoes not grow into high mountains? One would expect them to erupt explosively through the sulphur layers, like a submarine volcano building an island on Earth. The true answer may well be that some regions of the crust may be mixed, with silicate lava flows interspersed with enough sulphur that the crust still cannot support any great shields, while other regions may be essentially pure sulphur. Certainly the high temperature of one eruption detected from Earth (see above) implies that silicate lava does reach the surface.

If the flows are made of sulphur, what form is it in? It is probably simple yellow sulphur (S_8), although there has been spirited debate about the various other coloured forms that might exist. The colours of sulphur (Fig. 21.8) are due mainly to its strong, broad absorption in the ultraviolet, which extends a variable distance into the visible spectrum according to temperature. On seeing the colours portrayed in the Voyager pictures, some authors[8] suggested that they matched sulphur erupted at different temperatures. The flows from Ra Patera are particularly striking, as they change colour from 'dark brown' at the vent to 'orange' near the tips, as if the sulphur had cooled along the flows. In laboratory experiments, liquid sulphur ranges from black (520–715°K) through brown to red (≈450°K) and orange (≈420°K) to yellow (around the melting point which is 385–393°K). (The forms hotter than 430°K are very viscous and could not flow far.) If hot molten sulphur is quickly freeze-dried, it can retain its darker colours. However, these colours actually seem to be irrelevant to Io.[9] On the one hand, Io's true colours do not match the vivid laboratory sulphurs. On the other hand, the laboratory results are a poor guide to what might happen on Io. Largescale flows in industrial vats on Earth can be dusky red even close to the freezing point, probably due to trace impurities. Conversely Andrew Young (*1984*) noted that under ionian conditions all pure forms are likely to revert to the simple yellow variety within hours as they cool. Indeed, at Io's low temperatures, the yellow colour should be paler than on Earth. However, darker redder forms could persist in fallout from plumes, in microscopic droplets that have chilled very quickly (J. Moses and D. Nash, *1991*). In fact, all the 'reddish' colours of Io including

[8] Sagan (*1979*); Pieri *et al.*(*1984*); Clancy and Danielson (*1981*).
[9] Sill and Clark (*1982*); Young (*1984*); McEwen (*1988*); Greeley *et al.*(*1990*); Moses & Nash (*1991*).

Table 21.4. *Properties of sulphur and sulphur dioxide*

	Sulphur	SO_2
Colour (solid)	Yellow (etc.)	White
Density (solid)	$2.0 g/cm^3$	$1.6 g/cm^3$
Density (liquid)	$1.8 g/cm^3$	$1.5 g/cm^3$
Melting point	385–393°K	198°K
Boiling point	717°K [1 bar]	198°K [0.01 bar]
	1000°K [40 bar]	393°K [40 bar]

the Ra lava flows are quite good matches to ordinary yellow sulphur, and the different shades could be due to slight impurities.

Sulphur flows have been observed in a few volcanoes on Earth.[10] As on Io, they arise from sulphur deposits which are melted by renewal of silicate volcanism. The solidified flows are yellow or brown and resemble various lavas in texture. The largest, in Japan, ran for 1.4 km.

The most striking sulphur landscape was found in the crater of a volcano in Costa Rica. It was formed when an acidic water lake in the crater drained and evaporated away. The exposed crater floor developed sulphur 'lava ponds' about 20 metres across, as well as miniature sulphur volcanoes and flows. These ponds were at 389°K, dark brown, kept molten by bubbling hot gases. Gases in the crater included H_2S and SO_2. This may have been a miniature model of Io.

The plumes: sulphur and sulphur dioxide

The active plumes seem to be more like geysers than terrestrial volcanoes, but their fluid is not water, but sulphur (in the Pele class) or SO_2 (in the Prometheus class). The principles of their operation were outlined by B.A. Smith and colleagues (*1979c*), and worked out in detail by Susan Kieffer of the U.S. Geological Survey (*1982*).

Yellowstone's 'Old Faithful' geyser, if erupting into a vacuum in Io's low gravity, would produce a plume 35 km high. Such terrestrial geysers are driven by the sudden conversion of superheated water to steam; as the upper layers are blown off, the pressure below is reduced, so the boiling accelerates and a violent eruption results. The driving force for other types of eruption on Earth is also the formation of gas bubbles in a hot liquid – for instance, explosive volcanic eruptions are driven by formation of steam or carbon dioxide bubbles in silicate magma.

On Io, the two classes of plumes (Table 21.2) beautifully match the predictions for volcanoes propelled by sulphur and SO_2, respectively (Table 21.4). Sulphur boils above 717°K, and can produce a plume with the ejection speed and temperature of Pele. Boiling of liquid SO_2 at about 400°K can generate a SO_2 plume like Prometheus. In either case, the vapour rapidly expands and forces a mixture of vapour and liquid up the vent until it reaches the speed of sound. Expanding from a funnel-shaped vent, it can emerge with even faster speed. This mechanism can achieve ≈1 km/s, as seen in the real plumes, and is consistent with the apparent absence of smaller plumes with lower ejection speeds. As it

[10] For examples see Carr *et al.*(*1979*) and Cattermole (*1990*). The Costa Rica volcano was described by Oppenheimer and Stevenson (*1989*).

emerges, the vapour cools and condenses, forming sulphur or SO_2 'snow' which arches into the plume and falls back to the surface. (The plumes never achieve escape velocity, which is 2.5 km/s.)

Given Io's heat output, both sulphur and SO_2 should be molten at depths of only 1–2 km (Fig. 21.20). The total thickness of the sulphur layer ('thiosphere') should be less than 10 km from chemical abundance arguments, and less than 5 km because it would vaporise below that depth. The upper 1.5 km or so should be the solid crust. A Pele-type plume could result from a silicate magma eruption at ≈1000°K intruding into the sulphur ocean, boiling the sulphur and forcing it to erupt through its crust. Some silicate might well be entrained with these Pele-style eruptions, and could accumulate as a refractory coating to stabilise the walls of the vent.

The Loki complex seems to show a mixture of sulphur and SO_2 eruptions, and may be heated by a huge system of silicate magma, equivalent to hundreds of Hawaiis. Its heat flux is close to the maximum that can be sustained by silicate magma convection, so it has been likened to a boiling pot of water on a stove.

SO_2 is liquid at temperatures above 198°K (at 1 bar), which should occur at depths below a few hundred metres (estimates range from 100 m to 1 km). The liquid could flow in pores in the sulphur crust – a 'SO_2 aquifer'. Prometheus-style eruptions would occur when liquid sulphur (above 390°K) intrudes into the SO_2-rich crust and boils the SO_2. As the crust floats on the sulphur ocean, it forms a caldera in which liquid SO_2 bleeds out of the bottom of the walls and is vaporised – probably entraining sulphur in the plume (Fig. 21.20).

The Pele-type eruptions may be short-lived because they depend on episodic silicate eruptions below, and perhaps because sulphur is prone to solidify in the vent. Also, if the liquid sulphur exists only in 'magma chambers' rather than a widespread ocean, the eruptions will be limited by depletion of the available liquid. The Prometheus-type eruptions may be more prolonged because liquid SO_2 is extremely fluid (more so than water), and can diffuse horizontally in the crust to keep the eruption going. On a global scale, as SO_2 is liquid at lower pressures near the equator, it may tend to flow equatorwards, thus accounting for the preponderance of plumes at low latitudes.

SO_2 is also suspected of causing the bright white puffs that extend from many topographic features – calderas, lava flows, scarps, and mountains (Fig. 21.22). Around calderas and lava flows, they may be due to surface SO_2 vaporised by the heat of the eruptions. At the bases of scarps, the puffs may be due to springs of liquid SO_2 squirted out under pressure. These 'artesian' fountains would be due to SO_2 that is liquid at depths of little over 100 metres, and flashes into vapour and snow as it reaches the surface. The resulting force could propel it for ≈70 km across the plains, as observed.

The atmosphere of Io

We saw in Chapter 20.3 that there is abundant SO_2 frost on Io's surface. It covers ≈30% of the surface, mainly on the leading side, and is identified with the white equatorial zone on that side, and with the white aureoles around active volcanoes.

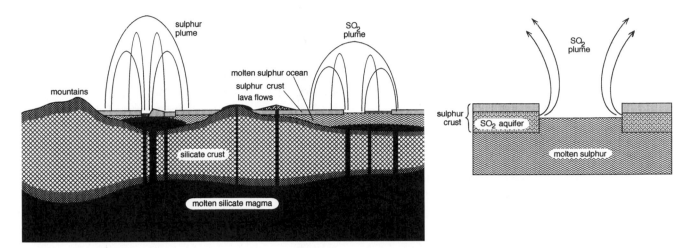

Figure 21.20. Model of Io's outer layers, showing how the two classes of plume are thought to be driven. The smaller diagram shows a SO_2 plume source on an enlarged scale. The sulphur crust may be a few km thick and the silicate crust a few tens of km thick (see text). For deeper layers see Fig. 20.24.

The first definite detection of Io's SO_2 atmosphere was by Voyager 1's IRIS. The Loki hotspot provided an infrared emitter against which absorption lines could be looked for, and a line at 7.4 μm was observed and identified with SO_2. The implied pressure was 10^{-7} bar, and as the observation was made near local noon through an erupting plume, it represented the densest atmosphere that is likely to occur. The field of view was wider than the plume, and the amount of gas observed was probably greater than the content of the plume, indicating an extended atmosphere. (The local atmosphere may have been supplemented by evaporation of SO_2 snow from the plume that was falling on the 'lava lake'; note the white Loki fallout was absent there in Fig. 21.11C.) Similar IRIS observations of the Amirani volcano at night, with a plume and hotspot surrounded by a surface colder than 100°K, showed no SO_2 in the plume; presumably it is all frozen out at night.

The IRIS Loki observation also excluded the presence of several other gases including CO_2, N_2O, CS_2, and H_2O (all less than 10^{-9} bar).

The SO_2 atmosphere has also been detected by an earthbound radio telescope, through its emission lines at 222 and 143 GHz (E. Lellouch and colleagues, *1990, 1992*). The deduced pressure ranged from 0.3 to 4×10^{-8} bar (10^{11} to 10^{12} mols/cm^3), but only over 5–20% of the sunlit surface. A few scans at the start of one session implied a lower pressure but over a larger area, and the researchers suggested that a large local source of SO_2 had just been switched off – perhaps a volcanic plume or a warm spot, rotating round to the night side. Surprisingly, the atmosphere appears to be very hot: ≈500–600°K at 40 km altitude.[11]

The variable SO_2 atmosphere implied by the observations can be understood by analysing the expected behaviour of SO_2 on Io. SO_2 is vaporised and frozen by several competing processes which are in precarious balance, and peak atmospheric pressures of SO_2 predicted by different theoretical treatments have ranged from 10^{-7} to 10^{-12} bar. Now it seems that all the predictions may be right!

The most detailed recent models[12] portray an SO_2 atmosphere that can reach 10^{-7} bar at mid-day or under the volcanic plumes, but is completely frozen out at night (Fig. 21.21). At any pressure above ≈10^{-11} bar, this is a true atmosphere – that is, dense enough to be controlled by molecular collisions so that it obeys the gas laws. Gas at lower pressures constitutes only an 'exosphere', and the magnetospheric plasma can penetrate down to the surface.

How much gas is emitted in the plumes is not known, but a plume like Loki can probably supply a local atmosphere of 10^{-7} bar (as observed by Voyager), which will form a supersonic shock wave around the perimeter of the plume, and should extend more thinly for tens of degrees away from the vent. A volcanic source can best explain the patchiness and high temperature of the atmosphere detected from Earth.

The other source of SO_2 gas is daily sublimation from the SO_2 surface frost. How much comes from this source is uncertain because it depends critically on the daily peak temperature. The mean subsolar temperature is 130°K, which would produce a SO_2 vapour pressure of 1.4×10^{-7} bar. However, the SO_2 frost is so reflective that regions covered by it are noticeably colder, probably only 120°K at midday, so the SO_2 vapour pressure will be only 10^{-8} bar or less. Pressures could be kept even lower if there are subsurface deposits of SO_2 ice, permeable to the atmosphere but hidden from the sun's warmth. It is possible (though unlikely) that such 'cold traps' could essentially eliminate any SO_2 atmosphere (pressure less than 10^{-12} bar).

The SO_2 vapour pressure depends so strongly on temperature that the equilibrium pressure will be much lower away from the subsolar region, and the models predict a supersonic wind blowing away from the subsolar region or the active plumes, towards the terminator where the SO_2 is freezing out. Voyager's IRIS showed temperatures less than 110°K at dawn and dusk, equivalent to SO_2 vapour pressures less than 2×10^{-10} bar, and consistent with the Pioneer

---

[11] An even higher temperature, ≈2000°K, was deduced for the very thin, unbound extension of the atmosphere above 700 km altitude (Schneider *et al.*, *1991b*). This is the inner part of the sodium cloud (Chapter 18.3), but sodium is a negligibly small fraction of Io's atmosphere.

[12] Fanale *et al.*(*1982*); Kumar and Hunten (*1982*); Johnson & Matson (*1989*); Ingersoll (*1989*); Moreno *et al.* (*1991*).

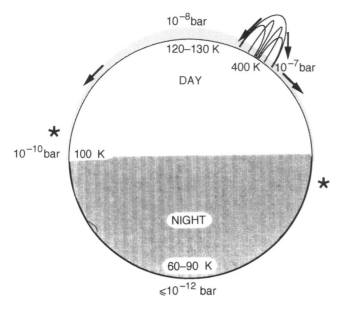

Figure 21.21. Model of Io's SO$_2$ atmosphere. Arrows indicate supersonic winds. Stars indicate where ionosphere was detected by Pioneer 10 occultations.

occultation limits (neutral atmosphere less than 10^{-8} to 10^{-9} bar at dawn and dusk). On the night side, when the temperature plummets to 90°K (or even 61°K in areas of thick SO$_2$ ice), it will all be frozen out (10^{-16} bar) – except in the active plumes. Similar temperatures are reached in eclipses, and SO$_2$ frost deposits formed then would explain 'post-eclipse brightening (Chapter 20.4). The sporadic occurrence of this brightening suggests that volcanic eruptions may be the main source of the SO$_2$ atmosphere.

At night, the frozen surface is unprotected against the magnetospheric plasma, which sputters atoms and molecules far into space. By day, similar sputtering occurs at the top of the atmosphere or of the volcanic plumes. Some atoms enter a corona around Io, while others escape into orbit around Jupiter to produce the Io clouds and torus (Chapter 18.3). The supply of gas to the orbital clouds, at ≈ 1 tonne/second, is no more than 0.1% of the output of the volcanoes.

The ionosphere detected by the Pioneer 10 occultation above Io's terminator (Chapter 20.4) has been a puzzle. It seems to require an atmosphere to support it, perhaps $\approx 10^{-9}$ bar, and all SO$_2$ should be frozen out there, even near a plume. However, the very hot SO$_2$ atmosphere reported by Lellouch and colleagues (*1992*) may provide the answer. Alternatively, there may be other gases; hydrogen sulphide (H$_2$S) and oxygen (O$_2$) have been proposed although neither has been detected.

H$_2$S could well be released from the volcanoes along with sulphur and SO$_2$. It has higher vapour pressures than SO$_2$, so if it is even 1% of the frost, it would maintain a significant atmosphere at the terminator. Small amounts of frozen H$_2$S are detected in the surface frosts (Chapter 20.3&4), and these deposits vary with time. However, no gaseous H$_2$S was detected by Lellouch and colleagues (*1990, 1992*) ($<10^{-10}$ bar).

Oxygen might be produced photochemically from SO$_2$, and being noncondensible at Io's temperatures, it might provide up to 10^{-10} bar of atmosphere even on the night side.

Another continuing puzzle is the nature of Io's polar regions. These regions must be colder than the equatorial zone, so why are they not covered with a white SO$_2$ polar cap – or at least a white circumpolar ring? (Moreno and colleagues (*1991*) predict that SO$_2$ will all be frozen out between 50° and 70° latitude so the poles should be clear.) Perhaps it is cold-trapped beneath the surface in the mid-latitudes, or is rapidly covered over by other deposits. But what is the dull grey material that does cover the poles? Solid H$_2$S, darkened by UV irradiation, is just one candidate that has been suggested.[13]

The resurfacing of Io

How fast is Io's surface renewed? Here we consider the global average; rates may be much faster at low latitudes than at high latitudes.

From the observed rate of change, the estimate is around 1 mm/yr. The absence of impact craters indicates that topography in the kilometre range is erased in ≤ 1 Myr, while the albedo changes between the Voyager missions suggest that most colour patterns will be painted over in less than a century. And if the equatorial SO$_2$ frost is warm enough by day to provide a substantial sublimation atmosphere, it too will need replenishing at ≈ 1 mm/yr.

The volcanoes are resurfacing Io in several ways, and are probably doing so at about the predicted rate. The fallout from the plumes is the most obvious source, but the actual rate is very uncertain because the size of the particles has not been measured. Estimates range from 10^3 to 10^6 tonnes/s over the whole moon, equivalent to between 0.01 and 10 mm/yr. Lava flows must be contributing a similar amount, in rarer but larger eruptions; if they resurfaced at a much lower rate, they would all be covered over. Explosive ash eruptions may also occur. Finally, the opening and closing of calderas may overturn portions of the crust at a similar rate. As for the hemispheric asymmetry: given that 'hot' eruptions have now been observed on the leading side, as well as the trailing side, the absence of visible lava flows and giant haloes on the leading side may be because they are even more rapidly covered over by the fallout from the Promethean plumes.

These estimates are interesting with regard to the possibility of variable tidal heating of Io. It has been proposed that the long-term average heat output may be ten times less than that observed by Voyager (Chapter 20.6), which has been maintained from 1972 to 1991. The present volcanic eruptions can probably account for the erasure of impact craters over the last million years, but it might be more difficult to do so if the volcanic output was an order of magnitude less. However, as a large proportion of the current heat output is simply radiated away from the Loki complex, it is contributing nothing to global resurfacing. Perhaps if the Loki calderas were sealed over, the local heat loss might decrease but productive volcanism might increase. More precise observations of the present resurfacing rates are obviously needed in order to make any definite statement on this question.

[13] Kuiper (*1973*); Nash and Howell (*1989*); Matson & Johnson (*1988*).

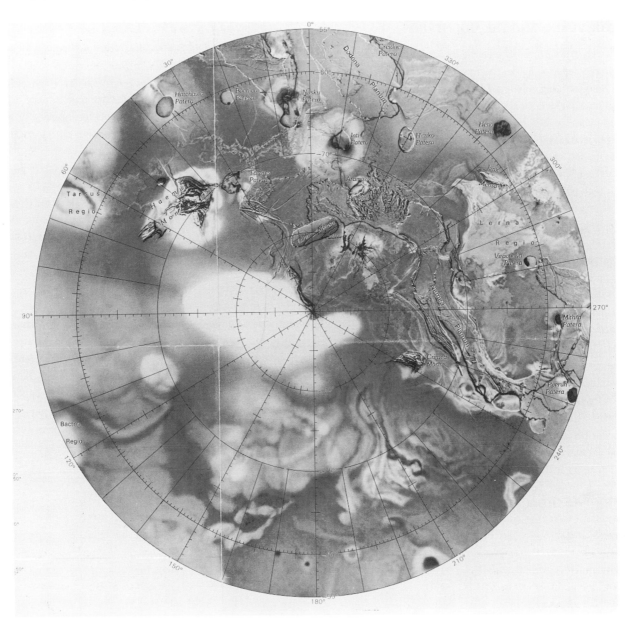

Figure 21.22. Map of the south polar region from Voyager 1, showing layered and eroded plates as well as several mountains and calderas. The irregular black 'fissures' at 81°S, 330°W, were a hotspot according to IRIS. From the U.S. Geological Survey, map I-1713, airbrush by P.G. Hagerty and P.M. Bridges.

Figure 21.23. Haemus Mons. The mountain is 9 km high. It is grooved and fissured, and surrounded by a white halo. South towards top left. (NASA image 16392.00; processed by Simon Mentha.)

21.5 THE PLAINS AND THE MOUNTAINS

The largely flat surface between the calderas is presumably made of volcanic deposits, masked by recent flows, fallout, and condensates. Some striking colour patches show no perceptible relief and are still unexplained. Near the north pole, some of the outstanding white patches are hills, rising above the grey plains (Fig. 21.11C).

In places the surface seems to form plates which are tilted, raised, faulted, or eroded at the edges. The scarps of the plates are typically a few hundred metres high. Especially large plates occur in the south polar region, bounded and crossed by faults. Some are stacked on top of each other in layers; their sharp edges range up to 1.7 km in height. Their edges are often serrated or even broken into mesas (Fig. 21.22), indicating some kind of erosion. The picture is strangely like the layered deposits at the poles of Mars, but the geological processes involved are unknown.

A possible candidate for the erosive agent is SO_2. White puffs attributed to springs of liquid SO_2 are common around these plates, and these springs may undercut the cliffs to produce the erosion.

The plains around the mountains are often stained by white patches or haloes, which again may be attributed to SO_2 venting at the base of the scarp. In this case, SO_2 could obviously be liberated if tides are flexing a sulphur crust up and down alongside the solid mountains. Like ice scraping against a rocky shore on Earth, the crust will break and release volatiles from below. Other signs of crustal mobility are the few mountains where faulted plates of plains material are tilted up against them.

The south polar region also contains the highest mountain, Haemus Mons. It is higher than Mt. Everest, towering 9 km above the plains (Fig. 21.23).

Most of the mountains are only 2–4 km high, and they occur at all latitudes, though mainly in the Pele-Aten region. They stick up ruggedly above the flat plains, and are obviously made of something much more rigid than sulphur, presumably silicate rocks, and must be supported on the underyling silicate crust. They do not look obviously like volcanoes, but more like mountains in the great ranges on Earth or Moon. Some of them are marked by closely-spaced striations, apparently embedded in the rock (e.g. Fig. 21.23). These mountains may be upended blocks of the silicate crust, revealing the lava layers of which it was formed. However, G. Schaber (1982) pointed out that some mountains (including Haemus Mons) have irregular off-centre hollows which could be large, old craters.

A trio of mountains on the north-west edge of the Pele 'hoof-print', Boösaule Montes, enclose a circular bay 170 km across that is filled with flat plains (Fig. 21.24). This may be a vast crater in the silicate crust. And on a tenfold larger scale, is it coincidence that these and other mountains form a loose ring surrounding Pele (Fig. 21.25)? Or is the Pele plume just the surface manifestation of a much larger and deeper volcanic centre? These mountains hint at great tectonic upheavals in the silicate crust which is otherwise hidden from us.

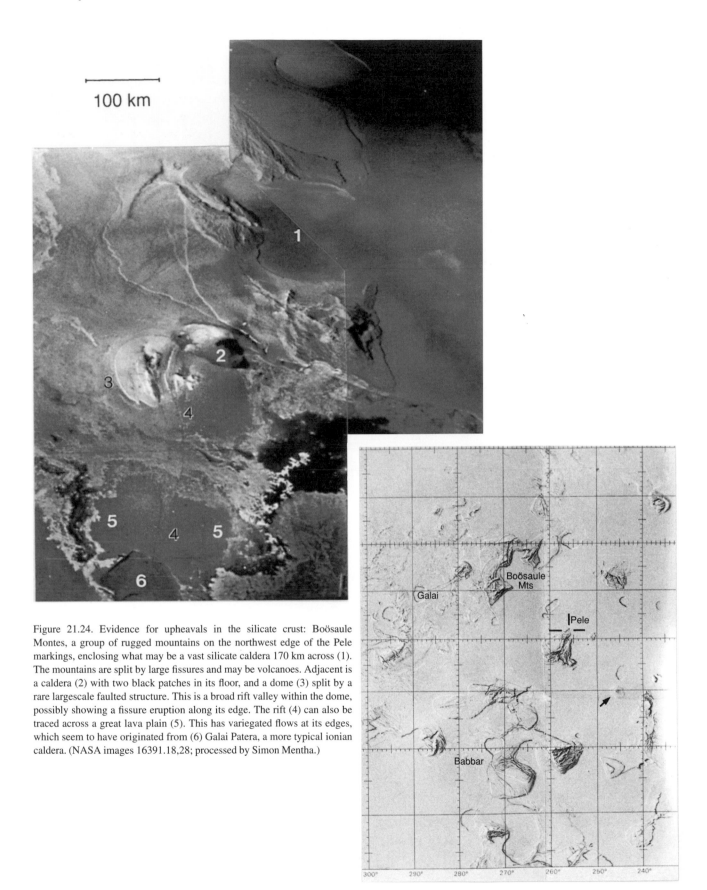

Figure 21.24. Evidence for upheavals in the silicate crust: Boösaule Montes, a group of rugged mountains on the northwest edge of the Pele markings, enclosing what may be a vast silicate caldera 170 km across (1). The mountains are split by large fissures and may be volcanoes. Adjacent is a caldera (2) with two black patches in its floor, and a dome (3) split by a rare largescale faulted structure. This is a broad rift valley within the dome, possibly showing a fissure eruption along its edge. The rift (4) can also be traced across a great lava plain (5). This has variegated flows at its edges, which seem to have originated from (6) Galai Patera, a more typical ionian caldera. (NASA images 16391.18,28; processed by Simon Mentha.)

Figure 21.25. Distribution of mountains in the Pele region. This map shows all discernible relief without the albedo markings. Note how Boösaule Mts and several other massifs form a rough ring around Pele. Arrow indicates a rare conical volcano, 2.5 km high. From the U.S. Geological Survey, map I-1713, airbrush by P.M. Bridges.

Figure 22.1. Map of Europa from Voyager images, from the U.S. Geological Survey; map I-1241, airbrush by J.L. Inge. This is a preliminary map with only approximate positions. For the south polar region, see Fig. 22.8.

INTERIOR—GEOLOGICAL SURVEY, RESTON, VA.—1979. REPRINTED, 1982—G82202

Prepared under JPL contract WO 8395

22: Europa

Europa is the smoothest object in the solar system, and one of the most reflective. Most of its surface is perfectly flat, and nothing exceeds 1 km in height. Its brilliant white surface is crisscrossed by thousands of dark lines. It has been likened to a crazed billiard ball, or to Percival Lowell's canal-infested renditions of Mars, or – most realistically – to the Arctic Ocean of Earth. The surface is indeed a cracked ice crust. Europa is probably the only world that has, or recently had, a water ocean like Earth's.

The Voyager maps of Europa (Fig. 22.1) show three main categories of feature. All have been named out of the classical myth of Europa, after people or places of the Mediterranean mythology. First, there are the sharp lines, which may be dark or bright or mixed; they are given Latin names ending in 'Linea' (dark line) or 'Flexus' (bright ridge). Second, there is brown mottling over large areas, which constitutes the markings seen from Earth. Third, there are a few circular features, either large dark patches (Latin singular 'Macula') or small craters.

We know that water-ice covers most and probably all of Europa's surface, mostly in the form of a deep fine frost or snow (Chapter 20). The contrast is low, only about 10%; the albedo is 0.71 for the bright plains, and 0.5–0.6 for dark bands and spots. So the 'dark' areas on Europa are still very bright.

Colours

There are two sources of dark reddish material that give Europa its (slight) tint. One is from outside, one from inside.

Figure 22.2. Europa in transit against Jupiter's South Temperate Belt. North is to top left. (NASA image 16317.38, processed by Simon Mentha.)

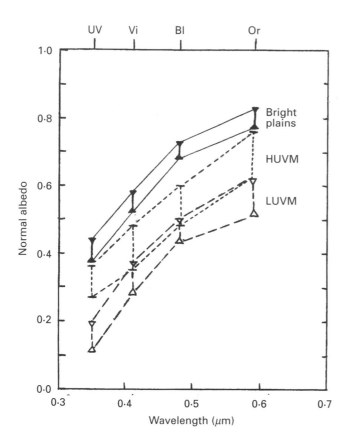

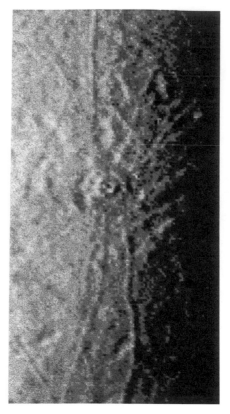

Figure 22.3. Low-resolution visible spectra of features on Europa, from the Voyager images. The four Voyager wavebands are indicated. 'HUVM' is broad grey-brown bands and large spots, while 'LUVM' is other lines and spots. (Data replotted from Johnson *et al.,1983*.)

Figure 22.4. The 'crater' Taliesin at 25°S, 135°W. This feature consists of two or three shallow rings; the inner ring, on a mound, is ≈30 km across. Is it an impact crater or some sort of ice volcano? (From NASA image 20649.46 [1219], processed by Simon Mentha.)

First, the smooth global variation of albedo and colour, following a cosine law from the leading to the trailing hemisphere, is attributed to sulphur ions implanted from the magnetosphere as described in Chapter 20.3.

Second, the local colours are interpreted as contamination of the ice with material from below – muddy ice that welled up to fill cracks and punctures in the pure ice crust. The raw Voyager photometry seemed to show two dark colour units, one encompassing the dark material in the lines, pits, and mottled areas, and the other occupying fewer but broader features (broad bands, wedge-shaped cracks, and 'maculae') (Fig. 22.3). But when the global cosine variation is accounted for, it is found that all the dark features have the same brown colour.[1]

Craters, palimpsests, and pits

The frequency of impact craters on a moon can indicate the age of the surface. On Europa, there are very few. Scrutiny of all the Voyager pictures revealed only five fresh craters, all 10–30 km across (B. Lucchitta and L. Soderblom, *1982*). One has dark rays. Two have central peaks. One of the latter (Fig. 22.4) has several concentric rings, is surrounded by a bright halo, and seems to be raised above the surrounding terrain, like some craters on Mars; if it is an impact crater, this suggests that there has been erosion of the surrounding surface. The density of these craters is comparable

to that on continental shields on Earth, and slightly less than that on the youngest lunar maria, implying an age of 0.1 to 1 billion years for the surface. However, this may apply only to select regions of more resistant crust, which must be tens of kilometres thick to support these craters.

There are also circular dark patches without any relief. The numerous small ones (less than 10 km across) seem to be unrelated to cratering. They are scattered across the surface, some on the lines and others not; perhaps they are traces of punctures in the ice crust? But the handful of large ones (80–130 km across) may be the scars of craters that have collapsed as the ice subsided. Such relics are called 'palimpsests'; a palimpsest is a parchment that has been erased and written over more than once. The one photographed most distinctly, Tyre Macula, shows traces of multiple rings (Fig. 22.5). Others are more irregular.

Closeups of the terminator region show numerous shallow, irregular pits (plus a few mounds). How the pits arose is unknown.

The canals of Europa

The whole surface is marked by dark grey-brown lines, which look like cracks in the crust, up to 30 km wide. The largest are thousands of kilometres long and one forms three-quarters of a great circle. They are straight or curved; they branch and cross each other. Global fracture systems coexist with much more intricate patterns of local fracturing, which include tortuous loops and wedges. Altogether the lines comprise 5–15% of the surface area.

[1] McEwen (*1986b*); Buratti and Golombek (*1988*).

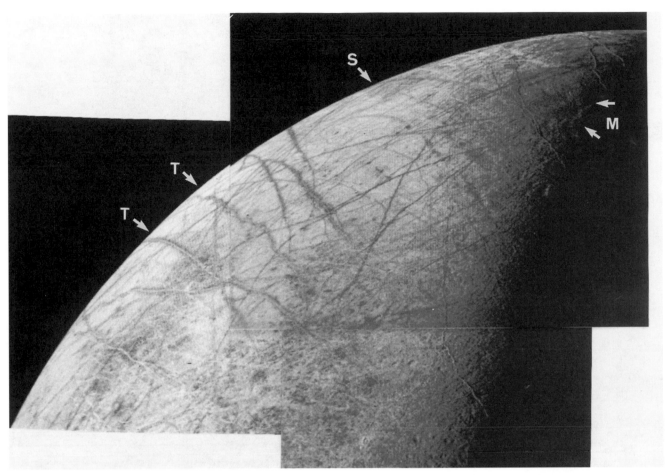

Figure 22.5. Part of the northern hemisphere. (M), Tyre Macula, showing shallow concentric polygonal 'moats' which confirm that it is a palimpsest like those on Ganymede. South of it is an irregular pit. (T), triple bands of the global fracture system. Triple bands and dark bands are evidently of various ages. The oldest are broken up into spots (S). (NASA images 20652.11 [1364] and 20651.55 [1348], processed by Simon Mentha.)

Most of the lines are simple dark strips. No shadows are cast in them, so they are less than 100 metres deep. But a change in the brightness of one of them with viewing angle implied that it might be a very gentle trough. There is no specular reflection from them so they do not constitute clear ice; they are probably muddy ice that has welled up from below in fractures. Many of the global-scale, great-circle bands are triple, as a white line runs down the centre of the broader dark line. In a few cases, the central bright line is definitely a ridge. From the ways in which lines intersect, one can tell that parts of some triple bands pre-date other fracture systems, whereas other parts are more recent (Fig. 22.5). So the global fractures forming the triple bands took place over a lengthy span of time. Fig. 22.5 shows how some old global-scale lines are broken up into chains of spots by younger, narrower lines; the older line is often obliterated by bright ice formed on each side of the younger one, but at the intersection there is sometimes a round dark 'oasis'. In a few places, isolated dark spots are aligned in rows (Fig. 22.6); these may mark lines which have disappeared except for a few remaining 'ponds'.

The most informative patterns are those of the wedge-shaped bands, as pointed out by Paul Schenk and colleagues (Figs. 22.6).[2]

Once they have been picked out from the tangle of other lines, it is obvious that they represent the drifting of pack-ice. A large section of the crust has clearly rotated, fraying at the edges, opening up the wedge-shaped cracks, and offsetting earlier lines by 25 km. To the west, the pattern of cracks suggests that the crust has been stretched. If one looks for compensating compression, the eye falls on Agenor Linea, a notable triple band whose white centre line may be a wrinkle ridge extruded within an ancient crack. The interior of the rotated region is a maze of tiny cracks, and some crust may perhaps have been piled up in this region. All this suggests that the crust was not much more than 10 km thick when the movements occurred.

Bright lines are also seen, near the terminator. They are low ridges caught by the sunlight, and are actually widespread over the surface. Most do not have high albedo, and many become faint dark lines at high sun angle – though a few bright lines can be seen under full sunlight, running parallel to the triple bands. The bright ridges are 5–10 km wide and at most a few hundred metres high. The most conspicuous set, in the southern hemisphere, has the remarkable form of arcs in a cycloidal pattern with a repeat length of 100 km or more (Figs. 22.7 & 8). Some of them cross each other. As they catch the light towards the terminator, they look like spider-webs strung across the surface.

[2] Schenk and Seyfert (1980); Schenk and McKinnon (1989).

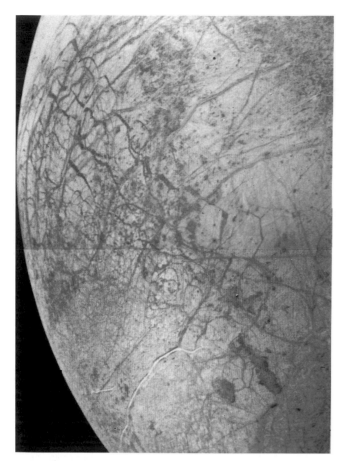

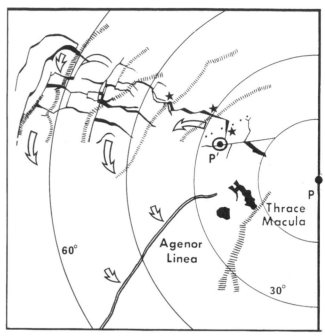

Figure 22.6. The area of wedge-shaped bands.
(A) Photograph of the whole area. Some of the wedge-shaped bands are crossed by just-visible bright striations parallel to the displacement; some of these extend into dark lines on either side. (NASA image 20650.22 [1255], processed by Simon Mentha.)

(B) Diagram of the wedge-shaped bands showing the apparent rotations (about point P or P') and offsets (marked by stars). Older, more diffuse bands, including triple bands, are shown striated. (Adapted from Schenk and McKinnon, *1989.*)

These bright lines are thought to be wrinkle ridges produced by compression of the surface. They cut across all other features, and so must be the youngest features on Europa.

The brown mottled areas

The mottled areas seem to be concentrations of small cracks or pits, filled with the same brownish material as the major lines. This is evident from the rough pitted texture of some mottled areas viewed near the terminator. The maximum relief is about 1 km. Another area, lit vertically in the photographs, seems to be a tangle of crooked lines branching smaller and smaller (in the region of the wedge-shaped cracks; Fig. 22.6). In general, the major lines do extend into the mottled areas, but tend to be narrower and more irregular within them.

So it is not clear whether the mottled areas are regions where the crust is more damaged (due to more internal activity?), or less rapidly coated by frost (due to less internal activity?), or where there is a greater flux of contaminants (due to more internal activity?), or where the silicate core is closer to the surface (and thus accessible via volcanism or via meteorite impacts).

Tectonics and resurfacing on Europa

According to the models of the galilean moons (Chapter 20.5), Europa consists mainly of rock, but has enough water to provide an outer layer ≈ 100 km thick, comprising 15% of the volume. Theorists are uncertain whether this should be mostly liquid or frozen (Chapter 20.6). The surface is certainly of ice, and its flatness and its fracture systems imply that it is quite a thin ice crust on a liquid interior, or has been so in the geologically recent past. But there are several unanswered questions about Europa, and there are two divergent models; the old and the young Europa. According to the 'old' model, the surface is many millions of years old, produced during a temporary episode of melting, and the water shell is now almost completely frozen, the ice being at least 40 km thick. According to the 'young' model, the surface is active today, with a global ocean covered by an ice crust less than 25 km thick.

The major open questions are as follows.

What produced the fracture systems? There are six theories, which are not mutually exclusive.

(i) Global expansion due to slow exudation of water from hydrated silicates in the core (Chapter 20.5). This could increase the surface area by 5–10%, similar to the area of the lines.

(ii) Global expansion due to freezing of the water shell. This could only increase the area by 0.7%.

(iii Stresses caused by uneven internal heating, producing regional convection in the ocean.

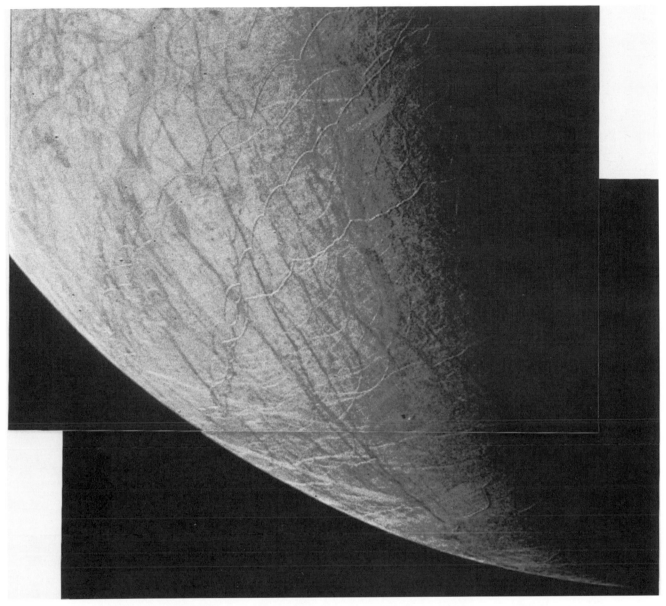

Figure 22.7. South polar region, showing arcuate ridges. The small bowl-shaped crater, called Rhiannon, is at 82°S, 220 km from the south pole. (NASA image 20652.19 [1372], processed by Simon Mentha.)

(iv) A global re-orientation of the crust, in response to uneven crustal thickness, caused by a previous history of uneven heating.

(v) Tidal stresses due to the present orbital resonance. This seems unlikely because present tidal stresses are too low to crack an ice crust; a much higher orbital eccentricity would be needed. Anyway, these stresses would produce fractures converging on the subjovian and antijovian points (longitudes 0° and 180°), which is clearly not the case (Fig. 22.9).

(vi) Tidal stresses due to non-synchronous rotation. This would produce fractures converging on the equator at longitudes 135° and 315°, which is not far off the actual pattern (Fig. 22.9).

The largest global line systems, including the triple bands, do converge vaguely around these points, although there is much criss-crossing. According to A. McEwen (*1986a*), the best fit is to points 25° further west. This suggests that the cracks were produced during a time of non-synchronous rotation that has now stopped. This could have occurred during the evolution of the orbital resonances, which might also have raised tides sufficient to crack the crust (Chapter 20.6). As the cracks do not all converge, they may have been produced at different times, and indeed the bright ridges and the western extensions of the dark lines do seem to have orientations more consistent with their being more recent. Of course, tidal fracturing may have been combined with global expansion, and regional heating could have contributed to regional fracture patterns.

How thick is the ice crust? From the observations, all one can infer is that it was not much more than 10 km thick when some of the fractures were produced. From theory, one can get widely divergent answers, as Europa appears to be near a critical state. A young Europa is possible if there is enough tidal heating (Table

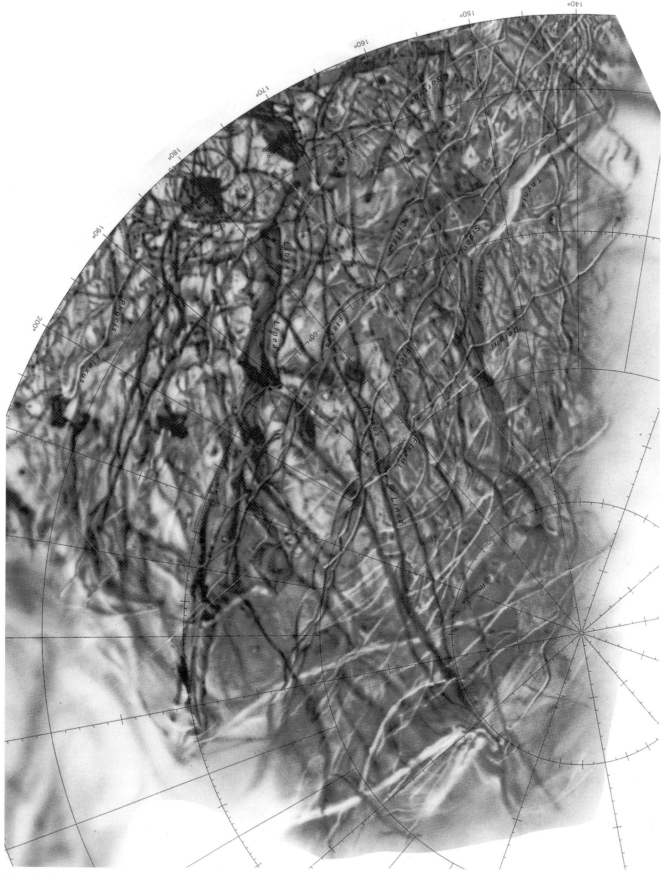

Figure 22.8. South polar region: map from the U.S. Geological Survey, I–1499, airbrush by S.L. Davis.

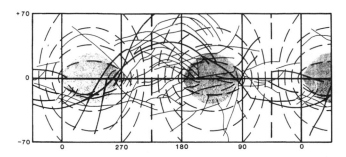

Figure 22.9. Map of global-scale lines on Europa. Superimposed on it (dashed lines) are the directions of tidal tensile stresses that would result if Europa were not quite in synchronous rotation now. Shaded areas would be under compression. These patterns would match the observed lines better if shifted 25° in longitude, suggesting that the fractures arose from non-synchronous rotation in the past. (From McEwen, *1986a*.)

20.3); thus S. Squyres and colleagues (*1983*) predict enough heating to maintain a global ocean with an ice crust only 16 km thick, and G. Ojakangas and D. Stevenson (*1989*) concur that it should be less than 25 km thick. In contrast, others such as M. Ross and G. Schubert (*1987*) calculate much less tidal heating, barely enough to maintain any liquid water. The difficulty is that if the ocean begins to freeze, the tidal heating is reduced further; and if the ice crust grows thicker than 30 km, glacial convection may occur within its lower layers (an asthenosphere), so that heat can be transported out by that route and the ocean may freeze completely. As the physical properties of an icy moon are really quite uncertain, this theoretical dichotomy is unresolved.

What would happen when a crack opens up in an ice crust? First, the exposed water would boil into the vacuum, but only briefly. Very soon the exposed water would freeze. It would not erupt over the surface because the crust is floating on the water. The darker colour of ice in the fractures suggests that the ocean below is contaminated with brown material: silicate particles from hot springs? sulphur? or organic compounds?

Geyser-like activity could occur only if there were volcanic activity on the floor of the ocean. A geyser could be driven by boiling from below, or possibly by bubbles of volcanic gases that made a froth light enough to reach the surface.

Arthur C. Clarke, in his novel *2010*, pictured Europa's oceans as an abode of life. The prospect is not impossible, at least for single-celled organisms, according to Ray Reynolds and Steven Squyres (*1982*). If the ocean does exist, it could well contain organic compounds. As for energy, there may be geothermal energy from volcanic vents, and the ice 'canals' may be transparent enough to allow sunlight into the ocean.

How is the frost renewed? The foregoing implies that Europa can never have large lakes of water at the surface. The bright surface frost probably comes from slow emissions of water vapour, or from occasional venting through cracks or geysers, or from magnetospheric sputtering of the surface ice (Chapter 20.3). (It cannot be due to vapour liberated by sunlight or by meteoroids, as these agents would act equally on Ganymede.) The fact that it does not cover the darker surfaces of the fractures suggests that the frost is quite old.

How old is the surface? This is the greatest unknown. The existence of several impact craters suggests an age of 100–500 million years for at least part of the crust, and comparison with the bright ray craters on Ganymede suggests that the 'fresh' frost surface could retain its brightness for this length of time. On the other hand, the analysis of magnetospheric colouring suggests that the frost may be renewed on a timescale of only a few years (Chapter 20.3, and McEwen, *1986b*).

Is Europa active today? The only positive hint has been one of the earliest Earth-based infrared measurements, which showed Europa warmer than in later observations, but this anomaly has not recurred (R. Howell and W. Sinton, *1989*). Perhaps the Galileo mission will show us whether any changes are occurring.

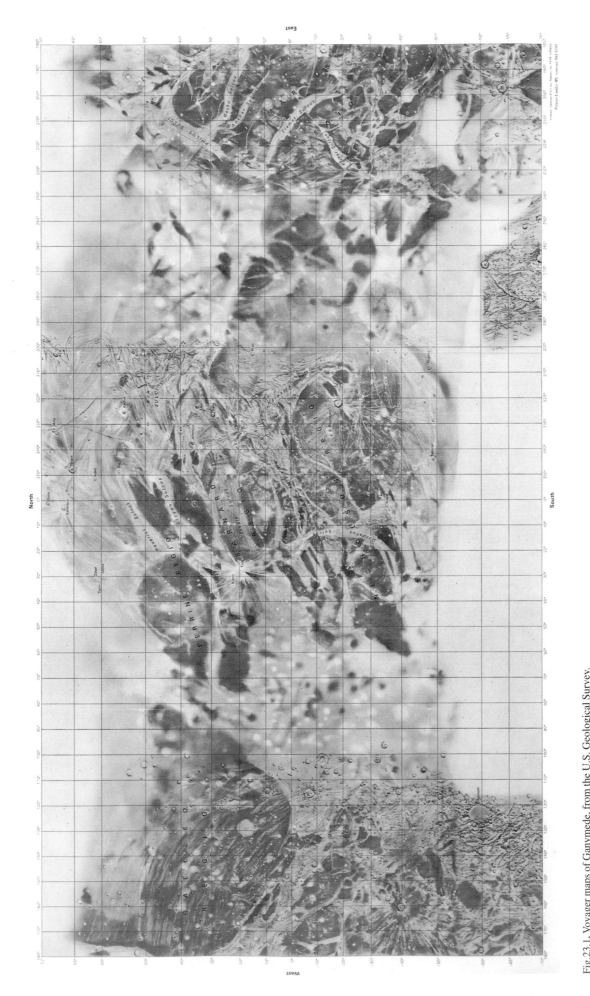

Fig. 23.1. Voyager maps of Ganymede, from the U.S. Geological Survey.

(A) USGS map I-1242, airbrush by J.L. Inge. The map is imprecise, being based on very preliminary navigation data, and has inaccuracies of up to 10° in positions of features. The overlap with the north polar region (map B) is especially bad as the north pole was misplaced by 10° in plotting map A. Neither does it represent albedo patterns very accurately, by comparison with full-disc photographs. Ganymede is now covered by a more detailed set of 15 USGS maps at a scale of 1 to 5 million, including maps B and C.

23: Ganymede

Ganymede is the largest moon in the solar system. It is very different from Io and Europa, and shows its own unique style of tectonics. About half its surface dates back to the early years of the solar system, as shown by the scars of intense cratering that it bears. The remainder, though also very old, was modified into massive bands of grooved terrain before it finally froze rigid. Over most of the moon, the relief does not exceed ≈1 km, but the freshest impact craters have depths of 3–5 km.[1]

The Voyager map of Ganymede (Fig.23.1) shows these two types of terrain. The dark regions are the ancient terrain, in which all the markings can be attributed to impacts, modified by ice creep over the aeons. These regions occupy 40% of the surface, and are named after astronomers who discovered satellites of Jupiter, with the Latin suffix 'Regio'. The lighter, less cratered areas are bands of parallel grooves, or more complex quilts of such bands. They occupy 60% of the surface, including most of the polar regions. They are given names from ancient mythologies of the Middle East, suffixed 'Sulcus' (meaning 'groove'; plural 'Sulci'). Craters are also named from Middle Eastern mythologies.

The grooved bands are often curved, and in several places they form neat arcs bounding the dark terrain (most notably the largest dark area, Galileo Regio, 4000 km across; Fig. 20.29). The resemblance to the forms of lunar maria is striking but illusory. Both dark and bright terrains (with mean visual albedoes 0.35 and 0.44

[1] This chapter is derived from the references listed in the preface to Part VI, and particularly from the comprehensive chapters by Passey and Shoemaker (*1982*), Shoemaker *et al.*(*1982*), and McKinnon and Parmentier (*1986*). For detailed views of grooved terrain see Lucchitta (*1980*).

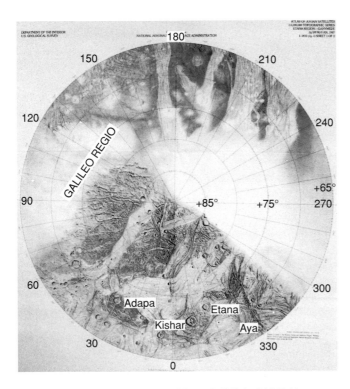

Fig. 23.1 (B). North polar region. USGS map I-1810, by P.M. Bridges.

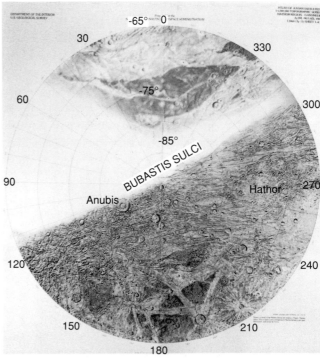

Fig. 23.1 (C). South polar regions. USGS map I-1860, by J.L. Inge and B.J. Hall.

respectively) are much brighter than the lunar surface, although not as bright as Europa (Fig. 23.3). Some of the fresher craters are surrounded by brilliant white ray systems that do match Europa (albedo up to 0.7).

The polar regions are overlaid by diffuse white polar caps, down to latitudes 40–45°N and S. These do not obscure the underlying topography, and they look like an atmospheric condensation, possibly water-frost (Chapter 20.4; Figs. 20.29 and 23.2). This does not imply that an atmosphere is present today; the caps could be millions of years old.

The surface of Ganymede consists mainly of water-ice, according to telescopic observations (Chapter 20.3). About 65% is covered in fairly pure frost – presumably the bright grooved terrain, polar caps, and crater rays. Most of the remainder, i.e. the dark terrain, consists of dirty ice, perhaps strewn with rocks or other dark material.

Beneath the visible surface, the outer half of the moon is believed to consist of water-ice (about 40–50% of the moon's mass; Chapter 20.5). The mechanical state of the ice will be considered later, but we should note that it probably consists of a rigid 'lithosphere' overlying a plastic 'asthenosphere', analogous to the layers in rocky planets. The asthenosphere is defined as a layer in which the ice can flow as in a glacier, particularly by convection; part of it may be melted, forming water/ice slush, or there may even be a liquid water layer. As nothing is known about any chemical layering of the ice, the lithosphere is often referred to simply as the crust.

Craters and palimpsests in the ancient terrain

The dark terrain is nearly saturated with craters a few tens of kilometres in diameter. As we know from the lunar highlands, this massive impact cratering ended just after 4 billion years ago, so Ganymede's dark terrain must have a similar age.

Most of the craters are shallow, with a bowed-up floor, and are thought to have subsided because the ice crust could not support high relief when they were formed. Craters with diameters less than 20 km have central bright spots or peaks, but larger craters usually have a central pit (Fig.23.4). These central pits are also seen in craters in the grooved terrain and on Callisto, so they must be a feature of cratering on icy planets (see next chapter). There may be a deficiency of craters more than 50 km across, as also on Callisto (see next chapter). The largest craters in the dark terrain are 100–150 km across.

There are larger circular features, but they are just light-coloured, roughened areas without walls, ≈100–300 km across and almost flat (Fig. 23.4). These are the 'palimpsests', and they occupy ≈25% of the area of the ancient terrain. Fig. 23.5 shows a dark area half filled with them! They are thought to be the remains of large ancient craters which have totally collapsed. Swarms of secondary craters are still present around some of them. From the distribution of these swarms, and from the fact that small younger craters in the outer parts of the palimpsests have dark floors, it is thought that the palimpsest covers the original crater plus its ejecta blanket. The inner region represents the original crater, and is

Fig. 23.2. Ganymede from Voyager 1; central meridian longitude ≈ 320°. The semicircular dark area in the lower (southern) half is Nicholson Regio. Note that white ray craters are more prominent in the polar regions; a colour version shows that these regions are generally whiter than lower latitudes. (NASA P-21207.) (For the opposite hemisphere, see Fig. 20.29.)

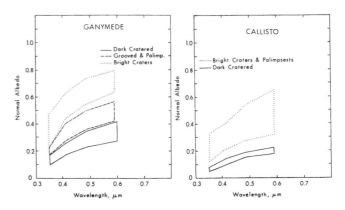

Fig. 23.3. Spectra of regions on Ganymede and Callisto, from the Voyager imaging data. (From Johnson et al., 1983.)

always light; sometimes there are traces of its original rim or multiple rings. It is likely that these craters collapsed soon after formation, when the ice lithosphere was perhaps thin, as there are plenty of fairly old craters in the same size range that show no sign of becoming palimpsests. The palimpsests may represent collapse over millions of years in a lithosphere only ≈10–15 km thick, or (perhaps more likely) they may have been created instantaneously, by impacts which smashed right through a lithosphere that was less than 10 km thick, allowing slush to well up in and around the hole. The lithosphere was probably growing thicker and stronger just as the terminal bombardment waned, ≈4 billion years ago, and so was retaining larger and larger craters. So the surviving surface is not completely saturated, but is still more densely cratered than anything younger.

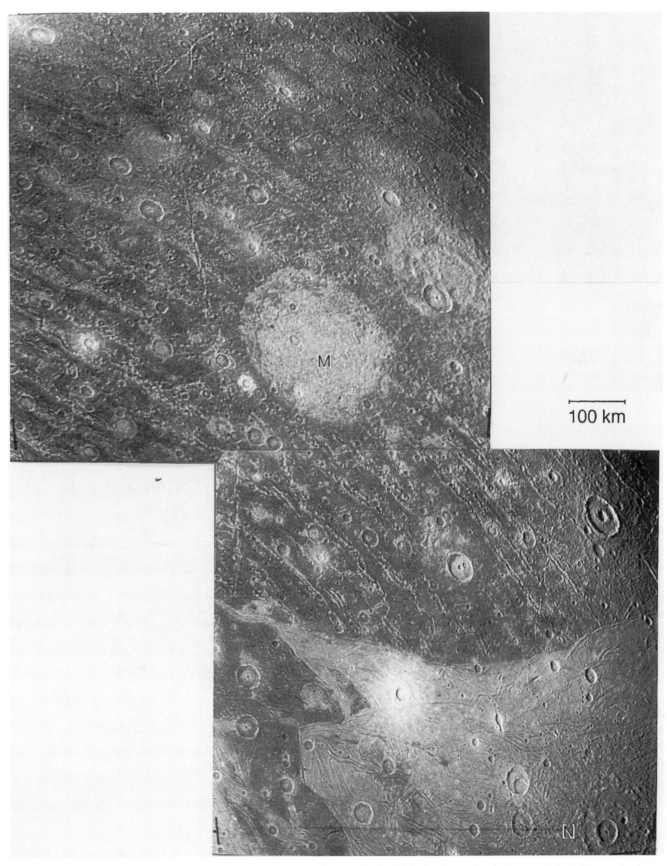

Fig. 23.4. Closeups of the southern part of Galileo Regio from Voyager 2. Note parallel furrows; one furrow taking an independent, north-south track; numerous craters (mostly post-dating the furrows). There are two large palimpsests of which the largest (Memphis; M) is 330 km across. At bottom is an area of smooth and grooved terrain, defaced by the Western Equatorial Basin (now called Ninki; N). Scale bar = 100 km. (NASA images 20637.02,14, processed by Simon Mentha.)

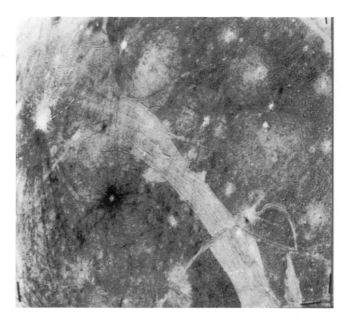

Fig. 23.5. Western Marius Regio. In the dark terrain at upper right, about half the area is occupied by four huge faint palimpsests 200–300 km across, and the remainder is marked by parallel furrows. The image includes several craters with dark rays or mixed bright and dark rays. The great grooved band is Tiamat Sulcus, named for the primeval turbulent ocean of Sumerian mythology. (NASA image 20635.45; processed by Simon Mentha.)

Furrows in the ancient terrain

Some of the dark areas contain curved, parallel 'furrows', in arcs that apparently form concentric patterns like those on Callisto (next chapter).[2] The most prominent set covers Galileo Regio (Fig. 23.4). If it originally belonged to a complete circular pattern, it must have covered most of one hemisphere! Each furrow is a valley with slightly raised walls and bowed floor, about 10 km from side to side and a few hundred metres deep, spaced about 50 km apart. They are likely to have originally been 'graben' (simple troughs bounded by pairs of near-vertical faults), produced by stretching of the crust, but have flexed as the ice relaxed. These furrows rarely cut across craters (unlike those on Callisto), but they are interrupted by craters and palimpsests, indicating that the ripple system pre-dates most of these impacts.

The centre of this system projects near 32°S, 190°W – a location which has been obliterated by later grooved bands. The neighbouring dark area, Marius Regio, has a less conspicuous and more closely-spaced set of furrows, almost but not quite aligned with those of Galileo Regio (Figs. 23.5&6). At the centre of this system (18°S, 165°W), there does seem to be a fragment of a giant palimpsest, ≈500 km across (Fig.23.8), with more furrows to its east. This could be the original centre of both furrow systems, if there has been more recent crustal movement. The simplest solution would be for Galileo Regio to have drifted ≈1300 km towards the centre, with crust disappearing in the intervening grooved band, Uruk Sulcus. Alternatively, S. Murchie and colleagues (1990) suggest that the furrows may have formed long after the original impact, along lines of weakness which might no longer be concentric with it, and they discern signs that dark material flowed like lava around the furrows.

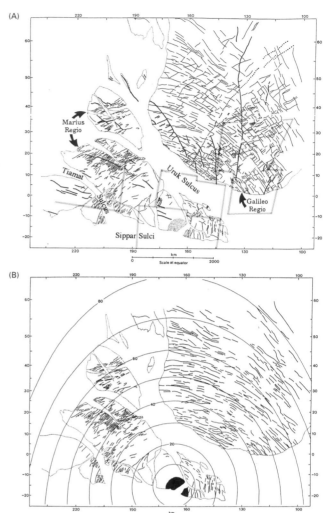

Fig. 23.6. Maps of furrow systems in ancient dark terrain. The shaded semi-circle is the giant palimpsest. Adapted from Murchie *et al.* (*1990*).
(A) Present positions of furrows. Shaded outlines mark the locations of Figs. 23.4–8.
(B) Positions of the concentric furrows relative to the giant palimpsest, with Galileo Regio rotated around its centre so as to produce a 500-km offset across Uruk Sulcus. This rotation produces local alignment of the furrows in Galileo Regio and Marius Regio, though it does not resolve the difference in radius. Concentric ovals on the map represent circles sentred on the giant palimpsest.

There may be a remnant of a similar system on the opposite (Jupiter-facing) hemisphere, in Nicholson, Barnard, and Perrine Regiones.[2] These furrows are less distinct but more extensive, centred somewhere near 60°N, 50°W.

There are a few isolated, deeper furrows or crater-chains that cut across other furrow systems. In Galileo Regio they mostly run north–south, in some cases for thousands of kilometres (Figs. 23.4&6). They form a radial pattern, projecting from ≈20°S, 140°W (well to the east of the giant palimpsest). According to Murchie and colleagues (*1990*), this falls in an area that was resurfaced late in the evolution of the dark terrain. They suggest that it

---

[2] For precise maps see Schenk and McKinnon (*1987*) and Murchie *et al.*(*1990*).

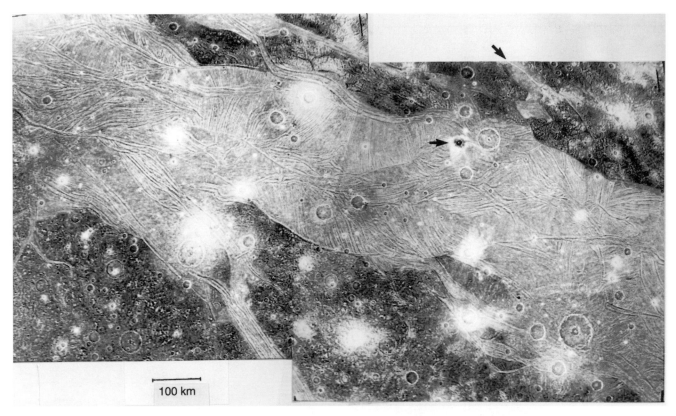

100 km

Fig. 23.7. A large band of grooved terrain, Uruk Sulcus. This area overlaps Fig. 23.8 to the south (below) and almost overlaps Fig.23.4 to the east (right); they are part of a mosaic by Voyager 2. Scale bar = 100 km.

In Uruk Sulcus, note details of the interlocking groove systems, which end at large grooves that separate them from the ancient dark terrain. Young craters have bright haloes and older ones have slightly dark haloes. Also note a crater with bright rays but very dark floor, possibly flooded with dark material (arrowed).

In Galileo Regio at top right, a narrow grooved band consists of a single central groove within a smooth band, which has apparently flowed over part of the dark terrain to the south (arrowed).
(NASA images 20637.17,20, processed by Simon Mentha.)

was a broad area of uplift, producing long radial cracks, 'volcanic' flows, and a region of early and very complex grooved terrain.

Craters in the grooved terrain

Different grooved bands have different crater densities; the average density is 1/10 that of dark terrain (Fig.23.9). If the densities are interpreted as relative ages, the oldest grooved terrain (in the south polar region) matches the youngest dark terrain, and the youngest grooved terrain matches the oldest lunar maria.

To get absolute ages, one would need to know what the impact rates were throughout geological history. These rates have been measured for the Moon, but can only be estimated for the Jupiter system. The rates may be similar for the Moon and Ganymede, as comets and asteroids would be 1–2 orders of magnitude less frequent so far from the sun, but would be concentrated by one order of magnitude by Jupiter's gravity.[3] If these very uncertain estimates are adopted, then the youngest grooved terrain is around 3 billion years old, contemporaneous with the oldest lunar maria and the oldest martian plains.

An extra factor is that most impacts should occur on the leading hemisphere – indeed, the frequency should be 15 times greater at

[3] Smith *et al.* (*1979b*). However, Shoemaker and Shoemaker (*1990*) suggest that the present rate for Ganymede is several times higher than for the Moon, mainly because of extinct short-period comets.

the apex than at the antapex of motion. Indeed, grooved terrain on the leading hemisphere is more cratered on average. (Also, as described below, the two large impact basins on Ganymede and the three great multi-ring structures on Callisto are all on the leading hemispheres.)

The grooved terrain does not have palimpsests, but it has some very decayed craters that have almost reached that state. The craters in the grooved terrain generally appear fresher than in the ancient terrain, implying that the grooved terrain formed after Ganymede had cooled somewhat and its crust had become more rigid. These craters often have central pits.

Younger craters with brilliant ray systems do occur in both types of terrain; presumably the rays are fresh ice flung out by the impact. The ray craters can be interpreted according to the simplest hypothesis: that the crust below the surface is fairly pure ice, that all impacts splash it across the surface, and that it darkens with age due to radiation and meteoroids. The most conspicuous ray crater is called Osiris, after the ancient Egyptian god (40°S, 160°W). It is 150 km across with rays extending for over 1000 km. The rays are noticeably less bright where they lie on polygons of dark ancient terrain. The same selectivity is shown by other ray systems such as in Fig. 23.10. This suggests that such large ray systems consist not only of ice flung out from the primary crater, but also of ice from secondary craters along the rays. But another factor is the darken-

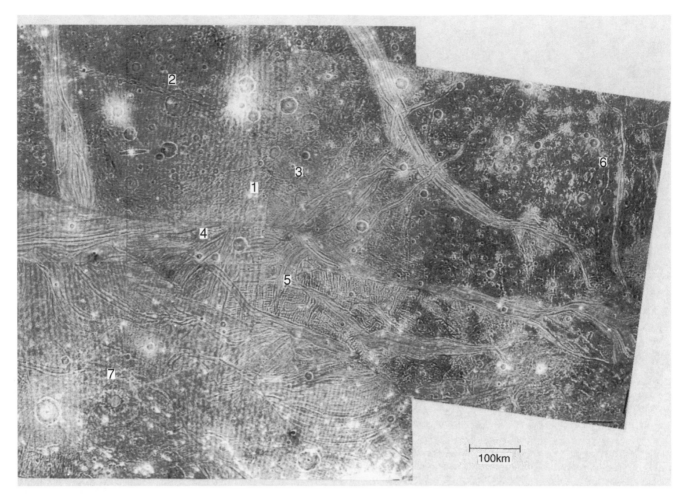

Fig. 23.8. A complex region centred around 17°S, 160°W. In centre is the giant palimpsest fragment close to the projected centre of the Galileo furrows system. To the north (upper) is ancient dark terrain, and to the south (lower) is complex grooved terrain. Scale bar = 100 km.

(1) The giant palimpsest. (2,3) Two rare examples of craters cut, stretched and offset by narrow bands of grooves. For crater 2 the displacement is a few km; for crater 3, it may be more, if the feature 40 km to the south is the other half of the same crater. (4) 'Escher pattern' of grooves. (5) A rare cross-hatched pattern where two orthogonal groove sets are superimposed. (6) Furrows running north-south, east of the giant palimpsest. These could be part of the same 'ripple pattern' seen in Galileo Regio; however, they seem to run straight to the north, and may once have been continuous with the mysterious isolated furrows that cross other features in Galileo Regio, in which case they were ≈2000 km long and unrelated to the giant palimpsest. (7) A local multi-ring structure on the grooved terrain, interpreted as a crater that has almost become a palimpsest; it has a dome in the centre. (NASA images 20636.02 and 20637.29, processed by Simon Mentha.)

ing of rays by micrometeoritic 'gardening'; this will darken rays more quickly if they lie on dark terrain. Ray craters are much less common on the leading side of the moon, and this is probably because such 'gardening' is more intense on that side.

Some of the older, larger craters in the bright terrain have very subtle dark haloes out to 2–3 times the radius of the crater (Fig. 23.7). The albedo is intermediate between the grooved and the dark terrain. These do not obscure the underlying grooves but may be old, thin ejecta blankets. P. Schenk and W. McKinnon (*1985*) concluded that these larger craters had excavated darker material lying 1–2 km deep, implying that the fresh ice of grooved terrain overlies darker crust. Although all crater ejecta initially looks bright thanks to its fresh ice, they argued that when it has had time to darken with aeons of exposure to radiation, the original albedo contrasts can be perceived. However, this theory does not explain the absence of radial striation nor gradation within the haloes. Could they rather mark the extent of the steam explosion produced by the original impact into ice? Perhaps the steam cloud would glaze the ice and

react with surface organic deposits left by meteoroids – literally cooking the surface.

Also, a few per cent of fresh craters have ejecta ending in a distinct scarp. V. Horner and R. Greeley (*1982*) interpreted these 'pedestal craters' as due to fluidised ejecta flow, perhaps caused by melting of ice on impact, as with similar craters on Mars. There are several other oddities that are unexplained, such as a few craters that have dark floors (e.g. Fig. 23.7).

A remaining oddity is that some fairly fresh craters have distinct *dark* ray patterns (Fig. 23.5). Although these could be due to impacts into subsurface pockets of dark material, J. Conca (*1981*) showed that they are equally abundant on cratered and grooved terrain. Schenk and McKinnon (*1991*) showed that they are strongly concentrated towards the antapex, and are 'redder' than typical dark terrain. So it is most likely that they may be caused not by impacts into dark subsurface layers, but by very dark, reddish meteorites – perhaps akin to the dark reddish 'type D' asteroids that populate the outer part of the asteroid belt. Dark rays are unex-

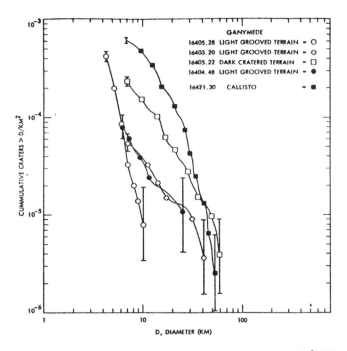

Fig. 23.9. Crater density plots, showing the number of craters per km² with diameter greater than D. Data are shown for Callisto (filled squares), Ganymede dark terrain (open squares), and examples of Ganymede grooved terrain (other symbols). (From Smith *et al.*(*1979a*). For further data see Smith *et al.*(*1979b*) and Murchie *et al.*(*1989*).)

pected because most ejecta must come from the surface, not the impactor. Schenk and McKinnon (*1991*) suggest that many craters have such dark-ray material, which is indeed masked by bright rays at first, but is revealed transiently because the bright rays fade away first. The greater frequency near the antapex may be partly because impacts there have lower velocities so the impactor contributes more to the ray material, and partly because ray systems persist longer on that side of the moon. No dark ray systems are seen on Callisto, but this is probably because the surface is too dark for them to show up.

The most impressive impact structure on Ganymede is a large basin called Gilgamesh, after the Sumerian hero of the earliest written epic (Fig.23.11). The central flat basin is 175 km across, but the most distinct scarp surrounding it, 1–2 km high and resembling the rims of lunar maria, has a radius of 275 km. Between the basin and the scarp is a chaotic mountainous zone of debris which is the most rugged terrain on Ganymede, with relief of 0.5–1.5 km. Outside the scarp, the debris grades outwards into a smooth ejecta blanket and radial strings of secondary craters – all familiar marks of impact cratering in a rigid crust like the Moon's.

Gilgamesh is emplaced on grooved terrain, but is itself cratered, so it probably formed soon after the grooved terrain. Its fresh appearance is probably due to its high latitude (60°S), where the crust must have been colder and more rigid than near the equator. A basin of similar size and age, called Ninki (formerly the Western Equatorial Basin), is centred at 7°S, 115°W, and it is much more decayed and crumpled than Gilgamesh (Fig. 23.4). As it is filled with hummocky terrain inside the inner basin as well as around it, it may have been partly transformed into a palimpsest.

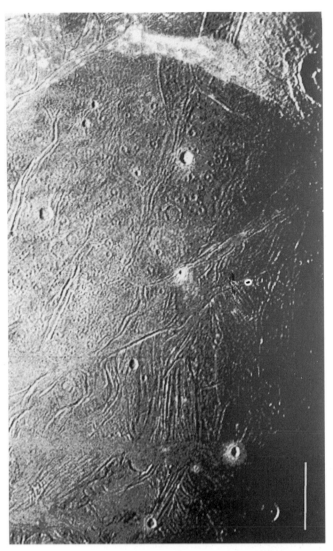

Fig. 23.10. Complex, unnamed region centred about 22°N, 308°W. It shows sinuous bands and grooves which all but obliterate a piece of ancient terrain. Also note a crater half-destroyed by a smooth band (near bottom). White splotches at top are ray deposits from the large sharp crater on the terminator (proposed name Ta-urt); note that they are only visible on the grooved bands. Scale bar = 100 km. (NASA image 16403.52, processed by Simon Mentha.)

The grooved bands

The grooved bands of Ganymede excited geologists because they were the first unmistakable sign of largescale plate tectonics on a world other than Earth. In fact they are similar in scale to Earth's Appalachian Mountains. But their geology is obviously very different, as they are made of ice. Now Voyager has revealed wrinkled or grooved bands on several smaller icy moons in the Saturn and Uranus systems, so this type of formation may be a natural outcome of internal activity in a world made of ice. But Ganymede's system is unique in its extent and regularity.

All the bands consist of parallel grooves 3–15 km apart and a few hundred metres deep (maximum 700 metres). The slopes are gentle, averaging ≈5° (maximum ≈20°). The ridges between the grooves are sometimes narrow, sometimes flat-topped. The grooved bands have sharp edges, usually parallel to the grooves, but sometimes bounded by a cross-cutting groove or fault instead.

Fig. 23.11. Gilgamesh (60°S, 125°W). Scale bar = 100 km. (NASA image 20639.15, processed by Simon Mentha.)

Such faults occur at some interfaces with dark terrain (Figs. 23.5&7), so must mark the original terminus of the groove system. Many of these faults are particularly deep grooves. The bands are often slightly lower than adjacent dark terrain. Some bands or isolated grooves extend into dark terrain, mostly near the edges, and sometimes taper to a point in it. Some bands are straight, others smoothly curved, and others weirdly kinked or tortuous. The narrowest bands are isolated grooves 10 km across, or pairs ('tramlines'). Much broader bands are in large light areas where multiple sets of grooves are fitted together like crazy paving, and many bands measure up to 100 km across. Broadest of all are Phrygia Sulcus, 450 km across, and Bubastis Sulci, a swathe of grooves near the south pole that is over 300 km across (Fig. 23.12).

Each band has clearly wiped out all previous topography – dark terrain, furrows, craters, or other bands, all are cleanly cut and erased (Figs. 23.8,10,13). The bands cut across each other at seemingly random angles. These interfaces suggest that some bands are younger than others, and the relative ages are generally confirmed by crater counts. However, the interfaces can be misleading, as a newer band can terminate at the edge of an older one. The degree of grooviness varies. In some crazy-paving areas, some of the youngest bands appear almost smooth. They usually have very slight lineations that suggest a very fine-scale grooved pattern, but these may be grooves introduced after the smooth band was emplaced. Some may be truly flat areas, and have been interpreted as due to 'lava flows' of water or icy slush. Although such eruptions seem physically improbable (see below), a few almost-smooth bands do look like flows (e.g. Fig. 23.7), including one which has overflowed into hollows or craters on either side (Fig. 23.14).

Has the crust moved across or along the grooves? There is evidence for movement in a few places, but they are rare and the offsets are always less than 100 km. Slight lateral motion seems indicated in Figs. 23.8&13. Fig. 23.8 also shows what may be a

crater cut in two by a band and pulled apart; this is the only such specimen. The fact that the primary structures are grooves, not ridges, has been taken as evidence that crust is pulled apart at the bands. But there is no strong evidence for such motion, and in many places the curvatures of the bands rule out any large movements. Moreover, massive crustal divergence at the bands should have been balanced by convergence somewhere else, of which there is no sign.

Rather, the bands look as if they have grown at the expense of preexisting crust, converting it into the parallel grooves – a sort of self-replicating parasitic pattern. How might this have happened?

Tectonics and resurfacing on Ganymede

The origin of the bands must lie in the first aeon of Ganymede's history, when the interior was warmer than it is now. (1 aeon is 10^9 years.) As we saw in Chapter 20, this moon probably consists of a rocky core surrounded by a massive shell of water-ice. The bands apparently formed when the ice lithosphere was much thinner, and could move over the asthenosphere or even over a liquid water mantle.

An important clue to the origin of the bands is that their shapes and relationships are very similar to those of the cracks on Europa. It is as if both moons suffered the same crustal cracking when their lithospheres were thin, but on Ganymede the cracks were broadened into the great grooved bands. On both moons, the cracking evidently went on for some time, and was intermittent; the old bands must have been completely rigid, and the underlying crust quite uniform, before new cracks could cut cleanly across them.

The prevailing theories of Ganymede's grooved terrain invoke a combination of three factors: progressive cooling of the crust, global expansion, and water/ice volcanism.

The evidence for progressive cooling of the lithosphere comes from studies of the forms of the craters on different types of terrain. The density of the craters gives the age of the terrain, and the degree of relaxation of the craters indicates the viscosity, and thus the temperature gradient, and thus the thickness, of the lithosphere when those craters formed. The physical parameters involved are very uncertain, but the rough deductions of Q.R. Passey and E.M. Shoemaker (*1982*) were as follows. The oldest dark terrain, with its furrow system and palimpsests, was formed about 4.0 aeons ago with a lithosphere ≈10 km thick. The oldest grooved terrain (including the south polar region) formed about 3.8 aeons ago with a lithosphere ≈35 km thick. The lithosphere thickened to become essentially rigid by the time Gilgamesh formed at a high southern latitude (3.5 aeons) and by the end of groove formation at lower latitudes (3.1 aeons). The thickness since then has been at least 80 km, as predicted from the current rate of radioactive heat flow, and perhaps much thicker.

Global expansion could well have occurred in the first aeon of Ganymede's history, due to exudation of water from the core, which would be much greater for Ganymede than for Europa.[4] Some expansion could also have resulted from freezing of this

[4] Shoemaker *et al.*(*1982*); Squyres (*1980*); Passey and Shoemaker (*1982*).

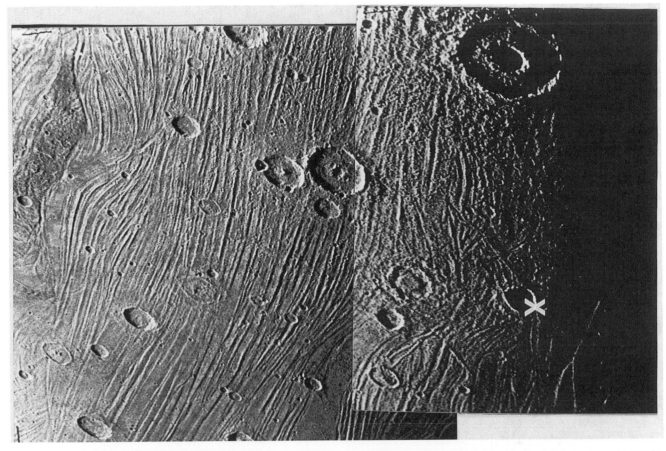

Fig. 23.12. South polar region. This shows one of the broadest sets of parallel grooves, over 300 km across, named Bubastis Sulci. The south pole is arrowed. The largest, two-ringed crater, mostly in darkness, 100 km across, is named Anubis after the Egyptian jackal-god of the underworld. (NASA images 20640.41,43, processed by Simon Mentha.)

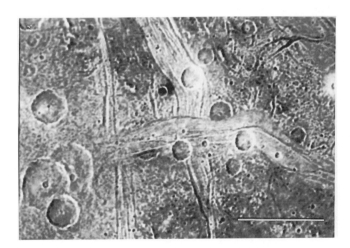

Fig.23.13. Closeup showing how a younger grooved band cuts across older grooved bands and craters. Three craters have been partly destroyed (not pulled apart), and the older grooved bands seem to be offset by 5–10 km. Scale bar =100 km. (NASA image 20637.38, processed by Simon Mentha.)

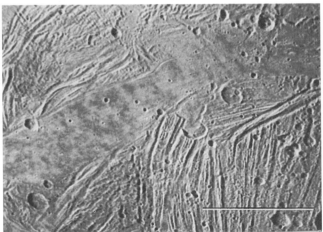

Fig. 23.14. An uncommon example of a smooth band that shows evidence of flow; it has overflowed into hollows or craters on either side. This is an arm of a larger almost-smooth band in the south polar region, which surrounds the great grooved band shown in Fig. 23.12 (slight overlap at bottom of this picture). Scale bar=100 km. (NASA image 20640.25, processed by Simon Mentha.)

water mantle, and from conversion between different forms of ice as the moon cooled, although this might have been offset by conversion into other, more compact forms of ice in the deeper layers. Overall, these expansive processes would naturally explain why the crust was cooling and thickening during the formation of the grooved bands, why they started in the colder polar regions, and perhaps also why such bands are not seen on Callisto (see below).

Global expansion can only have been a trigger: the bands occupy 60% of the surface area, and the expansion could not have been more than 2%, or Galileo Regio would have cracked under the strain. Indeed, the near-circular outline of Galileo Regio, and the tendency of grooved bands to taper into these dark regions, are

Fig. 23.15. North polar region. The lower half is part of a vast quilted area of smooth and grooved terrain. The upper half contains polygons of ancient dark terrain and, on the limb beyond the north pole, the edge of Galileo Regio. The north pole is marked. (NASA image 16407.19 processed by Simon Mentha.)

consistent with these regions being crustal remnants of a slightly smaller world.

'Water-ice volcanism' is the third factor invoked in explaining grooved bands. There is plenty of evidence for internal activity, and sometimes even surface flows, on Ganymede and several other icy satellites. The concept of water-ice volcanism had not been seriously studied before the Voyager discoveries, and its principles are still poorly understood. We said in the previous chapter that water should not erupt through an ice crust onto the surface. But this principle is less secure for Ganymede, whose ancient surface is not pure ice, and could be contaminated with something that makes it denser than water – perhaps meteoritic debris. Flows of water or an icy 'slush' could occur if the crust's density was increased by a 6–15% admixture of rocky material – or if the density of the water was reduced by volcanic gases from below. However, such flows should freeze before they could get very far. To spread for hundreds of kilometres without freezing, such a flow would have to be at least ten metres deep and all erupt within a few weeks – an almost unimaginable flood. Flows of ice, in the form of glaciers driven by forces from below, would seem more probable.

The prevailing theories call for global expansion, subsidence of rock-laden crust in the bands, and eruptions of water/slush to fill them. Two such theories are illustrated in Fig. 23.16(A&B). In (A),

a single band of crust subsides, slush flows up to fill the band, and grooves develop later by further extension and subsidence. (But why did the grooves develop so regularly in many bands?) In (B), each groove represents subsidence of a separate slice of crust; as the central slice subsides and is consumed in the plastic zone beneath, flanking strips break off and so the grooved band widens. (But why did the slush not cover the sinking slabs completely, and why are the edges of the bands so sharp?)

In Fig. 23.16(C) we add the possibility of compression, which the author has not seen advocated elsewhere. This model suggests that the grooved bands are analogous to the cracks on Europa, with the added feature of prolonged jostling between the crustal plates. As the plates separate, new crust forms, and as they converge, it is compressed into a ridge; the compression could be self-limiting and thus explain the very regular spacing of the grooves. As the ridges will also entail swellings below the crust, which will be melted away, the ridges will subside and new cracks develop flanking them. The dominance of grooves rather than ridges could be caused by slight, late global expansion as the moon cooled down. However, while the previous theories have difficulty explaining the bands filled with very regular grooves, the last theory has difficulty with the largely smooth bands in which there are only sparse grooves. Perhaps all these mechanisms operated to various degrees.

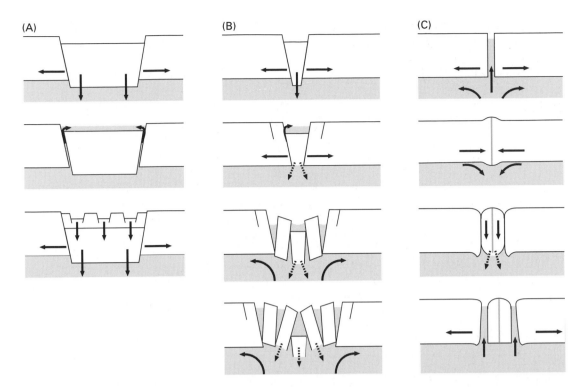

Fig. 23.16. Three models for the origin of the grooved bands. The almost-rigid ice 'lithosphere' lies on a more plastic ice 'asthenosphere', within which glacial flow can occur and local melting can occur.
(A) From Parmentier *et al.*(*1982*). The band represents a block of crust that subsides as the moon expands slightly. Water/ice slush flows up to fill it and freezes. Grooves develop later as expansion continues.
(B) From Shoemaker *et al.*(*1982*). Again, the model starts with global expansion and subsidence of a dense lithosphere, but this only creates a single groove. As it continues, and the sunken crust is dissolved away by a glacial convective cell, flanking strips of crust also break off and widen the grooved band.
(C) An alternative model which does not require the crust to sink. Small plates of crust repeatedly jostle to and fro, alternately creating grooves and ridges.

Whichever of these mechanisms was real, it must have operated episodically, to produce the cleanly cross-cutting patterns of the grooves. Possibly the heat flow rose and fell several times, due to tidal changes or to multiple stages of internal differentiation. Also, in any theory, the final stage must have involved some sort of surface flows to produce the smooth, flow-like areas.

Another difficulty with all these theories – particularly those invoking global expansion – is to explain why the same thing did not happen to Callisto. Its size and its calculated thermal history are very similar to those of Ganymede, and neither moon is likely to have suffered enormous tidal effects. One possibility is that Callisto did not fully differentiate, though most researchers think this unlikely. But Ganymede did get somewhat greater heating (double the accretional heat, and 55% more radioactive heat, for 37% greater mass; Table 20.3). It seems most likely that both moons underwent differentiation and tectonic activity, but that Ganymede's extra margin of heating was enough to prolong the activity for an extra few hundred million years after the great bombardment ceased – late enough to leave a visible record.

An unresolved question is what the dark material is, on both Ganymede and Callisto. Icy surfaces darken with age, but the ancient terrain on Ganymede is probably intrinsically darker than the grooved terrain: it is thought to be only 5–30% older, but 30–100% darker (relative to fresh ice). As there are bright ray craters only a few kilometres in diameter, the dark layer must be less than 0.5 km thick – indeed, no more than a few metres thick, if the major rays are indeed due to secondary cratering. However, the explanations given above for dark-floored and dark-haloed craters suggest a thickness of at least 1 km. Most likely, the ancient crust is uniformly slightly darker than the re-worked ice of the grooved terrain, but the top few metres is especially dark. This may be because the ice has been ablated, leaving a concentrate of the dark stuff.

The bright 'polar caps' may be due not to deposition of fresh ice, but to selective loss of surface ice from the equatorial regions by sublimation in the sunlight. Because the polar regions are colder, they may simply have retained their original frost (as well as more of subsequent crater rays), while the lower latitudes became enriched in a darker residue.

The dark residue that has accumulated on grooved terrain and later craters, since the end of geological activity, must come from the ubiquitous bombardment with radiation and/or meteoroids. Again, its nature is unknown. Its yellowish colour could be due to sulphur (unlikely so far from Io), to ferrous silicates (as in common rocks or meteorites), or to organics (as in carbonaceous chondrite meteorites and in comets). Perhaps the safest conclusion is that, if one leaves anything uncleaned for several billion years, it will get dirty.

(a)

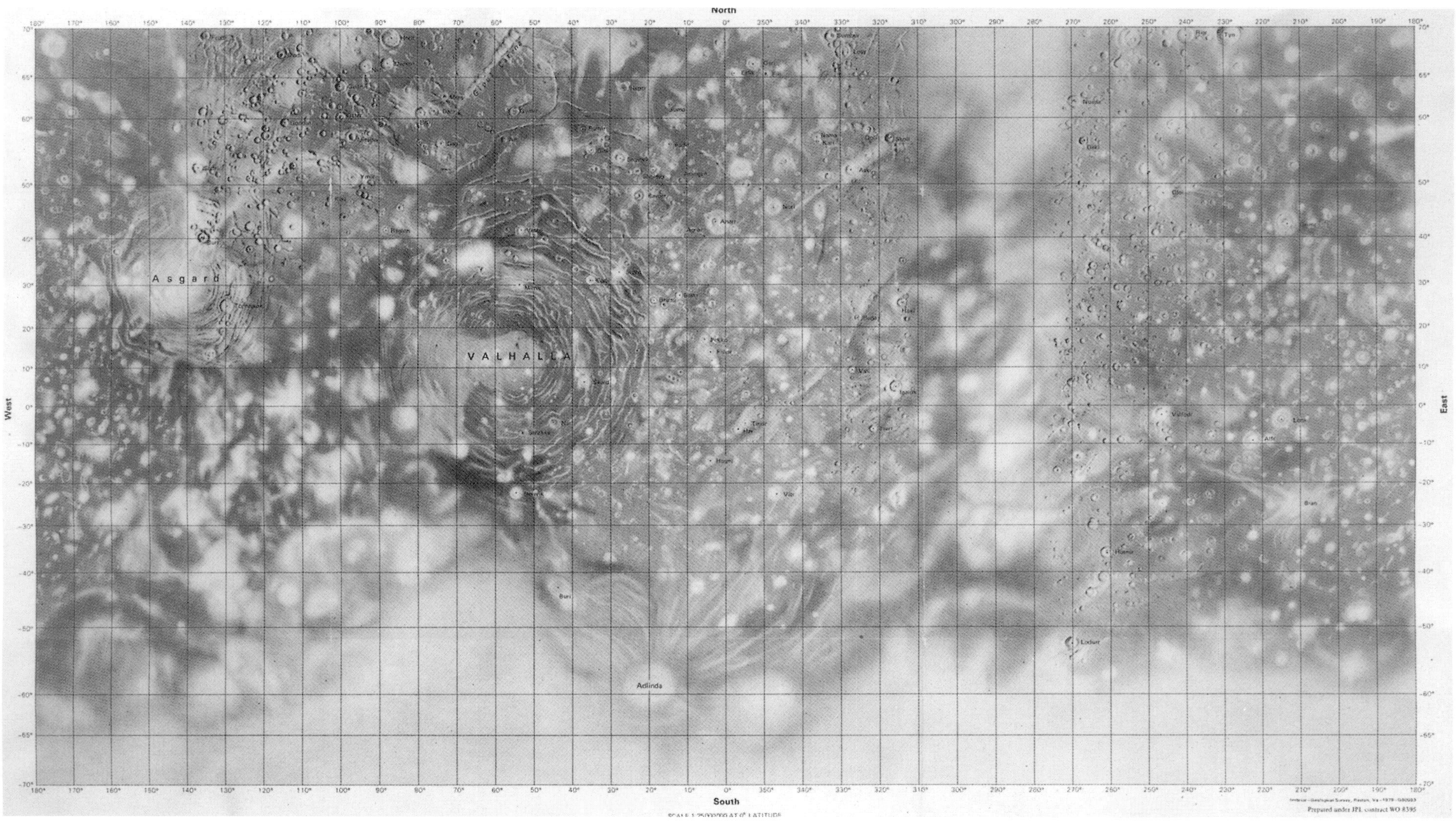

Figure 24.1. Map of Callisto from Voyager images, from the U.S. Geological Survey, map I-1239, airbrush by P.M. Bridges. This is a preliminary map and placement of features may be inaccurate by up to 10°. North polar region is on facing page. Callisto is now covered in greater detail in a set of 14 controlled photomosaic maps published by USGS.

24: Callisto

Callisto is the most cratered and least modified of all the planetary-sized bodies in the solar system. Its surface is entirely moulded by impacts. So it resembles the Moon or Mercury, but with fewer large late structures, and none of the mountains nor volcanism that arose on those rocky worlds. Some craters are 2–3 km deep, but most have subsided to less than 1 km.

Callisto is thought to have a massive ice crust and mantle like Ganymede, but its surface is darker. It has some fresh ice exposed, which may be identified with the fresher craters and the bright ray systems around them, but most of the surface is 'dirty ice' mixed with rocky or meteoritic debris (Chapter 20.3). Even so, the surface is still twice as bright as that of Earth's Moon.

So the only features on the Voyager map (Fig. 24.1) are impact craters and rings. They are given names befitting their frigid nature, from the myths of Scandinavia, Alaska, and Patagonia.

Craters and palimpsests

The surface is almost saturated with craters, which evidently date back to the end of the universal bombardment, about 4 aeons ago (Fig. 24.2). The most interesting aspect is the obvious rarity of craters greater than 50 km in diameter (Fig. 23.9). The ancient areas of Ganymede may show a similar deficiency. Most detailed studies of the craters have concluded that they must reflect a puzzling population of impactors which really did produce very few large craters.[1] In contrast, the moons of Saturn and Uranus are not at all short of large craters; they seem to have been bombarded by projectiles with a size distribution similar to that which afflicted the Moon and the inner planets (even though in the outer solar system they must have included more comets than asteroids). The moons of Saturn were also cratered by a swarm of debris orbiting the planet, but the crater distribution on Callisto does not seem consistent with such a swarm. Instead, the distribution seems to be unique to the jovian system, which is hard to explain.

[1] Including Strom *et al.*(*1981*), and Chapman and McKinnon (*1986*).

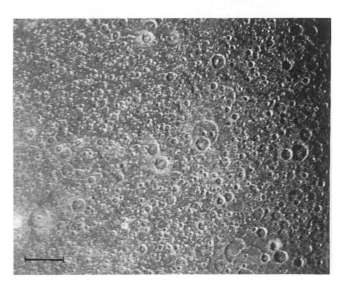

Fig. 24.2. A typical area of Callisto, almost saturated with craters less than 70 km in diameter. Scale bar=100 km. (NASA image 20617.09, processed by Simon Mentha.)

NORTH POLAR REGION

Fig. 24.1 (b). For legend see opposite

One possibility is suggested by the newly shattered comet Shoemaker-Levy 9 (next chapter). This comet was both captured and fragmented by Jupiter as it passed within the Roche limit. Could early impacts have been dominated by events of this sort, which broke comets into pieces and strewed them around the jovian system?

The alternative possibility is that craters larger than 50 km formed but collapsed. Palimpsests are not as visible on Callisto as on Ganymede, because they are mostly no brighter than their surroundings, but they are quite common. Perhaps many have been completely erased. This too would be hard to explain.

Turning to the craters that do cover the surface, we find that they are very like those of the dark terrain on Ganymede. Most have subsided by creep of the ice, and are now very shallow. Even the fresher ones are no more than 3 km deep. There are a few large ones (150 km across) which consist of two or three concentric rings.

Most of the craters wider than 20 km have central pits. As such pits are also seen in craters on both dark and bright terrain of Ganymede, they must be a feature of cratering on icy planets. A theory suggested by Q. Passey and E. Shoemaker (1982) is that they represent transient central peaks which, being fluidised by melted ice, quickly collapse. None are seen on Saturn's moons, which have much lower gravity; there, perhaps the peaks freeze before they can collapse.

Four straight chains of contiguous craters have been identified. The largest is Gipul Catena (named after a Norse river), in the north polar region, and another lies on the rings of Valhalla (Fig. 24.3). Their origin is unknown. Again, perhaps they record the impacts of newly shattered comets that had just flown through Jupiter's Roche zone.

Multi-ring structures

The most striking features on Callisto are two giant 'bullseye' structures. They are named after the two abodes of the Norse gods, Asgard and Valhalla. They appear to be the scars of large, late impacts in the ice crust.

Only the largest, Valhalla, was viewed in detail by Voyager (Fig. 24.3). At its centre is a roughly circular mottled bright patch, ≈600 km across, which is regarded as a giant palimpsest. The circular ripple pattern extends to a radius of 1200 km, and on the north side, incomplete concentric arcs go on out to a radius of 2000 km. There are no great scarps nor ejecta blankets, like those ringing the lunar basins; the Valhalla crater was apparently confined to what is now the central palimpsest. The crater density on Valhalla is two- to three-fold less than on the rest of Callisto, suggesting that this was one of the later impacts of the great bombardment.

The 'ripples' are somewhat irregular but many can be traced around arcs of more than 1000 km. They cut cleanly across some craters. There is an inner zone of continuous 'ripples' spaced ≈30 km apart, and a larger, outer zone of well-separated features ≈70–100 km apart. The exact forms would be better seen under a higher sun angle than Voyager had, but they mostly seem to be narrow ridges, although some are furrows up to 1 km deep with slightly raised walls – probably rift valleys ('graben') that have relaxed. The ridges are formed by tilted slabs of crust (Fig. 24.4), with bright material on their outer sides, possibly re-worked ice (H.J. Melosh, 1982).

Although they look like sets of ripples in a pond, the appearance is misleading. They cannot really be frozen tsunamis. The ice lithosphere must have been thick when they formed, and they probably developed gradually after the impact, as the ice lithosphere slowly flowed inwards to fill the central crater (Melosh, 1982).

The second-largest 'bullseye', Asgard, is very similar but only 1600 km in diameter (Fig. 24.5). A third such structure, so far unnamed, is only 900 km across (53°S, 36°W; Fig. 24.6), and is partly obscured by bright rays from the crater Adlinda.

In conclusion, Callisto is a long-dead world; and if we could drill 1 km down into its crust, we would find ice that has been perfectly preserved for 4 billion years.

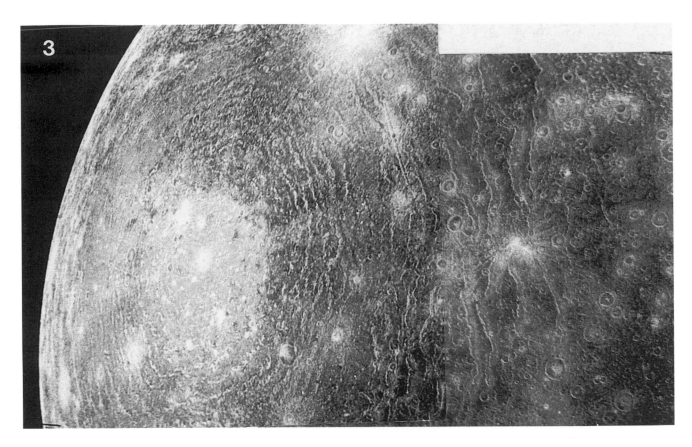

4 PALIMPSEST   GROOVES   RIDGES

CRATER

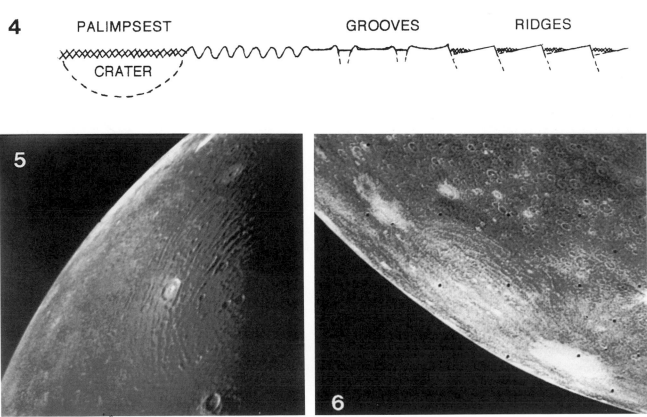

Figs 24.3–6. Multiring structures on Callisto, viewed by Voyager 1.

Fig. 24.3. Valhalla. (NASA images 16422.07,11, processed by Simon Mentha.)

Fig. 24.4. Valhalla: sketch of the topography, with dashed lines suggesting the original underlying forms. Cross-hatching indicates bright surface areas which have apparently filled in the crater and valleys. Vertical scale exaggerated, curvature of satellite not shown.

Fig. 24.5. Asgard. (NASA image 16428.09, processed by Simon Mentha.)

Fig. 24.6. The third large multiring system, unnamed, next to the bright ray crater Adlinda. (NASA image 16418.14, processed by Simon Mentha.)

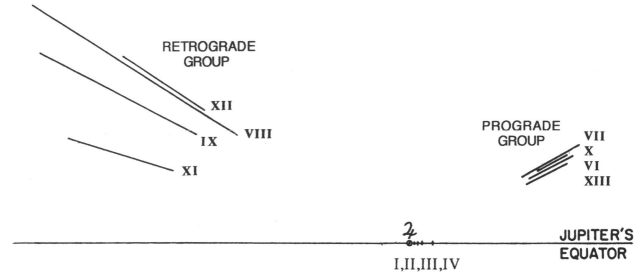

Fig. 25.1. Diagram of the orbits of the outer satellites, showing typical inclinations and orbital distances. (The orbits are, of course, not aligned in space but precess around the planet.)

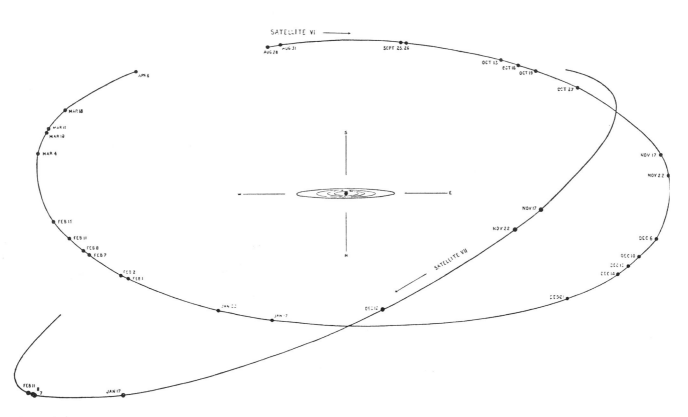

Fig. 25.2. The highly inclined orbits of satellites VI and VII, as observed from Greenwich in 1906/7. (From Christie, *1907*.)

25: The outer satellites

We have now reached the fringes of Jupiter's realm, whose only permanent occupants are a few scattered 'rocks' – the outer satellites.[1] They can only be detected from Earth by long photographic exposures, and no spacecraft have been anywhere near them. The largest (number VI, Himalia) is no more than 185 km in diameter; most of them measure less than 50 km. One group of four orbits 11–12 Mkm from the planet, with orbital inclinations of 27–29° (Table 19.1). The other group of four orbits 21–24 Mkm out, in a retrograde direction, with inclinations of 147–163°. For most of the time they are outside the magnetosphere. The outer group are so remote, with orbital radii 3% that of Jupiter itself, that they experience strong perturbations by solar gravity and have strongly non-elliptical orbits. The eccentricity of VIII's orbit, for example, varies between 0.16 and 0.66. It seems likely that each group comprises the fragments of a captured asteroid.

For many years they were known only by their numbers, given in order of discovery. Recently, the International Astronomical Union has given them mythological names – over the objections of their discoverers, dead or alive, who never saw any need for names. The names commemorate lesser-known lovers of Jupiter, ending in -a for members of the inner group, and in -e for members of the outer group (T. Owen, *1976*). For brevity, we will use the numbers.

The first ones discovered were the two brightest (satellites VI and VII), found in 1904 and 1905 by Charles Perrine at the Lick Observatory. Number VIII was found in 1908 by P. Melotte at Greenwich (though it was then lost until 1922, and lost again until 1938). The next four were found by Seth Nicholson, in a long career which started at Lick with satellite IX (1914), and resumed at the great telescope on Mt. Wilson with X and XI (1938) and XII (1951). The last, XIII, was found by Charles Kowal at Mt. Palomar (1974). Kowal also reported another possible satellite, 1975J1, but it has not been seen again. He was deliberately searching for smaller members of the groups, and the lack of any others is surprising if these objects are indeed fragments of collisions.

Astronomers can infer something about the compositions of these objects by measuring their colours and comparing them with

[1] This chapter is based on the review by Cruikshank *et al.*(*1982*), with more recent photometry by Tholen and Zellner (*1984*). For a photograph showing all eight of these satellites, see Kowal (*1976*).

Table 25.1 *Physical properties of the outer satellites*

| | | From Cruikshank *et al.*(*1982*) | | From Tholen & Zellner (*1984*) | | |
		Mean m_v	Diameter (km)	Mean m_v	Diameter (km)	B–V
XIII	Leda	20.2	15	–		
VI	Himalia	14.8	185	15.0	165	0.63
X	Lysithea	18.4	35	18.2	40	0.58
VII	Elara	16.8	75	16.6	80	0.64
XII	Ananke	18.9	30	–		
XI	Carme	18.0	40	17.9	45	0.64*
VIII	Pasiphae	17.1	50	16.9	70	0.74*
IX	Sinope	18.3	35	18.0	40	≈1.4*

m_v is the mean opposition visual magnitude. Diameter was deduced from this assuming albedo 0.03. *B–V* is the colour index. *Colour indices for the three outermost satellites show anomalies; see text. (For orbital data see Table 19.1.)

asteroids. In the inner group, the two largest members (VI and VII) have virtually identical grey spectra, and the less accurately measured spectrum of X is much the same. This supports the hypothesis that they arose from a single object. VI and VII are the only ones for which albedoes have been measured, by comparing their visual brightness with their thermal emission at 20 µm wavelength. They are extremely dark, with geometric albedoes of 0.02–0.03. Both the spectra and the albedoes are typical of asteroids classified as type C, which are believed to resemble the primitive meteorites known as carbonaceous chondrites. This is odd in that type C asteroids dominate the main asteroid belt, but are rare in its outer parts and in the 'Trojan clusters' in Jupiter's orbit. W. Hartmann (*1987*) suggested that they were among the asteroids flung out of the main belt by resonances with Jupiter, which later captured them.

In the outer group of satellites, the photometric results are peculiar and less precise (Tholen and Zellner, *1984*). VIII resembles a grey type C asteroid according to some results, but is redder according to others. XI is also grey in the visual range, but is unusually bright in the near-ultraviolet (intriguingly, at a wavelength where comets show emission). And IX seems to be as red as Mars, if these difficult measurements are correct. None of them matches a known type of asteroid, but more precise data are needed before any conclusions can be drawn.

VI (Himalia) fluctuates in brightness by 0.15 magnitudes, showing that it rotates in 9.5 (±0.3) hours. As the light-curve has two peaks, the satellite is probably elongated. These properties are again consistent with its being an asteroidal fragment. VII and VIII also show noticeable variations in brightness but they have not been observed well enough to establish their rotation periods.

If they were captured, they could not have been put into orbits as they are now by purely gravitational means. G. Kuiper (*1951*) suggested that the parent bodies were captured in the early days of the solar system, by gas drag in the proto-jovian nebula. This could

also have caused them to break apart, and eliminated smaller fragments. Alternatively, each parent body could have been captured later, and fragmented, by collision with another asteroid or comet.

If a human visitor could stand on one of these tiny bodies, the other outer moons would hardly ever come close enough to be seen with the naked eye. Cometary visitors would be even rarer, the sun would be distant and cold, and Jupiter would appear not quite as big and as bright as the full Moon does from Earth. In this frigid and empty place, the only signs of movement would be the circling of the galilean moons, and the ever-changing parade of clouds across the face of Jupiter.

Fig. 25.3. Jupiter is still capturing satellites today, though they do not last for long. As comets swing past, occasionally the interplay of jovian and solar gravity puts them into unstable orbits around the planet with periods of several years. At least three comets in the past century have been thus captured, for up to 11 years, performing one or two circuits around Jupiter before being flung back into orbits around the sun (e.g. Tancredi et al., *1990*).

This bizarre satellite of Jupiter is the last discovered and the soonest lost: comet Shoemaker-Levy 9 (1993e), discovered in 1993 March. Its misfor-

tune was to fly only 106 000 km from Jupiter's centre in 1992 July, well within the Roche limit. This both shattered it and swung it into a two-year orbit, which has taken it to an apogee of 49.5 million km at the time of writing. The train of fragments is now wider than Jupiter itself. (Image taken on 1993 March 27 with the 2.2 metre telescope on Mauna Kea; by courtesy of Jane Luu (University of California, Berkeley) and David Jewitt (University of Hawaii).)

When the shattered comet returns to Jupiter in 1994 July, it will plunge straight into the planet.

The Plates

Plates P1 to P24 show chronological sets of drawings, photographs, and colour images of Jupiter over a span of 160 years. They are chosen to illustrate typical aspects and important events on the planet, and include some of the best views ever obtained. Many of these pictures are reproduced by kind permission of the Royal Astronomical Society (RAS) or British Astronomical Association (BAA), as indicated. Illustrations credited only by the observer's name are from BAA archives and, if by living persons, are reproduced with their permission. Only a limited number of observers are represented, in the interests of uniformity, although many good drawings by others can be found in the original reports by the BAA and other organisations. Abbreviation: ω_2, System II longitude of central meridian.

PLATE P1

(1) 1831 Sep 5.   (2) 1840 Apr 26.   (3) 1851 May 10.

(1–3) Drawings by S. Schwabe. (From the archives of the RAS; no.2 copied by the author.) No.1 is the first known record of the Red Spot Hollow, which is also shown in nos.2 and 3.

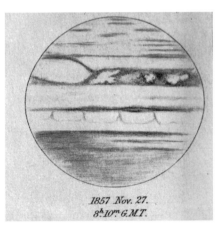

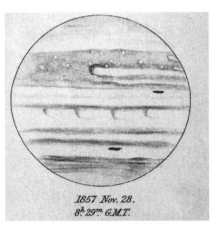

(4) 1857 Nov 27, 8h 10m.   (5) 1857 Nov 28, 6h 43m.   (6) 1857 Nov 28, 8h 29m.

(4–6) Drawings by W.R. Dawes (from Dawes, *1857*). No.4 shows the GRS as a ring; nos.5 and 6 show the p. and f. ends of a S. Tropical Disturbance (STropD), with many small bright ovals in S. Temperate region (compare Plate P22.6). NEBn faint with 'horse-tails'.

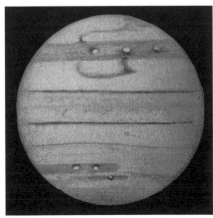
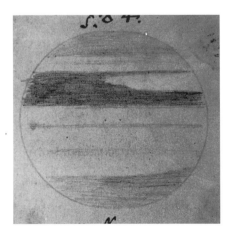

(7) 1859 Jan 15, 9h 40m.
W. Huggins.
GRS as ellipse, SEB and NEB faint.
(From RAS archives.)

(8) 1859 Nov 13, 10h 15m.
S. Schwabe.
Although this looks like the p. end of a STropD, it is actually the f. end of the Red Spot Hollow, following a SEB Revival. (From RAS archives.)

(9) 1860 Feb 26, 6h 30m.
S. Schwabe.
F. end of STropD.
(From RAS archives.)

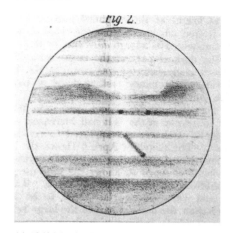

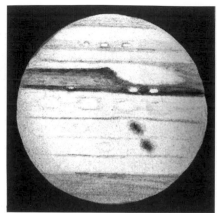

(1) 1860 Mar 2, 9h 24m.
Baxendell.
(From Long, *1860*.)

(2) 1860 Mar 2, 9h.
Huggins.
(From RAS archives; wrongly dated 1858 in some publications.)

(3) 1860 Apr 7, 8h
Huggins
(From RAS archives.)

(1–3) The NEB Revival of 1860. It was described as an oblique streak by J.W. Long (*1860*), but Huggins' simultaneous drawing resolves this into two unrelated spots, and his later drawing (from an unpublished series) shows vigorous turbulence. All drawings show the Red Spot Hollow.

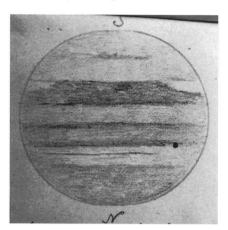

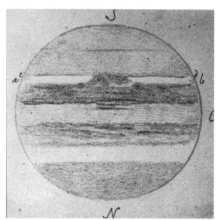

(4) 1861 Mar 19, 9h 29m.
Carpenter.
'Portholes' in the revived NEB(N); disturbance in SEB.

(5) 1862 Apr 2, 6h 50m.
Schwabe.
A long STropD.

(6) 1862 Apr 14, 6h 50m.
Schwabe.
The same STropD, much shorter.

(4–6) Aftermath of the SEB and NEB Revivals. (All from RAS archives.)

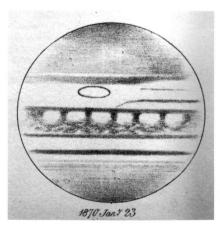

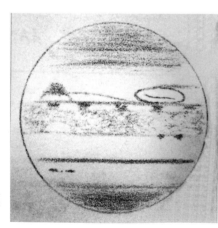

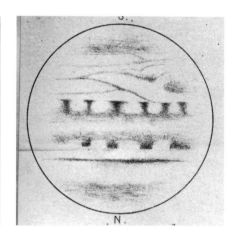

(7) 1870 Jan 23.
Gledhill.

(8) 1871 Dec 1.
Gledhill.

(9) 1872 Jan 6.
Pratt.

(7–9) The GRS before (1870) and during (1871/72) a SEB Revival. In no.9, the GRS is near the f. limb and its ring shape is disrupted. Throughout, EZ is shaded (coloured) with SEBn disturbed. No.9 also shows a NEB Revival in progress. (All from *Astronomical Register*, vol.8, p.209, and vol.10, pp.1, 42, 69, by courtesy of the RAS Library. Also see colour plate P12.1–2.)

(1) 1876 Jun 3.
G.D. Hirst (Australia).
GRS as 'the pink fish'. GRS and whole equatorial region are coloured orange-brown.

(2) 1878 Oct 20.
G.D. Hirst (Australia).
GRS as a vivid brick-red streak (on f. half), as SEB(S) fades.

(3) 1880 Nov 29.
G.T. Gwilliam.
GRS brick-red.
NTBs jetstream spots.

(1–3) These drawings are from coloured originals in the RAS archives. In nos.2, 3, and 4, the GRS is brick-red, NEB orange-brown, SEB(N)/EZ(S) either brown or grey, and NTB grey. (Also see colour plate P12.)

(4) 1880 Nov 29.
N.E. Green.
Simultaneous with (3).
(From Green, *1887*.)

(5) 1882 Dec 23.
N.E. Green.
After SEB Revival, GRS (on f. side) is pale orange, SEB(S) (on p. side) strong orange, SEB(N) still dark grey.
(From Green, *1887*.)

(6) 1886 Apr 11.
W.F. Denning.
Typical view of GRS, with SEB(S) faint p. it, and SEBn disturbed.
(From RAS archives.)

(7) 1890 Jul 30.
E.E. Barnard.
GRS; NEBn black spots with 'horse-tails'.
(From Barnard, *1891*.)

(8) 1891 Oct 22.
W.E. Jackson.
GRS; STB white spots.
(From BAA archives.)

(9) 1891 Oct 23.
W.E. Jackson.
Large dark streak on STB; NTBs jetstream spots.
(From BAA archives.)

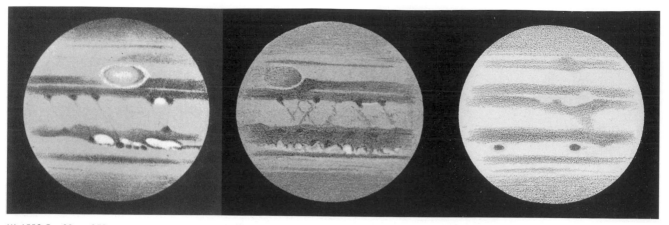

(1) 1893 Oct 30, ω_2 350.
Antoniadi.
GRS; NEB Revival active.

(2) 1894 Sep 6; ω_2 21.
Antoniadi.
GRS; NEB very broad.

(3) 1895 Oct 15; ω_2 270.
Antoniadi.
NEB receding, leaving two
dark 'little red spots'.

(1–7) Two successive NEB Revivals are covered by these beautiful early drawings by E.M. Antoniadi and T.E.R. Phillips. The originals are tinted. Nos.1 and 2 are from BAA Memoirs; nos.4–7, previously unpublished, are from BAA archives.

(4) 1896 Dec 14, ω_2 98.
Phillips.
Possibly the start of the NEB
Revival (compare Plate P2.1–3).

(5) 1897 Jan 27, ω_2 113.
Phillips.
NEB Revival underway.

(6) 1897 Feb 3, ω_2 56.
Phillips.
Red Spot Hollow on p. side.

(7) 1897 Mar 9, ω_2 178.
Phillips.
NEB very active. EZ ochre (also in nos.4–6).

(8) 1899 Jun 2, ω_2 282.
Molesworth.
EZ ochre.
(From BAA archives.)

(9) 1903 Aug 12, ω_2 207.
Phillips.
The great STropD. Much detail in far north.
(From Denning (1923), in Phillips' *Splendour of
the Heavens*.)

PLATE P5

All these drawings in Plates P5–P7 were published in BAA Memoirs, but are reproduced here either from BAA archives, or from Denning (*1923*) in Phillips' *Splendour of the Heavens* (which also contains other fine drawings from this period).

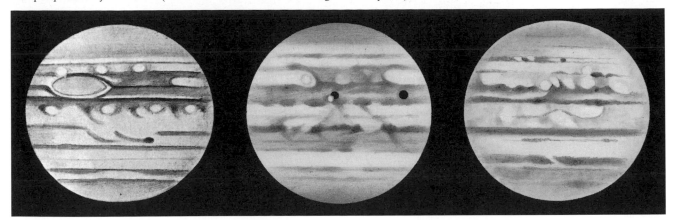

(1) 1906 Apr 15.
W.F. Denning.
STropD at f. edge of GRS. Another NEB Revival beginning. (From Denning, *1923*.)

(2) 1907 Feb 21, ω_2 241.
Scriven Bolton.
STropD. NTropZ dark. Shadow of II partly hidden by Io, whose own shadow follows it.

(3) 1911 Jun 17.
T.E.R. Phillips.
F. part of STropD, very disturbed. In 'interregnum' between SEBn and NEBs activity, EZ has only faint wisps.

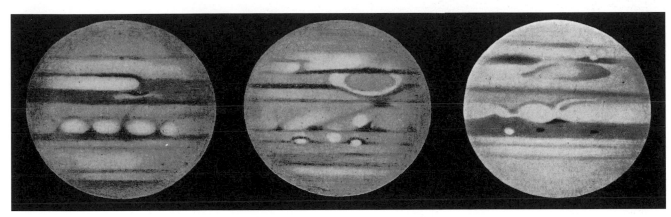

(4) 1913 Aug 28.
T.E.R. Phillips.
P. end of STropD.
Long-term NEBs activity has begun.
(From Denning, *1923*.)

(5) 1914 Aug 29.
T.E.R. Phillips.
F. end of STropD, p. GRS. 'Portholes' on NEBn with NTropZ dark.
(From Denning, *1923*.)

(6) 1919 Nov 29.
T.E.R. Phillips.
GRS as a huge oval with SEB faded. Broad NEB with barges and 'porthole'.
(From Denning, *1923*.)

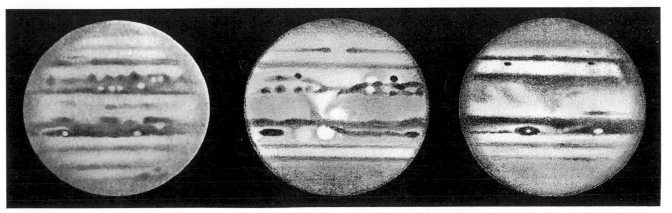

(7) 1920 Feb 13, ω_2 178.
T.E.R. Phillips. (From Denning, *1923*.)

(8) 1920 Feb 27, ω_2 47.
H. Thompson.

(9) 1920 Apr 13, ω_2 98.
T.E.R. Phillips.

(7–9) These show the great SEB Revival of 1919/20. In the STropZ, the two very dark spots which were retrograding on the SEBs jetstream in no.8 were returning on the STBn jetstream in no.9, having been reflected by the p. end of the STropD. EZ contains extensive ochre shadings. NEB is broad with 'portholes', and two very dark 'little red spots' on NEBn are shown in nos.8 and 9 respectively.

(1) 1921 Apr 25, ω_2 212.
Phillips.
S. Temperate Disturbance and f. end of STropD.

(2) 1925 Aug 7, ω_2 143.
Phillips.
P. end of STropD, and a rift opening through SEBn.

(3) 1926 Sep 19, ω_2 198.
Phillips. STB faded section; SEB faded, with residual streak ('Fragment') next to shadow of III; NEB disturbed; NTBs jetstream spots.

(4) 1927 Aug 2, ω_2 102.
Phillips.
STBn dark spots; SEB faint; NEB and NTropZ disturbed.

(5) 1927 Oct 6, ω_2 10.
Phillips.
GRS tapering into a ruddy S. Tropical Band; NEB and NTropZ disturbed.

(6) 1928 Aug 30, ω_2 120.
Peek.
Source region of the great SEB Revival. EZ shaded.

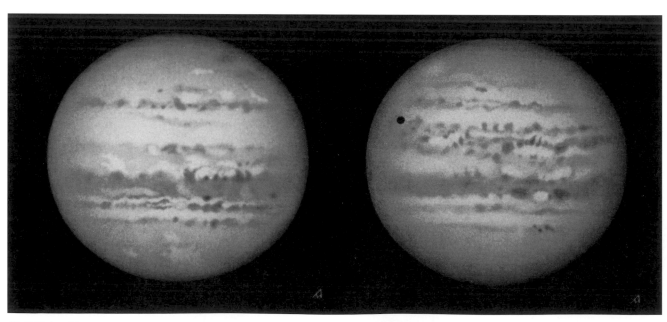

(7) 1928 Aug 11.
Antoniadi.
SEB faint before the Revival. NEB very broad.

(8) 1928 Dec 8.
Antoniadi.
The SEB Revival at its height. Prominent spot group on NEBn.

(1) 1928 Nov 5, ω_2 70.
Phillips.
SEB Revival at its height.

(2) 1928 Dec 5, ω_2 260.
Phillips.
SEB Revival continuing, with GRS on f. limb.
Prominent spot group on NEBn (also shown in
Plate P6.8, 3 days later).

(3) 1929 Nov 30, ω_2 311.
Peek.
GRS as a white oval. NTropZ disturbed, with
NTBs jetstream outbreak.

(1–3) The great SEB Revival of 1928 (continued) and its aftermath; drawings by T.E.R. Phillips and B.M. Peek.

(4) 1931 Jan 17, ω_2 14.
Phillips.
Mid-SEB outbreak.

(5) 1933 Jan 16, ω_2 84.
Phillips.
Curious STZB with tiny spots. SEBs jetstream
spots, with p. end of STropD as a dark bar. Broad
rifted NEB.

(6) 1934 Apr 19, ω_2 5.
Peek.
P. end of STropD, with spots on both sides of the
'circulating current'; extensive rifting in SEB.

(4–6) Long-running activity in the SEB and 'circulating current', in the 1930s.

(7) 1938 Jun 14, ω_2 277.
T.E.R. Phillips (incomplete).
P. end of STropD, with STBn jetstream spots p.
it.

(8) 1938 Jul 3, ω_2 184.
J.E. Phocas (redrawn by F.J. Hargreaves).
GRS on p. side, with SEBn and STBn spots
approaching it. Large 'barge' on NEBn.

(9) 1938 Jul 28, ω_2 330.
M.A. Ainslie.
Disturbance within the STropD.

(7–9) The SEB Revival of 1938, of which only the later stages were observed. There was intense spot activity on the SEBn and the STBn jetstream, and the
EZ was also disturbed and coloured.

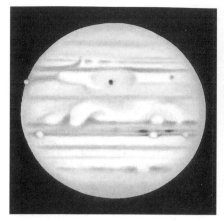

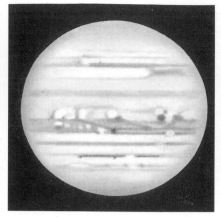

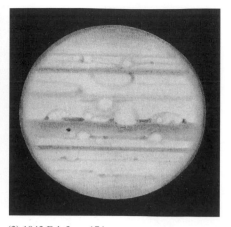

(1) 1942 Jan 6, ω_2 170.
F.J. Hargreaves.
II on p. limb, with its shadow on GRS. On p. side is the new STropD, and on f. side the STropZ oscillating spot. NEB broad with 'portholes'.

(2) 1942 Jan 10, ω_2 77.
F.J. Hargreaves.
P. end of the new STropD. NEB disturbed.

(3) 1943 Feb 2, ω_2 174.
B.M. Peek.
SEB faint; GRS a pale ring. NEB disturbed. NNTBs jetstream spots.

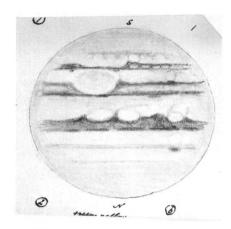

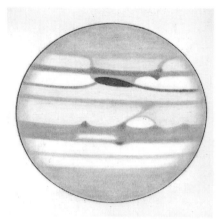

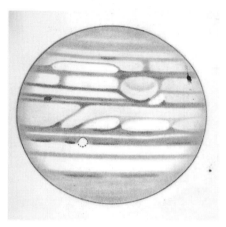

(4) 1945 Jun 25, ω_2 228.
J.E. Phocas.
Proto-oval FA on STBs alongside the GRS. (Unpublished, BAA archives.)

(5) 1947 Feb 11/12, ω_2 353.
E.J. Reese.
New STropD.

(6) 1947 Jun 9/10, ω_2 210.
E.J. Reese.
Proto-oval on STBs on p. side; GRS in a dark ring after SEB Revival.

(7) 1952 Aug 31, ω_2 282.
W.E. Fox.
GRS dark reddish; SEB faint; STBs white oval; dark belt segment in NTZ.
(Compare Plate P9.5&6.)

(8) 1952 Nov 28, ω_2 277.
W.E. Fox.
GRS impacted by S. and N. branches of SEB Revival. STBs white oval and NTZ belt segment shown again.

(9) 1953 Sep 17, ω_2 246.
E.J. Reese.
GRS as a white oval with dark STropZ p. it, after the SEB Revival.

(7–9) The GRS before, during, and after the SEB Revival of 1952. From 1943 to 1958, SEB fadings and Revivals occurred every three years.

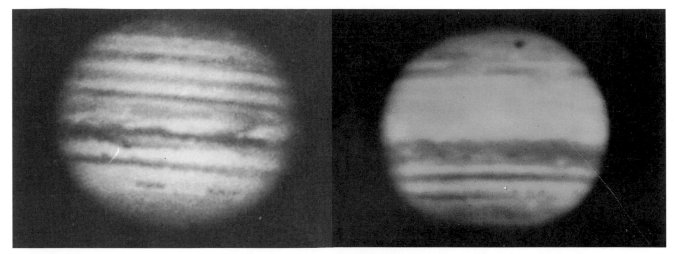

(1) 1939 Oct 13.
Lick Observatory.
Includes NTBs jetstream spot.

(2) 1941 Oct 22, ω_2 119.
Observatoire du Pic du Midi (38-cm refractor; Lyot, *1943*).
Dusky STZ divides proto-oval BC (on p. half) from DE (on f. half). SEB faint, but a small STropD is visible on f. half. NEB very broad and spotty.

(3) 1945 Apr 14, ω_2 331.
Observatoire du Pic du Midi (60-cm refractor; Lyot, *1953*).
NEB again very broad and disturbed (with satellite shadow). NNTBs has jetstream spots and a larger, slower spot on central meridian.

(4) 1950 Oct 5.
Mt. Palomar (5-metre reflector); Blue filter.
STBs white oval; GRS also a light oval; great white spot on SEBn, with dusky EZ. (Photograph from the Hale Observatories.)

(5, 6) 1952 Oct 24.
Mt. Palomar (5-metre reflector); (5) Red filter, (6) Blue filter. Another STBs white oval passing GRS, with shadow of III. GRS now a strongly red oval; SEB faint. NTZ belt segment is very dark. (Photograph from the Hale Observatories.)

(1) 1955 Feb 16, ω_2 221.
W.E. Fox.
Source of SEB Revival. Shadow of I N. of it, III (dark) Sf. it. Also STB oval DE.

(2) 1955 Dec 4, ω_2 209.
W.E. Fox.
New STropD.

(3) 1956 Nov 15, ω_2 287.
E.J. Reese.
STB oval FA, and GRS, with SEB faint again.

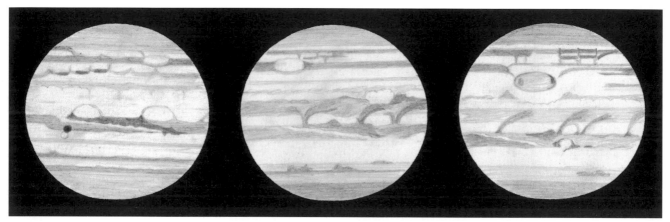

(4A) 1958 Apr 17, ω_2 58.
H. Sykes.
SEB Revival: retrograding SEB(S) spots. Large NEB rift.

(5A) 1961 Aug 22, ω_2 163.
H. Sykes.
SEB faint; massive 'EZs belt' with disturbed edges.

(6A) 1961 Aug 28, ω_2 18.
H. Sykes.
GRS a dark reddish 'doughnut'; SEB faint, with 'EZs belt'; NEB disturbed.

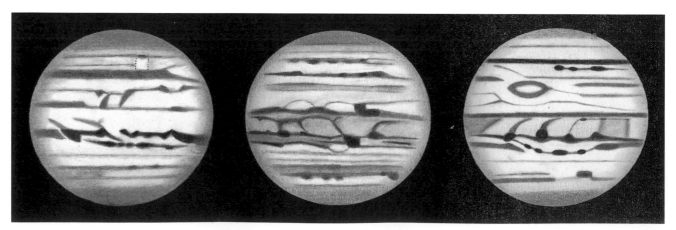

(4B) 1958 Apr 19, ω_2 35.
T. Sato.
SEB Revival: source region. Large NEB rift.

(5B) 1961 Jul 15, ω_2 257.
T. Sato.
SEB faint; massive 'EZs belt' with disturbed edges.

(6B) 1961 Aug 31, ω_2 23.
T. Sato.
GRS a dark reddish 'doughnut'; SEB faint, with 'EZs belt'; NEB disturbed.

(4–6) For 1958–1962, drawings by two observers show the same sets of features in different artistic styles. H. Sykes of Kuala Lumpur, Malaya, used a 30-cm reflector (except in no.4A, for which he used a 29-cm refractor in Sydney, Australia); drawings from BAA archives. T. Sato of Hiroshima, Japan, used a 25-cm reflector; drawings kindly provided by T. Sato. (Also see Fig. 9.8.)

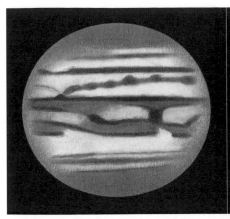

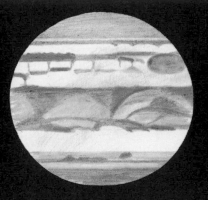

(1) 1962 Oct 24, ω_2 247.
T. Sato (Japan).
SEB Revival: source and S. branch. EZ mostly shaded. NTB absent.

(2) 1962 Oct 24, ω_2 337.
H. Sykes (Malaya).
SEB Revival: S. branch approaching GRS, with shadings in STropZ.
EZ heavily shaded. NTB absent.

(3) 1963 Aug 8.
J.H. Focas (J.E. Phocas; 63-cm refractor, Athens, Greece).
SEB again faint and GRS a dark reddish oval. EZ heavily shaded and NEB disturbed.

(4) 1964 Jul 25, ω_2 138.
T. Sato (Japan).
SEB Revival: central branch prograding, with bright spots.

(5) 1964 Aug 20, ω_2 204.
A.W. Heath (England).
SEB Revival: source region.

(6) 1966 Feb 6, ω_2 262.
J.B. Murray (England).
SEB double with SEBs jetstream spots. Dark NEB barge.

(4,5) These drawings, very early in the apparition, are among the few to show the SEB Revival of 1964. This was a weak affair which closely followed the incomplete Revival of 1962. EZ was still mostly shaded, and NTB absent. (See also Plates P12.12 and P13.1.)

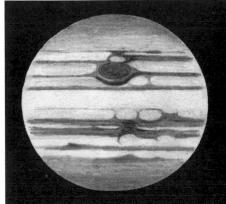

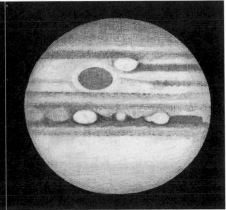

(7) 1967 Feb 25, ω_2 27.
A.R. and M.F. Pace (England).
GRS still has a dark red patch in ring. On S. edge, STB white oval with STB fading p. it.

(8) 1968 Jun 12, ω_2 108.
J.B. Murray (England).
STB Fade; SEBs jetstream spots; dark EB.

(9) 1969 Feb 28, ω_2 35.
M. Falorni (Italy).
STB white oval with Fade p. it; GRS dark red as SEB fades; dark EB. NTB absent.

PLATE P12. COLOUR DRAWINGS, 1872–1964. (Full captions can be found after Plate P24.)

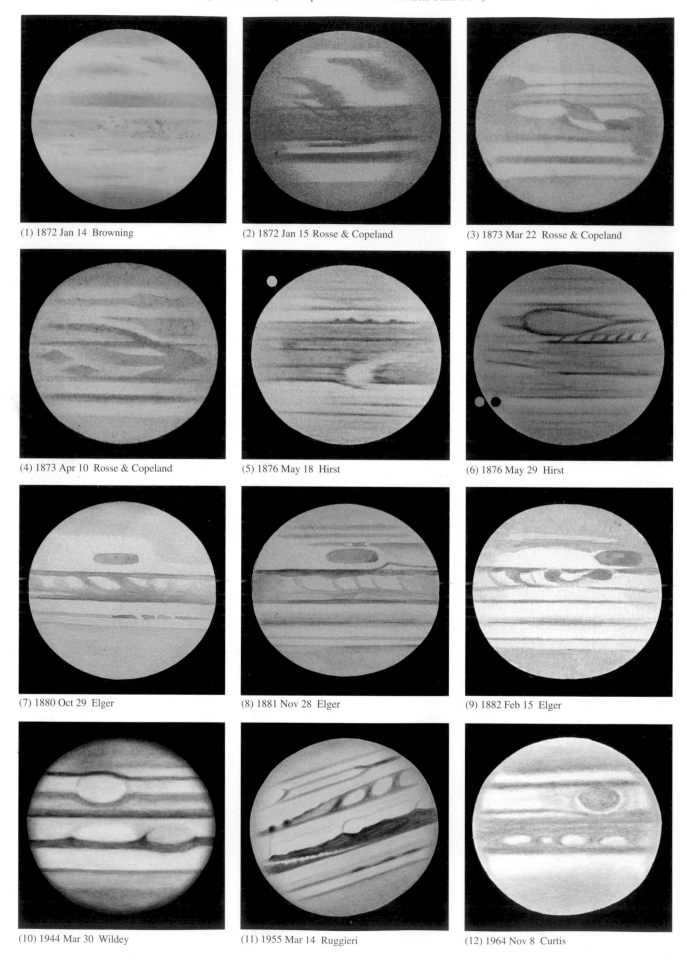

(1) 1872 Jan 14 Browning

(2) 1872 Jan 15 Rosse & Copeland

(3) 1873 Mar 22 Rosse & Copeland

(4) 1873 Apr 10 Rosse & Copeland

(5) 1876 May 18 Hirst

(6) 1876 May 29 Hirst

(7) 1880 Oct 29 Elger

(8) 1881 Nov 28 Elger

(9) 1882 Feb 15 Elger

(10) 1944 Mar 30 Wildey

(11) 1955 Mar 14 Ruggieri

(12) 1964 Nov 8 Curtis

PLATE P13. PHOTOGRAPHS FROM ARIZONA, 1964–1972. (Full captions can be found after Plate P24.)

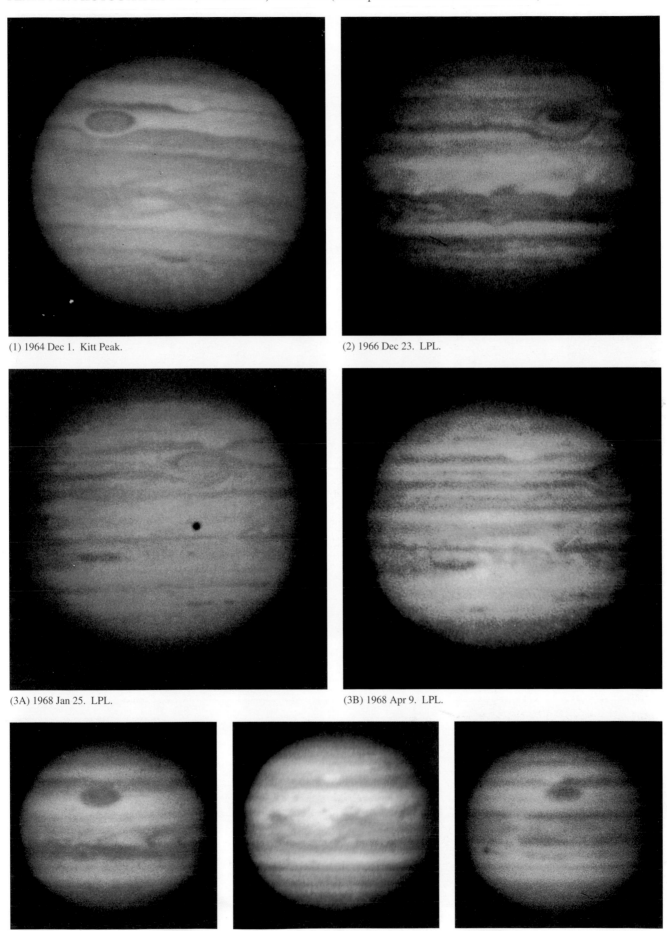

(1) 1964 Dec 1. Kitt Peak.

(2) 1966 Dec 23. LPL.

(3A) 1968 Jan 25. LPL.

(3B) 1968 Apr 9. LPL.

(4) 1970 May 16. LPL.

(5) 1971 Aug 7. LPL.

(6) 1972 Oct 3. LPL.

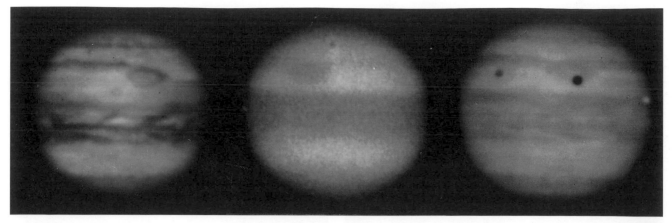

(1) 1961 Aug 4, ω_2 347.
Observatoire du Pic du Midi, France (60-cm refractor); yellow filter. GRS as 'doughnut'; STB white oval with STB faint f. it.

(2) 1962 Aug 26, ω_2 29.
J. Dragesco (France; 26-cm refractor); film used was evidently blue-sensitive. GRS and EZ very dark (reddish).

(3) 1962 Sep 14.
Observatoire du Pic du Midi (1-metre reflector); yellow filter. In transit are III (dark) and shadow (on GRS) and Io (on f. limb).

(1–3) These French photographs show the dark reddish GRS and EZ while the SEB faded in 1961–1962. NTB also absent. (From BAA archives. Also see Fig. 9.8.)

(4,5) 1968 Jan 25. LPL, Catalina Observatory, Arizona (154-cm reflector); (4) Red filter, (5) Blue filter.
GRS, EB, and Io (on NEBs towards p. limb) are strongly reddish. NEBs projections and N. Tropical Band are bluish.
(By courtesy of Lunar and Planetary Laboratory of the University of Arizona.)

(6) 1967 Jan 15.
G. Viscardy (Monte Carlo; 30-cm reflector).
GRS has a dark patch in a ring.

(7) 1969 May 30.
LPL, Catalina Observatory, Arizona (154-cm reflector), yellow filter.
Tiny dark spots on STBn jetstream. SEB faint.

(8) 1973 Sep 2.
J.B. Murray, Pic du Midi (1-metre reflector).
SEB faint. NEB rifts. Long-lived NTZB on p. side.

(1) 1971 Jul 27, ω_2 24. I. Hirabayashi (Japan).
Dark red GRS and source of the SEB Revival.
(See also Fig. 10.20.)

(2) 1971 Aug 10, ω_2 249. I. Hirabayashi (Japan).
The great SEB Revival: large, very bright spot on
central meridian.
(See also Plate P13.5 and Fig. 10.21.)

(3) 1972 Jul 13, ω_2 7. M. Foulkes (England).
SEB(S) fading again, with orange shading across
EZ and SEB, and GRS still dark red. Large light
spots spanning SEB(N) and NEB(S).

(4) 1973 Jun 7, ω_2 43. J.H. Rogers (England).
SEB(S) faint, GRS dark red, strong EB(S). North
of GRS are EZn plume *b* (very bright) and
NTropZ Little Red Spot (dark bar). NTB very
faint.

(5) 1975 Aug 27, ω_2 135. J.H. Rogers (England).
The great SEB Revival, with 3 sources visible.
EZn plume *a* bright in centre. III in transit in far
south, half dark and half bright.

(6) 1975 Oct 10, ω_2 47. P.J. Young (Texas).
SEB Revival: new SEB(S) crossing GRS. Start
of the long-term STB Fade p. oval BC and GRS.
NTBs jetstream outbreak including brilliant
white ovals.

(7) 1975 Oct 2, ω_2 186.
C. Boyer, Pic du Midi (France; 1-metre reflector, yellow filter).
The great SEB and NTBs outbreaks at their height; EZn plume *a* breaking
up, on f. side; STB oval DE.

(8) 1975 Dec 22, ω_2 23.
J. Dragesco, Pic du Midi (France; 1-metre reflector, yellow filter).
Aftermath of the 1975 disturbances. GRS as a ring on f. side; STB oval BC
near central meridian; STB Fade on p. side, with S. Tropical Dislocation.
NTropZ dusky.

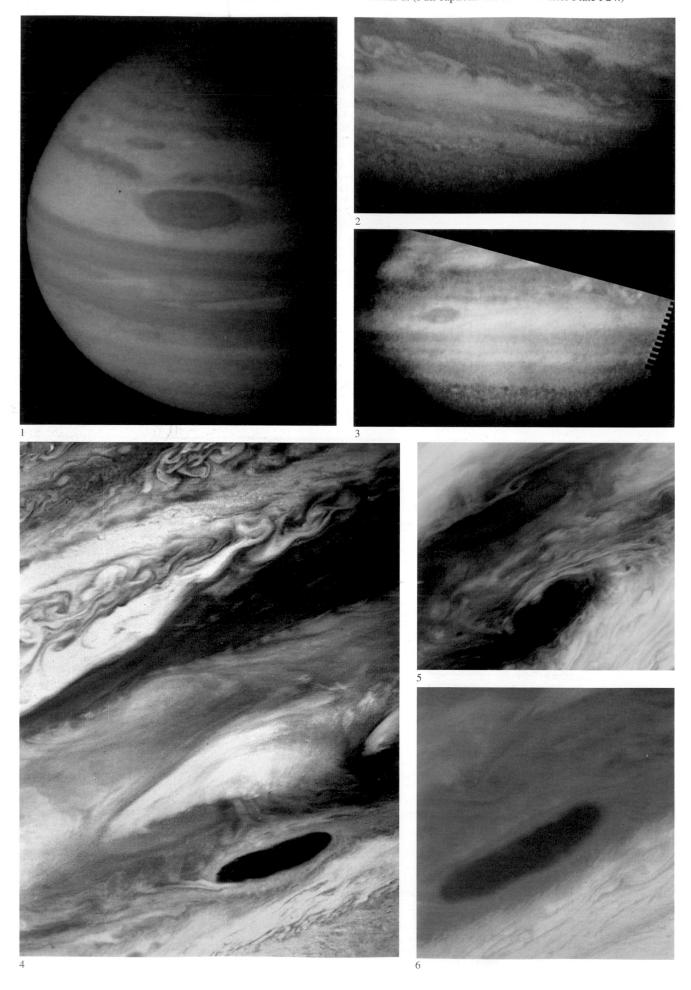

1

2

3

4

5

6

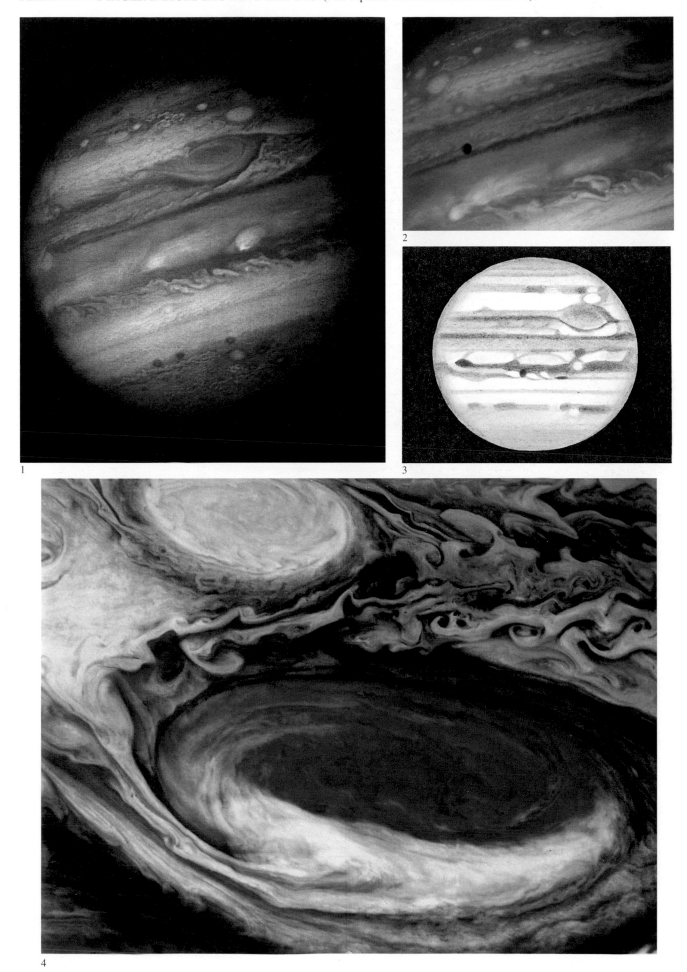

1

2

3

4

These drawings, all by J.H. Rogers, span the Voyager encounters of 1979 and show the evolution of the activity that appeared in 1978–1979. EZ was ochre from 1978 onwards. (Also see 1976 and 1979 planetary maps in Chapter 1.) Bright white spots are marked '+' or 'w'.

(1) 1976 Dec 8, ω_2 42.
GRS, with the long-lived EZs white spot (on f. side). Chaotic EZ(N) including the last record of plume a (just f. central meridian). Io and shadow on SEBs.

(2) 1978 Feb 1, ω_2 238.
STB Fade between ovals FA and BC. The brilliant white spot on SEBs was the start of long-running SEBs jetstream activity, while EZ has become shaded, yellowish.

(3) 1979 Jan 25, ω_2 57.
GRS with oval BC and STB Fade p. it. The long-lived EZs white spot re-erupting p. GRS. NEB rifted region. (Compare with Voyager images.)

(4) 1979 Mar 15, ω_2 129.
GRS on p. limb. New mid-SEB outbreak on central meridian. Dark barges on NEBn.

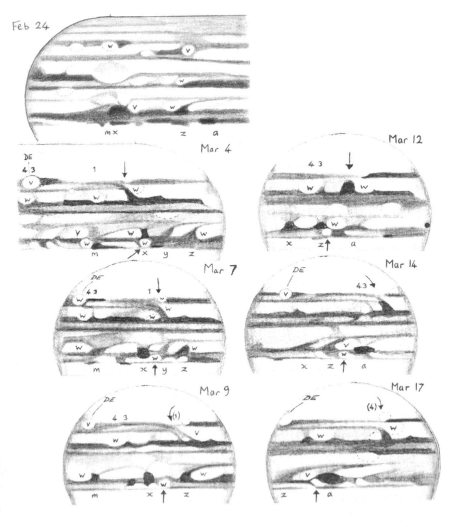

(6) 1980 Dec 27, ω_2 267.
The great dark complex at f. end of STropD, alongside oval FA, with STBn jetstream spots prograding p. it. Shadow of Io near p. limb.

(5) 1980 Feb 24 – Mar 17. Spots interacting with STropD and with NEB rifts.
Arrow from above marks the great dark complex at f. end of STropD, with SEBs jetstream spots running up to it and disappearing (1, 3, 4).
Arrow from below marks one of the white spots in a large mid-NEB rift ($\Delta\lambda_1 = +131$). NEBs projections (m,x,z,a) show remarkable interaction with this rift. A long NEBs complex splits into dark patches m and x; x retrogrades 20° in 10 days ($\Delta\lambda_1 \approx +58$), as though dragged along by the rift, but disappears before hitting projection z. Then a is intensified as the rift passes. (From Rogers, *1988*.)

(1) 1979 Feb 17, $\omega_2 \approx 101$.
Voyager 1, Violet filter (images 15920.44,50,56,62; spatially filtered).
Near p. limb are oval BC, GRS, and NEB rifted region. SEB turbulent f.
GRS. NNTBs jetstream spots. (NASA image.)

(2) 1979 May 12, $\omega_2 \approx 60$. *Voyager 2*, green filter (image 18915.28; spatially filtered, overlaid by grid of dots.)
GRS central; p. it are oval BC (near p. limb), the new STropD, and the long-lived EZs white spot. (NASA image.)

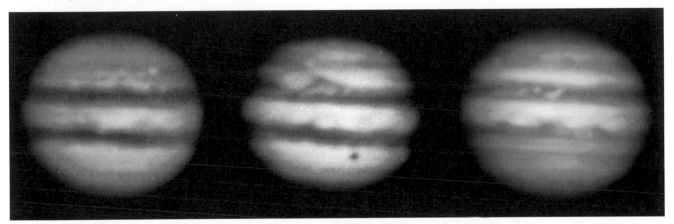

(3) 1984 May 24, ω_2 252. J. Dragesco (Benin).
STB between ovals BC and DE is disrupted, with S. Tropical Band; intense activity on SEBs jetstream.

(4) 1984 Jun 14, ω_2 49. J. Dragesco (Benin).
GRS and oval FA; S. Tropical Band near p. and f. limbs; SEBZ turbulence. Shadow of II on the faint NTB.

(5) 1985 Jul 21, ω_2 142. D. Parker (Florida).
New mid-SEB outbreak. Also a rift in NEB, interacting with NEBs projection. Dusky NTropZ with N. Tropical Band. Also see Fig. 4.1A (Aug 2), Fig. 20.3E (Aug 7), and Plate P22.3 (Aug 9).

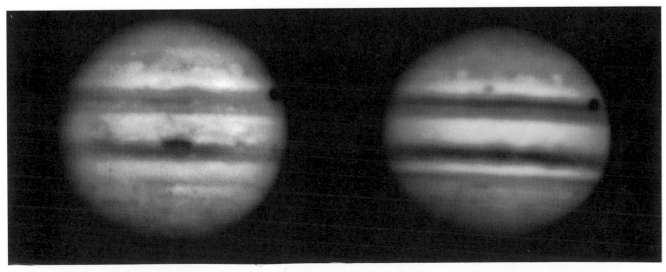

(6, 7) 1986 Jul 17, ω_2 261. J. Dragesco (Pic du Midi, 1-metre reflector); (6) Red filter, (7) Blue filter. Ovals BC and DE, with Little Red Spot in STropZ. Dark patch on NEBs is blue; NTB(S) red. Tiny NNTBs jetstream spots, with reddish strip N. of them. (See also colour plate P21.5.)

1

2

(1) 1979 Jun 29.
Voyager 2.

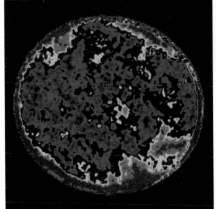

(2,3) Polarised light. B. Carlson & B. Lutz.

(4) Infrared light. D. Allen.

(5) 1986 Jul 10. J. Dragesco & J. Rogers.

(6) 1988 Dec 24. J. Bourgeois & J. Rogers.

PLATE P22

These drawings are by J.H. Rogers (Cambridge, England; 32-cm refractor; bright white spots marked '+' or 'w') and Isao Miyazaki (Okinawa, Japan; 20-cm reflector).

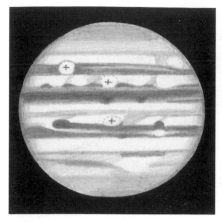

(1) 1982 Jul 19, ω_2 278.
Rogers.
Oval DE; intense SEBs jetstream activity; long NEB rift.

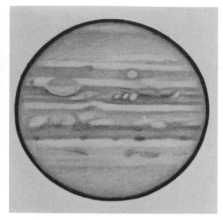

(2) 1983 Jun 20, ω_2 68.
Miyazaki.
GRS; oval DE; typical SEBZ turbulence f. GRS.

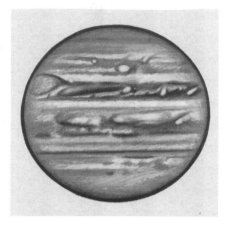

(3) 1985 Aug 9, ω_2 87.
Miyazaki.
Oval DE, with STB fading p. it; GRS on p. limb; SEB rifting supplemented by new mid-SEB outbreak on f. side; also a NEB rift.

(4) 1986 Sep 3, ω_2 247. Miyazaki.
Ovals BC and DE, with Little Red Spot in STropZ (compare Plate P19.7&8). The long-lived SEBn rift is alongside it. Unusual detail in far north.

(5) 1987 Oct 1, ω_2 320. Miyazaki.
STB very faint, replaced by dark SSTB with SSTBs white ovals. S. Tropical Band emerging p. GRS, in parallel with NNTB(S) advancing with NNTBs jetstream. NEB rifts.

(6) 1987 Oct 22, ω_2 101. Rogers.
Ovals BC (p.) and DE (f.) separated by two other STB white ovals, also with SSTB white ovals. Long-lived stationary white spot on SEBs.
NEB rifts. Small, very dark NTB(N) spots.

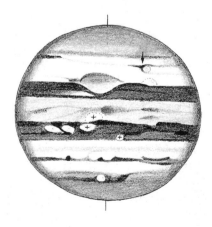

(7) 1989 Jan 22, ω_2 30.
Rogers.
GRS, with SEB now quiet. STB absent; arrow indicates SSTBn jetstream spot and oval FA. NEB very broad. NTB absent; NNTBs jetstream spots; still exceptional detail in far north. (Compare CCD image, Fig. 1.21.)

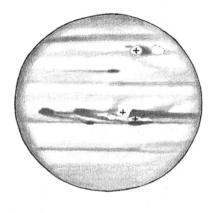

(8) 1989 Oct 18, ω_2 133.
Rogers.
STB, SEB, NTB all faint! Ovals BC and DE very close. Dark streak on SEBs is p. edge of long-lived STropZ white spot. Large rifts in NEB.

(9) 1990 Oct 13, ω_2 49.
Rogers.
SEB Revival. Cross-hatching indicates ochre colour in GRS and EZ. NTB also revived.

PLATE P23

These fine photographs record a complete activity cycle in the NEB: narrow with rifts in 1987, broad in 1988/89, with development of NEBn white spots and dark 'barges' that persisted as the belt receded in late 1989. Meanwhile, the STB and NTB disappeared. The dark southern belt is true SSTB with small white ovals at 40.5°S.

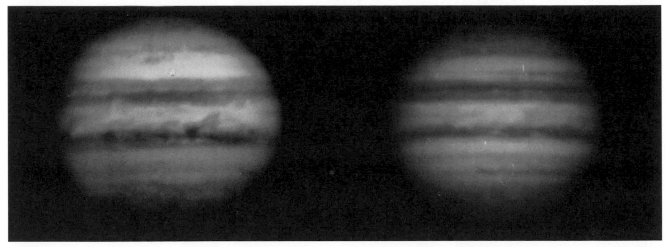

(1) 1986 Jul 21, ω_2 82. J. Dragesco (Pic du Midi, France; 1-m reflector), red filter.
STB 'Remnant'. The long-lived SEBn rift near f. limb with 'ripples' p. it. Rift in NEB.

(2) 1987 Oct 10, ω_2 281. T. Akutsu (Japan; 25-cm reflector), unsharp-masked. Dark S. Tropical Band and NNTB prograding in parallel (see also Plate P22.5). NEB narrow with small rifts.

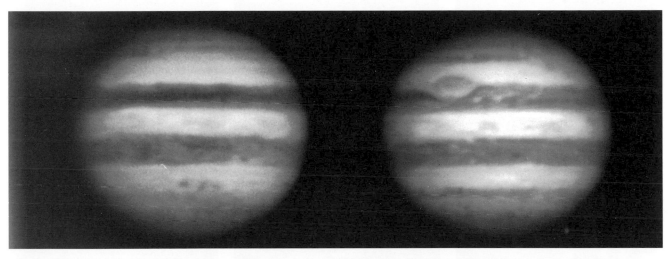

(3) 1988 Aug 27, ω_2 155. I. Miyazaki (Japan; 40-cm reflector).
NEB very broad with reticulate pattern. Small dark spots in NTZ are 'N. Temperate Disturbance'.

(4) 1988 Nov 12, ω_2 43. I. Miyazaki (Japan; 40-cm reflector).
SEBs jetstream spot is retrograding round N. edge of GRS. NEBs quiet; small mid-NEB rifts; white and dark spot appeared in NEB N. half.

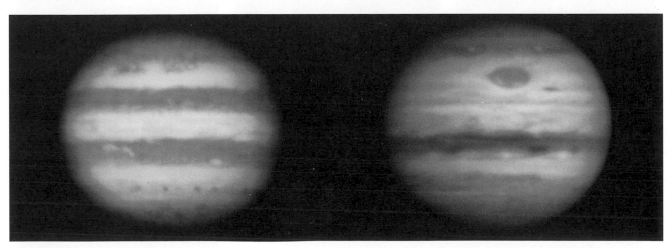

(5) 1988 Dec 31, ω_2 295. J. Bourgeois (Pic du Midi, France; 1-m reflector), yellow filter.
Ovals BC and DE on p. side with fragments of STB. NEB as in no.4. NNTBs jetstream spots. (See also Fig. 1.20.)

(6) 1989 Sep 28, ω_2 16. I. Miyazaki (Japan; 40-cm reflector).
SEB fading and GRS intensifying. NEBn receding, leaving array of dark barges and white spots in NTropZ. Main belt to north is N. Temp. Disturbance, much expanded since no.3.

PLATE P24. THE GLOBAL UPHEAVAL OF 1990. (Full captions are on following pages.)

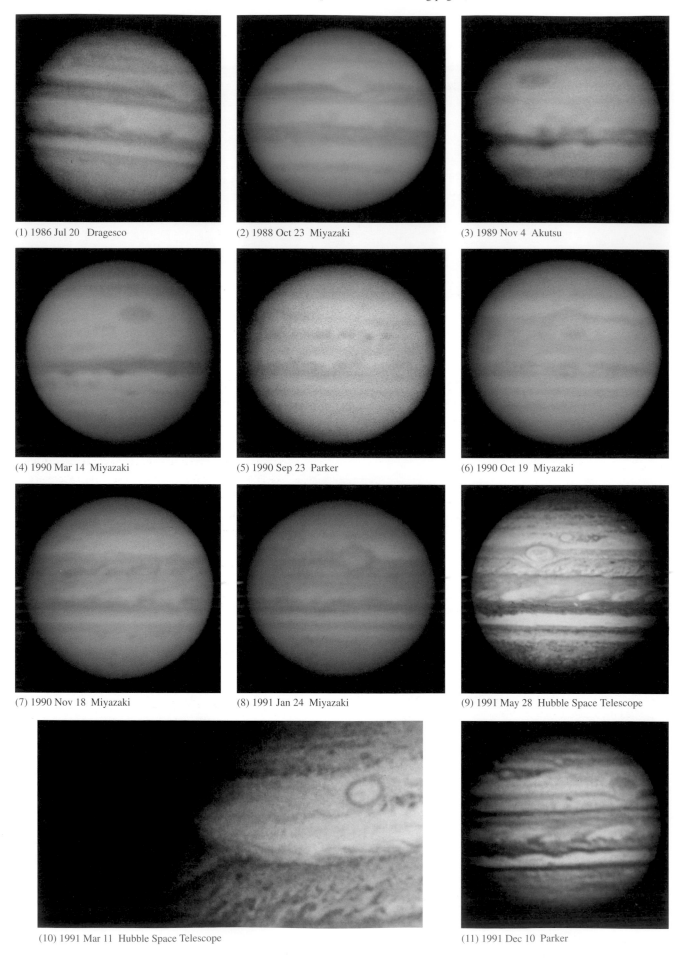

(1) 1986 Jul 20   Dragesco

(2) 1988 Oct 23   Miyazaki

(3) 1989 Nov 4   Akutsu

(4) 1990 Mar 14   Miyazaki

(5) 1990 Sep 23   Parker

(6) 1990 Oct 19   Miyazaki

(7) 1990 Nov 18   Miyazaki

(8) 1991 Jan 24   Miyazaki

(9) 1991 May 28   Hubble Space Telescope

(10) 1991 Mar 11   Hubble Space Telescope

(11) 1991 Dec 10   Parker

Captions to colour plates

PLATE P12. DRAWINGS FROM THE 19TH CENTURY AND MID-20TH CENTURY.

(1) 1872 Jan 14; J. Browning (England; 31-cm reflector).
EZ a striking tawny colour.
(From Browning, *1872*.)

(2) 1872 Jan 15; Lord Rosse & Dr. Copeland (Birr Castle, Ireland; 183-cm reflector).
EZ tawny (compare previous picture). The dusky mass at top right is the GRS, with SEB Revival p. it. (See also Plate P2.9.)
(From Rosse and Copeland, *1874*.)

(3) 1873 Mar 22, ω 150.
Nos. 3 and 4 are from the series of drawings by Rosse and Copeland in early 1873 that first revealed red colour in the GRS. Meanwhile, EZ partially cleared, with dark bluish spots. Here, GRS is near p. limb.
(From Rosse and Copeland, *1874*.)

(4) 1873 Apr 10, ω 135.
The reddish GRS in its Hollow spans the whole f. side of the disk.
(From Rosse and Copeland, *1874*.)

(5) 1876 May 18; G.D. Hirst (Sydney, Australia; 26-cm reflector).
EZ strongly tawny again. Spots on SEBs suggest jetstream activity.
(From RAS archives.)

(6) 1876 May 29; G.D. Hirst (Sydney, Australia; 26-cm reflector).
GRS large and reddish; it was named 'the pink fish' by the Australian observers. (Also see Plate P3.2.)
(From RAS archives.)

(7–9) Three drawings by T. Elger (England; 22-cm reflector), showing the GRS at its most conspicuous, with SEB(S) faint or absent. EZ is partly dusky, and SEBn shows the usual nineteenth-century array of spots. (From BAA archives.)

(7) 1880 Oct 29.
Dark grey spots moving on NTBs jetstream. (Also see Plate P3.3&4.)

(8) 1881 Nov 28.
NTB orange following the jetstream outbreak.

(9) 1882 Feb 15.

(10) 1944 Mar 30; H. Wildey (England; 32-cm reflector).
Following a SEB Revival, GRS is a white oval, SEB reddish, and EZ yellow.

(11) 1955 Mar 14; G. Ruggieri (Venezia, Italy; 25-cm reflector).
Advancing segment of reviving SEB is bluish; NEB reddish.

(12) 1964 Nov 8; A.C. Curtis (England; 30-cm reflector).
GRS red, and EZ deep orange-brown.

PLATE P13. PHOTOGRAPHS FROM PROFESSIONAL OBSERVATORIES IN ARIZONA, 1964–1972.

These were some of the highest-resolution colour images of Jupiter before spacecraft imagery. No.1 was taken by C.R. Lynds with the 2.1-metre reflector at NOAO Kitt Peak National Observatory, Arizona. Nos.2–6 are from the Lunar and Planetary Laboratory of the University of Arizona (LPL), taken with the 154-cm reflector at Catalina Observatory, Arizona. They are from Kuiper (*1972a*) and Larson *et al.*(*1973*), reproduced by permission.

(1) 1964 Dec 1 (Kitt Peak).
GRS red; EZ deep reddish (probably the most intense coloration this century); NTB also red after its revival. Following STBs white oval BC, STB Fade segment is cream-coloured. NEBn expanding and disturbed.

(2) 1966 Dec 23 (LPL).
GRS contains a shrunken inner red oval. SEBs jetstream spots. EZ normal, with blue NEBs projections.

(3A, *left*) 1968 Jan 25; (3B, *right*) 1968 Apr 9 (both LPL).
GRS a 'doughnut' with S. Tropical Band emerging p. it. EZ contains a new, orange EB. NEB largely faint with large NEBn 'barge'. On (A), Io is in transit following its shadow.

(4) 1970 May 16 (LPL).
SEB faint; GRS a red oval, overlapping orange STropZs coloration (which followed a STBn jetstream outbreak). Also orange colour over NEB and NTropZ.

(5) 1971 Aug 7 (LPL).
SEB Revival: massive disruption of EZ with huge orange, bluish, and white areas. NTB orange following a jetstream outbreak.

(6) 1972 Oct 3 (LPL).
Orange colour over the whole EZ and over the fading SEB; GRS still dark red though N. edge has receded.

PLATE P16. IMAGES FROM PIONEER AND VOYAGER SPACECRAFT.

In the NASA images in Plates P16, P17, and P20, south is normally to upper left. On this plate, small panels at right are close-ups of the NEB. Differences in processing of the images show up the range of colours that can be extracted.

(1) 1974 Dec 2 (Pioneer 11).

Oval DE passing the GRS, which is deep red while SEB is whitened. Fine details in S.S. Temperate region. (NASA image C8).

(2) 1973 Dec 3 (Pioneer 10).

Closeup of N. Tropical region, showing rifted region in the multicoloured NEB, NTB(S) is orange, and bluish streaks in NTropZ indicate the NTBs jetstream. (NASA image A9.)

(3) 1973 Dec 3 (Pioneer 10).

Closeup of N. Tropical region, showing Little Red Spot in NTropZ. (NASA image A16, reprocessed by Simon Mentha.)

(4) 1979 Jun 28 (Voyager 2).

Enhanced-colour image ranging from the turbulent STB (top) to the NTBs jetstream (bottom). The bright, turbulent SEBZ is f. the GRS. EZ(S) is all ochre. The swirling light structure in EZs is the long-lived white spot, apparently veiled by ochre material. (It habitually became obscure when approaching the GRS, as in this picture.) Several orange streaks lie N-S. across the NEBs edge. One of the great dark 'barges' in NEBn shows obvious signs of circulation. (NASA P-21731)

(5) 1979 Jul 6 (Voyager 2).

Closeup of NEB, with multicoloured cloud streaks. The very dark spot, showing obvious signs of cyclonic circulation, appears to be a new 'barge' forming in the wake of the NEB rifted region. (Images 20587.24,26,28, from NASA CD-ROM, processed by Simon Mentha.)

(6) 1979 Mar 2 (Voyager 1).

Closeup of NEB, showing multicoloured clouds, one of the great dark 'barges' in NEBn, and yellow and blue cloud streaks in NTropZ. The pale orange band at bottom is the faded NTB(S), marking the NTBs jetstream. (NASA P-21194c).

PLATE P17. VOYAGER 1 IMAGES OF THE GRS AND AN INTERACTION ON THE NEB.

(1) 1979 Feb 1 (Voyager 1).

STB Fade between ovals FA (on p. limb) and BC (passing GRS). S. Trop. Band p. GRS constitutes incipient S. Trop. Dislocation. Long-lived EZs white spot near p. limb. EZ(S) strongly ochre. NEB contains the long-lived rifted region; the very bright core of a NEBs plume (alongside the GRS) had appeared over the previous few days as the plume encountered the NEB rifted region. NNTBs jetstream spots passing an anticylonic white oval on NNTBn. (NASA image P-20993).

(2) 1979 Feb 9 (Voyager 1).

Similar view to no.1, showing motions in SEBs jetstream and in NEBs over 8 days. The bright NEBs plume is near the p. side. Following it, a large bluish area expanded and is now dissipating

(see Fig. 9.12) while an orange arc lying N-S. across the bluish area (presumably at higher altitude) represents the edge of white material which has blown in the f. direction from the plume core. (NASA image P-21085).

(Nos.1 and 2 are both standard NASA colour images, in which redness is exaggerated, although the exact balance differs in these two images.)

(3) 1979 Feb 13; J.H. Rogers (Los Angeles, California; 32-cm reflector).

This Earth-based drawing shows the same face of the planet as the Voyager 1 images, including the GRS and the NEB rifted region. The expanded bluish feature on NEBs is now near the p. limb.

(4) 1979 March 4 (Voyager 1).

Closeup of the GRS, with highly exaggerated colour. (South is to upper right.) A white patch that was orbiting round inside the GRS is in the N. half, and is drawing in white clouds from the STropZ. The whitish square of STropZ at top left is bounded by the circulations of GRS (f. side), SEBs (N. side), oval BC (S. side), and STB Fade (p. side). At the lower corner of this square, pinkish and bluish streaks can be seen crossing, presumably at different altitudes. (See Fig. 10.39 for a diagram and Figs. 10.2 and 12.4 for the larger setting.) In the turbulent STB f. oval BC, the cusps of eddies are all reddish (also in Plate P20.1). (NASA P-21431C.)

PLATE P20. VOYAGER IMAGES.

(1) 1979 Jul 3 (Voyager 2).

The GRS and surroundings. Intricate details of cyclonic turbulent regions in S³TB, SSTB, STB, and SEB, and of anticyclonic ovals in SSTB, STB (oval DE), and SEB (the GRS). A wavy white streamer connects the SEB rifts to the long-lived EZs white spot (just off bottom left corner). At bottom right, a sinuous orange band runs N-S. across the white and blue streaks of EZ(N). (NASA P-21742).

(2) 1979 Mar 2 (Voyager 1).

Closeup of N. Temperate region. In upper half, the pale orange band is the faint NTB(S), marking the fastest jetstream on the planet, which is indicated by oblique streaks on either side. In lower half, the two larger circles lie on the more moderately prograding NNTBs jetstream; these are typical jetstream spots, whose anticyclonic circulation is evident. Below is a turbulent 'folded-filament' region of NNTB, in which the cusps of eddies again show red spots. (NASA P-21193C).

PLATE P21. VOYAGER 2 IMAGE, AND MODERN EARTH-BASED IMAGES.

(1) 1979 Jun 29 (Voyager 2).

The new South Tropical Disturbance (STropD), with the GRS at f. side. Also see Fig. 10.59. (Images 20378.56,58, violet and orange images with interpolated green image, processed from NASA CD-ROM by Simon Mentha.)

(2,3) Polarised light: 1986 Nov 21 (Lowell Observatory, Arizona; 1.8-metre reflector; CCD polarimetry).

These images show the percentage polarisation of light, colour coded from 1% (blue) through 4–7% (red) to 11–14% (mauve); (no.2) in extreme red light near 750 nm, and (no.3) in the methane absorption band at 727 nm. Maximum polarisation is in the polar regions. The continuum image (no.2) also shows polarisation at the limb and in a dusky NEBs-EZ feature. North is to top right. (From Carlson and Lutz, *1989*, by courtesy of Barbara Carlson.)

(4) Infrared light: by David Allen (Anglo-Australian Telescope). Compiled from images at 1.74 μm (in blue), 2.16 μm (in green), and 2.34 μm (in red). This shows sunlight scattered from haze at very high altitudes. Haze over the EZ reflects at 1.74 μm, and the polar hoods reflect at longer wavelengths. (Copyright AAT Board; by courtesy of David Allen.)

(5,6) Colour composites made by the author from sets of monochrome images taken by J. Dragesco and J. Bourgeois, using the 1-metre reflector at the Pic du Midi, France. The component images were of high contrast for no.5, giving exaggerated colour, but of moderate contrast for no.6, giving normal colour.

(5) 1986 Jul 10; J. Dragesco & J. Rogers.

Ovals BC (near p. limb) and DE (near centre), with Little Red Spot in STropZ (on p. side). NTB(S) reddish. Much detail further N. including NNTBs jetstream spots and NNTBn white ovals. (Also see Plate P19.6 & 7.)

(6) 1988 Dec 24; J. Bourgeois & J. Rogers.

Jetstream spots on SEBs (including one retrograding round N. edge of GRS) and on NNTBs. Otherwise, no major activity nor colour, with STB and NTB absent.

PLATE P24. PHOTOGRAPHS COVERING THE GLOBAL UPHEAVAL OF 1990.

This global upheaval comprised the fading and revival of the SEB, coloration in the EZ which then migrated over the SEB, and a major NTBs jetstream outbreak.

All photos except nos.9 and 10 are by amateur astronomers: Jean Dragesco (Pic du Midi, France; 1-m reflector), Isao Miyazaki (Japan; 40-cm reflector), Tomio Akutsu (Japan; 32-cm reflector), and Donald C. Parker (Florida; 41-cm reflector). Nos.9 and 10 are NASA images from the Hubble Space Telescope. Times are in Universal Time. For other fine images from these years, see the BAA Reports.

(1) 1986 Jul 20, 03h 16m; Dragesco.

A normal aspect with pale GRS and strongly blue NEBs projections.

(2) 1988 Oct 23, 18h 18m; Miyazaki.

SEB and NEB still normal, with pale GRS and turbulent SEBZ f. it. STB and NTB absent.

(3) 1989 Nov 4, 16h 30m; Akutsu.

SEB has faded leaving bluish SEB(N) and red GRS. Yellow colour appearing in EZ(N). Contrast between bluish NEBs projections and brown NEBn barges, with white rift between.

(4) 1990 Mar 14, 13h 05m; Miyazaki.

EZ yellow colour is more intense. Chain of dark grey NTB(S) spots on p. side is the f. end of a major jetstream outbreak.

(5) 1990 Sep 23, 10h 10m; Parker.

SEB Revival under way, with GRS partly eroded (near p. limb). EZ colour has shifted southwards. NTB revived.

(6) 1990 Oct 19, 20h 18m; Miyazaki.

SEB Revival continuing, with dark bluish-grey spot trapped N. of GRS.

(7) 1990 Nov 18, 19h 16m; Miyazaki.

SEB Revival continuing, with dark grey and brown streaks embedded in colour which has now spread from equator to SEBs.

(8) 1991 Jan 24, 15h 12m; Miyazaki.

GRS now ringed by dark grey material which is extending into a S. Tropical Band p. it, contrasting with orange-brown colour across SEB. Oval DE is a ring on p. side.

(9) 1991 May 28; Hubble Space Telescope.

'Cold grey' S. Trop. Band now fills entire STropZ and encircles GRS. Oval FA is approaching GRS. Slight colour lingers in EZ(N), where bright areas are creamy in contrast to white zones. (NASA image.)

(10) 1991 Mar 11; Hubble Space Telescope.

The first Space Telescope image of Jupiter shows the GRS (on p. limb) and oval FA, with new dark rim. Continuing turbulence in SEB, with red spots on the cusps of eddies. This image, and no.9, were taken with the Wide Field/Planetary Camera, and sharpened to remove the effects of the telescope's spherical aberration. (NASA image.)

(11) 1991 Dec 10, 09h 32m; Parker (composite of CCD images).

Oval FA, with STB 'Remnant' f. it, is now p. the GRS. Ochre colour persists in SEB and EZ(S) but EZ(N) clear. NEB highly rifted. A solitary NTBs jetstream spot betokens a new outbreak, while N. Temp. Disturbance persists in NTZ.

Appendix 1
Measurements of longitude

Visual and photographic measurements of longitude

The principles of longitude measurement were described in Chapter 1.3.

Visual observers measure longitude by estimating the time of transit of a spot across the central meridian (CM). This is the transit time, and should be judged to the nearest minute. The observer should write down the transits as they occur, as in the following example:

1985 August 15/16
Seeing III

	UT	λ_1	λ_2
White oval in STB	23.50	–	221.9
Dark proj. NEBs	23.56	140.2	–
White spot in SEB	00.27	159.1	244.2

After the observing session, the observer must convert the transit times into longitudes, λ_1 or λ_2. Tables of the CM longitudes at a given time each day are published in almanacs (e.g. the BAA Handbook), and the rate of increase of longitude with time is given in Table A1.1. These tables allow one to calculate the CM longitude at each transit time. The λ_1 value should be used for equatorial features (NEBs, EZ, SEBn), and the λ_2 value for all other regions.

Sometimes a spot is not seen until it has passed the CM. In such cases, experienced observers may be able to estimate the number of minutes or degrees since the transit. Such estimates should be marked as such in the observing log. While estimates made more than 5° from the CM are not reliable enough to establish drift rates, they are often useful in confirming the continuous existence of a feature during intervals between accurate transits. A typical visual report form is reproduced in Fig. A1.1.

Photographs can of course yield accurate measurements of longitude, even for features far away from the CM. The most important condition is that the planet's limb must be clearly visible. This is not the case on white-light photos printed at high contrast, because of limb-darkening. To see the limb, one must use negatives or low-contrast prints, or prints which have been 'dodged' to make the periphery brighter, or blue-light photos. In difficult cases it is helpful to view the print against a bright light.

As the planet's tilt is very small, the distance east or west of the CM can be converted directly into a relative longitude. If a spot is $\alpha°$ from the CM, the distance from CM to spot divided by the distance from CM to limb is sin α. For speed and simplicity, the author measures photographs using a transparent grid, ruled with lines corresponding to 10° intervals of longitude (Fig. A1.2). This is placed over the photograph with the outermost lines touching the limb at the latitude of the spot to be measured.

Transit timings by a single good observer are typically consistent to within a minute or two. (One minute corresponds to 0.6° of longitude.) Comparison of longitudes measured by different observers and by photography may show systematic differences or 'personal equations' of a few degrees. This has been recognised for nearly a century. By comparing visual transits with longitudes measured from photographs, it has been found that many visual observers tend to transit early; visual personal equations (the mean of the observer's visual longitude minus photographic longitude) range from zero to about –5°. The BAA has consistently found such values in recent years and they were also documented by the ALPO in reports for 1961 and 1963/64. A personal equation may change during an apparition (perhaps in relation to the apparent orientation or phase of the planet), but significant changes are rare. For most observers, the personal equation is less than 3° and can be ignored.

Drift rates and rotation periods

Drift rates of spots are determined by plotting a graph of longitude against time. They can be expressed in degrees per month, or in metres per second, or as rotation periods (Table A1.2).

If the accuracy of the individual observations is ±1°, as from a single good visual observer or from photographs, the drift rate will be accurate to ±2°/month over one month or ±0.4°/month over 5 months. If the data from several visual observers are combined, personal equations may increase the scatter 2- to 3-fold. Even so, the accuracy is entirely adequate to distinguish the different jovian currents and variations within them.

The phase of the planet has a systematic effect on measured longitudes. Although the true phase is imperceptible, visual observers can see that the limb-darkening is enhanced on the side towards the

BRITISH ASTRONOMICAL ASSOCIATION

JUPITER SECTION REPORT FORM

Name_____ Location_____

Date_____ Start_____UT Finish_____UT

Telescope_____ Magnif'n_____ Seeing_____

GENERAL NOTES

TRANSITS

Feature	Time (UT)	λ_1	λ_2

LEFT:

Time (UT)_____

CM: ω_1 _____

CM: ω_2 _____

Seeing_____

RIGHT:

Time (UT)_____

CM: ω_1 _____

CM: ω_2 _____

Seeing_____

Figure A1.1. British Astronomical Association Jupiter section report form

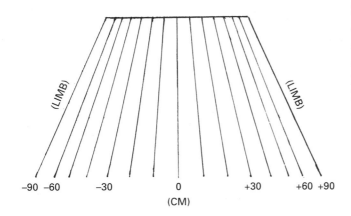

Fig. A1.2. Grid used for transparent overlay for measuring longitudes from photographs.

terminator, and this leads to a shift in the estimated position of the CM. Longitudes are therefore underestimated before opposition and overestimated after opposition; the greatest error (at quadrature) amounts to about 1.2° in visual transits, and a similar or smaller amount in photographs. As a result, drift rates during an apparition are too high by about +0.6°/month (corresponding to a rotation period 0.9 seconds too long). The effect is evident for long-lived features such as the GRS or STB white ovals, when drift rates between oppositions are compared with the drift rates within apparitions. This was first noticed by T.E.R. Phillips (see the 1904/5 BAA Memoir, p.87), and later measured by E.J. Reese and H.G. Solberg (*1966*) and by S. Cortesi (*1978*). Reese and Solberg estimated the shift on their photographs as 0.085 times the phase angle, which reaches 11.7° at quadrature, and all the published NMSUO measurements were already corrected by this amount. Otherwise, in published reports and in this book, no correction is made for the phase effect unless explicitly stated.

Sources of data on historical wind speeds

Drifts in longitude indicate the zonal wind speeds. All the published values that the author knows of and considers reliable, up to 1990/91, have been summarised within Part III of this book, in the form of tables or charts giving the average speed and range in each apparition. For most of the 9 slow currents and the equatorial current, the data are presented as charts. For some rarely-observed currents, and for the jetstreams, the data are listed in tables. When more than one value was published for a given apparition, I have plotted whatever seemed most appropriate: either the average, or the value that seemed most reliable, or both values if they probably referred to different spots. Some spots are also plotted individually if they had very divergent drifts.

The sources of these data are the reports listed in Appendix 3. I

have exercised some selection in the early years, before 1914, when all the drift rates were derived by individual observers. There was often disagreement between the values given by different observers, and this led Peek to omit many of these values from his book. However, the discrepancies over the period 1904–1910 appear to have been due to the results of one observer, Scriven Bolton, which often disagreed not only with those of P.B. Molesworth and other observers, but also with rates later found to be typical. For example, Bolton often gave identical drift rates for spots over a wide range of temperate latitudes, e.g. North Temperate Current for all the northern belts. Such a pattern contradicts all subsequent data, but can arise from an optical illusion that falsely connects high-latitude flecks with more distinct spots at lower latitudes. The 1908/9 BAA Memoir records that Phillips discussed with Bolton the numerous discrepancies in their charts and, perhaps charitably, agreed to differ about the N.N. Temperate region. I have therefore omitted all Scriven Bolton's drift rates. (A few high-latitude values by A.S. Williams are omitted for similar reasons).

This clears the way for us to accept most of the values by Molesworth and other observers. Some of Molesworth's drift rates have also been questioned, because he tended to plot an enormous number of small spots on his charts, and B.M. Peek suggested that this might have led to confusion. Also, he seems to have connected up some very sparse observations on the risky assumption that they followed the same drift as better-observed spots in the same latitude. Nevertheless, Molesworth's drift rates do generally agree with the steady currents that are now known to be permanent, even in the obscure high latitudes, where it was he who discovered several of them; and when he did find his values to be in error, he admitted it, which enhances his credibility as a scientist. I have only omitted some of Molesworth's values where the quoted mean intervals between observations were too long for reliable identification – e.g. values with mean intervals of >20 days in the SEBZ, which is now known to be a rapidly-changing region, where the observations must have been connected up largely by optimism.

For subsequent years, drift rates used in this book are mainly from the following sources (see Appendix 3 for references and further comments). All values from 1900 to 1948 are from the BAA reports. From 1948 to 1963, I mainly used the data from ALPO or BAA reports. From 1964 to 1974, I mainly used data from NMSUO or LPL reports when available, and also some from BAA or European reports. From 1975 to 1984, I preferentially used reports from the BAA or SAF. From 1985 to 1991, I used BAA reports. (The French group have also published excellent reports on these years, but as they were largely based on the same data as our BAA reports and agree with them, I have simply used the BAA values.)

Table A1.1. *Measurement of longitude by transits: Conversion of minutes into degrees*

| A. System I | | | | | | B. System II | | | | | |
h	°	m	°	m	°	h	°	m	°	m	°
1	36.6	10	6.1	1	0.6	1	36.3	10	6.0	1	0.6
2	73.2	15	9.1	2	1.2	2	72.5	15	9.1	2	1.2
3	109.7	20	12.2	3	1.8	3	108.8	20	12.1	3	1.8
4	146.3	25	15.2	4	2.4	4	145.1	25	15.1	4	2.4
5	182.9	30	18.3	5	3.0	5	181.3	30	18.1	5	3.0
6	219.5	35	21.3	6	3.7	6	217.6	35	21.2	6	3.65
7	256.1	40	24.4	7	4.3	7	253.8	40	24.2	7	4.25
8	292.7	45	27.4	8	4.9	8	290.1	45	27.2	8	4.85
9	329.2	50	30.5	9	5.5	9	326.4	50	30.2	9	5.45
10	365.8	55	33.5	10	6.1	10	362.6	55	33.2	10	6.05

Figures in the last column are given to the nearest 0.05° in order to reduce rounding errors when they are used to derive longitudes in a long series of timings.

Table A1.2. *Drift in longitude: Conversion of degrees/month into rotation periods*

$\Delta\lambda_1$ is rate of change in System I longitude in degrees per 30 days; $\Delta\lambda_2$ and $\Delta\lambda_3$, the same for Systems II and III. P is rotation period (sidereal). The longitude systems are defined by sidereal rotation of exactly 877.90°/day (System I), 870.27°/day (System II) and 870.536°/day (System III).

$$\Delta\lambda_1 = \Delta\lambda_2 + 228.9; \quad \Delta\lambda_3 = \Delta\lambda_2 + 8.0 = \Delta\lambda_1 - 220.9.$$

For conversion to wind speeds in metres per second:

$$u \text{ (metres / second)} = -\frac{\Delta\lambda_3 \times \cos\beta'}{2.080}$$

(where β' is the mean latitude; see Appendix 2).

The following tables are reproduced from Peek's book *The Planet Jupiter* (with permission from Faber and Faber Ltd.), with the addition of the first two columns.

A. System I

P	$\Delta\lambda_1$ 9ʰ 46ᵐ	$\Delta\lambda_1$ 9ʰ 47ᵐ	$\Delta\lambda_1$ 9ʰ 48ᵐ	$\Delta\lambda_1$ 9ʰ 49ᵐ	$\Delta\lambda_1$ 9ʰ 50ᵐ	$\Delta\lambda_1$ 9ʰ 51ᵐ
0ˢ		−156.7	−112.4	−67.5	−22.7	+21.9
1		−155.9	111.7	66.8	22.0	22.6
2		−155.2	110.9	66.0	21.3	23.3
3		−154.4	110.2	65.3	20.5	24.1
4		−153.7	109.4	64.5	19.8	24.8
5		−152.9	108.7	63.8	19.0	25.6
6		−152.1	107.9	63.0	18.3	26.3
7		−151.4	107.2	62.3	17.5	27.1
8		−150.6	106.4	61.5	16.8	27.8
9		−149.9	105.7	60.8	16.0	28.5
10		−149.1	104.9	60.0	15.3	29.3
11		−148.4	104.2	59.3	14.6	30.0
12		−147.6	103.4	58.5	13.8	30.8
13		−146.9	102.7	57.8	13.1	31.5
14		−146.1	101.9	57.0	12.3	32.2
15		−145.4	101.2	56.3	11.6	33.0
16		−144.6	100.4	55.5	10.8	33.7
17		−143.9	99.7	54.8	10.1	34.5
18	−188.3	−143.1	98.9	54.1	9.3	35.2
19	−187.5	−142.4	98.2	53.3	8.6	36.0
20	−186.8	−141.6	97.4	52.6	7.9	36.7
21	−186.0	−140.9	96.7	51.8	7.1	37.4
22	−185.3	−140.1	95.9	51.1	6.4	38.2
23	−184.5	−139.4	95.2	50.3	5.6	38.9
24	−183.8	−138.6	94.4	49.6	4.9	39.7
25	−183.0	−137.9	93.7	48.8	4.1	40.4
26	−182.3	−137.1	92.9	48.1	3.4	41.1
27	−181.5	−136.4	92.2	47.3	2.6	41.9
28	−180.8	−135.6	91.4	46.6	1.9	42.6
29	−180.0	−134.9	90.7	45.8	1.2	43.4
30	−179.2	−134.1	89.9	45.1	−0.4	44.1
31	−178.5	−133.4	89.2	44.4	+0.3	44.8
32	−177.7	−132.6	−88.4	−43.6	+1.1	+45.6
33	−177.0	−131.9	87.7	42.9	1.8	46.3
34	−176.2	−131.1	86.9	42.1	2.5	47.1
35	−175.5	−130.4	86.2	41.4	3.3	47.8
36	−174.7	−129.6	85.4	40.6	4.0	48.5
37	−174.0	−128.9	84.7	39.9	4.8	49.3
38	−173.2	−128.1	83.9	39.1	5.5	50.0
39	−172.5	−127.4	83.2	38.4	6.3	50.8
40	−171.7	−126.6	82.5	37.6	7.0	51.5
41	−171.0	−125.9	81.7	36.9	7.8	52.2
42	−170.2	−125.1	81.0	36.2	8.5	53.0
43	−169.5	−124.4	80.2	35.4	9.2	53.7
44	−168.7	−123.6	79.5	34.7	10.0	54.5
45	−167.9	−122.9	78.7	33.9	10.7	55.2
46	−167.2	−122.1	78.0	33.2	11.5	56.0
47	−166.4	−121.4	77.2	32.4	12.2	56.7
48	−165.7	−120.6	76.5	31.7	12.9	57.4
49	−164.9	−119.9	75.7	30.9	13.7	58.2
50	−164.2	−119.1	75.0	30.2	14.4	58.9
51	−163.4	−118.4	74.2	29.4	15.2	59.7
52	−162.7	−117.6	73.5	28.7	15.9	60.4
53	−161.9	−116.9	72.7	28.0	16.7	61.1
54	−161.2	−116.1	72.0	27.2	17.4	61.9
55	−160.4	−115.4	71.2	26.5	18.1	62.6
56	−159.7	−114.6	70.5	25.7	18.9	63.4
57	−158.9	−113.9	69.7	25.0	19.6	64.1
58	−158.2	−113.1	69.0	24.2	20.4	64.8
59	−157.4	−112.4	68.2	23.5	21.1	65.6
60	−156.7	−111.7	67.5	22.7	21.9	66.3
			−66.8	−22.0	+22.6	+67.0

B. System II

$\Delta\lambda_2$	P 9ʰ 52ᵐ	$\Delta\lambda_2$	P 9ʰ 53ᵐ	$\Delta\lambda_2$	P 9ʰ 54ᵐ	$\Delta\lambda_2$	P 9ʰ 55ᵐ
−162.6	0ˢ	−118.3	0ˢ	−74.1	0ˢ	−30.1	0ˢ
161.9	1	117.6	1	73.4	1	29.4	1
161.1	2	116.8	2	72.7	2	28.7	2
160.4	3	116.1	3	71.9	3	27.9	3
159.6	4	115.3	4	71.2	4	27.2	4
158.9	5	114.6	5	70.5	5	26.5	5
158.2	6	113.9	6	69.7	6	25.7	6
157.4	7	113.1	7	69.0	7	25.0	7
156.7	8	112.4	8	68.3	8	24.3	8
155.9	9	111.7	9	67.5	9	23.5	9
155.2	10	110.9	10	66.8	10	22.8	10
154.5	11	110.2	11	66.1	11	22.1	11
153.7	12	109.4	12	65.3	12	21.3	12
153.0	13	108.7	13	64.6	13	20.6	13
152.2	14	108.0	14	63.9	14	19.9	14
151.5	15	107.2	15	63.1	15	19.2	15
150.8	16	106.5	16	62.4	16	18.4	16
150.0	17	105.8	17	61.7	17	17.7	17
149.3	18	105.0	18	60.9	18	17.0	18
148.5	19	104.3	19	60.2	19	16.2	19
147.8	20	103.6	20	59.5	20	15.5	20
147.1	21	102.8	21	58.7	21	14.8	21
146.3	22	102.1	22	58.0	22	14.0	22
145.6	23	101.3	23	57.2	23	13.3	23
144.9	24	100.6	24	56.5	24	12.6	24
144.1	25	99.9	25	55.8	25	11.8	25
143.4	26	99.1	26	55.0	26	11.1	26
142.6	27	98.4	27	54.3	27	10.4	27
141.9	28	97.7	28	53.6	28	9.6	28
141.2	29	96.9	29	52.8	29	8.9	29
140.4	30	96.2	30	52.1	30	8.2	30
139.7	31	95.5	31	51.4	31	7.5	31
138.9	32	94.7	32	50.6	32	6.7	32
138.2	33	94.0	33	49.9	33	6.0	33
137.5	34	93.3	34	49.2	34	5.3	34
136.7	35	92.5	35	48.4	35	4.5	35
136.0	36	91.8	36	47.7	36	3.8	36
135.3	37	91.0	37	47.0	37	3.1	37
134.5	38	90.3	38	46.2	38	2.3	38
133.8	39	89.6	39	45.5	39	1.6	39
133.0	40	88.9	40	44.8	40	0.9	40
132.3	41	88.1	41	44.0	41	−0.1	41
131.6	42	87.4	42	43.3	42	+0.6	42
130.8	43	86.6	43	42.6	43	1.3	43
130.1	44	−85.9	44	41.9	44	+2.0	44
129.4	45	85.2	45	41.1	45	2.8	45
128.6	46	84.4	46	40.4	46	3.5	46
127.9	47	83.7	47	39.7	47	4.2	47
127.1	48	83.0	48	38.9	48	5.0	48
126.4	49	82.2	49	38.2	49	5.7	49
125.7	50	81.5	50	37.5	50	6.4	50
124.9	51	80.7	51	36.7	51	7.2	51
124.2	52	80.0	52	36.0	52	7.9	52
123.4	53	79.3	53	35.3	53	8.6	53
122.7	54	78.5	54	34.5	54	9.3	54
122.0	55	77.8	55	33.8	55	10.1	55
121.2	56	77.1	56	33.1	56	10.8	56
120.5	57	76.3	57	32.3	57	11.5	57
119.8	58	75.6	58	31.6	58	12.3	58
119.0	59	74.9	59	30.9	59	13.0	59
118.3	60	74.1	60	30.1	60	13.7	60
−117.6		−73.4		−29.4		+14.5	

B. System II (*continued*)

$\Delta\lambda_2$	P 9ʰ 56ᵐ	$\Delta\lambda_2$	P 9ʰ 57ᵐ	$\Delta\lambda_2$	P 9ʰ 58ᵐ	$\Delta\lambda_2$	P 9ʰ 59ᵐ
+13.7	0ˢ	+57.4	0ˢ	+101.0	0ˢ	+144.4	0ˢ
14.5	1	58.2	1	101.7	1	145.1	1
15.2	2	58.9	2	102.4	2	145.9	2
15.9	3	59.6	3	103.2	3	146.6	3
16.6	4	60.3	4	103.9	4	147.3	4
17.4	5	61.1	5	104.6	5	148.0	5
18.1	6	61.8	6	105.3	6	148.8	6
18.8	7	62.5	7	106.1	7	149.5	7
19.6	8	63.3	8	106.8	8	150.2	8
20.3	9	64.0	9	107.5	9	150.9	9
21.0	10	64.7	10	108.2	10	151.6	10
21.7	11	65.4	11	109.0	11	152.4	11
22.5	12	66.2	12	109.7	12	153.1	12
23.2	13	66.9	13	110.4	13	153.8	13
23.9	14	67.6	14	111.1	14	154.5	14
24.7	15	68.3	15	111.9	15	155.2	15
25.4	16	69.1	16	112.6	16	156.0	16
26.1	17	69.8	17	113.3	17	156.7	17
26.9	18	70.5	18	114.0	18	157.4	18
27.6	19	71.2	19	114.8	19	158.1	19
28.3	20	72.0	20	115.5	20	158.9	20
29.0	21	72.7	21	116.2	21	159.6	21
29.8	22	73.4	22	116.9	22	160.3	22
30.5	23	74.2	23	117.7	23	161.0	23
+31.2	24	+74.9	24	+118.4	24	+161.7	24
32.0	25	75.6	25	119.1	25	162.5	25
32.7	26	76.3	26	119.8	26	163.2	26
33.4	27	77.1	27	120.6	27	163.9	27
34.1	28	77.8	28	121.3	28	164.6	28
34.9	29	78.5	29	122.0	29	165.4	29
35.6	30	79.2	30	122.7	30	166.1	30
36.3	31	80.0	31	123.4	31	166.8	31
37.1	32	80.7	32	124.2	32	167.5	32
37.8	33	81.4	33	124.9	33	168.2	33
38.5	34	82.1	34	125.6	34	169.0	34
39.2	35	82.9	35	126.3	35	169.7	35
40.0	36	83.6	36	127.1	36	170.4	36
40.7	37	84.3	37	127.8	37	171.1	37
41.4	38	85.0	38	128.5	38	171.8	38
42.2	39	85.8	39	129.2	39	172.6	39
42.9	40	86.5	40	130.0	40	173.3	40
43.6	41	87.2	41	130.7	41	174.0	41
44.3	42	87.9	42	131.4	42	174.7	42
45.1	43	88.7	43	132.1	43	175.4	43
45.8	44	89.4	44	132.9	44	176.2	44
46.5	45	90.1	45	133.6	45	176.9	45
47.2	46	90.8	46	134.3	46	177.6	46
48.0	47	91.6	47	135.0	47	178.3	47
48.7	48	92.3	48	135.7	48	179.0	48
49.4	49	93.0	49	136.5	49	179.8	49
50.2	50	93.7	50	137.2	50	180.5	50
50.9	51	94.5	51	137.9	51	181.2	51
51.6	52	95.2	52	138.6	52	181.9	52
52.3	53	95.9	53	139.4	53	182.6	53
53.1	54	96.6	54	140.1	54	183.4	54
53.8	55	97.4	55	140.8	55	184.1	55
54.5	56	98.1	56	141.5	56	184.8	56
55.3	57	98.8	57	142.2	57	185.5	57
56.0	58	99.5	58	143.0	58	186.2	58
56.7	59	100.3	59	143.7	59	187.0	59
57.4	60	101.0	60	144.4	60	187.7	60
58.2		+101.7		+145.1		+188.4	

Appendix 2
Measurements of latitude

Micrometer and photographic measurements of latitude

Useful latitude measurements require a precision of ±1° or better, and this can only be achieved by multiple, careful micrometer measurements, or by measurements of high-resolution photographs.

Up to 1948, latitude measurements were done by micrometer at the telescope, as described in Peek's book. The method is only accurate if used by an experienced observer, and the measurements by Phillips and Peek typically had a scatter of ±1.0°.

Since 1950, latitudes have been measured from high-resolution photographs. These measurements can be more accurate, and can be made for faint spots and for the edges of minor belts, which was not possible using a micrometer because the measuring wire itself would obscure such features. Latitudes can only be measured from photographs on which the limb is clearly visible, as described above for longitudes. (One might hope that some belt edges might be sufficiently constant that they could be used as fixed reference points, but this does not seem to be so; even the NTBs and SEBs edges, which normally coincide with the fastest two jetstreams on the planet, show some variations in latitude.)

The method of calculating latitudes from the measurements, whether micrometric or photographic, is as follows (Fig. A2.1). Measure the distance of the feature from the south pole, s, and the polar radius, r. Calculate (r–s)/r. Call this sin θ and derive θ. Next we need the apparent tilt, δ'. (The figure usually given in almanacs is the zenocentric latitude of Earth, δ (or D_E), and δ' is obtained by: tan δ' = 1.07 tan δ.) Now we can calculate the 'mean latitude', β' = θ + δ'. In summary:

$$\beta' = \sin^{-1}\frac{(r-s)}{(r)} + \delta'$$

If Jupiter was a perfect sphere, there would be nothing more to do; but because it is oblate, there are two other definitions of latitude, and β' must be converted into one of them (Figure A2.1). Zenocentric latitude (β) is the angle subtended at the planet's centre between the feature and the equator. Zenographic latitude (β'') is the angle between the polar axis and the local horizon. They are interconverted having calculated their tangents:

$$\tan \beta'' = 1.07 \tan \beta',$$
$$\tan \beta' = 1.07 \tan \beta.$$

The number 1.07 is the ratio of the equatorial and polar diameters of the planet, and the most recent value is 1.0694. (A value of 1.0714 was used in the past and in the measurements listed here, but the difference is not significant.)

β and β'' would be identical on a sphere, but on Jupiter they differ by up to 4°. Throughout this book, zenographic latitude (β'') is used exclusively.

Sources of data on historical latitudes

In Part III of this book, the available latitude measurements for belts and zones are presented as charts (Figs. 6.2, 8.14, 9.17, 10.7, 12.7&8), with averages in tables (Tables 6.2, 8.1, 10.3, 12.1). The measurements for North and South Tropical Bands are given as tables (Tables 8.3 and 10.9). On all the charts, the symbols used are as follows: ●, centre of belt; ▲, N. edge; ▼, S. edge. (The base of the triangle marks the measurement, and I have sometimes connected them by lines to make it easier to see the extent of the belt; these are not error bars.) Numbers of high-precision observations are indicated by the size of the symbols: ., 1; ●, 2; ●, 3 or more. They are probably accurate to ±1° or better up to the temperate regions, and to ±2° for higher-latitude belts. Open symbols are medium-precision observations, given where high-precision data are lacking, and not included in statistics.

For higher-latitude belt edges and for spots, there are few published measurements, so the author has measured additional photographs.

The data used are as follows (see Appendix 3 for references).

From 1906 to 1948, the data are micrometer measurements from BAA Memoirs – mostly by Phillips and Peek, with some others for earlier years. Molesworth's values (1900–1903, converted to β'' from his β' values) are shown as medium-precision because they appear very scattered. For some years, the BAA data were supplemented with the author's measurements on professional photographs; as the limb was not clearly visible on most of the available prints, measurements were normalised to the BAA micrometer values for the major belts. These measurements for 1914, 1917/18, 1921, and 1938 were from Lowell Obs. photos published by Slipher (*1964*), and for 1941 and 1945 were from Pic du Midi photographs using prints in the BAA files (see Lyot, *1953*).

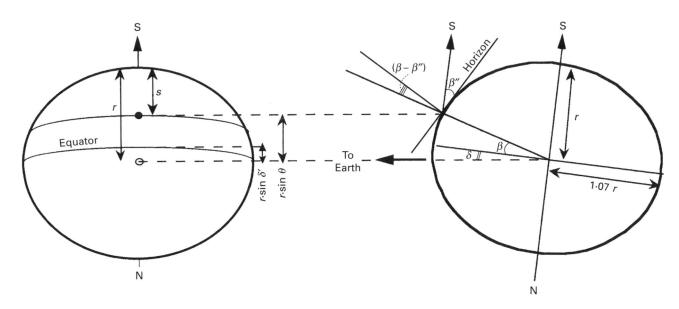

Fig. A2.1 Measurement of latitude.

From 1950 to 1989, high-precision values are measurements from photos, as follows. For 1950 to 1952, original prints of Mt. Palomar blue-light photos (Humason, *1961*; Reese, *1962a*) were measured by the author. For 1961 and 1962, NMSUO measurements were used, and for greater detail, Pic du Midi photographs in the BAA files were measured and normalised to the NMSUO measurements. For 1963 to 1974, the very accurate and detailed measurements published by NMSUO and LPL were used ; these have quoted accuracies of ±0.3° or better. For 1972 and 1973, measurements by J.B. Murray at Pic du Midi were also used (in BAA reports; averaged with LPL values for statistics). For 1975, values published by the SAF from Teide Observatory photos were used. For 1976 to 1979, measurements were by the author from published professional photos (in Sanchez-Lavega and Quesada, *1988*, and Beebe et al., *1989*). No professional data were available for the 1980s, but from 1984 onwards, several amateur photographers have been producing photos of sufficiently high resolution. For 1984 and 1985, the author measured fairly high-resolution photos by D. Parker in the BAA files. For 1986 onwards, measurements from high-resolution photos were used, as in the BAA reports.

It would not be possible to improve precision beyond ±0.3° or so because of the limitations of Earth-based resolution and the intrinsic diffuseness of the edges of jovian features.

For some apparitions, especially in the 1950s, only medium-precision latitudes could be obtained, perhaps accurate to ±2° to 3°, which are plotted as open symbols and omitted from statistical analysis. They are derived from measurements of medium-resolution amateur photographs, or from measurements of high-quality drawings normalised to high-precision photographic or micrometer measurements of the major belts. For 1953 to 1963, these were mostly from ALPO reports. For 1977 to 1985, the measurements published by the SAF (from photographs, converted here to zenographic coordinates) have been classified as medium-precision; some of them may be of higher precision, but the photographs used

before 1984 did have limitations due to resolution and limb-darkening.

Where two high-precision values were available for a single apparition, they were both plotted if significantly different, or averaged if not. They were entered separately in the statistics if they referred to different belt segments, but as a single average entry if they referred to the same belt.

Latitudes of spots

Latitudes for individual spots were not measured in the days of micrometers, and photographic measurements of them are fairly sparse. To supplement them, the author has measured more high-resolution photographs in the same way as for belts. The following tables list all the available measurements for the main classes of stable spots. All measurements are photographic unless otherwise stated.

Notes to Tables:

Entries in round brackets ( ) are imprecise; entries in square brackets [ ] look anomalous; these were not included in averages unless 'all features' is specified. Where two sources were given for an apparition, the average was used for the statistics. SL&Q is Sanchez-Lavega and Quesada (*1988*).

*Asterisk indicates photographs measured by the author, previously unpublished.

Abbreviations = d. or dk., dark; s., spot; ss., spots; v., very.

All latitudes are zenographic (β'').

Table A2.1 *Latitudes of white spots in the far north (presumed anti-cyclonic)*

Appar'n	Latitude in domain				Ref.
	NN	N³	N⁴	N⁵	
1950		47.8	[50.1]		Mt.Palomar*
1951		45.7			Mt.Palomar*
1952		44.0,46.5			Mt.Palomar*
1973	41.5	46.7			Pic (Murray-BAA)
1974		46.5			LPL
1975	41.4	46.5	55.2		SL&Q
	41.3	47.5			BAA
1976	39.9				NMSUO*
1979	40–41	46⁺	55	62	Voyager
	41.0	46.6⁺	55.7	62.2	Voyager*
1984	(≈40)				SAF
1986	41.9⁺				Pic(Dragesco–BAA)
1987/88		45⁺	52⁺	60,62	BAA
1988/89	41.2	48	51,52		BAA
1989/90	41.7				BAA
1990/91	41.3				BAA
Mean	41.2	46.4	53.0	61.4	
(±s.d)	±0.6	±1.0	±2.0	±1.2	

The mean is for 1973 onwards, excluding the approximate value for 1984. Values marked ⁺ referred to more than three spots and were given double weight.

Table A2.2 *Latitudes of NNTBn dark streaks (presumed cyclonic)*

Appar'n	Latitude	Ref.
1938	38.6	Lowell Obs.*
1939	(37.0)	Lick Obs.*
1964	38.6	Lick Obs.*
1968	(38.6)	LPL*
1971	38.1	NMSUO
1973	38.6	Pic (Murray)*
1975–77	38.5	SL&Q
1979	38.6	Voyager*
Mean	38.5 (*N*=6)	
(±s.d.)	(±0.2)	

Table A2.3 *Latitudes of NTBn dark streaks (presumed cyclonic)*

Appar'n	Latitude (n)	Ref.
1950	32.0 (1)	Mt. Palomar*
1951	30.3 (1)	Mt. Palomar*
1966/67	31.4 (4)	NMSUO
1971	31.6 (1)	NMSUO
1976/77	30.7 (2)	NMSUO*
1977/78	30.5 (1)	SL&Q*
1982	(32)	SAF
1985	30.3 (3)	Parker*
1986	31.0 ( )	BAA
1987/88	32.1 (5)	BAA
1988/89	(30) ( )	BAA
1989/90	29.4 (3)	BAA
Mean	30.9 (*N*=10)	
(±s.d.)	(±0.9)	

Table A2.4 *Latitudes of dark ovals in NTropZ (anticyclonic)*

This table includes all identified Little Red Spots (LRSs), plus other well-formed lozenges seen in the same latitude, which had various colours. They are identified by the numbers used in the original report. The accepted LRS's are entered in **bold** type.
**Measured from high-quality drawings, relative to belts which had been measured directly by micrometer.

Appar'n	Little Red Spots, etc.	β′	Ref.
1895/96	**'Violin'**	**[n.d.]**	**Antoniadi,**
	'Garnet'		**BAA**
1919/20	**[1] V.dk.oval**	**(19)**	**BAA****
	[3] V.dk.red oval,		
	w.s. in centre	**(22)**	
1951	Dusky ring	19.6	Mt.Palomar
1971	[B] Small dusky s.	19.7	NMSUO
1972	[BS1] Small bluish	19.2	LPL
	[BS2] spots NTropZ	19.0	
1973	**[RS1] Pioneer 10 LRS**	**19.2**	**LPL**
	[RS2] LRS (lozenge)	**19.2**	
	[RS3] Brown lozenge	17.8	
1974	Pale reddish s.(LRS?)	19.0	LPL
	Dark oval[see next table]	18.7	
1976/77	**[3] LRS**	**18.7**	**BAA, NMSUO**
	[6] LRS	**19.2**	
	[1] LRS?	**[n.d.]**	
1978/79	Pale brown oval	(20)	Voyager
Mean (all features):		19.3 ±0.95 (N=13)	
(precise, LRSs only):		19.1 ±0.25 (N=4)	

Table A2.5 *Latitudes of NEBn white ovals (anticyclonic) and dark spots (cyclonic)*

This lists only a small selection of the dark and white spots seen, from years where it was possible to estimate their latitudes. **Latitudes in the top half of the table were estimated from high-quality drawings or photographs, relative to the micrometer-measured latitudes of the edges of the NEB and NTB; all these values are from multiple spots except for 1965/66. Latitudes for later years were measured directly from photos. Spots in 1942/43 and 1974 apparently shifted in latitude.

Apparition	Dark spots or 'barges'	β''	White ovals β''	Ref.
(A) Estimates:				
1890	Small black ss.	14.7	–	(Barnard microm.)
1891	Dark spots	15.8	–	(Hough microm.)
1911	Dark streaks	16	–	**
1914	—		20.5	**
1917/18	Dark reddish barges	17	–	**
1918/19	—		19	**
1919/20	Dark barges	17	20	**
1922	—		19	**
1928/29	'Ruby' ss. NEBn	16	17	**
1929/30	Ditto (barges)	16.5	–	**
1938	Long dk.streaks	16	–	**
1941/42	—		18	**
1942/43	2 dark streaks	15→19		
	Other dark streaks	19	20	**
1965/66	V.dark barge	19	–	**
1966/67	D.brownish streaks	16–19	19.5	**
	Grey–brown lozenge	[20]		
1969	—		20.5	**
Mean		16.9	19.3	
(±s.d.)		±1.5 (*N*=15)	±1.2 (*N*=9)	

(B) Precise photo measurements:				
1967/68	Dark bars	15.5,16.3	19.3	NMSUO
1971	Dark bar NEBn	15.3	–	NMSUO
1972	D.grey s. NTropZ	16.0	–	LPL
	Reddish d.ss.NEBn	15.2	–	
1973	Reddish–brown seg.	15.0	–	LPL
1974	Dark & reddish ss. NEBn	15.7, 17.5,[19.0]	19.4	LPL
	D.s.NEBn→ d.oval NTropZ	16.3→ [18.7]	–	
1978/79	Dark brown barges	15.0	(19)	Voyager*
1987/88	Dark barge	16.3	–	BAA
1988/89	Barges on NEB(C)	15.8,	19.4,	BAA
(3 sectors)		16.0,17.6	19.1,20.2	
1989/90	Dk.barges NEBn	16.5	18.5	BAA
Mean		15.9	19.2	
		±0.6 (*N*=11)	±0.5 (*N*=4)	
	(all features:)	16.3		
		±1.2 (*N*=16)		

Table A2.6 *Latitudes of dark spots in SEB(S) (presumed cyclonic)*

Appar'n	Latitude	Ref.
1979	–16.3	Voyager (source of SEB outbreak)*
1989	–16.7	BAA
1991/92	–17.2	Miyazaki
Mean	–16.7 (±0.5) (*N*=3)	

Table A2.7 See next page

Table A2.8 *Latitudes of STBs white ovals (anticyclonic)*

Variable; see chart (Figs. 11.11&15).

Mean (1962–1991) = –32.9 (±0.6)

Table A2.9 *Latitudes of tiny white ovals in SSTB (anticyclonic)*

These are for tiny bright ovals; they do not include slightly larger, more northerly light ovals which Voyager showed to be cyclonic.
*Photos measured by the author. Those for 1914–1918 were from photos published by Slipher (1964), normalised to Phillips' micrometer measurements of belts as the limb was not visible on the photos; they are only of medium precision. For brevity, several apparitions have been grouped.

Apparition	β'' (n)	Ref.
1914	–42.8 (1)	Lowell Obs.*
1915	–39.1 (1)	Lowell Obs.*
1917/18	–39.6 (1)	Lowell Obs.*
Mean:	–40.5 (±2.0)	
1950–1951	–41.4 (2)	Mt.Palomar*
1965–1968	–40.1 (6)	NMSUO
1975–1977	–40.4 (2)	SAF,NMSUO*
1979	–40.5 (many)	Voyager
1984	–40.4 (3)	Parker*
1986–1989	–40.4 (many)	BAA
1989–1991	–41.0 (many)	BAA
Mean:	–40.6 (±0.4)	

Table A2.7 *Latitude of GRS (anticyclonic)*

Appar'n	Latitude			Ref.
	N.edge	Centre*	S.edge	
(a) Hough, micrometer measurements:				
1879/80		−23.6		
1880/81		−24.3		Peek book
1881/82		−25.2		
1891	−14.5	−22.1	−30.2	(MNRAS **52**, 410)
Mean:		−23.8 (±1.3)		
(b) Phillips, from his drawings, normalised to his micrometer measurements of belts: Phillips (1930)				
1914		−22.3		
1918/19		−20.8		
1919/20		−20.4		
1926		−23.0		
1927/28		−21.6		
Mean:		−21.6 (±1.1)		
(c) Photographs:				
1952	−16.2	−22.4	−28.5	Mt. Palomar*
1962/63	−16.2	−22.7	−29.2	
1963/64	−16.2	−22.5	−28.7	NMSUO
1964/65	−16.1	−22.1	−28.1	(Reese & Solberg,*1966*)
1965/66	−16.9	−22.3	−27.8	NMSUO
1966/67	*[−19.5]*	*[−23.1]*	*[−26.7]*	NMSUO
1967/68		−23.6		NMSUO
1969	−16.1	−22.3	−28.4	NMSUO
1970	−16.3	−22.5	−28.7	NMSUO
		−22.2		UAI (unpubl.)
1971	−16.3	−22.6	−28.8	NMSUO
		−21.8		UAI (unpubl.)
1972	*[−18.7]*	*[−23.6]*	*[−28.6]*	LPL (Minton)
	[−18.3]	*[−23.9]*	*[−28.5]*	BAA (Murray, Teide photos)
1973	−16.9	−22.4	−28.0	LPL (Minton)
		−21.7		BAA/UAI (Murray & Sette)
1974	−16.7	−22.2	−27.8	LPL (Minton)
		−21.6		UAI (Senigallesi & Sette, quoted in BAA Memoir)
1976/77	−17.1	−22.5	−27.9	BAA (NMSUO*)
1989/90	−16.7	−22.7	−28.7	BAA
(d) Observed circulation:				
1966–67	−16.2	−22.2	−28.3	NMSUO (Reese & Smith,*1968*)
1979	(−16)	−22.0	(−28)	Voyager (Mitchell et al.,*1981*; Dowling & Ingersoll,*1988*)
Mean (1952–90):	−16.5	−22.3	−28.4	
(±s.d.)	±0.4(*N*=13)	±0.3(*N*=19)	±0.5(*N*=13)	

The GRS data consist of: (a) micrometer measurements by G. Hough in the 19th century, of uncertain accuracy; (b) measurements by Phillips from his own drawings, normalised to micrometer measurements of belts, and subject to the inevitable uncertainties of this method; (c) precise measurements on high–resolution photographs from 1951 onwards; (d) high-resolution measurements of the GRS circulation.

Except for (d), latitudes are only entered when the dark oval outline of the GRS was clearly visible. Entries in square brackets are anomalous and not included in average; these were times when the red oval (1966/67, 1972) was smaller than the usual oval. Due to the curvature of the planet, the true centre (22.4°S) is 0.14° south of the apparent centre (22.3°S); this small difference has been ignored in compiling this table.

The conclusion is that the average latitude has been 22.3°S throughout, with no definite evidence for variation.

Appendix 3
Lists of apparitions and published reports

Table A3.1. *Dates of perihelion of Jupiter.*

1667	1856 Dec 22
1679	1868 Nov 16
1690	1880 Sep 25
1702	1892 Jul 24
1714	1904 Jun 1
1726	1916 Apr 17
1738	1928 Mar 15
1750	1940 Jan 23
1762	1951 Nov 20±
1773	1963 Sep 26
1785	1975 Aug 12
1797	1987 Jul 10
1809	1999
1821	2011
1833	2023
1845	2034

Dates from 1856 to 1987 are from the *Nautical Almanac* (HMSO). (The date for 1951 was not given exactly and has been interpolated.) Other dates were kindly provided by Mr. Neville Goodman of the BAA.

The dates of perihelion do not repeat exactly with Jupiter's orbital period of 11.862 years, but oscillate up to 14 days either side of it with a period of five jovian revolutions due to perturbations by Saturn.
In the year of perihelion, opposition falls in October (or only a few days outside it), and these are the most favourable oppositions for northern-hemisphere observers. This table therefore serves to indicate the approximate seasons of apparitions throughout telescopic history.

The solstice on Jupiter (maximum northern latitude of the sun) comes 11½ months after perihelion.

Table A3.2. *Dates of opposition, and published reports on the apparitions*

In the following tables, the abbreviations for organisations and journals are:

ALPO   Association of Lunar and Planetary Observers (*Journal of the ALPO*, also called *The Strolling Astronomer*)
BAA   British Astronomical Association (*Memoirs* and *Journal of the BAA*)
LPL   Lunar and Planetary Laboratory (University of Arizona) (*Communications of the LPL*)
NMSUO New Mexico State University Observatory
SAF   Société Astronomique de France (*L'Astronomie*, also called *Bulletin de SAF*)
SAI   Società Astronomica Italiana (*Memorie della SAI*)
SAS   Société Astronomique de Suisse (*Orion*)

(a) 1869–1890 Opposition	(b) 1891–1948 Opposition	BAA report	Opposition	BAA report
		Waugh WR, *Mem.BAA...*		Philips TER, *Mem. BAA...*
1869 Nov 7	1891 Sep 5	**1**, 73	1920 Feb 2	**26**(4), 69
1870 Dec 13	1892 Oct 12	**2**, 129	1921 Mar 4	**27**(4), 73
1872 Jan 15	1893 Nov 17	**3**, 121	1922 Apr 4	**29**(1)
1873 Feb 14	1894 Dec 22	**4**, 21	1923 May 5	*ibid.*
1874 Mar 17	1896 Jan 24	**5**, 51	1924 Jun 5	*ibid.*
1875 Apr 16	1897 Feb 23	**6**, 103	1925 Jul 9	**29**(3)
1876 May 17	1898 Mar 25	**7**, 71	1926 Aug 15	*ibid.*
1877 Jun 19		Cottam A, *Mem.BAA...*	1927 Sep 22	**30**(2)
1878 Jul 24	1899 Apr 25	**8**, 67	1928 Oct 29	**30**(4)
1879 Aug 31	1900 May 27	**10**, 87	1929 Dec 3	**32**(4)
1880 Oct 6		Phillips TER, *Mem.BAA...*	1931 Jan 6	*ibid.*
1881 Nov 12	1901 Jun 30	**12**, 73	1932 Feb 7	**34**(2)
1882 Dec 17	1902 Aug 5	*ibid.*	1933 Mar 9	*ibid.*
1884 Jan 19	1903 Sep 11	**14**, 68		Peek BM, *Mem.BAA...*
1885 Feb 18	1904 Oct 18	*ibid.*	1934 Apr 8	**34**(3)
1886 Mar 21	1905 Nov 23	**16**(1), 1	1935 May 10	*ibid.*
1887 Apr 20	1906 Dec 28	*ibid.*	1936 Jun 10	*ibid.*
1888 May 21	1908 Jan 29	**16**(2), 23	1937 Jul 15	**35**(1)
1889 Jun 24	1909 Feb 28	**16**(3), 37	1938 Aug 21	*ibid.*
1890 Jul 30	1910 Mar 30	**17**(4), 113	1939 Sep 27	*ibid.*
	1911 Apr 30	**19**(3), 57	1940 Nov 3	**35**(3)
	1912 May 31	**20**(1), 1	1941 Dec 8	*ibid.*
	1913 Jul 5	*ibid.*	1943 Jan 11	**35**(4)
	1914 Aug 10	**21**(1), 1		Alexander AFO'B, *JBAA...*
	1915 Sep 17	*ibid.*	1944 Feb 11	**63**, 251 & 328
	1916 Oct 23	**23**(3), 81	1945 Mar 13	*ibid.*
	1917 Nov 28	**24**(2), 29	1946 Apr 13	*ibid.*
	1919 Jan 1	**26**(3), 53	1947 May 14	*ibid.*
			1948 Jun 15	**64**, 72

(a) 1869–1890

In these years there were few comprehensive reports, but many individual reports in *Astronomical Register, The Observatory,* and *Monthly Notices of the Royal Astronomical Society (MNRAS)*. Also see: A.S. Williams (*1889*) *Zenographical Fragments, vol.I* (report on 1886/87); Williams (*1909*) *Zenographical Fragments, vol.II* (report on 1888); Williams (*1910*) *MNRAS* **71**, 145–156, 'The equatorial current of Jupiter in 1880.' Also see E.P. Sells and W.E. Cooke (*1913*) *Physical Observations of Jupiter made at the Adelaide Observatory.*

(b) 1891–1948

The principal record of the years 1891–1948 is the series of BAA *Memoirs*. Also, detailed reports by individual observers continued to appear in *MNRAS* up to 1907. The years 1904–1940 were extensively illustrated by Lowell Observatory photographs: E.C. Slipher (*1964*), *A Photographic Study of the Brighter Planets.*

(c) 1949–1964

Opposition	BAA reports	ALPO reports	European reports
1949 Jul 20	[i] McIntosh RA, *JBAA* **60**, 247	[ii]; [f] Reese EJ, *JALPO* **12**, 135 (*1958*)	–
1950 Aug 26	[unpubl.ms.]	[ii]	Du Martheray M, *Orion* (no. 30), 209
1951 Oct 3	Alexander AFO'B, *JBAA* **62**, 280	[ii]; [f] Both EE, *JALPO* **6**, 113 (*1952*)	Ruggieri G, *Mem.SAI* **23**, 259
1952 Nov 8	Alex. & Fox, *JBAA* **64**, 281&379	[ii]; [f] Reese EJ & Brookes RG, *JALPO* **7**, 125; Komoda K, *JALPO* **8**, 141	Ruggieri G, *Mem.SAI* **25**, 149
1953 Dec 13	–	[i] Brookes RG, *JALPO* **7**, 162; **8**, 6; [f] Brookes & Reese, *JALPO* **8**, 127	Ruggieri G, *Mem.SAI* **27**, 7
1955 Jan 15	[i] Alex., *JBAA* **66**, 124&208;	[i] Brookes RG, *JALPO* **9**, 30&58; [f] Brookes & Reese, *JALPO* **9**, 162	–
1956 Feb 16	[i] *ibid.*, & Fox WE, *JBAA* **67**, 307	[i] Brookes RG, *JALPO* **10**, 13; [f] Brookes & Reese, *JALPO* **11**, 22	Baldinelli L & Dall'Olmo U, *Mem.SAI*, **29**(1)
1957 Mar 17	[i] Lenham AP, *JBAA* **67**, 258; & Fox WE, *JBAA* **67**, 307; [f] Fox WE, *Mem.BAA* **39**(1)	[i] Squyres HP, *JALPO* **11**, 15 [f] Squyres & Reese, *JALPO* **12**, 37&88	Baldinelli L & Dall'Olmo U, *Coelum*, **26**(5)
1958 Apr 17	[i] Fox WE, *JBAA* **69**, 256; [f] Fox WE, *Mem.BAA* **39**(1)	[f] Reese EJ, *JALPO* **13**, 58	Dall'Olmo U, *Mem.SAI* **32**, 343
1959 May 18	[i] Fox WE, *JBAA* **69**, 269	[i] Budine PW, *JALPO* **13**, 101; **14**, 34 [f] Budine & Reese, *JALPO* **14**, 66; Sato T, *JALPO* **15**, 73	Dall'Olmo U, *Mem.SAI* **34**, 41
1960 Jun 20	–	[f] Glaser PR, *JALPO* **15**, 37; Reese EJ, *JALPO* **15**, 73	Dall'Olmo U, *Mem.SAI* **34**, 41 Hémeret J-L & Marin M, *L'Astr.* **76**, 171
1961 Jul 25	[f] Fox WE, *JBAA* **73**, 106 [i] Fox WE, *JBAA* **72**, 63	[i] Chapman CR, *JALPO* **15**, 212; **16**, 89&100; Sato T, *JALPO* **17**, 197 [f] Reese EJ, *JALPO* **16**, 193	Dall'Olmo U, *Mem.SAI* **35**, 147 Cortesi S, *Orion* (no. 75), 32
1962 Aug 31	[f] Fox WE, *JBAA* **75**, 35&108	[f] Reese EJ, *JALPO* **17**, 137; Wend RE, *JALPO* **18**, 209	Cortesi S, *Orion* (no. 80), 92 Marin, M, *L'Astr.* **77**, 105 (*1963*)
1963 Oct 8	[f] Fox WE, *JBAA* **75**, 187	[f] Reese EJ, *JALPO* **18**, 85; Wend RE, *JALPO* **19**, 109	Cortesi S, *Orion* (no. 87), 245 Marin M & Walbau M, *L'Astr.* **80**, 151 (*1966*)

(c) 1949–1964

From 1949 to 1964, some summer apparitions were reported briefly or not at all by the BAA, but the ALPO produced high-quality reports for most of these years. Summaries of drift rates for several preceding years were given by the ALPO in the 1955/56 report and by the BAA in the 1956/57 report. Useful reports were also produced by various European societies (a sample is listed) and by the Oriental Astronomical Association (published in Japanese in *The Heavens*; not listed). In this list: [f] final report; [i] interim report; [ii] numerous interim reports not listed.

(d) 1964–1974

Opposition	BAA reports	NMSUO reports:	European reports
1964 Nov 13	[f] Fox WE, *JBAA* **77**, 342	Reese EJ & Solberg HG, *Icarus* **5**, 266; Reese EJ & Smith BA, *Icarus* **5**, 248	Cortesi S, *Orion* (no.89),67; (no.92),208; [i] Marin M, *L'Astr.* **80**, 239 (*1966*)
1965 Dec 18	[f], *ibid.*; [i] Fox, *JBAA* **76**, 282	Solberg, *Icarus* **8**, 82; Reese & Smith, *Icarus* **9**, 474; Solberg & Reese, *NMSUO TN*-701-69-25	Cortesi S, *Orion* **12**(100), 37; Dragesco J, *L'Astr.***81**, 199 (*1967*)
1967 Jan 20	[f] Fox WE, *JBAA* **79**, 304	Solberg, *Icarus* **9**, 212; Reese & Smith, *Icarus* **9**, 474; Solberg & Reese, *NMSUO TN*-701-69-25	Cortesi S, *Orion* **13**, 57; Dragesco J, *L'Astr.* **82**, 59 (*1968*)
1968 Feb 20	[f] *ibid.*	Solberg, *Icarus* **10**, 412; Reese & Solberg, *NMSUO-TN*-701-69-28	Cortesi S, *Orion* **14**, 29; Dragesco J, *L'Astr.* **83**, 311 (*1969*)
1969 Mar 20	[f] Fox WE, *JBAA* **81**, 396	Reese, *Icarus* **12**, 249	Lecacheux J, *L'Astr.* **84**, 442 (*1970*); Falorni M *et al.*, *Mem.SAI* **42**, 33 (*1971*)
1970 Apr 21	[f] ibid.	Reese, *Icarus* **14**, 343 Reese, *NMSUO-TN*-701-70-302	Lecacheux J & A, *L'Astr.* **86**, 57 (*1972*)
1971 May 23	[i] Fox WE, *JBAA* **82**, 86&328	Reese, *Icarus* **17**, 57; Reese, *NMSUO-TN*-72-41	–

		LPL reports:	
		Minton RB, *JBAA* **83**, 263	
1972 Jun 24	[f] Murray JB & Rogers JH, *JBAA* **89**, 489 (*1979*)	Minton RB, *Icarus* **29**, 201	–
1973 Jul 30	[i] Rogers JH & Young PJ, *JBAA* **87**, 240; [f] McKim RJ & Murray JB, *Mem.BAA* **43**(1), 3	Minton RB, *Icarus* **29**, 211	Botton C, *L'Astr.* **89**, 94 (*1975*)
1974 Sep 5	[i] *ibid.*; [f] McKim RJ, *Mem.BAA* **43**(1), 10	Minton RB, *Icarus* **31**, 110	Botton C, *L'Astr.* **90**, 65 (*1976*)

(d) 1964–1974

From 1964 to 1974, the best data were from the professional observatories in New Mexico (NMSUO) and Arizona (LPL). Good reports were also published for some apparitions by the BAA and by several European groups.

In 1965–1971, the LPL produced no apparition reports but many colour-filter and true-colour photographs were published in *Communications of the LPL*, vols. 9 and 10. Full NMSUO reports were compiled from 1962 to 1971 but most were produced as preprints only. Some of their summaries published in *Icarus* concerned only the GRS and STB white ovals, but some were more comprehensive. Extensive summaries of NMSUO data from 1962 to 1970 were published by Reese (*1971a*), largely reprinted by Smith and Hunt (*1976*); and from 1970 to 1986 by Beebe *et al.* (*1989*) (references in Appendix 4).

ALPO reports also continued to be published regularly in their *Journal*. However, their reports since 1964 have often been at odds with observations by BAA members, as mentioned in BAA reports covering the 1970s. The discrepancies in 1975/76 were discussed (Rogers, *1979*; Mackal, *1979*). BAA data on equatorial current motions conflicted with ALPO reports for 1972, 1973, and 1974. Also, the ALPO reported a major SEB disturbance in spring, 1978, and STBn jetstream activity in 1979/80, largely on the basis of observations with a small telescope; neither of these events could be confirmed by our independent observations. In the present book, ALPO results from 1964 onwards are only referred to when the phenomena were independently confirmed.

(e) 1975–1992

Opposition	BAA reports (interim)	BAA reports (final)	European reports
1975 Oct 13	[ii]	Rogers JH, *Mem.BAA* **43**(1), 18 (*1990*)	Favero G & Ortolani S, *JALPO* **27**, 92 [UAI]; Dragesco et al., *L'Astr.* **94**, 115 (*1980*)
1976 Nov 18	[i] Fox WE, *JBAA* **87**, 330;	Foulkes M, *Mem.BAA* **43**(1), 37 (*1990*)	Favero G & Zatti P, *JALPO* **28**, 104 [UAI]; Néel R, *L'Astr.* **95**, 130 (*1981*)
1977 Dec 23	–	Foulkes M, *Mem.BAA* **43**(3), 3 (*1992*)	Néel R, *L'Astr.* **95**, 423 (*1981*)
1979 Jan 24	–	McKim RJ & Rogers JH, *Mem.BAA* **43**(3), 11	Néel R, *L'Astr.* **96**, 279 (*1982*)
1980 Feb 24	–	McKim RJ, *Mem.BAA* **43**(3), 39	Néel R, *L'Astr.* **97**, 31 (*1983*)
1981 Mar 26	–	McKim RJ, *Mem.BAA* **43**(3), 49	Néel R, *L'Astr.* **97**, 379 (*1983*)
1982 Apr 26	[i] Rogers, *JBAA* **93**, 164;	Rogers, *JBAA* **94**, 1	Néel R, *L'Astr.* **98**, 315 (*1984*)
1983 May 27	–	Rogers, *JBAA* **97**, 46&81	Néel R, *L'Astr.* **100**, 117 (*1986*)
1984 Jun 29	–	Rogers, *JBAA* **97**, 46&167	Néel R, *Obs. et Travaux SAF* p.5 (*1987*)
1985 Aug 5	[i] Rogers & Fox, *JBAA* **96**, 73&139;	Rogers, *JBAA* **98**, 151	[ii]; [f] Néel R, *Pulsar* **79**(no.664), 8 (*1988*)
1986 Sep 10	[i] Rogers, *JBAA* **97**, 195;	Rogers, *JBAA* **100**, 121	[ii]; [f] Néel R, *Pulsar* **80**(no.675), 205 (*1989*)
1987 Oct 18	[i] Rogers, *JBAA* **98**, 72	Rogers, *JBAA* **101**, 22	[ii]; [f] Néel R, *Pulsar* **82**(no.685), 119 (*1991*)
1988 Nov 23	[i] Rogers, *JBAA* **99**, 4	Rogers, *JBAA* **101**, 81	
1989 Dec 27	[i] Rogers, *JBAA* **99**, 271; **100**, 112	Rogers, *JBAA* **102**, 135	
1991 Jan 29	[i] Rogers, *JBAA* **100**, 270	Rogers, *JBAA* **102**, 324	
1992 Feb 29	[i] Rogers, *JBAA* **102**, 188	Rogers & Foulkes, *JBAA* (in prep.)	

(e) 1975–1992

From 1972 onwards, BAA reports have been comprehensive. The SAF have also produced excellent reports, which were more detailed than the BAA's in the early 1980s. The French reports from 1985 onwards have been by the Société d'Astronomie Populaire, published in *Pulsar* (Toulouse), including many interim reports by R. Néel. Most of the best data since 1985 were common to the SAF and BAA (including many high-quality photographs), and the reports agree in almost all details. Synopses of 1975–1985, based on amateur and professional photographs, were published by Sanchez-Lavega and Rodrigo (*1985*) and Sanchez-Lavega and Quesada (*1988*) (see Appendix 4).

Appendix 4
Bibliography
(The Planet)

A4.1. MAJOR BOOKS AND SPECIAL ISSUES OF JOURNALS

The following books were principally concerned with Jupiter, but are now out of date and/or out of print, with the exception of the first listed.

Belton MJS, West RA & Rahe J (eds.)(1989), *Time-variable Phenomena in the Jovian System*, NASA SP-494 (NASA, Washington DC, 1989).

Fimmel RO, Swindell W & Burgess E (1977), *Pioneer Odyssey* (NASA SP-349/396, Washington DC). [The book of the Pioneer mission.]

Gehrels T (ed.) (1976), *Jupiter* (University of Arizona Press, Tucson).

Hunt G & Moore P (1981), *Jupiter* (Mitchell Beazley, London).

Morrison D and Samz J (1980): *Voyage to Jupiter*, NASA SP-439 (Washington DC). [The book of the Voyager mission.]

Peek BM (1958) *The Planet Jupiter* (Faber & Faber, London).

The Pioneer results at Jupiter were described in the following special issues of journals:
> *Science* (1974 Jan 25) vol. **183**, pp.301–324;
> *Science* (1975 May 2) vol. **188**, pp.445–477;
> *Journal of Geophysical Research* (1974 Sep 1) vol. **79**, pp. 3487–3700.

The Voyager scientific results at Jupiter were described in the following special issues:
> *Science* (1979 June 1) vol. **204**, pp.913–1008;
> *Science* (1979 Nov 23) vol. **206**, pp.925–996;
> *Nature* (1979 August 30) vol. **280**, pp.725–806;
> *Geophysical Research Letters* (1980 Jan.) vol.**7**, pp.1–68;
> *Journal of Geophysical Research* (1981 Sep. 30) vol. **86** (no. A10) pp. 8121–8844;
> *Icarus* (1986 February) vol. **65** (no.2/3), pp.159–466.
> *Memoirs of the BAA* (1992 March) vol. **43** (no.3), pp. 23–38.

The Voyager and Galileo spacecraft were described in detail in:
> *Space Science Reviews*, vol. **21** (nos.2–3), pp.77–376 (1977): The Voyager missions to the outer solar system.
> *Space Science Reviews*, vol. **60** (nos.1–4) (1992): The Galileo mission. Reprinted as:
> Russell CT (ed.)(1992) *The Galileo Mission* (Kluwer Academic Publ., Dordrecht).

A4.2. OTHER REFERENCES

References are cited in the text by author(s) and date, and listed below in alphabetical order of first author. When a person is first author on more than one paper, any solo references are listed first in order of date, and then references with co-authors in order of date (not in alphabetical order of co-authors).

Some of the most commonly cited journals are abbreviated as follows:

JBAA: Journal of the British Astronomical Association

JALPO: Journal of the Association of Lunar and Planetary Observers (The Strolling Astronomer)
MNRAS: Monthly Notices of the Royal Astronomical Society
JGR: Journal of Geophysical Research
J.Atm.Sci.: Journal of the Atmospheric Sciences
Sky & Tel.: Sky and Telescope

Allison M (1990) *Icarus* **83**, 282–307. Planetary waves in Jupiter's equatorial atmosphere.

Antipov SV, Nezlin MV, Snezhkin EN & Trubnikov AS (1986) *Nature* **323**, 238–240. Rossby autosoliton and stationary model of the jovian Great Red Spot.

Antoniadi EM (1926) *Bull.Soc.Astron.de France (L'Astronomie)* **40**, 394–409. Le monde de Jupiter. [With many drawings.]

Armstrong KR, Minton RB, Riekc GH & Low FJ (1976) *Icarus* **29**, 287–297. Jupiter at 5 microns.

Ashbrook J (ed.) (1975) *Sky & Tel.* **49**(2), 72–78. Pioneer 11: through the dragon's mouth.

Atreya SK, Donahue TM, Sandel BR, Broadfoot AL & Smith GR (1979) *Geophys.Res.Lett.* **6**, 795–798. Jovian upper atmospheric temperature measurement by the Voyager 1 UV spectrophotometer.

Atreya SK, Donahue TM & Festou MC (1981) *Astrophys.J.Lett.* **247**, L43–L47. Jupiter: structure and composition of the upper atmosphere. [Voyager UVS occultation]

Atreya SK & Romani PN (1983) in: *Recent Advances in Planetary Meteorology* (ed. Hunt GE; Cambridge Univ. Press, 1985) pp.17–68. Photochemistry and clouds of Jupiter, Saturn, and Uranus.

Axel L (1972) *Astrophys. J.* **173**, 451–468. Inhomogeneous models of the atmosphere of Jupiter. [UV-absorbing haze]

BAA (1922) *JBAA* **22**, 289–294. (Meeting report.)

Baines KH, Hayden Smith W & Alexander C (1989), in: *Time-variable Phenomena in the Jovian System*, NASA SP-494 (eds. Belton MJS, West RA & Rahe J; NASA, Washington DC, 1989), pp. 363–370. Spatial and temporal variations of NH_3 abundance and cloud structure in the jovian troposphere derived from CCD/Coudé observations.

Barcilon A & Gierasch P (1970) *J.Atm.Sci.* **27**, 550–560. A moist Hadley cell model for Jupiter's cloud bands.

Barnard EE (1891) *MNRAS* **51**, 543–558. Observations of the planet Jupiter and his satellites during 1890 with the 12-inch equatoreal of the Lick Observatory.

Barnard EE (1892) *MNRAS* **52**, 6–7. Colour changes in the markings on the surface of the planet Jupiter.

Baron R, Joseph RD, Owen T, Tennyson J, Miller S & Ballester GE (1991) *Nature* **353**, 539–542. Imaging Jupiter's aurorae from H_3^+ emissions in the 3–4 μm band.

Basu D (1969) *Nature* **222**, 69. Relation between the visibility of Jupiter's Red Spot and solar activity.

Baum WA & Code AD (1953) *Astron. J.* **58**, 108, A photometric observation of the occultation of σ Arietis by Jupiter;

Beatty JK (1977 Aug.), *Sky & Tel.* **54** (2), 95. Voyaging to the outer planets.

Beebe RF (1990 Oct.) *Sky & Tel.* **80**, 359–364. Queen of the giant storms.

Beebe RF & Youngblood LA (1979) *Nature* **280**, 771–772. Pre-Voyager velocities, accelerations and shrinkage rates of jovian cloud features.

Beebe RF, Ingersoll AP, Hunt GE, Mitchell JL & Muller J-P (1980) Geophys. Res. Lett. **7**, 1–4, Measurements of wind vectors, eddy momentum transports, and energy conversions in Jupiter's atmosphere from Voyager 1 images.

Beebe RF and Hockey TA (1986) *Icarus* **67**, 96–105. A comparison of red spots in the atmosphere of Jupiter.

Beebe RF, Orton GS & West RA (1989), in: *Time-variable Phenomena in the Jovian System*, NASA SP-494 (eds. Belton MJS, West RA & Rahe J; NASA, Washington DC, 1989), pp.245–288. Time-variable nature of the jovian cloud properties and thermal structure: an observational perspective.

Bézard B, Baluteau JP & Marten A (1983) *Icarus* **54**, 434–455. Study of the deep cloud structure in the equatorial region of Jupiter from Voyager infrared and visible data.

Bjoraker GL, Larson HP & Kunde VG (1986a) *Icarus* **66**, 579–609. The gas composition of Jupiter derived from 5-μm airborne spectroscopic observations.

Bjoraker GL, Larson HP & Kunde VG (1986b) *Astrophys.J.* **311**, 1058–1072. The abundance and distribution of water vapor in Jupiter's atmosphere.

Browning J (1872) *MNRAS* **32**, 321–322. On some observations of Jupiter in 1871–72.

Budine PW (1968) *JALPO* **21**, 13–23 & 38–39. Jupiter in 1966–67: rotation periods.

Busse FH (1970) *Astrophys.J.* **159**, 629–639. Differential rotation in stellar convection zones.

Busse FH (1976) *Icarus* **29**, 255–260. A simple model of convection in the jovian atmosphere.

Busse FH (1983) *Geophys. Astrophys. Fluid Dynamics* **23**, 153 [review].

Caldwell J, Tokunaga AT & Gillett FC (1980) *Icarus* **41**, 667–675. Possible infrared aurorae on Jupiter.

Caldwell J, Tokunaga AT & Orton GS (1983) *Icarus* **53**, 133–140. Further observations of 8-μm polar brightenings of Jupiter.

Cameron AGW (1982) in: *Essays in Nuclear Astrophysics* (eds. Barnes CA, Clayton DD & Schramm DN; Cambridge Univ. Press, 1982), pp.23–44. Chapter 3: Elemental and nuclidic abundances in the solar system.

Campani G (1665) cited in *Phil.Trans.* **1**, 2–3. An accompt of the improvement of optick glasses.

Carlson BE, Prather MJ & Rossow WB (1987) *Astrophys.J.* **322**, 559–572. Cloud chemistry on Jupiter.

Carlson BE, Rossow WB & Orton GS (1988) *J.Atm.Sci.* **45**, 2066–2081. Cloud microphysics of the giant planets.

Carlson BE & Lutz BL (1989), in: *Time-variable Phenomena in the Jovian System*, NASA SP-494 (eds. Belton MJS, West RA & Rahe J; NASA, Washington DC, 1989), pp.289–296. Spatial and temporal variations in the atmosphere of Jupiter: polarimetric and photometric constraints.

Carlson BE, Lacis AA & Rossow WB (1992a) *Astrophys.J.* **388**, 648–668. The abundance and distribution of water vapor in the jovian troposphere as inferred from Voyager IRIS observations.

Carlson BE, Lacis AA & Rossow WB (1992b) *Astrophys.J.* **393**, 357–372. Ortho-para-hydrogen equilibration on Jupiter. (Also see *ibid.* **394**, L29.)

Cassini JD (1666) cited in *Phil.Trans.* **1**, 143–145. Of a permanent spot in Jupiter; by which is manifested the conversion of Jupiter about his own axis.

Cassini JD (1672) cited in *Phil.Trans.* **7**, 4039–4042. A relation of the return of a great permanent spot in the planet Jupiter, observed by Signor Cassini.

Chapman CR (1968) *Sky & Tel.* **35**, 276–278. The discovery of Jupiter's Red Spot.

Chapman CR (1969) *J.Atmos.Sci.* **26**, 986–990. Jupiter's zonal winds: variation with latitude.

Chapman CR & Reese EJ (1968) *Icarus* **9**, 326–335. A test of the uniformly rotating source hypothesis for the SEB disturbances on Jupiter.

Clarke AC (1982) *2010: Odyssey Two* (Granada, London, 1982).

Coffeen DL (1974) *JGR* **79**, 3645–. Optical polarisation measurements of the Jupiter atmosphere at 103° phase angle.

Combes M *et 8 al.* (1972) *L'Astronomie* (Bull.SAF) **86**, 1–14. L'occultation de β Scorpion par Jupiter le 13 Mai 1971.

Conrath BJ, Flasar FM, Pirraglia JA, Gierasch PJ & Hunt GE (1981a) *JGR* **86** (A10), 8769–8775. Thermal structure and dynamics of the jovian atmosphere: 2. Visible cloud features.

Conrath BJ, Gierasch PJ & Nath N (1981b) *Icarus* **48**, 256–282. Stability of zonal flows on Jupiter.

Conrath BJ, Gautier D, Hanel RA & Hornstein JS (1984) *Astrophys.J.* **282**, 807–815. The helium abundance of Saturn from Voyager measurements.

Cortesi S (1962), *Orion* **76**, 106–122. Contribution a l'étude de trois nouvelles formations persistantes de Jupiter.

Cortesi S (1978) *Icarus* **33**, 410–413. Determination quantitative de l'effect de phase dans la measure des longitudes sur Jupiter.

Dawes WR (1857) *MNRAS* **18**, 49–50. Further observations of the round bright spots on one of the belts of Jupiter.

De Callatay V & Dollfus A (1974) *Atlas of the Planets* (tr. Collon M; Heinemann, London).

Del Genio AD & McGrattan KB (1990) *Icarus* **84**, 29–53. Moist convection and the vertical structure and water abundance of Jupiter's atmosphere.

DeMarcus WC (1959) Astron.J. **63**, 2–28. The constitution of Jupiter and Saturn.

Deming D, Mumma MJ, Espenak F, Jennings DE, Kostiuk T, Wiedemann G, Loewenstein R & Piscitelli J (1989) *Astrophys. J.* **343**, 456–467. A search for p-mode oscillations of Jupiter: serendiptious observations of nonacoustic thermal wave structure.

Denning WF (1880) *MNRAS* **41**, 44–46. The motions and varieties of the jovian spots.

Denning WF (1883) *MNRAS* **44**, 63–73. Rotation period of Jupiter.

Denning WF (1885) *Nature* **32**, 31–34. Jupiter.

Denning WF (1898a) *MNRAS* **58**, 488–493. The Great Red Spot on Jupiter.

Denning WF (1898b) *MNRAS* **59**, 76–79. On a probable instance of periodically recurrent disturbance on the surface of Jupiter. [NTBs outbreaks.]

Denning WF (1899) *MNRAS* **59**, 574–584. Early history of the Great Red Spot on Jupiter.

Denning WF (1902) *JBAA* **12**, 121–125. Dark spot in Jupiter's South Temperate* region, 1901.

Denning WF (1904) *JBAA* **14**, 193–196. Dark spot in Jupiter's South Temperate* region.

Denning WF (1923), in: *Splendour of the Heavens* (ed. Phillips TER & Steavenson WH) (Hutchinson, London) vol.I, pp.333–356. Chapter VIII, Jupiter.

De Pater I (1986) *Icarus* **68**, 344–365. Jupiter's belt-zone structure at radio wavelengths, II: Comparison of observations with model atmospheric circulations.

Doel R (1977) *JALPO* **26**, 254–257. An update on SEB Disturbance analysis, Part I.

Doel R (1978) *JALPO* **27**, 105–115. An update on SEB Disturbance analysis, Parts II & III.

Dollfus A (1957) *Ann. d'Astrophysique Suppl.* 4. Etudes des planètes par la polarisation de leur lumière.

Dollfus A (1961) in: Kuiper GP & Middlehurst BM (eds.) (1961) *The Solar System, Vol. III: Planets and Satellites* (University of Chicago Press) pp.343–399. Polarization studies of planets.

Dollfus A (1990) *Comptes Rendues Acad. Sci. Paris* **311** (II), 1185–1190. Une nouvelle méthode d'analyse polarimétrique des surfaces planétaires.

Dowling TE & Ingersoll AP (1988) *J.Atm.Sci.* **45**, 1380–1396. Potential vorticity and layer thickness variations in the flow around Jupiter's Great Red Spot and white oval BC.

Dowling TE & Ingersoll AP (1989) *J.Atm.Sci.* **46**, 3256–3278. Jupiter's Great Red Spot as a shallow water system.

Dragesco J (1972), *JBAA* **82**, 200–201, Observations of occultations of β Scorpii on 1971 May 13 and 14.

Drake S (1957), *Discoveries and Opinions of Galileo* (Doubleday Anchor Books, New York)

Drossart P, Courtin R, Atreya S & Tokunaga A (1989), in: *Time-variable Phenomena in the Jovian System*, NASA SP-494 (eds. Belton MJS, West RA & Rahe J; NASA, Washington DC, 1989), pp.344–362. Variations in the jovian atmospheric composition and chemistry.

Drossart P, Lellouch E, Bézard B, Maillard J-P & Tarrago G (1990) *Icarus* **83**, 248–253. Jupiter: evidence for a phosphine enhancement at high northern latitudes.

Eshleman, von R (1975) *Science* **189**, 876–878, Jupiter's atmosphere: problems and potential of radio occultation;

Falorni M (1987) *JBAA* **97**, 215–219. The discovery of the Great Red Spot of Jupiter.

Favero G, Senigallesi P & Zatti P (1979) *JALPO* **27**, 240–245. Periodicity in the activity of Jupiter's atmosphere.

Flasar FM (1986) *Icarus* **65**, 280–303. Global dynamics and thermal structure of Jupiter's atmosphere.

Flasar FM, Conrath BJ, Pirraglia JA, Clark PC, French RG & Gierasch PJ (1981) *JGR* **86** (A10), 8759–8767. Thermal structure and dynamics of the jovian atmosphere, 1. The Great Red Spot.

Flasar FM & Gierasch (1986) *J.Atm.Sci.* **43**, 2683–2707. Mesoscale waves as a probe of Jupiter's deep atmosphere.

Focas JH (1962) *Mem.Soc.Roy.Liège* **7** = *Congrès et Colloques de l'Université de Liège* **24**, 535–540. Preliminary results concerning the atmospheric activity of Jupiter and Saturn.

Fountain JW (1972) *Commun.Lunar & Plan.Lab.* **9** (175), 327–337. Narrow-band photography of Jupiter and Saturn.

Fountain JW & Larson SM (1972) *Commun.Lunar & Plan.Lab.* **9** (174), 315–326. Multicolor photography of Jupiter.

Fountain JW, Coffeen DL, Doose LR, Gehrels T, Swindell W & Tomasko MG (1974) *Science* **184**, 1279–1281. Jupiter's clouds: equatorial plumes and other cloud forms in the Pioneer 10 images.

Fox WE (1967) *JBAA* **78**, 16–21. Presidential Address: The Great Red Spot on Jupiter.

Galileo G (1610) *Siderius Nuncius*; translation in Drake (1957).

Gautier D, Conrath B, Flasar M, Hanel R, Kunde V, Chedin A & Scott N (1981) *JGR* **86**(A10), 8713–8719. The helium abundance of Jupiter from Voyager.

Gautier D & Owen T (1983a) *Nature* **302**, 215–218. Cosmological implications of helium and deuterium abundances on Jupiter and Saturn.

Gautier D & Owen T (1983b) *Nature* **304**, 691–694. Cosmogonical implications of elemental and isotopic abundances in atmospheres of the giant planets.

Gautier D & Owen T (1989), in: *Origin and Evolution of Planetary and Satellite Atmospheres* (eds. Atreya SK, Pollack JB & Matthews MS; University of Arizona Press, Tucson, 1989), pp. 487–512. The composition of outer planet atmospheres.

Gehrels T (1974) *Sky & Tel.* **47**(2), 76–83. The fly-by of Jupiter.

Gehrels T, Herman BM & Owen T (1969) *Astron.J.* **74**, 190–199. Wavelength dependence of polarization: XIV. Atmosphere of Jupiter.

Gierasch PJ & Conrath BJ (1983) in: *Recent Advances in Planetary Meteorology* (ed. Hunt GE; Cambridge Univ. Press, 1985) pp.121–146. Energy conversion processes in the outer planets. [Esp. para-hydrogen]

Gierasch PJ, Conrath BJ & Magalhaes JA (1986) *Icarus* **67**, 456–483. Zonal mean properties of Jupiter's upper troposphere from Voyager infrared observations.

Gillett FC, Low FJ & Stein WA (1969), *Astrophys.J.* **157**, 925–934. The 2.8–14 micron spectrum of Jupiter.

Gillett FC & Westphal JA (1973), *Astrophys. J. Lett.* **179**, L153–154, Observations of 7.9-µm limb brightening on Jupiter.

Golitsyn GS (1970) *Icarus* **13**, 1–24. A similarity approach to the general circulation of planetary atmospheres.

Graf ER, Smith CE & McDevitt FR (1968) *Nature* **218**, 857. Correlation between solar activity and the brightness of Jupiter's GRS.

Green NE (1887) *Memoirs RAS* **49** (2), 259–270. On the belts and markings of Jupiter. [With many tinted drawings.]

Greene TF, Smith DW & Shorthill RW (1980) *Icarus* **44**, 102–115, Galilean satellite eclipse studies. I. Observations and satellite characteristics.

Griffith CA, Bézard B, Owen T & Gautier G (1992) *Icarus* **98**, 82–93. The tropospheric abundances of NH_3 and PH_3 in Jupiter's Great Red Spot from Voyager IRIS observations.

Grossman AS, Pollack JB, Reynolds RT, Summers AL & Graboske HC (1980) *Icarus* **42**, 358–379. The effect of dense cores on the structure and evolution of Jupiter and Saturn.

Gruithuisen FvP (1842) *Naturwissenschaftlich-astronomisches Jahrbuch* **5**, 86–91. Jupiter.

Gruithuisen FvP (1843) *Naturwissenschaftlich-astronomisches Jahrbuch* **6**, 106–113. Jupiter.

Haas WH (1953), *JALPO* **7** (1), 2–4, The occultation of the star σ Arietis by Jupiter on November 20, 1952.

Hanel R *et 12 al.* (1979a) *Science* **204**, 972–976. Infrared observations of the jovian system from Voyager 1.

Hanel R *et 13 al.* (1979b) *Science* **206**, 952–956. Infrared observations of the jovian system from Voyager 2.

Hanel RA, Conrath BJ, Herath LW, Kunde VG & Pirraglia JA (1981) *JGR* **86** (A10), 8705–8712. Albedo, internal heat, and energy balance of Jupiter: preliminary results of the Voyager infrared investigation.

Hargreaves FJ (1929) *MNRAS* **89**, 708–713. Further notes on the recent disturbance on Jupiter. [Irradiating spots in SEB.]

Hargreaves FJ (1939) *JBAA* **49** (9), 334–336. The Circulating Current in the STropZ of Jupiter.

Hargreaves FJ, Peek BM & Phillips TER (1928), *MNRAS* **89**, 209–215. The revival of activity in the southern hemisphere of Jupiter.

Harris DL (1961), in: Kuiper GP & Middlehurst BM (eds.) (1961) *The Solar System, Vol. III: Planets and Satellites* (University of Chicago Press) pp.272–342, Photometry and colorimetry of planets and satellites.

Hatzes A, Wenkert DD, Ingersoll AP & Danielson GE (1981) *JGR* **86** (A10), 8745–8749. Oscillations and velocity structure of a long-lived cyclonic spot.

Herschel W (1781) *Phil.Trans.* **71**, 115–137. Astronomical observations on the rotation of the planets round their axes, made with a view to determine whether the Earth's diurnal motion is perfectly equable.

Hess SL & Panofsky HA (1951) in: *Compendium of Meteorology* (ed. Malone TF; American Meteorological Society, Boston, 1951) pp.391–400. The atmospheres of the other planets.

Hide R (1961) *Nature* **190**, 895–896. Origin of Jupiter's Great Red Spot.

Hide R & Mason (1975) *Adv. in Physics* **24**, 47–100

Hockey TA (1989), Ph.D. dissertation: A historical interpretation of the study of visible cloud morphology on the planet Jupiter: 1610–1878 (New Mexico State University).

Hockey TA (1992) *J. History Astr.* **23**, 93–105. Seeing red: observations of colour in Jupiter's equatorial zone on the eve of the modern discovery of the Great Red Spot.

Hook R (1665) cited in *Phil.Trans.* **1**, 3. A spot in one of the belts of Jupiter.

Hord CW, West RA, Simmons KE, Coffeen DL, Sato M, Lane AL & Bergstrahl JT (1979) *Science* 206, 956–959. Photometric observations of Jupiter at 2400 Angstroms.

Hough GW (1905) *Popular Astronomy* **13** (no.121), 19–30. Our present knowledge of the condition of the surface of the planet Jupiter.

Houghton JT (1977, 1986) *The Physics of Atmospheres* (Cambridge University Press).

Hoyle F (1962) *Astronomy* (MacDonald & Co./Rathbone Books, London).

Hubbard WB (1977) *Icarus* **30**, 305–310. The jovian surface condition and cooling rate.

Hubbard WB (1989), in: *Origin and Evolution of Planetary and Satellite Atmospheres* (eds. Atreya SK, Pollack JB & Matthews MS; University of Arizona Press, Tucson, 1989), pp. 539–563. Structure and composition of giant planet interiors.

Hubbard WB, Nather RE, Evans DS, Tull RG, Wells DC, van Citters GW, Warner B & Vanden Bout P (1972) *Astron.J.* **77**, 41–59. The occultation of β Scorpii by Jupiter and Io.

Hubbard WB & Smoluchowski R (1973) *Space Sci. Rev.* **14**, 599–662. Structure of Jupiter and Saturn.

Hubbard WB, Hunten DM & Kliore A (1975) *Geophys. Res. Lett.* **12**, 265–268, Effect of jovian oblateness on Pioneer 10/11 radio occultations;

Hubbard WB & Slattery WL (1976) in: *Jupiter* (ed. Gehrels T; University of Arizona Press, Tucson, 1976), pp.176–194. Interior structure of Jupiter: theory of gravity sounding.

Humason ML (1961) in: Kuiper GP & Middlehurst BM (eds.) (1961) *The Solar System, Vol. III: Planets and Satellites* (University of Chicago Press) p.572 (& plates). Photographs of planets with the 200-inch telescope.

Hunt GE & Müller J-P (1979) *Nature* **280**, 778–780. Voyager observations of small-scale waves in the equatorial region of the jovian atmosphere.

Hunt GE, Conrath BJ & Pirraglia JA (1981) *JGR* **86** (A10), 8777–8781. Visible and infrared observations of jovian plumes during the Voyager encounter.

Hunt GE, Müller J-P & Gee P (1982) *Nature* **295**, 491–494. Convective growth rates of equatorial features in the jovian atmosphere.

Hunten DM (1976) in: *Jupiter* (ed. Gehrels T; University of Arizona Press, Tucson, 1976), pp.22–31. Atmospheres and ionospheres.

Hunten DM & Veverka J (1976) in: *Jupiter* (ed. Gehrels T; University of Arizona Press, Tucson, 1976), pp.247–283. Stellar and spacecraft occultations by Jupiter: a critical review of derived temperature profiles.

Ingersoll AP (1973) *Science* **182**, 1346–1348. Jupiter's Great Red Spot: a free atmospheric vortex?

Ingersoll AP (1976a, March) *Sci.Am.* **234** (no.3), 46–56. The atmosphere of Jupiter.

Ingersoll AP (1976b) *Icarus* **29**, 245–253. Pioneer 10 and 11 observations and the dynamics of Jupiter's atmosphere.

Ingersoll AP (1981 Dec.) *Sci.Am.* **245** (no.6), 66–80. Jupiter and Saturn.

Ingersoll AP (1990) *Science* **248**, 308–315. Atmospheric dynamics of the outer planets.

Ingersoll AP & Cuzzi JN (1969) *J.Atm.Sci.* **26**, 981–985. Dynamics of Jupiter's cloud bands.

Ingersoll AP, Munch G, Neugebauer G & Orton GS (1976) in: *Jupiter* (ed. Gehrels T; University of Arizona Press, Tucson, 1976), pp.197–205. Results of the infrared radiometer experiment on Pioneers 10 and 11.

Ingersoll AP & Porco CC (1978) *Icarus* **35**, 27–43. Solar heating and internal heat flow on Jupiter.

Ingersoll AP, Beebe RF, Collins SA, Hunt GE, Mitchell JL, Muller J-P, Smith BA & Terrile RJ (1979) *Nature* **280**, 773–775. Zonal velocity and texture in the jovian atmosphere inferred from Voyager images.

Ingersoll AP, Beebe RF, Mitchell JL, Garneau GW, Yagi GM & Muller J-P (1981), *JGR* **86**(A10), 8733–8744, Interaction of eddies and mean zonal flow on Jupiter as inferred from Voyager 1 and 2 images.

Ingersoll AP & Cuong PG (1981) *J.Atm.Sci.* **38**, 2067. Numerical model of long-lived jovian vortices.

Ingersoll AP & Pollard D (1982) *Icarus* **52**, 62–80. Motion in the interiors and atmospheres of Jupiter and Saturn: scale analysis, anelastic equations, barotropic stability criterion.

Ingersoll AP, Beebe RF, Conrath BJ & Hunt GE (1984), in: *Saturn* (eds. Gehrels T & Matthews MS; University of Arizona, Tucson, 1984), pp.195–238. Structure and dynamics of Saturn's atmosphere.

Ingersoll AP & Miller RL, (1986) *Icarus* **65**, 370–382. Motions in the interiors and atmospheres of Jupiter and Saturn: 2. Barotropic instabilities and normal modes of an adiabatic planet.

Johnson TV and Yeates CM (1983 Aug.) *Sky & Tel.* **66** (2), 99–106. Return to Jupiter: Project Galileo.

Keay CSL, Low FJ & Rieke GH (1972), *Sky & Tel.* **44** (5), 296–297, Infrared maps of Jupiter.

Keay CSL, Low FJ, Rieke GH & Minton RB (1973) *Astrophys.J.* **183**, 1063–1073. High-resolution maps of Jupiter at 5 µm.

Kiess CC, Corliss CH & Kiess HK (1960) *Astrophys.J.* **132**, 221–231. High-dispersion spectra of Jupiter.

Kim SJ, Caldwell J, Rivolo AR, Wagener R & Orton GS (1985) *Icarus* **64**, 233–248. Infrared polar brightening on Jupiter, III. Spectrometry from the Voyager 1 IRIS experiment.

Kim SJ, Drossart P, Caldwell J, Maillard JP, Goorvitch D, Moorwood A, Moneti A & Lecacheux J (1991) *Icarus* **91**, 145–153. The 2-µm polar haze of Jupiter.

Kliore AJ & Woiceshyn PM (1976) in: *Jupiter* (ed. Gehrels T; University of Arizona Press, Tucson, 1976), pp.216–237. Structure of the atmosphere of Jupiter from Pioneer 10 and 11 radio occultation measurements.

Kostiuk T, Espenak F & Mumma MJ (1989), in: *Time-variable Phenomena in the Jovian System*, NASA SP-494 (eds. Belton MJS, West RA & Rahe J; NASA, Washington DC, 1989), pp.234–241. Is ethane varying in the jovian north polar hot spot?

Kritzinger HH (ca.1912) *Uber die Bewegung des Roten Fleckes auf dem Planeten Jupiter* (Trowitsch und Sohn, Berlin).

Kritzinger HH (1914) *JBAA* **24**, 452–463. On the physical constitution of Jupiter.

Kuehn DM & Beebe RF (1993) *Icarus* **101**, 282–292. A study of the time variability of Jupiter's atmospheric structure.

Kuiper GP (1947) *Astron.J.* **52**, 147, cited in Report of Yerkes and McDonald Observatories.

Kuiper GP (1972a) *Sky & Telesc.* **43**, 4–8, 75–81. Lunar and Planetary Laboratory studies of Jupiter.

Kuiper GP (1972b) *Commun.Lunar & Plan.Lab.* **9** (173), 249–313. Interpretation of the Jupiter Red Spot. (Including Appendix by Larson SM.)

Kuiper GP & Middlehurst BM (eds.) (1961) *The Solar System, Vol. III: Planets and Satellites* (University of Chicago Press)

Larson H (1980) *An.Rev.Astron.Astrophys.* **18**, 43–75. Infrared spectroscopic observations of the outer planets, their satellites, and the asteroids.

Larson SM (1972a) *Commun.Lunar & Plan.Lab.* **9** (177), 353–359. Photographic observations of the occultation of β Scorpii by Jupiter.

Larson SM (1972b) *Commun.Lunar & Plan.Lab.* **9** (179), 371–383. Observations of the SEB disturbance on Jupiter in 1971.

Larson SM, Fountain JW & Minton RB (1973) *Commun.Lunar & Plan.Lab.* **10** (189), 40–41 (and Plates). Color photography of Jupiter.

Lellouch E, Drossart P, Encrenaz T, Guelachvili G, Lacombe N & Tarrago G (1989), in: *Time-variable Phenomena in the Jovian System*, NASA SP-494 (eds. Belton MJS, West RA & Rahe J; NASA, Washington DC, 1989), pp.371–373. A new analysis of the jovian 5-µm IRIS spectra.

Leovy CB, Friedson AJ & Orton GS (1991) *Nature* **354**, 380–382. The quasiquadrennial oscillation of Jupiter's equatorial stratosphere.

Lewis JS (1969) *Icarus* **10**, 365–378. The clouds of Jupiter and the NH_3–H_2O and NH_3–H_2S systems.

Lewis JS & Prinn RG (1970) *Science* **169**, 472–473. Jupiter's clouds: structure and composition.

Lewis JS & Fegley MB (1984) *Space Sci. Rev.* **39**, 163–192. Vertical distribution of disequilibrium species in Jupiter's troposphere.

Limaye SS (1986) *Icarus* **65**, 335–352. Jupiter: new estimates of the mean zonal flow at the cloud level.

Limaye SS, Revercomb HE, Sromovsky LA, Krauss RJ, Santek DA, Suomi VE, Collins SA & Avis CC (1982) *J.Atm.Sci.* **39**, 1413–1432. Jovian winds from Voyager 2. I: Zonal mean circulation.

Lindal GF *et 11 al.* (1981) *JGR* **86** (A10), 8721–8727. The atmosphere of Jupiter: an analysis of the Voyager radio occultation measurements.

Long JW (1860) *MNRAS* **20**, 243–245. On the appearance of Jupiter.

Lovelock JE (1965) *Nature* **207**, 568–570. A physical basis for life detection experiments.

Lunine JI (1989) *Science* **245**, 141–146. Origin and evolution of outer planet atmospheres.

Lyot B (1929) Récherches sur la polarisation de la lumiere des planetes et de quelque substances terrestres, *Ann. Obs. Paris* (Meudon) vol. 8; English translation NASA TT F-187 (1964).

Lyot B (1943), *Bull.Soc.Astron.de France* (*L'Astronomie*) **57**, 49–60, 67–72. Observations planétaires au Pic du Midi en 1941 par MM. Camichel, Gentili et Lyot.

Lyot B (1953), *Bull.Soc.Astron.de France* (*L'Astronomie*) **67**, 3–21. L'aspect des planetes au Pic du Midi dans une lunette de 60 cm d'ouverture.

Mackal PK (1966) *JALPO* **19**, 118–123. Latitude deviations of the NEBn and NEBs of Jupiter.

Mackal PK (1979) *JALPO* **27**, 245–246. An ALPO rejoinder to the BAA: Was there a NTrZ disturbance in 1975?

MacLow M-M & Ingersoll AP (1986) *Icarus* **65**, 353–369. Merging of vortices in the atmosphere of Jupiter: an analysis of Voyager images.

Magalhâes JA, Weir AL, Conrath BJ, Gierasch PJ & Leroy SS (1989) *Nature* **337**, 444–447. Slowly moving thermal features on Jupiter.

Magalhâes JA, Weir AL, Conrath BJ, Gierasch PJ & Leroy SS (1990) *Icarus* **88**, 39–72. Zonal motion and structure in Jupiter's upper troposphere from Voyager infrared and imaging obervations.

Marcus PS (1988) *Nature* **331**, 693–696. Numerical simulation of Jupiter's Great Red Spot.

Marth A (1892) *MNRAS* **52**, 462–467. Ephemeris for physical observations of Jupiter, 1892.

Martin AR (1972) *Spaceflight* **14**, 294–299 & 325–332. Missions to Jupiter.

Maxworthy T (1984) *Planet.Space Sci.* **32**, 1053–1058. The dynamics of a high-speed jovian jet.

Maxworthy T (1985) *Plan. Space Sci.* **33**, 987–991. Measurements and interpretation of a jovian near-equatorial feature.

Maxworthy T & Redekopp LG (1976a) *Nature* **260**, 509–511. New theory of the Great Red Spot from solitary waves in the Jovian atmosphere.

Maxworthy T & Redekopp LG (1976b) *Icarus* **29**, 261–271. A solitary wave theory of the Great Red Spot and other observed features in the jovian atmosphere.

Maxworthy T, Redekopp LG & Weidman PD (1978) *Icarus* **33**, 388–409. On the production and interaction of planetary solitary waves: applications to the jovian atmosphere.

Mayer AM (1870) J. Franklin Inst. 59, 136; reprinted without illustration in: *Astron. Register* **8**, 169–172. Observations on the planet Jupiter.

McDonald GD, Thompson WR & Sagan C (1992) *Icarus* **99**, 131–142. Radiation chemistry in the jovian stratosphere: laboratory simulations.

McIntosh RA (1936) *JBAA* **46** (8), 285–289. Colour variation in Jupiter's EZ.

McIntosh RA (1950) *JBAA* **60**, 247–250. Disturbance on Jupiter's South Equatorial region.

McKim RJ & Rogers JH (1992) *Mem.BAA* **43**(3), 23–38. A Voyager Atlas of Jupiter.

Mettig H-J (1991), cited in: *Sterne und Weltraum* **10**, 621–622.

Meyers SD, Sommeria J & Swinney HL (1989) *Physica D* **37**, 515–530. Laboratory study of the dynamics of jovian-type vortices.

Minton RB (1972a) *Commun.Lunar & Plan.Lab.* **9** (176), 339–351. Latitude measurements of Jupiter in the 0.89µ methane band.

Minton RB (1972b) *Commun.Lunar & Plan.Lab.* **9** (178), 361–369, *reprinted* (1973) *JBAA* **83**, 263–271. Initial development of the June 1971 SEB disturbance on Jupiter.

Minton RB (1973) *Commun.Lunar & Plan.Lab.* **9** (182), 397–402. Recent observations of Jupiter's NNTB current B.

Mitchell JL, Terrile RJ, Smith BA, Muller J-P, Ingersoll AP, Hunt GE, Collins SA & Beebe RF (1979) *Nature* **280**, 776–778. Jovian cloud structure and velocity fields.

Mitchell JL, Beebe RF, Ingersoll AP & Garneau GW (1981) *JGR* **86** (A10), 8751–8757. Flow fields within Jupiter's Great Red Spot and oval BC.

Molesworth PB (1905) *MNRAS* **65**, 691–706. Report on observations of Jupiter for 1903–4.

Münch G & Younkin RL (1964) Astron.J. **69**, 553. (Abstract)

Murray B (1989) *Journey into Space: The First Three Decades of Space Exploration* (Norton).

NASA (1971), *The Pioneer Mission to Jupiter*, NASA SP-268 (NASA, Washington DC).

Newburn RL & Gulkis S (1973) *Space Sci.Rev.* **14**, 179–271. A survey of the outer planets, Jupiter, Saturn, Uranus, Neptune, Pluto and their satellites.

Noll KS, Larson HP & Geballe TR (1990) *Icarus* **83**, 494–499. The abundance of AsH_3 on Jupiter.

Olivarez J (1984) *J.Assoc.Lunar Plan.Obs.***30**, 181–186. On the blue cloud features of Jupiter's NEBs-EZn region.

Orton GS (1975) *Icarus* **26**, 159–174. Spatially resolved absolute spectral reflectivity of Jupiter: 3390–8400 Å.

Orton GS & Ingersoll A (1976) in: *Jupiter* (ed. Gehrels T; University of Arizona Press, Tucson, 1976), pp.206–215. Pioneer 10 and 11 and ground-based infrared data on Jupiter: the thermal structure and He–H_2 ratio.

Orton GS, Ingersoll AP, Terrile RJ & Walton SR (1981) *Icarus* **47**, 145–158. Images of Jupiter from the Pioneer 10 and Pioneer 11 infrared radiometers: a comparison with visible and 5-µm images.

Orton GS et 14 al. (1991) *Science* **252**, 537–541. Thermal maps of Jupiter: spatial organization and time dependence of stratospheric temperatures, 1980 to 1990.

Owen T (1969) *Icarus* **10**, 355–364. The spectrum of Jupiter and Saturn in the photographic infrared.

Owen T & Terrile RJ (1981) *JGR* **86** (A10), 8797–8814. Colors on Jupiter.

Pagel BEJ (1989), in: *Evolutionary Phenomena in Galaxies* (eds. Beckman J & Pagel BEJ) (Cambridge Univ. Press).

Peek BM (1926) *JBAA* **37**, 62–64. Abnormal jovian spots in 1920.

Peek BM (1931) *MNRAS* **91**, 941–947. Rapidly-moving spots on Jupiter's NTB.

Peek BM (1937) *JBAA* **47** (3), 106–109; Jupiter Section, Interim Report, Apparition of 1936.

Peek BM (1939) *JBAA* **50** (1), 2–21. (Presidential Address.)

Pettit E & Richardson RS (1953) *Publ.Astr.Soc.Pacific* 65, 91–92. Motion pictures of the occultation of σ Arietis by Jupiter on Nov.20, 1952.

Phillips TER (1915) *JBAA* **26**, 1–16. Address of the President.

Phillips TER (1930) *Mem.BAA* 29(3), 79–86. Notes on the latitudes of the belts. [Synopsis of belt and GRS latitudes to date.]

Phillips TER *et al.*(1936), *JBAA* **47**, 57–58; (Meeting report)

Pilcher CB, Prinn RG & McCord T (1973) *J.Atmos.Sci.* **30**, 302–307. Spectroscopy of Jupiter: 3200–11200 Å.

Pirraglia JA, Conrath BJ, Allison MD & Gierasch PJ (1981) *Nature* **292**, 677–679. Thermal structure and dynamics of Saturn and Jupiter. [Voyager IRIS]

Pollack JB (1984) *An.Rev.Astron.Astrophys.* **22**, 389–424. Origin and history of the outer planets. [and their satellites]

Pollack JB, Atkinson DH, Seiff A & Anderson JD (1992) *Space Sci. Rev.* **60**, 143–178. Retrieval of a wind profile from the Galileo probe telemetry signal.

Ponnamperuma C (1976) *Icarus* **29**, 321–328. The organic chemistry and biology of the atmosphere of the planet Jupiter.

Prinn RG & Owen T (1976) in: *Jupiter* (ed. Gehrels T; University of Arizona Press, Tucson, 1976), pp. 319–371. Chemistry and spectroscopy of the jovian atmosphere.

Read PL (1992), *in:* Hopfinger EJ (ed.), *Rotating Fluids in Geophysical and Industrial Situations* (Springer, Vienna, in press). Long-lived eddies in the atmospheres of the major planets.

Read PL & Hide R (1983) *Nature* **302**, 126–129. Long-lived eddies in the atmospheres of Jupiter and Saturn.

Read PL & Hide R (1984) *Nature* **308**, 45–48. An isolated baroclinic eddy as a laboratory analogue of the Great Red Spot on Jupiter.

Reese EJ (1952), *JALPO (The Strolling Astronomer)* **6**, 33–35. Long-enduring brighter sections in the STZ of Jupiter. (Additional notes: *ibid.* **8**, 16 (1954); *ibid.* **12**, 54 (1958).

Reese EJ (1953a) *JBAA* **63**, 219–222. A possible clue to the rotation period of the solid nucleus of Jupiter.

Reese EJ (1953b) *JALPO (The Strolling Astronomer)* **7**, 88–92. The changeable aspect of Jupiter's Great Red Spot.

Reese EJ (1955) *JALPO* **9**, 64–68. Major SEB disturbances on Jupiter and an apparent clue to the true rotation of the giant planet.

Reese EJ (1960) *JALPO* **14**, 84–85. Jupiter: solar-induced changes.

Reese EJ (1962a) *Sky & Tel.* **24**, 70–74. Observing Jupiter.

Reese EJ (1962b) *JALPO* **16**, 260–263. Jupiter: a new disturbance and an 'old' theory.

Reese EJ (1971a), NMSUO preprint TN-71-36, Summary of jovian latitude and rotation period observations from 1898 to 1970.

Reese EJ (1971b), *Icarus* **14**, 343–354. Jupiter: its red spot and other features in 1969–1970.

Reese EJ (1972a) *Icarus* **17**, 57–72. Jupiter: its Red Spot and disturbances in 1970–1971.

Reese EJ (1972b), *Icarus* **17**, 704–706. An earlier generation of long-enduring south temperate ovals on Jupiter.

Reese EJ & Smith BA (1966) *Icarus* **5**, 248–257. A rapidly-moving spot on Jupiter's NTB.

Reese EJ & Solberg HG (1966) *Icarus* **5**, 266–273. Recent measures of the latitude and longitude of Jupiter's Red Spot.

Reese EJ & Smith BA (1968) *Icarus* **9**, 474–486. Evidence of vorticity in the Great Red Spot of Jupiter.

Reese E & Beebe R (1976) *Icarus* **29**, 225–230. Velocity variations of an equatorial plume throughout a jovian year.

Ridgway ST, Larson HP & Fink U (1976) in: *Jupiter* (ed. Gehrels T; University of Arizona Press, Tucson, 1976), pp.384–417. The infrared spectrum of Jupiter.

Rhines PB (1975) *J. Fluid Mech.* **69**, 417–443. Waves and turbulence on a beta plane.

Rogers JH (1976) *JBAA* **86**, 401–408. A high-velocity outbreak on the North Temperate Belt. Appendix: History of global upheavals. [Revised and reprinted in:] *Mem.BAA* **43**(3), 35–36 (1990).

Rogers JH (1978) in: *Proceedings of Astronomy West 78* (Western Amateur Astronomers). Patterns of activity in Jupiter's NEB.

Rogers JH (1979) *J.Assoc.Lunar & Plan.Obs.* **27**, 175–179. Jupiter in 1975–76: A postscript.

Rogers JH (1980) *JBAA* **90**, 132–147. Disturbances and dislocations on Jupiter.

Rogers JH (1983) *JBAA* **93**, 164–166. Recent events in the South Tropical Zone of Jupiter.

Rogers JH (1986) *JBAA* **97**, 46–51. Jupiter in 1983–84, I. Overview of the third South Tropical Dislocation.

Rogers JH (1988) *JBAA* **98**, 234–240. Strong interactions in the North Equatorial Belt of Jupiter.

Rogers JH (1989a) *JBAA* **99**, 95–100. Origins of outbreaks in the South Equatorial Belt of Jupiter.

Rogers JH (1989b) *JBAA* **99**, 135–143. Historical patterns of activity in the NEB and EZ of Jupiter.

Rogers JH (1990) *JBAA* **100**, 88–90. The patterns of jetstreams on Jupiter: correlation with Earth-based observations and consequences for belt nomenclature.

Rogers JH & Young PJ (1977) *JBAA* **87**, 240–251. Earth-based and Pioneer observations of Jupiter.

Rogers JH & Miyazaki I (1990) *Icarus* **87**, 193–197. First Earth-based observations of a south temperate jet stream on Jupiter.

Rogers J & Herbert D (1991) *JBAA* **101**, 351–360. The three white ovals and adjacent belts in the South Temperate region of Jupiter, 1940–1990.

Rosse, Earl of, and Copeland (1874) *MNRAS* **34**, 235–247. Notes to accompany chromolithographs from drawings of the Planet Jupiter…

Sagan C (1962) *Mem.Soc.Roy. Liege* **7**, 506–515. On the nature of the jovian Red Spot.

Sagan C (1973) *The Cosmic Connection* (Hodder & Stoughton, London), Chapters 3 & 4.

Sagan C and Miller SL (1960) *Astron.J.* 65, 499 (abstract). Molecular synthesis in simulated reducing planetary atmospheres.

Sagan C & Salpeter EE (1976) *Astrophys.J.Suppl.Ser.* **32**, 737–755. Particles, environments, and possible ecologies in the jovian atmosphere.

Sanchez-Lavega A (1985) *L'Astronomie* **99**, 375–385. La Grande Tache Rouge de Jupiter.

Sanchez-Lavega A & Rodrigo R (1985) *Astron. & Astrophys.* **148**, 67–78. Ground-based observations of synoptic cloud systems in southern equatorial to temperate latitudes of Jupiter from 1975 to 1983.

Sanchez-Lavega A & Quesada JA (1988) *Icarus* **76**, 533–557. Ground-based imaging of jovian cloud morphologies and motions: II. The northern hemisphere from 1975 to 1985.

Sanchez-Lavega A, Laques P, Parker DC & Viscardy G (1990) *Icarus* **87**, 475–483. Midscale dynamical features observed during 1987 in the NEB of Jupiter.

Sanchez-Lavega A, Miyazaki I, Parker D, Laques P & Lecacheux J (1991) *Icarus* **94**, 92–97. A disturbance on Jupiter's high-speed North Temperate jet during 1990.

Sato T (1969), *The Heavens (Japan)* **50** (no.530), 187, reprinted in *JBAA* **84**, 439–442 (1974). Statistical establishment of the repulsive force between the long-enduring white ovals in the STZ of Jupiter.

Sato T (1970) *JBAA* **81**, 30–33 (& p.328). A possible interpretation of the changeable aspect of the GRS on Jupiter.

Sato T (ed.) (1980) *Planet Guidebook 2* [in Japanese] (Seibundoshinkosha, Tokyo)

Schmidt JFJ (1865) *Astr.Nachr.* **65**, 81–86. Ueber die Bewegung dunkler und heller Flecken auf Jupiter.

Schoenberg von E & Heintz WD (1955) *Astr. Nachr.* **282**, 85–88. Uber Breitenänderungen der dunklen Jupiterstreifen.

Schröter JH (1788) *Beiträge zu den neuesten astronomischen Entdeckungen*, (Berlin).

Schröter JH (1798) *ibid.* vol.2 (Göttingen).

Sells EP & Cooke WE (1913) *Physical observations of Jupiter made at the Adelaide Observatory* (Adelaide).

Slipher EC (1964) *A Photographic Study of the Brighter Planets* (Lowell Observatory, Arizona).

Smith BA and Hunt GE (1976), in: *Jupiter* (ed. Gehrels T; University of Arizona Press), pp.564–585. Motions and morphology of clouds in the atmosphere of Jupiter.

Smith BA *et 21 al.* (1979a) *Science* **204**, 951–972. The Jupiter system through the eyes of Voyager 1.

Smith BA *et 21 al.* (1979b) *Science* **206**, 927–950. The galilean satellites and Jupiter: Voyager 2 imaging science results.

Smith DW (1980) *Icarus* **44**, 116–133, Galilean satellite eclipse studies. II. Jovian stratospheric and tropospheric aerosol content.

Smith DW, Greene TF & Shorthill RW (1977) *Icarus* **30**, 697–729. The upper jovian atmosphere aerosol content determined from a satellite eclipse observation.

Smith PH (1986) *Icarus* **65**, 264–279. The vertical structure of the jovian atmosphere.

Smith PH & Tomasko MG (1984) *Icarus* **58**, 35–73. Photometry and polarimetry of Jupiter at large phase angles: II. Polarimetry of the STropZ, SEB, and the polar regions from the Pioneer 10 and 11 missions.

Smoluchowski R (1975) *Am.Sci.* **63**, 638–648. Jupiter 1975.

Solberg HG (1969) *Planet.Space Sci.* **17**, 1573–1580. A 3-month oscillation in the longitude of Jupiter's Red Spot.

Solberg G (1972) *Commun.Lunar & Plan.Lab* **9** (180), 385–387. Rotation period for a subsurface source in the NNTB of Jupiter.

Sommeria J, Meyers SD & Swinney HL (1988) *Nature* **331**, 689–693. Laboratory simulation of Jupiter's Great Red Spot.

Sommeria J, Meyers SD & Swinney HL (1991), in: *Nonlinear Topics in Ocean Physics* (ed. Osborne AR) (North-Holland, Amsterdam), pp.227–269. Experiments on vortices and Rossby waves in eastward and westward jets.

Sromovsky LA, Revercomb HE, Suomi VE, Limaye SS & Krauss RJ (1982) *J.Atm.Sci.* **39**, 1433–1445: Jovian winds from Voyager 2. II: analysis of eddy transports.

Stevenson DJ (1982) *Planet. Space Sci.* **30**, 755–764. Formation of the giant planets.

Stevenson DJ & Salpeter EE (1976) in: *Jupiter* (ed. Gehrels T; University of Arizona Press, Tucson, 1976), pp.85–112. Interior models of Jupiter.

Stoker CR (1986) *Icarus* **67**, 106–125. Moist convection: a mechanism for producing the vertical structure of the jovian equatorial plumes.

Stoker CR & Hord CW (1985) *Icarus* **64**, 557–575. Vertical cloud structure of Jupiter's equatorial plumes.

Stone PH (1967) *J.Atm.Sci.* **24**, 642–652. An application of baroclinic stability theory to the dynamics of the jovian atmosphere.

Stone PH (1972) *J.Atm.Sci.* **29**, 419–426. On non-geostrophic baroclinic stability: Part III. The momentum and heat transports.

Stone PH (1976), in: Gehrels T (ed.), *Jupiter* (University of Arizona Press, Tucson), pp.586–618. The meteorology of the jovian atmosphere.

Streett WB, Ringermacher HI & Veronis G (1971) *Icarus* **14**, 319–342. On the structure and motions of Jupiter's Red Spot.

Swindell W & Doose LR (1974) *JGR* **79**, 3634–3644 . The imaging experiment on Pioneer 10.

Terrile RJ & Westphal JA (1977a) *Icarus* **30**, 274–281. The vertical cloud structure of Jupiter from 5 µm measurements.

Terrile RJ & Westphal JA (1977b) *Icarus* **30**, 730–735. Infrared imaging of Jupiter in the 8–14 µm spectral region.

Terrile RJ & Beebe RF (1979) *Science* **204**, 948–951. Summary of historical data.

Terrile RJ *et 7 al.* (1979a) *Science* **204**, 1007–1008. Infrared images of Jupiter at 5-µm wavelength during the Voyager 1 encounter.

Terrile RJ, Capps RW, Becklin EE & Cruikshank DP (1979b) *Science* **206**, 995–996. Jupiter's cloud distribution between the Voyager 1 and 2 encounters: results from 5-µm imaging.

Tokunaga AT, Knacke RF, Ridgway ST & Wallace L (1979) *Astrophys. J.* **232**, 603–615. High-resolution spectra of Jupiter in the 744–980 cm^{-1} spectral range.

Tokunaga AT, Beck SC, Geballe TR, Lacy JH & Serabyn E (1981) *Icarus* **48**, 283–289. The detection of HCN on Jupiter.

Tomasko MG, West RA & Castillo ND (1978) *Icarus* **33**, 558–592. Photometry and polarimetry of Jupiter at large phase angles: I. Analysis of imaging data of a prominent belt and a zone from Pioneer 10.

Tomasko MG, Karkoschka E & Martinek S (1986) *Icarus* **65**, 218–243. Observations of the limb darkening of Jupiter at ultraviolet wavelengths and constraints on the properties and distribution of stratospheric aerosols.

Trafton LM (1967) *Astrophys.J.* **147**, 765–781. Model atmospheres of the major planets.

Trento JJ (1987) *Prescription for Disaster* (Harrap, London), pp.212–225.

Veverka J, Wasserman LH, Elliot J, Sagan C & Liller W (1974) *Astron.J.* **79**, 73–84. The occultation of β Scorpii by Jupiter, I. The structure of the jovian upper atmosphere.

Wacker WK (1973) *JALPO* **24**, 136–138. The distribution of dark material in SEB disturbances on Jupiter.

Wacker WK (1975) *JALPO* **25**, 145–150 (& p.173). Large-scale disturbances on Jupiter.

Waff CB (1989 Sep) *Astronomy*, pp.44–52. The struggle for the outer planets.

Wallace L (1976) in: *Jupiter* (ed. Gehrels T; University of Arizona Press, Tucson, 1976), pp.284–303. The thermal structure of Jupiter in the stratosphere and upper troposphere.

Wallace L, Prather M & Belton MJS (1974) *Astrophys.J.* **193**, 481–493. The thermal structure of the atmosphere of Jupiter.

Waugh WR (1893) *JBAA* **4**, 88–89. Jupiter Section, 1893. Interim report on the disturbances in the NEB.

Weidenschilling SJ & Lewis JS (1973) *Icarus* **20**, 465–476. Atmospheric and cloud structures of the jovian planets.

West RA (1979a) *Icarus* **38**, 12–33. Spatially resolved methane band photometry of Jupiter: I.

West RA (1979b) *Icarus* **38**, 34–53. Spatially resolved methane band photometry of Jupiter: II.

West RA (1988) *Icarus* **75**, 381–398. Voyager 2 imaging eclipse observations of the Jovian high-altitude haze.

West RA and Tomasko MG (1980) *Icarus* **41**, 278–292. Spatially resolved methane band photometry of Jupiter: III.

West RA, Hord CW, Simmons KE, Coffeen DL, Sato M & Lane AL (1981) *JGR* **86** (A10), 8783–8792. Near-ultraviolet scattering properties of Jupiter.

West RA, Kupferman PN & Hart H (1985) *Icarus* **61**, 311–342. Voyager 1 imaging and IRIS observations of jovian methane absorption and thermal emission: implications for cloud structure.

West RA, Strobel DF & Tomasko MG (1986) *Icarus* **65**, 161–217. Clouds, aerosols and photochemistry in the jovian atmosphere.

Westphal JA (1969), *Astrophys.J.Lett.* **157**, L63–L64. Observations of localized 5-μm radiation from Jupiter.

Westphal JA, Matthews K & Terrile RJ (1974) *Astrophys.J.Lett.* **188**, L111–L112. Five-micron pictures of Jupiter.

Wildt R (1931) *Naturwissenschaft* **19**, 109.

Wildt R (1932) *Nachr. Ges. Akad. Wlss. Göttingen* **1**, 87–96. Absorptionsspektren und Atmosphären der grossen Planeten.

Wildt R (1939) *Proc.Am.Philos.Soc.* **81**, 135. [Theory of GRS]

Wildt R (1969) *J.Atm.Sci.* **26**, 795–797. The outer planets: some early history.

Williams AS (1889) *Zenographical Fragments, vol.I* (Mitchell & Hughes, London).

Williams AS (1896) *MNRAS* **56**, 143–151. On the drift of the surface material of Jupiter in different latitudes.

Williams AS (1897) *MNRAS* **58**, 10–14. The great equatorial current of Jupiter.

Williams AS (1899) *MNRAS* **59**, 376–381. Periodic variation in the colours of the two equatorial belts of Jupiter.

Williams AS (1909) *Zenographical Fragments, vol.II* (Taylor & Francis, London).

Williams AS (1910) *MNRAS* **71**, 145–156. The equatorial current of Jupiter in 1880.

Williams AS (1920) *MNRAS* **80**, 467–475. On the observed changes in the colour of Jupiter's EZ.

Williams AS (1922) *MNRAS* **82**, 417. On the tawny hue of Jupiter's EZ.

Williams AS (1930) *MNRAS* **90**, 696–700. Periodic variation in the colours of the two equatorial belts of Jupiter (II)

Williams AS (1936a) *MNRAS* **97**, 105–107. On the periodic variation in the colours of the two equatorial belts of Jupiter (III).

Williams AS (1936b) *JBAA* **47** (2), 68–70. The colour variations of Jupiter's EZ.

Williams GP (1975) *Nature* **257**, 778. Jupiter's atmospheric circulation.

Williams GP (1978) *J.Atm.Sci.* **35**, 1399–1426. Planetary circulations: 1. Barotropic representation of jovian and terrestrial turbulence.

Williams GP (1979) *J.Atm.Sci.* **36**, 932–968. Planetary circulations: 2. The jovian quasi-geostrophic regime.

Williams GP (1985) *Adv.Geophys.* **28A**, 381–429. Jovian and comparative atmospheric modelling.

Williams G & Robinson JR (1973). *J.Atm.Sci.* **30**, 684–717. Dynamics of a convectively unstable atmosphere: Jupiter?

Williams GP & Yamagata T (1984) *J.Atm.Sci.* **41**, 453–478. Geostrophic regimes, intermediate solitary vortices and Jovian eddies.

Williams GP & Wilson RJ (1988) *J.Atm.Sci.* **45**, 207–241. The stability and genesis of Rossby vortices.

Wright WH (1928) *MNRAS* **88**, 709–718. On photographs of the brighter planets by light of different colours.

Wright WH (1929) *MNRAS* **89**, 703–708. Photographic observations of certain jovian phenomena...

Young AT (1985) *Sky & Tel.* **69** (5), 399–403. What colour is the solar system?

Appendix 5
Bibliography
(Magnetosphere and satellites)

A5.1. MAJOR BOOKS AND SPECIAL ISSUES OF JOURNALS

The following books cover the magnetosphere thoroughly:

Gehrels T (ed.)(1976) *Jupiter* (University of Arizona Press, Tucson).
Dessler AJ (ed.)(1983) *Physics of the Jovian Magnetosphere* (Cambridge University Press).

The following books cover the jovian satellites extensively:

Beatty JK & Chaikin A (eds.) (1981, 1990) *The New Solar System* (Cambridge University Press & Sky Publishing Corp., Cambridge, Mass.).
Belton MJS, West RA & Rahe J (eds.)(1989) *Time-variable Phenomena in the Jovian System*, NASA SP-494 (NASA, Washington DC).
Burns JA & Matthews MS (eds.) (1986), *Satellites* (Univ of Arizona, Tucson).
Hamblin WK & Christiansen EH (1990) *Exploring the Planets* (Macmillan Publishing Co., New York). [A good textbook on the geology.]
Morrison D and Samz J (1980) *Voyage to Jupiter*, NASA SP-439 (NASA, Washington DC). [The book of the Voyager mission.]
Morrison D (ed.) (1982), *Satellites of Jupiter* (University of Arizona, Tucson).
Rothery DA (1992) *Satellites of the Outer Planets* (Oxford University Press).

The Pioneer results at Jupiter were described in the following special issues of journals:
Science (1974 Jan. 25) vol. **183**, pp.301–324;
Science (1975 May 2) vol. **188**, pp.445–477;
Journal of Geophysical Research (1974 Sep. 1) vol. **79**, pp. 3487–3700.
The Voyager results were described in the following special issues:
Science (1979 June 1) vol. **204**, pp.913–1008;
Science (1979 Nov. 23) vol. **206**, pp.925–996;
Nature (1979 August 30) vol. **280**, pp.725–806;
Icarus (1980 Nov.) vol. **44** no.2, The Satellites of Jupiter;
Journal of Geophysical Research (1981 Sep 30) vol. **86** no. A10, pp.8121–8844).
The Ulysses results were described in:
Science (1992 Sep.11) vol. **257**, pp.1487–1556.

A5.2. OTHER REFERENCES

(Conventions as in Appendix 4.)

Acuña MH & Ness NF (1976) *JGR* **81**, 2917–2922. The main magnetic field of Jupiter.
Acuña MH, Neubauer FM & Ness NF (1981) *JGR* **86**(A10), 8513–8521. Standing Alfvén wave current system at Io: Voyager 1 observations.
Antoniadi EM (1939) *J.Roy.Astron.Soc.Canada* **33**, 273–282. On the markings of the satellites of Jupiter in transit.
Atreya SK, Yung YL, Donahue TM & Barker ES (1977) *Astrophys.J.Lett.* **218**, L83–L87. Search for jovian auroral hot spots.
Atreya SK, Donahue TM & Waite JH (1979) *Nature* **280**, 795–796. An interpretation of the Voyager measurement of jovian electron density profiles.
Avigliano DP (1954) *JALPO* 8, 11–14. Jupiter's satellites, 1953–54.
Bagenal F (1989) in: Belton MJS, West RA & Rahe J (eds.)(1989) *Time-variable Phenomena in the Jovian System*, NASA SP-494 (NASA, Washington DC), pp.196–210. Torus-magnetosphere coupling.
Bagenal F and Sullivan JD (1981) *JGR* **86**(A10), 8447–8467. Direct plasma measurements in the Io torus and inner magnetosphere of Jupiter.
Bame SJ *et 9 al.* (1992) *Science* **257**, 1539–1543. Jupiter's magnetosphere: plasma description from the Ulysses flyby.
Barnard EE (1891) *MNRAS* **51**, 543–555, Observations of the planet Jupiter and his satellites during 1890 with the 12-inch equatoreal of the Lick Observatory; *ibid.* 556–558, On the phenomena of the transits of the first satellite of Jupiter.
Barnard EE (1892) *Astron. J.* **12**, 81–85. Discovery and observations of a fifth satellite to Jupiter.
Barnard EE (1894) *MNRAS* **54**, 134–136. On the dark poles and bright equatorial belt of the first satellite of Jupiter.
Baron R, Joseph RD, Owen T, Tennyson J, Miller S & Ballester GE (1991) *Nature* **353**, 539–542. Imaging Jupiter's aurorae from H_3^+ emissions in the 3–4 μm band.
Barton SG (1946) *Pop. Astron.* **54**, 122–131, The names of the satellites.
Baum R (1992) *JBAA* **102**, 316. A Barnard centenary: The finding of Jupiter V.
Becklin EE & Wynn-Williams CG (1979) *Nature* **279**, 400–401. Detection of Jupiter's ring at 2.2 μm.
Belcher JW (1987) *Science* **238**, 170–176. The Jupiter-Io connection: an Alfvén engine in space.
Bigg EK (1966) *Planet.Space Sci.* **14**, 741–758. Influence of the satellite Io on Jupiter's decametric emission.
Binder AB & Cruikshank DP (1964) *Icarus* **3**, 299–305. Evidence for an atmosphere on Io.
Binder AB & Cruikshank DP (1966) *Icarus* **5**, 7–9. Photometric search for atmospheres on Europa and Ganymede.
Boischot A, Lecacheux A, Kaiser ML, Desch MD, Alexander JK & Warwick JW (1981) *JGR* **86**(A10) 8213–8226. Radio Jupiter after Voyager: an overview of the planetary radio astronomy observations.
Borucki WJ & Magalhaes JA (1992) *Icarus* **96**, 1–14. Analysis of Voyager 2 images of jovian lightning.
Borucki WJ & Williams MA (1986) *JGR* **91**, 9893–9903. Lightning in the jovian water cloud.
Bridge HS *et 10 al.* (1979) *Science* **206**, 972–976. Plasma observations near Jupiter: initial results from Voyager 2.
British Astronomical Association (1901) *JBAA* **11**, 334–335. Meeting report.
British Astronomical Association (1954) *JBAA* **64**, 190–191. Meeting report.

Broadfoot AL *et 16 al.* (1979) *Science* **204**, 979–982. Extreme ultraviolet observations from Voyager 1 encounter with Jupiter.

Broadfoot AL *et 9 al.* (1981) *JGR* **86**(A10), 8259–8284. Overview of the Voyager ultraviolet spectrometry results through Jupiter encounter.

Brown RA (1974), in: Woszczyk A & Iwaniszewska C (eds.), *IAU Symposium 65: Exploration of the Planetary System* (D. Reidel, Hingman, Mass.) pp.527–531. Optical line emission from Io.

Brown RH, Cruikshank DP, Tokunaga AT, Smith RG & Clark RN (1988) *Icarus* **74**, 262–271. Search for volatiles on icy satellites: I, Europa.

Buratti BJ (1991) *Icarus* **92**, 312–323. Ganymede and Callisto: surface textural dichotomies and photometric analysis.

Buratti B & Golombek M (1988) *Icarus* **75**, 437–449. Geologic implications of spectrophotometric measurements of Europa.

Burke BF & Franklin FL (1955) *JGR* **60**, 213–217. Observations of a variable radio source associated with the planet Jupiter.

Caldwell J, Turgeon B & Hua X-M (1992) *Science* **257**, 1512–1515. Hubble Space Telescope imaging of the north polar aurora on Jupiter.

Carlson RW *et 8 al.* (1973) *Science* **182**, 53–55. An atmosphere on Ganymede from its occultation of SAO 186800 on 7 June 1972.

Carlson RW & Judge DL (1974), *JGR* **79**, 3623–3633. Pioneer 10 UV photometer observations at Jupiter encounter.

Carr MH (1986) *JGR* **91**, 3521–3532. Silicate volcanism on Io.

Carr MH, Masursky H, Strom RG and Terrile RJ (1979) *Nature* **280**, 729–733. Volcanic features of Io.

Carr TD, Desch MD & Alexander JK (1983), in: Dessler AJ (ed.) *Physics of the Jovian Magnetosphere* (Cambridge University Press), pp.226–284. Phenomenology of magnetospheric radio emission.

Cattermole P (1990 Dec.) *Astronomy Now* **4** (12), 21–26. Unlikely twins.

Chapman CR & McKinnon WB (1986), in: *Satellites* (ed. Burns JA & Matthews MS; University of Arizona, Tucson, 1986) pp. 492–580. Cratering of planetary satellites.

Christie WM (1907) *MNRAS* **67**, 561. Diagrams showing the positions of Jupiter's satellites VI and VII...

Clancy RT & Danielson GE (1981) *JGR* **86**(A10), 8627–8634. High resolution albedo measurements on Io from Voyager 1.

Clark RN (1980) *Icarus* **44**, 388–409. Ganymede, Europa, Callisto, and Saturn's rings: compositional analysis from reflectance spectroscopy.

Clark RN & McCord TB (1980) *Icarus* **41**, 323–339. The galilean satellites: new near-infrared spectral reflectance measurements (0.65–2.5 μm) and a 0.325–5.0 μm summary.

Clarke J, Caldwell J, Skinner T & Yelle R (1989) in: (NASA SP-494) pp.211–220. The aurora and airglow of Jupiter.

Cochran WD & Barker ES (1979) *Astrophys.J.Lett.* **234**, L151–L154. Variability of Lyman-α emission from Jupiter.

Conca J (1981) *Proc.Lunar & Plan.Sci.Conf.* **12**, 1599–1606. Dark-ray craters on Ganymede.

Consolmagno GJ & Lewis JS (1976) in: *Jupiter* (ed. Gehrels T; Univ. of Arizona, Tucson, 1976) pp.1035–1051. Structural and thermal models of icy galilean satellites.

Cook AF, Duxbury TC & Hunt GE (1979) *Nature* **280**, 794. First results on jovian lightning.

Cook AF & Duxbury TC (1981) *JGR* **86**(A10), 8815–8818. A fireball in Jupiter's atmosphere.

Cook AF, Jones AV & Shemansky DE (1981a) *JGR* **86**(A10), 8793–8796. Visible aurora in Jupiter's atmosphere?

Cook AF, Shoemaker EM, Smith BA, Danielson GE, Johnson TV & Synnott SP (1981b) *Science* **211**, 1419–1422. Volcanic origin of the eruptive plumes of Io.

Copeland R (1885) *MNRAS* **45**, 375. On the observation of the projection of Jupiter's first satellite on its own shadow.

Cruikshank DP, Degewij J & Zellner BH (1982) In: *Satellites of Jupiter* (ed. Morrison D; University of Arizona, Tucson), pp.129–146. The outer satellites of Jupiter.

Davies ME (1982) In: *Satellites of Jupiter* (ed. Morrison D; University of Arizona, Tucson), pp.911–936. Cartography and nomenclature for the galilean satellites.

Dawes WR (1860) *MNRAS* **20**, 245–247. On the appearance of Jupiter's satellites while transiting the disk of the planet.

De Pater I (1981) *Astron. Astrop.* **93**, 370–381. Radio maps of Jupiter's radiation belts and planetary disk at 6 cm.

De Pater I (1986) *Icarus* **68**, 344–365. Jupiter's belt-zone structure at radio wavelengths. II.

De Pater I (1990) *An.Rev.Astron.Astrop* **28**, 347–399. Radio images of the planets.

De Pater I & Dickel JR (1986) *Astrophys.J.* **308**, 459–471. Jupiter's belt-zone structure at radio wavelengths. I.

De Sitter W (1928) *Leiden Annals* **16** (Part 2). Orbital elements determining the longitudes of Jupiter's satellites, derived from observations.'

Dollfus A & Murray JB (1974) in: *IAU Symp. no.65, Exploration of the Planetary System* (eds. Woszczyk A & Iwanewska C) (IAU/Reidel, Dordrecht), pp.513–525. La rotation, la cartographie et la photometrie des satellites de Jupiter.

Dollfus A (1975) *Icarus* **25**, 416–431. Optical polarimetry of the galilean satellites of Jupiter.

Domingue DL, Hapke BW, Lockwood GW & Thompson DT (1991) *Icarus* **90**, 30–42. Europa's phase curve: implications for surface structure.

Douglass AE (1898) *MNRAS* **58**, 382–385. The markings on Venus.

Doyle LR & Borucki WJ (1989) in: (NASA SP-494) pp.384–389. Jupiter lightning locations.

Drossart P, Prangé R & Maillard J-P (1992) *Icarus* **97**, 10–25. Morphology of infrared H_3^+ emissions in the auroral regions of Jupiter.

Dutton D (1976 Dec) *Sky & Tel.* **52**, 482–484. Naked-eye observations of Jupiter's moons.

Eshleman VR, Tyler GL, Wood GE, Lindal GF, Anderson JD, Levy GS & Croft TA (1979a) *Science* **204**, 976–978. Radio science with Voyager 1 at Jupiter: preliminary profiles of the atmosphere and ionosphere.

Eshleman VR, Tyler GL, Wood GE, Lindal GF, Anderson JD, Levy GS & Croft TA (1979b) *Science* **206**, 959–962. Radio science with Voyager at Jupiter: initial Voyager 2 results and a Voyager 1 measure of the Io torus.

Eviatar A, Bar-Nun A & Podolak M (1985) *Icarus* **61**, 185–191. Europan surface phenomena.

Fanale FP, Hamilton Brown R, Cruikshank DP & Clake RN (1979) *Nature* **280**, 761–763. Significance of absorption features in Io's IR reflectance spectrum.

Fanale FP, Banerdt WB & Cruikshank DP (1981), *Geophys. Res. Lett.* **8**, 625–628. Io: could SO_2 condensation/sublimation cause the sometimes-reported post-eclipse brightening?

Fanale FP, Banerdt WB, Elson LS, Johnson TV & Zurek RW (1982) In: *Satellites of Jupiter* (ed. Morrison D; University of Arizona, Tucson, 1982), pp.756–781. Io's surface: its phase composition and influence on Io's atmosphere and Jupiter's magnetosphere.

Fillius RW, McIlwain CE & Mogro-Campero A (1975) *Science* **188**, 465–468. Radiation belts of Jupiter: a second look.

Fink U, Dekkers NH & Larson HP (1973) *Astrophys. J. Lett.* **179**, L155–L159. Infrared spectra of the galilean satellites of Jupiter.'

Fjeldbo G, Kliore A, Seidel B, Sweetnam D & Cain D (1975) *Astron. Astrophys.* **39**, 91–96. The Pioneer 10 radio occultation measurements of the ionosphere of Jupiter.

Fjeldbo G, Kliore A, Seidel B, Sweetnam D & Woiceshyn P (1976) in: *Jupiter* (ed.Gehrels T; University of Arizona, Tucson), pp.238–246. The Pioneer 11 radio occultation measurements of the jovian ionosphere.

Fox WE (1978) *JBAA* **88**, 360–361. Jupiter: double and triple satellite phenomena.

Fox WE (1986) *JBAA* **96**, 139. Observations of the mutual phenomens of Jupiter's satellites.

Frey H (1975) *Icarus* **25**, 439–446. Post-eclipse brightening and non-brightening of Io.

Gaskell RW, Synnott SP, McEwen AS & Schaber GG (1988) *Geophys.Res.Lett.* **15**, 581–584. Large-scale topography of Io: implications for internal structure and heat transfer.

Genova F & Aubier MG (1985) *Astron.Astrop.* **150**, 139–150. Io-dependent sources of the Jovian decameter emission.

Genova F, Zarka P & Lecacheux A (1989), in: Belton MJS, West RA & Rahe J (eds.) *Time-variable Phenomena in the Jovian System*, NASA SP-494 (NASA, Washington DC) pp.156–174. Jupiter decametric radiation.

Goguen JD *et 10 al* (1988) *Icarus* **76**, 465–484. Io hot spots: infrared photometry of satellite occultations.

Goldberg BA, Garneau GW & LaVoie SK (1984) *Science* **226**, 512– . Io's sodium cloud. [With colour images on cover.]

Goldstein RM & Green RR (1980) *Science* **207**, 179–180. Ganymede: radar surface characteristics.

Gorton S (ed.)(1867) *Astron.Reg.* **5**, 169, 191–194, 209–215. The planet Jupiter without his satellites.

Greeley R, Lee SW, Crown DA & Lancaster N (1990) *Icarus* **84**, 374–402. Observations of industrial sulfur flows: implications for Io.

Greenberg R (1982) in: *Satellites of Jupiter* (ed. Morrison D; University of Arizona, Tucson) pp. 65–92. Orbital evolution of the galilean satellites.

Greenberg R (1989) in (NASA SP-494) pp.100–115. Time-varying orbits and tidal heating of the galilean satellites.

Gurnett DA, Shaw RR, Anderson RR, Kurth WS & Scarf FL (1979) *Geophys.Res.Lett.* **6**, 511–514. Whistlers observed by Voyager 1: detection of lightning on Jupiter.

Gurnett DA & Goertz CK (1981) *JGR* **86**, 717–722. Multiple Alfvén wave reflections excited by Io: origin of the Jovian decametric arcs.

Hamilton DC, Gloeckler G, Krimigis SM & Lanzerotti LJ (1981) *JGR* **86**(A10), 8301–8318. Composition of nonthermal ions in the jovian magnetosphere.

Hanel R *et 13 al.* (1979b) *Science* **206**, 952–956. Infrared observations of the jovian system from Voyager 2.

Hapke B (1990) *Icarus* **88**, 407–417. Coherent backscatter and the radar characteristics of outer planet satellites.

Hartmann WK (1987) *Icarus* **71**, 57–68. A satellite-asteroid mystery and a possible early flux of scattered C-class asteroids.

Helfenstein P & Parmentier EM (1985) *Icarus* **61**, 175–184. Patterns of fracture and tidal streses due to non-synchronous rotation: implications for fracturing on Europa.

Horner VM & Greeley R (1982) *Icarus* **51**, 549–562. Pedestal craters on Ganymede.

Howell RR, Cruikshank DP and Fanale FP (1984) *Icarus* **57**, 83–92. Sulfur dioxide on Io: spatial distribution and physical state.

Howell RR & Sinton WM (1989), in: Belton MJS, West RA & Rahe J (eds.)(1989) *Time-variable Phenomena in the Jovian System*, NASA SP-494 (NASA, Washington DC), p.47–62. Io and Europa: the observational evidence for variability.

Ingersoll AP (1989) *Icarus* **81**, 298–313. Io meteorology: how atmospheric pressure is controlled locally by volcanos and surface frosts.

Jewitt DC (1982) in: *Satellites of Jupiter* (ed. Morrison D; University of Arizona, Tucson); pp.44–64. The rings of Jupiter.

Jewitt DC, Danielson GE & Synnott SP (1979) *Science* **206**, 951. Discovery of a new Jupiter satellite.

Jewitt DC & Danielson GE (1981) *JGR* **86** (A10), 8691–8697. The jovian ring.

Johnson JH (1931) *JBAA* **41**, 164–171. The discovery of the first four satellites of Jupiter.

Johnson TV, Soderblom LA, Mosher JA, Danielson GE, Cook AF & Kupferman P (1983) *J. Geophys. Res.* **88**, 5789–5805. Global multispectral mosaics of the icy Galilean satellites.

Johnson TV & Soderblom LA (1983 Dec.) *Sci. Am.* **249**, 60–71. Io.

Johnson TV, Veeder GJ, Matson DL, Brown RH, Nelson RM & Morrison D (1988) *Science* **242**, 1280–1283. Io: evidence for silicate volcanism in 1986.

Johnson TV & Matson DL (1989) in: *Atmospheres* (ed. Atreya S; University of Arizona, Tucson). Io's tenuous atmosphere.

Kieffer SW (1982) in: *Satellites of Jupiter* (ed. Morrison D; Univeristy of Arizona, Tucson) pp.647–723. Dynamics and thermodynamics of volcanic eruptions: implications for the plumes on Io.

Kim SJ, Drossart P, Caldwell J, Maillard J-P, Herbst T & Shure M (1991) *Nature* **353**, 536–539. Images of aurorae on Jupiter from H_3^+ emission at 4 μm.

Kim SJ, Caldwell J & Herbst TM (1992) *Icarus* **96**, 143–148. Locations of 4-μm hot spots on the poles of Jupiter.

Klein MJ, Thompson TJ & Bolton S (1989) In: Belton MJS, West RA & Rahe J (eds.)(1989) *Time-variable Phenomena in the Jovian System*, NASA SP-494 (NASA, Washington DC), pp.151–155. Systematic onservations and correlation studies of variation in the synchrotron radio emission from Jupiter.

Kliore AJ, Fjeldbo G, Seidel BL, Sweetman DN, Sesplaukis TT, Woiceshyn PM & Rasool SI (1975), *Icarus* **24**, 407–410. Atmosphere of Io from Pioneer 10 radio occultation measurements.

Kowal CT (1976) *Sky & Tel.* **51** (1976 April), 242–243. All the outer satellites of Jupiter.

Krimigis SM *et 10 al.* (1979) *Science* **206**, 977–984. Hot plasma environment at Jupiter: Voyager 2 results.

Krimigis SM *et 7 al.* (1981) *JGR* **86**(A10), 8227–8257. Characteristics of hot plasma in the jovian magnetosphere: results from the Voyager spacecraft.

Kuiper GP (1951) *Proc.Nat.Acad.Sci.USA* **37**, 717–721. On the origin of the natural satellites.

Kuiper GR (1973) *Commun. Lunar Plan. Lab.* **10** (no.187), 28–34. Comments on the galilean satellites.

Kumar S & Hunten DM (1982) In: *Satellites of Jupiter* (ed. Morrison D; University of Arizona, Tucson), pp.782–806. The atmospheres of Io and other satellites.

Kurth WS, Strayer BD, Gurnett DA & Scarf FL (1985) *Icarus* **61**, 497–507. A summary of whistlers observed by Voyager 1 at Jupiter.

Lane AL, Nelson RM & Matson DL (1981) *Nature* **292**, 38–39. Evidence for sulphur implantation in Europa's UV absorption band.

Laplace PS (1805), *Mecanique Celeste*, vol. 4 (Courcier, Paris).

Lellouch E, Belton M, de Pater I, Gulkis S & Encrenaz T (1990) *Nature* **346**, 639–641. Io's atmosphere from microwave detection of SO_2.

Lellouch E, Belton M, de Pater I, Paubert G, Gulkis S & Encrenaz T (1992) *Icarus* **98**, 271–295. The structure, stability, and global distribution of Io's atmosphere.

Lewis JS (1972) *Icarus* **16**, 241–252. Low-temperature condensation from the solar nebula.

Lewis JS (1973) *Space Sci. Rev.* **14**, 401–411. Chemistry of the outer solar system.

Lieske JH (1981) *Astron.Astrop.Suppl.Ser.* **44**, 209–216. Catalogue of eclipses of Jupiter's galilean satellites, 1610–2000.

Lieske JH (1987) *Astron.Astrophys.* **176**, 146–158. Galilean satellite evolution: observational evidence for secular changes in mean motions.

Livengood TA, Moos HW, Ballester GE & Prangé RM (1992) *Icarus* **97**, 26–45. Jovian ultraviolet auroral activity, 1981–1991.

Lucchitta BK (1980) *Icarus* **44**, 481–501. Grooved terrain on Ganymede.

Lucchitta BK & Soderblom LA (1982) In: *Satellites of Jupiter* (ed. Morrison D; University of Arizona, Tucson), pp.521–555. The geology of Europa.

Lunine JI & Stevenson DJ (1985) *Icarus* **64**, 345–367. Physics and chemistry of sulfur lakes on Io.

Lyot B (1943) *Bull.Soc.Astron.France (L'Astronomie)* **57**, 49–60, 67–72. Observations planétaires au Pic du Midi en 1941 par MM. Camichel, Gentili et Lyot.

Lyot B (1953) *Bull.Soc.Astron.France (L'Astronomie)* **67**, 3–21. Laspect des planetes au Pic du Midi dans une lunette de 60 cm d'ouverture.

Matson DL & Johnson TV (1988) EOS *Trans.AGU* **69**, 1227. Abstract: Io's atmosphere: evidence for H_2S.

Mallama A (1992) *Icarus* **95**, 309–318. Astrometry of the galilean satellites from mutual eclipses and occultations.

McEwen AS (1986a) *Nature* **321**, 49–51. Tidal reorientation and the fracturing of Jupiter's moon Europa.

McEwen AS (1986b) *JGR* **91**, 8077–8097. Exogenic and endogenic albedo and color patterns on Europa.

McEwen AS (1988) *Icarus* **73**, 385–426. Global color and albedo variations on Io.

McEwen AS, Johnson TV, Matson DL & Soderblom LA (1988) *Icarus* **75**, 450–478. The global distribution, abundance, and stability of SO_2 on Io.

McEwen AS, Lunine JI & Carr MH (1989), in: Belton MJS, West RA & Rahe J (eds.) *Time-variable Phenomena in the Jovian System*, NASA SP-494 (NASA, Washington DC), pp. 11–46. Dynamic geophysics of Io.

McKinnon WB & Parmentier EM (1986), in: *Satellites* (ed. Burns JA & Matthews MS; University of Arizona, Tucson, 1986) pp.718–763. Ganymede and Callisto.

McLeod BA, McCarthy DW & Freeman JD (1991) *Astron.J.* **102**, 1485–1489. Global high-resolution imaging of hotspots on Io.

McNutt RL, Belcher JW & Bridge HS (1981) *JGR* **86**(A10), 8319–8342. Positive ion observations in the middle magnetosphere of Jupiter.

Melosh HJ (1982) *JGR* **87** (B3), 1880–1890. A simple mechanical model of Valhalla basin, Callisto.

Mendillo M, Baumgardner J, Flynn B & Hughes WJ (1990) *Nature* **348**, 312–314. The extended sodium nebula of Jupiter.

Mendillo M, Flynn B & Baumgardner J (1992) *Science* **257**, 1510–1512. Imaging observations of Jupiter's sodium magneto-nebula during the Ulysses encounter.

Menietti JD & Gurnett DA (1980) *Geophys.Res.Lett.* **7**, 49–52. Whistler propagation in the jovian magnetosphere.

Metzger AE, Gilman DA, Luthey JL & Hurley KC (1983) *JGR* **88**, 7731. The detection of X-rays from Jupiter.

Mihalov JD, Collard HR, McKibben DD, Wolfe JH & Intriligator DS (1975) *Science* **188**, 448–451. Pioneer 11 encounter: preliminary results from the Ames Research Center plasma analyzer experiment.

Miller S (1990) *Astronomy Now* **4** (no.5), 20–21. Jovian moon gets new volcano.

Millis RL (1978) *Icarus* **33**, 319–321. Photoelectric photometry of JV.

Minton RB (1973) *Commun. Lunar & Plan. Lab* **10** (no.188), 35–36. The red polar caps of Io.

Molesworth PB (1891) *JBAA* **2**, 26–29. Report on variations of the satellites of Jupiter in colour and magnitude.

Molesworth PB (1899) *Mem.BAA* **7**, 207–210. Satellite observations and studies. [in 1898 Report]

Moore P (1965) *JBAA* **75**, 292–293; The colour of Io

Morabito LA, Synnott SP, Kupferman PN & Collins SA (1979) *Science* **204**, 972. Discovery of currently active extraterrestrial volcanism.

Moreno MA, Schubert G, Baumgardner J, Kivelson MG & Paige DA (1991) *Icarus* **93**, 63–81. Io's volcanic and sublimation atmospheres.

Morgan JS (1985 March) *Sky & Tel.* **69**, 202–203. Io and the "Jovian nebula".

Morrison D (1985 Mar.) *Sky & Tel.* **69**, 198–205. The enigma called Io.

Morrison D & Cruikshank DP (1974) *Space Sci.Rev.* **15**, 641–739. Physical properties of the natural satellites.

Morrison D & Burns JA (1976) in: *Jupiter* (ed. Gehrels T; Univ. of Arizona, Tucson) pp.991–1034. The jovian satellites.

Morrison D, Pieri D, Veverka J & Johnson TV (1979) *Nature* **280**, 753–755. Photometric evidence on long-term stability of albedo and colour markings on Io.

Morrison D (1982) *Ann.Rev.Astron.Astrop.* **20**, 469–493. The satellites of Jupiter and Saturn.

Moses JI & Nash DB (1991) *Icarus* **89**, 277–304. Phase transformations and the spectral reflectance of solid sulfur: Can metastable sulfur allotropes exist on Io?

Murchie SL, Head JW & Plescia JB (1989) *Icarus* **81**, 271–297. Crater densities and crater ages of different terrain types on Ganymede.

Murchie SL, Head JW & Plescia JB (1990) *JGR* **95** (B7), 10743–10768. Tectonic and volcanic evolution of dark terrain and its implications for the internal structure and evolution of Ganymede.

Murray JB (1975) *Icarus* **25**, 397–404. New observations of surface markings on Jupiter's satellites.

Nash DB & Nelson RM (1979) *Nature* **280**, 763–766. Spectral evidence for sublimates and adsorbates on Io.

Nash DB, Carr MH, Gradie J, Hunten DM & Yoder CF (1986) In: Burns JA & Matthews MS (eds.) *Satellites* (University of Arizona, Tucson), pp.629–688. Io.

Nash DB & Howell RR (1989) *Science* **244**, 454–457. Hydrogen sulfide on Io: evidence from telescopic and laboratory infrared spectra.

Nelson RM & Hapke BW (1978) *Icarus* **36**, 304–329. Spectral reflectivities of the Galilean satellites and Titan, 0.32–0.86 μm.

Nelson RM, Lane AL, Matson DL, Fanale FP, Nash DB & Johnson TV (1980) *Science* **210**, 784–786. Io: longitudinal distribution of sulfur dioxide frost.

Nelson ML, McCord TB, Clark RN, Johnson TV, Matson DL, Mosher JA & Soderblom LA (1986) *Icarus* **65**, 129–151. Europa: characterization and interpretation of global spectral units.'

Nelson RM et 7 al. (1993) *Icarus* **101**, 223–233. The brightness of Jupiter's satellite Io following emergence from eclipse: selected observations, 1981–1989.'

Ness NF, Acuña MH, Lepping RP, Burlaga LF, Behannon KW & Neubauer FM (1979) *Science* **206**, 966–972. Magnetic field studies at Jupiter by Voyager 2: preliminary results.

Neugebauer G, Becklin EE, Jewitt DC, Terrile RJ & Danielson GE (1981) *Astron.J.* **86**, 607–610. Spectra of the jovian ring and Amalthea.

Nicholson PD & Matthews K (1991) *Icarus* **93**, 331–346. Near-infrared observations of the jovian ring and small satellites.

Ojakangas GW & Stevenson DJ (1986) *Icarus* **66**, 341–358. Episodic volcanism of tidally heated satellites with application to Io.

Ojakangas GW & Stevenson DJ (1989) *Icarus* **81**, 220–241. Thermal state of an ice shell on Europa.

Oppenheimer C & Stevenson D (1989) *Nature* **342**, 790–793. Liquid sulphur lakes at Poás volcano.

Ostro SJ (1982) in: Morrison D (ed.), *Satellites of Jupiter* (University of Arizona, Tucson), pp.213–236. Radar properties of Europa, Ganymede, and Callisto.

Ostro SJ, Rosema KD, Campbell DB, Chandler JF & Shapiro II (1990) *Bull.AAS* **22**, 1109. Io's radar properties.

Ostro SJ & Shoemaker EM (1990) *Icarus* **85**, 335–345. The extraordinary radar echoes from Europa, Ganymede, and Callisto: a geological persepctive.

Owen T (1976), *Icarus* **29**, 159–163, Jovian satellite nomenclature

Owen T, Danielson GE, Cook AF, Hansen C, Hall VL & Duxbury TC (1979) *Nature* **281**, 442–446. Jupiter's rings.

Pagnini P (1931) *JBAA* **41**, 415–422. Galileo and Simon Mayer.

Parmentier EM, Squyres SW, Head JW and Allison ML (1982) *Nature* **295**, 290–293. The tectonics of Ganymede.

Passey QR & Shoemaker EM (1982), in: *Satellites of Jupiter* (ed. Morrison D; University of Arizona, Tucson) pp.379–434. Craters and basins on Ganymede and Callisto.

Peale SJ, Cassen P & Reynolds RT (1979) *Science* **203**, 892–894. Melting of Io by tidal dissipation.

Pearl J, Hanel R, Kunde V, Maguire W, Fox K, Gupta S, Ponnamperuma C & Raulin F (1979) *Nature* **280**, 755–758. Identification of gaseous SO_2 and new upper limits for other gases on Io.

Pearl JC & Sinton WM (1982) In: *Satellites of Jupiter* (ed. Morrison D; University of Arizona, Tucson), pp. 724–755. Hot spots of Io.

Phillips TER & Steavenson WH (1917) *JBAA* **28**, 56–59. A remarkable transit of Jupiter's third satellite.

Phillips TER (1921) *MNRAS* **82**, 96–100. The third satellite of Jupiter.

Pieri DC, Baloga SM, Nelson RM & Sagan C (1984) *Icarus* **60**, 685. Sulfur flows of Ra Patera, Io.

Pilcher CB, Ridgway ST & McCord TB (1972) *Science* **178**, 1087–1089. Galilean satellites: identification of water frost.

Poirier JP (1982) *Nature* **299**, 683–687. Rheology of ices: a key to the tectonics of the ice moons of Jupiter and Saturn.

Reynolds R & Squyres S (1982) cited in: *New Scientist* (1982 Dec.) p.783. Jupiter's icy moon could support life.

Riddle AC & Warwick JW (1976) *Icarus* **27**, 457–459. Redefinition of System III longitude.

Rogers J (1990) *JBAA* **100**, 113. The surface of Io.

Ross MN & Schubert G (1987) *Nature* **325**, 133–134. Tidal heating in an internal ocean model of Europa.

Sagan C (1979) *Nature* **280**, 750–753. Sulphur flows on Io.

Salama F, Allamandola LJ, Witteborn FC, Cruikshank DP, Sandford SA & Bregman JD (1990) *Icarus* **83**, 66–82. The 2.5–5.0 μm spectra of Io: evidence for H_2S and H_2O frozen in SO_2.

Sandel BR & 16 al. (1979) *Science* **206**, 962–966. Extreme ultraviolet observations from Voyager 2 encounter with Jupiter.

Sandel BR & Dessler AJ (1988) *JGR* **93**, 5487–5504. Dual periodicity of the jovian magnetosphere. [Definition of System IV]

Schaber GG (1982) In: *Satellites of Jupiter* (ed. Morrison D; University of Arizona, Tucson), pp.556–597. The geology of Io.

Schaeberle JM & Campbell WW (1891) *Publ.Astr.Soc.Pacific* **3**, 359–365. Observations of markings of Jupiter's third satellite.

Schenk PM & Seyfert CK (1980) *Eos* **61**, 286. Abstract.

Schenk PM & McKinnon WB (1985) *Proc. Lunar & Plan.Sci.Conf.16 (JGR Suppl.* **90**), C775–C783. Dark halo craters and the thickness of grooved terrain on Ganymede.

Schenk PM & McKinnon WB (1987) *Icarus* **72**, 209–234. Ring geometry on Ganymede and Callisto.

Schenk PM & McKinnon WB (1989) *Icarus* **79**, 75–100. Fault offsets and lateral crustal movement on Europa: evidence for a mobile ice shell.

Schenk PM & McKinnon WB (1991) *Icarus* **89**, 318–346. Dark-ray and dark-floor craters on Ganymede, and the provenance of large impactors in the jovian system.

Schneider NM, Hunten DM, Wells WK & Trafton LM (1987) *Science* **238**, 55–58. Eclipse measurements of Io's sodium atmosphere.

Schneider NM, Trauger JT, Wilson JK, Brown DI, Evans RW & Shemansky DE (1991a) *Science* **253**, 1394–1397. Molecular origin of Io's fast sodium.

Schneider NM, Hunten DM, Wells WK, Schultz AB & Fink U (1991b) *Astrophys.J.* **368**, 298–315. The structure of Io's corona.

Schubert G, Spohn T & Reynolds RT (1986) in: Burns JA & Matthews MS (eds.) *Satellites* (University of Arizona, Tucson, 1986), pp.224–292. Thermal histories, compositions and internal structures of the moons of the solar system.

Shemansky DE (1987) *JGR* **92** (A6), 6141–6146. Ratio of oxygen to sulfur in the Io plasma torus.

Shoemaker EM, Lucchitta BK, Plescia JB, Squyres SW & Wilhelms DE (1982), in: *Satellites of Jupiter* (ed. Morrison D; University of Arizona, Tucson) pp.435–520. The geology of Ganymede.

Shoemaker EM & Shoemaker CS (1990), in: Beatty KJ & Chaikin A (eds.) *The New Solar System*, 3rd edn. (Sky Publishing Corp./Cambridge Univ. Press), pp.259–274. The collision of solid bodies.

Showalter MR (1989) in: Belton MJS, West RA & Rahe J (eds.) *Time-variable Phenomena in the Jovian System*, NASA SP-494 (NASA, Washington DC), p.116–125. Anticipated time variations in (our understanding of) Jupiter's ring system.

Showalter MR, Burns JA, Cuzzi JN & Pollack JB (1985) *Nature* **316**, 526–528. Discovery of Jupiter's gossamer ring.

Sieveka EM & Johnson RE (1982) *Icarus* **51**, 528–548. Thermal- and plasma-induced molecular redistribution on the icy satellites.

Sill GT & Clark RN (1982) In: *Satellites of Jupiter* (ed. Morrison D; University of Arizona, Tucson, 1982), pp.174–212. Composition of the surfaces of the galilean satellites.

Simonelli DP & Veverka J (1984) *Icarus* **59**, 406–425. Voyager disk-integrated photometry of Io.

Sinton WM & Kaminski C (1988) *Icarus* **75**, 207–232; Infrared observations of eclipses of Io, its thermophysical parameters, and the thermal radiation of the Loki volcano and environs.

Skinner TE, Deland MT, Ballester GE, Coplin KA, Feldman PD & Moos HW (1988) *JGR* **93**, 29–34. Temporal variation of the jovian HI Lyman-α emission (1979–1986).

Smith BA *et 21 al* (1979a) *Science* **204**, 951–972. The Jupiter system through the eyes of Voyager 1.

Smith BA *et 21 al* (1979b) *Science* **206**, 927–950. The galilean satellites and Jupiter: Voyager 2 imaging science results.

Smith BA, Shoemaker EM, Kieffer SW and Cook AF (1979c) *Nature* **280**, 738–743. The role of SO_2 in volcanism on Io.

Smith EJ, Davis L, Jones DE, Colburn DS, Coleman PJ, Dyal P & Sonett CP (1974) *Science* **183**, 305–306. Magnetic field of Jupiter and its interaction with the solar wind.

Smith RA & Strobel DF (1985) *JGR* **90** (A10), 9469–9493. Energy partitioning in the Io plasma torus.

Smyth WH & McElroy MB (1978) *Astrophys.J.* **226**, 336–346. Io's sodium cloud: comparison of models and two-dimensional images.

Soderblom L *et 13 al.* (1980) *Geophys.Res.Lett.* **7**, 963–966. Spectrophotometry of Io: preliminary Voyager 1 results.

Spencer JR, Shure MA, Ressler ME, Goguen JD, Sinton WM, Toomey DW, Denault A & Westfall J (1990) *Nature* **348**, 618–621. Discovery of hotspots on Io using disk-resolved infrared imaging.

Spencer JR, Howell RR, Clark BE, Klassen DR & O'Connor D (1992) *Science* **257**, 1507–1510. Volcanic activity on Io at the time of the Ulysses encounter.

Squyres SW (1980) *Geophys. Res. Lett.* **7**, 593–596. Volume changes in Ganymede and Callisto and the origin of grooved terrain.

Squyres SW, Reynolds RT, Cassen PM & Peale SJ (1983) *Nature* **301**, 225–226. Liquid water and active resurfacing on Europa.

Staelin DH (1981) *JGR* **86**(A10), 8581–8584. Character of the jovian decametric arcs.

Steavenson WH (1915) *JBAA* **25**, 383–385. Ganymede.

Stebbins J & Jacobsen TS (1928) *Lick Obs. Bull.* No. **401**.

Strobel DF (1989) in: Belton MJS, West RA & Rahe J (eds.)(1989) *Time-variable Phenomena in the Jovian System*, NASA SP-494 (NASA, Washington DC), pp.183–195. Energetics, luminosity, and spectroscopy of Io's torus.

Strobel DF & Atreya SK (1986) in: *Physics of the Jovian Magnetosphere* (ed. Dessler AJ; Cambridge University Press) pp.51–67. Ionosphere.

Strom RG, Terrile RJ, Masursky H & Hansen C (1979) *Nature* **280**, 733–736. Volcanic eruption plumes on Io.

Strom RG, Woronow A & Gurnis M (1981) *JGR* **86**(A10), 8659–8674. Crater populations on Ganymede and Callisto.

Synnott SP (1980) *Science* **210**, 786–788. 1979J2: the discovery of a previously unknown jovian satellite.

Synnott SP (1981) *Science* **212**, 1392. 1979J3: discovery of a previously unknown satellite of Jupiter.

Synnott SP (1984) *Icarus* **58**, 178–181. Orbits of the small inner satellites of Jupiter.

Tancredi G, Lindgren M & Rickman H (1990) *Astron. & Astrophys.* **239**, 375–380. Temporary satellite capture and orbital evolution of comet P/Helin-Roman-Crockett.

Tholen DJ & Zellner B (1984) *Icarus* **58**, 246–253. Multicolor photometry of outer jovian satellites.

Thomas P & Veverka J (1982) in: *Satellites of Jupiter* (ed. Morrison D; University of Arizona, Tucson, 1982); pp.147–173. Amalthea.

Trauger JT (1984) *Science* **226**, 337–341. The jovian nebula: a post-Voyager perspective.

Van Allen JA *et al.*(1975) *Science* **188**, 459–462. Pioneer 11 observations of energetic particles in the jovian magnetosphere.

Veverka J (1975) in: *Planetary satellites* (ed. Burns JA; University of Arizona, Tucson, 1976) pp... Polarimetry of satellite surfaces.

Veverka J, Thomas P, Davies M & Morrison D (1981) *JGR* **86** (A10), 8675–8692. Amalthea: Voyager imaging results.

Vogt RE *et 10 al.* (1979) *Science* **204**, 1003–1007. Voyager 1: energetic ions and electrons in the jovian magnetosphere.

Westfall JE (1983–84) *JALPO* **30**, 45–53, 105–115, 145–154. Galilean satellite eclipse timings: 1975–82 report.

Westfall JE (1992) *JALPO* **36**, 63–74. Galilean satellite eclipse timings: the 1989/90 apparition.

Wolff RS & Mendis DA (1983) *JGR* **88**, 4749–4769. On the nature of the interaction of the jovian magnetosphere and the icy galilean satellites.

Yelle RV, McConnell JC, Sandel BR & Broadfoot AL (1987) *JGR* **92**, 15110–15124. The dependence of electroglow on the solar flux.

Yoder CF (1979) *Nature* **279**, 767–770. How tidal heating in Io drives the galilean orbital resonance locks.

Young AT (1984) *Icarus* **58**, 197–226. No sulfur flows on Io.

Yung YL and McElroy MB (1975) *Bull.Amer.Astron.Soc.* **7**, 387. Ganymede: possibility of an oxygen atmosphere.

Index